Lecture Notes in Mathematics

Edited by A. Dold and B. Eckmann

550

Proceedings of the Third Japan – USSR Symposium on Probability Theory

Edited by G. Maruyama and J. V. Prokhorov

Springer-Verlag
Berlin · Heidelberg · New York 1976

Editors
Gisiro Maruyama
Department of Mathematics
College of General Education
University of Tokyo
3–8–1 Komaba
Meguro-ku
Tokyo/Japan

Jurii V. Prokhorov
Mathematical Institute of the
Academy of Sciences of USSR
ul. Vavilova 42
Moskow 117 333/USSR

Library of Congress Cataloging in Publication Data

Japan-USSR Symposium on Probability Theory, 3d, Tashkend, 1975.
Proceedings of the third Japan-USSR Symposium on Probability Theory.

(Lecture notes in mathematics ; 550)
1. Probabilities--Congresses. I. Maruyama, Gishirō, 1916- II. Prokhorov, Iūriĭ Vasil'evich. III. Series: Lecture notes in mathematics (Berlin) ; 550.
QA3.L28 no. 550 [QA273.A1] 510'.8s [519.2] 76-49898

AMS Subject Classifications (1970): 60XX

ISBN 3-540-07995-5 Springer-Verlag Berlin · Heidelberg · New York
ISBN 0-387-07995-5 Springer-Verlag New York · Heidelberg · Berlin

Printed in Germany
Printing and binding: Beltz Offsetdruck, Hemsbach/Bergstr.

PREFACE

The Third USSR-Japan Symposium on Probability Theory was held in Tashkent, USSR, August 27 - September 2, 1975. 51 probabilists from Japan and 136 from the USSR attended the symposium.

This volume contains most of longer, one-hour and 45-minutes, papers presented to the symposium.

As well as the previous two, the third symposium was very fruitful and important for mathematicians of our countries, and we hope it will give rise to many interesting investigations in the future.

The Third Symposium was organized with the support of the Academy of Sciences of the USSR and the Academy of Sciences of the Uzbek SSR. It is a great pleasure for us to express our gratitude to the Organizing Committee and all those who have contributed to the success of the symposium and to the preparation of this volume.

Yu.V. Prokhorov G. Maruyama

TABLE OF CONTENTS

[x] Participants in paper

SOME LIMIT THEOREMS FOR A QUEUEING SYSTEM WITH ABSOLUTE PRIORITY IN HEAVY TRAFFIC

T.A.Azlarov, Ya.M.Husainov

1. We consider a queueing system with one server and with two arrival Poisson flows of customers with parameters λ_1 and λ_2. The serving times of the customers from the i -th $(i=1,2)$ flow are i.i.d.r.v.'s with distribution function $H_i(x)$ $(i=1,2)$ and finite expectation

$$\mu_i=\int_0^\infty [1-H_i(x)]\,dx < \infty , \qquad i=1,2.$$

Suppose that customers of the first flow (urgent demands) are served with absolute priority with respect to customers of the second flow (simple demands). Customers of each flow are served by the rule: "first arrived - first being served".

Let $\eta_1(t)$ - be the time interval between moment t and the moment when the serving of all the customers who had arrived by the moment t is over, i.e. the virtual waiting time of the urgent customer; $\eta_2(t)$ - be the time interval between moment t and the moment when the serving of all simple customers who had arrived by the moment t, under the condition that there are no urgent customers, is over; i.e. the virtual waiting time of the simple customer who arrives at moment t under the condition that the system has only simple customers.

Let

$$F(t,x,y)=P\{\eta_1(t)\le x;\ \eta_2(t)\le y\},\quad x,y\ge 0.$$

This queueing system was studied in [I] . There the following results were obtained.

If

$$\rho = \rho_1 + \rho_2 < 1 \qquad (\rho_i = \lambda_i \mu_i \ , \quad i = 1,2),$$

then the stationary distribution

$$F(x,y) = \lim_{t \to \infty} F(t,x,y)$$

exists, and the Laplace-Stieltjes transform of this distribution is obtained.

The asymptotical behaviour of one-dimensional distributions of the waiting time of urgent and simple customers in this system in heavy traffic was studied in [2] . It was proved that in both cases as well as in the system with one type arrival flow, limit distributions are exponential.

This paper deals with the asymptotical behaviour of two-dimensional distributions of the waiting times of the customers as $\rho \to 1$ is studied under any possible variations of ρ_1 and ρ_2 . The class of all possible limit distributions for $F(x,y)$ is described (these results were announced in [4]).

In what follows we suppose that

$$\beta_{i2} = \int_0^\infty x^2 dH_i(x) < \infty \ , \quad i = 1,2.$$

Denote

$$h_1(s) = \int_0^\infty e^{-sx} dH_1(x), \ \mathrm{Re}\, s \geq 0,$$

$$h_2(u)=\int_0^\infty e^{-ux}\,dH_2(x)\,,\quad \operatorname{Re} u \geq 0\,,$$

$$F(x)=\begin{cases} 0, & x\leq 0\,,\\ 1-e^{-x}, & x>0\,.\end{cases}$$

Theorem 1. If $\rho_1 \to a$, $\rho_2 \to 1-a$ $(0<a<1)$, then, for all $x,\ y \geq 0$,

$$\lim F(x;\ y M\eta_2) = A(x)\cdot E(y),$$

where $A(x)$ has the L-S transform

$$\alpha(s)=\int_0^\infty e^{-sx}\,dA(x)=(1-a)\cdot\left[1-\frac{a}{\mu_1}\cdot\frac{1-h_1(s)}{s}\right]^{-1}.$$

Note that $\alpha(s)$ is the Khinčin-Pollacek formula for the L-S transform of the average waiting time for M | G | 1 systems with one type of customers.

Thus, under the conditions of Theorem 1, heavy traffic does not affect the behaviour of the average waiting time expectation for the urgent customer.

Let $\{\xi_n\}$ be a sequence of i.i.d.r.v.'s with density

$$f(x)=\begin{cases} 0\,, & x\leq 0\,,\\ \dfrac{1}{2\sqrt{\pi x(1-b)}}\left[e^{-\frac{x}{4(1-b)}}-\dfrac{1}{2(1-b)}\left\{1-\int_0^x \exp\left(-\dfrac{u^2}{4x(1-b)}\right)du\right\}\right], & x>0.\end{cases}$$

Let $0<b<1$ and ν be lattice r.v.'s independent of ξ_i $(i=1,2,\dots)$, and let

$$P(\nu=k)=(1-b)b^k \qquad (k=0,1,2,\dots).$$

Denote

$$\zeta_\nu = \xi_1 + \xi_2 + \dots + \xi_\nu \ , \quad P(\zeta_\nu < y) = \mathcal{D}(y) \ ,$$

$$B(x,y) = \frac{1}{2\sqrt{\pi}} \int_0^x \int_0^y e^{-\frac{\theta}{2} - \frac{v}{4}} \frac{\theta}{v^{3/2}} e^{-\frac{\theta^2}{4v}} \, d\theta \, dv \ .$$

<u>Theorem 2.</u> Let $\rho_1 \to 1$, $\rho_2 \to 0$ such that $\frac{\rho_2}{1-\rho_1} \to \beta$, $0 \le \beta \le 1$. Then

1) if $\frac{\rho_2}{(1-\rho_1)^2} \to \infty$, then, for all x , $y \ge 0$,

$$\lim F(x M\eta_1 ; y M\eta_2) = \begin{cases} B(x,y) & , \quad \beta = 0 \ , \\ B(x, \frac{y}{1-\beta}) *^y \mathcal{D}(y) \ , & 0 < \beta < 1 \ , \\ E(x)\cdot E(y) & , \quad \beta = 1 \ . \end{cases}$$

where $*^y$ denotes the convolution with respect to y ;

2) if $\frac{\rho_2}{(1-\rho_1)^2} \to c$, $0 < c < \infty$, then, for all $x, y > 0$

$$\lim F(x M\eta_1 ; y) = C(x,y) \ ,$$

where $C(x,y)$ has the two-dimensional L-S transform

$$\tilde{C}(s,u) = \int_0^\infty \int_0^\infty e^{-sx} e^{-uy} dC(x,y) = \left\{ s + \frac{1}{2} \left[1 + \sqrt{1 + 2c \frac{\mu_1 \beta_{12}}{\mu_2} (1 - h_2(u))} \right] \right\}^{-1} .$$

<u>Remark</u>. We do not consider the case $\beta > 1$ because then a stationary regime does not exist.

<u>Theorem 3.</u> For $\rho_1 \to 0$, $\rho_2 \to 1$,

$$\lim F(x ; y M\eta_2) = E(y) \ .$$

The exponential limit theorem of [2] for the one-dimensional distribution of η_i as $\rho_i \uparrow 1$ is closely connected with our theorems.

The proof of Theorems 1 and 2 consists in asymptotical investigation of a formula for $\Phi(s,u)$, the L-S transform of the stationary distribution $F(x,y)$, obtained in [1] :

$$\Phi(s,u) = \frac{(s-u)\tilde{\Phi}(0,u)+u(1-\rho)}{s-(\lambda_1+\lambda_2)+\lambda_1 h_1(s)+\lambda_2 h_2(u)} \tag{I}$$

where $\tilde{\Phi}(0,u)$ - is determined from the condition that the L-S transform is an analytical function in the region

$$G = \{(s,u): s>0 \;,\; u>0\} \;.$$

2. Auxiliary results.

Lemma 1. The equation

$$s-\lambda_1-\lambda_2+\lambda_1 h_1(s)+\lambda_2 h_2(u) = 0 \tag{2}$$

has only one solution $s=P(u)$ in the region G .

Proof. Let us consider equation (2) for positive s and u

We have to prove that the equation

$$\frac{\lambda_1+\lambda_2[1-h_2(u)]-s}{\lambda_1} = h_1(s) \tag{3}$$

has only one solution $s=P(u)$ for $u>0$ which satisfies the following condition

$$\lambda_2[1-h_2(u)] \le s \le \lambda_1+\lambda_2[1-h_2(u)] \;.$$

On the figure below we can see the left and right hand sides of equality (3) for real s . If $s=\lambda_2[1-h_2(u)]$, then the value of the left hand side of (3) is larger than that of the right hand side. Since the straight line corresponding the left hand side of (3) has a negative angle coefficient and $h_1(s)$ is continuous and positive, there exists a point at which our curve is intersected by the line. This point is a solution of our equation.

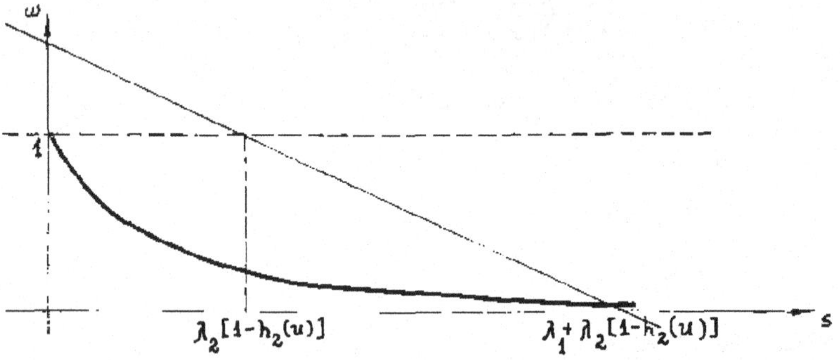

From the convexity of $h_1(s)$ it follows that there are no other real roots. The lemma is proved.

It is not difficult to prove that $P(u) \to 0$ as $u \to 0$. In fact, when $u \to 0$, the right hand side of equality (3) is not changed while the left hand side is transformed into the line

$$\omega = \frac{\lambda_1 - s}{\lambda_1} \tag{3'}$$

with the angle coefficient $-\frac{1}{\lambda_1}$. The derivative of $h_1(s)$ at $s=0$ is equal to $-\mu_1$. From the condition $\rho_1 < 1$ we conclude that the distance from the tangent to the curve is less than from the line (3') to the curve, what proves our assertion.

<u>Lemma 2.</u> For $u \to 0$ the solution of equation (2) is of the

following asymptotical form

$$P(u) = \frac{\rho_2}{1-\rho_1}\, u - \frac{\lambda_2 \beta_{22}(1-\rho_1)^2 + \lambda_1 \beta_{12} \rho_2^2}{2(1-\rho_1)^3} u^2 + o(u^2) .$$

Proof. Put $s = P(u)$ into equation (2), then it turns into an identity. Thus, if s is a function of u, then the left hand side of (2) is a function of u, which is identically zero. The derivative of the left hand side of (2) with respect to u is equal to zero, too. We obtain

$$P'(u) + \lambda_1 h_1'(P(u)) \cdot P'(u) + \lambda_2 h_2'(u) = 0$$

and

$$[1 + \lambda h_1'(P(u))] P''(u) + \lambda_1 h_1''(P(u)) \cdot (P'(u))^2 + \lambda_2 h_2''(u) = 0 .$$

From the first equation we find

$$P'(u) = - \frac{\lambda_2 h_2'(u)}{1 + \lambda_1 h_1'(P(u))} .$$

from the second one, taking into account the form of $P'(u)$, we obtain

$$P''(u) = \frac{\lambda_2 h_2''(u)[1+\lambda_1 h_1'(P(u))]^2 + \lambda_1 \lambda_2^2 (h_2'(P(u)))^2 h_1''(P(u))}{[1+\lambda_1 h_1'(P(u))]^3} .$$

Since $P(0) = 0$,

$$P'(0) = \frac{\lambda_2 \mu_2}{1 - \lambda_1 \mu_1} = \frac{\rho_2}{1 - \rho_1} ,$$

$$P''(0) = \frac{\lambda_2\beta_{22}(1-\lambda_1\mu_1)^2+\lambda_1(\lambda_2\mu_2)^2\beta_{12}}{(1-\lambda_1\mu_1)^3} = -\frac{\lambda_2\beta_{22}(1-\rho_1)^2+\lambda_1\beta_{12}\rho_2^2}{(1-\rho_1)^3}.$$

Thus, by the Taylor formula for small u we get

$$P(u) = \frac{\rho_2}{1-\rho_1}u - \frac{\lambda_2\beta_{22}(1-\rho_1)^2+\lambda_1\beta_{12}\rho_2^2}{2(1-\rho_1)^3}u^2 + o(u^2),$$

3. The proof of Theorem 1. Since the function $\Phi(s,u)$ is analytical in the region G, we have

$$\tilde{\Phi}(0,u) = \frac{u}{u-P(u)}(1-\rho).$$

Putting this expression into (1), we obtain

$$\Phi(s,u) = \frac{s-P(u)}{s-(\lambda_1+\lambda_2)+\lambda_1 h_1(s)+\lambda_2 h_2(u)} \cdot \frac{u}{u-P(u)}(1-\rho). \tag{4}$$

$$M\eta_{l_1} = \left[\frac{d\Phi(s,0)}{ds}\right]_{s=0} = \frac{\lambda_1\beta_{12}}{2(1-\rho_1)},$$

$$M\eta_{l_2} = -\left[\frac{d\Phi(0,u)}{du}\right]_{u=0} = \frac{\rho_2}{2(1-\rho)}\left[\frac{\beta_{22}}{\mu_2}+\frac{\rho_1\beta_{12}}{(1-\rho_1)\mu_1}\right]. \tag{5}$$

Now let $\delta = \delta(\rho) = [M\eta_{l_2}]^{-1}$. According to Lemma 2 and by (4) for $\rho_1 \to a$, $\rho_2 \to 1-a$, $0<a<1$, we have for all finite u:

$$\Phi(s,\delta u) = \frac{s-P(\delta u)}{s-\lambda_1[1-h_1(s)]-\lambda_2[1-h_2(\delta u)]} \times$$

$$\times \frac{1-\rho_1}{1+\left[\frac{\beta_{22}}{\mu_2} - \frac{\lambda_1\beta_{12}\rho_2}{(1-\rho_1)^2}\right]\cdot\left[\frac{\beta_{22}}{\mu_2}+\frac{\rho_1\beta_{12}}{(1-\rho_1)\mu_1}\right]^{-1} u+o(1)} \longrightarrow \frac{1-a}{1-\frac{a}{\mu_1}\cdot\frac{1-h_1(s)}{s}} \cdot \frac{1}{1+u} \cdot$$

The functions

$$\alpha(s) = (1-a)\left[1-\frac{a}{\mu_1}\cdot\frac{1-h_1(s)}{s}\right]^{-1} \qquad \text{and} \qquad \frac{1}{1+u}$$

are the L-S transforms of the distribution functions $A(x)$ and $F(x)$ respectively.

4. The proof of Theorem 2. I) Since $0<\beta_{12}<\infty$, we can write the following expansion for small s :

$$h_1(s) = 1-\mu_1 s + \frac{\beta_{12}}{2}s^2 + o(s^2).$$

Now (I) can be rewritten in the following way:

$$\Phi(s,u) = \frac{(s-u)\tilde{\Phi}(u,0)+u(1-\rho)}{\frac{\lambda_1\beta_{12}}{2}s^2+(1-\rho_1)s-\lambda_2[1-h_2(u)]+o(s^2)} =$$

$$= \frac{(s-u)\tilde{\Phi}(0,u)+u(1-\rho)}{\frac{\lambda_1\beta_{12}}{2}s^2+(1-\rho_1)s-\lambda_2[1-h_2(u)]}\left[1+\frac{o(s^2)}{\frac{\lambda_1\beta_{12}}{2}s^2+(1-\rho_1)s-\lambda_2(1-h_2(u))}\right] = \qquad (6)$$

$$= \frac{(s-u)\tilde{\Phi}(0,u)+u(1-\rho)}{\frac{\lambda_1\beta_{12}}{2}[s-P_1(u)][s-P_2(u)]}\left[1+\frac{o(s^2)}{\frac{\lambda_1\beta_{12}}{2}s^2+(1-\rho_1)s-\lambda_2(1-h_2(u))}\right],$$

where

$$P_\kappa(u) = \frac{-(1-\rho_1)-(-1)^\kappa\sqrt{(1-\rho_1)^2+2\lambda_1\lambda_2\beta_{12}[1-h_2(u)]}}{\lambda_1\beta_{12}} \qquad (\kappa=1,2).$$

The only root, which satisfies the condition $P(u)\to 0$ as $u\to 0$, is $P_1(u)$.

From (6) and Theorem 1 we obtain that, for small u,

$$\tilde{\Phi}(0,u) = \frac{u}{u-P_1(u)}(1-\rho).$$

Putting this equation into (6), we have

$$\Phi(s,u) = \frac{u(1-\rho)}{\frac{\lambda_1\beta_{12}}{2}\left[s-P_2(u)\right]\left[u-P_1(u)\right]}\left[1+\frac{o(s^2)}{\frac{\lambda_1\beta_{12}}{2}s^2+(1-\rho_1)s-\lambda_2(1-h_2(u))}\right]. \qquad (7)$$

It is easy to see that (5) holds in this case too.

Now let $\varepsilon=\varepsilon(\rho)=[M\eta_1]^{-1}$ and $\delta=\delta(\rho)=[M\eta_2]^{-1}$.

Then by (7)

$$\Phi(\varepsilon s,\delta u) = \frac{(1-\rho)u/\varepsilon}{\frac{\lambda_1\beta_{12}}{2}\left[s-\frac{P_2(\delta u)}{\varepsilon}\right]\left[u-\frac{P_1(\delta u)}{\delta}\right]}\cdot\left[1+\frac{o(\varepsilon^2)}{\frac{\lambda_1\beta_{12}}{2}\varepsilon^2s^2+(1-\rho_1)\varepsilon s-\lambda_2(1-h_2(\delta u))}\right]. \qquad (8)$$

If

$$h_2(q) = 1-\mu_2 q+\frac{\beta_{22}}{2}q^2+o(q^2),$$

then, for all finite u,

$$\frac{P_2(\delta u)}{\varepsilon} = \frac{-(1-\rho_1)-\sqrt{(1-\rho_1)^2+2\lambda_1\lambda_2\beta_{12}[1-h_2(\delta u)]}}{\varepsilon\lambda_1\beta_{12}} = \qquad (9)$$

$$= -\frac{1}{2}\left[1+\sqrt{1+4\rho_1\beta_{12}\left(1-\frac{\rho_2}{1-\rho_1}\right)\frac{u}{\mu_1\beta_{22}(1-\rho_1)+\rho_1\mu_1\beta_{12}}\left[\mu_2-\frac{1}{2}\beta_{22}\frac{u}{M\eta_2}+o\left(\frac{1}{M\eta_2}\right)\right]}\ \right],$$

$$\frac{P_1(\delta u)}{\delta} = -\frac{1-\rho_1}{\delta\lambda_1\beta_{12}}\left[1-\sqrt{1+\frac{2\lambda_1\lambda_2\beta_{12}}{(1-\rho_1)^2}\left[1-h_2(\delta u)\right]}\ \right] =$$

(IO)

$$= -\frac{\mu_1\beta_{22}(1-\rho_1)+\rho_1\mu_2\beta_{12}}{2\rho_1\beta_{12}\mu_2\left(\frac{1-\rho_1}{\rho_2}-1\right)}\left[1-\sqrt{1+4\beta_{12}\rho_1\left(1-\frac{\rho_2}{1-\rho_1}\right)\frac{u}{\mu_1\beta_{22}(1-\rho)+\rho_1\mu_2\beta_{12}}\left(\mu_2-\frac{1}{2}\beta_{22}\frac{u}{M\eta_2}+o\left(\frac{1}{M\eta_2}\right)\right)}\right]$$

From the conditions of the theorem and (9), (IO) and (8) we obtain

$$\lim\Phi(\varepsilon s;\delta u) = \frac{(1-b)\,u}{\left[s+\frac{1}{2}\left(1+\sqrt{1+4(1-b)u}\right)\right]\left[u+\frac{b}{2(1-b)}\left(1-\sqrt{1+4(1-b)u}\right)\right]} =$$

$$= \frac{1}{s+\frac{1}{2}\left(1\sqrt{1+4(1-b)u}\right)}\cdot\frac{1-b}{1-b\,\frac{2}{1+\sqrt{1+4(1-b)u}}}\ .$$

Thus, for all finite s and u,

$$\lim\Phi(\varepsilon s,\delta u) = \begin{cases} \dfrac{1}{s+\frac{1}{2}\left(1+\sqrt{1+4u}\right)} & \text{if } b=0 \\[2ex] \dfrac{1}{s+\frac{1}{2}\left(1+\sqrt{1+4(1-b)u}\right)}\ \dfrac{1-b}{1-b\,\dfrac{2}{1+\sqrt{1+4(1-b)u}}} & \text{if } 0<b<1 \\[2ex] \dfrac{1}{1+s}\cdot\dfrac{1}{1+u} & \text{if } b=1 \end{cases}$$

By using the inverse formula of the L-S transform we arrive at the first statement of the theorem (see [4]).

2) In this case $M\eta_{l_1} \nmid \infty$ and $M\eta_{l_2} < \infty$ (see (5)). Putting $s = s\varepsilon = s\,[M\eta_{l_1}]^{-1}$ into (7), we obtain

$$\Phi(s\varepsilon, u) = \frac{u\,(1-\rho)/\varepsilon}{\frac{\lambda_1\beta_{12}}{2}\left[s - \frac{P_2(u)}{\varepsilon}\right]\left[u - P_2(u)\right]}\left[1 + \frac{o(\varepsilon^2)}{\frac{\lambda_1\beta_{12}}{2}\varepsilon^2 s^2 + (1-\rho_1)\varepsilon s - \lambda_2[1-h_2(u)]}\right]. \quad (II)$$

Note that

$$\frac{P_2(u)}{\varepsilon} = -\frac{1}{2}\left[1 + \sqrt{1 + \frac{2\lambda_1\beta_{12}}{\mu_2}\cdot\frac{\rho_2}{(1-\rho_1)^2}(1-h_2(u))}\right] \quad . \quad (I2)$$

From (II) and (I2) we obtain the second statement of the theorem.

The proof of Theorem 3 is quite analogous to that of Theorem 2 and is omitted.

REFERENCES

I. Б р о д и С.М.,Кибернетика, 3, (I973).

2. Г н е д е н к о Б.В., Д а н и е л я н Э.А. и др., Приоритетные системы обслуживания, Москва,Издательство МГУ,I973,I7I-I73.

3. Д и т к и н В.А.,П р у д н и к о в А.П.,Операционное исчисление по двум переменным и его приложения, Москва, Физматгиз,I958, стр. 48, II2, I39.

4. А з л а р о в Т.А., Х у с а и н о в Я.М., Известия Академии наук Узбекской ССР, серия физико-математических наук, 6, (1974), 53-55.

Mathematical Institute
Academy of Sciences of the Uzbek SSR
Tashkent

ON CERTAIN PROBLEMS OF UNIFORM DISTRIBUTION OF REAL SEQUENCES

Yoshikazu Baba

Department of Mathematics
Shizuoka University
Shizuoka, Japan

1. Introduction.

Let a_n^1 be the first digit of 2^n, expressed in the base-ten, and consider the problem: What is the frequency of the occurrences of the digit $i(1\leq i\leq 9)$ in the sequence $\{a_n^1\}_{n\geq 1}$? Since $\log_{10}2$ is irrational, we can answer the question by making use of the ergodic property of the transformation $T_\theta\omega = \omega+\theta(\text{mod } 1)$ on the unit interval $[0,1)$ with Lebesgue measure where θ is any irrational number. The frequencies $P_i^1(1\leq i\leq 9)$ exist and $P_i^1 = \log_{10}(1+1/i)$ (cf. [1], Appendix 12). In this paper, we shall discuss some variants of the problem.

Let a_n^j, b_n^j, $B_n^j(n\geq 1, j\geq 1)$ be the jth digit, the jth digit from the end and the last j digits of r^n in the base-g expression $(r,g\geq 2)$, respectively. In case $r^n < g^j$, put $a_n^{j+1} = 0$, $b_n^{j+1} = 0$ and $B_n^{j+1} = \Sigma_{k=1}^{j+1} b_n^k g^{k-1}$. Evidently, the sequences $\{B_n^j\}_{n\geq 1}$ and $\{b_n^j\}_{n\geq 1}$ are periodic and we shall denote the period of $\{B_n^j\}_{n\geq 1}$ by $\pi(j) = \pi(j,g,r)$. Let P_i^j and Q_i^j be the frequencies of the occurrences of the digit $i(0\leq i\leq g-1)$ in the sequences $\{a_n^j\}_{n\geq 1}$ and $\{b_n^j\}_{n\geq 1}$, respectively. The existence of P_i^j ($= \lim_{n\to\infty} N(i,n)/n$, $N(i,n)$ = the number of occurrences of the digit i among $a_1^j, \ldots, a_n^j$) follows from the ergodicity of the transformation T_θ where θ is an irrational number $\log_g r$ which we assume to be irrational. As for Q_i^j, the existence is clear because of the periodicity of the sequence $\{b_n^j\}_{n\geq 1}$. The behavior of the sequence $\{B_n^j\}_{n\geq 1}$ in the interval $[0,g^j)$ is similar to that of the periodic orbit of the r-adic transformation on $[0,1)$, $T_r\omega=r\omega(\text{mod } 1)$, starting from $1/g^j$ and to discuss these is equivalent to the discussion of the occurrences of the digits in the periodic decimal $1/g^j = 0.x_1x_2\ldots x_n\ldots$ in the base-r expression.

We shall call mutually distinct dN numbers $A = \{a_1,a_2, \ldots ,a_{dN}\}$ in $[0,1)$, $d \geq 1$, to be 1/N-*uniformly distributed* if there exists a partition of A: $A = A_1+ A_2+ \ldots + A_N$ such that each A_k consists of d elements of A and is contained in the interval $I_k = [(k-1)/N, k/N)$, $1 \leq k \leq N$.

Theorem 1. *If* $\log_g r$ *is irrational, we have* $P_0^j > P_1^j > \dots > P_{g-1}^j$ *for any* $j \geq 1$ *and* $\lim_{j\to\infty} P_i^j = 1/g$ *for any* i $(0 \leq i \leq g-1)$.

Remarks. If $j = 1$, we omit the case of $i = 0$. $\log_g r$ is rational if and only if the prime factorizations of g and r are $g = p_1^{\alpha_1} \dots p_m^{\alpha_m}$ and $r = p_1^{\beta_1} \dots p_m^{\beta_m}$ with $\alpha_1/\beta_1 = \dots = \alpha_m/\beta_m$, and in this case the problem turns out trivial.

Theorem 2. *Suppose* g *and* r *are mutually prime,* $(g,r) = 1$. *Then, there exists an integer* $j_0 = j_0(g,r) \geq 1$ *such that if* $j > j_0$, $\pi(j) = \pi(j_0)g^{j-j_0}$ *and the fractions* $(r^n/g^j)_{1\leq n\leq\pi(j)}$ *are* $1/g^{j-j_0}$*-uniformly distributed in* $[0,1)$, *where* (x) *is the fractional part of a real number* x, $(x) = x - [x]$.

Next, we consider the case $(g,r) \neq 1$. If r contains all prime factors of g, it is evident that for any $j \geq 1$, there exists a number n_0 such that if $n > n_0$, then $g^j | r^n$ and $B_n^j = 0$, so that the behavior of the sequence $\{B_n^j\}_{n\geq 1}$ becomes trivial. This is the exclusive case. So, we shall assume that the prime factorizations of g and r take the forms of

$$(1) \qquad g = p_1^{\alpha_1} \dots p_m^{\alpha_m} p_{m+1}^{\alpha_{m+1}} \dots p_k^{\alpha_k}, \quad r = p_1^{\beta_1} \dots p_m^{\beta_m} q_{m+1}^{\beta_{m+1}} \dots q_l^{\beta_l}.$$

Put $g_0 = p_1^{\alpha_1} \dots p_m^{\alpha_m}$ and $g_1 = g/g_0 = p_{m+1}^{\alpha_{m+1}} \dots p_k^{\alpha_k}$, then we have

Theorem 3. *Suppose* g *and* r *are not mutually prime,* $(g,r) \neq 1$. *Then, there exists an integer* $j_0 = j_0(g,r) \geq 1$ *such that if* $j > j_0$, $\pi(j) = \pi(j_0)g_1^{j-j_0}$ *and the fractions* $(r^n/g^j)_n$ *in a period are* $1/g_1^{j-j_0}$ *-uniformly distributed in* $[0,1)$.

Remark. In this theorem d in the definition of the 1/N-uniform distribution is equal to $\pi(j_0)$. However, as is seen in examples of section 5, there are cases where we can make d smaller than $\pi(j_0)$.

We note here the well-known facts of periodic decimals. If the relation between g and r is exclusive, as is stated above, the corresponding decimal of $1/g^j$ is terminating. In ordinary cases, the decimal is pure recurring or mixed recurring according as the relation is $(g,r) = 1$ or $(g,r) \neq 1$. In the latter case, if the recurring part of the decimal starts at Nth digit, we may write the fractions in Theorem 3 as $(r^n/g^j)_{N\leq n\leq\pi(j)+N-1}$.

Some parts of our results on the 1/N-uniform distribution are contained in the works of Н.М.Коробов [2] and R.G.Stoneham [3],[4],[5] and [6] with different formulations from ours.

The next theorem is a direct corollary of Theorems 2 and 3.

Theorem 4. $\lim_{j\to\infty} Q_i^j = 1/g$ *for any* $i(0 \leq i \leq g-1)$. *If in particular* $(g,r) = 1$ *then we have for any* $j > j_0$

(2) $$Q_i^j = 1/g \ (0 \leq i \leq g-1)$$

where j_0 *is the number determined in Theorem 2.*

2. Proof of Theorem 1.

Since the situation is quite similar, we shall prove the theorem in the case of $g = 10$ and $r = 2$. For any $i(0 \leq i \leq 9)$, $a_n^2 = i$ if and only if the fractional part of $n\log_{10}2$ is contained in the disjoint union of intervals $\Sigma_{k=1}^{9}[\log_{10}(k+i/10), \log_{10}(k+(i+1)/10))$. So we have

$$P_i^2 = \Sigma_{k=1}^{9}\{\log_{10}(k+(i+1)/10) - \log_{10}(k+i/10)\}$$
$$= \Sigma_{k=1}^{9} \log_{10}(1+1/(10k+i)).$$

Similarly, for $j \geq 3$ we have

$$P_i^j = \log_{10}(1+1/(10^{j-1}+i)) + \log_{10}(1+1/(10^{j-1}+10+i)) + \log_{10}(1+1/(10^{j-1}+20+i)) + \ldots + \log_{10}(1+1/(10^j-(10-i))).$$

From this easily follows that $P_0^j > P_1^j > \ldots > P_9^j$. Next, simple calculation shows us

$$P_0^j - P_9^j < \log_{10}(1+1/10^{j-1}).$$

Since the right hand side of this inequality tends to zero as j tends to infinity, we have $\lim_{j\to\infty} P_i^j = 1/10$ for any $i(0 \leq i \leq 9)$.

3. Periods.

In this section, we shall determine the concrete form of the period $\pi(j)$. Clearly, we may obtain $\pi(j)$ as the minimal solution of the equation

(3) $$r^{n+\pi} \equiv r^n \pmod{g^j}.$$

If $(g,r) = 1$ this is equivalent to the equation

(4) $$r^\pi \equiv 1 \pmod{g^j}$$

and if $(g,r) \neq 1$, since for sufficiently large n we have $g_0^j | r^n$, the equation (3) becomes equivalent to

(5) $$r^\pi \equiv 1 \pmod{g_1^j}.$$

First, consider the case $(g,r) = 1$. For $j = 1$ the equality $r^{\pi(1)} = 1 + gu$ holds for some $u \geq 1$, and this may be rewritten as $r^{\pi(1)} = 1 + g^\delta u$ with $\delta \geq 1$ and $g \nmid u$.

Case 1. $(g,u) = 1$ with $g = 2$ and $\delta = 1$.

In this case, since r must be an odd number, $r = 1 + 2u$ with $(2,u) = 1$, so $\pi(1) = 1$ and $r^2 = 1 + 4(u + u^2) = 1 + 2^{j_0}u'$ for some $j_0 \geq 3$ and for some odd u'. Therefore we have $\pi(2) = \ldots = \pi(j_0) = 2$ and $\pi(j) = 2^{j-j_0+1} = \pi(j_0)2^{j-j_0}$ when $j > j_0$.

Case 2. $(g,u) = 1$ and not Case 1.

In this case, we easily obtain $\pi(1) = \ldots = \pi(j_0)$, $j_0 = \delta$, and $\pi(j) = \pi(j_0)g^{j-j_0}$ when $j > j_0$.

Case 3. $(g,u) \neq 1$.

Let $g = p_1^{\alpha_1} \ldots p_k^{\alpha_k}$ and $u = p_1^{\gamma_1} \ldots p_k^{\gamma_k}u_0$. Here, there is at least one γ_i which is positive and at least one $\gamma_{i'}$ less than $\alpha_{i'}$ $(1 \leq i,i' \leq k)$ and $(g,u_0) = 1$. First, put $t_1 = \Pi_{\alpha_i > \gamma_i} p_i^{\alpha_i-\gamma_i}$. Then, we have $r^{\pi(1)t_1} = (1 + ug^{\delta})^{t_1} = 1 + g^{\delta+1}u_1'$ with $u_1' \equiv u_1 (\mathrm{mod}\ g)$ where $u_1 = p_1^{\gamma_1'} \ldots p_k^{\gamma_k'}u_0$ with $\gamma_i > \gamma_i' (\equiv \gamma_i - \alpha_i) > 0$ (when $\alpha_i < \gamma_i$) and $\gamma_i' = 0$ (when $0 \leq \gamma_i \leq \alpha_i$). Next, put $t_2 = \Pi_{\alpha_i > \gamma_i'} p_i^{\alpha_i-\gamma_i'}$ and repeat the same argument. Then, after δ'-times of steps we have $(1+ug^{\delta})^{t_1 t_2 \ldots t_{\delta'}} = 1 + g^{\delta+\delta'}u_{\delta'}'$, with $u_{\delta'}' \equiv u_0$ (mod g), where δ' is equal to $\delta_0 \equiv \max(\gamma_1/\alpha_1, \ldots ,\gamma_k/\alpha_k)$ if δ_0 is an integer and is equal to $[\delta_0] + 1$ if δ_0 is not an integer. If we put $j_0 = \delta + \delta'$, then we have $\pi(j) = \pi(j_0)g^{j-j_0}$ when $j > j_0$, and we may write $\pi(j_0)$ as $\pi(1)t_1 t_2 \ldots t_{\delta'}$.

Lastly consider the case $(g,r) \neq 1$. In the equation (5) we have $(g_1,r) = 1$, so this case reduces to that of $(g,r) = 1$. We need only to rewrite g into g_1 in the results already obtained.

4. 1/N-uniform distribution.

We shall complete the proofs of Theorems 2 and 3 by making use of the concrete form of $\pi(j)$ obtained in the previous section.

Proof of Theorem 2. There exists a number $j_0 \geq 1$ such that if $j > j_0$ then, $\pi(j) = \pi(j_0)g^{j-j_0}$. Take this j_0. Since $r^{\pi(j_0)} \equiv 1$ (mod g^{j_0}), each of the g^{j-j_0} intervals in $[0,g^j)$: $I_1 = [0,g^{j_0})$, $I_2 = [g^{j_0},2g^{j_0})$, ... , $I_{g^{j-j_0}} = [g^j - g^{j_0},g^j)$ contains at most $\pi(j_0)$ numbers of $\{B_n^j\}_{1 \leq n \leq \pi(j)}$. But $\pi(j_0)g^{j-j_0}$ is equal to $\pi(j)$, each interval must contain just $\pi(j_0)$ numbers. If we interpret this into the unit interval, we obtain the result of Theorem 2.

Proof of Theorem 3. Take $g_0^j g_1^{j_0} = g^{j_0} g_0^{j-j_0} (\equiv \alpha)$ and devide the interval $[0, g^j)$ into $g_1^{j-j_0}$ intervals with length α: $I_1 = [0, \alpha)$, $I_2 = [\alpha, 2\alpha)$, ... , $I_{g_1^{j-j_0}} = [g^j - \alpha, g^j)$. Then, since the equation $r^{n+\pi} \equiv r^n \pmod{g_0^j g_1^{j_0}}$ is equivalent to the equation $r^\pi \equiv 1 \pmod{g_1^{j_0}}$ whose minimal solution is $\pi(j_0)$, each interval can contain at most $\pi(j_0)$ elements of $\{B_n^j\}_n$ in a period. Quite the same argument as in the proof of Theorem 2 enables us to have Theorem 3.

Remarks. By using Theorems 2 and 3, we can easily see that Theorem 4 holds. As is easily seen, $g | \pi(j)$ is a necessary condition for the relation (2) of Theorem 4 to hold for some j. In the case of $(g, r) \neq 1$, this is equivalent to $g_0 | \pi(1)$, as we can see in the calculations of $\pi(j)$.

5. Examples.

Ex. 1 If g is an odd prime, $(g, r) = 1$ and r is a primitive root of g^2, we have $j_0 = 1$, $\pi(j_0) = 1$ and $\pi(j) = (p-1)p^{j-1}$.

Ex. 2 $g = 10$ and $2 \leq r \leq 9$.

r	j_0	$\pi(1)$	$\pi(j_0)$	d_0	g_0	g_1
2	1	4	4	1	2	5
3	4	4	500	4	1	10
4	1	2	2	1	2	5
5	2	1	1	1	5	2
6	1	1	1	1	2	5
7	5	4	500	4	1	10
8	1	4	4	1	2	5
9	4	2	250	2	1	10

For convenience, we put $g_0 = 1$ and $g_1 = g$ when $(g, r) = 1$. d_0 in the table is the smallest number at which 1/N-uniform distribution holds. In cases of $r = 5$ and 6, as $g_0 \nmid \pi(1)$ the relation (2) of Theorem 4 does not hold. On the other hand, if $r = 2$, 4 and 8, using the formula for $\pi(j)$, we can show that (2) holds.

References

[1] V.I. Arnold and A. Avez: Problemes Ergodiques de la Mecanique Classique, Gauthier-Villars, Paris, 1967.

[2] Н.М. Коробов: О распределении знаков в периодических дробях, Мат. Сб. 89(131), (1972), 654-670.

[3] R.G. Stoneham: The reciprocals of integral powers of primes and

normal numbers, Proc. Amer. Math. Soc. 15 (1964), 200-208.

[4] R.G. Stoneham: On (j,ε)-normality in the rational fractions, Acta Arith. 16 (1970), 221-238.

[5] R.G. Stoneham: On absolute (j,ε)-normality in the rational fractions with applications to normal numbers, Acta Arith. 22 (1973), 277-286.

[6] R.G. Stoneham: On the uniform ε-distribution of residues within the periods of rational fractions with applications to normal numbers, Acta Arith. 22 (1973), 371-389.

NORMS OF GAUSSIAN SAMPLE FUNCTIONS

B.S.Cirel'son, I.A.Ibragimov and V.N.Sudakov

The object of this paper is a presentation and discussion of some results on the behavior of norms of Gaussian sample functions. Section 1 contains a general theorem concerning distributions of some nonlinear functionals on a linear space with a Gaussian measure. We suppose here that the functionals obey a variant of a "Lipschitz condition". For instance, norms of sample functions in Banach spaces or maximal values of sample functions (of course, if these norms and maximums are finite a.s. are among functionals satisfying this condition. For norms our inequalities may be considered as a further strengthening of Fernique's [2] and Marcus and Shepp's [5] results on almost Gaussian tails. The same inequalities, but in the case of more complicated functionals, are used in Section 2 to investigate in more detailes some analytical properties of distributions of norms and maximums. Theorem 4 of this section proves Dudley's hypothesis ([1] , p.197; see although Marcus and Shepp [5], p.435) which asserts that the distribution function of the norm has the density (except a possible atom at an essential infimum). Moreover, all possible discontinuities of this density are jumps; it has also Gaussian like tail near

As mentioned above, it is supposed, in all these theorems, that the norms under consideration are finite. Nevertheless, they are useful if one wishes to find some conditions for a norm to be finite. Namely, the inequalities of Section 1 are used in Section 3 to establish conditions under which sample functions of a stationary

Gaussian field belong to a functional space H_p^z of functions with prescribed differential properties.

Theorem 1 of Section 1 was proved by Cirel'son and Sudakov, other results of Sections 1 and 2 belong to Cirel'son; results of Section 3 were obtained by Ibragimov. But the paper in the whole is a result of collective work.

1. It is natural, from a general point of view, to consider a linear space with a Gaussian measure (E, γ) rather than a Gaussian process. To avoid unnecessary pathologic properties we restrict ourselves to linear spaces with a measure satisfying the following conditions (cf. Veršhic [7]):

(i) the measure γ is perfect;

(ii) there exists a countable system of linear γ - measurable functionals f_K on a linear subspace $E_1 \subset E$, $\gamma(E_1)=1$, which separates points of E_1 (i.e. $f_K(x)=0$ for all iff $x=0$).

Instead of (i) we could require (E, γ) to be a Lebesgue space (in the sence of Rohlin [9]).

Note that no topology in E is introduced.

Let F be the space of all linear measurable functionals on (E, γ). There exists a point $\bar{x} \in E$, barycentre of γ, such that, for all $f \in F$,

$$\int_E f(x)\gamma(dx) = f(\bar{x})$$

We may and will suppose $\bar{x}=0$. A centred Gaussian process (a parametric family of Gaussian random variables) defines a mapping of its parametric set into F.

The space F is a closed subspace of $L_2(\gamma)$ and so it inherits from $L_2(\gamma)$ the Hilbert space structure. The dual of F may be identified with the subspace $E_0 \subset E$ of vectors x_0 such that the shift operator $S_{x_0}: x \to x + x_0$ transforms γ into an equivalent measure. The space E_0 is a "coordinate-free" analogue of a Hilbert space with a reproducing kernel.

Definition 1. A measurable function φ on (E, γ) is said to satisfy Lipschitz condition with a constant C $(\varphi \in Lip(\gamma, C))$ if, for every $x_0 \in E_0$,

$$\operatorname*{ess\,sup}_{x \in E} |\varphi(x + x_0) - \varphi(x)| \leq C \|x_0\|_{E_0}$$

Define the distribution of a measurable functional φ as the measure μ on $\mathbb{R}$ such that $\mu(A) = \gamma\{x : \varphi(x) \in A\}$

Rather unexpectedly, it is possible to give a complete description of the class of the distributions of functionals $\varphi \in Lip(\gamma, C)$.

Definition 2. A probability measure μ on $\mathbb{R}$ is said to be of the class Lip $G(C)$ if may be represented as the image of the standard Gaussian measure on $\mathbb{R}$ under a mapping satisfying the usual Lipschitz condition with the constant C .

A visually equivalent definition can be given using the quantiles $J_\alpha(\mu)$, $0 < \alpha < 1$, defined by the condition

$$\mu(-\infty, J_\alpha(\mu)) \leq \alpha \leq \mu(-\infty, J_\alpha(\mu)$$

Clearly, $\mu \in Lip\, G(C)$ if and only if all its quantiles are correctly defined and

$$J_\alpha(\mu) - J_\beta(\mu) \leq C(J_\alpha(\gamma^1) - J_\beta(\gamma^1))$$

for $0<\beta<\alpha<1$; here γ^1 is a Gaussian measure on $\mathbb{R}$ with variance 1.

Theorem 1. Let (E,γ) be a linear space with a Gaussian measure and let $\varphi\in Lip(\gamma,C)$. Then the distribution of φ belongs to the class $Lip\,G(C)$.

Theorem 1 was proved in [10] . Here we give a few examples of functionals belonging to $Lip(\gamma,C)$.

1. Any linear functional $f\in F$ belongs to $Lip(\gamma,\|f\|_{L_2(\gamma)})$.

2. If $\varphi_1,\varphi_2,\ \dots\ ,\ \in Lip(\gamma,C)$ and $\varphi(x)=\sup_K f_K(x)<\infty$ a.s., then $\varphi\in Lip(\gamma,C)$.

3. If $f_1,f_2,\dots,\in F$ and $\varphi(x)=\sup_K f_K(x)<\infty$ a.s., then $\varphi\in Lip(\gamma,\sup_K\|f_K\|_{L_2(\gamma)})$ and the constant is the best possible.

4. Let $\xi(t,\omega)$ be a Gaussian process with continuous sample functions. Consider a Gaussian measure γ in the space $C[0,1]$, corresponding to the zero mean process $\xi(t,\omega)-m(t), m(t)=\mathbb{E}\xi(t,\cdot)$. The functional φ :

$$\varphi(x)=\sup_{t\in[0,1]}(x(t)+m(t)),\qquad x\in C[0,1]\ ,$$

belongs to the class $Lip(\gamma,C)$, $C^2=\sup_{t\in[0,1]}\mathbb{E}(\xi(t,\cdot)-m(t))^2$ (and hence by Theorem 1 the distribution of the random variable $\sup_{t\in[0,1]}\xi(t,\omega)$ belongs to the class $Lip\,G(C)$).

5. Let B be a separable Banach space and let γ be a Gaussian measure in this space, $\varphi(x)=\|x\|_B$ for $x\in B$; then

$$\varphi \in Lip(\gamma, C), \quad C^2 = \sup_{x^* \in B^*, |x^*| \leq 1} \int_B \langle x^*, x \rangle^2 \gamma(dx)$$

(B^* is the dual of B). Indeed, the norm in B can be represented as a supremum of a suitable countable set of linear functionals (this argument was used by Pettis [6]).

6. The previons assertion holds for some nonseparable Banach spaces: for $L_\infty[0,1]$, for spaces of Holder functions, for spaces H_p^z which will be used in Section 3. However, there exist (nonseparable) Banach spaces whose norms cannot be represented as the supremum of a countable set of linear functionals. Then the following fact may be useful.

<u>Lemma 1.</u> Let (E,γ) be a linear space with a Gaussian measure, N being a γ-measurable norm on E or, more generally, a convex positive gomogeneous (not necessarily positive) functional. Then there exist such $f_K \in F$ and $m_K \geqslant 0$ that $N(x) = \sup_K (f_K(x) + m_K)$ for almost all $x \in E$

<u>Sketch of the proof:</u> approximate N by "cylindrical" convex functionals f_K ; each of then approximate by the supremum of a finite set of functionals $f'_K(x) = f_K(x) + m_K$, $m_K \geqslant 0$ Joining these finite sets together, we obtain $N(x) = \overline{\lim_K} f'_K(x)$ a.s. To pass to then supremum starting from the upper limit we represent the latter as

$$\inf_{U \supset K_0} \sup_{t \in U} \xi(t, \omega),$$

where U runs through all relatively open subsets of the GB compact K , which is the closure of the set $\{f'_K : K = 1, 2, \dots\}$

in $L_2(\gamma)$; compact $K_0 \subset K$ is the set of all limit points of the sequence $\{f_K'\}$; ξ is the natural (isonormal) Gaussian process on K . Using the Itô and Nisio theorems about oscillation [4] , we can rewrite the last expression in the form

$$\sup_{t \in K_0} (\xi(t,\omega) + \frac{1}{2}\delta(t)),$$

where $\delta(\cdot)$ is the oscillation function. It is now easy to replace K_0 by a countable set.

Returning to Theorem 1, notice that its proof, given in [10], is not a purely probabilistic one; it is based on a known geometric theorem, giving the solution of the isoperimetric problem on an n - dimentional sphere. Theorem 1 hardly can be proved without using any geometrical considerations. But sometimes the following assertion, provable in a purely probabilistic way, can replace it.

Theorem 1 A. Let (E, γ) be a linear space with a Gaussian measure, $\varphi \in Lip(\gamma, C)$. Then the distribution of the functional φ coincides with that of the random veriable

$$m + w(\tau(\omega), \omega)$$

where $m = \int_E \varphi(x)\gamma(dx)$, ω are points of some probability space Ω , $w = w(t, \omega)$ is a Brownian motion adapted to a flow of σ -algebras $\{\mathcal{F}_t\}$; τ is an $\{\mathcal{F}_t\}$ - Markov time with the property $\tau(\omega) \le C^2$ a.s.

Proof. We shall consider only the finite-dimensional case. Let $E = \mathbb{R}^n$, and let γ be the standart n - dimensional Gaussian measure. Consider an n - dimensional Brownian motion

$W=W(s,\omega), s\geq 0, \omega\in\Omega$, and define its martingal

$$X(s,\cdot)=\mathbb{E}(\varphi(W(1,\cdot))|W(s',\cdot), s'\in[0,s])$$

for $s \in [0,1]$. Obsiously $X(0,\omega)=m$ and $X(1,\cdot)$ has the same distribution as φ . Let $\{T_s\}$ be the semigroup of operators, connected with the Markov process W . Then $X(s,\omega)=(T_{1-s}\varphi)(W(s,\omega))$. Ito's formula gives

$$dX_s=\langle dW_s,(grad\, T_{1-s}\varphi)(W_s)\rangle.$$

The function $T_{1-s}\varphi$ satisfies Lipschitz condition with the constant C , i.e. $|grad\, T_{1-s}\varphi|\leq C$ .Hence $(dX_s)^2\leq C^2ds$. Now we can introduce a new scale of time t such that $dt=|(grad\, T_{1-s}\varphi)(W_s)|^2ds$. With the new scale, $X(t,\omega)-m$ becomes a one-dimensional Brownian motion, and the nonrandom moment $s=1$ becomes a Markov time bounded by C^2.

<u>Corollary 1.</u> For every $u\geq 0$

$$\gamma\{x:\varphi(x)-m\geq uC\}\leq 2\int_u^\infty(2\pi)^{-1/2}\exp(-\frac{v^2}{2})dv.$$

Indeed,

$$\gamma\{x:\varphi(x)-m\geq uC\}=\mathbb{P}\{w(\tau)\geq uC\}\leq$$

$$\leq\mathbb{P}\{\max_{t\leq C^2} w(t)\geq uC\}=2\mathbb{P}\{w(C^2)\geq uC\}.$$

Of course, the same inequality holds for $\gamma\{x:\varphi(x)-m\leq -uC\}$.

<u>Corollary 2.</u> For every $u\geq 0$

$$\gamma\{x:\varphi(x)-M\geq uC\}\leq\int_u^\infty(2\pi)^{-1/2}\exp(-\frac{v^2}{2})dv$$

where M is the median.

This inequality will be used in Section 3. Note also that $|M-m| \leq (2\pi)^{-1/2} C$.

A question of interest is to investigate how essential is the Gaussian property of the measure γ for Theorem 1. We give a generalization of this theorem for a non-Gaussian case. We will not formulate here our results for infinite-dimensional spaces with measure; we will only give estimates which hold for every finite dimension.

Theorem 2. Let $\eta_1, \eta_2, \dots, \eta_n$ be independent random variables, each having the distribution from the class $Lip\, G(C_1)$, and let a function φ on $\mathbb{R}^n$ satisfy the Lipschitz condition with the constant C relative to the usual Euclidean metric on $\mathbb{R}^n$. Then the distribution of the random variable $\varphi(\eta_1, \dots, \eta_n)$ is of the class $Lip\, G(CC_1)$.

Proof. Represent η_K as $\chi_K(\zeta_K)$ where ζ_K are ortho-Gaussian, $\chi_K : \mathbb{R} \to \mathbb{R}$ have the Lipschitz property, and apply Theorem 1 to the composition $\varphi(\chi_1(\zeta_1), \dots, \chi_n(\zeta_n))$.

The uniform distribution on [0,1] belongs to the class $Lip\, G((2\pi)^{-1/2})$, hence Theorem 2 can be applied, and we arrive at the following purely geometric result.

Theorem 3. Let $G \in \mathbb{R}^n$ be an open subset with picewise smooth boundary ∂G, $(0,1)^n$ be the open -dimensional cube. Then

$$mes_{n-1}((0,1)^n \cap \partial G) \geq \exp(-\tfrac{1}{2} u^2),$$

where u is defined by the equality

$$mes_n((0,1)^n \cap G) = \Phi(u) = \int_{-\infty}^{u} (2\pi)^{-1/2} \exp(-\frac{v^2}{2}) dv,$$

mes_n denotes the n-dimensional volume, mes_{n-1} denotes the $(n-1)$ - dimensional surface area.(Using a more general definition of the "area", we can omit the requirement of smoothness).

Proof. Let us introduce, on $\mathbb{R}^n$, the function

$$\varphi(x) = dist(x, G) = \inf_{y \in G} |x-y|$$

satisfying the Lipschitz condition with constant 1. It follows from Theorem 2 that, for every $\varepsilon > 0$,

$$mes\{x \in (0,1)^n : \varphi(x) \leq \varepsilon\} \geq \Phi(u + \varepsilon\sqrt{2\pi}).$$

Subtracting the equality

$$mes\{x \in (0,1)^n : \varphi(x) = 0\} = \Phi(u),$$

dividing by ε and letting $\varepsilon \to +0$, we come to the conclusion of the theorem.

Remark. The expression $exp(-\frac{1}{2}u^2)$ on the right can probably be increased. But for no u it can be larger that

$$2\sqrt{3}(2\pi)^{-1/2} exp(-\frac{1}{2}u^2)$$

as the following example shows:

$$G = \{(x_1, \dots, x_n) : x_1 + \dots + x_n < \frac{1}{2\sqrt{3}} u\sqrt{n}\}$$

with n large enough.

Conditions of Theorem 2 demand that the distribution of every η_k belongs to $Lip\ G$. Roughly speaking, this means that the distribution decreases fast at infinity and has no downfalls in the finite domain. These conditions are essential as the following two examples show.

Example 1. Let η_k be independent and identically distributed random variables and let $\mathbb{P}\{\eta_1 > a\} = a^{-p}$ for every a large enough and for some $p \in (0,\infty)$. Then the distribution of the random variable $max(\eta_1,\dots,\eta_n)$ not only "leaves for" $+\infty$, but also disperses in the sence that, for every α, $\beta: 0<\beta<\alpha<1$, the distance between the α- quantile and the β - quantile tends to ∞ when $n\to\infty$.

Example 2. Let η_K be independent and identically distributed and $\mathbb{P}\{\eta_1 = 0\} = \frac{1}{2}$, $\mathbb{P}\{\eta_1 = 1\} = \frac{1}{2}$. Define

$$\varphi(x_1, \dots, x_n) = (x_1 + \dots + x_n - \frac{n}{2})^{1/3}$$

if every x_k is equal to 0 or 1, and let φ be not determined in other cases. The function φ is Lipschitz with constant $2^{2/3}$ but the distribution of the random variable $\varphi(\eta_1,\dots\eta_n)$ disperses as $n\to\infty$. Note that φ can be extended to the whole $\mathbb{R}^n$ without changing the Lipschitz constant. For instance, by putting

$$\varphi(y) = sup(\varphi(x) - 2^{2/3}|y-x|)$$

for $y \in R^n$, where the supremum is taken over all $x=(x_1,\dots,x_n)$ such that $x_K = 0$ or 1 .

Note also that the inequalities for the ratioes of quantiles of norms are much more stable in case of deviation from the Gaussian case than those for differences of quantiles of Lipschitzian functionals; see, for instance[3], [11].

2. In the present Section, we shall deal with functionals φ of the type

$$\varphi(x)=\sup_k(f_k(x)+m_k) \quad \text{on} \quad (E,\gamma) \tag{1}$$

only. Here f_k are linear measurable functionals and $m_k \geq 0$. It has been mentioned (see Lemma 1) that the norm of a Gaussian random element in a normed space may be represented in the form (1). In this case $m_k=0$ for separable spaces; in the opposite case one has to assume that the norm is measurable.

Other useful examples of functionals (1) are maximums (or supremums) of Gaussian pro cesses. As before, we suppose that all the functionals under concideration are finite a.s. Note that, adding if necessary a positive constant we can treat also the case $-\infty < \inf m_k < 0$ The case $\inf m_k = -\infty$ has not been investigated yet.

Let F denotes the distribution function of φ, i.e. $F(a) = \gamma\{x : \varphi(x) < a\}$.

Theorem 4. a) The distribution function F is continuous everywhere exept, may be, at the only point $a_0 = \inf\{a : F(a) > 0\}$ (it is possible that $a_0 = -\infty$ and then F is continuous on the whole real line). The function F is absolutely continuous on (a_0, ∞) and its derivative F' is defined and continuóu everywhere exept, may be, at a countable set of points where F' has jumps. If at F' has a jump, then $F'(a-0) > F'(a+0)$.

b) The derivative F' has bounded variation on intervals $[a,\infty)$, $a > a_0$ The upper second derivative

$$D^+F'(a) = \overline{\lim_{h\to 0}} \frac{F'(a+h) - F'(a)}{h}$$

is bounded from above on intervals $[a,\infty]$, $a>a_0$

c) For any $\varepsilon>0$ and a such that $F(a-\varepsilon)>0$,

$$F'(a)\le 6\,\frac{\max(a,\varepsilon)}{\varepsilon^2}\ln\frac{8}{F(a-\varepsilon)}\quad, \tag{2}$$

$$D^+F'(a)\le 30\left(\frac{\max(a,\varepsilon)}{\varepsilon^2}\ln\frac{8}{F(a-\varepsilon)}\right)^2 \tag{3}$$

d) If $\sigma=\sup\limits_{K}\|f_K\|_{L_2(\gamma)}>0$, then there exists $c\ge 0$ such that

$$1-F(a)=\exp\left(-\frac{a^2}{2\sigma^2}+ca+o(a)\right)\ , \tag{4}$$

$$F'(a)=\exp\left(-\frac{a^2}{2\sigma^2}+ca+o(a)\right)\ , \tag{5}$$

$$D^+F'(a)\le\exp\left(-\frac{a^2}{2\sigma^2}+ca+o(a)\right). \tag{6}$$

as $a\to+\infty$

e) If a function ε satisfies the conditions $\varepsilon(a)\to+0$ and

$a\varepsilon(a)\to+\infty$ as $a\to+\infty$, then

$$\begin{aligned}\frac{F'(a)}{1-F(a-\varepsilon(a))}&\le(1+o(1))\frac{\frac{1}{\sigma}\varphi'(\frac{a}{\sigma})}{1-\varphi(\frac{a-\varepsilon(a)}{\sigma})}=\\&=(1+o(1))\frac{a}{\sigma^2}\exp\left(-\frac{a\varepsilon(a)}{\sigma^2}\right),\qquad a\to+\infty\ ,\end{aligned} \tag{7}$$

$$\begin{aligned}\frac{D^+F'(a)}{1-F(a-\varepsilon(a))}&\le(1+o(1))\frac{\left|\frac{1}{\sigma^2}\varphi''(\frac{a}{\sigma})\right|}{1-\varphi(\frac{a-\varepsilon(a)}{\sigma})}=\\&=(1+o(1))\frac{a^2}{\sigma^4}\exp\left(-\frac{a\varepsilon(a)}{\sigma^2}\right),\qquad a\to+\infty\ .\end{aligned} \tag{8}$$

Here $\varphi(u)=(2\pi)^{-1/2}\int\limits_{-\infty}^{u}\exp\left(-\frac{v^2}{2}\right)dv$

f) If, in addition, for all K, $\|f_K\|_{L_2(\gamma)}=\sigma$ (it holds for the supremum of stationary processes), then F is continuous everywhere and differentiable on (a_0,∞). The derivative F' is continuous on (a_0,∞), has a finite limit as $a\to a_0+0$ and satisfies the Lipschitz

condition on intervals $[a,\infty), a>a_0$. In this last case one may put
condition on intervals .In this last case one may put
the sign " $=$ " instead of " $\leq$ " in (7).

The proof of Theorem 4 is published in [12] . Here is a very short sketch of it.

First, it is sufficient to find "good" estimates of F' and F'' when the space E is finite-dimensional and φ is a smooth convex function on E . Clearly, one can represent $F'(a)$ and $F''(a)$ as some integrals with the weight $exp(-\frac{1}{2}|x|^2)$ over the convex surface $S_a=\{x\in E : \varphi(x)=a\}$. Consider the partition of the domain $W_a=\{x\in E : \varphi(x)\geq a\}$ generated by half-lines drawn at any point of the surface S_a and orthogonal to it. The simple inequality

$$exp(-\tfrac{1}{2}u^2)\leq(1+max(u,0))\int_u^\infty exp(-\tfrac{v^2}{2})\,dv$$

gives us an estimate of the Gaussian density at the initial point of a half-line by the integral over the whole half-line and then an estimate of the integral over S_a by an appropriata integral over W_a The key moment of the proof is the estimation of the integral over W_a by $F(a-\varepsilon)$. This part of the proof is based on Theorem 1. Namely, let us consider the function

$$\psi(x)=dist(x,V_{a-\varepsilon}),\ V_{a-\varepsilon}=\{y\in E : \varphi(y)\leq a-\varepsilon\}.$$

This function satisfies the Lipschitz condition. If $F(a-\varepsilon)$ is not too small, Theorem 1 allows us to assert that the integral of ψ over W_a is not too large. Now, if ε is not too small, one estimates the integral in question over W_a by the integral of φ . That gives the necessary estimates of $F'(a)$. The way of estimating $F''(a)$ is analoqous but a little more complicated.

It is an open problem whether $1-F(a)$ and $F'(a)$ have an asymptotic (not only logarithmical) formulas near $+\infty$ depending on a finite number of parameters only. Does an example exist which containes infinitely many independent parameters and the behavior of $F'(a)$ at different points depends on different parameters (for example, the κ-th parameter has strong influence on $F'(2^\kappa)$ and very weakly affects $F'(2^\ell)$, $\ell \neq \kappa$) ?

3. Let $\xi(t,\omega), t \in \mathbb{R}^n$, be a stationary Guassian process with the correlation function $R(t)$ and the spectral measure F . In other words,

$$R(t-s) = \mathbb{E}\,\xi(t)\xi(s),$$

$$R(t) = \int_{\mathbb{R}^n} e^{i(t,\lambda)} F(d\lambda),$$

where $(t,\lambda) = \sum_1^n t_i \lambda_i$ if $t = (t_1, \dots, t_n)$, $\lambda = (\lambda_1, \dots, \lambda_n)$.

In this Section we shall present a result on sample behavior of ξ with a proof heavily based om Theorem 1.

Let us first define functional spaces $H_p^z(G)$ which will be used in stating the result.

Let G be an open subset of $\mathbb{R}^n$, $1 \le p \le \infty$, $z > 0$. A pair (κ, ρ) of nonnegative integers κ, ρ will be called admissible (with respect to z) if $\kappa > z - \rho > 0$.

For any $A \subset \mathbb{R}^n$ and $h \in \mathbb{R}^n$, denote by A_h the set

$\{x : x + \tau h \in A, \quad 0 \leq \tau \leq 1\}$.

Definition of the spaces $H_p^r(G)$ (see [8], p.189). The space $H_p^r(G)$ consists of all functions f on G such that

(i) $f \in L_p(G)$;

(ii) the generalized derivatives $f^{(s)}$ of order $s = (s_1, \dots, s_n), \sum s_i = \rho$ are defined and satisfy the following inequalities:

$$\|\Delta_h^k f^{(s)}\|_{L_p(G_{kh})} \leq C |h|^{r-\rho} \tag{9}$$

where C does not depend on h. Here (k, ρ) is an admissible pair and, for any φ,

$$\Delta_h^k \varphi(x) = \sum_{\ell=0}^{k} (-1)^{\ell+k} \binom{k}{\ell} \varphi(x + \ell h).$$

The $H_p^r(G)$ is a normed space with the norm

$$\|f\|_{H_p^r(G)} = \|f\|_p + C_f ,$$

where

$$\|f\|_p = \|f\|_{L_p(G)} = \left(\int_G |f(x)|^p dx\right)^{1/p} ,$$

$C_f = \inf C$ and the infimum is taken over all C from (9). This norm depends on the pair (k, ρ) but all such norms are equivalent for all admissible pairs.

It is well-known that the process ξ may be represented in the form

$$\xi(t) = \int_{R^n} e^{i(t,\lambda)} Z(d\lambda) ,$$

$$E|Z(d\lambda)^2| = F(d\lambda)$$

and the values of the random Gaussian measure $\mathcal{Z}$ on disjoint sets are independent. Let Γ_K denotes the cube $\{\lambda : \max_i |\lambda_i| \leq 2^K\}$ and

$$\Delta_K = \Gamma_{K+1} \setminus \Gamma_K \quad (K = 1, 2, \dots), \quad \Delta_0 = \Gamma_1 ,$$

Define the independent stationary Gaussian processes

$$\xi_K(t) = \int_{\Delta_K} e^{i(t,\lambda)} \mathcal{Z}(d\lambda).$$

Their correlation functions

$$R_K(t) = \int_{\Delta_K} e^{i(t,\lambda)} F(d\lambda).$$

Denote by Q the cube $\{t : 0 \leq t_i \leq 1\}$ and define the numbers $\alpha_{K,p}$ as follows:

$$\alpha_{K,p} = \sup_{\varphi} \int_Q \int_Q R_K(t-s)\varphi(t)\overline{\varphi(s)}\,dt\,ds ,$$

where the supremum is taken over all complex-valued functions φ on Q such that

$$\|\varphi\|_q = \|\varphi\|_{L_q(Q)} = 1 , \qquad q = \frac{p}{p-1} .$$

Now we state the main result of this Section.

Theorem 5. Almost all sample functions of the stationary Gaussian random process ξ belong to the space $H_p^z(Q) = H_p^z$, $1 \leq p < \infty$, if and only if

(i) $\sup_K 2^{zK} R_K^{1/2}(0) < \infty$;

(ii) $\sum_K \exp(-B \cdot 2^{-2zK} \alpha_{K,p}^{-1}) < \infty$

for all large B .

This theorem follows from Theorem 1 and the next result.

Theorem 6. Almost all sample functions of the stationary Gaussian random process ξ belong to the space H_p^{τ}, $1 \leq p \leq \infty$, if and only if

$$P\{\sup_{k} 2^{\tau k} \|\xi_k\|_p < \infty\} = 1$$

and hence if

$$\sum_{k} P\{2^{\tau k} \|\xi_k\|_p > B\} < \infty \qquad (10)$$

for all large B.

Will a function $f \in L_p(\mathbb{R}^n)$ belong to the space $H_p^{\tau}(\mathbb{R}^n)$ or not, depends on the behavior of its Dirichlet's or Vallee-Poussin sums (see [8], pp.367-378). The last theorem reminds these results. Its proof will be omitted and published elsewhere.

We show now how to derive Theorem 5 from the last result and results of Section 1.

To prove sufficiency, define

$$b_1^2 = \sup_{k} 2^{2\tau k} R_k(0) = \sup_{k} 2^{2\tau k} E|\xi_k(t)|^2,$$

$$b^p = 2 b_1^p \frac{2^{\frac{p}{2}+1}}{\sqrt{\pi}} \Gamma\left(\frac{p+1}{2}\right)$$

Then

$$P\{2^{\tau k} \|\xi_k\|_p \geq b\} \leq \frac{2^{\tau k p} E\|\xi_k\|_p^p}{b^p} \leq \frac{1}{2}.$$

Using Corollary 2 from Section 1 and taking into account Example 5 preceeded by Theorem 1, we conclude that, for all $B > 2b$,

$$P\{2^{2k}\|\xi_k\|_p > B\} \le 1 - \Phi\left(\frac{B}{2^{2k+1}\sqrt{\alpha_{k,p}}}\right),$$

where as above

$$\Phi(u) = (2\pi)^{-1/2}\int_{-\infty}^{u} \exp\left(-\frac{v^2}{2}\right) dv .$$

Hence

$$\sum_k P\{2^{2k}\|\xi_k\|_p > B\} \le \sum_k \exp\left(-\frac{1}{8}\,\frac{B^2}{2^{4k}\alpha_{k,p}}\right) < \infty$$

and the inclusion $\xi \in H_p^2$ follows from Theorem 6.

To prove necessity, note that

$$E\|\xi_k\|_1 = \int_Q E|\xi_k(t)|dt = \left(\frac{2}{\pi} R_k(0)\right)^{1/2}$$

and

$$(E(\|\xi_k\|_1 - E\|\xi_k\|_1)^2)^{1/2} =$$

$$= \left(E\left(\int_Q (|\xi_k(t)| - E|\xi_k(t)|)dt\right)^2\right)^{1/2} \le$$

$$\le \int_Q (E(|\xi_k(t)| - E|\xi_k(t)|)^2)^{1/2} dt = \left(\left(1 - \frac{2}{\pi}\right)R_k(0)\right)^{1/2}.$$

Using the simple inequality

$$P\{\|\xi_k\|_1 - E\|\xi_k\|_1 > -u\} \ge \frac{u^2}{u^2 + E(\|\xi_k\|_1 - E\|\xi_k\|_1)^2} \qquad (u > 0)$$

we obtain $P\{\|\xi_k\|_1 > \frac{1}{3}\sqrt{R_k(0)}\} \ge \frac{1}{3}$. It follows from the last inequality and (1o) that for large k.

$$P\{\|\xi_k\|_p > 2^{-2k}B\} < \frac{1}{3} \le P\{\|\xi_k\|_p > \frac{1}{3}\sqrt{R_k(0)}\}$$

and hence $\overline{\lim_{K}}\, 2^{2K}\sqrt{R_K(0)} \leq 3B$. To prove)ii), put

$$\alpha_K(\varphi) = \mathbb{E}\left|\int_Q \xi_K(t)\varphi(t)dt\right|^2 =$$

$$= \int_Q\int_Q R_K(t-s)\varphi(t)\overline{\varphi(s)}\,dt\,ds .$$

We have then for $B>0$:

$$\mathbb{P}\{2^{2K}\|\xi_K\|_p > B\} =$$

$$= \mathbb{P}\{\sup_{\|\varphi\|_q\leq 1} 2^{2K}\left|\int_Q \xi_K(t)\varphi(t)dt\right| > B\} \geq$$

$$\geq \sup_{\|\varphi\|_q\leq 1} \mathbb{P}\{2^{2K}\left|\int_Q \xi_K(t)\varphi(t)dt\right| > B\} =$$

$$= \sup_{\|\varphi\|_q\leq 1} 2\left(1-\Phi\left(\frac{B}{2^{2K}\sqrt{\alpha_K(\varphi)}}\right)\right) \geq$$

$$\geq \sup_{\|\varphi\|_q\leq 1} \frac{1}{2} exp\left(\frac{B^2}{2^{2\cdot 2K}\alpha_K(\varphi)}\right) =$$

$$= \frac{1}{2} exp\left(-\frac{B^2}{2^{2\cdot 2K}\alpha_{K,p}}\right).$$

Now (ii) follows from Theorem 6.

It is worth noting that the conditions (ii) for different p are generally speaking, not independent Namely, if $1\leq p\leq 2$, then

$$\left(\frac{\pi}{5}\right)^{n}\alpha_{K,2} \leq \alpha_{K,p} \leq \alpha_{K,2}$$

and hence ξ belongs or does not belong to H_p^2 for all $1\leq p\leq 2$ simultaneously.

The situation is different in the case $p\geq 2$. Denote by w a

Wiener process with $E|dw(\lambda)|^2 = d\lambda$ and consider the following stationary Gaussian process ξ on the line:

$$\xi(t) = \int_{-1}^{1} e^{i\lambda t}\, dw(\lambda) +$$

$$+ \sum_{K=0}^{\infty} \beta_K \Big(\int_{2^K}^{2^{K+1}} e^{i\lambda t}\, dw(\lambda) + \int_{-2^{K+1}}^{-2^K} e^{it\lambda}\, dw(\lambda) \Big),$$

$$\beta_K = \frac{1}{\sqrt{\ln K}}\, 2^{-K\tau}\, 2^{-(K+1)}\, 2^{\frac{2K}{p}}, \qquad p \geq 2.$$

An easy calculation shows that

$$\alpha_{K,p} \leq 2\pi \frac{2^{-2K\tau}}{\ln K}, \quad \alpha_{K,p'} \geq \frac{1}{8} \cdot \frac{2^{-2K\tau}}{\ln K} \cdot 2^{2K(\frac{1}{p} - \frac{1}{p'})}.$$

Hence, if $p' > p$, then

$$\sum \exp\Big(-\frac{B}{2^{2\tau K}\, \alpha_{K,p}}\Big) < \infty \qquad \text{for } B > 4\pi,$$

$$\sum \exp\Big(-\frac{B}{2^{2K\tau}\, \alpha_{K,p'}}\Big) = \infty \qquad \text{for all } B,$$

and $\xi \in H_p^{\tau}$ but $\xi \notin H_{p'}^{\tau}$.

REFERENCES

1. R.M.Dudley, Sample functions of the Gaussian proces, Ann.Probability, 1,1,1973,pp.66-103.
2. X.Fernique, Régularité de processus gaussiens, Invent.Math., 12, 4, 1971,pp.304-320.
3. J.Hoffmann-Jørgensen, Sums of independent Banach space valued random variables, Studia Math., 52,2,1974,pp.159-186.
4. K.Itô, M.Nisio, On the oscillation functions of Gaussian processes,Math.Scand., 22, 1, 1968,pp.209-223.
5. M.B.Marcus and L.A.Shepp, Sample behavior of Gaussian processes, Proc.Sixth Berkeley Symp.Math.Statist. Probability,Vol.2, 1972,pp.423-441,Univ.Calif.Press.
6. B.J.Pettis, On integration in vector spaces,Trans.Amer.Math. Soc., 44, 2, 1938,pp.277-304.
7. A.M.Veršik, The axiomatics of measure theory in linear spaces, Dokl.Akad.Nauk SSSR, 178, 2,1968,pp.278-281.English translation in:Soviet Math.Dokl., 9, 1,1968,pp.68-72.
8. S.M.Nikol'skii, Approximation of several variables and embedding theorems, Moscow, 1969(Russian)
9. V.A.Rohlin,On the fundamental ideas of measure theory,Mat.Sb., 25(67),1,1949,pp.107-150. English translation in:Amer. Math.Soc.Translation,71(1952).
10. V.N.Sudakov,B.S.Cirel'son.Extremal properties of half-spaces for spherically symmetrical measures, Zap.Naucn. Sem.Leningrad.Otdel.Mat.Inst.Steklov (LOMI),

41, 1974, pp.14–24. (Russian)

11. B.S.Cirel'son, Some properties of lacunary series and Gaussian measures connected with uniform variants of the Egoroff and Lusin properties, Teor.Verojatnost. i Primenen., 20, 3, 1975, pp.664–667 (Russian).

12. B.S.Cirel'son, The density of the distribution of the maximum of a Gaussian process, Teor.Verojatnost. i Primenen., 20, 4, 1975, pp.567–575 (Russian).

Steklov Mathematical Institute
Academy of Sciences of the USSR
Leningrad

ON A NEW APPROACH TO MARKOV PROCESSES

E.B.Dynkin

Foundations of the general theory of Markov processes are rather cumbersome. It seems that the situation can be improved by a new approach, proposed in [1]-[8]. We outline here the main ideas and consider examples of their applications to concrete problems.

1. General discussion

We begin with discussing some distinctive features of our approach.

1. The general non-homogeneous theory precedes the homogeneous one.

This is natural from the logical point of view: in the homogeneous case additional one-parameter groups of shifts are involved and the invariance with respect to a choice of time scale is destroyed. We treat the time-homogeneous case with the help of advanced inhomogeneous theory. In this way, many results are more simply and completely formulated; besides the intuitive picture, obscured by the technique of Laplace transforms in the homogeneous case, becomes more transparent.

2. All the theory is invariant with respect to time reversion.

Traditionally, it is assumed that, to each pair t, x, there corresponds a conditional probability distribution, $P_{t,x}$, of the future after t if the state at time t is x, but an analogous conditional probability distribution $P^{t,x}$ of the past before t is not usually considered. In our theory, both forward and backward transition propabilities play equal roles.

Usually processes with a fixed birth-time α and a random death-time β are considered. We assume that both α and β are random,

3. The regularity properties of a process are formulated not in topological terms but in terms of behaviour of some real functions along almost all the paths.

There exist two natural classes of real-valued functions associated with a process. The first one is defined with the help of the forward transition probabilities by the formula

$$f(t,x) = \begin{cases} P_{t,x}(A) & \text{for } t < u \\ 0 & \text{for } t \geq u \end{cases}$$

(here A is an event observable after time u). Functions of this class will be called right base functions. The dual class of left base functions is defined by the formula

$$f(t,x) = \begin{cases} P^{t,x}(A) & \text{for } t > s \\ 0 & \text{for } t \leq s \end{cases}$$

where A is an event observable before time s.

We call a process right regular if every right base function is right-continuous along almost all the paths[*)]. Replacing "right" with "left", we obtain the definition of left regularity. A process is two-sided regular if it is simultaneously right- and left-regular.

*) Note an obvious analogy between this concept and Meyer's "hypotèses droites".

All the probabilistic theory of potential can be developed on the basis of these definitions. In particular, a traditional pair of dual "nice" processes can be replaced with one two-sided regular process.

A remark on topologies. Using an appropriate countable family of right base functions, it is possible to provide the state space of a right regular process with a compact topology such that almost all the paths are continuous from the right. It can be done in many different ways with different exceptional sets of paths. It is reminiscent of the situation with coordinate systems: there exist many equivalent systems and we have no reason to prefer any special one.

Diffusion processes are an example of regular processes. Another example: Let x_t be a stationary right-continuous Markov process with two states at each time t ; then x_t is right-regular, x_{t-} is left-regular and (x_{t-}, x_t) is two-sided regular. A path of x_t and the corresponding path of (x_{t-}, x_t) are represented in Fig.1 (here $\sigma_1, \sigma_2, \ldots$ are

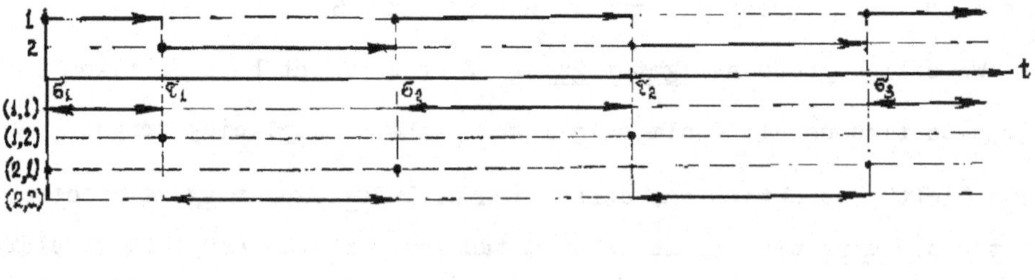

Figure 1

the times when x_t jumps from 2 to 1 and $\tau_1, \tau_2, \ldots$ are the times when it jumps from 1 to 2).

Generally, let x_t be a Markov process in a metric space and let

$P_{t,x}\{x_u \in \Gamma\}$ be continuous for $t < u$ and $P^{t,x}\{x_s \in \Gamma\}$ be continuous for $t > s$. The process x_t is regular if it is continuous. If x_t has one-side limits x_{t-} and x_{t+}, then the process (x_{t-}, x_{t+}) is regular.

2. Markov representations of stochastic systems

Our starting point is quite different from the traditional one. It is usual to start from the state x_t of a system at time t. But every real observation takes a certain time interval and "the state space at time t" is a mathematical abstraction which can be introduced for describing the same physical phenomenon in many different ways.

We start from the set $\mathcal{F}(I)$ of events observable in the time interval I. In precise mathematical terms, we assume that a σ-algebra $\mathcal{F}(I)$ on a fixed space Ω is associated with any open interval I of the real line R and that the following conditions are satisfied:

A. $\mathcal{F}(I_1) \subseteq \mathcal{F}(I_2)$ for $I_1 \subseteq I_2$.

B. If $I_n \uparrow I$, then $\mathcal{F}(I)$ is generated by the union of $\mathcal{F}(I_n)$

C. Let $I_n \downarrow I$. Let P_n be a probability measure on $\mathcal{F}(I_n)$ and $P_n = P_{n-1}$ on $\mathcal{F}(I_{n-1})$. Then there exists a measure P on $\mathcal{F}(I)$ which coincides with P_n on $\mathcal{F}(I_n)$, $n = 1, 2, \ldots$.

We say that $\{\mathcal{F}(I), P\}$ is a stochastic system if $\{\mathcal{F}(I)\}$ is a collection of σ-algebras satisfying the conditions A-C and P is a probability measure on a σ-algebra $\mathcal{F}$ which contains all the $\mathcal{F}(I)$ *)

*) A part of the theory can be developed without condition C.

The following construction explains the connection of this concept with the traditional theory. Let a Borel measurable space E_t **) be given for each $t \in R$ and let a probability measure $m_{t_1,\dots,t_n}$ on the product space $E_{t_1} \times \dots \times E_{t_n}$ be associated with every set $t_1, \dots, t_n$. Under the usual consistency conditions, a probability measure P on the space of all the functions $\omega(t) \in E_t$, $t \in R$, can be constructed with the help of the measures $m_{t_1,\dots,t_n}$. Let $\mathcal{F}(I)$ be a σ-algebra on Ω generated by the mappings $x_t(\omega) = \omega(t)$ for $t \in I$. The collection $\{\mathcal{F}(I), P\}$ is a stochastic system.

Starting from a stochastic system $\{\mathcal{F}(I), P\}$, we want to define state spaces E_t and paths $x_t(\omega)$ in the most proper way.

We require that:

a) E_t are Borel measurable spaces;

b) $x_t(\omega)$ is $\mathcal{F}(I)$ - measurable for $t \in I$;

c) σ- algebras $\mathcal{F}_{<t} = \mathcal{F}(-\infty, t)$ and $\mathcal{F}_{>t} = \mathcal{F}(t, +\infty)$ are conditionally independent given x_t, i.e.

$$P(AB \mid x_t) = P(A \mid x_t) P(B \mid x_t) \text{ a.s. } P \text{ for } A \in \mathcal{F}_{<t},\ B \in \mathcal{F}_{>t}\ .$$

Processes x_t satisfying conditions a)-c) are called <u>Markov representations</u> of the stochastic system $\{\mathcal{F}(I), P\}$. Our problem is to select, among all such representations, a representation with "the finest" paths. We shall summarize the solution of this problem which was obtained in [4]-[5].

**) We say that a measurable space $(E, \mathcal{B})$ is a Borel space if it is isomorphic to a space $(E_1, \mathcal{B}_1)$ where E_1 is a Borel subset in a complete separable metric space X and $\mathcal{B}_1$ is the collection of all Borel subsets of X contained in E_1.

A natural preliminary question is: which stochastic systems have Markov representations? The answer is: Markov representations exist if and only if :

(i) The σ - algebras $\mathcal{F}_{<t}$ and $\mathcal{F}_{>t}$ are conditionally independent relative to $\mathcal{F}\{t\} = \mathcal{F}(t-,t) \vee (\mathcal{F}(t,t+)$ *).

(ii) The σ - algebra $\mathcal{F}\{t\}$ is generated by a countable family of sets and by sets of measure 0 .

Let x_t be a Markov representation of $(\mathcal{F}(I),P)$ and let a probability measure $P_{t,x}$ on $\mathcal{F}_{>t}$ be given for every $t\in R, x\in E_t$ such that, for all $A\in\mathcal{F}$,

$$P(A|x_t)=P_{t,x_t}(A) \text{ a.s. } P$$

and

$$P_{s,x}(A|x_t)=P_{t,x_t}(A) \text{ a.s. } P_{s,x}, \quad s<t, x\in E_s .$$

The measures $P_{s,x}$ are called <u>forward transition probabilities</u> for $(\mathcal{F}(I),P)$. <u>Backward transition probabilities</u> are defined analogously. The existence of transition probabilities can be proved under the following condition:

AC. The two-dimensional distributions $m_{st}(B)=P\{(x_s,x_t)\in B\}$ are absolutely continuous with respect to the product $m_s\times m_t$ <u>of the corresponding</u> one-dimensional distributions**).

*) The σ - algebra $\mathcal{F}(t,t+)$ is the intersection of $\mathcal{F}(t,u)$ for all $u>t$; similarly $\mathcal{F}(t-,t)=\bigcap_{s<t}\mathcal{F}(s,t)$, $\mathcal{F}_{<t+}=\bigcap_{u>t}\mathcal{F}_{<u}$, etc.

**) In the case of the random birth-time α and death-time β , the measures m_{st} and m_t have to be defined as follows $m_{st}(B)=P\{\alpha<s,(x_s,x_t)\in B, t<\beta\}$ for $s<t$, $m_t(C)=P\{\alpha<t<\beta, x_t\in C\}$.

It is proved in [5] , that :

1. If a stochastic system has a Markov representation with forward (backward) transition probabilities, it has also a right (left) regular Markov representation.

2. If a stochastic system has a right regular and a left regular Markov representation, it has also a two-sided regular Markov representation.

3. A two-sided regular Markov representation is determined essentially uniquely. In particular, the class $\mathcal{C}_1$ of real-valued functions on $R \times \Omega$ which are indistinguishable[*)] from functions of the form $\xi_t(\omega) = f(t, x_t(\omega))$ with measurable $f(t,x)$ is uniguely determined. (The measurable structure in the state space $\mathcal{E} = \bigcup E_t$ is generated by the base functions. In the case of a two-sided regular x_t , $\mathcal{E}$ is a Borel space.)

Let $\xi_t(\omega)$ be real-valued function on $R \times \Omega$. We say that ξ is <u>solid</u> if it is measurable with respect to the σ-algebra generated by all evanescent sets and all products $(s,t) \times A$ where $s < t \in R$ and $A \in \mathcal{F}_{<s} \vee \mathcal{F}_{>t}$.

We call ξ <u>central</u> if it is solid and if the process ξ_t is well-measurable in the sence of Meyer with respect to $\mathcal{A}_t = \mathcal{F}^P_{<t+}$

*) We say that $\xi_t(\omega)$ and $\xi'_t(\omega)$ are <u>indistinguishable</u> if $P\{\xi_t(\omega) = \xi'_t(\omega)$ for all $t \in (\alpha, \beta)\} = 1$.

A function is called <u>evanescent</u> if it is indistinguishable from 0. The same word is applied to the set whose indicator is an evanescent function.

and with respect to $\mathcal{A}^t = \mathcal{F}^P_{>t-}$ (forward and backward well-measurability)[x)]

A class $\mathcal{C}$ of central functions is associated with every stochastic system. For a system which has a two-sided regular Markov representation, this class coincides with the class $\mathcal{C}_1$ described in n° 3.

3. Additive functionals

In this section, we assume that the conditions A, B and (i) of Section 2 are satisfied (the conditions C and (ii) are unnecessary).

A finite additive functional A of a stochastic system $(\mathcal{F}(I), P)$ is a measure $A(dt)$ on the real line R which depends on ω in such a way that:

a) for all ω, $A(dt)$ is concentrated on the interval (α, β);

b) for every I, $A(I)$ is solid and $\mathcal{F}^P(I)$ - measurable;

c) for all $s < t \in R$, $P\{\alpha < s < t < \beta,\ A(s,t) = \infty\} = 0$.

We say that A is a σ - functional if $A = \sum_1^\infty A_n$ where A_n are finite additive functionals.

A measure μ will de called a σ- measure if it can be represented as a denumerable sum of finite measures.

Starting from a σ-functional A, we define a measure μ on

---

x) We assume that the measure P is complete and denote by B^P the minimal σ- algebra which contains B and all the sets of measure 0 .

the σ-algebra C of all the central sets by the formula

(1) $$\mu(C) = P\int \mathbf{1}_C(t,\omega)\, A(dt) \quad . \qquad ^{*)}$$

We call μ the spectral measure of the σ- functional A . Obviously, μ charges no evanescent set. It is proved that μ is a σ- measure and that A is uniquely determined by μ .[**)] The functional A is continuous (i.e. $A\{t\}=0$ for all t a.s.) if and only if μ charges no set C with the property: $P\{(t,\omega)\in C$ for an uncountable set of $t\in(\alpha,\beta)\}=0$.

Now assume that the following condition (which is stronger than the condition (i) of Section 2) is satisfied:

(I) For every $s<t\in R$, σ -algebras $\mathcal{F}_{<t}$ and $\mathcal{F}_{>s}$ are conditionally independent relative to $\mathcal{F}(s,t)$.

Then each σ - measure on $\mathcal{C}$ which charges no evanescent set is a spectral measure of a σ- functional A .

If a stochastic system $(\mathcal{F}(I),P)$ has a two-sided regular Markov representation $(x_t, P_{t,x}, P^{t,x})$, then taking into account the result formulated at the end of Section 2 we can consider spectral measures as measures on the state space $\mathcal{E}$. Formula (1) can be rewritten in the form

(1a) $$\mu(\Gamma) = P\int \mathbf{1}_\Gamma(x_t)\, A(dt) \quad .$$

*) The integral of a function ξ with respect to a measure P is denoted by P_ξ . We denote by $\mathbf{1}_C$ an indicator function i.e.a function with the value 1 on C and the value 0 outside C .

**) Functionals, coinciding for almost all ω , are considered as indistinguishable.

Under the condition (I), this formula establishes a one-to-one correspondence between σ - functionals of $(\mathcal{F}(I),P)$ and σ -measures on $\mathcal{E}$ which charges no inaccessible set (a set Γ is called inaccessible if $P\{x_t \in \Gamma$ for some $t\in(\alpha,\beta)\} = 0$).*)

4. The split process

The expression "$(x_t(\mathcal{F}(I),P)$ is a Markov process" will be used as a synonym for the expression " x_t is a Markov representation of a stochastic system $(\mathcal{F}(I),P)$". The term "regular" will mean the same as "two-sided regular".

Let $(x_t\mathcal{F}(I),P)$ be a regular Markov process. We introduce a fine topology associated with the forward probabilities $P_{t,x}$: a set U is a fine neighborhood of (s,x) if $P_{s,x}\{x_t \in U$ for a time interval $[s, s+\varepsilon)$ with a positiv $\varepsilon\} = 1$. We say that a real-valued function f is fine on the set C if it is fine-continuous on a set $C\setminus B$ where B is inaccessible. Analogously, cofine topology and cofine functions are defined with the help of the backward transition probabilities $P^{t,x}$.

There exist measurable transformations $x\mapsto x^+$ and $x\mapsto x^-$ of $\mathcal{E}$ such that, for each fine f and cofine q , the function $f(x_t^+)$ is

*) The main part of these results was established in [6]. But the proof of Theorem 4.3 in [6] contains a mistake. For the case of a finite functional A , the assertion of Theorem 4.3 can be deduced easily from a result of M.Šur (Uspehi Matem.Nauk 29,6 (1974), 183-184).

a.s. a right-continuous modification of $f(x_t)$ and the function $q(x_t^-)$ is a leftcontinuous modification of $q(x_t)$. We denote x_t^+ by x_{t+} and x_t^- by x_{t-}.

Kuznecov proved [8] that every base function (see Section 1) is fine and cofine except for a single value of t. This implies easily that a metric can be introduced in $\mathcal{E}$ with the properties: a) $\mathcal{E}$ is compact; b) open sets are finely and cofinely open; c) x_{t+} and x_{t-} are right and left limits of x_t for all t a.s. P.

Let us associate three points $t-$, t and $t+$ with each real number t. The set V of all these points, completed by two points $-\infty$ and $+\infty$, will be called the <u>split real line</u>. Natural order and topology can be introduced in V. Combining the processes x_{t-}, x_t and x_{t+}, we define a stochastic process x_v with the parameter set V. We call it the <u>split process</u>.

σ - algebras $\mathcal{F}_{<v}$ and $\mathcal{F}_{>v}$ can be defined for each $v \in V$: $\mathcal{F}_{<t+}$ is the intersection of $\mathcal{F}_{<u}$ for all $u>t$, $\mathcal{F}_{<t-}$ is the minimal σ - algebra which contains $\mathcal{F}_{<s}$ for all $s<t$ etc. (according to condition B of Section 2, $\mathcal{F}_{<t-} = \mathcal{F}_{<t}$ and $\mathcal{F}_{>t+} = \mathcal{F}_{>t}$). The following Markov property is proved for (x_v, P): given x_v, the σ- algebras $\mathcal{F}_{<v}$ and $\mathcal{F}_{>v}$ are conditionally independent. If function f is fine and cofine, then it is continuous along almost all the paths of x_v.

The split process x_v has forward transition probabilities

$$P_{t+,x} = P_{t,x} \; (= P_{t,x^+}), \quad P_{t-,x} = P_{t,x^-}$$

and backward transition probabilities

$$P^{t+,x} = P^{t,x^+}, P^{t-,x} = P^{t,x} (= P^{t,x^-}) .$$

Note that, if $x^- = x^+$, then $P_{t,x^+} = P_{t,x}$, $P^{t,x^+} = P^{t,x^-} = P^{t,x}$ and therefore $x^+ = x$.

Let us call a point x of $\mathcal{E}$ a continuity point if

$$P_{t,x}\{P_{t,x_t} = P_{t,x}\} = 1, P^{t,x}\{P^{t,x_t} = P^{t,x}\} = 1 \tag{2}$$

and $x^+ = x^- = x$. The set $\mathcal{D}$ of all such points is called the continuity set. It is proved that conditions (2) are fulfilled q.e.*) Obviously, $x_{t-} = x_{t+} = x_t$ if $x_t \in \mathcal{D}$. This implies that a.s. $x_t \in \mathcal{D}$ except at most at a countable set of t .

For the regular Markov process described at the end of Section 1 (see Fig.1), the state space $\mathcal{E}$ consists of four real lines $R\times(1,1)$, $R\times(1,2)$, $R\times(2,1)$ and $R\times(2,2)$; the first and the fourth lines form the continuity set $\mathcal{D}$.

5. Capacity

Let $(x_t, \mathcal{F}(I), P)$ be a regular Markov process on a closed time interval. The value

$$\varphi(C) = P\{x_t \in C \quad \text{for some} \quad t\}$$

*) The abbreviation "q.e" means "quasi-everywhere", that is "outside an inaccessible set".

is called the capacity of C . It is defined for every analytic subset C of $\mathcal{E}$.

We want to establish the relation of this definition to the classical one based on the concept of the equilibrium potential.

We assume that condition AC of Section 2 is satisfied. Then there exists a measurable function $p(s,x;t,y)$ on $\mathcal{E}\times\mathcal{E}$ such that

$$m_{s,t}(dx,dy)=p(s,x;t,y)\,m_s(dx)\,m_t(dy)$$

and

$$\int_{E_t} p(s,x;t,y)\,m_t(dy)\,p(t,y;u,z)=p(s,x;u,z)$$

for all $s<t<u$, $x\in E_s$, $z\in E_u$ (see [5], [7]).

Theorem 1. Let C be an analytic subset of $\mathcal{D}$ and let $\bar{C}$ be its cofine closure. Then $\varphi(C)$ is equal to the maximum of $\nu(\mathcal{E})$ over all the measures ν which are concentrated on $\bar{C}$ and satisfy the following inequality

$$\int_{\mathcal{E}} p(s,x;t,y)\,\nu(dt,dy)\le 1 \quad \text{for all } s,x .$$

The maximum is attained for

$$\nu(\Gamma)=P\{x_\gamma\in\Gamma\}$$

where $\gamma=\sup\{t:x_t\in C\}$ is the last exit time from C. *)

*) The domain of γ is $\{\omega: x_t(\omega)\in C$ for some $t\}$. It coincides with the domain of x_γ since x_t is defined on a closed time interval.

The proof is based on the following propositions:

1. The split process x_v reaches every analytic subset of $\mathcal{D}$ with the same probability as the process x_t .

2. For each analytic $C \subset \mathcal{D}$ and for every $\varepsilon > 0$, there exists a finely and cofinely open set $U \supset C$ such that $\varphi(U) < \varphi(C) + \varepsilon$.

3. Backward transition probabilities $P^{s,x}$ of $(x_t, \mathcal{F}(I), P)$ can be selected in such a way that

$$P^{t,y}\{\tau < t\} = P\,\mathbf{1}_{\tau<t}\,p(\tau, x_\tau; t, y)$$

for each Markov time τ and all t, y .

Propositions 1 and 2 and the statement of Theorem 1 can be false if C is not a subset of $\mathcal{D}$.

6. Homogeneous case

In this case, two one-parameter groups of transformations

$$\theta_u : \Omega \to \Omega \quad , \quad i_u : \mathcal{E} \to \mathcal{E}$$

are given. We assume that the mapping $(u,x) \to i_u x$ of $R \times \mathcal{E}$ into $\mathcal{E}$ is measurable and $i_u E_t = E_{t+u}$. A collection $(x_t, \mathcal{F}(I), P, \theta_u, i_u)$ is called a homogeneous Markov process if $(x_t, \mathcal{F}(I), P)$ is a Markov process and if, for each u ,

(3) $$\theta_u \mathcal{F}(I) = \mathcal{F}(I+u) \quad ,$$

(4) $$i_u x_t(\theta_u \omega) = x_{t+u}(\omega) \text{ for all } t \text{ (a.s. } P) \quad ,$$

(5) $$P(\theta_{-u}A) = c(u)P(A).$$

Of course, the condition $P(\Omega)<\infty$ implies that the constant $c(u)$ should be equal to 1 .However,it is useful to extend the notion of a stochastic process and to admit infinite measures P for which all one-dimensional distributions are σ-finite. Obviously $c(u)=exp(\varkappa u)$ where $\varkappa$ is a constant independent of u .

The forward transition probabilities $P_{s,x}$ of a Markov process are called homogeneous if, for every $u\in R, A\in \mathcal{F}_{>t+u}$

$$P_{s,x}(\theta_{-u}A) = P_{s+u,\, i_u x}(A) \quad \text{q.e. on } \mathcal{E}_{<t} = \bigcup_{s<t} E_s$$

with i_u and θ_u satisfying conditions (3) and (4). An analogous definition can be introduced for the backward transition probabilities. Every homogeneous Markov process, subject to condition AC, has homogeneous forward and backward transition probabilities. The convers is true under a stronger condition: each measure m_{st} is equivalent to $m_s\times m_t$.

The following theorem allows us to reduce many problems concerning Markov processes with homogeneous forward transition probabilities to problems concerning homogeneous processes.

<u>Theorem 2.</u> Let a Markov process $(x_t, \mathcal{F}(I), P)$ have homogeneous forward transition probabilities $P_{s,x}$ which satisfy the condition:

AC$'$. For every t ,there exists a finite measure q_t on E_t such that $P_{s,x}\{x_t\in\Gamma\}=0$ for all $s<t, x\in E_s$ if $q_t(\Gamma)\leq 0$. Then there exists a homogeneous Markov process $(x_t, \mathcal{F}(I), \tilde{P})$ with the same forward transition probabilities and such that, for every

$t \in R$, $A \in \mathcal{F}_{>t}$, $A \subset \{\alpha < t\}$

(6) $$P(A) = \int_A \rho_t(x_t)\, d\tilde{P}$$

where $\rho_t(x)$ is a measurable function on $\mathcal{E}$ (independent of A).

It is convenient to identify all E_t with the space $E = E_0$ using the transformations i_t. We have a natural isomorphic mapping of $\mathcal{E}$ onto $R \times E$ which transforms x_t into $(t, \bar{x}_t)$ where $\bar{x}_t$ is a stochastic process in E *). For every $x \in E$, we set $P_x = P_{0,x}$.

To prove Theorem 2, we construct a finite measure q on E such that, if $q(\Gamma) = 0$, then $P_{s,x}\{\bar{x}_t \in \Gamma\} = 0$ for all $s < t, x \in E$. We set

(7) $$\eta(\Gamma) = \int_0^\infty dt\, e^{-æt} \int_E q(dx) P_x\{\bar{x}_t \in \Gamma\}$$

where $æ$ is a constant for which the integral has meaning. The function ρ_t is the density of the measure $m_t(\Gamma) = P\{\alpha < t, \bar{x}_t \in \Gamma\}$ with respect to $e^{æt}\eta$, and the measure $\tilde{P}$ satisfies condition (5) with $c(u) = e^{æu}$. Formula (6) can be rewritten as follows

$$\tilde{P}(A) = \int_{-\infty}^{u} dt\, e^{æt} \int_E q(dx) P_{t,x}(A) \quad (A \in \mathcal{F}_{>u},\ A \subset \{\alpha < u\}).$$

If a stochastic system has a Markov representation with homogeneous forward transition probabilities, then it has a right-regu-

---

*) Generally, the space $\mathcal{E}$ includes $E_{-\infty}$ and $E_{+\infty}$ which can not be identified with E_0. But in this section, it is more convenient to consider $\mathcal{E}$ as the union of E_t for $-\infty < t < +\infty$.

lar Markov representation with the same properties (see [5]). Analogous statements are true for left- and two-sided regular representations.

Now we have all the necessary tools to apply the general theory to the homogeneous case. Let us consider two examples.

Let a stochastic system $(\mathcal{F}(I),P)$ satisfy the homogeneity condition (3). We describe all additive functionals of $(\mathcal{F}(I)\,P)$ which are homogeneous in the following sense: $\theta_u A(\Lambda) = A(\Lambda + u)$ for every Borel subset Λ of R and every $u \in R$. Consider the regular Markov representation x_t of $(\mathcal{F}(I),P)$.

At first, we assume that the process $(x_t,\mathcal{F}(I),P)$ is homogeneous. Formula (1a) implies that, for each measurable subset Γ of E and each Borel subset Δ of R

(8)
$$\mu(\Gamma\times\Delta) = P\int_\Delta \mathbf{1}_\Gamma(\bar{x}_t)\,A(dt).$$

It follows from (3), (4) and (5) that

$$\mu((\Gamma+u)\times\Delta) = \bar{\bar{c}}(u)\,\mu(\Gamma\times\Delta)$$

with $\bar{\bar{c}}(u) = e^{\varkappa u}$. Hence

(9)
$$\mu(dt,dx) = e^{\varkappa t}\,dt\,\lambda(dx)$$

where λ is a σ - measure on E, and (8) implies that

$$\lambda(\Gamma)\int_\Delta e^{\varkappa t}\,dt = P\int_\Delta \mathbf{1}_\Gamma(\bar{x}_t)\,A(dt).$$

The last formula establishes a one-to-one correspondence between all the homogeneous σ-functionals of $(\mathcal{F}(I),P)$ and all the σ-measures λ on E which charge no sets inaccessible to $\bar{x}_t$. To prove this assertion, we need only show that, if λ charges no sets which are inaccessible to $\bar{x}_t$, then the measure μ defined by (9) has the same property relative to x_t. It can be checked with the help of the results of Section 5.

Now we consider an arbitrary process $(x_t, \mathcal{F}(I), P)$, satisfying the conditions of Theorem 2. It can be proved that there exists a measurable function $p(t,x,y)$ $(t>0, x,y\in E)$ such that

$$P_x\{\bar{x}_t\in\Gamma\}=\int_\Gamma p(t,x,y)\,\eta(dy)$$

and

$$\int_E p(s,x,y)\,\eta(dy)\,p(t,y,z)=p(s+t,x,z)$$

for all $s,t>0, x,y,z\in E$. Using this function, we can define functions ρ_t in (6) in such a way that

(10) $$\int_E \rho_s(x)\,\eta(dx)\,p(t-s,x,y)\uparrow\rho_t(y) \quad \text{as} \quad s\uparrow t \quad \text{for all } t,y.$$

If the process $(x_t,\mathcal{F}(I),P)$ is regular, then the homogeneous process $(x_t,\mathcal{F}(I),\widetilde{P})$ described in Theorem 2 is also regular. This follows from a result of Kusnecov.) Applying the results formulated above to $(x_t,\mathcal{F}(I),\widetilde{P})$, we prove that the formula

(11) $$\int_\Delta dt\int_\Gamma e^{\varkappa t}\rho_t(x)\,\lambda(dx)=P\int_\Delta \mathbf{1}_\Gamma(\bar{x}_t)\,A(dt)$$

defines a one-to-one correspondense between all homogeneous σ-functionals of $(\mathcal{F}(I), P)$ and all σ - measures λ on E which charge no sets inaccessible for $\bar{x}_t$.

Special classes of homogeneous additive functionals were investigated by Wentzell [12], McKean and Tanaka [9] and Revuz [10], [11] . In the case considered by Revuz, we have $\rho_t(x) \uparrow 1$ as $t \downarrow 0$, and formula (11) implies that

$$\lambda(\Gamma) = \lim_{u \downarrow 0} \frac{1}{u} P \int_0^u \mathbf{1}_\Gamma(\bar{x}_t) A(dt) \quad .$$

We wish now to find a homogeneous version of Theorem 1 concerning capacity. We consider a regular process which satisfies the conditions of Theorem 2 and we assume that the birth-time α is equal to 0 . Let Γ be a subset of E . We say that x belongs to the cofine closure $\bar{\Gamma}$ of Γ if $P^{u,x}\{\bar{x}_t \in \Gamma$ for some $t \in (s,u)\} = 1$ for all $u > s > 0$.

We prove that:

(i) the cofine closure of $R \times \Gamma$ relative to x_t is equal to $R \times \bar{\Gamma}$;

(ii) the continuity set $\mathcal{D}$ described in Section 4 is of the form $R \times \mathcal{D}_0$ where $\mathcal{D}_0$ is a meassurable subset of E .

Put

$$g_\varkappa(x,y) = \int_0^\infty e^{-\varkappa t} p(t,x,y)\, dt \quad ,$$

$$\varphi(\Gamma) = P\{\bar{x}_t \in \Gamma \text{ for some } t\} \quad .$$

Denote by $\varphi_\varkappa(\Gamma)$ the supremum of λ (E) over all measures λ which are concentrated on the cofine closure $\bar{\Gamma}$ of Γ and satisfy the inequality.

$$\int_E q_{\varkappa}(x,y)\,\lambda(dy) \le 1 \quad \text{for all } x .$$

Relying on Theorem 1, we prove that $\varphi_{\varkappa}(\Gamma) \le \varphi(\Gamma)$ for all analytic subsets Γ of $\mathcal{D}_0$ and all $\varkappa$. The equality can only hold in very special cases, but in quite a general situation $\varphi(\Gamma)=0$ if $\varphi_{\varkappa}(\Gamma)=0$ for some $\varkappa$. Namely, a sufficient condition for this last implication is that the conditional distribution of the last exit time,

$\gamma = \sup\{t : \bar{x}_t \in \Gamma\}$, satisfy the inequality

$$P\{s<\gamma<u \mid x_\gamma\} \ge C(x_\gamma)\int_s^u e^{-\gamma t}\rho_t(\bar{x}_t)\,dt \quad \text{a.s. for each } u>s>0 ,$$

where C is strictly positive measurable function and ρ_t satisfy condition (10).

References

[1] E.B.Dynkin, Initial and final behaviour of trajectories of Markov processes, Uspehi Mat.Nauk,26,4(1971),153-172; English translation: Russian Math.Surveys 26, 4(1973), 165-185.

[2] E.B.Dynkin, Integral representation of excessive measures and excessive functions, Uspehi Mat.Nauk 27,1(1972),43-80; English translation in: Russian Math.Surveys 27, 1(1972), 43-84.

[3] E.B.Dynkin, Regular Markov processes, Uspehi Mat.Nauk 28, 2(1973), 33-64 ; English translation in Russian Math. Surveys 28, 2 (1973), 33-64.

[4] E.B.Dynkin, Markov representations of stochastic systems,

Dokl.Akad.Nauk SSSR, 218, 5(1974),1009-1012; English translation in: Soviet Math.Dokl., 15,(1974),1442-1446.

[5] E.B.Dynkin,Markov representations of stochastic systems,Uspehi Mat.Nauk, 30, 1(1975),61-99(Russian).

[6] E.B.Dynkin, Additive functionals of Markov processes and stochastic systems, Annales de l'Institut Fourier,Grenoble, 25, fasc. 3/4 (1975).

[7] E.B.Dynkin and S.E.Kuznecov,Determining functions of Markov processes and corresponding dual regular classes, Dokl. Akad.Nauk SSSR 214, 1(1974), 25-28; English translation in: Soviet Math.Dokl. 15(1974), 20-23.

[8] S.E.Kuznecov, Excessive functions connected with a regular Markov split process, Teoriya Veroyatnostei i ee Primem., 20, 1(1975), 144-150(Russian).

[9] H.P.McKean and H.Tanaka, Additive functionals of the Brownian path, Mem. Coll.Sci.Univ.Kyoto, A 33,3(1961),479-506.

[10] D.Revuz, Mesures associées aux fonctionnelles additives de Markov I , Trans.Amer.Math.Soc., 148(1970),501-531.

[11] D.Revuz, Mesures associées aux fonctionnelles additives de Markov 11, Z.Wahrscheinlichkeitstheorie verv.Geb.,16(1970) 336-344.

[12] A.D.Wentzell, Nonnegative additive functionals of Markov processes ,Dokl.Akad.Nauk SSSR, 137(1961),17-20,(Russian).

Central Economical Mathematical Institute

Academy of Sciences of the USSR, Moscow

Limit theorems for linear combinations of order statistics

V.A.Egorov, V.B.Nevzorov

The aim of this paper is to investigate some asymptotic properties of linear combinations of order statistics (l.c.o.s.). The asymptotic normality of l.c.o.s. under different conditions was proved by various methods ([I]-[5]). W.Rosenkrantz and N.O'Reilly ([6]) were the first to obtain estimates for the rate l.c.o.s. convergence to the normal law. They used the Skorokhod representation theorem and noted that this method can not give the rate of convergence better than $n^{-1/4}$.

Estimates for trimmed means, optimal up to a logarithmic multiplier, were obtained by the authors of this paper in [7]. Similar estimates were also proved for linear combinations of absolute order statistics ([8],[9]). Some asymptotic expansions for trimmed means were given in [IO].

In the first part of this paper we consider linear combinations of uniform order statistics and give some optimal estimates for their rate of convergence to the normal law. The results for order statistics from continuous distributions are proved in the second part.

I. Let $U_1, U_2, \ldots, U_n$ be a sample from the uniform distribution on the interval $[0,1]$ and $U_{1,n} \leq U_{2,n} \leq \ldots \leq U_{n,n}$ be the corresponding order statistics. The random variable (r.v.) $U_{k,n}$ is known to have the B-distribution with parameters k and $n-k+1$, i.e.

$$P\{U_{k,n} < x\} = B(x) = \frac{n!}{(k-1)!(n-k)!} \int_0^x u^{k-1}(1-u)^{n-k} du \quad (0 \leq x \leq 1) \qquad (I)$$

Lemma I. (H.Cramér [II]). The r.v. $U_{K,n}$ has the following representation

$$U_{K,n} = \frac{\eta_1^2 + \eta_2^2 + \cdots + \eta_{2K}^2}{\eta_1^2 + \eta_2^2 + \cdots + \eta_{2(n+1)}^2}$$

with independent identically distributed (i.i.d.) r.v.

$$\eta_i \in \mathcal{N}(0,1) \quad (i = 1, 2, \dots)$$

The next result follows from Lemma I.

Lemma 2.

$$U_{K,n} = \frac{\xi_1 + \xi_2 + \dots + \xi_K}{\xi_1 + \xi_2 + \dots + \xi_{n+1}} \quad , \tag{2}$$

where i.i.d. r.v. $\xi_1, \xi_2, \dots$ have the exponential distribution, i.e.

$P\{\xi_i < x\} = max\, [0, 1 - e^{-x}]$.

Lemma 3. (H.Cramér [II]).

$$EU_{K,n} = \frac{K}{n+1} \quad , \qquad B_2 = \mathfrak{D}U_{K,n} = \frac{K(n-K+1)}{(n+1)^2\,(n+2)} \quad . \tag{3}$$

Now let us consider r.v. $T_n = \sum_{i=1}^{n} c_i u_{i,n}$ with some constants $c_1, c_2, \dots$ and put

$$b_i = \sum_{K=i}^{n} c_K \quad (i = 1, 2, \dots, n), \quad b_{n+1} = 0 \ .$$

It follows, from Lemma 2, that

$$T_n = \sum_{K=1}^{n+1} \frac{b_i \xi_i}{\xi_1 + \xi_2 + \dots + \xi_{n+1}} \tag{4}$$

It is not difficult to verify that

$$ET_n = \frac{1}{n+1} \sum_{i=1}^{n} i c_i = \frac{1}{n+1} \sum_{i=1}^{n+1} b_i \tag{5}$$

and

$$\mathfrak{D}T_n = \frac{1}{(n+1)^2(n+2)}\Big[\sum_{i=1}^{n} i(n-i+1)c_i^2 + 2\sum_{1\le i<j\le n} i(n-j+1)c_i c_j\Big] = \tag{6}$$

$$= \frac{n}{(n+1)^2(n+2)}\sum_{i=1}^{n+1} b_i^2 - \frac{2}{(n+1)^2(n+2)}\sum_{1\le i<j\le n+1} b_i b_j \quad .$$

Let $$\Phi(x) = \frac{1}{\sqrt{2\pi}}\int_{-\infty}^{x} e^{-\frac{t^2}{2}}\,dt \quad .$$

Theorem I. The inequality

$$\sup_x \left| P\left\{ \frac{T_n - ET_n}{\sqrt{\mathfrak{D}T_n}} < x \right\} - \Phi(x) \right| \le C \, \frac{\sum_{i=1}^{n+1} | b_i - \bar{b} |^3}{\{\sum_{i=1}^{n+1} (b_i - \bar{b})^2\}^{3/2}} \tag{7}$$

holds with $\bar{b} = ET_n = \frac{1}{n+1}\sum_{i=1}^{n+1} b_i$ and an absolute constant C .

The next result follows from Theorem I.

Theorem 2.

$$\sup_x \left| P\left\{ \frac{U_{K,n} - \frac{K}{n+1}}{\beta_2^{1/2}} < x \right\} - \Phi(x) \right| \le C\left(\frac{1}{\sqrt{K}} + \frac{1}{\sqrt{n-K+1}}\right) \tag{8}$$

where C is an absolute constant.

In fact, we can choose the constants in the definition of r.v. T_n in such a way: $c_K = 1$, $c_i = 0\ (i \ne K)$ In this case $b_1 = b_2 = \dots$ $\dots = b_K = 1,\ b_{K+1} = b_{K+2} = \dots = b_{n+1} = 0$ and $T_n = U_{K,n}$. Then

$$ET_n = \frac{K}{n+1} \quad ,$$

$$\mathfrak{D}T_n = \frac{K(n-K+1)}{(n+1)^2(n+2)} = \beta_2 \ ; \quad \sum_{i=1}^{n+1} | b_i - \bar{b} |^3 = K\left(1 - \frac{K}{n+1}\right)^3 + (n-K+1)\,\frac{K}{(n+1)^3} =$$

$$= \frac{K(n-K+1)}{(n+1)^3}\left(K^2 + (n-K+1)^2\right); \ \Big\{\sum_{i=1}^{n+1} (b_i - \bar{b})^2\Big\}^{3/2} = \Big[\frac{K(n-K+1)}{n+1}\Big]^{3/2}$$

and

$$\frac{\sum_{i=1}^{n+1} |b_i-\bar{b}|^3}{\{\sum_{i=1}^{n+1}(b_i-\bar{b})^2\}^{3/2}} \le \frac{K^{3/2}}{(n-K+1)^{1/2}(n+1)^{3/2}} + \frac{(n-K+1)^{3/2}}{K^{1/2}(n+1)^{3/2}} \le \frac{1}{\sqrt{n-K+1}} + \frac{1}{\sqrt{K}} .$$

Now one can see that (8) follows from (7).

Remark I. The coefficients $\{C_i\}$ in the definition of r.v. T_n are often taken as values of a certain continuous function $J(x)(0\le x\le 1)$ at different points: $C_i = J(\frac{i}{n+1})$. In this case the rate of convergence in Theorem I has the order $n^{-1/2}$ under the condition

$$0 < \int_0^1 t(1-t)J^3(t)\,dt < \infty$$

for the function $J(x) \ge 0$. In fact, we have in this situation:

$$\sum_{i=1}^{n+1}|b_i-\bar{b}|^3 = \sum_{i=1}^{n+1}\Big|\sum_{j=i}^{n} J(\tfrac{j}{n+1}) - \tfrac{1}{n+1}\sum_{j=1}^{n} j\,J(\tfrac{j}{n+1})\Big|^3 = \sum_{j=1}^{n+1}\Big|\sum_{j=i}^{n+1} J(\tfrac{j}{n+1})(1-\tfrac{j}{n+1}) -$$

$$-\sum_{j=1}^{i-1}\tfrac{j}{n+1}J(\tfrac{j}{n+1})\Big|^3 \le 4\sum_{i=1}^{n+1}\Big|\sum_{j=i}^{n} J(\tfrac{j}{n+1})(1-\tfrac{j}{n+1})\Big|^3 + 4\sum_{i=1}^{n+1}\Big|\sum_{j=1}^{i}\tfrac{j}{n+1}J(\tfrac{j}{n+1})\Big|^3 \le$$

$$\le 4\sum_{i=1}^{n+1}(n-i+1)^2\sum_{j=i}^{n}(1-\tfrac{j}{n+1})^3 J^3(\tfrac{j}{n+1}) + 4\sum_{i=1}^{n+1} i^2\sum_{j=1}^{i}\tfrac{j^3}{(n+1)^3}J^3(\tfrac{j}{n+1}) \le \tag{9}$$

$$\le 4\sum_{j=1}^{n+1} J^3(\tfrac{j}{n+1})(1-\tfrac{j}{n+1})^3 j\,(n+1)^2 + 4\sum_{j=1}^{n+1} J^3(\tfrac{j}{n+1})\tfrac{j^3}{(n+1)^3}(n+1)^2(n+1-j) \le$$

$$\le 8(n+1)^3\sum_{j=1}^{n+1} J^3(\tfrac{j}{n+1})(1-\tfrac{j}{n+1})\tfrac{j}{n+1} \le$$

$$\le C_1(n+1)^4\int_0^1 J^3(t)\,t(1-t)\,dt$$

with a certain constant C_1 . We can also find that

$$\sum_{i=1}^{n+1}(b_i-\bar{b})^2 \ge C_2(n+1)^3 \tag{10}$$

where a constant C_2 depends on the function $J(t)$ only. The inequalities (9) and (10) prove our remark.

Now we give the proof of Theorem 1. Let us note that

$$P\{T_n - ET_n < x\} = P\{\sum_{i=1}^{n+1} b_i \xi_i < (x+ET_n)\sum_{i=1}^{n+1} \xi_i\} = \tag{11}$$

$$= P\{\sum_{i=1}^{n+1}(b_i - x - ET_n)\xi_i < 0\} = P\{\Sigma < 0\} = P\{\frac{\Sigma - E\Sigma}{(\mathcal{D}\Sigma)^{1/2}} < -\frac{E\Sigma}{(\mathcal{D}\Sigma)^{1/2}}\}$$

with r.v. $\xi_1, \xi_2, \ldots, \xi_{n+1}$ from the representation (2),

$$E\Sigma = \sum_{i=1}^{n+1}(b_i - x - ET_n) = \sum_{i=1}^{n+1} b_i - (n+1)ET_n - x(n+1) = -x(n+1)$$

and

$$\mathcal{D}\Sigma = \sum_{i=1}^{n+1}(b_i - x - ET_n)^2 = \sum_{i=1}^{n+1}(b_i - ET_n)^2 + (n+1)x^2 = \sum_{i=1}^{n+1} b_i^2 - \frac{1}{n+1}(\sum_{i=1}^{n+1} b_i)^2 + (n+1)x^2 .$$

It is not difficult to show that

$$\Gamma_n = \sum_{i=1}^{n+1} E|(b_i - x - ET_n)\xi_i|^3 \leq C\{\sum_{i=1}^{n+1}|b_i - ET_n|^3 + (n+1)|x|^3\} \tag{12}$$

where C is an absolute constant. With the help of Bikelis' nonuniform estimate ([12]) we can obtain the following inequalities:

$$|P\{\frac{\Sigma - E\Sigma}{(\mathcal{D}\Sigma)^{1/2}} < -\frac{E\Sigma}{(\mathcal{D}\Sigma)^{1/2}}\} - \Phi(-\frac{E\Sigma}{(\mathcal{D}\Sigma)^{1/2}})| \leq$$

$$\leq \frac{C\{\sum_{i=1}^{n+1}|b_i - ET_n|^3 + (n+1)|x|^3\}}{(\mathcal{D}E)^{3/2}(1 + \frac{|E\Sigma|^3}{(\mathcal{D}\Sigma)^{3/2}})} \leq \frac{C\{\sum_{i=1}^{n+1}|b_i - ET_n|^3 + (n+1)|x|^3\}}{(\mathcal{D}\Sigma)^{3/2} + |E\Sigma|^3} \leq$$

$$\leq \frac{C\{\sum_{i=1}^{n+1}|b_i - ET_n|^3 + (n+1)|x|^3\}}{\{\sum_{i=1}^{n+1}(b_i - ET_n)^2\}^{3/2} + (n+1)^3|x|^3} \leq C\{\frac{\sum_{i=1}^{n+1}|b_i - ET_n|^3}{\{\sum_{i=1}^{n+1}(b_i - ET_n)^2\}^{3/2}} + \frac{1}{(n+1)^2}\} \leq \tag{13}$$

$$\leq C\frac{\sum_{i=1}^{n+1}|b_i - ET_n|^3}{\{\sum_{i=1}^{n+1}(b_i - ET_n)^2\}^{3/2}} ,$$

taking into account that

$$\frac{\sum_{i=1}^{n+1} | b_i - ET_n |^3}{\{\sum_{i=1}^{n+1} (b_i - ET_n)^2\}^{3/2}} \geq \frac{1}{\sqrt{n+1}} \quad . \tag{I4}$$

In fact, $\frac{1}{n+1}\sum_{i=1}^{n+1} | b_i - ET_n|^3$ is the third absolute central moment for the r.v. which takes values $b_1, b_2, \ldots, b_{n+1}$ with equal probabilities and $\frac{1}{n+1}\sum_{i=1}^{n+1} (b_i - ET_n)^2$ is the corresponding variance. Then (I4) follows from the inequality

$$E|Z-EZ|^3 \geq \{E(Z-EZ)^2\}^{3/2}.$$

We obtain from (II) and (I3) that

$$|P\{T_n - ET_n < x\} - \Phi(\frac{x(n+1)}{\sqrt{\sum_{i=1}^{n+1}(b_i - ET_n)^2 + (n+1)x^2}})| \leq C \frac{\sum_{i=1}^{n+1} | b_i - \bar{b}|^3}{\{\sum_{i=1}^{n+1} (b_i - \bar{b})^2\}^{3/2}} \quad . \tag{I5}$$

Let us note that

$$\sum_{i=1}^{n+1} (b_i - ET_n)^2 = \frac{n}{n+1}\sum_{i=1}^{n+1} b_i^2 - \frac{2}{n+1}\sum_{1\leq l<j\leq n+1} b_l b_j = (n+2)(n+1)\mathcal{D}T_n$$

and introduce

$$y = \frac{x(n+1)}{\sqrt{(n+2)(n+1)\mathcal{D}T_n}} = \frac{x}{\sqrt{\mathcal{D}T_n}} \sqrt{\frac{n+1}{n+2}}$$

and

$$B_n^2 = \frac{n+2}{n+1}\mathcal{D}T_n = \frac{n}{(n+1)^3}\sum_{i=1}^{n+1} b_i^2 - \frac{2}{(n+1)^3}\sum_{1\leq l<j\leq n+1} b_l b_j$$

We can rewrite inequality (I5) in such a way:

$$\sup_y |P\{\frac{T_n - ET_n}{B_n} < y\} - \Phi(\frac{y}{\sqrt{1+\frac{y^2}{n+1}}})| \leq \frac{C\sum_{i=1}^{n+1} | b_i - \bar{b} |^3}{\{\sum_{i=1}^{n+1} | b_i - \bar{b}|^2\}^{3/2}} \quad . \tag{I6}$$

It easy to prove that

$$\sup_y |\Phi(y) - \Phi(\frac{y}{\sqrt{1+\frac{y^2}{n+1}}})| \leq \frac{C}{n+1} \tag{I7}$$

where C is an absolute constant. Now Theorem I is a consequence of inequalities (I6), (I7) and (I4).

Remark 2. It is interesting to note that the distribution of the r.v. T_n depends on the set $\{b_i\}_{i=1}^n$ only and the order of the arrangement of these coefficients is not essential. So chosing the coefficients $b_1, b_2, \dots, b_n$ in another order $(b_1', b_2', \dots, b_n')$ and taking $c_i' = b_i' - b_{i+1}'$ $(i=1,2,\dots,n-1)$, $c_n' = b_n'$, we obtain that the r.v. $T_n' = \sum_{i=1}^n c_i' U_{i,n}$ has the same distribution as the r.v. T_n. In particular, if $b_1' > b_2' > \dots > b_n'$, we have a linear combination of uniform order statistics with non-negative (except, may be, $c_n' = \min_{1 \le i \le n} b_i$) coefficients.

Remark 3. The method, used in the proof of Theorem I, is also suitable for obtaining asymptotic expansions for linear combinations of uniform order statistics.

2. A number of results for order statistics corresponding to continuous distribution functions (d.f.) can be obtained by means of Smirnov's transformation from the results for uniform order statistics. For example, using this transformation we can obtain from Theorem 2 the following

Theorem 3. Let $X_1, X_2, \dots, X_n$ be n i.i.d. r.v. each having a continuous d.f. $F(x)$ with the density $p(x)$ such that $\sup_{x: p(x)>0} |p'(x)| \le M$, and let $X_{1,n} \le X_{2,n} \le \dots \le X_{n,n}$ denote the ordered X's. Then the following inequality

$$\sup_x \left| P\left\{ \frac{X_{K,n} - G(\frac{K}{n+1})}{\sqrt{\beta^2}} < x \right\} - \Phi(x) \right| \le C\left(\frac{1}{\sqrt{K}} + \frac{1}{\sqrt{n-K+1}} + \frac{M\sqrt{\beta_2}}{p^2(G(\frac{K}{n+1}))} \right) \qquad (I8)$$

holds for any $1 \le K \le n$. Here $G(x) = F^{-1}(x)$, C is an absolute constant and $\beta_2 = \dfrac{K(n-K+1)}{(n+1)^2(n+2)}$.

In fact,

$$P\{X_{K,n} < x\} = P\{F(X_{K,n}) < F(x)\} = P\{U_{K,n} < F(x)\}$$

and

$$P\{X_{K,n} < x + G(\tfrac{K}{n+1})\} = P\{U_{K,n} < F(x + G(\tfrac{K}{n+1}))\} = \qquad (19)$$

$$= P\{U_{K,n} < \frac{K}{n+1} + x\rho(G(\tfrac{K}{n+1})) + \frac{x^2}{2}\theta\}$$

with $|\theta| \le M$.

It follows from (19) that

$$P\{\frac{(X_{K,n} - G(\frac{K}{n+1}))\rho(G(\frac{K}{n+1}))}{\sqrt{\beta_2}} < x\} =$$

$$= P\{\frac{U_{K,n} - \frac{K}{n+1}}{\sqrt{\beta_2}} < x + \frac{x^2\sqrt{\beta_2}\,\theta}{2\rho^2(G(\frac{K}{n+1}))}\} = \Phi(x + \frac{x^2\sqrt{\beta_2}\,\theta}{2\rho^2(G(\frac{K}{n+1}))} + O(\frac{1}{\sqrt{K}} + \frac{1}{\sqrt{n-K+1}},) =$$

$$= \Phi(x) + O(\frac{1}{\sqrt{K}} + \frac{1}{\sqrt{n-K+1}} + \frac{M\sqrt{\beta_2}}{\rho^2(G(\frac{K}{n+1}))} .$$

and (16) holds true.

Note that Theorem 3 is a generalization of Reiss' result for quantiles ([13]).

Smirnov's transformation is also useful in investigation of linear combinations of order statistics.

Let $X_{1,n} \le X_{2,n} \le \dots \le X_{n,n}$ be order statistics from a sample with d.f. $F(x)$, and let

$$S_n = \sum_{i=\alpha_n}^{\beta_n} J(\frac{i}{n+1}) X_{i,n} \qquad (20)$$

where $\mathcal{g}(t) \geq 0$ is a continuous on (0,I) function and $0 < \alpha_n < \beta_n \leq n$ are some integers.

Let us assume that $F(x)$ has two continuous derivatives and $p(x) = F'(x) > 0$ and set $G(x) = F^{-1}(x)$. Using Smirnov's transformation, we can obtain the following representation

$$S_n = \sum_{i=\alpha_n}^{\beta_n} \mathcal{g}(\tfrac{i}{n+1}) G(\tfrac{i}{n+1}) + \sum_{i=\alpha_n}^{\beta_n} G'(\tfrac{i}{n+1}) \mathcal{g}(\tfrac{i}{n+1})(U_{i,n} - \tfrac{i}{n+1}) + \tag{2I}$$

$$+ \frac{1}{2} \sum_{i=\alpha_n}^{\beta_n} \mathcal{g}(\tfrac{i}{n+1}) G''(\theta_{i,n})(U_{i,n} - \tfrac{i}{n+1})^2 = I_1 + I_2 + I_3 \quad ,$$

where $\{U_{i,n}\}_{i=1}^{n}$ are uniform order statistics, $\theta_{i,n} \in [U_{i,n}, \frac{i}{n+1}]$.

Let

$$\Gamma = \{ \max_{\alpha_n \leq i \leq \beta_n} \left| \frac{U_{i,n} - \frac{i}{n+1}}{\sqrt{i(n-i+1)}} \right| > \frac{C\sqrt{\ln n}}{n^{3/2}} \} \tag{22}$$

with a certain positive constant C.

<u>Lemma 4.</u> Let $\alpha_n > A \ln n$ and $n - \beta_n > A \ln n$. Then

$$P\{\Gamma\} = o(\tfrac{1}{\sqrt{n}}) \quad ,$$

if

$$A > c^2 \geq 96 \quad . \tag{23}$$

Really we have from Lemma 2 that

$$U_{i,n} - \frac{i}{n+1} = \frac{(n-i+1)\sum_{j=1}^{i} \xi_j - i \sum_{j=i+1}^{n+1} \xi_j}{(n+1)\sum_{j=1}^{n+1} \xi_j} \quad ,$$

where $\{\xi_i\}_{i=1}^{n}$ are independent r.v.'s with the same density function

$$p(x) = \begin{cases} e^{-x}, & \text{if } x > 0, \\ 0, & \text{if } x \leq 0. \end{cases}$$

It follows from the Chebyshev inequality that

$$P\{\sum_{j=1}^{n+1} \xi_j < \frac{n}{2}\} = O(\frac{1}{n}) \quad .$$

The last assertion helps us to obtain the following estimate

$$P\{\Gamma\} \leq P\Big\{\max_{\alpha_n \leq i \leq \beta_n} \Big| \frac{(n-i+1)\sum_{j=1}^{i} \xi_j - i\sum_{j=i+1}^{n+1} \xi_j}{\sqrt{i(n-i+1)}} \Big| > \frac{c}{2}\sqrt{n \ln n}\Big\} +$$

$$+ O(\frac{1}{n}) \leq P\{\max_{\alpha_n \leq i \leq \beta_n} |\sum_{j=1}^{i} \frac{(\xi_j - 1)}{\sqrt{i}}| > \frac{c}{4}\sqrt{\ln n}\} +$$

$$+ P\{\max_{\alpha_n \leq i \leq \beta_n} |\frac{\sum_{j=i+1}^{n} (\xi_j - 1)}{\sqrt{n-i+1}}| > \frac{c}{4}\sqrt{\ln n}\} + O(\frac{1}{n}) \leq$$

$$\leq \sum_{i=\alpha_n}^{\beta_n} [P\{|\frac{\sum_{j=1}^{i} (\xi_j - 1)}{\sqrt{i}}| > \frac{c}{4}\sqrt{\ln n}\} + P\{|\frac{\sum_{j=i+1}^{n+1} (\xi_j - 1)}{\sqrt{n-i+1}}| > \frac{c}{4}\sqrt{\ln n}\}] +$$

$$+ O(\frac{1}{n}) \leq 4n \exp\{-\frac{c^2 \ln n}{64}\} + O(\frac{1}{n}) = O(\frac{1}{\sqrt{n}}) \quad . \tag{24}$$

Here we used Petrov's exponential estimates for the distributions of sums of independent r.v.'s ([I4]).

Remark 4. Acting somewhat more accurately we can prove Lemma 4 with the condition

$$\mathcal{A} > c^2 > 24 \tag{23'}$$

instead of (23).

By virtue of Lemma 4 we obtain that, on $\bar{\Gamma}$,

$$|J_3| \leq c \frac{\ln n}{n^3} \sum_{i=\alpha_n}^{\beta_n} J(\frac{i}{n+1}) |G''(\theta_{i,n})| i(n-i+1) \quad , \tag{25}$$

$$\frac{i}{n+1}\left(1-\frac{c}{\sqrt{A}}\right) < 1-\theta_{i,n} < \frac{1}{n+1}\left(1+\frac{c}{\sqrt{A}}\right)$$

and

$$\frac{n-i+1}{n+1}\left(1-\frac{c}{\sqrt{A}}\right) < 1-\theta_{i,n} < \frac{n-i+1}{n+1}\left(1+\frac{c}{\sqrt{A}}\right) \quad .$$

Now using condition (23) or (23') we have that, on $\bar{\Gamma}$,

$$\theta_{i,n} \asymp \frac{i}{n+1} \quad , \quad 1-\theta_{i,n} \asymp \frac{n-i+1}{n+1} \qquad {}^{*)}$$

For the case $\liminf_{n\to\infty} \frac{\alpha_n}{n} = 0$ ($\limsup_{n\to\infty} \frac{\beta_n}{n} = 1$) we shall suppose that

$G''(x)$ ($G''(1-x)$) is a regularly varying function for $x \to 0$. If this condition is fulfilled, then the inequality

$$|\mathfrak{I}_3| \le C \ln n \sum_{i=\alpha_n}^{\beta_n} \frac{1}{n} \mathfrak{I}\left(\frac{i}{n+1}\right) \left|G''\left(\frac{i}{n+1}\right)\right| \frac{i}{n+1}\left(1-\frac{i}{n+1}\right) \le \tag{26}$$

$$\le C \ln n \int_{\frac{\alpha_n}{n}}^{\frac{\beta_n}{n}} \mathfrak{I}(t)|G''(t)|t(1-t)\,dt \le C \ln n \int_0^1 \mathfrak{I}(t)|G''(t)|t(1-t)\,dt$$

holds on $\bar{\Gamma}$. It follows from (26) that

$$|\mathfrak{I}_3| = O(\ln n) \tag{27}$$

on $\bar{\Gamma}$ if

$$\int_0^1 \mathfrak{I}(t)\,|G''(t)|\,t(1-t)\,dt < \infty \quad . \tag{28}$$

If the following condition

---

*) The assertain $\alpha_n \asymp \beta_n$ means that $0 < \liminf_{n\to\infty} \frac{\alpha_n}{\beta_n} \le \limsup_{n\to\infty} \frac{\alpha_n}{\beta_n} < \infty$

$$0<\int_0^1 J^3(t)[G'(t)]^3 t(1-t)\,dt<\infty \tag{29}$$

is fulfilled, then we have from Theorem I and Remark I that

$$\sup_x \left| P\left\{ \frac{I_2}{\sqrt{\mathcal{D} I_2}} < x \right\} - \Phi(x) \right| = O\left(\frac{1}{\sqrt{n}}\right) . \tag{30}$$

In (30) $\mathcal{D} I_2$ can be obtained by the substitution $c_i = J(\frac{i}{n+1})\, G'(\frac{i}{n+1})$ $(\alpha_n \leq i \leq \beta_n)$ and $c_i = 0$ $(i < \alpha_n,\ i > \beta_n)$ in (6). It is not difficult to prove that

$$\mathcal{D} I_2 > Cn , \tag{31}$$

where C is a certain positive constant, if

$$\beta_n - \alpha_n \asymp n ,\ p(x) = F'(x) > 0 \qquad (x \in (F^{-1}(\tfrac{\alpha_n}{n})))$$

and $J(x) \not\equiv 0$ is a continuous non-negative function.

Now we remind a lemma due to V.V.Petrov.

Lemma 5. ([15]). Let X and Y be r.v.'s with the d.f.'s $F(x)$ and $G(x)$ and let $H(x)$ be the d.f. of their sum $Z = X + Y$. If for all x $(-\infty < x < \infty)$ $|F(x) - \Phi(x)| \leq \mathcal{K}$, where $\mathcal{K}$ is a non-negative constant, then, for any $\varepsilon > 0$ and for all x ,

$$|H(x) - \Phi(x)| \leq \mathcal{K} + P\{|Y| > \varepsilon\} + \frac{\varepsilon}{\sqrt{2\pi}} .$$

It follows from Lemma 5 and formulas (21), (24), (27), (30) and (31) that

$$\sup_x \left| P\left\{ \frac{S_n - I_1}{\sqrt{\mathcal{D} I_2}} < x \right\} - \Phi(x) \right| = O\left(\frac{\ln n}{\sqrt{n}}\right) . \tag{32}$$

Theorem 4. Let the r.v. S_n be defined by (20) with a continuous non-negative function $\mathcal{J}(x) \not\equiv 0$ and let the d.f. have two continuous derivatives and $p(x) = F'(x) > 0$. Let also $G(x) = F^{-1}(x)$. If $G''(x)$ and $G''(1-x)$ are regularly varying functions as $x \to 0$ and the following conditions are fulfilled

$$\alpha_n > 24 \ln n, \quad n - \beta_n > 24 \ln n, \quad \beta_n - \alpha_n \asymp n, \tag{33}$$

$$\int_0^1 \mathcal{J}^3(t)(G'(t))^3 t(1-t)\,dt < \infty \quad , \tag{34}$$

$$\int_0^1 \mathcal{J}(t) G''(t) | t(1-t)\,dt < \infty \quad , \tag{35}$$

then

$$\Delta_n = \sup_x | P\{\frac{S_n - A_n}{B_n} < x\} - \Phi(x)| = O(\frac{\ln n}{\sqrt{n}}), \tag{36}$$

where $A_n = I_1$, $B_n^2 = \mathcal{D} I_2$ were defined in (21).

Theorem 5 can be proved with the same technique.

Theorem 5. Let

$$0 < \alpha = \liminf_{n \to \infty} \frac{\alpha_n}{n} < 1, \quad 0 < \beta = \limsup_{n \to \infty} \frac{\beta_n}{n} < 1 \quad ,$$

$\mathcal{J}(t) \not\equiv 0$ be a continuous non-negative function, and let $F(x)$ have two continuous derivatives on $(G(\alpha), G(\beta))$ and $F'(x) > 0$ on $(G(\alpha), G(\beta))$. Then

$$\Delta_n = O(\frac{\ln n}{\sqrt{n}}) \tag{37}$$

provided the constants A_n and B_n are properly chosen.

Now we consider the r.v. $\hat{S}_n = \sum_{j=1}^{n} \mathcal{J}(\frac{i}{n+1}) X_{i,n}$

Theorem 6. Let all the conditions of Theorem 4, except (33), be fulfilled and let $J(x)$ increase in a neighbourhood of 0 and decrease in a neighbourhood of 1. If the following inequalities

$$J(x)\,G(x^{3/2}\,|\ln x|^{-1/2}) \geq -C$$
$$J(1-x)\,G(1-x^{3/2}\,|\ln x|^{-1/2}) \leq C \tag{38}$$

hold with a positive constant C in a neighbourhood of 0, then

$$\sup_x \left| P\left\{ \frac{S_n - A_n}{B_n} < x \right\} - \Phi(x) \right| = O\left(\frac{\ln n}{\sqrt{n}}\right), \tag{39}$$

constants A_n and B_n being properly chosen.

Let us prove this theorem. Let $m = 24 \ln n$, $S_n = \sum_{m < i < n-m} J\left(\frac{i}{n+1}\right) X_{i,n}$ and $R_n = \hat{S}_n - S_n$.

We can assume, without loss of generality, that $0 < F(0)$. Then

$$P\{X_{m,n} > 0\} = \tag{40}$$

$$= n C_{n-1}^{m-1} \int_0^\infty [F(x)]^{m-1} [1-F(x)]^{n-m} p(x)\,dx \leq n^m (1-F(0))^{n-m} = O\left(\frac{1}{\sqrt{n}}\right)$$

and

$$P\{X_{n-m,n} < 0\} = O\left(\frac{1}{\sqrt{n}}\right). \tag{41}$$

Using (38), (40) and (41), we obtain the following inequality

$$P\{|R_n| > C \ln n\} \leq P\left\{X_{1,n} < -\frac{C}{J(\frac{m}{n})}\right\} + P\left\{X_{n,n} > \frac{C}{J(1-\frac{m}{n})}\right\} + O\left(\frac{1}{\sqrt{n}}\right) =$$

$$= 1-(1-F(-\frac{C}{\mathcal{J}(\frac{m}{n})}))^n + 1 - F^n(\frac{C}{\mathcal{J}(1-\frac{m}{n})}) + O(\frac{1}{\sqrt{n}}) \leq$$

$$\leq 2[1-(1-(\frac{m}{n})^{3/2}|\ln\frac{m}{n}|^{-1/2})^n] + O(\frac{1}{\sqrt{n}}) = O(\frac{m^{3/2}}{\sqrt{n}\,\ln^{1/2} n}) + O(\frac{1}{\sqrt{n}}) = O(\frac{\ln n}{\sqrt{n}}) \tag{42}$$

where the constant C coincides with that from (38).

We have $\frac{\hat{S}_n - A_n}{B_n} = \frac{S_n - A_n}{B_n} + \frac{R_n}{B_n}$, where $B_n^2 = \mathcal{D}I_2 \times n$, $A_n = I_1$ and S_n are taken from Theorem 4, and $P\{\frac{|R_n|}{B_n} > \frac{C\ln n}{\sqrt{n}}\} = O(\frac{\ln n}{\sqrt{n}})$. Now Theorem 6 follows from Theorem 4 and Lemma 5.

Now we show several examples when the conditions of Theorem 6 are fulfilled.

a) Let the r.v. X_1 have a symmetric distribution with d.f. $F(x) = 1 - \frac{c}{x^\alpha}$ for $x < x_0 > 0$ and $\alpha > 0;\ c > 0$. In this case it easy to verify that we can apply Theorem 6 with $\mathcal{J}(x) = x^\beta$ $(0 \leq x \leq 1)$ if $\beta > max(\frac{3}{2\alpha}, \frac{1}{3} + \frac{1}{\alpha})$.

b) Let us now consider symmetric r.v.'s $X_1, X_2, \dots$ with d.f. $F(x) = 1 - c_1 e^{-c_2 x^\alpha}$ for $x > x_0 > 0$ and positive constants c_1, c_2 and α. The conditions of Theorem 6 are fulfilled if we take $\mathcal{J}(x) = x^\beta$ $(0 \leq x \leq 1)$ with the parameter $\beta > \frac{1}{3}$.

c) Let symmetric r.v.'s $X_1, X_2, \dots$ have d.f. $F(x) = 1 - \frac{c}{(\ln x)^\alpha}$ $(c > 0, \alpha > 0)$ for $x > x_0 > 1$. Then the function $G''(x)$ is not a regularly varying one as $x \to 0$, and therefore we can not apply Theorem 6.

References

I. Stigler M.S., (I969), Linear functions of order statistics, Ann. Math.Statist., 40, 770-787.

2. Bickel P.L. (I967), Some contributions to the theory of order statistics, Proc. Fifth Berkeley Symp.Math.Statist.Prob., I, 575-59I, Univ. California Press.

3. Moore D.S. (I968), An elementary proof of asymptotic normality of linear functions of order statistics, Ann.Math. Statist., 39, 263-265.

4. Shorack G. (I969), Asymptotic normality of linear combinations of functions of order statistics, Ann.Math. Statist.,40, 204I-2050.

5. Stigler M.S. (I973), The asymptotic distribution of the trimmed means, Ann.Statist., I, 472-477.

6. Rosenkrantz W., O'Reilly N. (I972), Application of the Skorokhod representation theorem to rates of convergence for linear combinations of order statistics, Ann. Math. Statist., 43, I204-I2I2.

7. Egorov V.A., Nevzorov V.B. (I974), Some estimates for the rate of convergence of the sum of order statistics to the normal law, Problems of probabilitic distribution II, I05-I28, Notes of Sci. Semin. of Leningrad Branch of Math.Inst., 4I, Leningrad (in Russian).

8. Egorov V.A., Nevzorov V.B. (I974), Sums of order statistics and the normal law, Vestnik Leningrad University, N I, 5-II (in Russian).

9. Egorov V.A., Nevzorov V.B. (I975), On the rate of convergence of linear combinations of absolute order statistics to the normal law, Theoriya Veroyatnostei i Primenen. XX, 207-2I5 (in Russian).

IO.Egorov V.A., Nevzorov V.B. (I975), Asymptotic expansions for the distribution functions of sums of absolute order statistics, Vestnik Leningrad University, N I9, I8-25 (in Russian).

II.Cramér H. (I946), Mathematical methods of statistics. Prinseton Univ.Press.

I2. Bikelis A. (I966), Estimates for the remainder term in the central limit theorem, Litovsky mat.sborn., VI, 32I-346 (in Russian).

I3. Reiss P.D. (I974), On the accuracy of the normal approximation for quantiles, Ann.Prob., 2, 74I-744.

I4. Petrov V.V. (I972), Sums of independent random variables, Moscow (in Russian).

I5. Petrov V.V (I958), On the central limit theorem for m-dependent variables, Selected Transl.Math. Statist. and Prob.,vol.9 (I970), Amer.Math.Soc., 53-67.

Department of Mathematic and Mechanics

Leningrad State University

Leningrad

Some estimates of the rate of convergence in multidimensional limit theorems for homogeneous Markov processes

Š. K. Formanov

0. Let Ω be the set of points ω, $\mathcal{F}_\Omega$ be a σ-algebra of its subsets and $p(\omega,A)$, $\omega\in\Omega$, $A\in\mathcal{F}_\Omega$, be a transition probability: for fixed A, $p(\omega,A)$ is an $\mathcal{F}_\Omega$-measurable function and for fixed ω, is a probability measure on $\mathcal{F}_\Omega$.

Given an initial distribution $\pi(\cdot)$, probabilistic properties of a sequence of random variables are defined by $x_1, x_2, \dots, x_n$

$$P(x_1\in A) = \pi(A),$$
$$P(x_n\in A) = \int_\Omega p^{(n-1)}(\omega,A)\,\pi\,d\omega$$

where $p^{(l)}(\omega,A)$ is the transition probability after l steps:

$$p^{(n)}(\omega,A) = \begin{cases} p(\omega,A) & (n=1), \\ \int_\Omega p^{(n-1)}(\eta,A)\,p(\omega,d\eta) & (n>1). \end{cases}$$

In this paper, we shall assume that $P(\cdot,\cdot)$ satisfies the following condition

$$\sup_{\omega,\bar\omega,A} |p(\omega,A)-p(\bar\omega,A)| = \rho < 1;\ \omega,\bar\omega\in\Omega,\ A\in\mathcal{F}_\Omega. \qquad (0.1)$$

If condition (0.1) is fulfilled, there exists the stationary distribution $p(\cdot)$:

$$\sup_{\omega,A} |p^{(n)}(\omega,A)-p(A)| \leq \rho^{n-1},\quad \omega\in\Omega,\ A\in\mathcal{F}_\Omega.$$

Assume, for the sake of simplicity, that $\pi(\cdot)=p(\cdot)$, although all the below statements, with slight changes in the definitions, hold true for any initial distribution.

Let

$$f(\omega)=(f_1(\omega), f_2(\omega), \dots, f_k(\omega))$$

where $f_i(\omega)$ $(i=\overline{1,k})$ are $\mathcal{F}_\Omega$ -measurable real functions on Ω. Without loss of generality, we shall suppose that

$$\int_\Omega f_i(\eta)\,P(d\eta)=0, \qquad i=\overline{1,k}.$$

Suppose that the matrix $\Lambda=\|\sigma_{ij}\|_1^k$, where

$$\sigma_{ij}=Mf_i(x_1)f_j(x_1)+\sum_{s=1}^{\infty} Mf_i(x_1)f_j(x_{s+1})+\sum_{s=1}^{\infty} Mf_i(x_{s+1})f_j(x_1)$$

is positive definite.

Let

$$S_n=\frac{1}{\sqrt{n}}\sum_{s=1}^{n+1} Bf(x_s)$$

where the matrix B satisfies the condition $B^T B=\Lambda^{-1}$ (T denotes transposition), $\Phi(A)$ is the k -dimensional normal distribution with zero expectation and unit covariance matrix.

Let

$$P_n(A)=P(S_n\in A), \quad \Delta_n(A)=P_n(A)-\Phi(A),$$

$$\Delta_n(g)=\int_{R^k} g\,d(P_n-\Phi)$$

where $A\in\mathcal{B}^k$, the class of all Borel sets on $\mathcal{R}^k$, g is a bounded measurable real function on $\mathcal{R}^k$.

In this paper, we shall give several theorems on estimates of the rate of convergence to zero of $\Delta_n(A)$ and $\Delta_n(g)$. These estimates generalize the results of [5], [6], [7], [8] to the case of homogeneous Markov processes. Besides, they are new for the one-dimensional case too (compare [2]).

1. Main Results. Let (x, y) be the inner product of vectors x and y from $\mathcal{R}^k$, $|x|$ is the norm of vector x,

$$S(x,\varepsilon)=\{y: |x-y|<\varepsilon\}, \qquad x\in\mathcal{R}^k.$$

Let, for any real function g on R^k,

$$W_g(A) = \sup_{x,y\in A} |g(x)-g(y)|, \quad A \subseteq R^k,$$

and
$$W_g(x,\varepsilon) = W_g(s(x,\varepsilon)), \quad g_u(x) = g(x+u), \quad u \in R^k,$$

$$\beta_3 = \sup_{\omega} \int_{\Omega} |Bf(\eta)|^3 \, p(\omega, d\eta),$$

and let $h(\cdot)$ denote a positive constant, not the same at different places, which depends on the arguments in the parentheses.

Theorem 1. For every bounded measurable function g on R^k.

$$|\Delta_n(g)| \le h(k,\rho) W_g(R^k)\beta_3 n^{-1/2} + 2 \sup_{u\in R^k} \int_{R^k} W_{g_u}(\cdot, \varepsilon_n)\, d\Phi$$

where

$$\varepsilon_n = h(k,\rho)\beta_3 n^{-1/2}.$$

Theorem 2. For every bounded measurable function g on R^k,

$$|\Delta_n(g)| \le h(k,\rho) W_g(R^k)\beta_3 n^{-1/2} + \int_{R^k} W_g(\cdot, \varepsilon_n)\, d\Phi$$

where

$$\varepsilon_n = h(k,\rho)\, n^{-1/2} \log n \cdot \beta_3$$

The quantity

$$L_{\pi}(P,Q) = \inf\{\varepsilon : \varepsilon > 0,\ Q(A) \le P(A^{\varepsilon}) + \varepsilon,\ P(A) \le Q(A^{\varepsilon}) + \varepsilon,\ A \in B^k\},$$

where $A^{\varepsilon} = \{y : |x-y| < \varepsilon$ in some $x \in A\}$, is the Lévy-Prokhorov distance between two distributions P and Q.

The estimation of the rate of convergence to zero of $L_{\pi}(P_n, \Phi)$ is of certain interest since the convergence in this metric is equivalent to weak convergence P_n to Φ. Using the lemma on "smoothing" for Lévy-Prokhorov distance [11], we can prove the following theorem.

Theorem 3. The inequality

$$L_{\pi}(P_n, \Phi) \le \frac{h(k,\rho)}{\sqrt{n}} \sum_{\nu=1}^{k} \frac{1}{\lambda^3} \{\sup_{\omega} \int_{\Omega} |(\theta_{\nu}, f(\eta))|^3 \, p(\omega, d\eta)\}$$

holds true, where θ_ν, $\nu=\overline{1,k}$, are orthonormal eigenvectors of Λ and λ_ν^2 are the corresponding eigenvalues. As noted in [7], when proving Theorem 2, we can get the estimate

$$L_\pi(P_n,\Phi)=O\left(\frac{\ln n}{\sqrt{n}}\right).$$

It should be mentioned that the estimates of Theorems 1 and 2 are of universal character and allow us to obtain uniform for some classes of sets, estimates of the remainder term in the multidimensional limit theorem. E.g.,

$$E(d,\varepsilon_0)=\{A: A\in\mathcal{B}^k,\ \Phi((\partial A)^\varepsilon)\leq d\varepsilon \quad \text{for any } \varepsilon, 0<\varepsilon\leq\varepsilon_0\}$$

where ∂A is the boundary of the set A, d and ε_0 are certain positive numbers. This class is a Φ-uniform class of sets in the sense of [4].

From Theorem 2 it follows:

$$\sup_{A\in E(d,\varepsilon_0)}|\Delta_n(A)|\leq h(k,\rho)\beta_3 n^{-1/2}\left(1+\left(d+\frac{1}{\varepsilon_0}\right)\log n\right).$$

If we consider the shift invariant subclass of $E(d,\varepsilon_0)$,

$$E^*(d,\varepsilon_0)=\{A: A+u\in E(d,\varepsilon_0)\} \quad \text{for any} \quad u\in R^k\}$$

we see that the logarithmic factor in the right side of the preceding estimate may be omitted according Theorem 1. It is easy to construct an example of a set from $E(d,\varepsilon_0)$, not belonging to the class $E^*(d,\varepsilon_0)$ (see [8]).

One should note that, according Lemma 1 in Sazonov V.V. [5], the class of all convex Borel sets $\mathcal{C}$ is in $E^*(d(k),\varepsilon_0)$ for any $\varepsilon_0>0$ where

$$d(k)=\frac{2\sqrt{2}}{\sqrt{\pi}}\,k^{3/2}.$$

In the author's paper [9] the estimate

$$\sup_{A\in E^*(d,\varepsilon_0)} |\Delta_n(A)| = O\left(\frac{1}{\sqrt{n}}\right)$$

was obtained under the condition that the moment of order $3+\delta$ $(0<\delta\leq 1)$ with respect to the transition probability exists.

Note that, if

$$\beta_s = \sup_{\omega} \int_{\Omega} |B f(\eta)|^s p(\omega, d\eta), \quad s\geq 3,$$

exists, we can study the asymptotic behaviour of

$$\Delta_n^*(g) = \int_{R^k} g\, d\left(P_n - \sum_{j=1}^{s-2} n^{-j/2} P_j(-\Phi)\right)$$

where $P_j(-\Phi)$ are the functions known from limit theorems and defined by the moments up to order $j+2$. If the multidimensional analogue of the condition of Theorem 5 from [2] (the Cramér type condition) is fulfilled, we can show that

$$|\Delta_n(g)| \leq W_g(\mathcal{R}^k)\delta(n) + (1+\delta_1(n)) \int_{R^k} W_g(\cdot, e^{-\gamma n})\, d\Phi$$

where

$$\delta(n) = O(n^{-(s-2)/2}), \quad \delta_1(n) = O(1)$$

and γ is a positive constant independent of g. Here, instead of the Cramér type condition, one can assume the following condition to be satisfied: let $\psi(t)$ be the Fourier transform of g, i.e.

$$\Psi(t) = \int_{R^k} e^{i(t,x)} g(x)\, dx,$$

and

$$\int_{R^k} |t|^{s-2} |\Psi(t)|\, dt < \infty.$$

Then

$$\Delta_n^*(g) = o(n^{-(s-2)/2})$$

(see [8]).

Now let us consider non-uniform estimates of $\Delta_n(A)$. According to [6], for any $A\subset \mathcal{R}^k$, assume

$$\beta(A) = \begin{cases} \inf\limits_{x\in A} |x| & \text{if} \quad 0 \bar{\in} A, \\ \inf\limits_{x\in R^k\setminus A} & \text{if} \quad 0 \in A. \end{cases}$$

Theorem 4. If $|f(\omega)| \le N < \infty$, then, for any $A \in \mathfrak{C}$ and any integer $q > 0$,

$$|\Delta_n(A)| \le \frac{h(k,\rho,N,q)}{\sqrt{n}\,(1+\beta^q(A))}.$$

It is known that the random vectors

$$f(x_1),\ f(x_2), \dots,\ f(x_n), \dots$$

may form a non-Markov process. To prove this theorem, it is important this sequence should be a Markov process. Therefore we assume that $\Omega = R^k$, $f(x) = (x_1, \dots, x_k)$, $x \in R^k$; it would be sufficient to assume $f(\cdot)$ to be one-to-one.

Theorem 5. For any $A \in \mathfrak{C}$,

$$|\Delta_n(A)| \le \frac{h(k,\rho)}{\sqrt{n}} \cdot \frac{max(\beta_4^{3/4}, \beta_4^4)}{1+\beta^3(A)}.$$

<u>2. Proof of Theorems 1 - 3.</u> Suppose that

$$\Lambda = J \tag{2.1}$$

where J is the unit matrix; consequently $B = J$ and also

$$\frac{\beta_3}{\sqrt{n}} \le h(k,\rho). \tag{2.2}$$

For Theorems 1, 2 and 3 assumptions (2.1) and (2.2) do not bound the generality.

Let

$$f_n(\omega) = (f_{n1}(\omega), \dots, f_{nk}(\omega)) = \begin{cases} f(\omega) & \text{if} \quad |f(\omega)| \le \sqrt{n}, \\ 0 & \text{if} \quad |f(\omega)| > \sqrt{n}, \end{cases}$$

$$\bar{S}_n = \frac{1}{\sqrt{n}} \sum_{\nu=1}^{n+1} f_n(x_\nu), \qquad \Psi_n(t) = Me^{i(t,\bar{S}_n)}.$$

Put

$$M(\rho)=\frac{9}{1-\rho}, \qquad \bar{M}(\rho)=136\,M^3(\rho)+\frac{32}{3}M^2(\rho)+1,$$

$$|\tilde{t}|=max(1,|t|), \qquad T=\frac{\sqrt{n}}{80k\,\bar{M}(\rho)\cdot\beta_3}.$$

In the sequel we shall need estimates of the derivatives of the function $\Psi_n(t)$. Let us fix a direction θ and let t_θ be the projection onto θ. In the author's paper [10] the following lemma is proved with the use of the general spectral theory of linear operators acting in the Banach space of all bounded functions with the uniform norm.

Lemma 2.1. For any $\nu \geq 0$ and $|t| \leq T_n$,

$$\sup_\theta \left| \frac{\partial^\nu}{\partial t_\theta^\nu}\left(\Psi_n(t)-e^{-|t|^2/2}\right)\right| \leq h(k,\rho,\nu)\,|\tilde{t}|^{\nu^*}\cdot e^{-|t|^2/3\nu}\cdot\frac{\beta_3}{\sqrt{n}}$$

where $\nu^* = max(2\nu, \nu+3)$.

We note that in [10] in the given estimate, $\frac{\beta_3}{\sqrt{n}}+\frac{\beta_2}{\sqrt{n}}$ stands instead of $\frac{\beta_3}{\sqrt{n}}$. But since $\Lambda = J$,

$$1=\sigma_{ii}\leq h(\rho)\cdot\beta_2 .$$

Proof of Theorem 1. Let

$$\bar{P}_n(A)=P\{\bar{S}_n\in A\},$$

and let a vector X have the distribution K from Lemma 3.10 of [7] with $s=2$, k_ε is the distribution of the vector εX. The characteristic function of K_ε

$$\zeta_\varepsilon(t)=\zeta(t,\varepsilon)=0 \qquad (2.3)$$

for $|t|\geq h(k)/\varepsilon$. Here $\zeta(t)$ is the characteristic function of K. Let us prove that

$$\|(P_n-\Phi)*K_\varepsilon\|\leq h(k,\rho)\cdot\beta_3\cdot n^{-\frac{1}{2}}, \qquad (2.4)$$

where $\|Q\|$ denotes the full variation of the generalized measure Q, $*$ is the sign of composition.

Let $p_n(x)$ and $\xi_n(t)$ denote the density and the Fourier-Stieltjes transform of the generalized measure $(P_n - \Phi) * K_3$. Then it is easy to make sure of the fact that (see [7])

$$\|(\bar{P}_n - \Phi) * K_3\| \leq h(k) \prod_{m=1}^{k} (I + I_m)^{1/2k} , \qquad (2.5)$$

where

$$I = \int_{R^k} p_n^2(x)\,dx, \qquad I_m = \int_{R^k} x_m^{2(k+2)} p_n^2(x)\,dx ,$$

x_m is the m-th coordinate of the vector x. According to the Parseval formula,

$$I = (2\pi)^{-k} \int_{R^k} |\xi_n(t)|^2 dt, \quad I_m = (2\pi)^{-k} \int_{R^k} \left| \frac{\partial^{k+2}}{\partial t_m^{k+2}} \xi_n(t) \right|^2 dt.$$

Let $\varepsilon = \frac{1}{T_n}$. By applying Lemma 2.1 with $\nu = 0$, we have

$$I \leq \frac{h(k,\rho)}{n} \beta_3^2 . \qquad (2.6)$$

We note that, for $0 \leq r \leq k+2$,

$$\left| \frac{\partial^{k+2-r}}{\partial t_m^{k+2-r}} \zeta_\varepsilon(t) \right| \leq h(k,\rho). \qquad (2.7)$$

The next inequality follows from (2.3), (2.7) and from Lemma 2.1:

$$I_m \leq h(k,\rho) \frac{\beta_3^2}{n} \qquad (2.8)$$

where

$$h(k,\rho) = \max_{1 \leq \nu \leq k+2} \sum_{r=0}^{k+2} C_{k+2}^{r} h(k,\rho,r,\nu).$$

The inequality (2.4) follows from (2.5), (2.6) and (2.8). The further proof of the theorem is the same as in (7).

Theorem 2 may be proved as Theorem 4.1) of [7] by application of Corollary 2.1 of this paper.

We shall give a short proof of Theorem 3.

For any $A \in B^k$, we have the inequalities

$$|P(S_n \in A) - P(\bar{S}_n \in A)| \leq n P(|f(x_1)| \geq \sqrt{n}) \leq \frac{1}{\sqrt{n}} M |f(x_1)|^3.$$

Hence

$$L_{\pi}(P_n, \bar{P}_n) \leq \frac{\beta_3}{\sqrt{n}} . \qquad (2.9)$$

Let $\|\alpha\| = \alpha_1 + \dots + \alpha_k$ where α_i, $i = \overline{1,k}$, are integer positive numbers. For any complex function $W(t)$ on R^k

$$\mathcal{D}^{\ell} W(t) = \Big(\sum_{\|\alpha\|=\ell} \Big| \frac{\partial^{\|\alpha\|}}{\partial t_1^{\alpha_1} \dots \partial t_k^{\alpha_k}} W(t) \Big|^2 \Big)^{1/2}.$$

In the Lemma 4 of [11] we put

$$T = T_n, \quad F = \bar{P}_n, \quad \varphi(t) = \Psi_n(t), \quad G = \Phi, \quad \gamma(t) = e^{-|t|^2/2}.$$

Then, according to this lemma,

$$L_\pi(\bar{P}_n, \Phi) \leq h(k) \Big\{ \frac{1}{T_n} + \Big(\int_{|t| \leq T_n} \big(|\Psi(t) - \gamma(t)|^2 + \sum_{j=1}^{\ell_0} |\mathcal{D}^j (\Psi_n(t) - \gamma(t))|^2 \big) dt \Big)^{1/2} \Big\}. \quad (2.10)$$

where $\ell_0 = [\frac{k}{2}] + 1$ ($[\cdot]$ - is the integer part). From Lemma 2.1 it follows that

$$|(\Psi_n(t) - \gamma(t))| \leq h(k,\rho) \frac{\beta_3}{\sqrt{n}} (1 + |t|^3) e^{-\frac{|t|^2}{30}},$$

$$\mathcal{D}^{\ell} (\Psi_n(t) - \gamma(t)) \leq h(k,\rho) \frac{\beta_3}{\sqrt{n}} (1 + |t|^{k+4}) e^{-\frac{|t|^2}{30}}$$

where $\ell \leq [\frac{k}{2}] + 1$. From inequality (2.10) we have

$$L_\pi(\bar{P}_n, \Phi) \leq h(k,\rho) \frac{\beta_3}{\sqrt{n}}. \quad (2.11)$$

Then Theorem 3 follows from (2.9) and (2.11).

3. Proof of Theorem 4. Let ξ and η be such vectors that

$$P(\xi \in A) = P(A), \quad M|\xi|^{q-1} = b, \quad M|\xi|^q < \infty,$$
$$M|\eta|^\nu \leq \alpha, \quad \nu = \overline{1,q}, \quad q \geq 3.$$

Let

$$T > 0, \quad \eta^{(T)} = T^{-1}\eta, \quad P'_T(A) = P\{\eta^{(T)} \in A\},$$
$$\Delta(A) = \Phi(A) - P(A), \quad \Delta_T(A) = (\Delta * P'_T)(A),$$
$$V(A) = \rho^q(A)\Delta(A), \quad V_T(A) = \rho^q(A)\Delta_T(A), \quad \delta = \sup_{A \in \mathfrak{C}} |V(A)|,$$
$$\delta_T = \sup_{A \in B^k} |V_T(A)|.$$

Lemma 3.1. For all $T \geq 1$,

$$\delta \leq 4\delta_T + h(k,b,\alpha) T^{-1},$$

where $h(k,b,\alpha) \leq \alpha h(k) \max(b, b^{\frac{1}{q-1}})$.

This lemma is proved in [6] for the case $q=3$. In our case, its proof goes along the lines of the proof of Lemma 4 in [6] .

Let vector $\eta^{(1)}$ has the distribution with the density

$$\mathcal{U}(x) = \alpha_{k,\tau} \left[\frac{J_{k/2}\left(\frac{|x|}{2\tau}\right)}{|x|^{k/2}} \right]^{2\tau},$$

where $J_k(z)$ is the Bessel function of order $k/2$, the constant $\alpha_{k,\tau}$ is such that $\int_{R^k} \mathcal{U}(x)\,dx = 1$, τ is some positive number. It is easy to see, that $M|\eta^{(1)}|^\nu \leq h(k,\tau,\nu) < \infty$ for all $\nu \leq k+q$ if $\tau \geq 2 + \frac{q-1}{k+1}$.

Let

$$W_T(t) = M e^{i(t,\eta^{(T)})}.$$

It is easy to check that $W_T(t) = 0$ for $|t| > T$ and

$$\sup_\theta \left| \frac{\partial^\nu}{\partial t_\theta^\nu} W_T(t) \right| \leq M|\eta(T)|^\nu \leq \frac{h(k,\nu)}{T^\nu} \qquad (3.1)$$

for all $\nu = k+q$. First we suppose that

$$\Lambda = J, \quad \frac{1}{\sqrt{n}} \leq h(k,\rho,N,q). \qquad (3.2)$$

Let

$$T = T_n = h(k,\rho,N,q)\cdot\sqrt{n}, \quad P = P_n(A).$$

Let

$$H_n(t) = (Me^{i(t,S_n)} - e^{-|t|^2/2})\, W_{T_n}(t).$$

Then from Lemma 2.1 and inequality (3.2) we have

$$\sup_\theta \int_{R^k} \left| \frac{\partial^\nu}{\partial t_\theta^\nu} H_n(t) \right| dt \leq \frac{h(k,\rho,N,q)}{\sqrt{n}} \qquad (3.3)$$

for all $\nu \leq k+q$.

Note that the generalized measure $\Delta_{T_n}(\cdot)$ with the characteristic function $H_n(t)$ satisfies all the conditions of Lemma 1 of Rotar's paper [6]. According to this (3.3) it follows that

$$\delta_{T_n} \leq \frac{h(k,\rho,N,q)}{\sqrt{n}}.$$

Now Theorem 4 with condition (3.2) follows from Lemma 3.1.

It is evident that

$$\sup_{\eta} |Bf(\eta)| \leq h(k,\rho,\mathcal{N}).$$

Therefore, the first part of condition (3.2) is unessential.

If $\frac{1}{\sqrt{n}} \geq h(k,\rho,\mathcal{N},q)$, then

$$|\Delta_n(A)| \leq \frac{h(k,\rho,\mathcal{N},q)}{\sqrt{n}}. \tag{3.4}$$

For any distribution P, put

$$P^*(A) = \begin{cases} 1-P(A) & \text{if } 0 \in A, \\ -P(A) & \text{if } 0 \bar{\in} A. \end{cases}$$

It is evident that $\Delta_n(A) = |P^*(A) - \Phi^*(A)|$ and

$$|\beta^q(A) P_n^*(A)| \leq M|S_n|^q, \quad |\beta^q(A)\Phi^*(A)| \leq h(k,q).$$

Hence

$$|\Delta_n(A)| \leq |P_n^*(A)| + |\Phi^*(A)| \leq \beta^{-q}(A)(M|S_n|^q + h(k,q)). \tag{3.5}$$

According to the elementary inequality

$$(|x_1|^2 + \ldots + |x_k|^2)^{q/2} \leq (\sqrt{k})^{q-2}(|x_1|^q + \ldots + |x_k|^q), \quad q \geq 2,$$

we have

$$M|S_n|^q \leq (\sqrt{k})^{q-2} \sum_{j=1}^{k} M|S_{n_j}|^q, \tag{3.6}$$

where S_{n_j} is the j-th component of the vector S_n.

According to Lemma 7.4 of [1], page 205, there exists a constant h, independent of n and such that

$$M|S_{n_j}|^q \leq h$$

for any $j = 1, 2, \ldots, k$. Therefore, from (3.5) and (3.6) it follows that

$$|\Delta_n(A)| \leq \beta^{-q}(A)\, h(k,\rho,\mathcal{N},q) \frac{1}{\sqrt{n}}. \tag{3.7}$$

Theorem 4 follows now from (3.4) and (3.7).

4. Proof of Theorem 5. Let $\mu(\omega, A)$ be some kernel in the sense in which it is used in the theory of Markov processes, i.e.,

for fixed $\omega \in \mathcal{R}^k$, $\mu(\omega, A)$ is a countably additive set function on $\mathcal{B}^k$ and, for fixed $A \in \mathcal{B}^k$, $\mu(\omega, A)$ is a measurable function on $\mathcal{R}^k$.

Define the operator Q_p acting to kernels $\mu(\omega, A)$ as follows:

$$(Q_p \mu)(\omega, A) = \int_{\mathcal{R}^k} \mu(u, A-u)\, p(\omega, du)$$

where $A-u = \{a-u, a \in A\}$.

It is evident that this operator is linear. Put

$$\mu_0(u, A) = \begin{cases} 1 & \text{if} \quad 0 \in A, \\ 0 & \text{if} \quad 0 \bar{\in} A. \end{cases}$$

It is easy to check that

$$(Q_p \mu_0)(\omega, A) = p(\omega, A)$$

and

$$P(x_2 + \ldots + x_{n+1} \in A-u \mid_{x_1 = u}) = (Q_p^n \mu_0)(u, A-u).$$

Hence

$$W_n(A) = P(x_1 + \ldots + x_{n+1} \in A) = \int_{\mathcal{R}^k} (Q_p^n \mu_0)(u, A-u) p du. \tag{4.1}$$

This representation will play an important role in further reasoning.

Put

$$S = \{x : x \in \mathcal{R}^k, |x| \leqslant \sqrt{n}\}, \quad \bar{S} = \mathcal{R}^k \setminus S$$

$$\tilde{p}(A) = \begin{cases} p(A \cap S) & \text{if} \quad 0 \bar{\in} A, \\ p((A \cap S) \cup \bar{S}) & \text{if} \quad 0 \in A, \end{cases}$$

$$\tilde{p}(\omega, A) = \begin{cases} p(\omega, A \cap S) & \text{if} \quad 0 \bar{\in} A, \\ p(\omega, (A \cap S) \cup \bar{S}) & \text{if} \quad 0 \in A. \end{cases}$$

Consider the following non-negative measures

$$p_q(A) = \begin{cases} \tilde{p}(A \cap S) & \text{if } 0 \bar{\in} A, \\ p(A \cap S) - \int_{\bar{S}} p(dx) & \text{if } 0 \in A, \end{cases}$$

$$q(\omega, A) = \begin{cases} \tilde{p}(\omega, A \cap S) & \text{if } 0 \bar{\in} A, \\ \tilde{p}(\omega, A \cap S) - \int_{\bar{S}} p(\omega, dx) & \text{if } 0 \in A. \end{cases}$$

It is easy to check that $p(A) - p_q(A)$, $\tilde{p}(A) - p_q(A)$, $p(\omega, A) - q(\omega, A)$ and $\tilde{p}(\omega, A) - q(\omega, A)$ are non-negative measures.

Consider a sequence

$$\tilde{x}_1, \tilde{x}_2, \dots, \tilde{x}_n \dots$$

of random vectors, forming a Markov process with the initial distribution $\tilde{p}(A)$ and the transition probability $\tilde{p}(\omega, A)$, with the state space S. Let

$$\tilde{S}_n = \frac{\tilde{x}_1 + \dots + \tilde{x}_{n+1}}{\sqrt{n}}, \qquad \tilde{P}_n(A) = P(\tilde{S}_n \in A), \quad A \in \mathcal{B}^k.$$

According to (4.1), we have

$$\tilde{W}_n(A) = P(\tilde{x}_1 + \dots + \tilde{x}_{n+1} \in A) =$$
$$= \int_{R^k} (Q_{\tilde{p}}^n \mu_0)(u, A-u)\, p_1(du). \qquad (4.2)$$

Define the measure

$$\Gamma_n(A) = \int_{R^k} (Q_q^n \mu_0)(k, A-u)\, p_q(du). \qquad (4.3)$$

Let us prove that $W_n(A) - \Gamma_n(A)$ and $\tilde{W}_n(A) - \Gamma_n(A)$ are non-negative measures on $\mathcal{B}^k$. Note that operators Q_p preserve non-negativeness. From (4.1) and (4.3) we obtain

$$W_n(A) - \Gamma_n(A) = \int_{R^k} ((Q_p^n - Q_q^n)\mu_0)(u, A-u)\, p(du) +$$

$$+ \int_{R^k} (Q_q^n \mu_0)(u, A-u)(p(du) - p_q(du)) = I + I'.$$

I' is hence non-negative. For I we have the following representation

$$I = \int_{R^k} ((Q_p^n - Q_q^n)\mu_0)(u, A-u)\, p(du) =$$

$$=\sum_{m=0}^{n-1}\int_{R^k}((Q_p^{n-m-1}(Q_p-Q_q)Q_q^m)\mu_0)(u,A-u)p(du)$$

where $Q_p^0=Q_q^0=E$ is the unit operator. Since the operator Q_p-Q_q is generated by the kernel $p(\cdot,\cdot)-q(\cdot,\cdot)$, I is non-negative. Analogously we can prove that the measure

$$\widetilde{W}_n(A)-T_n(A)$$

is non-negative. As in the case of Theorem 4, it is sufficient to prove Theorem 5 for

$$\Lambda=J,\quad \frac{max(\beta_4^{3/4},\beta_4^4)}{\sqrt{n}}\leqslant h(k,\rho)=\frac{1-\rho}{36k^2} \tag{4.4}$$

Lemma 4.1. For any $A\in\mathfrak{B}^k$

$$|P_n(A)-\bar{P}_n(A)|\leqslant\frac{h(k,\rho)max(\beta_4^{3/4},\beta_4^4)}{\sqrt{n}(1+\beta^3(A))}.$$

In what follows by $\sqrt{n}A$ we denote the sets

$$\{x:\frac{x}{\sqrt{n}}\in A\}.$$

Assume that we have managed to show that

$$|P_n(A)-T_n(\sqrt{n}A)|\leqslant h(k,\rho)\frac{max(\beta_4^{3/4},\beta_4^4)}{\sqrt{n}\beta^3(A)}; \tag{4.5}$$

$$|\widetilde{P}_n(A)-T_n(\sqrt{n}A)|\leqslant h(k,\rho)\frac{max(\beta_4^{3/4},\beta_4^4)}{\sqrt{n}\beta^3(A)}. \tag{4.6}$$

Then from this the lemma would immediately follow. Note that, without loss generality, one may assume that $0\bar{\in}A$. Since

$$P_n(A)=W_n(\sqrt{n}A),$$

$$P_n(A)-T_n(\sqrt{n}A)=\int_A W_n(\sqrt{n}dx)-T_n(\sqrt{n}dx)\leqslant$$

$$\leqslant\beta^{-3}(A)\int_A|x|^3[W_n(\sqrt{n}dx)-T_n(\sqrt{n}dx)]\leqslant$$

$$\leqslant\sqrt{k}\beta^{-3}(A)\sum_{s=1}^{k}\int_{-\infty}^{\infty}|\frac{y}{\sqrt{n}}|^3d(W_{n,s}(y)-T_{n,s}(y)), \tag{4.7}$$

where

$$W_{n,s}(y)=\int_{u_s<y}W_n(du),\quad T_{n,s}(y)=\int_{u_s<y}T_n(du),$$

u_s is the s-th coordinate of the vector $u=(u_1,\dots,u_k)$.

To estimate the integrals in the right hand side of (4.7), we shall use the ideas of [3], [6] for the case of independent random variables. As is clear from further application, in the case of Markov chains there appear great difficulties.

We have

$$\int_{-\infty}^{\infty}\left|\frac{y}{\sqrt{n}}\right|^3 dW_{n,s}(y)=\left|\int_{-\infty}^{0}\frac{y^3}{n^{3/2}}\,dW_{n,s}(y)\right|+\int_{0}^{\infty}\frac{y^3}{n^{3/2}}\,dW_{n,s}(y), \qquad (4.8)$$

and

$$\int_{-\infty}^{0}\left|\frac{y}{\sqrt{n}}\right|^3 dW_{n,s}(y)=\frac{1}{n^{3/2}}\int\limits_{y_1^{(s)}+\dots+y_{n+1}^{(s)}<0}\cdots\int (y_1^{(s)}+\dots+y_{n+1}^{(s)})^3\,p(dy_1)\dots p(y_n,dy_{n+1})=$$

$$=\frac{1}{n^{3/2}}\sum_{i=1}^{n+1}\int\limits_{y_1^{(s)}+\dots+y_{n+1}^{(s)}<0}\cdots\int y_i^{(s)}(y_1^{(s)}+\dots+y_{n+1}^{(s)})^2\,p(dy_1)\dots p(y_n,dy_{n+1})=$$

$$=\frac{1}{n^{3/2}}\sum_{i=1}^{n+1}\int\limits_{z_1^{(i)}+z_{i+1}^{(n+1)}}\cdots\int\sum_{m=0}^{2}C_2^m(z_1^{(i)})^m y_i^{(s)}(z_{i+1}^{(n+1)})^{2-m}\,p(dy_1)\dots p(y_n,dy_{n+1})$$

where $y_i^{(s)}$ is the s-th component of the vector y_i,

$$z_1^{(i)}=y_1^{(s)}+\dots+y_i^{(s)},\quad z_{i+1}^{(n+1)}=y_{i+1}^{(s)}+\dots+y_{n+1}^{(s)},\quad z_i^{(i)}=0.$$

Put

$$P^{(j)}(u,A)=(Q_p^j\mu_0)(u,A),\qquad Q^{(j)}(u,A)=(Q_q^j\mu_0)(u,A),$$

$$P_s^{(j)}(u,y)=\int\limits_{x_s<y}P^{(j)}(u,dv),\quad Q_s^{(j)}(u,y)=\int\limits_{x_s<y}Q^{(j)}(u,dx).$$

Then

$$\int_{-\infty}^{0}\frac{y^3}{n^{3/2}}\,dW_{n,s}(y)=J_0^{(1)}+J_0^{(2)}+J_1^{(1)}+J_1^{(2)}+J_2^{(1)}+J_2^{(2)}, \qquad (4.9)$$

where

$$J_0^{(1)}=\int_{R^k}p(dy_1)\frac{y_1^{(s)}}{\sqrt{n}}\int_{-\infty}^{-y_1^{(s)}}\frac{u^2}{n}\,dP_s^{(n)}(y_1,u),$$

$$J_0^{(2)}=\sum_{i=2}^{n}\underbrace{\int\cdots\int}_{i}p(dy_1)p(y_1,dy_2)\dots p(y_{i-1},dy_i)\frac{y_i^{(s)}}{\sqrt{n}}\int_{-\infty}^{-z_1^{(i)}}\frac{u^2}{n}\,dP_s^{(n+1-i)}(y_i,u),$$

$$J_1^{(1)}=2\int_{R^k}p(dy_1)\frac{(y_1^{(s)})^2}{n}\int_{-\infty}^{-y_1^{(s)}}\frac{u}{\sqrt{n}}\,dP_s^{(n)}(y_1,u),$$

$$J_1^{(2)}=\sum_{i=2}^{n}\underbrace{\int\cdots\int}_{i}p(dy_1)p(y_1,dy_2)\dots p(y_{i-1},dy_i)\frac{z_1^{(i)}}{\sqrt{n}}\frac{y_i^{(s)}}{\sqrt{n}}\int_{-\infty}^{-z_1^{(i)}}\frac{u}{\sqrt{n}}\,dP_s^{(n+1-i)}(y_i,u),$$

$$J_2^{(1)} = \int_{\mathcal{R}^k} p(dy_1) \frac{(y_1^{(s)})^3}{n^{3/2}} \int_{-\infty}^{-y_i^{(s)}} dP_s^{(n)}(y_1, u),$$

$$J_2^{(2)} = \sum_{i=2}^{n} \underbrace{\int \dots \int}_{i} p(dy_1) p(y_1, dy_2) \dots p(y_{i-1}, dy_i) \frac{(z_1^{(i)})^2}{n} \frac{y_i^{(s)}}{\sqrt{n}} \int_{-\infty}^{-z_1^{(i)}} dP_s^{(n+1-i)}(y_i, u).$$

Similarly to (4.9), we have

$$\int_{-\infty}^{0} \frac{y^3}{(\sqrt{n})^3} \, d\Gamma_{n,s}(y) = \bar{J}_0^{(1)} + \bar{J}_0^{(2)} + \bar{J}_1^{(1)} + \bar{J}_1^{(2)} + \bar{J}_2^{(1)} + \bar{J}_2^{(2)} \tag{4.10}$$

where $\bar{J}_i^{(j)}$ is obtained from the expression for $J_i^{(j)}$ by substitutioning $p_q(\cdot)$, $q(\cdot,\cdot)$ for $p(\cdot)$, $p(\cdot,\cdot)$ respectively. From (4.3) - (4.10) it follows that

$$\left| \int_{-\infty}^{0} \frac{y^3}{(\sqrt{n})^3} \, d(W_{n,s}(y) - T_{n,s}(y)) \right| \le |J_0^{(1)} - \bar{J}_0^{(1)}| + |J_0^{(2)} - \bar{J}_0^{(2)}| +$$

$$+ |J_1^{(1)} - \bar{J}_1^{(1)}| + |J_1^{(2)} - \bar{J}_1^{(2)}| + |J_2^{(1)} - \bar{J}_2^{(1)}| + |J_2^{(2)} - \bar{J}_2^{(2)}|. \tag{4.11}$$

Estimate now $|J_0^{(1)} - \bar{J}_0^{(1)}|$. It is easy to see that

$$|J_0^{(1)} - \bar{J}_0^{(1)}| \le \left| \int_{\mathcal{R}^k} (p(dy_1) - p_q(dy_1)) \right| \frac{y_1^{(s)}}{\sqrt{n}} \int_{-\infty}^{-y_1^{(s)}} \frac{u^2}{n} \, dP_s^{(n)}(y_1, u) \Big| +$$

$$+ \int_{\mathcal{R}^k} P_q(dy_1) \frac{y_1^{(s)}}{\sqrt{n}} \int_{-\infty}^{-y_1^{(s)}} \frac{u^2}{n} \, d(P_s^{(n)}(y_1, u) - Q_s^{(n)}(y_1, u)) \Big| = R_0^{(1)} + R_0^{(2)}. \tag{4.12}$$

Further,

$$R_0^{(1)} \le \int_{S} p(dy_1) \frac{|y_1^{(s)}|}{\sqrt{n}} \int_{-\infty}^{\infty} \frac{u^2}{n} \, dP_s^{(n)}(y_1, u). \tag{4.13}$$

It is easy to check that

$$\int_{-\infty}^{\infty} \frac{u^2}{n} \, dP_s^{(n+1)}(y_1, u) = \frac{1}{n} \sum_{\nu=2}^{n+1} \int_{\mathcal{R}^k} (y_\nu^{(s)})^2 p^{(\nu)}(y_1, dy_\nu) +$$

$$+ \frac{2}{n} \sum_{j=2}^{n+1} \sum_{\nu=1}^{n+1-j} \int_{\mathcal{R}^k} y_1^{(s)} (p^{(j)}(y_1, dy_j) - p(dy_j)) \int_{\mathcal{R}^k} y_\nu^{(s)} (p^{(\nu)}(y_j, dy_\nu) - p(dy_\nu)) +$$

$$+ \frac{2}{n} \sum_{j=2}^{n+1} \sum_{\nu=1}^{n+1-j} \int_{\mathcal{R}^k} y_j^{(s)} p(dy_j) \int_{\mathcal{R}^k} y_\nu^{(s)} (p^{(\nu)}(y_j, dy_\nu) - p(dy_\nu)).$$

Hence

$$\int_{-\infty}^{\infty} \frac{u^2}{n} \, dP_s^{(n)}(y_1, u) \le h(\rho) \sup_x \int_{\mathcal{R}^k} y^2 p(x, dy). \tag{4.14}$$

Now estimate $R_0^{(2)}$. One can check that

$$\int_{-\infty}^{\infty} \frac{u^2}{n} \, dQ_s^{(n+1)} = \sum_{j=2}^{n+1} \int_{\mathcal{R}^k} \frac{(y_j^{(s)})}{n} q^{(n-1-j)}(y_j, \mathcal{R}^k) q^{(j)}(y_1, dy_j) +$$

$$+2\sum_{j=2}^{n+1}\sum_{\nu=1}^{n+1-j}\int_{R^k}\frac{y_j^{(s)}}{\sqrt{n}}q^{(j)}(y_1,dy_j)\int_{R^k}\frac{y_\nu^{(s)}}{\sqrt{n}}q^{(n-1-j-\nu)}(y_\nu,R^k)q^{\nu}(y_j,dy_\nu)$$

where $q^{\nu}(u,A)$ is the ν -th iteration of the measure $q(\cdot,\cdot)$.

One can check by induction that, for any j and A,

$$p^{(j)}(\omega,A)-q^{(j)}(\omega,A)\leq j\cdot\sup_x\int_{\bar{S}}p(x,dy). \qquad (4.15)$$

After some transformation, we get that, for $\tau=1,2,3$,

$$\left|\int_{R^k}\left(\frac{y_i^{(s)}}{\sqrt{n}}\right)^{\tau}q^{(j)}(y_1,dy_i)-\int_{R^k}\left(\frac{y_i^{(s)}}{\sqrt{n}}\right)^{\tau}p^{(j)}(y_1,dy_i)\right|\leq$$

$$\leq(j-1)\frac{\sup_x\int_{R^k}|y|^{\tau}p(x,dy)}{(\sqrt{n})^{\tau}}\sup_u\int_{\bar{S}}p(u,dy)+\sup_u\int_{\bar{S}}\left(\frac{|y|}{\sqrt{n}}\right)^{\tau}p(u,dy). \qquad (4.16)$$

According to (4.15) and (4.16),

$$\left|\int_{R^k}\frac{y_i^{(s)}}{\sqrt{n}}q^{(j)}(y_1,dy_i)\int_{R^k}\frac{y_\nu^{(s)}}{\sqrt{n}}(q^{(n-i-j-\nu)}(y_\nu,R^k)-1)(q^{(\nu)}(y_j,dy_\nu)-1)q^{\nu}(y_j,dy_\nu)-\right.$$

$$\left.-\int_{R^k}\frac{y_i^{(s)}}{\sqrt{n}}p^{(j)}(y_1,dy_i)\int_{R^k}\frac{y_\nu^{(s)}}{\sqrt{n}}(q^{(n-i-j-\nu)}(y_\nu,R^k)-1)p^{\nu}(y_i,dy_\nu)\right|\leq\frac{8}{n^3}\max(\beta_4^2,\beta_4^{5/2}). \qquad (4.17)$$

It is easy to check that

$$\left|\int_{R^k}\frac{y_i^{(s)}}{\sqrt{n}}p^{(j)}(y_1,dy_i)\int_{R^k}\frac{y_\nu^{(s)}}{\sqrt{n}}(q^{(n-i-j-\nu)}(y_\nu,R^k)-1)p^{\nu}(y_i,dy_\nu)\right|\leq$$

$$\leq 2\frac{\beta_4^{3/2}}{n^2}(\rho^{2j/3}+\rho^{2\nu/3}), \qquad (4.18)$$

$$\left|\int_{R^k}\frac{y_i^{(s)}}{\sqrt{n}}q^{(j)}(y_1,dy_i)\int_{R^k}\frac{y_\nu^{(s)}}{\sqrt{n}}q^{(\nu)}(y_i,dy_\nu)-\int_{R^k}\frac{y_i^{(s)}}{\sqrt{n}}p^{(j)}(y_1,dy_i)\int_{R^k}\frac{y_\nu^{(s)}}{\sqrt{n}}p^{\nu}(y_i,dy_\nu)\right|\leq$$

$$\leq\left(\frac{3}{n^2}\rho^{3j/4}+\frac{6}{n^3}+8\frac{\rho^{3\nu/4}}{n^2}\right)\max(\beta_4,\beta_4^{5/2}), \qquad (4.19)$$

$$\left|\int_{R^k}\frac{(y_i^{(s)})^2}{n^2}(q^{(n-i-j)}(y_i,R^k)-1)\,q^{(j)}(u,dy_i)\right|\leq\frac{\beta_4^{3/2}}{n^2} \qquad (4.20)$$

Put, for $\tau=0,1,2$

$$a_{n,\tau}=\int_{-\infty}^{\infty}\left|\frac{u}{\sqrt{n}}\right|^{\tau}d\,(P_s^{(n)}(y,u)-Q_s^{(n)}(y,u)).$$

From (4.15) it follows that

$$a_{n,0}\leq\frac{\beta_4}{n}. \qquad (4.21)$$

From (4.16) - (4.20) we get

$$a_{n,2}\leq h(\rho)\frac{\max(\beta_4,\beta_4^{5/2})}{n}. \qquad (4.22)$$

It is evident that

$$a_{n,1} \leq \int_{|u|\leq\sqrt{n}} d\left(P_s^{(n)}(y,u) - Q_s^{(n)}(y,u)\right) + \int_{|u|>\sqrt{n}} \left|\frac{u}{\sqrt{n}}\right|^2 d\left(P_s^{(n)}(y,u) - Q_s^{(n)}(y,u)\right).$$

Therefore from (4.21) - (4.22) it follows that

$$a_{n,1} \leq a_{n,0} + a_{n,2} \leq h(\rho)\frac{\max(\beta_4, \beta_4^{5/2})}{n}. \qquad (4.23)$$

Finally, from (4.21) - (4.23) we get

$$R_0^{(2)} \leq h(\rho)\frac{\max(\beta_4, \beta_4^{11/4})}{\sqrt{n}}. \qquad (4.24)$$

So (4.12) - (4.14) and (4.24) imply

$$|J_0^{(1)} - \bar{J}_0^{(1)}| \leq h(\rho)\frac{\max(\beta_4, \beta_4^{11/4})}{\sqrt{n}}. \qquad (4.25)$$

After similar reasoning and cumbersome transformations, we arrive at the following estimates:

$$|J_0^{(2)} - \bar{J}_0^{(2)}| \leq h(\rho)\frac{\max(\beta_4^{5/4}, \beta_4^{4})}{\sqrt{n}}, \qquad (4.26)$$

$$|J_1^{(1)} - \bar{J}_1^{(1)}| \leq h(\rho)\frac{\max(\beta_4^{3/4}, \beta_4^{13/4})}{\sqrt{n}}, \qquad (4.27)$$

$$|J_1^{(2)} - \bar{J}_1^{(2)}| \leq h(\rho)\frac{\max(\beta_4^{5/4}, \beta_4^{29/8})}{\sqrt{n}}, \qquad (4.28)$$

$$|J_2^{(2)} - \bar{J}_2^{(2)}| \leq h(\rho)\frac{\max(\beta_4^{5/4}, \beta_4^{9/4})}{\sqrt{n}}. \qquad (4.28)$$

From (4.25) - (4.29) inequality (4.5) follows. Inequality (4.6) is proved similarly.

Lemma 4.1 is proved.

Now the proof of Theorem 5 goes along the lines of the proof in the case of independent random vectors with application of Lemmas 2.1 and 4.1.

Acknowledgement. The author is grateful to V.I. Rotar' for useful and stimulating discussions.

References

1. G.L. Doob, Stochastic processes, Moscow, 1956 (Russian translation).

2. S.V. Nagaev , Teoriya Veroyatnostei i ee primeneniya, 5, 1 (1961), 62-81.

3. L.V. Osipov, Vestnik Leningradskogo Universiteta, 19, 4(1967), 45-62.

4. P. Billingsley & O. Topsøe, Z.Wahrscheinlichkeitstheorie verw. Geb., 7, 1(1967), 1-16.

5. V.V. Sazonov, Synkhya, ser. A, 30(1968), 203-225.

6. V.I. Rotar', Teoriya Veroyatnostei i ee Primeneniya, 15, 4(1970), 647-665.

7. R.W. Bhattacharya, Proc. 6th Berkeley Sympos. Math.Statist.and Probability, vol.II, 1971, 453 - 484.

8. R.W.Bhattacharya, Ann.Math.Statist., 42(1971), 241 - 259.

9. Š.K.Formanov, Doklady Akademii Nauk SSSR 204, 1(1972), 29 - 31.

10.Š.K.Formanov,Izvestiya Akademii Nauk UzSSR,ser.fiz.-mat.3(1972), 33 - 37; 6(1972), 35 - 52.

11.V.V.Yurinskii,Teoriya Veroyatnostei i ee Primeneniya,20, 1(1975), 3 - 12.

Mathematical Institute

Academy of Sciences of the Uzbek SSR

Tashkent

EXPECTATION SEMIGROUP OF A CASCADE PROCESS AND A LIMIT THEOREM

Tetsuo Fujimagari

Introduction

A mathematical study on an electron-photon cascade was given by T.E.Harris in his book [2], and after that, a similar stochastic process was studied by M.Motoo and the author [1] as a class of branching Markov processes on a certain space of measures. In [1], the stochastic process was called as a cascade process and was shown to be characterized by a pair of a nonnegative constant and a σ-finite measure. The characterization was carried out by using a system of nonlinear integral equations. We are now concerned with the investigation of asymptotic properties of the cascade process.

In §1 we shall present the definition of a cascade process and its fundamental equations without proof from [1]. In §2 we shall consider the expected number of particles of the cascade process existing in some set of states, which are regarded as energies of particles, at a given time and the corresponding semigroup of operators, which will be called an expectation semigroup. A system of linear integral equations as well as the infinitesimal generator of the expectation semigroup will be given there. In the last section, we shall consider the expected number of particles with energies above a given level. It will be given a limit theorem of a normal distribution character on this expected number, a similar result of which was originally given in [2]. The theorem will mean that, if we take $-\log u$ instead of u as a scale of energy, then the energy distribution of the cascade process is asymptotically normally distributed.

§1. Description of a cascade process and its fundamental equations

In this section we are devoted to presenting several notions and a definition of a cascade process and its fundamental equations from the work of M.Motoo and the author [1]. Detailed properties and proofs are to be seen in [1].

Let S be an interval $(0,1]$ and set

$$M_1 = \{ \mu;\ \mu \text{ is a measure on } S \text{ such that } \mu=0 \text{ or } \mu=\sum_i x_i\delta_{x_i}\ (x_i \in S)$$
$$\text{and } \|\mu\| = \sum_i x_i \leqq 1 \},$$

where $x_i\delta_{x_i}$ is a measure which is concentrated at a point x_i and has a mass x_i at the point, and where $\sum_i$ denotes a finite or countably infinite sum. The set M_1 is shown to be a compact metrizable space with the weak*-topology. Set

$$B_0^* = \{ f \,;\, f \text{ is a Borel function on } S \text{ such that } 0 \leqq f \leqq 1 \text{ and } f = 1 \text{ in some neighborhood of } 0 \}$$

and

$$C_0^* = \{ f \in B_0^* \,;\, f \text{ is a continuous function on } S \text{ such that } f > 0 \}$$

and define a function $\hat{f}(\mu)$ on M_1 for any $f \in B_0^*$ by

$$\hat{f}(\mu) = \exp\left(\int_S \frac{1}{x} \log f(x)\, \mu(dx) \right) ,$$

where $\log 0 = -\infty$ and $e^{-\infty} = 0$. Then $\hat{f}$ is a continuous function on M_1 if $f \in C_0^*$. We denote by $C(M_1)$ the set of all continuous functions on M_1. In addition, set

$$\theta_a f(x) = f(ax) , \qquad x \in S$$

for any Borel function f on S and $a \in S$.

Now a *cascade process* $(\mu_t) = \{ \mu_t, P_\mu \,;\, \mu \in M_1\}$ is defined as a Markov process on the state space M_1 which corresponds in a usual way to a *cascade semigroup* $\{T_t \,;\, t \geqq 0\}$, which in turn is defined as such a semigroup that satisfies : (a) $\{T_t \,;\, t \geqq 0\}$ is a strongly continuous and contraction semigroup of nonnegative linear operators on $C(M_1)$ and $T_t 1 = 1$ $(t \geqq 0)$, (b) $T_t\hat{f}(\mu+\nu) = T_t\hat{f}(\mu)\, T_t\hat{f}(\nu)$ for any $f \in C_0^*$ and for $\mu, \nu \in M_1$ such that $\mu+\nu \in M_1$ also, and (c) $T_t\hat{f}(a\delta_a) = T_t\widehat{\theta_a f}(\delta_1)$ for any $f \in C_0^*$ and for $a \in S$. In a cascade process (μ_t), a state $\mu = \sum_i x_i\delta_{x_i} \in M_1$ may be interpreted as the existence of particles with energies (or masses, etc.) $x_1, x_2, \cdots$, and the property (b) which will be called a *branching property* is an abstruction of the independence of each particle of the cascade process, and moreover the property (c) may represent a *homogeneity* of the process.

It is known from the results of [1] that a cascade process can be characterized by a nonnegative constant m and a nonnegative Borel measure $\Pi(d\mu)$ on $M_1 - \{\delta_1\}$ which satisfies

$$(1.1) \qquad \int_{M_1-\{\delta_1\}} (1-M(\mu))\, \Pi(d\mu) < +\infty ,$$

where $M(\mu)$ is a function on M_1 defined by

$$M(\mu) = \begin{cases} \max_i x_i & \text{if} \quad \mu = \sum_i x_i \delta_{x_i} \\ 0 & \text{if} \quad \mu = 0 \end{cases} .$$

Let $(x_t) = \{x_t, P_x^\circ ; x \in [0,1]\}$ be a right continuous strong Markov process on $[0,1]$ generated by an infinitesimal generator A° such that the domain $D(A^\circ)$ contains $C^1[0,1]$ the set of all continuously differentiable functions on $[0,1]$ and for $f \in C^1[0,1]$

$$(1.2) \qquad A^\circ f(x) = -mx \frac{df}{dx} + \int_{M_1 - \{\delta_1\}} \Pi(d\mu)(f(xM(\mu)) - f(x)) .$$

This process (x_t) and the measure $\Pi(d\mu)$ will be called an *underlying process* and a *branching measure*, respectively, of the cascade process.

For any d ($\frac{2}{3} < d < 1$), set $S_d = (\frac{1-d}{d} , 1]$ and

$$B_d^* = \{f \in B_\circ^*; \ f=1 \ \text{ on } \ S - S_d \} .$$

Define a multiplication of $a \in S$ and $\mu \in M_1$ by

$$a \cdot \mu = \begin{cases} \sum_i ax_i \delta_{ax_i} & \text{if} \quad \mu = \sum_i x_i \delta_{x_i} \\ 0 & \text{if} \quad \mu = 0 \end{cases}$$

and a Markov time τ_d° ($\frac{2}{3} < d < 1$) of the underlying process (x_t) by

$$\tau_d^\circ = \begin{cases} \inf \{ t; \ \frac{x_t}{x_0} \leqq d\} & \\ + \infty & \text{if } \{ \cdots \} = \phi \end{cases} .$$

It is shown in [1] that $u_t(x) = T_t \hat{f}(x\delta_x)$ for $f \in B_d^*$ is a solution of the following (S_d)-equation :

$$(S_d) \quad u_t(x) = E_x^\circ[f(x_t); \ t < \tau_d^\circ] + E_x^\circ[u_{t-\tau_d^\circ}(xd); \ x_{\tau_d^\circ} = xd, \ \tau_d^\circ \leqq t]$$

$$+ E_x^\circ[\int_0^t ds \int_{M_1 - \{\delta_1\}} \Pi(d\mu) \ \chi(x_s M(\mu) < xd < x_s) \hat{u}_{t-s}(x_s \cdot \mu)],$$

where $E_x^\circ[\ \cdot \ ]$ stands for the expectation by the probability measure P_x° and $\chi(\ \cdot \ )$ for the indicator function of the event $\{x_s M(\mu) < xd < x_s\}$.

§2. Expectation semigroup and its infinitesimal generator

We define in this section an expectation semigroup of a cascade process and deduce its infinitesimal generator. Similar treatments about this section are seen also in the paper by N.Ikeda, M.Nagasawa, and S.Watanabe [3].

Set

$$B_d = \{f \; ; \; f \text{ is a bounded Borel function on } S \text{ and } f=0 \text{ on } S-S_d\}$$

for any d ($\frac{2}{3} < d < 1$), $B_0 = \bigcup_{\frac{2}{3}<d<1} B_d$, and $C_0 = B_0 \cap C(0,1]$.

Define

$$\|f\| = \sup_{0<x\leq 1} |f(x)|$$

and

$$\check{f}(\mu) = \int_S \frac{1}{x} f(x)\mu(dx), \qquad \mu \in M_1 \tag{2.1}$$

for any $f \in B_0$. Then we have the following properties of the function $\check{f}$ on M_1.

<u>Lemma 2.1</u>. (i) *For* $f, g \in B_0$ *and for real numbers* α, β,

$$(\alpha f+\beta g)^{\vee}(\mu) = \alpha\check{f}(\mu) + \beta\check{g}(\mu), \qquad \mu \in M_1 .$$

(ii) *For* $f \in B_0$ *and* $a \in S$,

$$(\theta_a f)^{\vee}(\mu) = \check{f}(a\cdot\mu), \qquad \mu \in M_1 .$$

(iii) *For* $f \in B_0$ *and for* $\mu, \nu \in M_1$ *such as* $\mu+\nu \in M_1$ *also,*

$$\check{f}(\mu+\nu) = \check{f}(\mu) + \check{f}(\nu) .$$

(iv) *For* $f \in B_d$,

$$\|\check{f}\| = \sup_{\mu \in M_1} |\check{f}(\mu)| \leq \frac{d}{1-d} \|f\|$$

and $\check{f}(\mu) = 0$ *if* $M(\mu) \leq \frac{1-d}{d}$.

(v) $\check{f} \in C(M_1)$ *for any* $f \in C_0$ *and if* $f \in B_0$, $\check{f}$ *is a bounded measurable function on* M_1.

<u>Proof:</u> It is clear for (i), (ii), and (iii). As for (iv),

$$\check{f}(\mu) = \int_{S_d} \frac{1}{x} f(x)\mu(dx), \qquad \mu \in M_1$$

if $f \in B_d$, and since $\|\mu\| \leq 1$ and so

$$(2.2) \qquad \int_{S_d} \frac{1}{x}\,\mu(dx) \leqq \frac{d}{1-d}\,,$$

it follows

$$|\check{f}(\mu)| \leqq \|f\| \int_{S_d} \frac{1}{x}\,\mu(dx) \leqq \frac{d}{1-d}\|f\|\,, \qquad \mu \in M_1\,.$$

The second part of (iv) is also clear. For the first part of (v) it is obvious because M_1 has the weak*-topology, and for the second part it follows from the standard arguement.

Now we define

$$(2.3) \qquad M_t f(x) = E_x[\,\check{f}(\mu_t)]\,, \qquad x \in S \quad \text{and} \quad t \geqq 0\,,$$

for all $f \in B_\circ$, where $E_x[\,\cdot\,]$ stands for the expectation by the probability measure $P_{x\delta_x}$ of a cascade process (μ_t, P_μ).

Lemma 2.2. (i) *If* $f \in B_d$, $M_t f \in B_d$ *also and*

$$\|M_t f\| \leqq \frac{d}{1-d}\,\|f\|\,.$$

(ii) *For* $f \in B_\circ$,

$$(2.4) \qquad \widecheck{M_t f}(\mu) = E_\mu[\,\check{f}(\mu_t)\,]\,, \qquad \mu \in M_1 \quad \textit{and} \quad t \geqq 0\,.$$

(iii) *For* $f \in B_\circ$ *and* $a \in S$,

$$(2.5) \qquad M_t f(ax) = M_t \theta_a f(x)\,, \qquad x \in S \quad \textit{and} \quad t \geqq 0\,.$$

Proof: (i) Let $f \in B_d$. Then, since $M_t f(x) = E_1[\,\check{f}(x\cdot\mu_t)\,]$ by the homogeneity property of (μ_t) (see, Proposition 2.3 of [1]) and $M(x\cdot\mu_t) \leqq x$, it follows $M_t f(x) = 0$ for $x \leqq \frac{1-d}{d}$ from Lemma 2.1 (iv) and hence $M_t f \in B_d$. Moreover,

$$|M_t f(x)| \leqq E_x\,|\check{f}(\mu_t)| \leqq \frac{d}{1-d}\|f\|\,, \qquad x \in S$$

by Lemma 2.1 (iv).

(ii) For $f \in B_\circ$ and for $\mu, \nu \in M_1$ such as $\mu+\nu \in M_1$ also,

$$\begin{aligned}(2.6) \qquad E_{\mu+\nu}[\,\check{f}(\mu_t)\,] &= E_\mu^{(1)} \otimes E_\nu^{(2)}[\,\check{f}(\mu_t(w_1) + \mu_t(w_2))\,] \\ &= E_\mu^{(1)} \otimes E_\nu^{(2)}[\,\check{f}(\mu_t(w_1)) + \check{f}(\mu_t(w_2))\,] \\ &= E_\mu[\,\check{f}(\mu_t)\,] + E_\nu[\,\check{f}(\mu_t)\,]\end{aligned}$$

by the branching property of (μ_t) (see, Proposition 2.2 of [1]) . Since $M_t f \in B_d$ for $f \in B_d$, putting the restriction of $\mu \in M_1$ on S_d by

$\mu|_{S_d} = x_1\delta_{x_1} + \cdots + x_n\delta_{x_n}$ ($n \leqq \frac{d}{1-d}$) , we have

$$\widetilde{M_t f}(\mu) = \widetilde{M_t f}(x_1\delta_{x_1} + \cdots + x_n\delta_{x_n}) = M_t f(x_1) + \cdots + M_t f(x_n)$$
$$= E_{x_1}[\ \check{f}(\mu_t)\ ] + \cdots + E_{x_n}[\ \check{f}(\mu_t)\ ]$$
$$= E_{x_1\delta_{x_1} + \cdots + x_n\delta_{x_n}}[\ \check{f}(\mu_t)\ ] = E_\mu[\ \check{f}(\mu_t)\ ]$$

from (2.6).

(iii) $M_t f(ax) = E_x[\ \check{f}(a\cdot\mu_t)\ ] = E_x[\ \widetilde{\theta_a f}(\mu_t)\ ] = M_t\theta_a f(x)$.

Theorem 2.1. $\{M_t\ ;\ t \geqq 0\}$ *is a strongly continuous semigroup of nonnegative linear operators on* C_o *and* $M_0 = I$ (= *identity*). *Moreover, if* $f \in C_o^r(0,1] = \{f \in C_o\ ;\ f$ *is r-times continuously differentiable on* $S\}$, *it holds* $M_t f \in C_o^r(0,1]$ *also* ($r=0,1,2,\cdots$) .

Proof: Since $\{T_t\ ;\ t\geqq 0\}$ is a strongly continuous semigroup on $C(M_1)$ and $\check{f} \in C(M_1)$ for $f \in C_o$, $M_t f(x) = T_t\check{f}(x\delta_x)$ implies $M_t f \in C_o$ for $f \in C_o$ and $\|M_t f - f\| \leqq \|T_t\check{f} - \check{f}\| \longrightarrow 0$ as $t \longrightarrow 0$. In addition, by Lemma 2.2(ii),

$$M_{t+s}f(x) = T_{t+s}\check{f}(x\delta_x) = T_t(T_s\check{f})(x\delta_x)$$
$$= T_t(\widetilde{M_s f})(x\delta_x) = M_t M_s f(x) .$$

Thus, $\{M_t\ ;\ t\geqq 0\}$ is a strongly continuous semigroup of nonnegative linear operators on C_o and $M_0 f(x) = T_0\check{f}(x\delta_x) = f(x)$. The last part of the assertion is easily shown because, if $f \in B_d \cap C_o^r(0,1]$,

$$M_t f(x) = M_t\theta_x f(1) = E_1[\int_S \frac{1}{y}\,\theta_x f(y)\mu_t(dy)\ ]$$
$$= E_1[\int_{S_d} \frac{1}{y}\, f(xy)\mu_t(dy)\ ]$$

and $E_1[\int_{S_d} \frac{1}{y}\,\mu_t(dy)\ ] \leqq \frac{d}{1-d}$, so that $M_t f(x)$ can be differentiated w.r.t. x within the integral sign of the right-hand side.

Definition. The strongly continuous semigroup $\{M_t;\ t\geqq 0\}$, $M_0 = I$ of the nonnegative linear operators on C_o is called an *expectation semigroup* of a cascade process (μ_t).

Now, let $f \in B_d^+ = \{f \in B_d\ ;\ f \geqq 0\}$ and put $f(\alpha,x) = e^{-\alpha f(x)}$, $\alpha \geqq 0$, then $f(\alpha,\cdot) \in B_d^*$ and

(2.7) $\widehat{f(\alpha,\cdot)}(\mu) = \exp(\int \frac{1}{x}\log f(\alpha,x)\mu(dx))$

$$= \exp(-\alpha\int \frac{1}{x} f(x)\mu(dx))$$
$$= \exp(-\alpha\check{f}(\mu)) .$$

Lemma 2.3. *For* $f \in B_d^+$,

$$(2.8) \qquad M_t f(x) = -\frac{d}{d\alpha} T_t\widehat{f(\alpha,\cdot)}(x\delta_x)\Big|_{\alpha=0}$$

and moreover

$$(2.9) \qquad \widecheck{M_t f}(\mu) = -\frac{d}{d\alpha} T_t\widehat{f(\alpha,\cdot)}(\mu)\Big|_{\alpha=0} , \qquad \mu \in M_1 .$$

Proof:

$$T_t\widehat{f(\alpha,\cdot)}(x\delta_x) = E_x[\ \widehat{f(\alpha,\cdot)}(\mu_t)\ ] = E_x[\ e^{-\alpha\check{f}(\mu_t)}\ ] ,$$

where $|\check{f}(\mu_t)| \leqq \frac{d}{1-d}\|f\|$ by Lemma 2.1(iv). Therefore, $T_t\widehat{f(\alpha,\cdot)}(x\delta_x)$ is differentiable w.r.t. $\alpha \geqq 0$ and

$$\frac{d}{d\alpha} T_t\widehat{f(\alpha,\cdot)}(x\delta_x)\Big|_{\alpha=0} = -E_x[\ \check{f}(\mu_t)\ ] = -M_t f(x)$$

which gives (2.8). Further, since $T_t\widehat{f(\alpha,\cdot)}|_S \in B_d^*$ for $f \in B_d^+$ and $f(0,x) = 1$, putting $\mu|_{S_d} = x_1\delta_{x_1} + \cdots + x_n\delta_{x_n}$ ($n \leqq \frac{d}{1-d}$) for $\mu \in M_1$,

$$T_t\widehat{f(\alpha,\cdot)}(\mu) = T_t\widehat{f(\alpha,\cdot)}(\ x_1\delta_{x_1} + \cdots + x_n\delta_{x_n}\ )$$
$$= T_t\widehat{f(\alpha,\cdot)}(\ x_1\delta_{x_1}\ ) \cdots T_t\widehat{f(\alpha,\cdot)}(\ x_n\delta_{x_n}\ )$$

by the branching property and then by differentiating both sides w.r.t. α at $\alpha = 0$, we have

$$\frac{d}{d\alpha} T_t\widehat{f(\alpha,\cdot)}(\mu)\Big|_{\alpha=0} = \frac{d}{d\alpha} T_t\widehat{f(\alpha,\cdot)}(x_1\delta_{x_1})\Big|_{\alpha=0} + \cdots + \frac{d}{d\alpha} T_t\widehat{f(\alpha,\cdot)}(x_n\delta_{x_n})\Big|_{\alpha=0}$$
$$= -M_t f(x_1) - \cdots - M_t f(x_n)$$
$$= -\widecheck{M_t f}(\ x_1\delta_{x_1} + \cdots + x_n\delta_{x_n}\ )$$
$$= -\widecheck{M_t f}(\mu)$$

which gives (2.9).

Theorem 2.2. *For* $f \in B_d$ ($\frac{2}{3} < d < 1$), $v_t(x) = M_t f(x)$ *is a solution of the following linear integral equation :*

$$(2.10) \qquad v_t(x) = E_x^\circ[f(x_t)\ ;\ t < \tau_d^\circ] + E_x^\circ[v_{t-\tau_d^\circ}(xd)\ ;\ x_{\tau_d^\circ} = xd,\ \tau_d^\circ \leqq t]$$
$$+ E_x^\circ[\int_0^t ds \int_{M_1-\{\delta_1\}} \Pi(d\mu)\ \chi(x_s M(\mu) < xd < x_s)\check{v}_{t-s}(x_s\cdot\mu)] .$$

Proof: Since M_t is a linear operator and the equation (2.10) is also a linear equation, it suffices to prove the equation (2.10) for all $f \in B_d^+$. Thus, let $f \in B_d^+$, then

$$v_t(x) = -\frac{d}{d\alpha} T_t\widehat{f(\alpha,\cdot)}(x\delta_x)\Big|_{\alpha=0}$$

by Lemma 2.3, and by setting $f_t(\alpha,x) = T_t\widehat{f(\alpha,\cdot)}(x\delta_x)$ where $f(\alpha,\cdot) \in B_d^*$, $f_t(\alpha,x)$ satisfies the (S_d)-equation :

$$f_t(\alpha,x) = E_x^\circ[f(\alpha,x_t) \; ; \; t<\tau_d^\circ] + E_x^\circ[f_{t-\tau_d^\circ}(\alpha,xd) \; ; \; x_{\tau_d^\circ} = xd, \; \tau_d^\circ \leqq t]$$

$$+ E_x^\circ[\int_0^t ds \int_{M_1-\{\delta_1\}} \Pi(d\mu) \; \chi(x_s M(\mu) < xd < x_s)\widehat{f_{t-s}(\alpha,\cdot)}(x_s\cdot\mu)]$$

$$= I + II + III \;, \quad \text{say.}$$

We shall differentiate I, II, and III w.r.t. α at $\alpha = 0$.

$$-\frac{d}{d\alpha} I\Big|_{\alpha=0} = E_x^\circ[\, f(x_t) \; ; \; t < \tau_d^\circ \,]$$

and

$$-\frac{d}{d\alpha} II\Big|_{\alpha=0} = E_x^\circ[\, v_{t-\tau_d^\circ}(xd) \; ; \; x_{\tau_d^\circ} = xd, \; \tau_d^\circ \leqq t \,] .$$

As for the third term III, put $\nu = x_s\cdot\mu$ and $\nu|_{S_d} = y_1\delta_{y_1} + \cdots + y_n\delta_{y_n}$ $(n \leqq \frac{d}{1-d})$. Then,

$$\widehat{f_{t-s}(\alpha,\cdot)}(\nu) = \widehat{f_{t-s}(\alpha,\cdot)}(\, y_1\delta_{y_1} + \cdots + y_n\delta_{y_n} \,)$$

$$= f_{t-s}(\alpha,y_1) \cdots f_{t-s}(\alpha,y_n)$$

and

$$\Big|\frac{d}{d\alpha}\widehat{f_{t-s}(\alpha,\cdot)}(\nu)\Big| \leqq \Big|\frac{d}{d\alpha} f_{t-s}(\alpha,y_1)\Big| + \cdots + \Big|\frac{d}{d\alpha} f_{t-s}(\alpha,y_n)\Big|$$

$$\leqq \left(\frac{d}{1-d} \right)^2 \| f \|$$

since $|f_{t-s}(\alpha,y)| \leqq 1$ and $|\frac{d}{d\alpha} f_{t-s}(\alpha,y)| \leqq \frac{d}{1-d} \|f\|$.

Besides,

$$-\frac{d}{d\alpha}\widehat{f_{t-s}(\alpha,\cdot)}(\nu)\Big|_{\alpha=0} = -\frac{d}{d\alpha} T_{t-s}\widehat{f(\alpha,\cdot)}(\nu)\Big|_{\alpha=0}$$

$$= \widecheck{M_{t-s}f}(\nu) = \check{v}_{t-s}(\nu)$$

by Lemma 2.3 and

$$E_x^\circ[\int_0^t ds \int_{M_1-\{\delta_1\}} \Pi(d\mu) \; \chi(x_s M(\mu) < xd < x_s)] \leqq 1 \quad ,$$

so that

$$-\frac{d}{d\alpha}\, III\Big|_{\alpha=0} = E_x^\circ[\int_0^t ds \int_{M_1-\{\delta_1\}} \Pi(d\mu)\ \chi(x_s M(\mu) < xd < x_s)\check{v}_{t-s}(x_s\cdot\mu)] .$$

Thus it has been proved that $v_t(x)$ satisfies the equation (2.10).

We now deduce the infinitesimal generator N of the expectation semigroup $\{M_t\ ;\ t\geqq 0\}$. Denote by $D(N)$ the domain of N : for $f \in D(N)$,

$$\lim_{t\to 0} \left\|\frac{M_t f - f}{t} - Nf\right\| = 0 .$$

Theorem 2.3. *If* $f \in C_\circ^1(0,1]$ *, then* $f \in D(N)$ *and*

$$(2.11)\quad Nf(x) = -mx\frac{df}{dx} + \int_{M_1-\{\delta_1\}} \Pi(d\mu)(\ \check{f}(x\cdot\mu) - f(x)\ ) .$$

Proof: To begin with, if $f \in C_\circ^1(0,1]$, then f belongs to the domain $D(A^\circ)$ of the infinitesimal generator of the semigroup $\{T_t^\circ\ ;\ t\geqq 0\}$ corresponding to the underlying process $(\ x_t\ )$ and (1.2) holds :

$$A^\circ f(x) = -mx\frac{df}{dx} + \int_{M_1-\{\delta_1\}} \Pi(d\mu)(\ f(xM(\mu)) - f(x)\ ) .$$

Moreover, as in the proof of Lemma 7.2 of [1],

$$(2.12)\quad T_t^\circ f(x) = E_x^\circ[f(x_t)\ ;\ t<\tau_d^\circ] + E_x^\circ[T_{t-\tau_d^\circ}^\circ f(xd)\ ;\ x_{\tau_d^\circ} = xd,\ \tau_d^\circ \leqq t]$$
$$+ E_x^\circ[\int_0^t ds \int_{M_1-\{\delta_1\}} \Pi(d\mu)\ \chi(x_s M(\mu) < xd < x_s) T_{t-s}^\circ f(x_s M(\mu))]$$

for all $f \in B_d$. Thus, for $f \in B_d \cap C_\circ^1(0,1]$, we have, by making use of the equation (2.10),

$$\frac{M_t f(x) - f(x)}{t} = \frac{T_t^\circ f(x) - f(x)}{t} + I + II ,$$

where

$$I = \frac{1}{t} E_x^\circ[\int_0^t ds \int \Pi(d\mu)\ \chi(x_s M(\mu) < xd < x_s)(\check{f}(x_s\cdot\mu) - f(x_s M(\mu)))]$$

and

$$II = \frac{1}{t} E_x^\circ[\ M_{t-\tau_d^\circ} f(xd) - T_{t-\tau_d^\circ}^\circ f(xd)\ ;\ x_{\tau_d^\circ} = xd,\ \tau_d^\circ \leqq t\ ] +$$

$$+ \frac{1}{t} E_x^\circ [\int_0^t ds \int \Pi(d\mu)\ \chi(x_s M(\mu) < xd < x_s) \{ (\widetilde{M_{t-s} f}(x_s\cdot\mu) - \check{f}(x_s\cdot\mu))$$

$$+ (f(x_s M(\mu)) - T_{t-s}^\circ f(x_s M(\mu)))] .$$

Now, let t go to 0 in the above equalities. It is already known

$$\lim_{t\to 0} \frac{T_t^\circ f(x) - f(x)}{t} = A^\circ f(x) \quad \text{uniformly in } x.$$

Since $M(\mu) \geqq d$ implies $\check{f}(x\cdot\mu) = f(xM(\mu))$, $x \in S$,

$$I = \frac{1}{t} E_x^\circ [\int_0^t ds \int \Pi(d\mu)\ \chi(M(\mu)<d)\ \chi(x_s M(\mu)<xd<x_s)(\check{f}(x_s\cdot\mu) - f(x_s M(\mu)))]$$

and $\int \Pi(d\mu)\ \chi(M(\mu) < d) < +\infty$, so that

$$\lim_{t\to 0} I = \int \Pi(d\mu)\ \chi(M(\mu) < d)(\check{f}(x\cdot\mu) - f(xM(\mu))) .$$

Further, to see the uniform convergence of $\lim_{t\to 0} I$, we estimate the following :

$$|E_x^\circ[\ \chi(x_s M(\mu)<xd<x_s)(\check{f}(x_s\cdot\mu) - f(x_s M(\mu))) - (\check{f}(x\cdot\mu) - f(xM(\mu)))]|$$

$$\leqq E_x^\circ|\ \chi(x_s M(\mu)<xd<x_s)(\check{f}(x_s\cdot\mu) - f(x_s M(\mu))) - (\check{f}(x_s\cdot\mu) - f(x_s M(\mu)))|$$

$$+ E_x^\circ|\ (\check{f}(x_s\cdot\mu) - f(x_s M(\mu))) - (\check{f}(x\cdot\mu) - f(xM(\mu)))|$$

$$\leqq (\frac{d}{1-d} + 1)\ \|f\|\ P_x^\circ(x_s M(\mu) \geqq xd, \text{ or } xd \geqq x_s)$$

$$+ E_1^\circ\ |\check{f}(xx_s\cdot\mu) - \check{f}(x\cdot\mu)| + E_1^\circ|\ f(xx_s M(\mu)) - f(xM(\mu))|$$

where

$$|f(xx_s y) - f(xy)| = |f'(xv + cxy(x_s - 1))(xx_s y - xy)| \qquad (0 < c < 1)$$

$$\leqq \|f'\| \cdot |x_s - 1|$$

and putting $\mu|_{S_d} = y_1\delta_{y_1} + \cdots + y_n\delta_{y_n}$ $\quad (n \leqq \frac{d}{1-d})$,

$$|\check{f}(xx_s\cdot\mu) - \check{f}(x\cdot\mu)| = |(f(xx_s y_1)+\cdots+f(xx_s y_n)) - (f(xy_1)+\cdots+f(xy_n))|$$

$$\leqq |f(xx_s y_1) - f(xy_1)| + \cdots + |f(xx_s y_n) - f(xy_n)|$$

$$\leqq n\ \|f'\|\cdot |x_s - 1| \leqq \frac{d}{1-d} \|f'\| \cdot |x_s - 1| .$$

Therefore,

$$|I - \int \Pi(d\mu)\ \chi(M(\mu) < d)\ \ (\check{f}(x\cdot\mu) - f(xM(\mu)))|$$

$$\leqq \frac{1}{t}\int_0^t ds \int \Pi(d\mu)\ \chi(M(\mu)<d)\,(\frac{d}{1-d}+1)\{\|f\|P_1^\circ(x_sM(\mu) \geqq d, \text{or } d \geqq x_s) + \|f'\|E_1^\circ|x_s - 1|\}$$

$\longrightarrow 0$ uniformly in x as $t \longrightarrow 0$.

For the first term II, it is carried out as the following:

$$|II| \leqq \frac{1}{t} E_x^\circ[|M_{t-\tau_d^\circ} f(xd) - f(xd)| + |f(xd) - T_{t-\tau_d^\circ}^\circ f(xd)|;\ x_{\tau_d^\circ} = xd,\ \tau_d^\circ \leqq t]$$

$$+ \frac{1}{t} E_x^\circ[\int_0^t ds \int \Pi(d\mu)\ \chi(x_sM(\mu)<xd<x_s)\{|\widecheck{M_{t-s}f}(x_s\cdot\mu) - \check{f}(x_s\cdot\mu)|$$

$$+ |f(x_sM(\mu)) - T_{t-s}^\circ f(x_sM(\mu))|\}\ ]$$

$$\leqq \frac{1}{t} P_1^\circ(\ \tau_d^\circ \leqq t\ )\{\ \sup_{0\leqq s\leqq t} \|\widecheck{M_s f} - \check{f}\| + \sup_{0\leqq s\leqq t}\|T_s^\circ f - f\|\ \}$$

since it holds by setting $f = 1$ in the (S_d)-equation:

$$P_1^\circ(\ \tau_d^\circ \leqq t\ ) = P_x^\circ(\ \tau_d^\circ \leqq t\ ) = P_x^\circ(\ x_{\tau_d^\circ} = xd,\ \tau_d^\circ \leqq t\ )$$

$$+ E_x^\circ[\int_0^t ds \int \Pi(d\mu)\ \chi(x_sM(\mu) < xd < x_s)].$$

Here

$$\|\widecheck{M_s f} - \check{f}\| \leqq \frac{d}{1-d}\|M_s f - f\|\ .$$

Thus, from the strong continuity of $\{T_t^\circ\}$ and $\{M_t\}$ on $C_\circ$ and from Lemma 7.1 of [1], it follows

$$\lim_{t\to 0} II = 0 \quad \text{uniformly in } x.$$

Therefore, we have shown that $f \in C_\circ^1(0,1]$ implies $f \in D(N)$ and that it holds for $f \in B_d \cap C_\circ^1(0,1]$

$$Nf(x) = A^\circ f(x) + \int \Pi(d\mu)\ \ \chi(M(\mu) < d)\,(\check{f}(x\cdot\mu) - f(xM(\mu)))$$

$$= -\,mx\frac{df}{dx} + \int \Pi(d\mu)\,(\check{f}(x\cdot\mu) - f(x))\ .$$

This completes the proof.

If $f \in D(N)$, then $M_t f \in D(N)$ also and it satisfies

$$\frac{\partial}{\partial t} M_t f = N\ M_t f$$

by the general semigroup theory. Therefore we have

Corollary. *If* $f \in C_0^1(0,1]$, *it holds*

(2.13) $$M_t f(x) - f(x) = -mx \int_0^t \frac{\partial}{\partial x} M_s f(x) ds + \int_0^t ds \int_{M_1 - \{\delta_1\}} \Pi(d\mu) (\widetilde{M_s f(x\cdot\mu)} - M_s f(x)) .$$

§3. Limit theorem related to the expectation semigroup

We consider an asymptotic behavior of the expected number of particles having energies not less than a given level. For this, let u be $0 \leq u \leq 1$ and set

$$M(t,u) = E_1[\check{\chi}_{(u,1]}(\mu_t)]$$

where $\chi_{(u,1]}(x)$ is an indicator function of the set $(u,1]$. Then, $M(t,u)$ is a nonincreasing function of u and satisfies $M(t,1) = 0$ and

(3.1) $$M(t,u) \leq \frac{1}{u} .$$

We define a nonnegative measure $p(t,dx)$ on $[0,\infty)$ for all $t \geq 0$ by

$$p(t,dx) = e^{-x} d_x M(t,e^{-x}) ,$$

then we have

Theorem 3.1. *It holds*

(3.2) $$\int_0^\infty e^{-\alpha x} p(t,dx) = e^{-t(m+m\alpha-\lambda(\alpha))}$$

for all $\alpha > 0$ *and* $t \geq 0$, *where*

$$\lambda(\alpha) = \int_{M_1-\{\delta_1\}} \Pi(d\mu) (\int_S x^\alpha \mu(dx) - 1) .$$

Before we proceed to the proof, we shall give two lemmas. Let h_ε $(0 < \varepsilon < 1)$ be a function such that $h_\varepsilon(x)$ is a nondecreasing and continuously differentiable function on $(0,\infty)$,

$$h_\varepsilon(x) = \begin{cases} 0 & \text{if } 0 < x \leq 1 \\ 1 & \text{if } 1+\varepsilon \leq x < \infty \end{cases} ,$$

and for all $x>0$, $h_\varepsilon(x)$ is nondecreasing and converges to $\chi_{(1,\infty)}(x)$ as ε decreases to zero. Setting

$$v_\varepsilon(t,x,u) = M_t\ \theta_{\frac{1}{u}}\ h_\varepsilon(x)\ , \qquad x \in S$$

for $0<u<1$, we have

Lemma 3.1. $v_\varepsilon(t,x,u)$ *satisfies*

$$\text{(3.3)}\quad v_\varepsilon(t,x,u) - \theta_{\frac{1}{u}}\ h_\varepsilon(x) = -\ mx \int_0^t \frac{\partial}{\partial x} v_\varepsilon(t,x,u)ds$$
$$+ \int_0^t ds \int_{M_1-\{\delta_1\}} \Pi(d\mu)\,(\widetilde{v_\varepsilon(s,\cdot,u)}(x\cdot\mu) - v_\varepsilon(s,x,u)).$$

Moreover, $v_\varepsilon(t,x,u)$ *is nondecreasing in* x, $v_\varepsilon(t,x,u)=0$ *if* $x \leqq u$,

$$0 \leqq v_\varepsilon(t,x,u) \leqq \frac{1}{u}\ ,$$

and

$$0 \leqq \frac{\partial}{\partial x} v_\varepsilon(t,x,u) \leqq \frac{1}{u}\left\|\frac{dh_\varepsilon}{dx}\right\| .$$

Proof: Since $\theta_{\frac{1}{u}}\ h_\varepsilon(x)$ restricted on S is a function belonging to $C^1_\circ(0,1]$ for $0<u<1$, (3.3) follows directly from (2.13). The other properties of $v_\varepsilon(t,x,u)$ are easily seen by the following :

$$v_\varepsilon(t,x,u) = M_t\ \theta_{\frac{x}{u}}\ h_\varepsilon(1) = E_1[\ \widetilde{\theta_{\frac{x}{u}} h_\varepsilon}(\mu_t)\ ]$$
$$= E_1[\int_{(u,1]} \frac{1}{y}\, h_\varepsilon(\frac{xy}{u})\ \mu_t(dy)]$$

and

$$\frac{\partial}{\partial x} v_\varepsilon(t,x,u) = E_1[\int_{(u,1]} \frac{1}{u}\frac{d}{dx} h_\varepsilon(\frac{xy}{u})\ \mu_t(dy)] .$$

We set

$$\psi_\varepsilon(t,x,\alpha) = (\alpha+1)\int_0^1 u^\alpha\ v_\varepsilon(t,x,u)du$$

for all $\alpha>0$, which is finite because $0 \leqq v_\varepsilon(t,x,u) \leqq \frac{1}{u}$.

Lemma 3.2. $\psi_\varepsilon(t,1,\alpha)$ *satisfies the equation :*

$$(3.4)\qquad \psi_\varepsilon(t,1,\alpha) = (\alpha+1)\int_0^1 u^\alpha h_\varepsilon\left(\frac{1}{u}\right)du - (m + m\alpha - \lambda(\alpha))\int_0^t \psi_\varepsilon(s,1,\alpha)ds$$

for all $\alpha > 0$.

<u>Proof</u>: From (3.3),

$$\psi_\varepsilon(t,1,\alpha) - (\alpha+1)\int_0^1 u^\alpha \, \theta_{\frac{1}{u}} h_\varepsilon(1)du$$

$$= -(\alpha+1)m\int_0^1 u^\alpha du \int_0^t \frac{\partial}{\partial x} v_\varepsilon(s,x,u)\Big|_{x=1} ds$$

$$+ (\alpha+1)\int_0^1 u^\alpha du \int_0^t ds \int \Pi(d\mu)\,(\widetilde{v_\varepsilon(s,\cdot,u)}(\mu) - v_\varepsilon(s,1,u))$$

$= I + II$, say.

$$I = -(\alpha+1)m\int_0^t ds \int_0^1 u^\alpha \frac{\partial}{\partial x} v_\varepsilon(s,x,u)\Big|_{x=1} du$$

$$= -m\int_0^t ds \, \frac{\partial}{\partial x}\left\{(\alpha+1)\int_0^1 u^\alpha v_\varepsilon(s,x,u)du\right\}\Big|_{x=1} ,$$

where

$$(\alpha+1)\int_0^1 u^\alpha v_\varepsilon(s,x,u)du = (\alpha+1)\int_0^1 u^\alpha v_\varepsilon(s,1,\tfrac{u}{x})du$$

$$= x^{\alpha+1}(\alpha+1)\int_0^1 u^\alpha v_\varepsilon(s,1,u)du$$

$$= x^{\alpha+1}\psi_\varepsilon(s,1,\alpha) .$$

Therefore,

$$I = -(\alpha+1)m\int_0^t \psi_\varepsilon(s,1,\alpha)ds .$$

As for the second term II,

$$II = (\alpha+1)\lim_{r\to 0}\int_r^1 u^\alpha du \int_0^t ds \int \Pi(d\mu)\{(\widetilde{v_\varepsilon(s,\cdot,u)}(\mu) - v_\varepsilon(s,M(\mu),u)) - (v_\varepsilon(s,1,u) - v_\varepsilon(s,M(\mu),u))\}.$$

Since, here

$$\widecheck{v_\varepsilon(s,\cdot,u)}(\mu) - v_\varepsilon(s,M(\mu),u) = \widecheck{v_\varepsilon(s,\cdot,u)}(\ \mu - M(\mu)\,\delta_{M(\mu)}\ )$$

$$= \chi(M(\mu) < 1-u)\ \widecheck{v_\varepsilon(s,\cdot,u)}(\ \mu - M(\mu)\,\delta_{M(\mu)}\ ) \geqq 0$$

and $v_\varepsilon(s,1,u) - v_\varepsilon(s,M(\mu),u) \geqq 0$,

$$II = (\alpha+1)\lim_{r\to 0}\int_r^1 u^\alpha du\int_0^t ds\int \Pi(d\mu)\,\chi(M(\mu)<1-r)\{\widecheck{v_\varepsilon(s,\cdot,u)}(\mu)-v_\varepsilon(s,M(\mu),u)\}$$

$$-(\alpha+1)\lim_{r\to 0}\int_r^1 u^\alpha du\int_0^t ds\int \Pi(d\mu)\{v_\varepsilon(s,1,u)-v_\varepsilon(s,M(\mu),u)\}$$

$$= (\alpha+1)\lim_{r\to 0}\int_0^t ds\int \Pi(d\mu)\int_r^1 u^\alpha\ \chi(M(\mu)<1-r)\{\widecheck{v_\varepsilon(s,\cdot,u)}(\mu)-v_\varepsilon(s,M(\mu),u)\}du$$

$$-(\alpha+1)\lim_{r\to 0}\int_0^t ds\int \Pi(d\mu)\int_r^1 u^\alpha\ \{v_\varepsilon(s,M(\mu),u)-v_\varepsilon(s,1,u)\}du$$

$$= (\alpha+1)\int_0^t ds\int \Pi(d\mu)\int_0^1 u^\alpha\ \{\widecheck{v_\varepsilon(s,\cdot,u)}(\mu)-v_\varepsilon(s,M(\mu),u)\}du$$

$$-(\alpha+1)\int_0^t ds\int \Pi(d\mu)\int_0^1 u^\alpha\ \{v_\varepsilon(s,M(\mu),u)-v_\varepsilon(s,1,u)\}du$$

$$= (\alpha+1)\int_0^t ds\int \Pi(d\mu)\int_0^1 u^\alpha\ \{\widecheck{v_\varepsilon(s,\cdot,u)}(\mu)-v_\varepsilon(s,1,u)\}du$$

$$= \int_0^t ds\int \Pi(d\mu)\ \{(\alpha+1)\int_0^1 u^\alpha\ \ \widecheck{v_\varepsilon(s,\cdot,u)}(\mu)du - \psi_\varepsilon(s,1,\alpha)\}\ ,$$

where

$$\int_0^1 u^\alpha\ \widecheck{v_\varepsilon(s,\cdot,u)}(\mu)du = \lim_{r\to 0}\int_r^1 u^\alpha\ du\int_S \frac{1}{x}\,v_\varepsilon(s,x,u)\,\mu(dx)$$

$$= \lim_{r\to 0}\int_r^1 u^\alpha\ du\int_{(r,1]}\frac{1}{x}\,v_\varepsilon(s,1,\tfrac{u}{x})\,\mu(dx)$$

$$= \lim_{r\to 0}\int_{(r,1]}\frac{1}{x}\,\mu(dx)\int_r^x u^\alpha\ v_\varepsilon(s,1,\tfrac{u}{x})\,du$$

$$= \lim_{r\to 0}\int_{(r,1]}\frac{1}{x}\,\mu(dx)\cdot x^{\alpha+1}\int_{r/x}^1 u^\alpha\ v_\varepsilon(s,1,u)\,du$$

$$= \lim_{r\to 0}\int_{(r,1]}x^\alpha\ \mu(dx)\{\int_0^1 u^\alpha\ \ v_\varepsilon(s,1,u)du - \int_0^{r/x}u^\alpha\ v_\varepsilon(s,1,u)du\}$$

$$= \int_S x^\alpha \mu(dx) \cdot \int_0^1 u^\alpha v_\varepsilon(s,1,u)du$$

$$- \lim_{r\to 0} \int_{(r,1]} x^\alpha \mu(dx) \int_0^{r/x} u^\alpha \; v_\varepsilon(s,1,u)du \; .$$

Since $0 \leqq v_\varepsilon(s,1,u) \leqq \frac{1}{u}$,

$$0 \leqq \int_{(r,1]} x^\alpha \mu(dx) \int_0^{r/x} u^\alpha \; v_\varepsilon(s,1,u)du \leqq \int_{(r,1]} x^\alpha \mu(dx) \int_0^{r/x} u^{\alpha-1} du$$

$$= \frac{r^\alpha}{\alpha} \int_{(r,1]} \mu(dx) \leqq \frac{r^\alpha}{\alpha}$$

and so

$$\lim_{r\to 0} \int_{(r,1]} x^\alpha \mu(dx) \int_0^{r/x} u^\alpha \; v_\varepsilon(s,1,u)du = 0 \; .$$

Therefore,

$$II = \int_0^t ds \int \Pi(d\mu) \left(\int_S x^\alpha \mu(dx) - 1 \right) \psi_\varepsilon(s,1,\alpha)$$

$$= \lambda(\alpha) \int_0^t \psi_\varepsilon(s,1,\alpha)ds \; .$$

Thus, we have had (3.4).

Proof of Theorem 3.1. By the definition of $p(t,dx)$,

$$\int_0^\infty e^{-\alpha x} \; p(t,dx) = \int_0^\infty e^{-(\alpha+1)x} \; d_x M(t,e^{-x})$$

$$= (\alpha+1) \int_0^\infty e^{-(\alpha+1)x} \; M(t,e^{-x})dx$$

$$= (\alpha+1) \int_0^1 u^\alpha \; M(t,u)du$$

$$= \psi(t,\alpha) \; , \quad \text{say.}$$

On the other hand, since $v_\varepsilon(t,1,u)$ is nondecreasing and converges to $M_t \, \chi_{(u,1]}(1) = M(t,u)$ as ε decreases to zero, $\psi_\varepsilon(t,1,\alpha)$ is also nondecreasing and converges to

$$(\alpha+1)\int_0^1 u^\alpha M(t,u)du = \psi(t,\alpha)$$

as ε decreases to zero. Therefore, by letting $\varepsilon \downarrow 0$ in (3.4), $\psi(t,\alpha)$ satisfies

$$\psi(t,\alpha) = 1 - (m + m\alpha - \lambda(\alpha))\int_0^t \psi(s,\alpha)ds ,$$

so that

$$\psi(t,\alpha) = e^{-t(m + m\alpha - \lambda(\alpha))} , \quad .$$

which prove the theorem.

Now, let $\sigma_\mu(du)$ be a measure on $(0,\infty)$ defined by

$$\sigma_\mu(du) = \int_{\{x;-\log x \in du\}} \mu(dx) ,$$

then

(3.5) $$\int_S x^\alpha \mu(dx) = \int_{(0,\infty)} e^{-\alpha u} \sigma_\mu(du) .$$

We denote the support of the branching measure Π by $\mathrm{Supp}(\Pi)$.

Lemma 3.3. *If*

(3.6) $$\mathrm{Supp}(\Pi) \subset \{\mu \in M_1-\{\delta_1\} ; \|\mu\| = 1\} ,$$

then

(3.7) $$\lambda(\alpha) = \int_{(0,\infty)} (e^{-\alpha u} - 1)\nu(du) , \quad \alpha > 0 ,$$

where $\nu(du)$ *is a measure on* $(0,\infty)$ *defined by*

$$\nu(du) = \int_{M_1-\{\delta_1\}} \Pi(d\mu)\, \sigma_\mu(du)$$

and satisfies

(3.8) $$\int_{(0,\infty)} \frac{u}{1+u}\, \nu(du) < +\infty .$$

Proof: Since

$$\int_{(0,\infty)} \sigma_\mu(du) = \|\mu\| = 1 , \qquad \mu \in \mathrm{Supp}(\,\Pi\,)$$

from the assumption, we have

$$\lambda(\alpha) = \int \Pi(d\mu)\,(\int_{(0,\infty)} e^{-\alpha u}\,\sigma_\mu(du) - \int_{(0,\infty)} \sigma_\mu(du))$$

$$= \int_{(0,\infty)} (e^{-\alpha u} - 1) \int \Pi(d\mu)\ \sigma_\mu(du)$$

$$= \int_{(0,\infty)} (e^{-\alpha u} - 1)\ \nu(du) \quad .$$

(3.8) is shown as the following.

$$\int_{(0,\infty)} \frac{u}{1+u}\,\nu(du) = \int \Pi(d\mu) \int_{(0,\infty)} \frac{u}{1+u}\,\sigma_\mu(du) = \int \Pi(d\mu) \int_{(0,1)} \frac{-\log x}{1-\log x}\,\mu(dx)$$

$$= \int \Pi(d\mu)\,\chi(M(\mu) \leqq \tfrac{1}{2}) \int_{(0,1)} \frac{-\log x}{1-\log x}\,\mu(dx)$$

$$+ \int \Pi(d\mu)\,\chi(M(\mu) > \tfrac{1}{2}) \int_{(0,1)} \frac{-\log x}{1-\log x}\,\mu(dx)$$

where the first term of the right-hand side is finite because

$$\int \Pi(d\mu)\ \ \chi(M(\mu)\ \leqq \tfrac{1}{2}\,) \quad < \quad +\,\infty \quad ,$$

while, for the second term, if we set $\mu(dx) = M(\mu)\,\delta_{M(\mu)} + \bar{\mu}(dx)$, then $\|\bar{\mu}\| = 1 - M(\mu)$ and the second term is equal to

$$\int \Pi(d\mu)\ \ \chi(M(\mu) > \tfrac{1}{2})\ \{\frac{-\log M(\mu)}{1-\log M(\mu)} \cdot M(\mu) + \int_{(0,\frac{1}{2})} \frac{-\log x}{1-\log x}\ \ \bar{\mu}(dx)\}$$

$$\leqq \int \Pi(d\mu)\ \ \chi(M(\mu) > \tfrac{1}{2})\,(1-M(\mu)) \cdot \frac{-\log M(\mu)}{1-M(\mu)} + \int \Pi(d\mu)\ \|\bar{\mu}\|$$

$$\leqq 2 \int \Pi(d\mu)\,(1-M(\mu)) + \int \Pi(d\mu)\,(1-M(\mu))' < \infty$$

by (1.1), so that the integral in (3.8) is finite.

Now we assume $\mathrm{Supp}(\,\Pi\,) \subset \{\,\mu \in M_1 - \{\delta_1\}\ ;\ \ \|\mu\| = 1\}$, then, from

Lemma 3.3, there exists a stochastic process $(X(t), P)$, $X(0) = 0$, with independent nonnegative increments whose Laplace transform has the form:

$$(3.9) \qquad E[\, e^{-\alpha X(t)}\,] = e^{t\lambda(\alpha)} \,, \qquad \alpha > 0 \,.$$

If, in addition, we assume

$$(3.10) \qquad \int_{(0,\infty)} u^2 \,\nu(du) < +\infty \,,$$

then, the expectation $E[X(t)]$ and the variance $Var(X(t))$ of $X(t)$ exist and are given by

$$(3.11) \quad E[X(t)] = t \int_{(0,\infty)} u \,\nu(du) = t \int \Pi(d\mu) \int_{(0,1]} (-\log x)\, \mu(dx)$$

and

$$(3.12) \quad Var(X(t)) = t \int_{(0,\infty)} u^2 \,\nu(du) = t \int \Pi(d\mu) \int_{(0,1]} (\log x)^2\, \mu(dx) \,.$$

Moreover, we know it holds the normal limit theorem for the process $(X(t), P)$ under the condition (3.10), that is

$$(3.13) \quad \lim_{t\to\infty} P\left(x_1 < \frac{X(t)-E[X(t)]}{\sqrt{Var(X(t))}} \leqq x_2 \right) = \int_{x_1}^{x_2} \frac{1}{\sqrt{2\pi}} e^{-\frac{x^2}{2}} dx$$

for any x_1 and x_2 $(-\infty < x_1 < x_2 < \infty)$.

Thus, if we assume $m = 0$ as well as (3.6), we have

$$(3.14) \quad p(t,dx) = P(\, X(t) \in dx\,)$$

by Theorem 3.1 and (3.9). Therefore, we have the following limit theorem on $M(t,u)$ from (3.10), (3.11), (3.12), (3.13), and (3.14).

Theorem 3.2. *If it is assumed* $m = 0$,

$$Supp(\,\Pi\,) \subset \{\, \mu \in M_1 - \{\delta_1\} \,;\, \|\mu\| = 1 \}$$

and

$$\sigma^2 = \int_{M_1-\{\delta_1\}} \Pi(d\mu) \int_{(0,1]} (\log x)^2\, \mu(dx) < +\infty \qquad (\sigma > 0)\,,$$

then, for any x_1 *and* x_2 $(-\infty < x_1 < x_2 < \infty)$,

$$(3.15) \quad \lim_{t\to\infty} \int_{e^{-\gamma t - x_1\sigma\sqrt{t}}}^{e^{-\gamma t - x_2\sigma\sqrt{t}}} u \, d_u M(t,u) = \int_{x_1}^{x_2} \frac{1}{\sqrt{2\pi}} e^{-\frac{x^2}{2}} dx \,,$$

where

$$\gamma = \int_{M_1-\{\delta_1\}} \Pi(d\mu) \int_{(0,1]} (-\log x)\ \mu(dx) \quad .$$

From Theorem 3.2, we see that *the energy distribution of the cascade process is asymptotically normally distributed, if we take* $-\log u$ *instead of* u *as a scale of energy.*

References

[1] Fujimagari, T., and M. Motoo, *Cascade semigroup and their characterization.* Kōdai Math. Sem. Rep. 23 (4) (1971), 402-472.

[2] Harris, T.E., *The theory of branching processes.* Springer (1963).

[3] Ikeda, N., M. Nagasawa, and S. Watanaba, *Branching Markov processes* I; II; III. J. Math. Kyoto Univ. 8 (1968), 233-278; 365-410; 9 (1969), 95-160.

POTENTIAL THEORY OF SYMMETRIC MARKOV PROCESSES AND ITS APPLICATIONS

M. Fukushima

We give a systematic study of a general symmetric standard process $\underset{\sim}{M}$ in connection with its Dirichlet form $\mathcal{E}$ and apply it to a characterization of a family of symmetric diffusion processes on an Euclidean domain.

Some potential theoretic notions relevant to $\underset{\sim}{M}$ and $\mathcal{E}$ were identified in [5] under the two restrictions that $\mathcal{E}$ is regular and that $\underset{\sim}{M}$ is properly associated with $\mathcal{E}$. Further study of the relation has been made in M.L.Silverstein[11] under the same conditions. These restrictions are quite effective in the consideration of construction problem. In fact, given a regular Dirichlet form, a properly associated Hunt process was constructed in [5], [7], [11], and this has been generalized to a non-symmetric regular Dirichlet form by S. Carrillo Menendez[3]. It is still important however to clarify the true scope of these restrictions in the relvant potential theory.

In the first three sections of this paper, we develop the theory without above two restrictions. Theorem 1 is a generalization of a result of M.Takano[12] who assumed the absolute continuity of the resolvent. Theorem 3 enables us to give a straightforward proof of the theorem of [6] concerning the absolute continuity of the transition function. In §3, we introduce an equivalence relation among all symmetric standard processes.

$\mathcal{E}$ is assumed to be regular in §4 but it turns out that the proper association of $\underset{\sim}{M}$ with $\mathcal{E}$ is rather a consequence of the simple association. This fact, combined with the results of [9] and the representation theorem of Beurling-Deny[1], makes it possible to single out in §5 the family of those symmetric diffusions on an Euclidean domain D whose Dirichlet forms possess $C_0^\infty(D)$ as their cores. The equivalence classes(in the sense of §3) of this family are shown to be in one to one correspondence with those closable integro-differential forms of the type

$$\mathcal{E}(u, v) = \sum_{i,j=1}^{n} \int_D \frac{\partial u(x)}{\partial x_i} \frac{\partial v(x)}{\partial x_j} \nu_{ij}(dx) + \int_D u(x)v(x)k(dx).$$

A closability criterion for such form is given in §6.

§ 1. Almost polar sets and the fine capacity

Let X be a locally compact separable Hausdorff space and m be an everywhere dense positive Radon measure on X. $L^2(X;m)$ denotes the real L^2-space with the usual inner product $(\, , \,)$. The family of all Borel sets of X is denoted by $\mathcal{B}(X)$. $C_0(X)$ (resp. $C_\infty(X)$) stands for the space of all continuous functions on X with compact support (resp. vanishing at infinity).

We assume throughout this and the next sections that we are given a standard Markov process $\underset{\sim}{M} = \{\Omega, \mathcal{M}, \mathcal{M}_t, X_t, P_x\}_{x \in X_\Delta}$ on $(X_\Delta, \mathcal{B}(X_\Delta))$ which is <u>m-symmetric</u> in the following sense :

$$\int_X p_t f(x) g(x) m(dx) = \int_X f(x) p_t g(x) m(dx), \quad t > 0,$$

for any non-negative Borel functions f and g. Here X_Δ is the one-point compactification of X, Δ playing the role of the cemetery. Any function f on X is extended to X_Δ by setting $f(\Delta) = 0$. $\{ p_t, t > 0 \}$ is the transition function of $\underset{\sim}{M}$. $\{p_t, t > 0\}$ then determines uniquely a strongly continuous contraction semi-group $\{T_t, t > 0\}$ of symmetric Markov operators on $L^2(X;m)$, which in turn defines a Dirichlet form $\mathcal{E}$ on $L^2(X;m)$ by the following formula [7] :

$$(1) \quad \begin{cases} \mathcal{D}[\mathcal{E}] = \{u \in L^2(X;m) \; ; \; \lim\limits_{t \downarrow 0} \frac{1}{t} (u - T_t u, u) < \infty \} \\ \mathcal{E}(u, v) = \lim\limits_{t \downarrow 0} \frac{1}{t} (u - T_t u, v), \quad u, v \in \mathcal{D}[\mathcal{E}]. \end{cases}$$

We say that $\mathcal{E}$ <u>is the Dirichlet form of</u> $\underset{\sim}{M}$ or alternatively $\underset{\sim}{M}$ <u>is associated with</u> $\mathcal{E}$.

A Dirichlet form is, by definiton, a closed Markov symmetric form. Here we adopt those notiòns formulated in [7] but the definition of Markovity of a symmetric form is slightly strengthened for future convenience : a symmetric form $\mathcal{E}$ on $L^2(X;m)$ is called <u>Markov</u> if there exists, for any $\delta > 0$, a function $\phi_\delta(t)$, $t \in R^1$, satisfying the next two conditions :

$$(2) \quad \begin{cases} \phi_\delta(t) = t, \; t \in [0,1], \quad -\delta \leq \phi_\delta(t) \leq 1 + \delta, \; t \in R^1. \\ 0 \leq \phi_\delta(t') - \phi_\delta(t) \leq t' - t \quad \text{whenever} \quad t < t'. \end{cases}$$

Moreover

$$(3) \quad u \in \mathcal{D}[\mathcal{E}] \Rightarrow \phi_\delta(u) \in \mathcal{D}[\mathcal{E}], \; \mathcal{E}(\phi_\delta(u), \phi_\delta(u)) \leq \mathcal{E}(u,u).$$

Let $\mathcal{E}$ be the Dirichlet form of $\underset{\sim}{M}$. The underlying space X is now endowed with two kinds of topology : the original topology and the fine topology relative to the standard process $\underset{\sim}{M}$, the latter being stronger than the former [2], [4]. We put, for a finely open Borel set A, $\mathcal{L}_A = \{u \in \mathcal{D}[\mathcal{E}] \ ; \ u \geqq 1 \text{ m-a.e. on } A\}$. The <u>fine (1-) capacity</u> $Cap_f(A)$ is defined by $Cap_f(A) = \inf_{u \in \mathcal{L}_A} \mathcal{E}_1(u, u)$ for any finely open Borel set A and by $Cap_f(A) = \inf_{A'} Cap_f(A')$ for any set $A \subset X$, where A' ranges over all finely open borel sets containing A. Just as in the case of the ordinary (1-) capacity Cap(A) [5], it can be shown that the fine capacity is a non-decreasing and countably subadditive set function.

For a finely open Borel set A with $Cap_f(A) < \infty$, there exists a unique element $e_A \in \mathcal{L}_A$ which minimizes $\mathcal{E}_1(u, u)$ on the closed convex set $\mathcal{L}_A$. e_A is said to be the (1-) <u>equilibrium potential</u> of A. $e_A = 1$ m-a.e. on A and e_A is <u>almost</u> (1-) <u>excessive</u> in the following sense (cf. [7]) : a function $u \in L^2(X;m)$ is called almost (1-) excessive if $u \geqq 0$ m-a.e. and $e^{-t}T_t u \leqq u$ m-a.e., $t > 0$. The next lemma due to M.L.Silverstein is an immediate consequence of the formula (1).

<u>Lemma 1</u>. If u_1 and $u_2 \in L^2(X;m)$ are almost excessive, $u_1 \leqq u_2$ m-a.e. and $u_2 \in \mathcal{D}[\mathcal{E}]$, then $u_1 \in \mathcal{D}[\mathcal{E}]$ and $\mathcal{E}_1(u_1, u_1) \leqq \mathcal{E}_1(u_2, u_2)$.

<u>Corollary</u>. If an almost excessive function u satisties $u \leqq e_A$ m-a.e. and $u = 1$ m-a.e. on A, then $u = e_A$ m-a.e.

We now consider the hitting time and (1-) hitting probability of $B \in \mathcal{B}(X)$ defined by $\sigma_B(\omega) = \inf\{t > 0 \ ; \ X_t(\omega) \in B\}$ and $p_B(x) = E_x(e^{-\sigma_B}) = \int_\Omega \exp(-\sigma_B(\omega)) P_x(d\omega)$ respectively. By convention we put $\inf \phi = \infty$ and $e^{-\infty} = 0$.

Lemma 2. Let A be a finely open Borel set with $Cap_f(A) < \infty$, then p_A is a version of the equilibrium potential e_A.

Proof. Since p_A is 1-excessive with respect to p_t and $p_A(x) = 1$ for any $x \in A$, it suffices to show in view of Corollary to Lemma 1 that

(4) $p_A \leqq e_A$ m-a.e.

Following a reasoning of M.L.Silverstein, we take a Borel modification $\tilde{e}_A$ of e_A such that $\tilde{e}_A(x) = 1$, $x \in A$, and put $Y_t(\omega) = e^{-t}\tilde{e}_A(X_t(\omega))$, $t > 0$, $\omega \in \Omega$. Let g be any non-negative Borel function on X such that $\int_X g(x)m(dx) = 1$. Then the process $\{Y_t, \mathcal{M}_t, P_{g\cdot m}\}_{t\geqq 0}$ is a supermartingale. Indeed the Markovity of $\underset{\sim}{M}$ implies, for $0 \leqq s < t$, $E_{g\cdot m}(e^{-t}\tilde{e}_A(X_t)|\mathcal{M}_s) = e^{-s}e^{-(t-s)}p_{t-s}\tilde{e}_A(X_s)$, $P_{g\cdot m}$-almost surely, which in turn is not greater than $e^{-s}\tilde{e}_A(X_s)$, $P_{g\cdot m}$-a.s., because the set $\{x \in X ; e^{-(t-s)}p_{t-s}\tilde{e}_A(x) > \tilde{e}_A(x)\}$ is m-negligible.

Let D be a finite subset of $(0, \infty)$ with $\min D = a$ and $\max D = b$ and put $\sigma(D;A) = \min\{t \in D ; X_t \in A\}$. If the set in the braces is empty, we set $\sigma(D;A) = b$. Doob's optional sampling theorem applied to $\{Y_t, \mathcal{M}_t, P_{g\cdot m}\}_{t\in D}$ yields

$$E_{g\cdot m}(e^{-\sigma(D;A)};\sigma(D;A) < b) \leqq E_{g\cdot m}(Y_{\sigma(D;A)}) \leqq E_{g\cdot m}(Y_a) \leqq (g, \tilde{e}_A).$$

Letting D increase to a countable dense subset of $(0,b)$ and then b tend to infinity, we arrive at $(g, p_A) \leqq (g, e_A)$, which means (4).

According to [8], a set A is called (m-) almost polar if there exists a Borel set $B \supset A$ such that $P_x(\sigma_B < \infty) = 0$ for m-a.e.$x \in X$.

Theorem 1. A subset of X is almost polar if and only if its fine capacity is zero.

Proof. Suppose $Cap_f(N) = 0$, then there exists a decreasing sequence $\{A_n\}$ of finely open Borel sets such that $A_n \supset N$ and

$Cap_f(A_n) \to 0$, $n \to \infty$. Since $Cap_f(A_n) = \mathcal{E}_1(e_{A_n}, e_{A_n}) \geq (e_{A_n}, e_{A_n})$, we have from Lemma 2

(5) $\lim_{n\to\infty} p_{A_n}(x) = 0$ m-a.e.

and consequently $p_B(x) = 0$ m-a.e. for $B = \bigcap_{n=1}^{\infty} A_n (\supset N)$. Hence N is almost polar.

In order to prove the converse, take an everywhere positive function $h \in C_\infty(X) \cap L^2(X\,;\,m)$ and observe $X = \bigcup_{\ell=1}^{\infty} A_\ell$ where $A_\ell = \{x \in X\,;\, \ell R_1 h(x) > 1\}$, $\{R_\alpha,\ \alpha > 0\}$ being the resolvent of $\underset{\sim}{M}$.

Now, in view of the countable subadditititity of the fine capacity, it suffices to get $Cap_f(N) = 0$ under the assumption that N is a Borel almost polar set such that $N \subset A_\ell$ for a fixed ℓ. Note that A_ℓ is finely open Borel and $Cap_f(A_\ell) < \infty$ since $R_1 h$ is a finely continuous Borel function belonging to $\mathcal{D}[\mathcal{E}]$.

Let g be an everywhere positive Borel function such that $\int_X g(x)m(dx) = 1$. Since any semi-polar set is of potential zero and consequently m-negligible according to [8 ; §2(i)], we can use the approximation theorem [2 ; (11.2)] to select a decreasing sequence $\{G_n\}$ of open sets containing N such that $\lim_{n\to\infty} \sigma_{G_n} \wedge \zeta = \sigma_N \wedge \zeta = \zeta$, $P_{g\cdot m}$-almost surely, ζ being the life time. Put $G_n' = G_n \cap A_\ell$. Evidently $\sigma_{G_n'} \wedge \zeta \to \zeta$, $n \to \infty$, and $E_{g\cdot m}(e^{-\sigma_{G_n'}} R_1 h(X_{\sigma_{G_n'}})) = (g, R_1 h) - E_{g\cdot m}(\int_0^{\sigma_{G_n'}\wedge\zeta} e^{-t} h(X_t)dt) \to 0$, $n \to \infty$. But the first term dominates $\frac{1}{\ell}(g, p_{G_n'})$ because $X_{\sigma_{G_n'}}$ is located on the fine closure of A_ℓ where $1 \leq \ell R_1 h(x)$. Therefore

(6) $\lim_{n\to\infty} p_{G_n'}(x) = 0$ m-a.e.

On the other hand, $\{G_n'\}$ is a decreasing sequence of finely open Borel sets with finite fine-Capacity and hence it is easy to see that $e_{G_n'}$ converges to some $e_0 \in \mathcal{D}[\mathcal{E}]$ with metric $\mathcal{E}_1$. Lemma 2 and (6) mean $e_0 = 0$ m-a.e. We arrive at $Cap_f(N) \leq \lim_{n\to\infty} Cap_f(G_n') = \lim_{n\to\infty} \mathcal{E}_1(e_{G_n'}, e_{G_n'}) = \mathcal{E}_1(e_0, e_0) = 0$. q.e.d.

§2. q.e. fine continuity of transition function

"q.e." will mean "except for an almost polar set". A function u defined q.e. on X is called q.e. finely continuous or finely continuous q.e. if there exists a nearly Borel almost polar set B such that X - B is finely open and u is nearly Borel and finely continuous on X - B.

Let B be a Borel set. X - B is said to be $\underset{\sim}{M}$-invariant if

(7) P_x(either X_t is in B or X_{t-} exists and is in B for some $t \geq 0$) = 0 whenever $x \in X - B$.

The present definition of $\underset{\sim}{M}$-invariance is stronger than that given in [8] but we know from [2 ; (10.20)] and [8 ; §2(ii)] that a Borel set B is almost polar if and only if (7) holds for m-a.e. $x \in X$. Therefore we can see that every assertion of [8 ; §2] is valid with the notion of $\underset{\sim}{M}$-invariance being strengthened in the above sense. Particularily we have the following.

Lemma 3. (i) A set N is almost polar if and only if there exists a Borel set $B \supset N$ such that $m(B) = 0$ and X - B is $\underset{\sim}{M}$-invariant. (ii) A function u is q.e. finely continuous if and only if there exists a Borel set B with the property that $m(B) = 0$, X - B is $\underset{\sim}{M}$-invariant and u is Borel and finely continuous on X - B.

It follows that

(8) $\mathrm{Cap}_f(\{x ; |u(x)| > \varepsilon\}) \leq \frac{1}{\varepsilon^2} \mathcal{E}_1(u, u)$

for any q.e. finely continuous function $u \in \mathcal{D}[\mathcal{E}]$. In fact let B be the set for u in Lemma 3 (ii). Then the left hand side of (6) equals in view of Theorem 1 to the fine capacity of the finely open Borel set $\{x \in X - B ; |u(x)| > \varepsilon\}$ on which $\frac{1}{\varepsilon}|u(x)|$ is greater then 1.

Theorem 2. Let $\{u_n\}$ be a sequence of q.e. finely continuous function in $\mathcal{D}[\mathcal{E}]$ such that $\mathcal{E}_1(u_n-u_m, u_n-u_m) \to 0$, $n,m \to \infty$. Then a subsequence converges q.e. on X to a q.e. finely continuous function u. Moreover $\mathcal{E}_1(u_n-u, u_n-u) \to 0$, $n \to \infty$.

Proof. We can take a Borel set B_0 which possesses the properties of Lemma 3 (ii) for all u_n. Put $X_0 = X - B_0$. (8) implies that there exists a subsequence $\{u_{n_k}\}$ and a decreasing sequence $\{A_m\}$ of finely open Borel sets such that $A_m \subset X_0$, $\lim_{m\to\infty} \mathrm{Cap}_f(A_m) = 0$ and the restrictions to each set $X_0 - A_m$ of $\{u_{n_k}\}$ are uniformly convergent. Then the limit function $u(x) = \lim_{n_k\to\infty} u_{n_k}(x)$ has the property that its restriction to each $X_0 - A_m$ is Borel and relatively finely continuous.

On the other hand, we saw in the proof of Theorem 1 that the relation (5) holds, which combined with [8 ; §2(X)] can be strengthened as

$$(9) \qquad \lim_{m\to\infty} p_{A_m}(x) = 0 \qquad \text{q.e. on } X_0.$$

Denote by C the exceptional Borel set in (9) and put $B = \bigcap_{m=1}^{\infty} (C \cup A_m)$. B is almost polar.

Now take any $x \in X_0 - B$. $P_x(\sigma_{C\cup A_m} > 0) = 1$ for sufficiently large m by Blumenthal zero-one law. In view of [4 ; (4.13)], $X_0 - C \cup A_m$ is a fine neighbourhood of x and hence u is finely continuous at x. q.e.d.

As consequences of Theorem 2, we have

Theorem 3. (i) Each element of $\mathcal{D}[\mathcal{E}]$ has a q.e. finely continuous version.

(ii) For any non-negative Borel function $u \in L^2(X ; m)$, $p_t u$ is a q.e. finely continuous version of $T_t u$, $t > 0$.

Proof. (i) This is because $R_1 u$, $u \in C_0(X)$, are Borel, finely continuous and $\mathcal{E}_1$-dense in $\mathcal{D}[\mathcal{E}]$. (ii) Just as in [6 ; Theorem 2], it suffices to use the inequality in [7 ; Lemma 3.2].

Finally we can get

Theorem 4. The next three conditions are mutually equivalent:

(a) Any almost polar set is polar.

(b) $R_\alpha(x, \cdot)$ is absolutely continuous with respect to m for each $\alpha > 0$ and $x \in X$.

(c) $p_t(x, \cdot)$ is absolutely continuous with respect to m for each $t > 0$ and $x \in X$.

Proof. The equivalence of (a) and (b) was proven in [8 ; §2(viii)]. The proof of (a) $\Rightarrow$ (c) is now the same as that of [6 ; Theorem 3]. Indeed let B be an m-negligible Borel set. Then

(10) $\quad p_t(x, B) = 0$ for m-a.e. $x \in X$,

by the symmetry of p_t. By virtue of Theorem 3(ii) and [8 ; §2(vi)], (10) holds for q.e. $x \in X$ and hence up to a polar set N by our hypothesis (a). Then $p_{2t}(x, B) = E_x(p_t(X_t, B), X_t \notin N) = 0$, $t > 0$.

q.e.d.

§ 3. Equivalence of symmetric standard processes

Let us introduce an equivalence relation $\sim$ among the family of all m-symmetric standard processes on $(X, \mathcal{B}(X))$: $\underset{\sim}{M}_1 \sim \underset{\sim}{M}_2$ if there exists an m-negligible Borel set B such that

(a) $X - B$ is both $\underset{\sim}{M}_1$-invariant and $\underset{\sim}{M}_2$-invariant,

(b) the restrictions to $X - B$ of $\underset{\sim}{M}_1$ and $\underset{\sim}{M}_2$ have a common transition function.

The next theorem particularily implies that the relation $\sim$ is indeed an equivalence relation.

Theorem 5. m-symmetric standard processes $\underset{\sim}{M}_1$ and $\underset{\sim}{M}_2$ on $(X, \mathcal{B}(X))$ have a common semi-group on $L^2(X ; m)$ (in other words they are associated with a common Dirichlet form on $L^2(X ; m)$) if and only if $M_1 \sim M_2$.

To prove this, we need a lemma.

Lemma 4. Let $\underset{\sim}{M}_1$ and $\underset{\sim}{M}_2$ be m-symmetric standard processes on $(X, \mathcal{B}(X))$ possessing a common semi-group on $L^2(X ; m)$.

(i) A set is almost polar for $\underset{\sim}{M}_1$ if and only if so it is for $\underset{\sim}{M}_2$.

(ii) A function is q.e. finely continuous for $\underset{\sim}{M}_1$ if and only if so it is for $\underset{\sim}{M}_2$.

Proof. (i) Trivially $\underset{\sim}{M}_1$ and $\underset{\sim}{M}_2$ are in duality in the sense of [8]. Hence they have the common notion of almost polarity [8 ; §2 (iv)]. (ii) Since $\underset{\sim}{M}_1$ is m-symmetric, any semi-polar set is almost polar for $\underset{\sim}{M}_1$ in view of [8 ; §2(xiv)]. But $\underset{\sim}{M}_2$ is in duality with $\underset{\sim}{M}_1$. Hence using [8 ; §2(xiv)] again we are led to the conclusion (ii).

Proof of Theorem 5. Under the assumption of Lemma 4, the notions of almost polarity and q.e. fine continuity have the same meanings for $\mathcal{M}_1$ and $\mathcal{M}_2$. In view of Theorem 3(ii) and [8 ; §2, (vi)], there exists an almost polar Borel set B_0 such that

$$(11) \qquad p_t^{(1)}u(x) = p_t^{(2)}u(x), \qquad x \in X - B_0 ,$$

for all rationals $t > 0$ and all $u \in C_1$. Here $p_t^{(i)}$ is the transition function of $\mathcal{M}_i$, $i = 1,2$, and C_1 is a countable uniformly dense subset of $C_0(X)$. (11) then holds for all $t \geq 0$ by the right continuity. Applying Lemma 3(i) to $\mathcal{M}_1$ and $\mathcal{M}_2$ alternatively, we can find an increasing sequence $\{B_n\}$ of m-negligible Borel sets containing B_0 such that $X - B_{2n-1}$(resp. $X-B_{2n}$) is $\mathcal{M}_1$-invariant (resp. $\mathcal{M}_2$-invariant). Then $\mathcal{M}_1 \sim \mathcal{M}_2$ with $B = \bigcup_n B_n$. q.e.d.

§ 4. Regularity and the proper association

From now on we are concerned with regular Dirichlet forms. Given a symmetric form $\mathcal{E}$ on $L^2(X ; m)$, a subcollection $\tilde{C} \subset \mathcal{D}[\mathcal{E}] \cap C_0(X)$ is said to be a core of the form $\mathcal{E}$ if $\tilde{C}$ is both uniformly dense in $C_0(X)$ and $\mathcal{E}_1$-dense in $\mathcal{D}[\mathcal{E}]$. $\mathcal{E}$ is called regular if $\mathcal{E}$ possesses a core.

The above definition of the regularity of a symmetric form is the same as the original one of Beurling-Deny [1]. Apparently it is stronger than the previous one of author's [5],[7], where a symmetric form $\mathcal{E}$ was said to be regular if $\mathcal{D}[\mathcal{E}] \cap C_\infty(X)$ is both uniformly dense in $C_\infty(X)$ and $\mathcal{E}_1$-dense in $\mathcal{D}[\mathcal{E}]$. However the previous definition reduces to the present one when $\mathcal{E}$ is a Dirichlet form. To see this, take any non-negative function $f \in \mathcal{D}[\mathcal{E}] \cap C_\infty(X)$ and put $f_n = f - f \wedge (\frac{1}{n})$. Then f_n is in $\mathcal{D}[\mathcal{E}] \cap C_0(X)$ and converges uniformly to f on X. Furthermore $\mathcal{E}_1(f_n, f_n)$ is dominated uniformly by $4\mathcal{E}_1(f, f)$ and hence the Cesaro sum of a subsequence of $\{f_n\}$ is $\mathcal{E}_1$-convergent to f. In this connection, we refer the reader to M.L. Silverstein [11 ; Lemma 10.2(i)] where it is shown that the part on an open set of a regular Dirichlet form is again a regular one in the present sense.

We start this section with a probabilistic description of those notions of capacity and quasi-continuity in [5],[7]. The first assertion of the next theorem is a counterpart of Theorem 1.

Theorem 6. Let $\underset{\sim}{M}$ be an m-symmetric Hunt process on $(X, \mathcal{B}(X))$ whose Dirichlet form $\mathcal{E}$ is regular.

(i) A set is almost polar if and only if its capacity is zero.

(ii) Any quasi-continuous function is q.e. finely continuous.

(iii) Any q.e. finely continuous function of $\mathcal{D}[\mathcal{E}]$ is quasi-continuous.

Proof. (i) "If" part follows from the same part of Theorem 1 because $Cap_f(A) \leq Cap(A)$ for any $A \subset X$. Conversely suppose N is a compact almost polar set. Let $\{G_n\}$ be a decreasing sequence of open sets such that $\bar{G}_n$ is compact, $G_n \supset \bar{G}_{n+1}$ and $\bigcap_{n=1}^{\infty} G_n = N$

Then the quasi-left continuity of $\underset{\sim}{M}$ on $(0, \infty)$ and the same remark as in the proof of Theorem 1 yield $\lim_{n\to\infty} p_{G_n}(x) = p_N(x) = 0$ m-a.e.

Since $Cap(G_n)$ is finite, Corollary to Lemma 1 leads us to $Cap(N)$ $\leq \lim_{n\to\infty} Cap(G_n) = \lim_{n\to\infty} \mathcal{E}_1(p_{G_n}, p_{G_n}) = 0$. Now it suffices to note that Cap is a Choquet capacity [5].

(ii) Similar to the proof of Theorem 2.

(iii) Let $u \in \mathcal{D}[\mathcal{E}]$ be q.e. finely continuous.

We can take its quasi-continuous version $\tilde{u}$. By virtue of (ii), $\tilde{u}$ is again q.e. finely continuous. Therefore $u = \tilde{u}$ q.e. in view of [8 ; § 2(vi)] and hence $u = \tilde{u}$ up to a set of capacity zero by (i), which means that u is quasi-continuous. q.e.d.

Next we restate the existence theorem in [5] and [7] in the following simpler manner.

Theorem 7. Let $\mathcal{E}$ be a regular Dirichlet form on $L^2(X ; m)$. Then there exists an m-symmetric Hunt process on $(X, \mathcal{B}(X))$ whose Dirichlet form is $\mathcal{E}$.

In [5] and [7], a Hunt process was constructed outside some Borel set B of zero capacity. It suffices to make each point of B trap to get the process on $(X, \mathcal{B}(X))$ of Theorem 7. Theorem 7 combined with Lemma 4 immediately leads us to the following extension of Theorem 6.

Corollary. Theorem 6 is valid with the Hunt process there being replaced by a standard process.

The Hunt process constructed in [5] and [7] was not only associated with a regular Dirichlet form but also properly associated in the sense that $p_t u$ is a quasi-continuous version of $T_t u$ for any $t > 0$ and any non-negative Borel function $u \in L^2(X ; m)$. We now see however that the proper association is rather a class property, namely, it is a consequence of the simple association. Indeed combining Corollary to Theorem 7 with Theorem 3, we get

Theorem 8. Let $\underset{\sim}{M}$ be an m-symmetric standard process on $(X, \mathcal{B}(X))$ whose Dirichlet form $\mathcal{E}$ is regular. Then, for any non-negative Borel function $u \in L^2(X ; m)$,

(i) $p_t u$ is a quasi-continuous version of $T_t u$, $t > 0$,

(ii) $R_\alpha u$ is a quasi-continuous version of $G_\alpha u$, $\alpha > 0$.

Corollary to Theorem 7 was first proved in [5] for a Hunt process properly associated with a regular Dirichlet form. Now all results of [5 ; § 3] remain true for any m-symmetric standard process whose Dirichlet form is regular because of Corollary to Theorem 7 and Theorem 8. Some arguments of [5 ; § 3] can be simplified by making use of reasonings similar to § 1 (cf. M.L. Silverstein [11 ; § 7]).

§ 5. A family of symmetric diffusions

A standard process $\underset{\sim}{M}$ on $(X, \mathcal{B}(X))$ is called a diffusion process if

$$P_x(X_t \text{ is continuous in } t \in [0, \zeta)) = 1, \quad x \in X.$$

Theorem 8 and the remark following it make it possible to restate a theorem of [9] in a simpler way :

Theorem 9. Let $\mathcal{E}$ be a regular Dirichlet form on $L^2(X ; m)$. Then the next two statements are equivalent :

(i) $\mathcal{E}$ has the local property,

(ii) there exists an m-symmetric diffusion process $\underset{\sim}{M}$ on $(X, \mathcal{B}(X))$ whose Dirichlet form is $\mathcal{E}$.

Let us consider a concrete case that X is a domain D of the Euclidean n-space R^n. As before m is an everywhere dense positive Radon measure on D. $C_0^\infty(D)$ stands for the space of all infinitely differentiable functions on D with compact support.

We are concerned with the family $\mathfrak{M}$ of symmetric diffusions on D defined by $\mathfrak{M} = \{\underset{\sim}{M} ; \underset{\sim}{M}$ is an m-symmetric diffusion process on $(D, \mathcal{B}(D))$ whose Dirichlet form possesses $C_0^\infty(D)$ as its core$\}$. Next we introduce an <u>integro-differential form of local type</u> $\mathcal{E}$ by

$$(12) \quad \begin{cases} \mathcal{D}[\mathcal{E}] = C_0^\infty(D) \\ \mathcal{E}(u, v) = \sum_{i,j=1}^{n} \int_D \frac{\partial u(x)}{\partial x_i} \frac{\partial v(x)}{\partial x_j} \nu_{ij}(dx) + \int_D u(x)v(x)k(dx) \end{cases}$$

where ν_{ij}, $1 \leqq i, j \leqq n$, are Radon measures on D such that $\nu_{ij} = \nu_{ij}$ $(i \neq j)$ and $\sum_{i,j=1}^{n} \nu_{ij}(K)\xi_i\xi_j \geqq 0$ for any compact set $K \subset D$ and any $\xi \in R^n$, and k is a positive Radon measure on D.

<u>Theorem 10</u>. There is a one-to-one correspondence between the equivalence classes $\mathfrak{M}/\sim$ and the family of those integro-differential forms of local type (12) which are closable on $L^2(D ; m)$. Here $\sim$ is the equivalence relation defined in §3.

<u>Proof</u>. Let $\underset{\sim}{M}$ be a member of $\mathfrak{M}$ and $\mathcal{E}$ be its Dirichlet form on $L^2(D ; m)$. $\mathcal{E}$ is then regular and has the local property by virtue of Theorem 9. Hence on account of Beurling-Deny representation theorem [1], $\mathcal{E}$ can be expressed on its core $C_0^\infty(D)$ as an integro-differential form of local type (12).

Conversely let $\mathcal{E}$ be an integro-differential form of local type (12) which is closable on $L^2(D ; m)$. Since $\mathcal{E}$ is Markov in the sense of (1) and possesses the local property, its smallest closed extension $\overline{\mathcal{E}}$ on $L^2(D ; m)$ is a regular Dirichlet form still possessing the local property according to [9 ; Theorem 1]. Therefore there exists an m-symmetric diffusion process $\underset{\sim}{M}$ on $(D, \mathcal{B}(D))$ whose Dirichlet form is $\overline{\mathcal{E}}$ by Theorem 9. Of course, $\underset{\sim}{M} \in \mathfrak{M}$. q.e.d.

§ 6. <u>Superposition of closable symmetric forms</u>

Theorem 10 reduces our study of the family $\mathfrak{M}$ of symmetric diffusions on D to the following analytic question : under what conditions on the data $\{m, \nu_{ij}, k\}$, is the symmetric form (12) closable on $L^2(D ; m)$? In this connection several examples were already considered in [7 ; § 6]. Now we give a method of superposition of symmetric forms to get new closable symmetric forms inductively.

Let X be R^n and m be the n-dimensional Lebesgue measure dx.

Denote by $\underset{\sim}{D}(u, v)$ the Dirichlet integral $\sum_{i=1}^{n} \int_D \frac{\partial u}{\partial x_i} \frac{\partial v}{\partial x_j} dx$. We are concerned with a closable symmetric form $\mathcal{E}$ on $L^2(R^n)$ such that $\mathcal{D}[\mathcal{E}] = C_0^\infty(R^n)$ and $\mathcal{E}(u, u) \geq c \underset{\sim}{D}(u, u)$, $u \in C_0^\infty(R^n)$, for some positive constant c.

More specifically difine $\mathcal{E}$ by

$$(13) \quad \begin{cases} \mathcal{D}[\mathcal{E}] = C_0^\infty(R^n) \\ \mathcal{E}(u, v) = \int_{\underset{\sim}{\Theta}} \mathcal{E}^\theta(u^\theta, v^\theta)\mu(d\theta) + c \underset{\sim}{D}(u, v) . \end{cases}$$

Here we are given an auxiliary σ-finite measure space $(\underset{\sim}{\Theta}, \mathcal{B}(\Theta), \mu)$, a collection $\{F_\theta \ ; \ \theta \in \underset{\sim}{\Theta}\}$ of $(n-1)$-dimensional hyperplanes F_θ and, for each $\theta \in \underset{\sim}{\Theta}$, a symmetric form $\mathcal{E}^\theta$ on $L^2(F_\theta)$ with $\mathcal{D}[\mathcal{E}^\theta] = C_0^\infty(F_\theta)$, $L^2(F_\theta)$ being relevant to the $(n-1)$-dimensional Lebesgue measure on F_θ. In order to make sense of the form (13), we suppose

(14) $\mathcal{E}^\theta(u^\theta, u^\theta)$ is a μ-integrable function of $\theta \in \underset{\sim}{\Theta}$ for each $u \in C_0^\infty(R^n)$,

where u^θ is the restriction to F_θ of u.

<u>Theorem 11</u>. (i) Suppose $\mathcal{E}^\theta$ is closable on $L^2(F_\theta)$ for each $\theta \in \underset{\sim}{\Theta}$, then so is the form (13) on $L^2(R^n)$.

(ii) Suppose $\mathcal{E}^\theta$ is a closable integro-differential form of local type on $L^2(F_\theta)$ for each $\theta \in \underset{\sim}{\Theta}$, then so is the form (13) on $L^2(R^n)$.

<u>Proof</u>. (i) Consider $u_k \in C_0^\infty(R^n)$ such that $\mathcal{E}(u_k - u_\ell, u_k - u_\ell)$ 0, $k, \ell \to \infty$, and $u_k \to 0$ in $L^2(R^n)$. Since the Dirichlet integral is closable, we have then

$$(15) \quad \underset{\sim}{D}(u_k, u_k) \to 0, \quad k \to \infty.$$

It is well known that the trace operator from $H^1(R^n)$ to $L^2(F_\theta)$ is continuous. Hence (15) implies

$$(16) \quad u_k^\theta \to 0 \quad \text{in} \quad L^2(F_\theta) \quad \text{for each} \quad \theta \in \underset{\sim}{\Theta}.$$

Choose a subsequence k_j such that

$$\sum_{j=1}^{\infty} \sqrt{\mathcal{E}(u_{k_j} - u_{k_{j+1}}, u_{k_j} - u_{k_{j+1}})} < \infty.$$

Since μ is σ-finite, this means that

$$\sum_{j=1}^{\infty} \sqrt{\mathcal{E}^\theta(u_{k_j}^\theta - u_{k_{j+1}}^\theta, u_{k_j}^\theta - u_{k_{j+1}}^\theta)} < \infty$$

for μ-a.e. $\theta \in \Theta$. In view of (16), we now have $\lim_{k_j\to\infty} \mathcal{E}^\theta(u_{k_j}^\theta, u_{k_j}^\theta) = 0$ for such θ because of the closability of $\mathcal{E}^\theta$.

By making use of Fatou's lemma,

$$\mathcal{E}(u_m, u_m) \leq \int_{\Theta} \lim_{k_j\to\infty} \mathcal{E}^\theta(u_m^\theta - u_{k_j}^\theta, u_m^\theta - u_{k_j}^\theta)\mu(d\theta)$$

$$+ c\, \mathcal{D}(u_m, u_m) \leq \lim_{k_j\to\infty} \mathcal{E}(u_m - u_{k_j}, u_m - u_{k_j})$$

$$+ c\, \mathcal{D}(u_m, u_m),$$

which can be made arbitrarily small.

(ii) Under the assumption of (ii), $\mathcal{E}$ is a closable Markov-symmetric form possessing the local property. Hence the same arguments as in the proof of Theorem 10 leads us to the conclusion. q.e.d.

Theorem 11 gives us another proof of the closability of the example [7 ; §6,2]. Applying Theorem 11 inductively, we can also prove the closability of Example 1.1 of N. Ikeda-S. Watanabe [10] where the corresponding diffusion is constructed by the method of skew-product.

References

[1] A. Beurling and J. Deny, Dirichlet spaces, Proc. Nat. Acad. Sci. U.S.A. 45 (1959), 208-215.

[2] R.M. Blumenthal and R.K. Getoor, Markov processes and potential theory, Academic press, New York and London, 1968.

[3] S. Carrillo Menendez, Processus de Markov associé a une forme de Dirichlet non symétrique, Z. Wahrscheinlichkeitstheorie verw. Gebiete, 33(1975), 139-154.

[4] E.B. Dynkin, Markov processes, vol.1, vol.2, Springer-Verlag, Berlin-Heidelberg-New York, 1965.

[5] M. Fukushima, Dirichlet spaces and strong Markov processes, Trans. Amer. Math. Soc., 162 (1971), 185-224.

[6] M. Fukushima, On transition probabilities of symmetric strong Markov processes, J. Math. Kyoto Univ., 12 (1972), 431-450.

[7] M. Fukushima, On the generation of Markov processes by symmetric forms, Proc. 2nd Japan-USSR symp. on probability theory, Lecture Notes in Math., vol. 330, Springer-Verlag, Berlin-Heidelberg-New York, 1973.

[8] M. Fukushima, Almost polar sets and an ergodic theorem, J. Math. Soc. Japan, 26 (1974), 17-32.

[9] M. Fukushima, Local property of Dirichlet forms and continuity of sample paths, Z. Wahrscheinlichkeitstheorie verw. Gebiete, 29 (1974), 1-6.

[10] N. Ikeda and S. Watanabe, The local structure of a class of diffusions and related problems, Proc. 2nd Japan-USSR symp. on probability theory, Lecture Notes in Math., vol 330, Springer-Verlag, Berlin-Heidelberg-New York, 1973.

[11] M.L. Silverstein, Symmetric Markov processes, Lecture Notes in Math., vol.426, Springer-Verlag, Berlin-Heidelberg-New York, 1974.

[12] M. Takano, On a fine capacity related to a symmetric Markov process, Proc. Japan Acad., 48 (1972), 599-602.

Masatoshi Fukushima
Department of Mathematics
Osaka University
Toyonaka, Osaka, Japan

HILBERT SPACE METHODS IN CLASSICAL PROBLEMS OF MATHEMATICAL STATISTICS

O.V.Gerlein, A.M.Kagan

I. Introduction and summary. The main object of the paper is a family $\mathcal{S}=\{(\cdot,\cdot)_\theta, \theta\in\Theta\}$ of scalar products (sc.prod.) on a Hilbert space $\mathcal{H}$ depending on a parameter θ . The aim of the paper is to demonstrate that, for $\mathcal{S}$, it is possible to define natural analogues of fundamental concepts of classical statistics. Remind that the latter deals with a parametric family $\mathcal{P}=\{P_\theta, \theta\in\Theta\}$ of probability measures (distributions) defined on a measurable space $(\mathfrak{X},\mathfrak{A})$.

More precisely, we shall define, in terms of $\mathcal{S}$, analogues of sufficient statistics, the Neyman-Fisher factorization theorem, the Fisher information, the maximum likelihood method, the likelihood ratio test, and we show that our approach preserves many essential results of the classical theory.

The authors' viewpoint is that the linear theory of statistical inference is as interesting and rich as the classical one.

In Section 2 we construct, on the base of $\mathcal{S}$, the family of Radon-Nikodym operators (analogues of probability densities in the classical case) in terms of which all the subsequent results are formulated. Then a criterion of sufficiency of a subspace is given which is quite analogous to the classical Neyman-Fisher factorization theorem. In Section 3 the Fisher information contained in $\mathcal{H}$ is introduced and investigated. In Section 4 the case of a finite-dimensional $\mathcal{H}$ is considered which leads to many simplifications. Section 5 deals with the most important case when the elements of $\mathcal{H}$ are functions on a measurable space $(\mathfrak{X},\mathfrak{A})$ and the sc.prod.'s$(\cdot,\cdot)_\theta$ are defined with the help of distributions on $(\mathfrak{X},\mathfrak{A})$:

$$(\varphi,\psi)_\theta = E_\theta\varphi\psi = \int_{\mathfrak{X}} \varphi(x)\psi(x)\,dP_\theta \tag{I.I}$$

In this situation estimators of the so called maximum likelihood $\mathcal{H}$ -method are defined. Their behaviour appears tobe completely analogous to that of classical maximum likelihood estimators. In Section 6 again the case of a finite-dimensional $\mathcal{H}$ is specially considered, the special choice of $\mathcal{H}$ leads to a generalization of the classical method of moments. Note that the method of estimation of $\theta\in\Theta\subset R^s$ proposed in Section 6 allows to effectively make use of an arbitrarily large number of sample and population moments in contrast to the classical method of moments when the number of utilized moments has to be equal to the dimension of the parameter (this defect of the method of moments is well known; see [I] . In Section 7 a measure of divergence (called W -divergency) between two sc.prod's. is introduced. It can be considered as a discrete version of the Fisher information and is a useful tool for constructing in Section 8 certain goodness-of-fit tests which are of some interest. The results of Sections 2,3,5 have been obtained jointly by the authors [2]; the special case considered in Sections 4,6 was investigated earlier by one of the authors (A.M.Kagan). Introducing W-divergency (see Section 7) is due to A.M.Kagan and the results of Section 8 based on it were obtained jointly by O.V.Gerlein and R.Pincus (Mathematical Institute, GDR, Berlin) during stay of the latter at the Mathematical Institute of the Academy of Sciences of the USSR in

Leningrad.

2. Radon-Nikodym operators. Sufficient subspaces and the factorization theorem. Let $\mathcal{H}$ be a Hilbert space with a sc.prod. $(\cdot,\cdot)$ and let $\mathcal{S}=\{(\cdot,\cdot)_\theta, \theta\in\Theta\}$ be a family of sc.prod.'s on $\mathcal{H}$. By $\|\cdot\|$ and $\|\cdot\|_\theta$ we shall denote the norms generated by the sc.prod.'s $(\cdot,\cdot)$ and $(\cdot,\cdot)_\theta$ respectively.

Definition I. We say that $\mathcal{S}$ is dominated by the sc.prod. $(\cdot,\cdot)$ if

$$\|\varphi\|=0 \longrightarrow \|\varphi\|_\theta \equiv 0\ , \quad \theta\in\Theta \quad . \tag{2.I}$$

In what follows we consider only dominated families of sc.prod!s without special mentioning this fact.

It is natural to associate with a sc.prod. $(\cdot,\cdot)$ the operator A_θ in $\mathcal{H}$ defined by the formula

$$(\varphi,\psi)_\theta = (\varphi, A_\theta\psi) \ . \tag{2.2}$$

Evidently, A_θ is a non-negative linear symmetric operator. For simplicity we shall assume the operators $A_\theta, \theta\in\Theta$, to bounded, al though some of the results hold without this assumption.

The operators $A_\theta, \theta\in\Theta$, play in the linear theory the same role as the Radon-Nikodym derivatives $\frac{dP_\theta}{d\mu}=p_\theta$ do in the classical theory which deals with a family $\mathcal{P}=\{P_\theta, \theta\in\Theta\}$ of measures dominated by a measure μ. Note in this connection that if the sc.prod.'s are defined by (I.I) and $\mathcal{P}$ is dominated by a measure μ then the operators A_θ are of the form

$$A_\theta\varphi = p_\theta\varphi \quad . \tag{2.3}$$

Hence, if $\underset{\mu}{\text{ess sup}}\, p_\theta = C_\theta < \infty$, the operators A_θ from (2.3) are bounded and $\| A_\theta \| = C_\theta$.

Thus it is natural to formulate the subsequent results of the theory of linear statistical inference in terms of the family $\mathcal{A} = \{ A_\theta, \theta \in \Theta \}$ of self-adjoint operators.

Let now $\mathcal{M}$ be a (closed) subspace of $\mathcal{H}$ and $\mathcal{M}_\theta$ be a completion of $\mathcal{M}$ with respect to the norm $\|\cdot\|_\theta$.

Definition 2. $\mathcal{M}$ is said to be a sufficient subspace for the family $\mathcal{S} = \{(\cdot,\cdot)_\theta, \theta \in \Theta\}$ of sc.prod.'s if, for any $\varphi \in \mathcal{H}$, there exists a $\hat{\varphi} \in \mathcal{M}_\theta$ which is independent of θ and such that

$$\hat{E}_\theta (\varphi | \mathcal{M}_\theta) = \hat{\varphi} , \tag{2.4}$$

where $\hat{E}_\theta(\cdot|\mathcal{M})$ denotes the projecting operator into $\mathcal{M}_\theta$ (conditional expectation in a wide sense) when $\mathcal{H}$ is endowed with the sc.prod. $(\cdot,\cdot)_\theta$.

If $\mathcal{H}$ is the space of all functions on $(\mathcal{X}, \mathcal{O}\!\mathcal{L})$ squareintegrable with respect to a (probability) measure μ , $\mathcal{P} = \{P_\theta, \theta \in \Theta\}$ is a family of distributions on $(\mathcal{X}, \mathcal{O}\!\mathcal{L})$ dominated by μ, $\mathcal{b}$ is a σ-subalgebra of $\mathcal{O}\!\mathcal{L}$ and $\mathcal{M}$ is the subspace of $\mathcal{H}$ formed by all $\mathcal{b}$-measurable functions, then definition 2 coincides with the well-known definition of sufficiency of the subalgebra $\mathcal{b}$ for $\mathcal{P}$ (see, for example, [3]).

Theorem I (factorization theorem). Suppose $\mathcal{H}$ to be separable. Then $\mathcal{M}$ is a sufficient subspace for the family $\mathcal{S}$ of sc.prod.'s iff the operators A_θ admit the following factorization:

$$A_\theta = Q R_\theta , \quad \theta \in \Theta \tag{2.5}$$

where Q is a bounded non-negative self-adjoint operator in $\mathcal{H}$ and the operators R_θ, $\theta\in\Theta$, leave $\mathcal{M}$ invariant: $R_\theta\mathcal{M}\subset\mathcal{M}$.

It is interesting to compare Theorem I with the classical Neyman-Fisher factorization theorem (see [4]) stating that a statistic $T:(\mathcal{X},\mathcal{O}\!\mathcal{L})\to(\mathcal{T},\mathcal{B})$ is sufficient for a dominated family $\mathcal{P}=\{P_\theta,\theta\in\Theta\}$ iff the densities $\frac{dP_\theta}{d\mu}=p_\theta$ are factorized in the form:

$$p_\theta(x)=R_\theta(T)q(x) \quad . \tag{2.6}$$

The analogy between (2.5) and (2.6) is clear.Factorization of A_θ in the form (2.5) and not in the inverse form $A_\theta=R_\theta Q$ is implied by non-commutativity of the operators A_θ in a general case while in case of (2.3) they commutate.

<u>3. Informant and Fisher information in $\mathcal{H}$.</u> In the first part of this section we suppose Θ to be an interval of the real line R^1.

<u>Definition 3.</u> The operator J_θ in $\mathcal{H}$ defined by the formula

$$(J_\theta\varphi,\varphi)_\theta=\frac{d}{d\theta}(\varphi,\varphi)_\theta \tag{3.1}$$

is called the <u>information operator of the family S .</u>

Evidently, J_θ is a linear operator (maybe unbounded);we denote its domain by $\mathcal{D}(J_\theta)$. It is easily to verify that J_θ is an extension of the operator $A_\theta^{-1}A_\theta'$ where A_θ' means the weak derivative.

Suppose now that there exists an element $\mathbb{1}\in\mathcal{H}$ with the following property:

$$\|\mathbb{1}\|_\theta=1 \quad , \quad \theta\in\Theta \quad . \tag{3.2}$$

Let us fix this element and suppose moreover that

$$\mathbb{1} \in \mathcal{D}(J_\theta) , \quad \theta \in \Theta \quad . \tag{3.3}$$

Definition 4. $J_\theta \mathbb{1}$, $\theta \in \Theta$, is called the informant of the family $\mathcal{S}$ of sc.prod.'s defined on the space $\mathcal{H}$ with fixed element $\mathbb{1}$.

The informant is an analogue of the logarithmic derivative $\frac{\partial}{\partial\theta} \log p_\theta$ of the density and we preserve for it the term "informant" proposed by Yu.V.Linnik for the logarithmic derivative.

Definition 5. We call

$$I(\theta) = I(\theta; \mathcal{H}) = \| J_\theta \mathbb{1} \|_\theta^2 \quad . \tag{3.4}$$

the Fisher information contained in $\mathcal{H}$ about the parametric element $\theta \cdot \mathbb{1}$.

Note that, if $\mathcal{H}$ is the space of all functions on $(\mathfrak{X}, \mathfrak{A})$ square-integrable with respect to a (probability) measure μ , $\mathcal{P} = \{P_\theta, \theta \in \Theta\}$ is a family of distributions on $(\mathfrak{X}, \mathfrak{A})$ dominated by μ , the densities $\frac{dP_\theta}{d\mu} = p_\theta$ satisfy the condition $\operatorname{ess\,sup}_\mu p_\theta < \infty$, the logarithmic derivative $\frac{\partial}{\partial\theta} \log p_\theta$ is square integrable with respect to P_θ and, for $\varphi \in \mathcal{H}$, we have $\frac{d}{d\theta} \int_{\mathfrak{X}} \varphi p_\theta \, d\mu = \int_{\mathfrak{X}} \varphi \frac{d}{d\theta} p_\theta \, d\mu$ (these regularity type conditions are usual in the estimation theory; see [6,Ch.32]), then, for the element $\mathbb{1}(x) \equiv 1, x \in \mathfrak{X}$, we have

$$J_\theta \mathbb{1} = \frac{\partial}{\partial\theta} \log p_\theta \quad .$$

$$I(\theta) = \| J_\theta \mathbb{1} \|_\theta^2 = \int_{\mathfrak{X}} \left(\frac{\partial}{\partial\theta} \log p_\theta \right)^2 p_\theta \, d\mu \quad .$$

We shall show now that $I(\theta; \mathcal{H})$ preserves almost all essential properties of the classical Fisher information (for the proofs see

[2]).

I) Non-negativity of the information. If $\mathbb{1} \in \mathfrak{D}(J_\theta)$, then $I(\theta;\mathcal{H}) \geq 0$ and the equality holds for a given θ iff

$$\frac{d}{d\theta}(\mathbb{1},\varphi)_\theta = 0 \quad , \quad \varphi \in \mathcal{H} \quad . \tag{3.5}$$

For elements $\varphi \in \mathcal{H}$ different from $\mathbb{1}$, the sc.prod.'s $(\varphi,\varphi)_\theta$ may depend on θ even if $I(\theta;\mathcal{H}) = 0$.

2) Monotonicity of the information. Let $\hat{\mathcal{H}}$ be a subspace of $\mathcal{H}$ such that $\mathbb{1} \in \hat{\mathcal{H}}$; if $\mathbb{1} \in \mathfrak{D}(J_\theta)$ and $\mathbb{1} \in \mathfrak{D}(\hat{J}_\theta)$, where $\hat{J}_\theta$ is the information operator defined in $\hat{\mathcal{H}}$ by the formula (3.I), then

$$I(\theta;\hat{\mathcal{H}}) \leq I(\theta;\mathcal{H}) , \quad \theta \in \Theta \quad . \tag{3.6}$$

3) Invariance of the information with respect to reduction to a sufficient subspace. If $\hat{\mathcal{H}}$ is a sufficient subspace and the conditions of 2) are satisfied, then

$$I(\theta;\hat{\mathcal{H}}) = I(\theta;\mathcal{H}) , \quad \theta \in \Theta \quad . \tag{3.7}$$

4) Additivity of the information with respect to tensor multiplication of spaces. Let $\mathcal{H} = \mathcal{H}_1 \otimes \mathcal{H}_2$, $\mathbb{1} = \mathbb{1}_1 \otimes \mathbb{1}_2$, where $\mathbb{1}$, $\mathbb{1}_1$, $\mathbb{1}_2$ are fixed elements in $\mathcal{H}, \mathcal{H}_1, \mathcal{H}_2$ and let J_θ , $J_{\theta,1}$, $J_{\theta,2}$ denote the information operators in $\mathcal{H}$, $\mathcal{H}_1$, $\mathcal{H}_2$ respectively. If $\mathbb{1}_i \in \mathfrak{D}(J_{\theta,i})$, $i=1,2$, then $\mathbb{1} \in \mathfrak{D}(J_\theta)$ and

$$I(\theta;\mathcal{H}) = I(\theta;\mathcal{H}_1) + I(\theta;\mathcal{H}_2) \quad . \tag{3.8}$$

In particular,

$$I(\theta;\underbrace{\mathcal{H}_1 \otimes \cdots \otimes \mathcal{H}_1}_{n}) = n\, I(\theta;\mathcal{H}_1) \quad . \tag{3.9}$$

5) An alternative representation of the information. Suppose that $1 \in \mathfrak{D}(A_\theta^{-1} A_\theta')$, $1 \in \mathfrak{D}(J_\theta')$ where the prime stands for the weak derivative and moreover

$$\|(A_{\theta+\Delta\theta} - A_\theta)1\| \leq C(\theta)|\Delta\theta| , \tag{3.10}$$

$$\|(J_{\theta+\Delta\theta} - J_\theta)1\| \longrightarrow 0 , \quad \Delta\theta \longrightarrow 0 . \tag{3.11}$$

Then

$$I(\theta;\mathcal{H}) = -(J_\theta' 1, 1)_\theta . \tag{3.12}$$

(3.12) is an analogue of the formula

$$\int_{\mathfrak{X}} \left(\frac{\partial}{\partial\theta} \log p_\theta\right)^2 p_\theta d\mu = -\int_{\mathfrak{X}} \frac{\partial^2}{\partial\theta^2} \log p_\theta \cdot p_\theta d\mu . \tag{3.13}$$

which is very useful in the estimation theory.

We pass now to analogues of the Rao-Cramér and Bhattacharya inequalities.

Definition 6. An element $\varphi \in \mathcal{H}$ is called an unbiased estimator of the parametric element $g(\theta)\cdot 1$, $g(\theta):\Theta \longrightarrow R^1$, if

$$(\varphi, 1)_\theta = (g(\theta)\cdot 1, 1)_\theta = g(\theta), \quad \theta \in \Theta . \tag{3.14}$$

Introduce the operators $J_{i,\theta}$ tramsforming $\mathcal{H}$ according to the formula

$$(J_{i,\theta}\varphi, \psi)_\theta = \frac{d^i}{d\theta^i}(\varphi, \psi)_\theta . \tag{3.15}$$

One can easily see that $J_{i,\theta}$ is an extension of $A_\theta^{-1} A_\theta^{(i)}$, where $A_\theta^{(i)}$ denotes the -th weak derivative.

If for some κ the condition

$$\mathbb{1}\in\mathcal{D}(J_{i,\theta})\,,\quad i=1,\dots,\kappa\,,\quad \theta\in\Theta\quad . \tag{3.16}$$

is satisfied, we denote by $\Gamma(\theta)$ the Gram matrix of the elements $J_{1,\theta}\mathbb{1}=J_\theta\mathbb{1},\ J_{2,\theta}\mathbb{1},\ \dots\ .\ J_{\kappa,\theta}\mathbb{1}$:

$$\Gamma(\theta)=\|(J_{i,\theta}\mathbb{1},\,J_{\ell,\theta}\mathbb{1})_\theta\|_{i,\ell=1}^{\kappa}\ ,$$

For a κ times differentiable function $g(\theta)$, we set $\gamma(\theta)=\|g'(\theta)\dots g^{(\kappa)}(\theta)\|^T$, where T means transposition.

6) The Bhattacharya inequality. Let the condition (3.16) hold, the matrix $\Gamma(\theta)$ be non-singular and $\varphi\in\mathcal{H}$ be an unbiased estimator of the parametric element $g(\theta)\cdot\mathbb{1}$. Then $g(\theta)$ is κ times differentiable and

$$\|\varphi-g(\theta)\cdot\mathbb{1}\|_\theta^2\geq\gamma(\theta)^T\Gamma(\theta)^{-1}\gamma(\theta)\quad . \tag{3.17}$$

For $\kappa=1$ (3.17) becomes

$$\|\varphi-g(\theta)\cdot\mathbb{1}\|_\theta^2\geq\frac{[g'(\theta)]^2}{I(\theta;\mathcal{H})} \tag{3.18}$$

which is a straightforward analogue of the Rao-Cramér inequality.

In the remaining part of the section we suppose θ vector-valued, $\theta=\|\theta_1,\dots,\theta_s\|^T$, Θ is an open set in R^s.

<u>Definition 7.</u> The vector $J(\theta)$ whose components $J_1(\theta),\dots,J_s(\theta)$ are the operators transforming $\mathcal{H}$ according to the formula

$$(J_r(\theta)\varphi,\varphi)_\theta=\frac{\partial}{\partial\theta_r}(\varphi,\varphi)_\theta \tag{3.19}$$

is called the <u>information vector- operator of the family</u> S .

Let $\mathbb{1}$ be a fixed element of $\mathcal{H}$ satisfying the condition (3.2) such that $\mathbb{1} \in \mathcal{D}(J_r(\theta))$, $r = 1, \dots, s$.

Definition 8. The vector $J(\theta)\mathbb{1}$ with components $J_1(\theta)\mathbb{1}, \dots, J_s(\theta)\mathbb{1}$ is called the vector-informant of the family S of sc.prod.'s on $\mathcal{H}$ with fixed element $\mathbb{1}$.

Definition 9. We call

$$I(\theta;\mathcal{H}) = \| I_{rq}(\theta:\mathcal{H}) \|_{rq=1}^{s}, \quad I_{rq}(\theta;\mathcal{H}) = (J_r(\theta)\mathbb{1}, J_q(\theta)\mathbb{1})_\theta \tag{3.20}$$

the matrix of the Fisher information contained in $\mathcal{H}$ about the parametric element $\theta \cdot \mathbb{1}$.

The properties I)–5) mentioned above hold also for the information matrix if the inequalities in I)–2) have the usual matrix meaning: $A \leq B$ means that $B - A$ is a non-negatively definite matrix.

Certain comments are necessary in connection with analogues of the Rao-Cramér and Bhattacharya inequalities. Elements $\Phi \in \underbrace{\mathcal{H} \times \dots \times \mathcal{H}}_{m}$, $\Phi = \| \varphi_1 \dots \varphi_m \|^T$ will be used for estimating parametric elements $g(\theta) \cdot \mathbb{1}$, $g(\theta) = \| g_1(\theta) \dots g_m(\theta) \|^T$, $g: \Theta \longrightarrow R^m$. Unbiasedness of Φ as an estimator of $g(\theta)$ means unbiasedness of φ_i as an estimator of $g_i(\theta)$, $i = 1, \dots, m$, in sense of Definition 6.

Let

$$B(\Phi, \theta) = \| (\varphi_i - g_i(\theta) \cdot \mathbb{1}, \varphi_j - g_j(\theta) \cdot \mathbb{1})_\theta \|_{i,j=1}^{m} .$$

If, for some K,

$$\mathbb{1} \in \mathcal{D}\left(A_\theta^{-1} \frac{\partial^{r_1 + \dots + r_s}}{\partial \theta_1^{r_1} \dots \partial \theta_s^{r_s}} A_\theta\right), \; r_1 \geq 0, \dots, r_s \geq 0, \; 1 \leq r_1 + \dots + r_s \leq K , \tag{3.21}$$

(all the derivatives are weak), then let us write the elements

$A_\theta^{-1}\frac{\partial^{r_1+\dots+r_s}}{\partial\theta_1^{r_1}\dots\partial\theta_s^{r_s}}A_\theta\mathbb{1}$, $1\le r_1+\dots+r_s\le K$, as a sequence

$$\lambda^{(1)}(\theta)\ ,\ \dots\ ,\ \lambda^{(\ell)}(\theta)\quad . \tag{3.22}$$

The order of enumerating $\lambda^{(1)}(\theta),\dots,\lambda^{(\ell)}(\theta)$ is of no importance.

Set

$$\Lambda_\theta^T=\|\lambda^{(1)}(\theta)\dots\lambda^{(\ell)}(\theta)\|\ ,\quad \Gamma(\theta)=\|(\lambda^{(i)}(\theta),\ \lambda^{(j)}(\theta))_\theta\|_{i,j=1}^{\ell}\ .$$

For a parametric vector-function $g(\theta)=\|g_1(\theta)\dots g_m(\theta)\|^T$ whose components have all possible derivatives up to the order K , we write the derivatives $\frac{\partial^{r_1+\dots+r_s}}{\partial\theta_1^{r_1}\dots\partial\theta_s^{r_s}}g_i(\theta)$, $r_1\ge 0,\dots\ r_s\ge 0$, $1\le r_1+\dots+r_s\le K$, as a sequence

$$\gamma_i^{(1)}(\theta)\ ,\ \dots\ ,\ \gamma_i^{(\ell)}(\theta) \tag{3.23}$$

following the enumeration chosen in (3.22). Define the matrix

$$G(\theta)=\begin{Vmatrix}\gamma_1^{(1)}(\theta)\dots & \gamma_1^{(\ell)}(\theta)\\ \dots\quad\dots & \dots\\ \gamma_m^{(1)}(\theta)\dots & \gamma_m^{(\ell)}(\theta)\end{Vmatrix}\quad .$$

The multidimensional Bhattacharya inequality. Let the condition (3.21) be satisfied, the matrix $\Gamma(\theta)$ be non-singular and $\Phi\in\underbrace{\mathcal{H}\times\dots\times\mathcal{H}}_{m}$ be an unbiased estimator of the element $g(\theta)\cdot\mathbb{1}$. Then the components of $g(\theta)$ have all possible derivatives up to the order K and

$$B(\Phi,\theta)\ge G(\theta)^T\Gamma(\theta)G(\theta)\quad . \tag{3.24}$$

For $K=1$, $m=s$ and $g(\theta)=\theta$, the inequality (3.24) is reduced to a straightforward analogue of the multidimensional Rao-Cramér inequality:

$$B(\Phi,\theta)\ge I(\theta;\mathcal{H})^{-1}\ . \tag{3.25}$$

<u>4. A finite-dimensional space $\mathcal{H}$.</u> Let $\mathcal{H}$ be generated by elements $\varphi_1, \dots, \varphi_m$. For the matrix $A(\theta) = \| a_{ij}(\theta) \|_{i,j=1}^{m}$ determining the operator A_θ by the relation

$$A_\theta \varphi_i = \sum_{k=1}^{m} a_{ki} \varphi_k \quad ,$$

we obtain from (2.2):

$$\Pi A(\theta) = \Pi(\theta)$$

where

$$\Pi(\theta) = \| \pi_{ij}(\theta) \|_{i,j=1}^{m} = \| (\varphi_i, \varphi_j)_\theta \|_{i,j=1}^{m}, \quad \Pi = \| \pi_{ij} \|_{i,j=1}^{m} = \| (\varphi_i, \varphi_j) \|_{i,j=1}^{m} .$$

The Gram matrix may be always assumed nonsingular so that

$$A(\theta) = \Pi^{-1} \Pi(\theta) \quad .$$

If $\Pi(\theta)$ is non-singular and the functions $\pi_i(\theta) = (\varphi_i, \mathbb{1})_\theta$, $i = 1, \dots, m$, are differentiable with respect to the components $\theta_1, \dots, \theta_s$ of the parameter $\theta = \| \theta_1 \dots \theta_s \|^{T}$, then it can be easily seen that $\mathbb{1} \in \mathcal{D}(A_\theta^{-1} \frac{\partial}{\partial \theta_r} A_\theta)$, $r = 1, \dots, s$, and the components of the vector-informant are of the form

$$J_r(\theta) \cdot \mathbb{1} = - \sum_{j=1}^{m} \frac{\Lambda_{1,j+1}^{(r)}(\theta)}{\det \Pi(\theta)} \tag{4.1}$$

where $\Lambda_{1,j+1}^{(r)}(\theta)$ is the cofactor of the element with the indices 1, $j+1$ of the matrix

$$\Lambda^{(r)}(\theta) = \left\| \begin{array}{c|ccc} 0 & \frac{\partial}{\partial \theta_r} \pi_1(\theta) & \cdots & \frac{\partial}{\partial \theta_r} \pi_m(\theta) \\ \hline \frac{\partial}{\partial \theta_r} \pi_1(\theta) & & & \\ \cdots & & \Pi(\theta) & \\ \frac{\partial}{\partial \theta_r} \pi_m(\theta) & & & \end{array} \right\| \quad . \tag{4.2}$$

The information matrix is

$$I(\theta; \mathcal{H}) = \| I_{rq}(\theta; \mathcal{H}) \|_{r,q=1}^{s}, \quad I_{rq}(\theta; \mathcal{H}) = - \frac{\det \Lambda^{(r,q)}(\theta)}{\det \Pi(\theta)} \tag{4.3}$$

where

$$\Lambda^{(r,q)}(\theta) = \begin{Vmatrix} 0 & \frac{\partial}{\partial\theta_r}\pi_1(\theta) \cdots \frac{\partial}{\partial\theta_r}\pi_m(\theta) \\ \frac{\partial}{\partial\theta_q}\pi_1(\theta) & \\ \cdots & \Pi(\theta) \\ \frac{\partial}{\partial\theta_q}\pi_m(\theta) & \end{Vmatrix} \qquad (4.4)$$

As an illustration, we give here the result of calculating the matrix of the Fisher information (about the s-dimensional parameter $\theta = \|\theta_1,...,\theta_s\|^T$) contained in the space $\mathcal{H}$ of all polynomials of degree $\leq K$ of the observations $x_1,...,x_n$ in the linear regression model

$$x_1 = c_{11}\theta_1 + \ldots + c_{1s}\theta_s + \varepsilon_1 \quad , \qquad (4.5)$$

$$x_n = c_{n1}\theta_1 + \ldots + c_{ns}\theta_s + \varepsilon_n \quad ,$$

with the design matrix $C = \|c_{ij}\|$. We assume the errors $\varepsilon_1,...,\varepsilon_n$ to be independent with the distribution functions $P(\varepsilon_i < x) = F_i(x) = F(\frac{x}{\sigma_i})$ satisfying the conditions

$$\int x\,dF = 0 \quad , \quad \int x^{2K}\,dF < \infty$$

Let rank $C = s \leq n$ and

$$P = \begin{Vmatrix} P_1 & 0 & \ldots & 0 \\ 0 & P_2 & \ldots & 0 \\ \ldots & \ldots & \ldots & \ldots \\ 0 & 0 & \ldots & P_n \end{Vmatrix}$$

denote the weight matrix of the observations so that, for some σ^2,

$$\sigma_i^2 = \frac{\sigma^2}{P_i} \quad , \qquad i = 1, \ldots, n \ .$$

Simple calculation shows that

$$I(\theta;\mathcal{H}) = \frac{I_K}{\sigma^2}\, C^T P C \qquad (4.6)$$

where I_K has the meaning of the Fisher information about the (scalar) parameter a contained in polynomials of degree $\leq K$ of the observations $x_1,...,x_n$ obtained in the set-up of direct measurements:

$$x_1 = a + \zeta_1$$

$$\ldots\ldots$$

$$x_n = a + \zeta_n$$

with $\zeta_1,\dots,\zeta_n$ independent identically distributed random variables with the distribution function $F(x)$. The following expression for I_K was obtained in [5]:

$$I_K = \frac{R_{11}}{\det R} \tag{4.7}$$

where

$$R = \|\rho_{ij}\|_{i,j=1}^{K}, \ \rho_{11} = \alpha_2, \ \rho_{1j} = \rho_{j1} = \alpha_{j+1} - j\alpha_2\alpha_{j-1} \qquad \text{for } j>1,$$

$$\rho_{ij} = \rho_{ji} = \alpha_{i+j} - j\alpha_{j-1}\alpha_{i+1} - i\alpha_{j+1}\alpha_{i-1} - \alpha_i\alpha_j + ij\alpha_2\alpha_{i-1}\alpha_{j-1} \qquad \text{for } i>1,\ j>1;$$

$\alpha_i = \int x^i dF$ and R_{11} denotes the cofactor of the element ρ_{11} of R.

5. An analogue of the maximum likelihood method. Suppose in this section that elements of $\mathcal{H}$ are functions on a measurable space $(\mathfrak{X},\mathfrak{A})$, the sc.prod.'s $(.,.)_\theta$ are defined by a family $\{P_\theta, \theta\in\Theta\}$ of distributions on $(\mathfrak{X},\mathfrak{A})$:

$$(\varphi,\psi)_\theta = E_\theta\varphi\psi = \int_{\mathfrak{X}} \varphi(x)\psi(x)dP_\theta , \tag{5.1}$$

and the function $\mathbb{1}(x)\equiv 1$, $x\in\mathfrak{X}$, belongs to $\mathcal{H}$ and plays the role of the fixed element $\mathbb{1}$.

We consider here the case of a vector-valued parameter $\theta = \|\theta_1 \dots \theta_s\|^T$ belinging to an open set $\Theta\subset R^s$ and, for the sake of convenience, we introduce here a new notation for the vector-informant, which is also a function on $(\mathfrak{X},\mathfrak{A})$:

$$J(\theta)\mathbb{1}(x) = g_\theta(x), \quad g_\theta = \|g_\theta^{(1)} \dots g_\theta^{(s)}\|^T .$$

Let $x_1,\dots,x_n$ be independent random variables (observations) taking values in $(\mathfrak{X},\mathfrak{A})$ and distributed each according to the probability P_θ, θ being unknown. In fact we shall deal not with observations $x_1,\dots,x_n$ themselves but only with observations $\varphi(x_1),\dots,\varphi(x_n)$ of elements

$\varphi \in \mathcal{H}$. Thus elements of $\mathcal{H}$ can be considered as admissible statistics in the problem in question.

As it was mentioned in Section 3, components of $g_\theta(x)$ are analogues of logarithmic derivatives of the density with respect to $\theta_1,\dots,\theta_s$ That is why it is natural to investigate the following analogue of the classical maximum likelihood estimator.

Definition 10. A statistic $\hat{\theta}_n = \hat{\theta}_n(x_1,\dots,x_n)$ taking values in R^s is called an estimator of the $\mathcal{H}$-method of maximum likelihood if it is a solution of the system of equations

$$\sum_{i=1}^{n} g_\theta^{(1)}(x_i) = 0 \ , \qquad (5.2)$$
$$\cdots\cdots\cdots\cdots$$
$$\sum_{i=1}^{n} g_\theta^{(s)}(x_i) = 0 \ .$$

The estimator $\hat{\theta}_n$ (for brevity we call it the $\mathcal{H}$-estimator of m.l.) depends on $x_1,\dots,x_n$ only through observations of elements $\varphi \in \mathcal{H}$. We state now results describing asymptotic behaviour of $\hat{\theta}$ as $n \to \infty$. Let θ_0 denote the true value of the parameter and U be a vicinity of θ_0 .

Theorem 2. (Measurability, strong consistency and asymptotic normality of the $\mathcal{H}$-estimator of m.l.) Suppose that the following conditions are satisfied.

1) The functions $g_\theta^{(r)}(x), r=1,\dots,s$, are continuous and continuously differentiable with respect to cach component of $\theta \in U$ for all $x \in \mathfrak{X}$.

2) The functions $h_0^{(r)}(\theta) = (J_r(\theta)\mathbb{1},\mathbb{1})_{\theta_0}$, $r=1,\dots,s$. are continuous for $\theta \in U$.

3) The functions $M_{1,q}^{(r)}(x) = \sup \left| \frac{\partial}{\partial \theta_q} g_\theta^{(r)}(x) \right|$, $r,q=1,\dots,s$, are integrable with respect to the measure P_{θ_0} .

4) $I_{rq}(\theta_0;\mathcal{H}) = -(\frac{\partial}{\partial\theta_q} J_r(\theta_0)\mathbb{1},\mathbb{1})_{\theta_0}$, $r,q = 1,\dots,s$, where $\frac{\partial}{\partial\theta_q} J_r(\theta)$ means the weak derivative, and the matrix $I(\theta;\mathcal{H})$ is nonsingular.

Then there exists a statistic $\hat{\theta}_n = \hat{\theta}_n(x_1,\dots,x_n)$ with values in R^s which satisfies (5.2) such that

$$P_{\theta_0}\{|\hat{\theta}_n - \theta_0| \to 0\} = 1 \quad , \quad n \to \infty \quad .$$

Let now, in addition to conditions 1)-4), the following conditions be satisfied.

5). The second derivatives $\frac{\partial^2}{\partial\theta_p\partial\theta_q} g_\theta^{(r)}(x)$, $r,p,q = 1,\dots,s$, exist and are continuous functions of $\theta \in U$ for all $x \in \mathcal{X}$.

6). The functions $M_{2,pq}^{(r)}(x) = \sup_{\theta\in U}\left|\frac{\partial^2}{\partial\theta_p\partial\theta_q} g_\theta^{(r)}(x)\right|$, $r,p,q = 1,\dots,s$, are integrable with respect to P_{θ_0} .

7) $-E_{\theta_0}(\frac{\partial}{\partial\theta_0} g_\theta^{(r)}(x)) = I_{rq}(\theta_0;\mathcal{H})$, $r,q = 1,\dots,s$.

Then the random vector $\sqrt{n}(\hat{\theta}_n - \theta_0)$ is distributed asymptotically (as $n \to \infty$) according to the s-dimensional normal law $\mathcal{N}(0, I(\theta_0;\mathcal{H})^{-1})$.

We make now a few remarks concerning the conditions of Theorem 2.

In connection with condition 4) we note that $-(\frac{\partial}{\partial\theta_q} J_r(\theta)\mathbb{1},\mathbb{1})_\theta$ is an alternative representatuon of the element $I_{rq}(\theta;\mathcal{H})$ of the information matrix; this was established, under certain conditions in Section 3 (the dimension of Θ here irrelevant). Condition 7) is a regularity type condition which means that for $\theta = \theta_0$

$$(\frac{\partial}{\partial\theta_q} g_\theta^{(r)}, \mathbb{1})_\theta = (\frac{\partial}{\partial\theta_q} J_r(\theta)\mathbb{1},\mathbb{1})_\theta$$

and hence it is automatically satisfied if

$$\frac{\partial}{\partial\theta_q} J_r(\theta)\mathbb{1}(x) = \frac{\partial}{\partial\theta_q} g_\theta^{(r)}(x) \quad \text{for } \theta=\theta_0 \text{ , all } x\in\mathfrak{X}, r,q=1,\dots,s \text{ .}$$

To demonstrate an optimal property of the $\mathcal{H}$ -estimator of m.l. let us consider in $\underbrace{\mathcal{H}\times\dots\times\mathcal{H}}_{s}$ a family of elements $\{\varphi_\theta, \theta\in\Theta\}$, $\varphi_\theta = \|\varphi_\theta^{(1)}, \dots, \varphi_\theta^{(s)}\|^T$, satisfying the following conditions:

$$(\varphi_\theta^{(r)}, \mathbb{1})_\theta \equiv 0 \text{ , } \theta\in\Theta \text{ , } r=1,\dots,s \text{ ,} \tag{5.3}$$

$$\Phi(\theta) = \|(\frac{\partial}{\partial\theta_q}\varphi_\theta^{(r)}, \mathbb{1})_\theta\|_{r,q=1}^{s} \text{ , } \quad \theta\in\Theta \quad \text{, is non-singular} \tag{5.4}$$

(by $\frac{\partial}{\partial\theta_q}\varphi_\theta^{(r)}$ we mean the weak derivative in $\mathcal{H}$).

For independent observations $x_1, \dots, x_n$ we define the estimator $\hat\theta_n = \hat\theta_n(x_1, \dots, x_n)$ as a solution of the system of equations

$$\sum_{i=1}^{n} \varphi_\theta^{(1)}(x_i) = 0 \text{ ,} \tag{5.5}$$
$$\dots\dots\dots\dots\dots$$
$$\sum_{i=1}^{n} \varphi_\theta^{(s)}(x_i) = 0 \text{ .}$$

The following theorem is a complete analog of Theorem 2.

<u>Theorem 3.</u> (Measurability, strong consistency and asymptotic normality of $\hat\theta_n$) Suppose that conditions (5.3), (5.4) are satisfied as well as the following ones.

1) The functions $\varphi_\theta^{(r)}(x)$; $r=1,\dots,s$; are continuous and continuously differentiable with respect to each component of $\theta\in U$ for all $x\in\mathfrak{X}$.

2) The functions $h_1^{(r)}(\theta) = (\varphi_\theta^{(r)}, \mathbb{1})_{\theta_0}$; $r=1,\dots,s$; are continuous for $\theta\in U$

3) The functions $M_{3,q}^{(r)}(x) = \sup_{\theta\in U}\left|\frac{\partial}{\partial\theta_q}\varphi_\theta^{(r)}(x)\right|$; $r,q=1,\dots,s$; are integrable with respect to the measure P_{θ_0} .

Then there exists a statistic $\hat\theta_n = \hat\theta_n(x_1,\dots,x_n)$ with values in R^s which satisfies (5.5) and such that

$$P_{\theta_0}\{\widehat{\theta}_n \to \theta_0\} = 1 \quad , \quad n \to \infty \quad .$$

Let, additionally to 1)-3) the following conditions be satisfied.

4) The second derivatives $\frac{\partial^2}{\partial\theta_p\partial\theta_q}\varphi^{(r)}_\theta(x)$; $r,p,q = 1,\dots,s$; exist and are continuous functions of $\theta \in U$ for all $x \in \mathfrak{X}$.

5) The functions $M^{(r)}_{4,pq}(x) = \sup_{\theta\in U}\left|\frac{\partial^2}{\partial\theta_p\partial\theta_q}\varphi^{(r)}_\theta(x)\right|$; $r,p,q = 1,\dots,s$; are integrable with respect to the measure P_{θ_0} ,

6) $E_{\theta_0}\left(\frac{\partial}{\partial\theta_q}\varphi^{(r)}_{\theta_0}(x)\right) = \left(\frac{\partial}{\partial\theta_q}\varphi^{(r)}_{\theta_0}, \mathbb{1}\right)_{\theta_0}$; $\quad r,q = 1, \dots, s$.

Then the random vector $\sqrt{n}(\widehat{\theta}_n - \theta_0)$ is distributed asymptotically (as $n \to \infty$) according to the normal law $\mathcal{N}(0, \Phi(\theta_0)^{-1}\Gamma(\theta_0)(\Phi(\theta_0)^T)^{-1})$, where $\Gamma(\theta)$ is the Gram matrix of the elements $\varphi^{(1)}_\theta, \dots, \varphi^{(s)}_\theta$ and $\Phi(\theta)$ was defined in (5.4).

The next theorem shows that, under certain conditions, the concentration of the limit distribution of the $\mathcal{H}$-estimator of m.l.is greater than that for the estimator determined by (5.5).

Theorem 4. Let the operators A_θ , $\theta \in \Theta$, be such that

$$\|(A_{\theta+\Delta\theta} - A_\theta)\mathbb{1}\| \le C(\theta)|\Delta\theta| \quad ,$$

$$\mathbb{1} \in \mathcal{D}\left(A^{-1}_\theta \frac{\partial}{\partial\theta_r} A_\theta\right) ; \quad r = 1, \dots, s ;$$

where $\frac{\partial}{\partial\theta_r} A_\theta$ means the weak derivative, and let the family $\{\varphi_\theta, \theta \in \Theta\}$ satisfy conditions (5.3),(5.4) and moreover be strongly continuous in $\theta \in \Theta$, i.e.

$$\|\varphi_{\theta+\Delta\theta} - \varphi_\theta\| \to 0 \quad , \quad \Delta\theta \to 0 \quad .$$

If the matrix $I(\theta;\mathcal{H})$ is non-singular then

$$\Phi(\theta)^{-1}\Gamma(\theta)(\Phi(\theta)^T)^{-1} \ge I(\theta;\mathcal{H})^{-1} \quad . \tag{5.6}$$

6. Estimators of the $\mathcal{H}$-method of maximum likelihood in case of a finite-dimensional space $\mathcal{H}$ and a correct version of the method of moments. Here we consider the case when $\mathcal{H}$ is generated by functions $\varphi_1(x), \dots, \varphi_m(x)$ on a measurable space $(\mathcal{X}, \mathcal{O}\!\iota)$, the sc.prod.'s are defined by (5.I) and the function $\mathbb{1}(x) \equiv 1$, $x \in \mathcal{X}$, belongs to $\mathcal{H}$ and is chosen as $\mathbf{1}$. In this situation the conditions of Theorems 2,3,4 can be essentially simplified.

Taking (4.I) into account, the system (5.2) can be written as follows:

$$\begin{aligned} &\sum_{j=1}^{m} \lambda_j^{(1)}(\theta)\bar{\varphi}_j = 0 \;, \\ &\dots\dots\dots\dots\dots \\ &\sum_{j=1}^{m} \lambda_j^{(s)}(\theta)\bar{\varphi}_j = 0 \;, \end{aligned} \qquad (6.I)$$

where

$$\bar{\varphi}_j = \frac{1}{n}\sum_{i=1}^{n} \varphi_j(x_i) \;, \quad \lambda_j^{(r)}(\theta) = \frac{\Lambda_{1,j+1}^{(r)}(\theta)}{\det \Pi(\theta)} \;;\; j=1,\dots,m;\; r=1,\dots,s;$$

and $\Pi(\theta)$ and $\Lambda_{1,j+1}^{(r)}(\theta)$ were defined in Section 4.

Theorem 5. Suppose that for $\theta \in \Theta$ the following conditions are satisfied.

1) The matrix $\Pi(\theta)$ is non-singular.

2) The functions $\pi_{11}(\theta), \pi_{12}(\theta), \dots, \pi_{mm}(\theta)$, $\frac{\partial}{\partial\theta_1}\pi_1(\theta), \frac{\partial}{\partial\theta_2}\pi_1(\theta), \dots, \frac{\partial}{\partial\theta_s}\pi_m(\theta)$ defined in Section 4 are differentiable with respect to the components $\theta_1, \dots, \theta_s$.

3) The information matrix $I(\theta; \mathcal{H})$ is non-singular.

Then there exists a statistic $\hat{\theta}_n = \hat{\theta}(\bar{\varphi}_1, \dots, \bar{\varphi}_m)$, $\hat{\theta}_n : R^m \to R^s$, depending on n only through $\bar{\varphi}_1, \dots, \bar{\varphi}_m$ and satisfying (6.I) such that

$$P_\theta\{|\hat{\theta}_n - \theta| \to 0\} = 1 \;, \quad n \to \infty \;.$$

Let conditions 1)-3) be satisfied and, moreover,

4) The functions $\pi_{11}(\theta), \pi_{12}(\theta), \dots, \pi_{mm}(\theta)$ and $\frac{\partial}{\partial\theta_1}\pi_1(\theta)$,

$\frac{\partial}{\partial\theta_2}\pi_1(\theta)$, ..., $\frac{\partial}{\partial\theta_s}\pi_m(\theta)$ are twice continuously differentiable with respect to $\theta_1, \dots, \theta_s$. Then the random vector $\sqrt{n}(\hat{\theta}_n - \theta)$ is distirbuted asymptotically (as $n \to \infty$) according to the s-dimensional normal distribution $\mathcal{N}(0, I(\theta; \mathcal{H})^{-1})$.

Let us consider specially the case when $\mathcal{X} = R^1$, $\mathcal{O}$ is the σ-algebra of Borel sets, the distributions P_θ, $\theta \in \Theta$, have all the moments up to order 2κ:

$$\alpha_1(\theta) = \int x \, dP_\theta \ , \ \dots \ , \ \alpha_{2\kappa}(\theta) = \int x^{2\kappa} dP_\theta$$

and the space $\mathcal{H}$ is generated by the functions $1, x^1, \dots, x^\kappa$.

Then (6.I) turns into

$$\sum_{j=1}^{\kappa+1} \lambda_j^{(1)} a_{j-1} = 0 \ , \ \dots \ , \ \sum_{j=1}^{\kappa+1} \lambda_j^{(s)} a_{j-1} = 0 \qquad (6.2)$$

where $a_j = \frac{1}{n}\sum_{i=1}^{n} x_i^j$; $j = 0, \dots, \kappa$; are the sample moments,

$$\Pi(\theta) = \begin{Vmatrix} 1 & \alpha_1(\theta) & \dots & \alpha_\kappa(\theta) \\ \alpha_1(\theta) & \alpha_2(\theta) & & \alpha_{\kappa+1}(\theta) \\ \dots & \dots & \dots & \dots \\ \alpha_\kappa(\theta) & \alpha_{\kappa+1}(\theta) & \dots & \alpha_{2\kappa}(\theta) \end{Vmatrix}, \quad \Lambda^{(r)}(\theta) = \begin{Vmatrix} 0 & 0 \ \frac{\partial}{\partial\theta_r}\alpha_1(\theta) \dots \frac{\partial}{\partial\theta_r}\alpha_\kappa(\theta) \\ 0 & \\ \frac{\partial}{\partial\theta_r}\alpha_1(\theta) & \\ \dots & \Pi(\theta) \\ \frac{\partial}{\partial\theta_r}\alpha_\kappa(\theta) & \end{Vmatrix}$$

The method of estimating θ by the statistic $\hat{\theta}_n = \hat{\theta}(a_1, \dots, a_\kappa)$ satisfying (6.2) can be considered as a correct version of the method of moments because in (6.2) the numbers κ of utilized sample moments and 2κ of population moments are connected with the dimension s of the parametric set Θ only by the natural condition $\kappa \geq s$. This fact essentially distinguishes the method of estimating θ by (6.2) from the classical method of moments (see [6, ch.33]) prescribing to utilize for estimating an s-dimensional parameter θ just s (not more!) moments and to find the estimator $\tilde{\theta}_n = \tilde{\theta}(a_1, \dots, a_\kappa)$ from the following system of equations:

$$\alpha_1(\theta) = a_1 \quad , \qquad (6.3)$$
$$\dots\dots\dots\dots$$
$$\alpha_s(\theta) = a_s \quad .$$

It is known that, if $\alpha_{2s}(\theta) < \infty$, the functions $\alpha_1(\theta), \dots, \alpha_s(\theta)$ are twice continuously differentiable, the matrix $B(\theta) = \| \frac{\partial}{\partial \theta_q} \alpha_r(\theta) \|_{q,r=1}^{s}$ is non-singular, then the random vector $\sqrt{n}(\hat{\theta}_n - \theta)$ is distributed asymptotically according to a normal law with zero mean vector and the covariance matrix determined by $B(\theta)$ and the moments $\alpha_1(\theta), \dots, \alpha_{2s}(\theta)$.

The estimator of the method of moments is a special case of estimators determined by (5.5). Making use of this fact and of simplifications implied by finite dimension of $\mathcal{H}$, one obtains the following result.

<u>Theorem 6.</u> Suppose that $K \geq S$, the matrices $\Pi(\theta)$, $B(\theta)$, $I(\theta; \mathcal{H})$ are non-singular, the functions $\alpha_{K+1}(\theta), \dots, \alpha_{2K}(\theta)$ are twice and the functions $\alpha_1(\theta), \dots, \alpha_K(\theta)$ are theree times continuously differentiable. Then the $\mathcal{H}$-estimator of m.l. determined by (6.2) is asymptotically better in the sense of Theorem 4 than the estimator of the classical method of moments.

Note that, under more restrictive conditions, the estimator (6.2) was investigated by Shenton [7,8] who obtained it in a little different way.

<u>7. W-divergency between two scalar products.</u> Let $\Theta = \{\theta_1, \theta_2\}$ so that $\mathcal{S}$ consists of two sc.prod.'s $(\cdot\cdot\cdot)_1$ and $(\cdot\cdot\cdot)_2$. Denote by A_1 and A_2 the operators corresponding to these sc.prod.'s according to the formula (2.2). Let $\|\cdot\|_i$ be the norm generated by the sc.prod. $(\cdot\cdot\cdot)_i$, $i = 1, 2$, and suppose that

$$\| 1 \|_i = 1 \quad , \quad i = 1, 2 \quad . \qquad (7.1)$$

Definition II. If $1 \in \mathfrak{D}(A_1^{-1} A_2)$, we call the number

$$W(1:2) = \log\{1 + \|(I - A_1^{-1} A_2)1\|_1^2\} \quad , \qquad (7.2)$$

W-divergency between the sc.prod.'s $(\cdot,\cdot)_1$ and $(\cdot,\cdot)_2$ on the space $\mathcal{H}$ with the fixed element 1, i being the identity operatir. Similarly, if $1 \in \mathfrak{D}(A_2^{-1} A_2)$, the number

$$W(2:1) = \log\{1 + \|(I - A_2^{-1} A_1)1\|_2^2\} \qquad (7.3)$$

is called W-divergency between $(\cdot,\cdot)_2$ and $(\cdot,\cdot)_1$.

Consider as an example the following special case. Let P_1, P_2 be probability measures on a space $(\mathfrak{X}, \mathfrak{A})$, $\mu = \frac{1}{2}(P_1 + P_2)$ and let $\mathcal{H}$ be the space of all functions $\varphi(x)$ on $(\mathfrak{X}, \mathfrak{A})$ with $\int_{\mathfrak{X}} \varphi^2 d\mu < \infty$, the scalar product being defined as $(\varphi, \psi) = \int_{\mathfrak{X}} \varphi(x)\psi(x)\, d\mu$.

Set

$$\frac{dP_i}{d\mu} = p_i \quad , \quad (\varphi, \psi)_i = (\varphi, A_i \psi) = \int_{\mathfrak{X}} \varphi \psi p_i \, d\mu$$

sothat $A_i \varphi = p_i \varphi$, $i = 1,2$. If $1(x) \equiv 1$, $x \in \mathfrak{X}$, and $1 \in \mathfrak{D}(A_1^{-1} A_2)$ (what means that $\|\frac{p_2}{p_1}\|_1^2 = \int_{p_1 > 0} \frac{p_2^2}{p_1} d\mu < \infty$), then (7.2) implies that

$$W(1:2) = \log\{1 + E_1(1 - \frac{p_2}{p_1})^2\}$$

where E_1 denotes the expectation corresponding to P_1.

The following properties of W-divergency are analogous to the properties of Fisher information formulated in Section 3.

I) Non-negativity:

$$W(1:2) \geq 0$$

the eqyality holding iff, for all $\varphi \in \mathcal{H}$,

$$(\varphi, 1)_1 = (\varphi, 1)_2 \quad .$$

2) Monotonicity. Let $\hat{\mathcal{H}}$ be a subspace of $\mathcal{H}$ such that $1 \in \hat{\mathcal{H}}$ and

let the operators $\hat{A}_i$; $i=1,2$; be defined in $\hat{\mathcal{H}}$ by (2.2). If $\mathbb{1}\in\mathfrak{D}(\hat{A}_1^{-1}A_2)$ and

$$\hat{W}(1:2)=\log\{1+\|(I-\hat{A}_1^{-1}\hat{A}_2)\mathbb{1}\|_1^2\} ,$$

then

$$\hat{W}(1:2)\leq W(1:2) .$$

3) Invariance with respect to reduction to a sufficient subspace. If, in the above situation, $\hat{\mathcal{H}}$ is a sufficient subspace for the family $\mathcal{S}$, then

$$\hat{W}(1:2)=W(1:2) .$$

4) Additivity with respect to tensor multiplication of spaces. Let $\mathcal{H}=\mathcal{H}_1\otimes\mathcal{H}_2$, $\mathbb{1}=\mathbb{1}_1\otimes\mathbb{1}_2$, the elements $\mathbb{1}\in\mathcal{H}_i$ $(i=1,2)$ satisfying the condition (7.I). If $W(1:2)$, $W_1(1:2)$, $W_2(1:2)$ denote W-divergencies between the sc.prod.'s $(\cdot,\cdot)_1$ and $(\cdot,\cdot)_2$ in $\mathcal{H}$ with the fixed element $\mathbb{1}$, in $\mathcal{H}_1$ with $\mathbb{1}_1$, and in $\mathcal{H}_2$ with $\mathbb{1}_2$ respectively, then

$$W(1:2)=W_1(1:2)+W_2(1:2) .$$

5) Relation to the Fisher information. Let $\mathcal{S}$ be a family of sc.prod.'s $(\cdot,\cdot)_\theta$ on $\mathcal{H}$ indexed by an s-dimensional parameter $\theta=\|\theta_1,\dots,\theta_s\|^T$, $\theta\in\mathcal{H}$, Θ being an open set in R^s. Suppose that the operators A_θ, $\theta\in\Theta$ which correspond to $(\cdot,\cdot)_\theta$ according to (2.2) satisfy the following conditions:

(i) $$\|A_\theta^{-1}(A_{\theta+\Delta\theta}-A_\theta)\mathbb{1}\|< C(\theta)\,|\Delta\theta| ,$$

(ii) the function $A_\theta\mathbb{1}$ has strongly continuous strong derivatives $A_\theta^{(r)}\mathbb{1}$ with respect to θ_r: $r=1,\dots,s$. Then

$$W(\theta:\theta+\Delta\theta)=\sum_{r,q=1}^{s} I_{rq}(\theta:\mathcal{H})\Delta\theta_r\Delta\theta_q+o(|\Delta\theta|^2) .$$

The next section is devoted to some applications of W-divergency.

8. Goodness-of-fit tests based on W-divergency. We consider here the same as in Section 6, when $\mathcal{H}$ is generated by functions $\varphi_0(x)\equiv 1, \varphi_1(x), \ldots, \varphi_m(x)$ on a measurable space $(\mathcal{X},\mathcal{O}\!\ell)$ and square-integrable with respect to probability measures P_θ on $(\mathcal{X},\mathcal{O}\!\ell)$, $\theta\in\Theta\subset R^s$ (the function φ_0 is singled out for the convenience), the sc.prod.'s being defined by (5.I). we suppose the matrix $\Pi(\theta)=\|(\varphi_i,\varphi_j)_\theta\|_{i,j=0}^{m}$ to be non-singular.

Let $x_1,\ldots,x_n$ be independent random variables each distributed according to P_θ. We shall need not the x_i's themselves but only values of $\varphi_i(x_i)$, $j=1,\ldots,m$ (moreover only the values of $\bar{\varphi}_j=\frac{1}{n}\sum_{i=1}^{n}\varphi_j(x_i)$, $j=1,\ldots,m$).

We begin with the hypothe sis

$$H_0:\ E_\theta\varphi_1=(\varphi_1,\mathbb{1})_\theta=c_1(\theta),\ldots,E_\theta\varphi_m=(\varphi_m,\mathbb{1})_\theta=c_m(\theta) \qquad (8.I)$$

where $c_1(\theta),\ldots,c_m(\theta)$ are given functions, E_θ denotes the expectation with respect to the measure P_θ.

Let us first construct the empirical measure on $(\mathcal{X},\mathcal{O}\!\ell)$ corresponding to a sample $(x_1,\ldots,x_n)$ by assigning the mass $\frac{1}{n}$ to every point $x_1,\ldots,x_n$ and define the empirical sc.prod.:

$$(\varphi,\psi)_n=\int_{\mathcal{X}}\varphi(x)\psi(x)\,dP_n=\frac{1}{n}\sum_{i=1}^{n}\varphi(x_i)\psi(x_i) \quad .$$

$\|\varphi\|_\theta=0$ implies that $\|\varphi\|_n=(\varphi,\varphi)_n=0$ with P_θ-probability 1 .

Define now the "random" operator A_n by the formula

$$(\varphi_i,\varphi_j)_n=(\varphi_i,A_n\varphi_j)$$

and consider the W-divergency between the sc.prod.'s $(\cdot\cdot\cdot)_n$ and $(\cdot\cdot\cdot)_\theta$ denoting it for convenience by $W(A_\theta,A_n)$. Let us introduce the statistic

$$W_n^* = \min_\theta (A_\theta, A_n) \; . \qquad (8.2)$$

<u>Theorem 7.</u> Suppose that $m \geq s$, the functions $c_1(\theta), \ldots, c_m(\theta)$ are continuously differentiable with respect to all components θ_r $(r=1,\ldots,s)$ and rank $\| \frac{\partial}{\partial \theta_r} c_j(\theta) \|_{j,r=1,\ldots,m} =$. The nullhypothesis (8.I) being true, the statistic nW_n^* is asymptotically (as $n \to \infty$) distributed according to the chi-squre distribution with $m-s$ degrees of freedom.

Hence the test rejecting the hypothesis (8.I) when

$$nW_n^* > \chi^2_{m-s,\alpha}$$

where $\chi^2_{m-s,\alpha}$ denotes the $100(1-\alpha)$ -percent point of the chi-square distribution with $m-s$ degrees of freedom, has asymptotically the significance level α .

Consider now the hypothesis

$$H_0: \theta_1 = \theta_{10} \, , \ldots , \theta_r = \theta_{r_0} \qquad (r \leq s) \qquad (8.3)$$

where $\theta_{10}, \ldots, \theta_{r_0}$ are given and the components $\theta_{r+1}, \ldots, \theta_s$ are the nuisance parameters.

Let $\hat{\theta}_n = \hat{\theta}(\bar{\varphi}_1, \ldots, \bar{\varphi}_m)$ be the $\mathcal{H}$ -estimator of m.l. for the whole parameter $\theta = \| \theta_1 \ldots \theta_s \|^T$ and $\hat{\hat{\theta}}_n = \hat{\hat{\theta}}(\bar{\varphi}_1, \ldots, \bar{\varphi}_m)$ be the $\mathcal{H}$ -estimator of m.l. for the $(s-r)$ -dimensional parameter $\eta = \| \theta_{r+1} \ldots \theta_s \|$ when the values $\theta_1, \ldots, \theta_s$ are determined by the hypothesis (8.3). Denote by $W(A_{\hat{\theta}}, A_{\hat{\hat{\theta}}})$ the W -divergency between the sc.prod.'s $(\cdot,\cdot)_{\hat{\theta}}$ and $(\cdot\,\cdot\,\cdot)_{\hat{\hat{\theta}}}$.

<u>Theorem 8.</u> Suppose that the conditions of Theorem 5 are satisfied which guarantee the asymptotic normality of $\hat{\theta}_n$ and $\hat{\hat{\theta}}_n$. Then, the null hypothesis (8.3) deing true, the statistic $nW(A_{\hat{\theta}}, A_{\hat{\hat{\theta}}})$ is distributed asymptotically (as $n \to \infty$) according to the chi-square distribution with r degrees of freedom.

In similar ways one can construct tests for other hypotheses naturally formulated within the linear theory.

REFERENCES

I Chernoff H., Sequential analysis and optimal design, Philadelphia, SIAM, 1972.

2 Gerlein O.V. and Kagan A.M., Hilbert space methods in classical problems of mathematical statistics (in Russian), Zapiski nauchnyh seminarov Leningrad.otdelenija Matemat. Instituta, Vol.53(1975), 64-100.

3 Bahadur R.R., Sufficiency and statistical decision functions, Ann.Math.Stat., 25,3 (1954), 423-462.

4 Halmos P. and Savage L., Application of the Radom-Nikodym theorem to the theory of sufficient statistics, Ann.Math. Stat., 20,2(1949), 225-241.

5 Kagan A.M., Klebanov L.B. and Fintushal S.M., Asymptotic behaviour of the polynomial Pitman estimators (in Russian), Zapiski nauchnyh seminarov Leningrad.otdeleniya Matemat. Instituta, Vol.43(1974), 30-39

6 Cramér H., Mathematical methods of statistics, Princeton Univ. Press, 1946.

7 Shenton L.R., Maximum likelihood and efficiency of the method of moments, Biometrika, 37, I, (1950), III-II6.

8 Shenton L.R., Moment estimators and maximum likelihood, Biometrika, 45,3, (1958), 3II-320.

Steklov Mathematical Institute
Akademy of Sciences of the USSR
Leningrad

ON THE MARTINGALE APROACH TO STATISTICAL PROBLEMS FOR STOCHASTIC PROCESSES WITH BOUNDARY CONDITIONS

B. Grigelionis

Introduction

After fundamental papers of W.Feller there were many investigations concerning construction and classification of diffusion processes on an m-dimensional differentiable manifild with a boundary. For these aims analytical methods as well as purely probabilistic ones were developed (for bibliography see e.g. [1]).

As is well known, these processes behave themselves near an inner point of a manifold as solutions of Itô stochastic differential equations in m-dimensional Euclidean spaces and, near a boundary point, they behave themselves as solutions of stochastic aquations in semispaces of m-dimensional Euclidean spaces with certain boundary conditions. Thus in both cases they can be characterized, locally, in terms of martingales as a special subclass of local semimartingales. From this point of view it seems to be important to investigate a general subclass of local semimartingales of this type in order to apply them, for example, in statistical problems for diffusion processes with boundary conditions.

In the paper, a subclass of local semimartingales is investigated which have certain local characteristics with respect to a given family of σ-algebras and a probability measure and take values in a semispace of the m-dimensional Euclidean space. The main attention is payed to transformations of local characteristics when we change the probability measure by absolute continuity and the family of σ-

-algebras and to the structure of functionals which are non-negative local martingales or square-integrable martingales adapted to the family of σ-algebras generated by the considered semimartingale. The results obtained are used in the non-linear filtering theory and in problems of absolute continuity of probability measures. The proofs of theorems are outlined, the details can be restored without difficulty or can be found in [I] and [2] . Note that analogous results are obtained for local semimartingales taking values in a finite one-dimensional interval (see [3]).

§ I. Notation and definitions

Let $(\Omega, \mathcal{F}, P)$ be a complete probability space and $\{\mathcal{F}_t, t \geq 0\}$ be an increasing right continuous family of σ-subalgebras $\mathcal{F}$, containing all null subsets. We shall use the standard terminology and notation of the martingale theory (see [4],[5]). Let $\mathcal{M}$ be the class of all right continuous square-integrable martingales with respect to the family $\{\mathcal{F}_t, t \geq 0\}$; $\mathcal{M}_c$ be the subclass of $\mathcal{M}$ of continuous elements; $\mathcal{M}^{loc}$ be the class of stochastic processes $M = \{M_t, t \geq 0\}$ such that $M^{(n)} = \{M_{t \wedge T_n}, t \geq 0\} \in \mathcal{M}$ for all $n \geq 1$ and some sequence of stopping times $\{T_n\}$, $T_n \uparrow \infty$ as $n \to \infty$ $\mathcal{A}^+$ be the class of all natural increasing stochastic processes $A = \{A_t, t \geq 0\}$, $A_0 = 0$, adapted to the family $\{\mathcal{F}_t, t \geq 0\}$ and $EA_t < \infty$ for all $t \geq 0$; $\mathcal{A} = \{A : A = A_1 - A_2, A_1, A_2 \in \mathcal{A}^+\}$. Analogously the classes of stochastic processes $\mathcal{M}_c^{loc}$, $\mathcal{A}_c$, $\mathcal{A}_c^{loc}$, $\mathcal{A}^{+loc}$ and $\mathcal{A}_c^{+loc}$ are defined.

Denote by $R_+^m = \{x : x = (x_1, \dots, x_m),\ x_1 \geq 0$, the semi-space of the Euclidean space R^m , by $\mathcal{B}_+^m$ the σ-algebra of Borel subsets of R^m,

$$G=\{x: x\in R_+^m, x_1>0\},\ \partial G=\{x: x\in R_+^m,\ x_1=0\}$$

$$U_\varepsilon=\{x: |x|\geq\varepsilon,\ x\in R_+^m\},\ \hat{x}=(x_2,\ldots,x_m).$$

Further we shall need the notion of a stochastic point measure adapted to the family $\{\mathcal{F}_t,\ t\geq 0\}$ (see [6]).

Definition I. We say that a random sequence (S_κ, X_κ), $\kappa\geq 1$, such that $0<S_\kappa\leq\infty$, $S_\kappa\neq S_\ell$ for $\kappa\neq\ell$ a.e. on the set $\{S_\kappa<\infty\}$, X_κ taking its values in a measurable space $(\mathfrak{X},\mathcal{A})$ defines a stochastic point measure p on $\mathcal{B}_+^1\times\mathcal{A}$ adapted to the family $\{\mathcal{F}_t, t\geq 0\}$ if, for all $t\geq 0$, $\Gamma\in\mathcal{A}$,

$$\{S_\kappa\leq t,\ X_\kappa\in\Gamma\}\in\mathcal{F}_t$$

and there exists a sequence $\Gamma^{(n)}\in\mathcal{A}$, $\Gamma^{(n)}\subset\Gamma^{(n+1)}$, $n\geq 1$, such that $\bigcup_{n\geq 1}\Gamma^{(n)}=\mathfrak{X}$ and, for all $t\geq 0$, $n\geq 1$,

$$P\{p([0,t]\times\Gamma^{(n)})<\infty\}=1,$$

where

$$p([s,t]\times\Gamma)=\sum_{s<S_\kappa\leq t}\chi_\Gamma(X_\kappa);$$

$\chi_\Gamma(x)$ is the indicator of the set Γ.

Denote by Π the measure on $\mathcal{B}_+^1\times\mathcal{A}$ such that $\Pi([0,t]\times\Gamma)\in\mathcal{O}\!\mathcal{L}^{+\,loc}$ and $q([0,t]\times\Gamma)=p([0,t]\times\Gamma)-\Pi([0,t]\times\Gamma)\in\mathcal{M}^{loc}$ for all $\Gamma\in\mathcal{A}\cap\Gamma^{(n)}$, $n\geq 1$.

Let us consider now random functions $\delta(t)$, $\gamma(t)$, $a(t)=(a_1(t),\ldots,a_m(t))$, $\beta(t)=(\beta_2(t),\ldots,\beta_m(t))$, $A(t)=\|a_{ij}(t)\|_1^m$, $B(t)=\|\beta_{ij}(t)\|_2^m$, $\pi(t,\Gamma)$, $t\geq 0$, $\Gamma\in\mathcal{B}_+^m$, adapted to the family $\{\mathcal{F}_t, t\geq 0\}$, (t,ω)-measurable and such that, for all $t\geq 0$, $\delta(t)\geq 0$, $\gamma(t)\geq 0$, the matrices $A(t)$, $B(t)$ are symmetric and non-negatively defined and $\pi(t,\Gamma)$ is a measure on $\mathcal{B}^m_+$. Further we shall assume that $m \geq 2$; simplifications of the formulations for the case $m=1$ will be evident.

<u>Definition 2.</u> A stochastic process $X = \{X_t = (X^1_t, \dots, X^m_t), t \geq 0\}$ taking values in R^m_+ is said to have local characteristics $(a, A, \delta, \gamma, \beta, B, \pi)$ with respect to the family $\{\mathcal{F}_t, t \geq 0\}$ and the measure P if it has the following structure:

$$X^1_t = X^1_0 + \int_0^t \chi_G(X_s) a_1(s) ds + \int_0^t \gamma(s) d\varphi_s + M^1_t + \int_0^t \int_G x_1 p(ds, dx) , \tag{I}$$

$$X^i_t = X^i_0 + \int_0^t \chi_G(X_s) a_i(s) ds + \int_0^t \beta_i(s) d\varphi_s + M^i_t + N^i_t + \int_0^t \int_{R^m_+} \chi(x) x_i q(ds, dx) + \int_0^t \int_{R^m_+} (1 - \chi(x)) x_i p(ds, dx), \quad i = 2, \dots, m , \tag{2}$$

where $X^1_0 \geq 0$, $\varphi \in \mathcal{O}\!\mathcal{L}^{+loc}_c$, $M^i, N^i \in \mathcal{M}^{loc}_c$, the measure $p(\cdot, \cdot)$ is a stochastic point measure on $\mathcal{B}^1_+ \times \mathcal{B}^m_+$ (with $\Gamma^{(n)} = \mathcal{U}_{1/n}$); adapted to $\{\mathcal{F}_t, t \geq 0\}$, moreover, for all $t \geq 0$, $\Gamma \in \mathcal{B}^m_+ \cap \mathcal{U}_{1/n}$, $n \geq 1$, a.e.

$$\Pi([0,t] \times \Gamma) = \int_0^t \pi(s, \Gamma) d\varphi_s , \tag{3}$$

$$\int_0^t \chi_{\partial G}(X_s) d\varphi_s = \varphi_t , \quad \int_0^t \delta(s) d\varphi_s = \int_0^t \chi_{\partial G}(X_s) ds , \tag{4}$$

$$\langle M^i, M^j \rangle_t = \int_0^t \chi_G(X_s) a_{ij}(s) ds, \quad i, j = 1, \dots, m , \tag{5}$$

$$\langle N^i, N^j \rangle_t = \int_0^t \beta_{ij}(s) d\varphi_s ; \quad i, j = 2, \dots, m ; \tag{6}$$

$\chi(x) = 1$ for $|x| \leq 1$ and $= 0$ for $|x| > 1$.

It is naturally assumed that, for all $t \geq 0$, a.e.

$$\int_0^t \chi_G(X_s)|a_i(s)|ds < \infty, \quad \int_0^t |\beta_i(s)|d\varphi_s < \infty, \quad \int_0^t \gamma(s)d\varphi_s < \infty,$$

$$\int_0^t \chi_G(X_s)|a_{ij}(s)|ds < \infty, \quad \int_0^t |\beta_{ij}(s)|d\varphi_s < \infty,$$

$$\int_0^t \int_{R_+^m} (|x|^2 \wedge 1)\pi(s,dx)d\varphi_s < \infty \text{ and } \int_0^t \int_G (x_1 \wedge 1)\pi(s,dx)d\varphi_s < \infty .$$

Note that p is the jump measure of X, φ plays the role of the local time on the boundary ∂G for X, and the local characteristics $(\delta, \gamma, \beta, B, \pi)$ determine the behaviour of X on ∂G. It is also important to note that in the case when

$$X_0^1 = 0, \, a(t) \equiv 0, \, A(t) \equiv 0, \, \delta(t) \equiv 0, \, \gamma(t) \equiv 0, \, \pi(t, G) \equiv 0,$$

we shall have that $X_t \in \partial G$ for all $t \geq 0$, $\varphi_t \equiv t$, and thus the stochastic process $\hat{X}_t = (X_t^2, \dots, X_t^m), t \geq 0$, will be an (m-1)-dimensional locally infinitely divisible process having local characteristics (β, B, π) with respect to the family of σ-algebras $\{\mathcal{F}_t, t \geq 0\}$ and the measure p, i.e. the stochastic process of the following structure:

$$\hat{X}_t = \hat{X}_0 + \int_0^t \beta(s)ds + \mathcal{N}_t + \int_0^t \int_{|\hat{x}| \leq 1} \hat{x}\, q(ds, d\hat{x}) + \int_0^t \int_{|\hat{x}| > 1} \hat{x}\, p(ds, d\hat{x}),$$

where $\mathcal{N}^i \in \mathcal{M}_c^{loc}$, $i = 2, \dots, m$, p is a stochastic point measure on $\mathcal{B}_+^1 \times \mathcal{B}^{m-1}$ adapted to $\{\mathcal{F}_t, t \geq 0\}$ such that, for all $t \geq 0$ a.e.

$$\langle \mathcal{N}^i, \mathcal{N}^j \rangle_t = \int_0^t \beta_{ij}(s)ds ; \quad i,j = 2, \dots, m ;$$

$$\Pi([0,t]\times\Gamma)=\int_0^t \pi(s,\Gamma)\,ds\ ,\quad \Gamma\in\mathcal{B}^{m-1}\ ,$$

and

$$\int_0^t\int_{R^{m-1}}(|\hat{x}|^2\wedge 1)\,\pi(s,dx)\,ds<\infty\ .$$

Lemma 1. If, for all $t\geq 0$, a.e.

$$\delta(t)+\gamma(t)+\int_{R^m_+}(|x|^2\wedge 1)\,\pi(t,dx)>0\ , \tag{7}$$

then φ, M and $\mathcal{N}$ are determined uniquely for a stochastic process X with local characteristics $(a, A, \delta, \gamma, \beta, B, \pi)$.

Proof of this statement easily follows from (1)-(7) if we use the following known results. First, if $M\in\mathcal{M}_c^{loc}\cap\mathcal{O}\mathcal{l}^{loc}$, then $M_t\equiv M_0$ a.e. (see [4]). Secondly, if $M, \mathcal{N}\in\mathcal{M}^{loc}$ and the measures generated on $\mathcal{B}^1_+$ by the functions $\langle M\rangle_t$ and $\langle\mathcal{N}\rangle_t$, $t\geq 0$, are singular, then $M\perp\mathcal{N}$ (see [7]).

In what follows we shall assume the condition (7) to be fulfilled.

Later on we shall also need the following notation:

F_M^{loc} is the class of m-dimensional (t,ω)-measurable functions $g(t)=(g_1(t),\dots,g_m(t)),\ t\geq 0$, such that $g(T)$ is $\mathcal{F}_T$-measurable for all stopping times T and, for all $t\geq 0$, a.e.

$$\int_0^t \chi_G(X_s)\,(g(s),\,g(s)A(s))\,ds<\infty\ ;$$

$$F_M=\{g: g\in F_M^{loc},\ E[\int_0^t\chi_G(X_s)(g(s),g(s)A(s))\,ds]<\infty,\ t\geq 0\}\ ;$$

$F_{\mathcal{N}}^{loc}$ is the class of (m-1)-dimensional (t,ω)-measurable functions $\psi(t)=(\psi_2(t),\dots,\psi_m(t)),t\geq 0$ such that $\psi(T)$ is $\mathcal{F}_T$ -measurable for all stopping times T and, for all $t \geq 0$, a.e.

$$\int_0^t (\psi(s),\psi(s)B(s))\,d\varphi_s < \infty \ ;$$

$$F^{\mathcal{N}} = \{\psi : \psi \in F_{\mathcal{N}}^{loc},\ E[\int_0^t (\psi(s),\psi(s)B(s)d\varphi_s] < \infty, t \geq 0\} \ ;$$

F is the class of (t,x,ω) -measurable functions $\eta(t,x), t\geq 0, x\in R_+^m$, such that, for all x and stopping times T , $\eta(x,T)$ is $\mathcal{F}_T$ -measurable;

F_Q^{loc} is the subclass of functions $\eta \in F$ such that, for all $t \geq 0$, a.e.

$$\int_0^t \int_{R_+^m} \eta^2(s,x)\pi(s,dx)\,d\varphi_s < \infty \ ;$$

$$F^Q = \{\eta : \eta \in F_Q^{loc},\ E[\int_0^t \int_{R_+^m} \eta^2(s,x)\pi(s,dx)\,d\varphi_s] < \infty, t \geq 0\} \ ;$$

F_P is the subclass of functions $\eta \in F$ for which the sums $\sum_{0\leq s\leq t} \eta(s,X_s - X_{s-})$ converge a.e. for all $t \geq 0$. The value of this sum is denoted by

$$P_t(\eta) = \int_0^t \int_{R_+^m} \eta(s,x)\,p(ds,dx) .$$

Further let

$$M_t(g) = \sum_{i=1}^m \int_0^t g_i(s)\,dM_s^i \ , \quad g \in F_M^{loc} \ ,$$

$$\mathcal{N}_t(\psi) = \sum_{i=2}^m \int_0^t \psi_i(s)\,d\mathcal{N}_s^i \ , \quad \psi \in F_{\mathcal{N}}^{loc} \ ,$$

$$Q_t(\eta)=\int_0^t\int_{R_+^m}\eta(s,x)\,q(ds,dx),\quad \eta\in F_Q^{loc}.$$

§ 2. Transformations of the local characteristics

Let $\mathcal{F}_t^X=\hat{\mathcal{F}}_{t+0}$, $t\geq 0$, where $\hat{\mathcal{F}}_t$ is the σ-algebra generated by stochastic vectors X_s, $0\leq s\leq t$, and completed with respect to the measure P. The classes $\mathfrak{M}_X$, $\mathfrak{M}_{sX}$ etc. are defined analogously to $\mathfrak{M}$, $\mathfrak{M}_c$ etc. by substituting the family $\{\mathcal{F}_t, t\geq 0\}$ with $\{\mathcal{F}_t^X, t\geq 0\}$. We shall denote by $E^t Z(t)$ (t,ω)-measurable modifications of $E(Z(t)\,|\,\mathcal{F}_t^X)$, $t\geq 0$, for stochastic functions $Z(t)$, $t\geq 0$.

(I) Assume that the functions $A(t)$, $\gamma(t)$, $\delta(t)$, $B(t)$ and φ_t, $t\geq 0$, are adapted to the family of σ-algebras $\{\mathcal{F}_t^X, t\geq 0\}$ and, for all $t\geq 0$,

$$E\Big[\int_0^t\int_G(x_1\wedge 1)\pi(s,dx)\,d\varphi_s\Big]<\infty,\quad E\Big[\int_0^t\int_{R_+^m}(|x|^2\wedge 1)\pi(s,dx)\,d\varphi_s\Big]<\infty,$$

$$E\Big[\int_0^t\chi_G(X_s)|a_{ij}(s)|\,ds\Big]<\infty,\quad E\Big[\int_0^t|\beta_{ij}(s)|\,d\varphi_s\Big]<\infty,$$

$$E\Big[\Big|\int_0^t\chi_G(X_s)(a(s)-\bar a(s))\,ds\Big|^2\Big]<\infty,\quad E\Big[\Big|\int_0^t(\beta(s)-\bar\beta(s))\,d\varphi_s\Big|^2\Big]<\infty,$$

$$E\Big[\Big|\int_0^t\int_{R_+^m}\chi(x)\hat x(\pi(s,dx)-\bar\pi(s,dx))\,d\varphi_s\Big|^2\Big]<\infty,$$

where $\bar a(t)=E^t a(t)$, $\bar\beta(t)=E^t\beta(t)$, $\bar\pi(t,\Gamma)=E^t\pi(t,\Gamma)$.

Theorem 1. Under assumption (I) the stochastic process X has the local characteristics $(\bar a, A, \delta, \gamma, \bar\beta, B, \bar\pi)$ with respect to the family of σ-algebras $\{\mathcal{F}_t^X, t\geq 0\}$ and the measure P.

P r o o f. Put

$$\bar{q}(t,\Gamma) = p([0,t]\times\Gamma) - \int_0^t \bar{\pi}(s,\Gamma)\, d\varphi_s \, . \tag{8}$$

$$\bar{M}_t^1 = X_t^1 - X_0^1 - \int_0^t \chi_G(X_s)\bar{a}_1(s)ds + \int_0^t \gamma(s)d\varphi_s - \int_0^t\int_G x_1\, p(ds,dx) \, , \tag{9}$$

$$\bar{M}_t^i = \int_0^t \chi_G(X_s)\, d\bar{L}_s^i \, . \tag{10}$$

$$\bar{N}_t^i = \int_0^t \chi_{\partial G}(X_s)\, d\bar{L}_s^i \, , \quad i = 2, \dots, m \, , \tag{11}$$

where

$$\bar{L}_t^i = X_t^i - X_0^i - \int_0^t \chi_G(X_s)\bar{a}_i(s)ds - \int_0^t \bar{\beta}_i(s)d\varphi_s - \int_0^t\int_{R_+^m} \chi(x)x_i\bar{q}(ds,dx) - \int_0^t\int_{R_+^m} (1-\chi(x))x_i\, p(ds,dx) \, . \tag{12}$$

Under assumption (I) it is easy to check that, for all $\Gamma \in B_+^m \cap \mathcal{U}_{1/n}$ $n \geq 1$, $\bar{q}(\cdot,\Gamma) \in \mathcal{M}_X$, $\bar{M}^1, \bar{L}^i \in \mathcal{M}_{c,X}$, $i = 2, \dots, m$.
Using the well known theorem on quadratic variation of martingales from $\mathcal{M}_c$ (see [5]), we obtain that a.e.

$$\langle \bar{M}^1 \rangle_t = \int_0^t \chi_G(X_s)\, a_{11}(s)\, ds \, , \tag{13}$$

$$\langle \bar{M}^1, \bar{L}^i \rangle_t = \int_0^t \chi_G(X_s)\, a_{i1}(s)\, ds \, , \quad i = 2, \dots, m \, , \tag{14}$$

and

$$\langle \bar{L}^i, \bar{L}^j \rangle_t = \int_0^t \chi_G(X_s)\, a_{ij}(s)\, ds + \int_0^t \beta_{ij}(s)\, d\varphi_s ; \quad i,j = 2, \dots, m \, . \tag{15}$$

The theorem follows now from (8)-(15).

<u>Remark 1.</u> Using (9)-(12), it is not difficult to verify that

a.e.

$$\bar{M}_t^i = M_t^i + \int_0^t \chi_G(X_s)(a_i(s) - \tilde{a}_i(s))ds \; ; \quad i = 1, \dots, m \; ; \tag{I6}$$

and

$$\bar{\mathcal{N}}_t^i = \mathcal{N}_t^i + \int_0^t (\beta_i(s) - \bar{\beta}_i(s)) d\varphi_s + \int_0^t \int_{R_+^m} \psi(x) x_i (\pi(s,dx) - \tag{I7}$$

$$- \bar{\pi}(s,dx)) d\varphi_s \; ; \quad i = 2, \dots, m.$$

Further wa put

$$\alpha_t = \alpha_t(g,\psi,\eta) = \exp\{M_t(g) + \mathcal{N}_t(\psi) + Q_t(\chi_{\{|\eta|<1\}}\eta) +$$

$$+ P_t(\chi_{\{|\eta|>1\}}\eta) - \tfrac{1}{2}\langle M(g)\rangle_t - \tfrac{1}{2}\langle \mathcal{N}(\psi)\rangle_t - \tag{I8}$$

$$- \int_0^t \int_{R_+^m} \chi_{\{|\eta|<1\}}(s,x)(e^{\eta(s,x)} - 1 - \eta(s,x))\pi(s,dx)d\varphi_s -$$

$$- \int_0^t \int_{R_+^m} \chi_{\{|\eta|>1\}}(s,x)(e^{\eta(s,x)} - 1)\pi(s,dx)\, d\varphi_s \}$$

for functions $g \in \Gamma_M^{loc}$, $\psi \in \Gamma_{\mathcal{N}}^{loc}$ and $\eta \in \Gamma$ such that $\chi_{\{|\eta|<1\}}\eta \in \Gamma_Q^{loc}$ and, for all $t \geq 0$, a.e.

$$\left[\int_0^t \int_{R_+^m} \chi_{\{|\eta|>1\}}(s,x)(e^{\eta(s,x)} - 1)^2 \pi(s,dx)\, d\varphi_s\right] < \infty,$$

$$\left|\int_0^t \chi_G(X_s) g(s) A(s)\, ds\right| < \infty, \quad \left|\int_0^t \psi(s) B(s)\, d\varphi_s\right| < \infty.$$

Under these conditions,it is easy to vertify that $\alpha \in \mathcal{M}^{loc}$.

(II) Assume that, for all $t \geq 0$, $E\alpha_t = 1$.

Define the measure $\tilde{P}$ on the σ-algebra $\mathcal{F}_\infty = \bigvee_{t \geq 0} \mathcal{F}_t$ by putting

$$\tilde{P}(A) = \int_A \alpha_t \, dP \, , \quad A \in \mathcal{F}_t \, , \quad t \geq 0 \, .$$

<u>Theorem 2.</u> Under assupmtion (II) the stochastic process X has the local characteristics $(\tilde{a}, A, \delta, \gamma, \tilde{\beta}, B, \tilde{\pi})$ with respect to the family of σ-algebras $\{\mathcal{F}_t, t \geq 0\}$ and the measure $\tilde{P}$, where

$$\tilde{a}(t) = a(t) + g(t) A(t) \, ,$$

$$\tilde{\beta}(t) = \beta(t) + \psi(t) B(t) + \int_{R_+^m} \chi(x) \hat{x} (e^{\eta(t,x)} - 1) \pi(t, dx),$$

$$\tilde{\pi}(t, \Gamma) = \int_\Gamma e^{\eta(t,x)} \pi(t, dx) \, .$$

Proof of this theorem follows from the known properties of stochastic integrals after standard calculations using the generalized Ito formula.

<u>§ 3. Absolute continuity of measures corresponding to stochastic processes with boundary conditions</u>

In deriving conditions for absolute continuity of the measures, corresponding to semimartingales, and explicit formulas for the Radon-Nikodym densities in terms of local characteristics, we make use of the following results. (1) Formulas expressing the structure of non-negative local martingales adapted to a given family of σ-algebras. (2) Transformation formulas for local characteristics considered in the preceding section (3). Uniqueness conditions for solutions of the corresponding martingale problems.

Let $\mathcal{D}$ be the space of all right continuous and having left limits functions $\omega(t)$ defined on R_+^1 and taking values in R_+^m,

$\mathcal{D}_t$ be the σ-algebra generated by cylindrical sets

$$C_s(\Gamma)=\{\omega:\omega\in D,\ \omega(s)\in\Gamma\},\ 0\le s\le t,\ \Gamma\in\mathcal{B}_+^m,\ \mathcal{D}=\bigvee_{t\ge 0}\mathcal{D}_t,$$

$$Y(t,\omega)=\omega(t),\ t\ge 0,\ \omega\in D.$$

Definition 3. We say that a probability measure μ on the σ-algebra $\mathcal{D}$ is a solution of the martingale problem for $(a, A, \delta, \gamma, \beta, B, \pi)$ with the initial distribution $\mu_0(\Gamma)$, $\Gamma\in\mathcal{B}_+^m$, if $\mu(C_0(\Gamma))=\mu_0(\Gamma)$, $\Gamma\in\mathcal{B}_+^m$, and the stochastic process Y on the probability space $(D,\mathcal{D},\mu)$ has the local characteristics $(a, A, \delta, \gamma, \beta, B, \pi)$ with respect to the family $\{\mathcal{D}_t, t\ge 0\}$ and the measure μ.

Now consider a right continuous and having left limits stochastic process $X'=\{X'_t,\ t\ge 0\}$ taking values in R_+^m such that the measure μ', corresponding to X' on $(D,\mathcal{D})$, is a solution of the martingale problem for $(a', A', \delta', \gamma', \beta', B', \pi')$ with the initial distribution μ_0'.

Let a stochastic process X satisfy the assumptions of § I and 2.

(III) Assume that, for all $t\ge 0$, a.e.

I) $A(t)=A'(t,X)$, $\delta(t)=\delta'(t,X)$, $B(t)=B'(t,X)$, $\gamma(t)=\gamma'(t,X)$;

2) $a'(t,X)=\bar{a}(t)+\hat{h}(t)A(t)$;

3) $\pi'(t,X,\Gamma)=\int_\Gamma \hat{\rho}(t,x)\bar{\pi}(t,dx)$, $\Gamma\in\mathcal{B}_+^m$;

4) $\beta'(t,X)=\bar{\beta}(t)+\hat{\xi}(t)B(t)+\int_{R_+^m}\chi(x)\hat{x}(\hat{\rho}(t,x)-1)\bar{\pi}(t,dx)$;

5) $\mu_0'(\Gamma)=\int_\Gamma \rho_0(x)\mu_0(dx)$, $\Gamma\in\mathcal{B}_+^m$,

where the functions $\hat{h}(t)$, $\hat{\rho}(t,x)$ and $\hat{\zeta}(t)$ are measurable in all the variables and adapted to the family $\{\mathcal{F}_t^X, t\geq 0\}$; μ_0 is a probability distribution of X_0.

(IV) Assume that $\hat{h}\in F_{M,X}^{loc}$, $\hat{\zeta}\in F_{N,X}^{loc}$, $\chi_{\{|\ln\hat{\rho}|<1\}}\ln\hat{\rho}\in F_{q,X}^{loc}$ and, for all $t\geq 0$, a.e.

$$|\int_0^t \chi_G(X_s)\hat{h}(s)A(s)ds|<\infty,\quad |\int_0^t \hat{\zeta}(s)B(s)d\varphi_s|<\infty ,$$

$$[\int_0^t\int_{R_+^m} \chi_{\{|\ln\hat{\rho}|\geq 1\}}(s,x)(\hat{\rho}(t,x)-1)^2\bar{\pi}(s,dx)d\varphi_s]<\infty$$

and

$$E\,\mathcal{L}_t(\hat{h},\hat{\zeta},\ln\hat{\rho})=1 ,$$

where the classes $F_{M,X}^{loc}$, $F_{M,X}$ etc. are defined analogously to F_M^{loc}, F_M etc. by substituting the family $\{\mathcal{F}_t, t\geq 0\}$ and π with $\{\mathcal{F}_t^X, t\geq 0\}$ and $\bar{\pi}$, $\bar{\mathcal{L}}_t(g,\psi,\eta)$ is defined by substituting (M,N,q,π) with $(\bar{M},\bar{N},\bar{q},\bar{\pi})$ in (I8).

(V) Assume that the martingale problem for $(a',A',\delta',\gamma',\beta',B',\pi')$ with the initial distribution μ_0' has a unique solution.

Let μ_T and μ_T' be the measures, corresponding to X and X', on the σ-algebra $\mathcal{D}_T$, P_T be the restriction of the measure P to the σ-algebra $\mathcal{F}_T^X$, and define P_T' by the formula:

$$P_T'(A)=\mu_T'(B),\quad A\in\mathcal{F}_T^X ,$$

if $A=\{X\in B\}$, $B\in\mathcal{D}_T$. Note that P_T' will be a measure, if, for example, $\mu_T'\ll\mu_T$.

<u>Theorem 3.</u> Under assumptions (I),(III) and (IV)

$$\mu'_T \ll \mu_T \ , \quad P'_T \ll P_T$$

and

$$\frac{dP'_T}{dP_T}(X) = \rho_0(X_0)\,\bar{\alpha}_T(\hat{h},\hat{\xi},\ln\hat{\rho}) \ . \tag{I9}$$

Moreover, if

$$P\{\rho_0(X_0)\bar{\alpha}_T(\hat{h},\hat{\xi},\ln\hat{\rho})>0\}=1 \ ,$$

then

$$\mu'_T \sim \mu_T , \quad P'_T \sim P_T \ .$$

P r o o f. From assumptions (I), (III), (IV) and Theorems I and 2 we derive that the stochastic process X has the initial distribution μ'_0 and local characteristics $(a', A', \delta', \gamma', \beta', B', \pi')$ with respect to the family $\{\mathcal{F}^X_t, t \geq 0\}$ and the measure $\tilde{P}$ defined by the formula

$$\tilde{P}(A) = \int_A \rho_0(X_0)\,\bar{\alpha}_t(\hat{h},\hat{\xi},\ln\hat{\rho})\,dP, \ A \in \mathcal{F}^X_t, \ t \geq 0 \ . \tag{20}$$

Using assumption (V) and (20), we obtain directly that

$$\mu'_T \ll \mu_T , \quad P'_T \ll P_T$$

and formula (I9). The remaining statement is evident.

<u>Remark 2.</u> Note that general conditions of absolute continuity are obtained for Markov processes in terms of their generators in [8]. The case of locally infinitely divisible processes was considered in [9]-[II].

<u>§ 4. Stochastic equations for non-linear filtering</u>

In deriving stochastic filtering equations for a posteriori distributions of the unobservable components of a stochastic process with the observable component X being a semimartingale, we use the following results. (1) Transformation formulas for the local characteristics of X when we substitute the family of σ-algebras. Representation formulas for martingales from $\mathcal{M}_X$. In the case of diffusion type process X, such equations were obtained in [I2](see also [I3]). The case of locally infinitely divisible processes was investigated in [I4].

Let the observable process X satisfy assumption (I).

(III') Assume that, for all $t \geq 0$, a.e.

I) $A(t)=A'(t,X)$, $\delta(t)=\delta'(t,X)$, $B(t)=B'(t,X)$, $\gamma(t)=\gamma'(t,X)$;

2) $a(t)=a'(t,X)+h(t)A'(t,X)$;

3) $\pi(t,\Gamma)=\int_\Gamma \rho(t,x)\pi'(t,X,dx)$;

4) $\beta(t)=\beta'(t,X)+\zeta(t)B(t)+\int_{R_+^m} \chi(x)\hat{x}(\rho(t,x|-1)\pi'(t,X,dx)$,

where the functions $h(t)$, $\zeta(t)$ and $\rho(t,x)$ are measurable in all the variables and adapted to the family $\{\mathcal{F}_t, t \geq 0\}$.

Put $\hat{h}(t)=E^t h(t)$, $\bar{\zeta}(t)=E^t\zeta(t)$, $\bar{\rho}(t,x)=$
$E^t\rho(t,x)$, $t\geq 0$, $x\in R_+^m$.

(VI) Assume that every martingale $\hat{M}\in\mathcal{M}_X$ can be represented as the sum

$$\hat{M}_t=\hat{M}_0+\bar{M}_t(g)+\bar{N}_t(\psi)+\bar{Q}_t(\eta)$$

for some $g\in F_{M,X}$, $\psi\in F_{N,X}$ and $\eta\in F_{Q,X}$, where the stochastic integrals $\bar{M}(g)$, $\bar{N}(\psi)$ and $\bar{Q}(\eta)$ are defined analogously to the integrals $M(g)$, $\mathcal{N}(\psi)$ and $Q(\eta)$.

We shall not discuss here conditions under which assumption (VI) is satisfied (see [I5],[I6]).

Let θ_t, $t \geq 0$, be an unobservable stochastic process with values in a measurable space $(\mathcal{Y},\mathcal{B})$ adapted to the family of σ-algebras $\{\mathcal{F}_t, t \geq 0\}$. Denote by $D(\tilde{A})$ the space of real $\mathcal{B}$-measurable functions $f(\theta)$, $\theta \in \mathcal{Y}$, for which $E|f(\theta_t)|^2 < \infty$, $t \geq 0$, and there exists a (t,ω)-measurable function $\tilde{A}_t f$, $t \geq 0$, adapted to $\{\mathcal{F}_t, t \geq 0\}$ such that, for all $t \geq 0$,

$$E\left[\int_0^t |\tilde{A}_s f|^2 ds\right] < \infty$$

and

$$M_t^f = f(\theta_t) - \int_0^t \tilde{A}_s f \, ds \in \mathfrak{M} .$$

Let

$$\mathfrak{M}_M = \{M(g), g \in F_M\} ,$$
$$\mathfrak{M}_{\mathcal{N}} = \{\mathcal{N}(\psi), \psi \in F_{\mathcal{N}}\} ,$$
$$\mathfrak{M}_Q = \{Q(\eta), \eta \in F_Q\} .$$

From the known properties of stochastic integrals $M(g)$, $\mathcal{N}(\psi)$ and $Q(\eta)$, it follows that $\mathfrak{M}_M$, $\mathfrak{M}_N$ and $\mathfrak{M}_Q$ are orthogonal subspaces of $\mathfrak{M}$. Let us denote the projections of the martingale M^f to $\mathfrak{M}_M$, $\mathfrak{M}_N$ and $\mathfrak{M}_Q$ by M'^f, M''^f and M'''^f, correspondingly. There exist, unique up to equivalence, functions $\tilde{D}f \in F_M$, $\tilde{L}f \in F_{\mathcal{N}}$ and $\tilde{F}f \in F_Q$ such that, for all $t \geq 0$ a.e.

$$M_t'^f = M_t(\tilde{D}f),\ M_t''^f = \mathcal{N}_t(\tilde{L}f),\ M_t'''^f = Q_t(\tilde{F}f) .$$

Recall that we consider the functions $g_1, g_2 \in F_M$ to be equivalent if, for all $t \geq 0$,

$$E\left[\int_0^t (g_1(s)-g_2(s),(g_1(s)-g_2(s))A(s))\chi_G(X_s)\,ds\right]=0 \quad .$$

Equivalence is defined analogously in the classes F_N and F_Q.

(VII) Assume that, for all $t \geq 0$,

$$E\left[\int_0^t |f(\theta_s)|^2 \chi_G(X_s)(h(s)-\bar{h}(s),(h(s)-\bar{h}(s))A(s))\,ds\right]<\infty \ ,$$

$$E\left[\int_0^t |f(\theta_s)|^2(\zeta(s)-\hat{\zeta}(s),(\zeta(s)-\bar{\zeta}(s))B(s))\,d\varphi_s\right]<\infty$$

and

$$E\left[\int_0^t \int_{R_+^m} \left(|f(\theta_s)|^2 \frac{|\rho(s,x)-\bar{\rho}(s,x)|^2}{\rho(s,x)\,\bar{\rho}(s,x)} + \right.\right.$$

$$\left.\left. + |\tilde{F}_{s,x} f|^2 \frac{\rho(s,x)}{\bar{\rho}(s,x)}\right)\pi(s,dx)\,d\mathfrak{f}_s\right]<\infty \quad .$$

<u>Theorem 4.</u> Under assumptions (I), (III'), (VI) and (VII) for $f \in D(\tilde{A})$ the following relation is true (the stochastic non-linear filtering equation):

$$E^t f(\theta_t) = E^0 f(\theta_0) + \int_0^t E^s \tilde{A}_s f\,ds + \bar{M}_t(g_f) + \tag{21}$$

$$+ \bar{N}_t(\psi_f) + \bar{Q}_t(\eta_f)$$

for all $t \geq 0$ a.e., where

$$g_f(t) = E^t[f(\theta_t)(h(t)-\bar{h}(t)) + \tilde{D}_t f] \ ,$$

$$\psi_f(t) = E^t[f(\theta_t)(\xi(t) - \bar{\xi}(t)) + \tilde{L}_t f] \quad ,$$

$$\eta_f(t,x) = E^t[f(\theta_t)(\frac{\rho(t,x)}{\bar{\rho}(t,x)} - 1) + \tilde{F}_{t,x} f \frac{\rho(t,x)}{\bar{\rho}(t,x)}] \; .$$

P r o o f. Let

$$\hat{M}_t^f = E^t f(\theta_t) - \int_0^t E^s \tilde{A}_s f \, ds \, , \; t \geq 0 \; .$$

It is easy to verify that, for all $f \in D(\tilde{A})$, $\hat{M}^f \in \mathcal{M}_X$. The assertion of the theorem follows after standard reasoning from assumptions (III), (VI) (I6) and (I7) (cf. [I4]).

Equation (2I) describing the evolution of the a posteriori distributions of the unobservable component of a stochastic process is nonlinear. It turned out that this equation can be linearized by means of some nonlinear transformation (see [2] and the bibliography there.

Assuming that $\bar{h} \in F^{loc}_{M,X}$, $\bar{\xi} \in F^{loc}_{N,X}$, $\chi_{\{|ln\bar{\rho}|<1\}} ln\bar{\rho} \in F^{loc}_{a,X}$ and, for all $t \geq 0$, a.e.

$$|\int_0^t \chi_G(X_s)\bar{h}(s)A(s)ds| < \infty, \quad |\int_0^t \bar{\xi}(s)B(s)d\varphi_s| < \infty$$

and

$$[\int_0^t \int_{R_+^m} \chi_{\{|ln\bar{\rho}|\geq 1\}}(s,x)(\frac{1}{\bar{\rho}(s,x)} - 1)^2 \bar{\pi}(s,dx)\, d\varphi_s] < \infty \; ,$$

we put

$$\hat{E}^t Z(t) = [\alpha_t(-\bar{h}, -\bar{\xi}, -ln\bar{\rho})]^{-1} E^t Z(t) \; .$$

If the assumptions of Theorem 4 are satisfied we put

$$\hat{M}_t(g) = \bar{M}_t(g) + \int_0^t \chi_G(X_s)(g(s), \bar{h}(s)A(s))\,ds\ , \quad g \in F_{M,X}\ ,$$

$$\hat{\mathcal{N}}_t(\psi) = \bar{\mathcal{N}}_t(\psi) + \int_0^t (\psi(s), \bar{\xi}(s)B(s))\,d\varphi_s\ , \quad \psi \in F_{\mathcal{N},X}$$

and

$$\hat{Q}_t(\eta) = \bar{Q}_t(\eta) + \int_0^t \int_{R_+^m} \eta(s,x)(\bar{\rho}(t,x)-1)\,\pi'(s,X,dx)\,d\varphi_s\ , \quad \eta \in F_{Q,X}\ .$$

It is particularly important to underline that it is not necessary to know the a posteriori distributions of the process θ_t, $t \geq 0$, for the construction of the stochastic integrals $\hat{M}(g)$, $\hat{\mathcal{N}}(\psi)$ and $\hat{Q}(\eta)$.

Theorem 5. Under assumptions (I), (III'), (VI) and (VII) for $f \in D(\tilde{A})$ the following relation is true (the reduced stochastic nonlinear filtering equation):

$$\hat{E}^t f(\theta_t) = \hat{E}^0 f(\theta_0) + \int_0^t \hat{E}^s(\tilde{A}_s f)\,ds + \hat{M}_t(g_f) + \tag{22}$$
$$+ \hat{\mathcal{N}}_t(\hat{\psi}_f) + \hat{Q}_t(\hat{\eta}_f)$$

for all $t \geq 0$ a.e., where

$$\hat{g}_f(t) = \hat{E}^t[f(\theta_t)h(t) + \tilde{D}_t f]\ ,$$

$$\hat{\psi}_f(t) = \hat{E}^t[f(\theta_t)\zeta(t) + \tilde{L}_t f]\ ,$$

$$\hat{\eta}_f(t,x) = \hat{E}^t[f(\theta_t)(\rho(t,x)-1 + \tilde{F}_{t,x} f\,\rho(t,x)]\ .$$

P r o o f. Relation (22) follows immediately from (2I) and the expression for $\mathcal{L}_t$ by using the generalized Itô formula.

References

I. Б. Григелионис, О статических задачах случайных процессов с граничными условиям, Liet.matem.rink., XVI, 1(I976).

2. Б. Григелионис, О редуцированных стохастических уравнениях нелинейной фильтрации случайных процессов, Liet.matem.rink., XVI, 3(I976).

3. Б. Григелионис, О случайных процессах в конечном пространственном интервале, Liet.matem.rink., XVI, 2(I976).

4. H.Kunita, S.Watanabe, On square integrable martingales, Nagoya Math.J., 30(I967), 209-245.

5. P.A. Meyer, Intégrales stochastiques, Séminaire de Probabilités I, Lecture Notes in Math., Springer-Verlag, 39(I967), 72-I62.

6. Б. Григелионис, Случайные точечные процессы и мартингалы, Liet.matem.rink., XV, 3(I965), IOI-II4.

7. Д. Сургайлис, "Нервенство Шварца" и некоторые другие результаты для интегрируемых в квадрате мартингалов, Liet.matem. rink., XIII, 3(I973), 2II-2I7.

8. H.Kunita, Absolute continuity of Markov processes and generators, Nagoya Math.J., 36(I969), I-26.

9. Б. Григелионис, Об абсолютно непрерывной замене меры и марковском свойстве случайных процессов, Liet.matem.rink., IX, 1(I969), 57-7I.

IO. Б. Григелионис, Об абсолютной непрерывности мер, соответствующих случайным процессам, Liet.matem.rink.,XI,4(I97I), 783-794.

II. Б. Григелионис, О структуре плотностей мер, соответствующих случайным процессам, Liet.matem.rink., XIII,1(I973), 7I-78.

I2. M. Fujisaki, G. Kallianpur, H. Kunita, Stochastic differential equations for the nonlinear filtering problem, Osaka J. Math.,9, 1(I972),I9-40.

I3. Р.Ш.Липцер, А.Н. Ширяев, Статистика случайных процессов, М., "Наука", I974.

I4. Б. Григелионис, О стохастических уравнениях нелинейной фильтрации случайных процессов, Liet.matem.rink.,XII, 4(I972), 37-5I.

I5. Б. Григелионис, О представлении стохастическими интегралами мартингалов, интегрируемых с квадратом, Liet.matem.rink., XIV, 4(I974), 45-6I.

I6. Б. Григелионис, Структура функционалов от случайных процессов, Труды школы-семинара по теории случайных процессов, ч.I, 7I-I06. Вильнюс, I975.

Institute of Physics and Mathematics
Academy of Sciences of the Lituan.SSR

Vilnius

PROBABILITIES OF THE FIRST EXIT FOR CONTINUOUS PROCESSES WITH INDEPENDENT INCREMENTS ON A MARKOV CHAIN

D.V.Gusak

Let $\xi(t)$ $(t \geq 0, \xi(0)=0)$ be a continuous homogeneous process with independent increments which is defined on an homogeneous Markov chain $x(t)$ by the characteristic function (ch.f.)

$$M e^{i\alpha\xi(t)} = \| M(e^{i\alpha\xi(t)}, x(t)=z \mid x(0)=\kappa\} \| =$$

$$= \exp\{t(\mathbb{Q} + i\alpha\mathbb{A} - \frac{\alpha^2}{2}\mathbb{B}^2)\} \quad (1 \leq \kappa, z \leq n),$$

$$\mathbb{A} = \| a_\kappa \delta_{\kappa z} \|, \ \mathbb{B}^2 = \| \delta_\kappa^2 \delta_{\kappa z} \| \ (-\infty < a_\kappa < \infty, \ 0 \leq \delta_\kappa^2 < \infty)$$

$$\mathbb{Q} \equiv \mathcal{N}[\mathbb{P} - \mathbb{I}] = \frac{1}{t} \ln \mathbb{P}(t),$$

$$\mathbb{P}(t) = \| \mathcal{P}\{x(t)=z \mid x(0)=\kappa\} \|,$$

$$\mathbb{I} = \| \delta_{\kappa z} \|_{\kappa,z=1}^{n} .$$

The distributions of some boundary functionals for this process are described by equations with the operator

$$\mathbb{L}_t = \frac{1}{2}\mathbb{B}^2 \frac{\partial^2}{\partial x^2} - \mathbb{A}\frac{\partial}{\partial x} + \mathbb{Q} - \mathbb{I}\frac{\partial}{\partial t}$$

to which the following kernel function corresponds

$$\mathbb{K}(s,\alpha) = -\frac{1}{2}\alpha^2 \mathbb{B}^2 + i\alpha\mathbb{A} + \mathbb{Q} - s\mathbb{I} .$$

We shall be concerned only with limit characteristics of $\xi(t)$ which are described by equations with the operator

$$\mathbb{L}_0 = \frac{1}{2}\mathbb{B}^2\frac{\partial^2}{\partial x^2} - \mathbb{A}\frac{\partial}{\partial x} + \mathbb{Q} \tag{2}$$

and its kernel function

$$\mathbb{K}_0(\alpha) = -\frac{1}{2}\alpha^2\mathbb{B}^2 + i\alpha\mathbb{A} + \mathbb{Q} \quad , \tag{3}$$

especially with the probabilities of the first exit of $\xi(t)$ from $[x-c, x]$ $(0 \leq x \leq c)$ (or $[y, y+c]$, $-c \leq y \leq 0$). The characteristics of such a kind are the distributions of the absolute extrema and the probabilities of the first exit from a strip.

For analogous characteristics of a compound Poisson process and other functionals, V.S.Korolyuk developed an aproach based on the potential theory (see [1]) . For processes which are defined on a Markov chain, we shall consider only the continuous case and try to obtain analogous results for this case. This separate consideration of the continuous case is justified by the possibility to apply a specific aproach connected with an algebric factorization for $\mathbb{K}(s, \alpha)$.

We shall begin with an ordinary Wiener process

$$\xi(t) = at + \sigma w(t) \quad ,$$

$$M e^{i\alpha\xi(t)} = \exp\{t\ K_0(\alpha)\} \quad ,$$

$$K_0(\alpha) = -\frac{1}{2}\alpha^2\sigma^2 + i\alpha a \quad .$$

Let us introduce the following boundary functionals:

$$\xi^+(t) = \sup_{0\le u\le t} \xi(u), \quad \xi^-(t) = \inf_{0\le u\le t} \xi(u),$$

$$\tau^x = \inf\{s : \xi(s) > x\} \quad (x \ge 0),$$

$$\tau_x = \inf\{s : \xi(s) < x\} \quad (x \le 0), \tag{4}$$

$$T(t) = \inf\{s : \xi(s) = \xi^+(t)\} \quad ;$$

$$\xi^+ = \sup_{0\le u<\infty} \xi(u) \quad (a < 0),$$

$$Q_c(x) = \mathcal{P}\{\tau_{x-c} < \tau^x\} \quad (0 \le x \le c), \tag{5}$$

$$Q^c(x) = \mathcal{P}\{\tau^x < \tau_{x-c}\} \quad (0 \le x \le c).$$

The distributions of the first group of finctionals (4) are described by the components of the factorization for $K(s,\alpha) = s - K_0(\alpha)$ ($\Im m\, \alpha = 0$)

$$\frac{s}{s - K_0(\alpha)} = \frac{\alpha_s^-}{\alpha_s^- + i\alpha} \cdot \frac{\alpha_s^+}{\alpha_s^+ - i\alpha} \quad \left(\alpha_s^\pm = \frac{\sqrt{a^2 + 2s\sigma^2} \mp a}{\sigma^2}\right). \tag{6}$$

If $\theta_s > 0$ ($s > 0$) is an exponetial random variable ($\mathcal{P}\{\theta_s > t\} = e^{-st}$), then it is easy to show that

$$M e^{i\alpha \xi^\pm(\theta_s)} = \frac{\alpha_s^\pm}{\alpha_s^\pm \mp i\alpha},$$

$$M e^{-s\tau^x} = e^{-\alpha_s^+ x}, \tag{7}$$

$$Me^{-\mu T(\theta_s)} = \frac{\alpha_s^+}{\alpha_{s+\mu}} = \frac{\sqrt{a^2+2s\sigma^2}-a}{\sqrt{a^2+2(s+\mu)\sigma^2}-a} \quad .$$

For the second group (5) of limit characteristics connected with $K_0(\alpha)$, we have the relations

$$\begin{aligned} Q_c(x) &= 1 - Q^c(x) = \frac{1-e^{2ax\sigma^{-2}}}{1-e^{2ac\sigma^{-2}}} \quad (a \neq 0,\ 0 \leq x \leq c), \\ Me^{i\alpha\xi^+} &= \frac{2|a|}{\sigma^2}\left(\frac{2|a|}{\sigma^2} - i\alpha\right)^{-1} \quad (a \leq 0), \\ Me^{-\mu T(\infty)} &= \frac{2|a|}{\sqrt{a^2+2\mu\sigma^2}-a} \quad (a < 0). \end{aligned} \tag{8}$$

These results for the probabilities $Q_c(x)$, $Q^c(x)$ and $\mathcal{P}\{\xi^+ < x\}$ can be casy obtained by resolving the equation

$$L_0 F(x) \equiv \frac{1}{2}\sigma^2 \frac{d^2F}{dx^2} - a\frac{dF}{dx} = 0 \quad ,$$

which, together with the corresponding boundary conditions, determines these probabilities uniquely.

To begin with the consideration of processes on a Markov chain we introduce additional notation:

$$E_\pm = \{r : (b_r^2 > 0) \cap (b_r^2 = 0, \pm a_r > 0\}, \quad m_\pm = \sum_{k=1}^{n} \delta(k \in E_\pm),$$

$$\mathbb{I}_\pm = \| \delta_{kr}\, \delta(k \in E_\pm) \|, \quad \mathfrak{D}^2 = \frac{1}{2}\mathbb{B}^2 + (\mathbb{I} - \mathbb{I}_+ \mathbb{I}_-),$$

$$\mathbb{M}e^{-\alpha\xi^+(\theta_s)} = \| M\{e^{-\alpha\xi^+(\theta_s)},\ x(\theta_s) = r \mid x(0) = k\} \| \quad .$$

$$\mathbb{M}e^{-s\tau^z} = \| M\{ e^{-s\tau^z}, x(\tau^z) = r \mid x(0) = k \} \| ,$$

$$\mathbb{M}e^{-\mu T(\theta_s)} = \| M\{ e^{-\mu T(\theta_s)}, x(\theta_s) = r \mid x(0) = k \} \| ,$$

$$\mathbb{Q}^c(z) = \| \mathcal{P}\{ \tau_{z-c} < \tau^z, x(\tau_{z-c}) = r \mid x(0) = k \} \| ,$$

$$\mathbb{Q}^c(z) = \| \mathcal{P}\{ \tau^z < \tau_{z-c}, x(\tau^z) = r \mid x(0) = k \} \| .$$

We suppose $x(t)$ to be a regular and invertible Markov chain. On this account

$$\mathbb{P}(\theta_s) = \| \mathcal{P}\{ x(\theta_s) = r \mid x(0) = k \} \| \underset{s \to 0}{\longrightarrow} \mathbb{P}_0 ,$$

$$\mathbb{P}_0 = \| p_{kr} \| , \quad p_{kr} = \rho_r > 0 \quad (1 \le k, r \le n), \tag{9}$$

$$\mathbb{P} \mathbb{R}^{-1} = \mathbb{R}^{-1} \mathbb{P}^T \quad (\mathbb{R} = \| \delta_{kr} \rho_k \|) .$$

To formulate the needed auxiliary propositions which are established in [2], we note that condition (9) permits to transform $\mathbb{Q}$ into a symmetric matric by multiplying it by a diagonal one $\mathbb{C} = \mathcal{N}\mathbb{R}^{-1}$. Therefore we suppose $\mathbb{Q}$ to be a symmetric matrix. In this case $\mathbb{R} = \mathcal{N}$.

P r o p o s i t i o n 1. If $x(t)$ satisfies conditions (9) with $\mathbb{R} = \mathcal{N}$, the following identity holds true ($\mathrm{Im}\,\alpha = 0$)

$$s\mathbb{I} - \mathbb{Q} + \frac{\alpha}{2}\mathbb{B}^2 - i\alpha\mathbb{A} = (\mathbb{X}_-(s) + i\alpha\,\mathbb{I}_-)\mathbb{D}^2(\mathbb{X}_+(s) - i\alpha\,\mathbb{I}_+) , \tag{10}$$

where the components are equivalent to diagonal matrices:

$$\mathbb{X}_\pm(s) \mp i\alpha \mathbb{I}_\pm \sim \Lambda_s^\pm \mp i\alpha \mathbb{I}_\pm \qquad (|\mathbb{X}_\pm(s)| \neq 0,\ s>0),$$

$$\Lambda_s^\pm = \| \delta_{k r} [\lambda_k^\pm(s)\, \delta(k \in E_\pm) + \delta(k \bar{\in} E_\pm)] \| \quad ,$$

$\pm\lambda_k^\pm(s) > 0$ $(s>0)$ being the roots of the characteristic equation

$$\Delta(\lambda, s) \equiv \left| \frac{1}{2} \mathbb{B}^2 \lambda^2 + \mathbb{A}\lambda + \mathbb{Q} - s\mathbb{I} \right| = 0 \quad .$$

P r o p o s i t i o n 2. Under the conditions of Proposition 1,

$$\mathrm{M} e^{i\alpha \xi^+(\theta_s)} = (\mathbb{X}_+(s) - i\alpha \mathbb{I}_+)^{-1} \mathbb{X}_+(s)\, \mathbb{P}(\theta_s) \, ,$$

$$\mathrm{M} e^{i\alpha \xi^+} = \lim_{t\to\infty} \mathrm{M} e^{i\alpha \xi^+(t)} = (\mathbb{I} - i\alpha \mathbb{X}_+^{-1} \mathbb{I}_+)^{-1} \mathbb{P}_0 \, (a_0 < 0), \tag{11}$$

where $\mathbb{X}_\pm = \lim_{s\to\infty} \mathbb{X}_\pm(s)$, $|\mathbb{X}_+| \neq 0$ and $a_0 = \sum_{k=1}^{n} a_k \rho_k < 0 \Leftrightarrow |\mathbb{X}_+| \neq 0$.
It is easy to show that the distribution of ξ^+

$$\mathbb{F}(x) = \lim_{t\to\infty} \| \mathcal{P}\{\xi^+(t) < x,\, x(t) = r \mid x(0) = k\} \| \;(\mathbb{F}(\infty) = \mathbb{P}_0),$$

determined by its ch.f. in (11), satisfies the equation

$$\mathrm{L}_0 \mathbb{F}(x) \equiv \frac{1}{2} \mathbb{B}^2 \frac{d^2 \mathbb{F}}{dx^2} - \mathbb{A} \frac{d\mathbb{F}}{dx} + \mathbb{Q}\mathbb{F} \quad (x>0) \quad . \tag{12}$$

To find $\mathbb{Q}_c(x)$ and $\mathbb{Q}^c(x)$ (they satisfy this equation too) we need a solution of (12), which does not contain $\mathbb{P}_0$ (just $\mathbb{F}(x)$ contains it). To this end we consider the function

$$(\mathbb{X}_+(s) + \lambda \mathbb{I}_+)^{-1} \mathbb{X}_+(s) = \mathbb{I} - \lambda (\mathbb{X}_+(s) + \lambda \mathbb{I}_+)^{-1} \mathbb{I}_+ \, , \tag{13}$$

which determines the generating function for τ^x ; to be precise,

$$\int_0^\infty e^{-\lambda x} M e^{-s\tau^x} dx = \frac{1}{\lambda} M e^{-s\tau^{\theta_\lambda}} = (X_+(s) + \lambda I_+)^{-1} I_+ . \qquad (14)$$

Hence the matrix $Me^{-s\tau^x}$ contains 0 -columns with index $\kappa \in E_+$ for $I_+ \neq I$. We can specify the structure of it, by supposing for simplicity

$$I_+ = \begin{Vmatrix} I^+ & 0 \\ 0 & 0 \end{Vmatrix} , \quad I^+ = \| \delta_{\kappa z} \|_{\kappa, z=1}^{m_+} , \quad m_+ \leq n .$$

In fact, since $\tau^0 > 0$ with a positive probability for the initial states $x(0) = \kappa \bar{\in} E_+$, then

$$\tau^x \doteq \tau^0 + \tau_+^x .$$

τ_+^x is the time necessary for reaching $x > 0$ after the passage through 0 . The last stochastic relation implies

$$Me^{-s\tau^x} = \begin{Vmatrix} I^+ & 0 \\ M_+ e^{-s\tau^0} & 0 \end{Vmatrix} \cdot \begin{Vmatrix} M^+ e^{-s\tau_+^x} & 0 \\ 0 & 0 \end{Vmatrix} , \qquad (15)$$

$M^+ (M_+)$ is the expectation sign corresponding to the initial states $\kappa \in E_+ \ (\kappa \bar{\in} E_+)$.

Let us denote by $\hat{\xi}(t)$ the process defined on the converted chain $\hat{x}(t)$ with the ch.f.

$$Me^{i\alpha\hat{\xi}(t)} = exp\{ K_0^*(\alpha) \} .$$

The corresponding functionals for $\hat{\xi}(t)$ will be denoted by $\hat{\tau}^z$, $\hat{\xi}^+(t)$, ξ^+.

L e m m a. The functions $R_+(x)$ and $\hat{R}_+(-x)$ ($R_+(x)$ and $\hat{R}_+(x)$ are defined by $\tilde{R}_1^+(\lambda) = (X_+ + \lambda I_+)^{-1} I_+$ and $\tilde{R}_2(\lambda) = (X_-^T + \lambda I_-)^{-1} I_-$ respectively) satisfy equation (12). Both of them have the following probabilistic interpretation for $z \geq 0$:

$$\mathbb{R}_+(z) = \begin{cases} \| \mathcal{P}\{\tau^z < \infty, x(\tau^z) = \iota \mid x(0) = \kappa\} \| & \text{if } |X_+| \neq 0, \\ \mathbb{P}(\tau^z) = \| \mathcal{P}\{x(\tau^z) = \iota \mid x(0) = \kappa\} \| & \text{if } |X_+| = 0; \end{cases}$$

$$\hat{\mathbb{R}}_+(z) = \begin{cases} \| \mathcal{P}\{\hat{\tau}^z < \infty, \hat{x}(\hat{\tau}^z) = \iota \mid \hat{x}(0) = \kappa\} \| & \text{if } |X_-| \neq 0, \\ \hat{\mathbb{P}}(\hat{\tau}^z) = \| \mathcal{P}\{\hat{x}(\hat{\tau}^z) = \iota \mid \hat{x}(0) = \kappa\} \| & \text{if } |X_-| = 0. \end{cases}$$

To prove lemma we shall begin with a non-degenerate case, when $\mathbb{I}_\pm = \mathbb{I}$. In this case it is easy to realize the inversion of the functions $\widetilde{\mathbb{R}}_1^+(\lambda)$ and $\widetilde{\mathbb{R}}_2^+(\lambda)$ (with respect to λ), and as a consequence of this inversion we find:

$$\mathbb{R}_+(z) = e^{-zX_+}, \quad \hat{\mathbb{R}}_+(z) = e^{-zX_-^T}.$$

It is obvious that

$$\begin{aligned} \mathbb{L}_0\mathbb{R}_+(z) &= (\tfrac{1}{2}B^2X_+^2 + AX_+ + \mathbb{Q})e^{-zX_+}, \\ \mathbb{L}_0\hat{\mathbb{R}}_+(-z) &= (\tfrac{1}{2}B^2(X_-^T)^2 - AX_-^T + \mathbb{Q})e^{zX_-^T}. \end{aligned} \tag{16}$$

By comparing the corresponding coefficients of powers of $i\alpha$ in the factorization identity (10), we obtain the two equations for $X_\pm$:

$$\begin{aligned} \tfrac{1}{2}B^2X_+^2 + AX_+ + \mathbb{Q} &= 0, \\ \tfrac{1}{2}X_-^2B^2 - X_-A + \mathbb{Q} &= 0. \end{aligned} \tag{17}$$

On this account we conclude, from (16) and the symmetry of $\mathbb{Q}$,that

$$\mathbb{L}_0 \mathbb{R}_+(z) = 0 , \quad \mathbb{L}_0 \hat{\mathbb{R}}_+(-z) = 0 \quad . \tag{18}$$

In the degenerate case ($\mathbb{I}_\pm \neq \mathbb{I}$) it is difficult to verify the realizability of (18) in the same way as for $\mathbb{I}_\pm = \mathbb{I}$. But in this case it is easy to prove that the equations obtained after integral transformations of (18) are valid for

$$\mathbb{S}_1(\alpha) = \int_0^\infty e^{i\alpha z} \mathbb{R}_+(z) dz = (\mathbb{X}_+ - i\alpha \mathbb{I}_+)^{-1} \mathbb{I}_+ ,$$

$$\mathbb{S}_2(\alpha) = \int_{-\infty}^0 e^{i\alpha z} \hat{\mathbb{R}}_+(-z) dz = (\mathbb{X}_-^T + i\alpha \mathbb{I}_-)^{-1} \mathbb{I}_- .$$

Since $\mathbb{K}_0^*(\alpha) = \mathbb{K}_0(\alpha)$, both of them satisty the equations corresponding to (18). To be precise,

$$\mathbb{K}_0(\alpha)\, \mathbb{S}_1(\alpha) = -\frac{i\alpha}{2} \mathbb{B}^2 + \mathbb{C}_1 \quad ,$$
$$\mathbb{K}_0^*(\alpha)\, \mathbb{S}_2(\alpha) = -\frac{i\alpha}{2} \mathbb{B}^2 + \mathbb{C}_2 \quad . \tag{19}$$

The right-hand sides of (19) (depending on the initial conditions) differ only by a constant matrix. Hence we conclude that $\mathbb{R}_+(z)$ and $\hat{\mathbb{R}}_+(-z)$ satisfy equations (18).

It is not difficult to derive equation (12) for $\mathbb{Q}_c(z)$ and $\mathbb{Q}^c(z)$ and to show that their boundary conditions depend on fixed values of $\mathbb{R}_+$ and $\hat{\mathbb{R}}_+$. On the basis of this fact we arrive at the following two theorems.

T h e o r e m 1. If $\mathbb{Q}$ is a symmetric matrix, $\mathbb{B}^2 > 0$ and $a_0 = \sum a_K \rho_K \neq 0$, then the probabilities of the first exit from the interval $[x - c, x]$ $(0 \leqslant x \leqslant c)$, which satisfy equation (12) and the boundary conditions

$$\mathbb{Q}_c(0) = \mathbb{Q}^c(c) = 0 \ , \ \mathbb{Q}^c(0) = \mathbb{Q}_c(c) = \mathbb{I} \ , \tag{20}$$

are determined by the following relations

$$\begin{aligned} \mathbb{Q}_c(x) &= (e^{x\mathcal{X}_-^T} - e^{-x\mathcal{X}_+})(e^{c\mathcal{X}_-^T} - e^{-c\mathcal{X}_+})^{-1}, \\ \mathbb{Q}^c(x) &= (e^{(x-c)\mathcal{X}_-^T} - e^{(c-x)\mathcal{X}_+})(e^{-c\mathcal{X}_-^T} - e^{c\mathcal{X}_+})^{-1}. \end{aligned} \tag{21}$$

The proof follows from Lemma the above which permits to establish that relations (21) for $\mathbb{Q}_c(x)$ and $\mathbb{Q}^c(x)$ satisfy equation (12). These relations are combined from solutions $\mathbb{R}_+(x)$ and $\hat{\mathbb{R}}_+(x)$ in such a way that boundary conditions (20) are fulfilled.

If $|\mathcal{X}_+| \neq 0 \quad (a_0 < 0 \ , \ |\mathcal{X}_-| = 0 \ )$, it is easy to show that

$$\mathbb{Q}^c(y+c) \xrightarrow[c \to \infty]{} 0 \ , \ \mathbb{Q}_c(y+c) \xrightarrow[c \to \infty]{} e^{y\mathcal{X}_-^T} \ (y < 0) \ , \tag{22}$$

where

$$e^{y\mathcal{X}_-^T} = \mathbb{P}(\tau_y) \qquad (\ y \leq 0\ ) \ .$$

By analogy with (22), we can found limit relations for the case

$$|\mathcal{X}_-| \neq 0 \ \ (a_0 > 0, \ |\mathcal{X}_+| = 0)$$

$$\mathbb{Q}_c(x) \xrightarrow[c \to \infty]{} 0 \ , \quad \mathbb{Q}^c(x) \longrightarrow e^{-x\mathcal{X}_+} = \mathbb{P}(\tau^x) \ .$$

In the degenarate case $(\ \mathbb{I}_{\pm} \neq \mathbb{I}\ )$ boundary conditions for $\mathbb{Q}_c(z)$ and $\mathbb{Q}^c(z)$ appear more complicated that (20). They depend on the values $\mathbb{R}_+(0)$, $\hat{\mathbb{R}}_+(0)$, $\mathbb{R}_+(\pm c)$, $\hat{\mathbb{R}}_+(\pm c)$, where

$$\mathbb{R}_+(0) = \lim_{\alpha\to\infty} \alpha\,(X_+ + \alpha\,\mathbb{I}_+)^{-1}\,\mathbb{I}_+ \neq \mathbb{I}\ ,$$

$$\hat{\mathbb{R}}_+(0) = \lim_{\alpha\to\infty} \alpha\,(X_-^T + \alpha\,\mathbb{I}_-)^{-1}\,\mathbb{I}_- \neq \mathbb{I},\ \ \mathbb{R}_+(0) \neq \hat{\mathbb{R}}_+(0)\ .$$

Introducing notation

$$\mathbb{P}_c(\tau_0) = \|\mathcal{P}\{\tau_0 < \tau^c,\ x(\tau_0) = z\,|\,x(0) = \kappa\}\| = \mathcal{P}\{\xi^+(\tau_0) < c\}\ ,$$

$$\mathbb{P}^0(\tau_{-c}) = \|\mathcal{P}\{\tau_{-c} < \tau^0,\ x(\tau_{-c}) = z\,|\,x(0) = \kappa\}\| = \mathbb{P}(\tau_{-c}) - \mathcal{P}\{\xi^+(\tau_{-c}) > 0\}\ ,$$

$$\mathbb{P}_{-c}(\tau_0) = \|\mathcal{P}\{\tau^0 < \tau_{-c},\ x(\tau^0) = z\,|\,x(0) = \kappa\}\| = \mathcal{P}\{\xi^-(\tau^0) > -c\}\ ,$$

$$\mathbb{P}_0(\tau^c) = \|\mathcal{P}\{\tau^c < \tau_0,\ x(\tau^c) = z\,|\,x(0) = \kappa\}\| = \mathbb{P}(\tau_c) - \mathcal{P}\{\xi^-(\tau_c) < 0\}\ ,$$

we can write boundary conditions for $\mathbb{Q}_c(z)$ and $\mathbb{Q}^c(z)$:

$$\mathbb{Q}_c(0) = \mathbb{P}^0(\tau_{-c}) \geq 0,\ \ \mathbb{Q}_c(c) = \mathbb{P}_c(\tau_0) \neq \mathbb{I}\,(\mathbb{I}_- \neq \mathbb{I})\,;$$

$$\mathbb{Q}^c(0) = \mathbb{P}_{-c}(\tau^0) \neq \mathbb{I}\ \ (\mathbb{I}_+ \neq \mathbb{I}),\ \ \mathbb{Q}^c(c) = \mathbb{P}_0(\tau^c) \geq 0\ . \tag{23}$$

By analogy with Theorem 1 we can prove Theorem 2 for the case $\mathbb{I}_\pm \neq \mathbb{I}$.

T h e o r e m 2 . If $\mathbb{Q}$ is a symmetric matrix , $\mathbb{I}_\pm \neq \mathbb{I}$ and $a_0 \neq 0$, then the probabilities $\mathbb{Q}_c(z)$ and $\mathbb{Q}^c(z)$ are determined by the following relations

$$\mathbb{Q}_c(z) = (\hat{\mathbb{R}}_+(-z) - \mathbb{R}(z))(\hat{\mathbb{R}}_+(-c) - \mathbb{R}_+(c))^{-1}\,\mathbb{P}_c(\tau_0), \tag{24}$$

$$\mathbb{Q}^c(z) = (\hat{\mathbb{R}}_+(c-z) - \mathbb{R}_+(z-c))(\hat{\mathbb{R}}_+(c) - \mathbb{R}_+(-c))^{-1}\,\mathbb{P}_{-c}(\tau_0).$$

According to the above Lemma, the right-hand sides of (24) satisfy equation (12) and conditions (23), which determine $Q_c(z)$ and $Q^c(z)$.

It should de ramarked that the relations (21) and (24) remain valid even in the case $a_0 = 0$ if

$$\Delta(c) \neq 0, \Delta(-c) \neq 0 \quad (\Delta(x) = \det[\hat{R}_+(-x) - R_+(x)]) .$$

But, if $\Delta(c) = \Delta(-c) = 0$, then we must take into account a linear solution of (12),

$$R_+^0(x) = x P_0 + \text{const} .$$

Such a situation arises in the case when $A = 0$, $B^2 > 0$.
In this case $(\Delta(-c) = \Delta(c) = 0)$, instead of (21) and (24), we have the following relations for the probabilities in question:

$$Q_c(x) = (\hat{R}_+(-x) - R_+(x) + x P_0)(\hat{R}_+(-c) - R_+(c) + c P_0)^{-1}, \quad (25)$$

$$Q^c(x) = (\hat{R}_+(c-x) - R_+(x-c) + (x-c) P_0)(\hat{R}_+(c) - R_+(-c) - c P_0)^{-1}.$$

As for $T(\theta_s)$ and $T(\infty)$, we restrict ourselves by the remark, that their distribution depend on a ratio of the matrices in the factorisation identity (10) (see [2] , Theorem 3).
For the case $I_+ = I$,

$$M e^{-\mu T(\theta_s)} = X_+^{-1}(s+\mu) X_+(s) P(\theta_s) , \quad (26)$$

$$M e^{-\mu T(\infty)} = X_+^{-1}(\mu) X_+(0) P_0 \quad (\text{if } a_0 < 0) .$$

References

1 В.С.Королюк, Граничные задачи для сложных пуассоновских процессов, Киев, 1975.

2 Д.В.Гусак, О непрерывных однородных процессах с независимыми приращениями на цепи Маркова.Сб.Теория вероятностей и математическая статистика (to appear).

Mathematical Institute Academy
of Sciences of the Ukrainian SSR
Kiev

NONCOMMUTATIVE ANALOGUES OF THE CRAMÉR-RAO INEQUALITY IN THE QUANTUM MEASUREMENT THEORY

A.S.Holevo

I. Introduction. Recently deep connections between the classical mathematical statistics and the quantum measurement theory were established. Mathematical statistics, particularly estimation theory, may be interpreted as statistical measurement theory for classical systems, establishing limits of the accuracy of parameter measurements and giving prescriptions for the choice of optimal measuring procedure. This range of problems is at least equally interesting in quantum theory in which fundamental limitations to the accuracy of measurements are internally inherent [I]-[3] Apart from their practical significance, e.g. in quantum communications theory which recently attracted attention to these problems, they shed light on some important points of quantum theory and are of general interest for its foundations and interpretation (see [4]).

In general terms the problem may be formulated as follows. Given a family of states $\{S_{\vec{\theta}}\}$ of the system under consideration, parametrized by a vector $\vec{\theta} = (\theta_1, \dots, \theta_n)$, and a criterion of accuracy of the parameter measurement, one has to: (I) estimate the best possible accuracy of measurements, (II) find a measurement, optimizing the given criterion of accuracy.

To formulate these problems mathematically we remind here the definitions of principal concepts of quantum statistical decision theory (see [5], [6] for more detail). To avoid some technical difficulties we shall proceed partly heuristically, assuming all the necessary regularity conditions to be fulfilled.

A quantum system is described by a Hilbert space H. The reader is recommended to keep in mind a finite dimensional unitary space since most of regularity conditions are automatically fulfilled in this case. A state of the system is described by a density operator S in H, i.e. an Hermitian positive operator with the unit trace [2]. Writing the spectral decomposition $S = \sum_{\kappa} s_{\kappa} |\kappa)(\kappa|$, (where $|\kappa)(\kappa|$ is, in Dirac's notation [I], the projection operator onto the κ-th eigenvector $|\kappa)$ of the operator S), we have $s_{\kappa} \geq 0$, $\sum_{\kappa} s_{\kappa} = 1$. States on a finite dimensional H may be considered as noncommutative analogues of the classical probability distributions on a finite space of events. An observable is a self-adjoint operator in H; the mean value of an observable X with respect to a state S is equal to $\sum_{\kappa} s_{\kappa} (\kappa|X|\kappa) = m(S \cdot X)$, where m is the trace of an operator in H. We shall denote this mean value by $\langle S, X \rangle$. Let a family of states $\{S_{\vec{\theta}}\}$ be given, where $\vec{\theta}$ is a parameter running through a set $\Theta \subset \mathbb{R}^n$.

A decision rule is any mapping $S \to \mu_S$, transforming states S into probability distributions $\mu_S(d\theta_1 \dots d\theta_n)$ on Θ, and satisfying the condition

$$\mu_{\lambda S_1 + (1-\lambda) S_2} = \lambda \mu_{S_1} + (1-\lambda) \mu_{S_2}, \quad 0 \leq \lambda \leq 1. \tag{I}$$

This definition is based on the following consideration. If one performs an experiment with the system to determine the true values of $\theta_1, \dots, \theta_n$, one obtains, in general, random variables

$\hat{\theta}_1, \dots, \hat{\theta}_n$, the distribution of which is determined by the state S of the system before the experiment. Condition (I) means that a mixture of states results in the corresponding mixture of distributions.

The weighted mean square error (m.s.e.) of the decision rule $S \to \mu_S$ is equal to

$$\Sigma_{\vec{\theta}} = \int \sum_{k,j} g_{kj} (\hat{\theta}_k - \theta_k)(\hat{\theta}_j - \theta_j) \mu_{S_{\vec{\theta}}}(d\hat{\theta}_1 \dots d\hat{\theta}_n), \tag{2}$$

where $G = \| g_{kj} \|$ is a real positive weight matrix. Since, as in classical statistics, we cannot expect a uniformly optimal decision rule minimizing $\Sigma_{\vec{\theta}}$ for all $\vec{\theta} \in \Theta$ to exist, we introduce the condition of unbiasedness

$$\int \hat{\theta}_k \mu_{S_{\vec{\theta}}}(d\hat{\theta}_1 \dots d\hat{\theta}_n) = \theta_k , \quad k = 1, \dots, n. \tag{3}$$

In what follows we shall study the problem of minimizing the functional $\Sigma_{\vec{\theta}}$ over all decision rules, satisfying condition (3).

The decision rules introduced above are analogous to randomized procedures of classical statistics. It is well-known, however, that in classical estimation theory only deterministic procedures (i.e. point estimates) are of real interest, since they constitute a complete class under very general regularity assumptions (see, e.g. [8]). It is remarkable that this fact has no analogy in the quantum case [5] . We must, however, explain at this point what plays the role of deterministic procedures in quantum estimation theory.

By a Θ - measurement* we shall mean a measure $X(d\theta_1 \dots d\theta_n)$ on Θ with values in the cone of positive Hermitian operators, normalized by the condition $X(\Theta) = I$, where I is the identity operator in H. There is one-to-one correspondence between decision rules and measurements described by the relation ([6])

$$\mu_S (d\theta_1 \dots d\theta_n) = \langle S, X(d\theta_1 \dots d\theta_n) \rangle , \quad \forall S.$$

A measurement is called simple if the measure $X(d\theta_1 \dots d\theta_n)$ is projection-valued, i.e. it is a spectral measure. In this and only in this case the measurement is uniquely determined by the estimates, i.e. by the family of commuting observables

$$T_k = \int \hat{\theta}_k \, X(d\hat{\theta}_1 \dots d\hat{\theta}_k), \quad k = 1, \dots, n.$$

Simple measurements are thus analogous to classical deterministic procedures. The fact that several observables commute and thus have a common spectral resolution means physically that these observables are "compatible", i.e. admit a joint measurement. It can be shown that an arbitrary measurement $X(d\theta_1 \dots d\theta_n)$ can be reduced to a joint measurement of some commuting observables $T_1', \dots, T_n'$ over an extension of the initial system including an auxiliary independent quantum system (quantum randomization) [5] , [6] .

---

* Although generally accepted, this term appeares to be not very apt since the measure $X(d\theta_1 \dots d\theta_n)$, giving a complete statistical description of possible outcomes of the measurement, tells nothing about the actual mechanism of the measurement.

Now we shall discuss briefly the relation between the concept of decision rule and physical reality. It is clear that any real physical experiment in principle can be associated with some decision rule as a complete statistical description of possible outcomes of the experiment. However the converse problem of realizing a given decision rule is a nontrivial one even in the classical case (see in this connection [8]). In quantum theory the problem of realizing a decision rule is closely related to justification of the fundamental quantum postulate that any observable (i.e. a self-adjoint operator in H) describes a real physical quantity [I] , [2] . Thus even if an optimal decision rule, i.e. a mathematical solution of problem (II), is found (what is rarely the case) the question of its practical realization arises in applications. Such questions are out of the scope of mathematical theory and are in fact problems of experimental physics. At the same time any theoretical bound for the accuracy of measurements, obtained via the mathematical concept of decision rule, will definitely hold for the results of real measuring procedures.

In what follows we shall concentrate our attention on problem(I).

2. A survey of noncommutative analogues of the Cramér-Rao inequality. A characteristic feature of the quantum case is the existence of two different types of inequalities giving lower bounds for the m.s.e. of an unbiased decision rule.

Fix a point $\vec{\theta}\in\Theta$ and put $S=S_{\vec{\theta}}$, $S'_k=\frac{\partial}{\partial\theta_k}S_{\vec{\theta}}$. Following Helstrom [9] , consider the symmetrized logarithmic derivatives (s.l.d.), which are defined by the equation

$$S'_k = L_k \circ S \equiv \frac{1}{2}(L_k S + S L_k). \tag{4}$$

In the case $\dim H < \infty$ a sufficient condition for the existence and uniqueness of a solution of (4) is that the operator S should be non-degenerate.

Let us introduce a noncommutative analogue of the Fisher information matrix*

$$\mathbb{A} = \| \langle S, L_k \circ L_j \rangle \|_{k,j=1}^{n}$$

and the correlation matrix of the estimation errors

$$\mathbb{B} = \| \int (\hat{\theta}_k - \theta_k)(\hat{\theta}_j - \theta_j) \mu_S (d\hat{\theta}_1 \dots d\hat{\theta}_n) \|_{k,j=1}^{n}$$

For unbiased measurements the inequality $\mathbb{B} \geq \mathbb{A}^{-1}$ may be proved quite similarly to the classical Cramér-Rao inequality [9] (see also § 5). For the m.s.e. $\Sigma = \mathrm{Tr}\, G \mathbb{B}$, it follows

$$\Sigma \geq \mathrm{Tr}\, G \mathbb{A}^{-1} \equiv \Sigma_1 \tag{5}$$

where Tr denotes the trace of a matrix.

---

* In classical statistics the Fisher information tensor corresponds to the natural Riemannian metric on the variety of all probability distributions with the Hellinger distance arccos $\int \sqrt{dP_1 \, dP_2}$ [8]. Note, without going into details, that in the noncommutative case an analogous role is played by the distance arccos $m(|S_1^{1/2} \cdot S_2^{1/2}|)$, where $|X| = \sqrt{X^* X}$.

Another inequality for $\mathbb{B}$ was obtained by Yuen and Lax [10], where the left logarithmic derivative V_k, defined from the equation $S_k' = V_k \cdot S$, was used. If S is non-degenerate, then $V_k = S_k' \cdot S^{-1}$.

Let us introduce the matrix

$$\widetilde{\mathbb{A}} = \| \langle S, V_j^* \cdot V_k \rangle \| = \| m(S_k' \cdot S^{-1} \cdot S_j') \|.$$

It is important to understand that, unlike the real matrix $\mathbb{A}$, the matrix $\widetilde{\mathbb{A}}$ is a complex Hermitian one. The inequality $\mathbb{B} \geq \widetilde{\mathbb{A}}^{-1}$ holds and is to be understood as an inequality for complex Hermitian matrices.

Let us show how this inequality may be used to obtain a lower bound for the m.s.e. $\Sigma = \operatorname{Tr} G \mathbb{B}$. We have $\mathbb{B} \geq \operatorname{Re} \widetilde{\mathbb{A}}^{-1} + i \operatorname{Im} \widetilde{\mathbb{A}}^{-1}$, where $\operatorname{Re} \widetilde{\mathbb{A}}^{-1}$ is a real symmetric, $\operatorname{Im} \widetilde{\mathbb{A}}^{-1}$ is a real skew-symmetric matrix. Introducing the real symmetric matrix $X = \mathbb{B} - \operatorname{Re} \mathbb{A}^{-1}$, we have $X \geq i \operatorname{Im} \widetilde{\mathbb{A}}^{-1}$, hence

$$\Sigma \geq \operatorname{Tr} G \operatorname{Re} \widetilde{\mathbb{A}}^{-1} +$$

$$+ \min \{ \operatorname{Tr} G X : X \text{ is real symmetric and } X \geq i \operatorname{Im} \widetilde{\mathbb{A}}^{-1} \} \quad (6)$$

To evaluate the minimum we use the following device [11]. For each real normal (in particular, for a skew-symmetric) matrix M the polar decomposition $M = |M| \cdot O = O \cdot |M|$ holds, where $|M| = \sqrt{M^T M}$ and $O = \lim_{\varepsilon \downarrow 0} (|M| + \varepsilon \mathbb{1})^{-1} \cdot M$ is a skew-symmetric partially isometric matrix. If N is similar to M, $N = T \cdot M \cdot T^{-1}$, then one may write

$$N = \operatorname{abs} N \cdot \operatorname{sgn} N = \operatorname{sgn} N \cdot \operatorname{abs} N,$$

where

$$\mathrm{abs}\, N = T\cdot|M|\cdot T^{-1}, \quad \mathrm{sgn}\, N = T\cdot O\cdot T^{-1}.$$

We emphasize that abs $N \neq |N|$.

<u>Lemma I.</u> Let S be a real skew-symmetric matrix and G be a strictly positive matrix, then

$$\min\left\{\mathrm{Tr}\, G X : X \text{ real symmetric and } X \geq iS\right\} = \mathrm{Tr}\, \mathrm{abs}\,(G\cdot S),$$

and the minimum is attained for $X = S\cdot \mathrm{sgn}\,(G\cdot S)$.

<u>Proof:</u> Putting $Y = \sqrt{G}\, X \sqrt{G}$ and taking into account that $Y^T = Y,\ S^T = -S$, we have $Y \geq \pm i \sqrt{G}\, S \sqrt{G}$, whence

$$\mathrm{Tr}\, Y \geq \mathrm{Tr}\,|i\sqrt{G}\, S\sqrt{G}| = \mathrm{Tr}\,|\sqrt{G}\, S\sqrt{G}|.$$

Since $GS = \sqrt{G}\,(\sqrt{G}\, S\sqrt{G})\sqrt{G}^{-1}$ is similar to the skew-symmetric matrix $\sqrt{G}\, S\sqrt{G}$, we have

$$\mathrm{Tr}\, G X = \mathrm{Tr}\, Y \geq \mathrm{Tr}\,|\sqrt{G}\, S\sqrt{G}| = \mathrm{Tr}\, \mathrm{abs}\,(G\cdot S).$$

The second statement of the lemma is easily verified.

From inequality (6) and Lemma I we obtain the following inequality for the m.s.e. (See Holevo [I2]):

$$\Sigma \geq \mathrm{Tr}(G\cdot \mathrm{Re}\,\tilde{A}^{-1}) + \mathrm{Tr}\, \mathrm{abs}\,(G\cdot \mathrm{Im}\,\tilde{A}^{-1}) \equiv \Sigma_2.$$

In the case of a one-dimensional parameter $(n=1)$ the bound Σ_1 is always better than Σ_2 $(\Sigma_1 \geq \Sigma_2)$, the equality being attained if and only if $\frac{d}{d\theta} S_\theta$ commutes with S_θ . If $n>1$, then the bounds are incompatible in the sense that $\Sigma_1 \gtrless \Sigma_2$ depending on the choice of the weight matrix $\mathcal{G}$.

In what follows we shall try to disclose the mechanism of the situation and show how these two bounds follow from a more general inequality.

To conclude this section we remark, that bound (7), unlike (5), takes into account "incompatibility" of several parameters (the second term in (7)), and in a number of cases it may be used to obtain nontrivial limits provided the computation of $\tilde{\mathbb{A}}$ does not require solving operator equation (4) for the s.l.d.

Example I. Let $\dim H = 2$ and $|1)$, $|2)$ be a basis in H. Let

$$S_\theta = \frac{1}{2}\begin{Vmatrix} 1+\theta_1 & \theta_2 - i\theta_3 \\ \theta_2 + i\theta_3 & 1-\theta_1 \end{Vmatrix}$$

be the matrix of a density operator in this basis, $\vec{\theta} = (\theta_1, \theta_2, \theta_3)$ being Stokes parameters satisfying $\theta_1^2 + \theta_2^2 + \theta_3^2 \leq 1$.

For example, such a matrix describes polarization of a photon, θ_1, θ_2 characterizing the degree of polarization along the axes $|1)$, $|2)$ correspondingly, θ_3 the degree of the circular polarization [7].

We have $\frac{\partial}{\partial \theta_i} S_{\vec{\theta}} = \frac{1}{2}\sigma_i$, where σ_i are Pauli matrices. A straightforward computation gives

$$\tilde{\mathbb{A}}^{-1} = \left\{ \begin{Vmatrix} 1-\theta_1^2 & \theta_1\theta_2 & \theta_1\theta_3 \\ \theta_1\theta_2 & 1-\theta_2^2 & -\theta_2\theta_3 \\ \theta_1\theta_3 & -\theta_2\theta_3 & 1-\theta_3^2 \end{Vmatrix} - i \begin{Vmatrix} 0 & -\theta_3 & \theta_2 \\ \theta_3 & 0 & \theta_1 \\ -\theta_2 & -\theta_1 & 0 \end{Vmatrix} \right\}.$$

Taking the unit matrix for the weight and applying (7), we find the lower bound for the m.s.e. of measurement of parameters $\theta_1, \theta_2, \theta_3$:

$$\sum_{i=1}^{3} \overline{\Delta\theta_i^2} \geq \left[\left(3 - \sum_{i=1}^{3}\theta_i^2\right) + 2\sqrt{\sum_{i=1}^{3}\theta_i^2}\right] = 4 - 4\left[\frac{1-\sqrt{\sum_{i=1}^{3}\theta_i^2}}{2}\right]^2.$$

Note that the expression on the right-hand side attains its maximum, equal to 4, for a "completely polarized" state $\left(\sum_{i=1}^{3}\theta_i^2 = 1\right)$ and its minimum, equal to 3, for the "unpolarized" state: $\theta_1 = \theta_2 = \theta_3 = 0$.

The family of states $S_\theta^{(n)} = S_\theta \otimes \ldots \otimes S_\theta$ which are tensor products of n copies of the S_θ may be considered as a noncommutative analogue of the Bernoulli scheme. It follows that the m.s.e. of a measurement of $\theta_1, \theta_2, \theta_3$ in this scheme satisfies the inequalities

$$\sum_{i=1}^{3} \overline{\Delta\theta_i^2} \geq \frac{4}{n}\left[1 - \left(\frac{1 - \sqrt{\sum\theta_i^2}}{2}\right)^2\right] \geq \frac{3}{n}.$$

Example 2. Let $\dim H < \infty$, $S = \sum s_{kj} |k)(j|$ be the matrix representation of a density operator in a fixed basis $\{|j)\}$. Matrix elements s_{kj} may be referred to as unknown parameters satisfying the condition $s_{kj} = \overline{s}_{jk}$.

Fix k, j $(k \neq j)$ and find the lower bound for the error of joint estimation of the real and imaginary parts of s_{kj}. We have

$$\frac{\partial S}{\partial \operatorname{Re} s_{kj}} = |k)(j| + |j)(k|, \quad \frac{\partial S}{\partial \operatorname{Im} s_{kj}} = i\left[|k)(j| - |j)(k|\right].$$

It follows easily that

$$\tilde{A}^{-1} = \frac{1}{4}\begin{Vmatrix} a+b & -i(a-b) \\ i(a-b) & a+b \end{Vmatrix}, \quad a = (k|S^{-1}|k)^{-1}, \quad b = (j|S^{-1}|j)^{-1}.$$

For $G = 1$ we obtain from (7)

$$\overline{(\Delta \operatorname{Re} s_{kj})^2} + \overline{(\Delta \operatorname{Im} s_{kj})^2} \geq \frac{1}{2}\left(a + b + |a - b|\right) = \\ = \max\left\{ (k|S^{-1}|k)^{-1}, (j|S^{-1}|j)^{-1} \right\}.$$

3. Symplectic structure determined by a quantum state. Just as in probability theory the space $L^2(P)$ is canonically associated with the probability distribution P, in quantum theory we can associate, with a state S, the Euclidean space $\mathfrak{h}(S)$ of Hermitian operators $X, Y, \dots$ with the inner product

$$\sigma(X, Y) = \mathrm{Re}\langle S, X\cdot Y\rangle = \langle S, X\circ Y\rangle.$$

In the case $\dim H < \infty$ this definition has to be made more precise since X, Y may be unbounded operators; however, we have agreed not to dwell on such subtleties here. A precise general definition of $\mathfrak{h}(S)$ will be given elsewhere. For simplicity we assume S to be nondegenerate.

Unlike the classical case, the space $\mathfrak{h}(S)$, corresponding to a quantum state, possesses additional symplectic structure. Namely, there is a bilinear skew-symmetric form on $\mathfrak{h}(S)$

$$\vartheta(X, Y) = 2\,\mathrm{Im}\langle S, X\cdot Y\rangle = i^{-1}\langle S, [X,Y]\rangle,$$

where $[X, Y] = XY - YX$ is the commutator of the operators X, Y. Note that this form is degenerate: $\vartheta(X, Y) = 0$ for all Y iff $[X, S] = 0$. Thus the null subspace of the form ϑ consists of operators commuting with S; in particular, $\vartheta \equiv 0$ iff $\dim H < \infty$ and $S = (\dim H)^{-1}\cdot I$.

The forms σ and ϑ obey the uncertainty relation (cf. [3])

$$\sigma(X, X)\cdot\sigma(Y, Y) \geq \tfrac{1}{4}\vartheta(X, Y)^2, \qquad X, Y \in \mathfrak{h}(S).$$

Indeed, this is equivalent [12] to the inequality

$$\sigma(X, X) + \sigma(Y, Y) \geq \vartheta(X, Y), \qquad X, Y \in \mathfrak{h}(S), \tag{8}$$

which in turn follows from $\langle S, (X + iY)^*(X + iY)\rangle \geq 0$.

We define the commutation operator $\mathcal{D}$ of S in $\mathfrak{h}(S)$ by the relation

$$\vartheta(X,Y) = \sigma(X, \mathcal{D}Y), \qquad X,Y \in \mathcal{h}(\mathcal{S}).$$

The operator $\mathcal{D}$ is a skew-symmetric operator in $\mathcal{h}(\mathcal{S})$. Putting $Y = \frac{1}{2}\mathcal{D}X$ in (8) we get $1 + \frac{1}{4}\mathcal{D}^2 \geq 0$ in $\mathcal{h}(\mathcal{S})$. From the definition of the operator $\mathcal{D}$ and the forms ϑ and α it follows directly that the element $Y = \mathcal{D}X \in \mathcal{h}(\mathcal{S})$ is the solution of the equation

$$Y \circ \mathcal{S} = i^{-1}[X, \mathcal{S}]. \tag{9}$$

We shall give an explicit description of action of the operator $\mathcal{D}$. Let $\mathcal{S} = \sum_{\kappa} s_\kappa |\kappa)(\kappa|$ be a spectral resolution of the density operator. Since we assumed $\mathcal{S}$ to be nondegenerate, $s_\kappa > 0$. Then, using (9), we can see that

$$\mathcal{D}: \quad X \to 2i \sum_{\kappa,j} \frac{s_\kappa - s_j}{s_\kappa + s_j} |\kappa)(\kappa|X|j)(j|. \tag{I0}$$

Thus the action of $\mathcal{D}$ consists of multiplication of the matrix elements $(\kappa|X|j)$ in the matrix representation $X = \sum_{\kappa,j} |\kappa)(\kappa|X|j)(j|$ by the numbers $2i\,\frac{s_\kappa - s_j}{s_\kappa + s_j}$. Since the components of the representation are orthogonal in $\mathcal{h}(\mathcal{S})$, relation (I0) gives the "diagonal form" of $\mathcal{D}$. Thus, for an arbitrary function f, we have

$$f(\mathcal{D})X = \sum_{\kappa,j} f\left(2i\,\frac{s_\kappa - s_j}{s_\kappa + s_j}\right)|\kappa)(\kappa|X|j)(j|. \tag{II}$$

Being skew-symmetric, the operator $\mathcal{D}$ is normal and thus admits the polar decomposition in $\mathcal{h}(\mathcal{S})$, $\mathcal{D} = |\mathcal{D}|\mathcal{J} = \mathcal{J}|\mathcal{D}|$, where $|\mathcal{D}|$ is a positive operator in $\mathcal{h}(\mathcal{S})$, $\mathcal{J}$ is a skew-symmetric partially isometric operator. The action of these operators is given by the formulas

$$|\vartheta| : \quad X \to 2 \sum_{k,j} \frac{s_k - s_j}{s_k + s_j} |k)(k|X|j)(j| ;$$

$$\mathcal{J} : \quad X \to i \sum_{k,j} \operatorname{sgn}(s_k - s_j) |k)(k|X|j)(j| .$$

Now we shall briefly describe an important class of quantum states which are natural noncommutative analogues of Gaussian distributions. Our presentation here is deliberately heuristic since in the Gaussian case $\dim H = \infty$. For a more careful treatment see [6], [12].

Let Z be a <u>symplectic space</u>, i.e. a real linear space with a nondegenerate skew-symmetric form Δ and let $\{ R(z) ; z \in Z \}$ be an irreducible family of self-adjoint operators in a Hilbert space H, satisfying the <u>canonical commutation relation</u> (c.c.r.)

$$[R(z_1), R(z_2)] = -i \Delta (z_1, z_2) \cdot I , \quad z_1, z_2 \in Z. \tag{12}$$

In physics the symplectic space Z is the "phase space" of a classical system, and the mapping $z \to R(z)$ defines "quantization" of the system, i.e. assignes to it the corresponding quantum system determined by the "canonical observables" $R(z)$.

Each trace-class operator T in H defines the function

$$\mathcal{F}_z [T] = \operatorname{m} (T \cdot \exp i R(z)) , \quad z \in Z.$$

The correspondence $T \to \mathcal{F}_z [T]$, which we call a <u>Weyl transform</u>, is in many respects analogous to the classical Fourier transform, its usefulness lying in the reduction of the operator algebra to a functional calculus. Each state S is uniquely determined by its

<u>characteristic function</u>

$$\mathcal{F}_z[S] = \langle S, \exp i R(z) \rangle$$

The state S is called <u>Gaussian</u> if

$$\mathcal{F}_z[S] = \exp[i\mu(z) - \tfrac{1}{2}\alpha(z,z)], \tag{I3}$$

where μ is linear function, α is a scalar product on $\mathcal{Z}$.

Let us denote by $\mathcal{R}$ the space of operators of the form $c \cdot I + R(z)$, $c \in \mathbb{R}$, $z \in \mathcal{Z}$. For any state S, for which $\mathcal{R} \subset \mathfrak{h}_y(S)$, the <u>mean value</u>

$$\mathcal{V}(R(z), I),$$

and the <u>correlation function</u>

$$\mathcal{V}(R(z_1), R(z_2)) - \mathcal{V}(R(z_1), I) \cdot \mathcal{V}(R(z_2), I)$$

are defined. For a Gaussian state they are equal correspondingly to $\mu(z)$ and $\alpha(z_1, z_2)$.

We have the following characterization of Gaussian states:

<u>A state S such that $\mathcal{R} \subset \mathfrak{h}_y(S)$ is Gaussian iff the subspace $\mathcal{R}$ is invariant under the commutation operator ϑ.</u>

Indeed, let S be a Gaussian state. Define the operator $\mathfrak{D}$ in $\mathcal{Z}$ by the relation

$$\Delta(z, w) = \alpha(z, \mathfrak{D} w), \qquad z, w \in \mathcal{Z},$$

and show that

$$\vartheta\left(R(z)-\mu(z)\cdot I\right)=R(\mathfrak{D}z)-\mu(\mathfrak{D}z)\cdot I.$$

Since $\vartheta(I)=0$, it will follow that $\vartheta(\mathcal{R})\subset\mathcal{R}$.

In view of (9) it is sufficient to check that

$$\{R(\mathfrak{D}z)-\mu(\mathfrak{D}z)\cdot I\}\circ S=i^{-1}[R(z)-\mu(z)I, S]. \tag{14}$$

We shall make use of properties of the Weyl transform established in [12] :

$$\mathcal{F}_w[R(z)\circ S]=-i\nabla_z\mathcal{F}_w[S];\ \mathcal{F}_w[[R(z), S]]=\Delta(z,w)\cdot\mathcal{F}_w[S],$$

where ∇_z is the derivative in the direction z at the point w .
Putting $z=\mathfrak{D}^{-1}z'$ write (14) in terms of Weyl transform:

$$(\nabla_{z'}-i\mu(z'))\mathcal{F}_w[S]=-\alpha(z', w)\,\mathcal{F}_w[S]. \tag{15}$$

The characteristic function (13) is easily checked to satisfy this equation.

Conversely, let $\mathcal{R}$ be invariant under ϑ . Then

$$\vartheta\left(R(z)-\mu(z)\cdot I\right)=R(\mathfrak{D}z)-f(z)\cdot I,$$

where μ is the mean value of S, $\mathfrak{D}$ is a linear operator, f is a linear function on Z . To determine f consider the equality

$$\langle S, R(\mathfrak{D}z)\rangle-f(z)=\langle S, \vartheta(R(z)-\mu(z)\cdot I)\rangle=0,$$

implying that $f(z)=\mu(\mathfrak{D}z)$. Denoting by α the correlation function of S , we have

$$\alpha(w, \mathfrak{D}z) = \sigma(R(w) - \mu(w)\cdot I, R(\mathfrak{D}z) - \mu(\mathfrak{D}z)\cdot I) =$$
$$= \sigma(R(w) - \mu(w)\cdot I, \vartheta(R(z) - \mu(z)\cdot I)) =$$
$$= \vartheta(R(w) - \mu(w)\cdot I, R(z) - \mu(z)\cdot I) =$$
$$= \Delta(z, w).$$

Thus $\mathfrak{D}$ is the same as in the first part of the proof, and the characteristic function of the state is a solution of the equation (15). Choosing an orthonormal basis $\{z_j\}$ in $\mathcal{Z}$ (with respect to α) and rewriting (15) in coordinates $w = (w_1, \dots, w_k)$, we have

$$\left[\frac{\partial}{\partial w_j} - i\mu(z_j)\right] \mathcal{F}_{w_1 \dots w_k}[S] = -w_j \mathcal{F}_{w_1 \dots w_k}[S], \; j = 1, \dots, k.$$

The unique solution of this equation, satisfying $\mathcal{F}_{0\dots0}[S] = 1$, is the characteristic function of the Gaussian state (13). The result is proved.

4. <u>Operator characteristic function and moments of a measurement.</u> With each Θ - measurement $X(d\theta_1 \dots d\theta_n)$ we associate the <u>operator characteristic function</u>

$$\Phi(t_1, \dots t_n) = \int_{\Theta} \exp[i \sum_j t_j \theta_j] X(d\theta_1 \dots d\theta_n),$$

which uniquely determines the measurement. In order that $\Phi(\vec{t})$, $\vec{t} \in \mathbb{R}^n$, should be an operator characteristic function of a measurement, it is necessary and sufficient that $\Phi(0) = I$, $\Phi(\vec{t})$ should be weakly continuous at $\vec{t} = 0$ and one of the following equivalent conditions hold

$$\sum_{k,j} c_k \bar{c}_j \Phi(\vec{t}^k - \vec{t}^j) \geq 0, \ \forall c_j \in \mathbb{C}; \quad \sum_{k,j} (\psi_k | \Phi(\vec{t}^k - \vec{t}^j) | \psi_j) \geq 0, \ \forall \psi_j \in H. \tag{I6}$$

Operator moments of a measurement (if they exist) are defined by the relation

$$M_{j_1 \dots j_s} = \int \theta_{j_1} \dots \theta_{j_s} X(d\theta_1 \dots d\theta_n);$$

they are related to the characteristic function by the formula

$$M_{j_1 \dots j_s} = i^{-s} \partial^s / \partial t_{j_1} \dots \partial t_{j_s} \Phi(t_1, \dots t_n) \Big|_{\vec{t}=0}.$$

The first moments M_j and second central moments $M^\circ_{kj} = M_{kj} - M_k \circ M_j$ are especially interesting. They satisfy the inequality

$$\| M^\circ_{kj} \|_{k,j=1}^N \geq \frac{1}{2} \| [M_k, M_j] \|_{k,j=1}^N , \tag{I7}$$

where in both sides block mattices stand, elements of which are operators in H [I2] .

It is interesting whether condition (I7) is sufficient for the operators M_k, M°_{kj} to be correspondingly the first and the second operator moments of a measurement. One can construct an operator-valued function $\Phi(\vec{t})$ for which

$$M_j = \frac{\partial \Phi}{\partial t_j}\Big|_{\vec{t}=0}; \quad M_{kj} = M^\circ_{kj} + M_k \circ M_j = -\frac{\partial^2 \Phi}{\partial t_k \partial t_j}\Big|_{\vec{t}=0}.$$

Such a function is, for example,

$$\Phi(\vec{t}) = \exp[i \sum_j t_j M_j - \frac{1}{2} \sum_{k,j} t_j t_k M^\circ_{kj}].$$

It is not known, however, whether this function is positive-definite in the sense of (I6).

Consider an important particular case in which positive answers to these questions are known. Let $R(z),\ z\in Z$ be the canonical observables of the c.c.r. introduced in the previous section. We call a measurement $X(d\theta_1\dots d\theta_n)$ canonical if its operator characteristic function has the form

$$\exp\left[i\sum_{j=1}^{n} t_j R(z_j) - \frac{1}{2}\sum_{k,j=1}^{n} æ_{kj} t_k t_j \right],$$

where $z_j \in Z$, $\| æ_{kj} \|$ is a real symmetric matrix. We have $M_j = R(z_j)$, $\overset{\circ}{M}_{kj} = æ_{kj}\cdot I$ and condition (I6) takes the form

$$\| æ_{kj} \| \geq \frac{i}{2} \| \Delta(z_k, z_j) \|, \tag{I9}$$

where both the matrices have real elements. The following proposition holds:

<u>Expression (I8) determines the operator characteristic function of a measurement iff the condition (I9) is fulfilled.</u>

<u>5. Minimization of the weighted mean square error.</u> We return now to the problem of minimization of the m.s.e. Σ.

Differentiating unbiasedness condition (3) at point $\vec{\theta}$, we obtain

$$\langle S'_k, M_j \rangle = \delta_{jk}, \quad j,k = 1,\dots,n \tag{20}$$

We call a measurement locally-unbiased (at point $\vec{\theta}$) if its operator moments satisfy (20). Using s.l.d. the equations (20) may be rewritten as

$$\mathfrak{a}(L_k, M_j) = \delta_{jk}, \quad j,k = 1,\dots n.$$

Elements of the correlation matrix $\mathbf{B}$ have the form

$$\int (\hat{\theta}_k - \theta_k)(\hat{\theta}_j - \theta_j)\mu_S(d\hat{\theta}_1 \dots d\hat{\theta}_n) =$$

$$= \langle S, M^{\circ}_{kj} \rangle + \langle S, (M_k - \theta_k) \circ (M_j - \theta_j) \rangle .$$

Putting $\varkappa_{kj} = \langle S, M^{\circ}_{kj} \rangle$, $X_k = M_k - \theta_k$, we obtain the m.s.e.:

$$\sum_{k,j} g_{kj} [\varkappa_{kj} + \mathfrak{a}(X_k, X_j)]. \tag{2I}$$

The variables $\varkappa_{kj} \in \mathbb{R}$, $X_j \in \mathfrak{h}(S)$ satisfy the conditions

(a) $\mathfrak{a}(L_k, X_j) = \delta_{kj}$; (b) $\| \varkappa_{kj} \|_{k,j=1}^{n} \geq \frac{i}{2} \| \vartheta(X_k, X_j) \|_{k,j=1}^{n}$

The condition (a) follows from the local unbiasedness and the equalities $\mathfrak{a}(L_k, I) = 0$, which are obtained by differentiating the identity $\langle S_{\vec{\theta}}, I \rangle \equiv 1$. The condition (b) follows from (I7) by taking mean values of both sides of the inequality with respect to the state S, and taking into account that $\vartheta(M_k, M_j) = \vartheta(X_k, X_j)$.

We shall obtain a lower bound for the Σ, minimizing expression (2I) with respect to all $X_j \in \mathfrak{h}(S)$ and all real matrices $\| \varkappa_{kj} \|_{k,j=1}^{n}$, satisfying the conditions (a), (b). This bound will be accessible if there exists a measurement $X(d\theta_1 \dots d\theta_n)$ for which $M_j - \theta_j = X_j^{\circ}$, $\langle S, M^{\circ}_{kj} \rangle = \varkappa^{\circ}_{kj}$, where X_j°, $\varkappa^{\circ}_{kj}$ are obtained via minimization of (2I) subject to the conditions (a),(b).

In addition to the real space $\mathfrak{h}(S)$ we introduce the complex space $\mathfrak{h}_{\mathbb{C}}(S)$ of operators $X, Y, \ldots$ (in general, non-self-adjoint) with the inner product $\sigma(X,Y)=\langle S, X^{*}\circ Y\rangle$. The following result, proved in [12], allows to reduce the minimization problem.

Lemma 2. The condition (b) holds iff there exist elements $Y_k \in \mathfrak{h}(S)$ and a symmetric operator $\mathcal{F}$ in $\mathfrak{h}(S)$ satisfying

(c) $X_k = \mathcal{F} Y_k$;

(d) $\varkappa_{kj} = \sigma(Y_k, \mathcal{F}(1-\mathcal{F})Y_j)$;

(e) $0 \leq \mathcal{F} \leq (1+\frac{i}{2}\vartheta)^{-1}$ in $\mathfrak{h}_{\mathbb{C}}(S)$.

In the last condition we mean by $\mathcal{F}, \vartheta$ the complex linear extensions of these operators onto $\mathfrak{h}_{\mathbb{C}}(S)$.

Now substituting (c), (d) into the expression (21), we get

$$\sum_{k,j} g_{kj}[\varkappa_{kj} + \sigma(X_k, X_j)] = \sum_{k,j} g_{kj}\,\sigma(Y_k, \mathcal{F} Y_j).$$

Thus the problem is reduced to the evaluation of

$$\min \sum_{k,j} g_{kj}\,\sigma(Y_k, \mathcal{F} Y_j),$$

over all $\mathcal{F}$, satisfying (e) and all Y_j satisfying

$$\sigma(L_j, \mathcal{F} Y_k) = \delta_{jk}.$$

Fix $\mathcal{F} \geq 0$ and consider minimization over Y_j. This is a classical multiple regression problem. The minimum is attained for

$$\vec{Y}^{\circ} = \| \sigma(L_k, \mathcal{F} L_j)\|^{-1} \vec{L},$$

where

$$\vec{Y}^{\circ} = \| Y_1^{\circ}, \dots, Y_n^{\circ} \|^T, \quad \vec{L} = \| L_1, \dots, L_n \|^T,$$

and it is equal to $\operatorname{Tr} G \| \alpha(L_k, \mathcal{F} L_j) \|^{-1}$. Thus we may write

$$\Sigma \geq \min \left\{ \operatorname{Tr} G \| \alpha(L_k, \mathcal{F} L_j) \|^{-1} : \ 0 \leq \mathcal{F} \leq \left(1 + \tfrac{i}{2} \mathcal{D}\right)^{-1} \right\} = \Sigma_0 . \quad (22)$$

Denote by $\mathcal{F}_0$ the solution of the problem on the right-hand side of (22) and introduce the matrix $\mathbb{F}_0 = \| \alpha(L_k, \mathcal{F} L_j) \|_{k,j=1}^{n}$. Then the optimal vector $\vec{X}^{\circ} = \| X_1^{\circ}, \dots, X_n^{\circ} \|^T$ and the matrix $\mathbb{K}^{\circ} = \| æ_{kj}^{\circ} \|$ are

$$\vec{X}^{\circ} = \mathbb{F}_0^{-1} \cdot \vec{L}^{\circ}; \qquad \mathbb{K}^{\circ} = \mathbb{F}_0^{-1} \cdot \mathbb{H} \cdot \mathbb{F}_0^{-1}, \quad (23)$$

where $\vec{L}^{\circ} = \| \mathcal{F}_0 L_1, \dots, \mathcal{F}_0 L_n \|^T$, $\mathbb{H} = \| \alpha(L_j, \mathcal{F}(1 - \mathcal{F}) L_k) \|$.

We can find explicitly the optimal $\mathcal{F}_0$ in some important particular cases. However, before doing it, we make an obvious but useful remark. Let $\mathfrak{M}$ be a subspace of $\mathfrak{h}(S)$ such that:

(f) $\mathfrak{M}$ is an invariant subspace of the commutation operator $\mathcal{D}$;

(g) $L_j \in \mathfrak{M}$, $j = 1, \dots, n$.

Then the quantity Σ_0 does not change if the operators $\mathcal{F}, \mathcal{D}$ in definition (22) of Σ_0 are considered as acting in $\mathfrak{M}$.

Using this remark we can easily prove that <u>in the Gaussian case the best unbiased measurement of the mean value is a canonical one</u> (this was proved earlier in a different way [12] under a strong extraneous restriction).

In fact, assume that the mean value of the Gaussian state is of the form $\mu(z) = \theta_1 \mu_1(z) + \dots + \theta_n \mu_n(z)$. Then, for an arbitrary $\vec{\theta}$, s.l.d. are $R(z_j) - \mu(z_j) \cdot I$, where z_j is defined by the equation [6]

$$\mu_j(z) = \alpha(z_j, z), \quad z \in \mathcal{Z}.$$

If we take the space of canonical observables $\mathcal{R}$ for $\mathfrak{M}$ in the previous remark, then conditions (f), (g) are fulfilled, so that

$$\Sigma_0 = \min\{\mathrm{Tr}\, G\, \|\alpha(z_j, \mathcal{F} z_k)\|^{-1} : 0 \leq \mathcal{F} \leq (1 + \tfrac{i}{2}\mathcal{D})^{-1}\},$$

where $\mathcal{F}, \mathcal{D}$ act in $\mathcal{Z}$. If $\mathcal{F}_0$ is a solution of this problem, then $M_j = R(z_j^0)$, where $\|z_j^0\| = \|\alpha(z_j, \mathcal{F}_0 z_k)\|^{-1} \cdot \|\mathcal{F} z_k\|$.

Since condition (b) coincides with (19), there exists a canonical measurement with the first moments $R(z_j^0)$ and the second central moments $\varkappa_{kj}^0 \cdot I$, which is an optimal measurement. Moreover, since the parameters of the measurement turned to be independent of the point $\vec{\theta}$ under consideration, this canonical measurement is uniformly optimal in the class of unbiased measurements.

We can calculate the bound Σ_0 in the case when <u>the subspace $\mathcal{L} \subset \mathfrak{h}(S)$ generated by the operators L_j is invariant under the commutation operator $\mathcal{D}$</u>.

Introduce the matrices

$$\mathbf{D} = \|\mathcal{D}(L_k, L_j)\|, \quad \mathbf{A} = \|\alpha(L_k, L_j)\|, \quad \mathbf{F} = \|\alpha(L_k, \mathcal{F} L_j)\|.$$

According to the preceding remark, we may consider the operator $\mathcal{F}$ as acting in $\mathcal{L}$. Then the condition $0 \leq \mathcal{F} \leq (1 + \frac{i}{2}\mathcal{D})^{-1}$, in the matrix form, turns into

$$0 \leq \mathbb{F} \leq \mathbb{A}(\mathbb{1} + \tfrac{i}{2}\mathbb{A}^{-1}\mathbb{D})^{-1} \qquad \text{or} \qquad \mathbb{F}^{-1} \geq \mathbb{A}^{-1} + \tfrac{i}{2}\mathbb{A}^{-1}\mathbb{D}\mathbb{A}^{-1}$$

and the minimized functional is equal to $\mathrm{Tr}\, \mathcal{G}\mathbb{F}^{-1}$. Introducing the real matrix $\mathbb{Y} = \mathbb{F}^{-1} - \mathbb{A}^{-1}$, we have $\mathbb{Y} \geq \frac{i}{2}\mathbb{A}^{-1}\mathbb{D}\mathbb{A}^{-1}$. It follows from Lemma I that the minimum is equal to

$$\mathrm{Tr}\left[\mathcal{G}\mathbb{A}^{-1} + \tfrac{1}{2}\, abs\,(\mathcal{G}\mathbb{A}^{-1}\mathbb{D}\mathbb{A}^{-1})\right],$$

and it is attained for

$$\mathbb{F}_0^{-1} = \mathbb{A}^{-1} + \mathbb{A}^{-1}\mathbb{D}\mathbb{A}^{-1} \cdot sgn\,(\mathcal{G}\mathbb{A}^{-1}\mathbb{D}\mathbb{A}^{-1}).$$

Using (23), we obtain

$$\vec{X}^{\circ} = \mathbb{A}^{-1}\vec{L}\,; \quad \mathbb{K}^{\circ} = \mathbb{A}^{-1}\mathbb{D}\mathbb{A}^{-1} \cdot sgn\,(\mathcal{G}\mathbb{A}^{-1}\mathbb{D}\mathbb{A}^{-1}).$$

Note that the expression for $\vec{X}^{\circ}$ has the same form as the classical expression for the vector of the best unbiased estimate. In the general case an optimal $\mathcal{F}_0$ will not leave $\mathcal{L}$ invariant; according to (23) the vector $\vec{X}^{\circ}$ has the form $\mathbb{A}^{-1}\vec{L} + \vec{\Psi}$, where the components of $\vec{\Psi}$ lie in the complement of $\mathcal{L}$.

Now we shall show that

$$\underline{\Sigma}_0 \geq \underline{\Sigma}_1 \;; \quad \underline{\Sigma}_0 \geq \underline{\Sigma}_2 ,$$

the equality being attained, in the first case, when $\mathcal{V}(L_k, L_j) \equiv 0$, and, in the second case, when $\mathcal{L}$ is invariant under $\mathcal{D}$.

The first inequality is most easily obtained by considering expression (21). From (b) it follows that $\| \varkappa_{kj} \| \geq 0$, thus

$$\Sigma_0 \geq \min \sum_{kj} g_{kj}\, \sigma(X_k, X_j),$$

where the minimum is taken over all X_j, satisfying (a). From the classical regression theory this minimum is equal to Σ_1. The equality is attained for $\mathbb{K}^\circ = 0,\ \vec{X}^\circ = \mathbb{A}^{-1}\cdot\vec{\mathbb{L}}$.

To prove the second inequality it is sufficient to show that $\mathbb{F} \leq \widetilde{\mathbb{A}}$ for all $\mathcal{F}$, satisfying $0 \leq \mathcal{F} \leq (1+\frac{i}{2}\vartheta)^{-1}$.

From the last inequality we shall have

$$\mathbb{F} = \| \sigma(L_j, \mathcal{F} L_k) \| \leq \| \sigma(L_j, (1+\tfrac{i}{2}\vartheta)^{-1} L_k) \|, \tag{24}$$

in the sense of the ordering of complex Hermitian matrices. We shall show that

$$\widetilde{\mathbb{A}} = \| \sigma(L_j, (1+\tfrac{i}{2}\vartheta^{-1}) L_k) \|. \tag{25}$$

Taking into account that

$$(1+\tfrac{i}{2}\vartheta)^{-1} = (1+\tfrac{1}{4}\vartheta^2)^{-1}\cdot(1-\tfrac{i}{2}\vartheta),$$

we have

$$\sigma(L_j, (1+\tfrac{i}{2}\vartheta)^{-1} L_k) = \sigma(\hat{L}_j, (1-\tfrac{i}{2}\vartheta)\hat{L}_k) = \langle S, \hat{L}_k \cdot \hat{L}_j \rangle,$$

where $\hat{L}_j = (1+\frac{1}{4}\vartheta^2)^{-1/2} L_j$. Using (II), we find

$$(1+\tfrac{1}{4}\vartheta^2)^{-1/2} X = \tfrac{1}{2}(S^{1/2} X S^{-1/2} + S^{-1/2} X S^{1/2}) = S^{-1/2}(X \circ S) S^{-1/2},$$

so that $\hat{L}_j = S^{-1/2} S_j' S^{-1/2}$ *. It follows that

$$\langle S, \hat{L}_k \cdot \hat{L}_j \rangle = m(S_k' \cdot S^{-1} \cdot S_j'),$$

and the inequality is proved. If $\mathcal{L}$ is invariant under $\mathcal{D}$, then by the preceding remark we may consider $\mathcal{F}$ as acting in $\mathcal{L}$. Then condition (24) is equivalent to $\mathcal{F} \leq (1+\frac{i}{2}\mathcal{D})^{-1}$, so that $\Sigma_2 = \Sigma_0$ in this case.

6. Gauge-invariant case. Assume that the dimension of the unknown parameter is even, $\vec{\theta} = (\xi_1, \eta_1, \ldots, \xi_m, \eta_m)$. We shall denote the corresponding s.l.d.'s by $L_j^\xi, L_j^\eta; j = 1, \ldots, m$. Let the following conditions be fulfilled

(h) $L_j^\eta = \mathcal{J} L_j^\xi, \quad L_j^\xi = -\mathcal{J} L_j^\eta,$

where $\mathcal{J}$ is the operator from the polar decomposition of the commutation operator $\mathcal{D}$ (see § 3);

(i) the weight matrix is such that the corresponding quadratic loss function is of the form

$$\sum_{k,j=1}^{m} g_{kj}' [(\hat{\xi}_k - \xi_k)(\hat{\xi}_j - \xi_j) + (\hat{\eta}_k - \eta_k)(\hat{\eta}_j - \eta_j)].$$

We shall call this case gauge-invariant since the situation is invariant under the action of the "gauge group of the first kind"

---

* The $\hat{L}_j$ is apparently one more noncommutative analogue of the logarithmic derivative. The "Cramér-Rao inequality", resulting from it, coincides with the inequality obtained via left logarithmic derivative.However, unlike the latter, the $\hat{L}_j$ is a self-adjont operator.

$\exp(t\mathcal{J}), t \in \mathbb{R}$. We prove the following proposition:

<u>In the gauge-invariant case the solution of problem (22) is</u> $\mathcal{F}_0 = (1+\frac{1}{2}|\vartheta|)^{-1}$. <u>Moreover,</u> $\Sigma_0 = \Sigma_2$ <u>in this case.</u>

To prove it we recall that $\Sigma_2 = \min\{\mathrm{Tr}\, G\widetilde{\mathbb{B}} : \widetilde{\mathbb{B}}$ is real symmetric matrix, $\widetilde{\mathbb{B}} \geq \widetilde{\mathbb{A}}^{-1}\}$ (cf.(6)). Put

$$\widetilde{\mathbb{B}} = \begin{Vmatrix} \widetilde{B}_{\xi\xi} & \widetilde{B}_{\xi\eta} \\ \widetilde{B}_{\eta\xi} & \widetilde{B}_{\eta\eta} \end{Vmatrix}, \quad \widetilde{\mathbb{A}} = \begin{Vmatrix} \widetilde{\mathbb{A}}_{\xi\xi} & \widetilde{\mathbb{A}}_{\xi\eta} \\ \widetilde{\mathbb{A}}_{\eta\xi} & \widetilde{\mathbb{A}}_{\eta\eta} \end{Vmatrix},$$

where all blocks are $m \times m$ - matrices. We have $\widetilde{B}^T_{\xi\xi} = \widetilde{B}_{\xi\xi}$, $\widetilde{B}^T_{\xi\eta} = \widetilde{B}_{\eta\xi}$, $\widetilde{B}^T_{\eta\eta} = \widetilde{B}_{\eta\eta}$. Taking into account (25), we have

$$\widetilde{\mathbb{A}}_{\xi\xi} = \| \sigma(L^{\xi}_j, (1+\tfrac{i}{2}\vartheta)^{-1} L^{\xi}_k) \|,$$

$$\widetilde{\mathbb{A}}_{\xi\eta} = \| \sigma(L^{\xi}_j, (1+\tfrac{i}{2}\vartheta)^{-1} L^{\eta}_k) \|,$$

$$\widetilde{\mathbb{A}}_{\eta\eta} = \| \sigma(L^{\eta}_j, (1+\tfrac{i}{2}\vartheta)^{-1} L^{\eta}_k) \|.$$

From (h) it follows that

$$\widetilde{\mathbb{A}}_{\xi\eta} = -\| \sigma(L^{\xi}_j, (1+\tfrac{i}{2}\vartheta)^{-1} L^{\eta}_k) \|, \quad \widetilde{\mathbb{A}}_{\eta\eta} = \widetilde{\mathbb{A}}_{\xi\xi}$$

(we use also the fact that $\mathcal{J}$ and ϑ commute). Since $(1+\frac{i}{2}\vartheta)$ is Hermitian, and $\mathcal{J}$ is skew-Hermitian in $\mathcal{H}_c(S)$, we obtain

$$\widetilde{\mathbb{A}}^*_{\xi\xi} = \widetilde{\mathbb{A}}_{\xi\xi}, \quad \widetilde{\mathbb{A}}^*_{\xi\eta} = -\widetilde{\mathbb{A}}_{\xi\eta} = \widetilde{\mathbb{A}}_{\eta\xi}.$$

The inequality $\widetilde{\mathbb{B}} \geq \widetilde{\mathbb{A}}^{-1}$ may be rewritten in the form

$$b\widetilde{\mathbb{B}}b^* + a\widetilde{\mathbb{A}}a^* \geq ab^*, \quad \forall a, b,$$

where $a = \|a_\xi, a_\eta\|$, $b = \|b_\xi, b_\eta\|$ are $2m$-dimensional rows with complex elements. Putting here $a_\eta = i a_\xi$, $b_\eta = i b_\xi$ and taking into account the relations between the blocks of $\tilde{A}$ and $\tilde{B}$, we get

$$b_\xi \left(\frac{\tilde{B}_{\xi\xi} + \tilde{B}_{\eta\eta}}{2} \right) b_\xi^* + a_\xi (\tilde{A}_{\xi\xi} + i \tilde{A}_{\eta\xi}) a_\xi^* \geq a_\xi b_\xi^*,$$

or, equivalently

$$\frac{1}{2} (\tilde{B}_{\xi\xi} + \tilde{B}_{\eta\eta}) \geq (\tilde{A}_{\xi\xi} + i \tilde{A}_{\eta\xi})^{-1}. \tag{26}$$

We have

$$\tilde{A}_{\xi\xi} + i \tilde{A}_{\eta\xi} = \| \sigma(L_j^\xi, (1 + \tfrac{i}{2} \vartheta)^{-1} (1 + i \mathcal{J}) L_k^\xi) \|.$$

Making use of the polar decomposition of the operator ϑ, we have

$$(1 + \tfrac{i}{2} \vartheta)^{-1} (1 + i \mathcal{J}) = (1 + \tfrac{1}{4} \vartheta^2)^{-1} (1 - \tfrac{i}{2} \vartheta)(1 + i \mathcal{J}) =$$

$$= (1 - \tfrac{1}{4} |\vartheta|^2)^{-1} (1 - \tfrac{1}{2} |\vartheta|)(1 + i \mathcal{J}) =$$

$$= (1 + \tfrac{1}{2} |\vartheta|)^{-1} (1 + i \mathcal{J}),$$

whence

$$\tilde{A}_{\xi\xi} + i \tilde{A}_{\eta\xi} =$$

$$= \| \sigma(L_j^\xi, (1 + \tfrac{1}{2} |\vartheta|)^{-1} L_k^\xi) \| + i \| \sigma(L_j^\xi, (1 + \tfrac{1}{2} |\vartheta|)^{-1} L_k^\eta) \|,$$

where both matrices on the right-hand side are real, the first being symmetric and the second skew-symmetric. Thus the right-hand side gives the decomposition of $\tilde{A}_{\xi\xi} + i \tilde{A}_{\eta\xi}$ into the real and the imaginary parts. It is known that the mapping

$$A \to \begin{Vmatrix} \mathrm{Re}\,\tilde{A} & \mathrm{Im}\,\tilde{A} \\ -\mathrm{Im}\,\tilde{A} & \mathrm{Re}\,\tilde{A} \end{Vmatrix},$$

transforming complex $m \times m$ -matrix $\tilde{A}$ into real $2m \times 2m$ -matrix, preserves algebraic operations and partial ordering of matrices, so that (26) is equivalent to

$$\left\| \begin{matrix} \frac{1}{2}(\tilde{\mathbb{B}}_{\xi\xi}+\tilde{\mathbb{B}}_{\eta\eta}) & 0 \\ 0 & \frac{1}{2}(\tilde{\mathbb{B}}_{\xi\xi}+\tilde{\mathbb{B}}_{\eta\eta}) \end{matrix} \right\| \geq \tilde{\mathbb{F}}^{-1} \equiv \left\| \begin{matrix} \tilde{\mathbb{F}}_{\xi\xi} & \tilde{\mathbb{F}}_{\xi\eta} \\ \tilde{\mathbb{F}}_{\eta\xi} & \tilde{\mathbb{F}}_{\eta\eta} \end{matrix} \right\|^{-1},$$

where $\tilde{\mathbb{F}}_{\xi\xi} = \| \alpha(L_j^{\xi}, (1+\frac{1}{2}|\vartheta|^{-1}L_k^{\xi})\|$, $\tilde{\mathbb{F}}_{\xi\eta} = \|\alpha(L_j^{\xi}, (1+\frac{1}{2}|\vartheta|)^{-1}L_k^{\eta})\|$, and so on.

Thus for the weight matrix $\mathcal{G}$, satisfying (i), we have $\Sigma_2 \geq \mathrm{Tr}\,\mathcal{G}\tilde{\mathbb{F}}^{-1}$. As shown below, the operator $\mathcal{F} = (1+\frac{1}{2}|\vartheta|)^{-1}$ satisfies condition (e) of Lemma 2. then it follows from (22) that $\mathrm{Tr}\,\mathcal{G}\tilde{\mathbb{F}}^{-1} \geq \Sigma_0$. Since always $\Sigma_0 \geq \Sigma_2$, we have $\Sigma_0 = \Sigma_2$. It follows also that $(1+\frac{1}{2}|\vartheta|)^{-1} = \mathcal{F}_0$, $\tilde{\mathbb{F}} = \mathbb{F}_0$, and the proposition is proved.

It remains to check that $(1+\frac{1}{2}|\vartheta|)^{-1} \leq (1+\frac{i}{2}\vartheta)^{-1}$ in $\mathfrak{h}_{\mathbb{C}}$. Since the function $x \mapsto x^{-1}$ is operator monotone, this is equivalent to the fact that $1+\frac{1}{2}|\vartheta| \geq 1+\frac{i}{2}\vartheta$ or $|\vartheta| \geq i\vartheta$ in $\mathfrak{h}_{\mathbb{C}}$, which follows directly from the definition of $|\vartheta|$.

If the best locally unbiased measurement exists, then its first and second operator moments satisfy relation (23) with

$$\tilde{\mathbb{F}}_0 = \left\| \begin{matrix} \mathbb{F}_{0\xi\xi} & \mathbb{F}_{0\xi\eta} \\ \mathbb{F}_{0\eta\xi} & \mathbb{F}_{0\eta\eta} \end{matrix} \right\|, \qquad \mathbb{H} = \left\| \begin{matrix} \mathbb{H}_{\xi\xi} & \mathbb{H}_{\xi\eta} \\ \mathbb{H}_{\eta\xi} & \mathbb{H}_{\eta\eta} \end{matrix} \right\|,$$

$$\mathbb{F}_{0\alpha\beta} = \| \alpha(L_j^{\alpha}, (1+\tfrac{1}{2}|\vartheta|)^{-1}L_k^{\beta})\|, \qquad \alpha, \beta = \xi, \eta;$$

$$\mathbb{H}_{\alpha\beta} = \tfrac{1}{2}\| \alpha(L_j^{\alpha}, (1+\tfrac{1}{2}|\vartheta|)^{-2}|\vartheta|\, L_k^{\beta})\|;$$

$$\vec{\mathbb{L}}^{0} = \| (1+\tfrac{1}{2}|\vartheta|)^{-1}L_1^{\xi}, \dots, (1+\tfrac{1}{2}|\vartheta|)^{-1}L_m^{\xi}, \\ (1+\tfrac{1}{2}|\vartheta|)^{-1}L_1^{\eta}, \dots, (1+\tfrac{1}{2}|\vartheta|)^{-1}L_m^{\eta} \|.$$

These formulas have an interesting application in the quantum communications theory which will be considered elsewhere.

References

I. Dirac P.A.M., The principles of quantum mechanics, 4th edition, Calderon Press, Oxford, 1959.

2. von Heumann J., Mathematical foundations of quantum mechanics, Princeton University Press, 1955.

3. Robertson H.P., The uncertainty principle, Physical Review, 34, 163-164, 1929.

4. Helstrom C.W., "Simultaneous measurement" from the standpoint of quantum estimation theory. Foundations of physics, 4, 453--463, 1974.

5. Holevo A.S., Statistical problems in quantum physics, Proc. of the Second Japan-USSR Sympos. on Probability Theory (Kyoto,1972); Lecture Notes in Mathematics, 33o, Springer-Verlag, 1973.

6. Holevo A.S., Statistical decision theory for quantum systems. J. Multivariate Analysis, 3, 337-394, 1973.

7. Klauder J.R., Sudarshan E.C.G. Fundamentals of quantum optics, W.A. Benjamin, Inc. New York-Amsterdam 1968.

8. Čentcov N.N., Statistical decision rules and optimal inference, Moscow, 1972 (in Russian).

9. Helstrom C.W., Quantum detection and estimation theory. J.Statist. Phys. I, 231-252, 1969.

10. Yuen H.P.H. and Lax M., Multiple-parameter quantum estimation and measurement of non-self-adjoint observables, IEEE Trans. inf. theory, IT-19, 74o-75o, 1973.

11. Belavkin V.P., Linear estimation of noncommuting quantum observables under their indirect measurement, Radiotechnika i Electronika, 17, 2533-254o, 1972 (In Russian).

12. Holevo A.S., Some statistical problems for quantum Gaussian states, IEEE Trans. inf. theory, IT-21, 533-543, 1975.

Steklov Mathematical Institute
Academy of Sciences of the USSR

Moscow

Test of hypotheses for distributions with monotone likelihood ratio: case of vector valued parameter.

by

Goro Ishii
Osaka City University

1. Introduction.

To tests of statistical hypotheses with regard to a vector valued parameter, the concept of monotone likelihood ratio is applied here. The concept for a family of distributions with a real parameter is introduced by Karlin [2] and Rubin [3], and is used to derive admissible decision procedures. Since then S. Karlin wrote many papers about so called Polya type distributions which are an extension of the notion of distributions with monotone likelihood ratio. But all such extensions fall in the case of real parameter and do not include the vector valued parameter case. Here a generalization of the concept to the case of vector valued parameter is given. In terms of partial orders on sample space and parameter space derived from convex cones in respective spaces, a concept of monotone likelihood ratio for vector-valued parameters is defined. When the null hypotheses and the alternative hypotheses can be expressed as partial order relations between parameter points, some admissible test procedures are derived. Examples given in the section 3 deal with exponential family and distribution of matrix variates derived from multivariate normal samples (cf. James [1]), where the convex cones are either the positive orthant of R^n or the set of all semi-positive definite matrices.

2. Monotone Likelihood Ratio.

Let $\mathfrak{X} \subset R^n$ be a sample space and $\Delta \subset R^n$ a convex cone where $0 \notin \Delta$. A partial order of $X \in \mathfrak{X}$ is defined as follows:

$$X_1 < X_2 \text{ if } X_2 - X_1 \in \Delta.$$

It is called that X_1 is samller than X_2 in the sense Δ.

Let $\Theta \subset R^m$ be a parameter space and $\Gamma \subset R$ a convex cone where $0 \notin \Gamma$. A partial order of $\theta \in \Theta$ is defined as follows:

$$\theta_1 < \theta_2 \text{ if } \theta_2 - \theta_1 \in \Gamma.$$

It is called that θ_1 is samller than θ_2 in the sense Γ.

Let $f(x,\theta)$ be a probability density function of a random vector $X \in \mathfrak{X}$ with respect to a certain σ-finite measure μ. It is assumed that the carrier of $f(x,\theta)$ is independent of θ and $f(x,\theta)$, $\theta \in \Theta$ are dominated by each other.

Definition. A family of distributions $P = \{f(x,\theta) : \theta \in \Theta\}$ is said to have monotone likelihood ratio in the sense Γ and Δ, if for any $\theta_2 > \theta_1$ and $X_2 > X_1$, the density functions satisfy

$$\frac{f(x_2 , \theta_2)}{f(x_2 , \theta_1)} > \frac{f(x_1 , \theta_2)}{f(x_1 , \theta_1)}. \tag{1}$$

At first we shall consider the problem of testing a hypothesis

$$H_1 : \theta \leqq \theta_0$$

against an alternative hypothesis

$$K_1 : \theta > \theta_1$$

where $\theta_0 < \theta_1$.

Let ϕ be the most powerfull test of level α for the simple hypothesis θ_0 against the simple alternative hypothesis θ_1 derived from Neyman-Pearson's Lemma. That is,

$$\phi(x) = \begin{cases} 1 & \text{when} \quad x \in C \\ \gamma & \text{when} \quad x \in D \\ 0 & \text{when} \quad x \in E \end{cases} \tag{2}$$

where

$$C = \{x : \frac{f(x,\theta_1)}{f(x,\theta_0)} > d\},$$

$$D = \{x : \frac{f(x,\theta_1)}{f(x,\theta_0)} = d\},$$

and

$$E = \{x : \frac{f(x,\theta_1)}{f(x,\theta_0)} < d\}.$$

For any $\theta_3 > \theta_2 \in H_1 \cup K_1$, define a function

$$h(x;\theta_3 ,\theta_2) = \frac{f(x,\theta_3)}{f(x,\theta_2)}.$$

The function h is an increasing function of X in the sense Δ.
It is assumed here that

$$\begin{aligned} &\operatorname*{ess\,sup}_{X \in \mathfrak{X}} h(x;\theta_3 ,\theta_2) > \operatorname*{ess\,sup}_{X \in D} h(x;\theta_3 ,\theta_2) \\ &\quad > \operatorname*{ess\,inf}_{X \in D} h(x;\theta_3 ,\theta_2) > \operatorname*{ess\,inf}_{X \in \mathfrak{X}} h(x;\theta_3 ,\theta_2). \end{aligned} \tag{3}$$

Where ess sup is the essentially supremum with respect to the measure μ.

We also make the following assumption:

(4) If $h(x_a ,\theta_3 ,\theta_2) > h(x_b ,\theta_3 ,\theta_2)$, then there is some $x_c \in \{x : h(x,\theta_3 ,\theta_2) = h(x_b ,\theta_3 ,\theta_2)\}$ such that $x_a > x_c$.

<u>Theorem</u> 1. Let P have monotone likelihood ratio, and satisfy the conditions (3) and (4).
For testing H_1 against K_1, the test ϕ given by (2) is
(i) level α : $E_\theta \phi(x) \leqq \alpha$ for all $\theta \in H_1$,
(ii) the power function $\beta(\theta) = E_\theta \phi(x)$ is monotone increasing in the sense Γ, that is, if $\theta_3 > \theta_2$, then $\beta(\theta_3) > \beta(\theta_2)$ unless $\beta(\theta) = 1$ for $\theta \in H_1 \cup K_1$. and
(iii) admissible, if $P_{\theta_0}(x \in D) = 0$.

<u>Proof</u>.
(a) If the value of h is greater than 1 for all $x \in D \cup C$, then

$$\beta(\theta_3) = \int \phi(x) f(x,\theta_3) d\mu(x) = \int \phi(x) h(x;\theta_3,\theta_2) f(x,\theta_2) d\mu(x)$$
$$> \int \phi(x) f(x,\theta_2) d\mu(x) = \beta(\theta_2).$$

(b) If the value of h is smaller than 1 for all $X \in D \cup E$, then

$$1-\beta(\theta_3) = \int (1-\phi(x)) f(x,\theta_3) d\mu(x) = \int (1-\phi(x)) h(x;\theta_3,\theta_2) f(x,\theta_2) d\mu(x)$$
$$< \int (1-\phi(x)) f(x,\theta_2) d\mu(x) = 1-\beta(\theta_2).$$

That is,

$$\beta(\theta_3) > \beta(\theta_2).$$

(c) When the value of h is greater than 1 for some part of D and smaller than 1 for another part of D, the sample space $\mathfrak{X}$ is divided into four parts:

$$\mathfrak{X}_1 = \{x : h(x;\theta_3,\theta_2) > M\}$$
$$\mathfrak{X}_2 = \{x : m \leqq h(x;\theta_3,\theta_2) \leqq M\} \cap (C \cup D)$$
$$\mathfrak{X}_3 = \{x : m \leqq h(x;\theta_3,\theta_2) \leqq M\} \cap E$$
$$\mathfrak{X}_4 = \{x : h(x;\theta_3,\theta_2) < m\}$$

where

$$M = \operatorname*{ess\,sup}_{X \in D} h(x;\theta_3,\theta_2)$$

and

$$m = \operatorname*{ess\,inf}_{X \in D} h(x;\theta_3,\theta_2).$$

Then by the assumption (4)

$$\mathfrak{X}_1 \subset C,\ \mathfrak{X}_4 \subset E,\ \mathfrak{X}_1 \cup \mathfrak{X}_2 = C \cup D,\ \mathfrak{X}_3 \cup \mathfrak{X}_4 = E, \text{ and } \mathfrak{X} = \mathfrak{X}_1 \cup \mathfrak{X}_2 \cup \mathfrak{X}_3 \cup \mathfrak{X}_4.$$

Set

$$a_1 = \int_{\mathfrak{X}_1} \phi(x) f(x,\theta_2) d\mu(x), \qquad b_1 = \int_{\mathfrak{X}_1} \phi(x) f(x,\theta_3) d\mu(x),$$

$$a_2 = \int_{\mathfrak{X}_2} \phi(x) f(x,\theta_2) d\mu(x), \qquad b_2 = \int_{\mathfrak{X}_2} \phi(x) f(x,\theta_3) d\mu(x),$$

$$a_3' = \int_{\mathfrak{X}_3} (1-\phi(x)) f(x,\theta_2) d\mu(x), \qquad b_3' = \int_{\mathfrak{X}_3} (1-\phi(x)) f(x,\theta_3) d\mu(x),$$

$$a_4 = \int_{\mathfrak{X}_4} (1-\phi(x)) f(x,\theta_2) d\mu(x), \qquad b_4 = \int_{\mathfrak{X}_4} (1-\phi(x)) f(x,\theta_3) d\mu(x),$$

$$a_3'' = \int_D (1-\phi(x)) f(x,\theta_2) du(x), \qquad b_3'' = \int_D (1-\phi(x)) f(x,\theta_3) d\mu(x),$$

$$a_3 = a_3' + a_3'' . \qquad b_3 = b_3' + b_3'' .$$

Then

$$\beta(\theta_2) = a_1 + a_2 , \qquad \beta(\theta_3) = b_1 + b_2$$

$$a_1 + a_2 + a_3 + a_4 = 1, \qquad b_1 + b_2 + b_3 + b_4 = 1.$$

And from the assumption (3)

$$a_1 , b_1 , a_4 , b_4 > 0.$$

From the definition of $\mathfrak{X}_i$, $i = 1, 2, 3$ and 4, we have

$$\frac{b_1}{a_1} > \frac{b_2}{a_2} , \qquad \frac{b_1}{a_1} > \frac{b_3}{a_3}$$

$$\frac{b_2}{a_2} > \frac{b_4}{a_4} , \qquad \frac{b_3}{a_3} > \frac{b_4}{a_4}$$

If $a_i = 0$, then $b_i = 0$, $i = 3, 4$ and in such case we shall omit b_i/a_i in the followings. Then there are two possible cases:

Case 1.

$$\frac{b_1}{a_1} > \frac{b_2}{a_2} \geqq \frac{b_3}{a_3} > \frac{b_4}{a_4} ,$$

and

Case 2.

$$\frac{b_1}{a_1} > \frac{b_3}{a_3} > \frac{b_2}{a_2} > \frac{b_4}{a_4} .$$

In the case 1

$$\frac{b_1 + b_2}{a_1 + a_2} > \frac{b_3 + b_4}{a_3 + a_4} \quad \text{and} \quad \frac{b_1 + b_2 + b_3 + b_4}{a_1 + a_2 + a_3 + a_4} = 1.$$

Then

$$b_1 + b_2 > a_1 + a_2 .$$

In the case 2

$$\frac{b_1 + b_3}{a_1 + a_3} > \frac{b_1 + b_2}{a_1 + a_2} > \frac{b_3 + b_2}{a_3 + a_2} > \frac{b_3 + b_4}{a_3 + a_4} .$$

$$b_1 + b_2 > a_1 + a_2$$

Hence, in both cases,

$$\beta(\theta_3) > \beta(\theta_2).$$

This shows the part (ii) and the part (i). If $P_{\theta_0}(X \in D) = 0$, then the part (iii) is clear because at the point θ_1, the test ϕ is the best and unique.

Next we shall consider the test of

$$H_2 : \theta \leq \theta_0$$

against

$$K_2 : \theta > \theta_0.$$

Theorem 2. Under the same assumptions as in Theorem 1, for testing H_2 against K_2, the test ϕ given by (2) for any $\theta_1 \in K_2$ is

(1) level α,

(ii) the power function $\beta(\theta)$ is monotone increasing in the sense Γ, and

(iii) admissible, if $P_{\theta_0}(X \in D) = 0$.

Proof. In the proof of $\beta(\theta_3) > \beta(\theta_2)$ in Theorem 1, the relation $\theta_3 > \theta_2$ is used. (i), (ii), and (iii) of this theorem follow by the same way.

A set in $\mathfrak{X}$ or Θ is called type A if any two points on the set are not greater than or smaller than each other in the sense Δ or Γ. For a type A set $A \subseteq \Theta$, the notation $\theta < A$ means $\theta < \theta'$ for some $\theta' \in A$. For the test ϕ given by (2), set

$$\Xi = \{\theta : E_\theta \phi(x) = \alpha\}.$$

The set Ξ is of type A and $\theta_0 \in \Xi$. We shall consider the test of

$$H_3 : \theta \leq \Xi,$$

against

$$K_3 : \theta > \Xi.$$

Theorem 3. Under the same assumptions as in Theorem 1, for testing H_3 against K_3, the test ϕ is

(i) level α,

(ii) the power function $\beta(\theta)$ is monotone increasing,

and

(iii) admissible, if $P_{\theta_0}(X \in D) = 0$.

The proof is easy.

In the inequality (1), if > turns to $\geq$, then the family is said to have monotone likelihood ratio in the weak sense. In this case the part (ii) of Theorems is altered to : (ii') if $\theta_3 > \theta_2$ then $\beta(\theta_3) \geq \beta(\theta_2)$.

In the above theorems, under the condition $P_{\theta_0}(X \in D) = 0$, the admissibility is proved. If $P_{\theta_0}(X \in D) \neq 0$, there are many equivalent tests for testing θ_0 against θ_1. For the problems H_i against K_i , i= 1, 2, 3, in some cases all equavalent tests are admissible and in some other cases some tests are not admissible.
We can construct such a case in the example 3.

3. Examples.

Example 1. A special case of noncentral Wishart distribution. If the m × n matrix Y has independent normally distributed columns with covariance σ and E(Y) = M, then the distribution of $X = YY^T$ is

$$f = \exp\{-\tfrac{1}{2}\mathrm{tr}(\sigma^{-1}MM^T)\}\,{}_0F_1(\tfrac{n}{2}, \tfrac{1}{4}\sigma^{-1}MM^T\sigma^{-1}X)$$

$$\frac{1}{2^{\frac{mn}{2}}\Gamma_m(\frac{n}{2})|\sigma|^{\frac{n}{2}}}\exp\{-\tfrac{1}{2}\mathrm{tr}\sigma^{-1}X\}|X|^{\frac{1}{2}(n-m-1)}.$$

If rank M = 1,

$${}_0F_1 = \sum_{k=0}^{\infty}(\mathrm{tr}\tfrac{1}{4}\sigma^{-1}MM^T\sigma^{-1}X)^k \quad (K!(\tfrac{n}{2})_k)$$

and f is written as

$$f = \sum_k h(k,\theta)g(x,k)$$

where $\theta = MM^T$,

$$h = \frac{(\mathrm{tr}\frac{1}{2}\sigma^{-1}\theta)^k}{k!}\exp\{-\tfrac{1}{2}\mathrm{tr}\sigma^{-1}\theta\}$$

$$g = \frac{(\mathrm{tr}\frac{1}{4}\sigma^{-1}\theta\sigma^{-1}X)^k|X|^{\frac{1}{2}(n-m-1)}}{(\frac{n}{2})_k(\mathrm{tr}\frac{1}{2}\sigma^{-1}\theta)^k 2^{\frac{mn}{2}}\Gamma_m(\frac{n}{2})|\sigma|^{\frac{n}{2}}}\exp\{-\tfrac{1}{2}\sigma^{-1}X\}.$$

The following lemma is a special case of a result in [3].

Lemma 1. If $f(x,\theta)$ has a form

$$f = \sum_k h(k,\theta)g(x,k)$$

and $h(k,\theta)$ has monotone likelihood ratio with respect to θ and $g(x,k)$ has monotone likelihood ratio with respect to k, then f has monotone likelihood ratio.

In this example, the sample space $\mathcal{X}$ ={ all positive definite m × m matrices} and the parameter space Θ ={all semi-positive definite m × m matrices}. We shall take $\Delta = \Gamma$ = the set of all semi-positive definite symmetric matrices. Using the partial orders derived from Δ and Γ and Lemma 1, when rank M = 1 noncentral Wishart distribution has monotone likelihood ratio.

For the test of

$$H : \theta = 0 \text{ or } M = 0$$
$$\text{vs } K : \theta > 0 \text{ or } M \neq 0$$

or "each column of M lies on a given line which passes through the origin" the critical region is a set

$$\operatorname{tr}\sigma^{-1}MM^T\sigma^{-1}X > \text{constant.}$$

Example 2. A special case of noncentral multivariate F. If the matrix variate X is m × p and Y is m × n with $p \leq m \leq n$, if the columns are all independently normally distributed with covariance σ, and if $E(X) = M$, $E(Y) = 0$, then the distribution of

$$F = X^T(YY^T)^{-1}X$$

depends upon $\theta = M^T\sigma^{-1}M$, and is

$$f = \exp\{-\tfrac{1}{2}\operatorname{tr}\theta\}\,{}_1F_1(\tfrac{1}{2}(p+m):\tfrac{1}{2}m\theta(1+F^{-1})^{-1})$$
$$\cdot\frac{\Gamma_p(\frac{1}{2}(p+m))}{\Gamma_p(\frac{1}{2}m)\Gamma_p(\frac{1}{2}(p+n-m))}\;\frac{|F|^{\frac{1}{2}(m-p-1)}}{|1+F|^{\frac{1}{2}(p+n)}}$$

If rank M = 1,

$${}_1F_1 = \sum_{k=0}^{\infty}\frac{(\frac{p+m}{2})_k(\operatorname{tr}\frac{1}{2}\theta(1+f^{-1})^{-1})^k}{k!\,(\frac{m}{2})_k}.$$

By the same reason as in noncentral Wishart distribution, when rank M = 1, f has monotone likelihood ratio.

Example 3. Independent Poisson distribution. Let $X = (X_1,\cdots,X_n)$ be a set of n random variables which follow independently Poisson distributions with parameters

$$\theta = (\theta_1,\cdots,\theta_n).$$

Set

$$\Delta = \{X : X > 0\}$$
$$\Gamma = \{\theta : \theta \geq 0\}$$

where $Y > 0$ and $Y \geq 0$ mean that all components of Y are positive and nonnegative and $Y \neq 0$ respectively. By the partial orders derived from Δ and Γ, the family has monotone likelihood ratio.

Example 4. If $X = (X_1, \cdots, X_n)$ follows the multinomial distribution

$$f = \frac{m!}{x_1! \cdots x_m! (m-\sum x_i)!} \theta_1^{x_1} \cdots \theta_m^{x_m} (1-\sum \theta_i)^{m-\sum x_i}$$

then f has monotone likelihood ratio where Δ and Γ are the same as in example 3.

In the above two examples and the next example, the density function is an exponential type

$$f = \exp\{\theta_1 x_1 + \cdots + \theta_n x_n + b(\theta) + k(x)\}.$$

Then

$$h(x;\theta_3,\theta_2) = \text{constant } \exp\{x_1(\theta_{12}-\theta_{11}) + \cdots + x_n(\theta_{n2}-\theta_{n1})\}$$

and the critical region has a form

$$\{X : x_1 a_1 + \cdots + x_n a_n \geqq \text{constant}\}$$

Example 5. Multivariate normal with known covariance. Let $x = (x_1, \cdots, x_n)$ be a random variable which follows the multivariate normal distribution $N(\theta, \sigma)$ where σ is known. Set

$$\Delta = \{X : X\sigma^{-\frac{1}{2}} > 0\},$$

$$\Gamma = \{\theta : \theta\sigma^{-\frac{1}{2}} \geqq 0\}.$$

By the partial orders derived from Δ and Γ, the family has monotone likelihood ratio.

In this example the condition (3) is not satisfied. But by the special form of h and critical region and the fact that the set of θ with constant power has a form $c_1\theta_1 + \cdots + c_n\theta_n$ = constant, the results of the theorems still hold true.

Example 6. (K. Takeuchi) Double exponential distribution. Let X_i, $i = 1, 2, \cdots, n$ be a sample from a distribution $f(x,\theta) = \frac{1}{2}e^{-|x-\theta|}$. The family has monotone likelihood ratio in the weak sense where $X \in R^n$, $\theta \in R$, Δ = positive orthant and $\Gamma = (0, \infty)$. By Theorem 2 we can derive some admissible test procedures for testing $H : \theta \leqq \theta_0$ vs $K : \theta > \theta_0$.

References

1. James, A. T. (1964) Distributions of matrix variates and latent roots derived from normal samples. Ann. Math. Statist. 35. 475-501.

2. Karlin, S. and H. Rubin. (1956) The theory of decision procedures

for distributions with monotone likelihood ratio. Ann. Math. Statist. 27. 272-299.

3. Karlin, S. (1957) Decision theory for Polya type distributions. Case of two actions, 1. Proceedings of the Third Berkeley Symposium on Probability and Statistics. 115-128.

4. Lehmann, E. L. (1959) Testing Statistical Hypotheses. Wiley.

CRITERIA OF ABSOLUTE CONTINUITY OF MEASURES CORRESPONDING TO MULTIVARIATE POINT PROCESSES

Yu.M.Kabanov, R.S.Lipcer, A.N.Širyaev

1. Introduction. Main result.

1. Let $(\Omega, \mathcal{F})$ be a measurable space with an increasing right-continuous family $(\mathcal{F}_t)$, $t \geq 0$, of σ - subalgebras of $\mathcal{F}$.

According to Jacod [1], we define a multivariate point process as a sequence (T_n, X_n), $n \geq 1$ where T_n are stopping times with respect to the family $(\mathcal{F}_t)$, $t \geq 0$, such that

1) $T_1 > 0$,

2) $T_n < T_{n+1}$ if $T_n < \infty$,

3) $T_n = T_n$ if $T_n = \infty$,

4) $\lim_n T_n = T_\infty \leq \infty$,

X_n are $\mathcal{F}_{T_n}$ - measurable with values in a measurable space $(E, \mathcal{E})$.

A multivariate point process (T_n, X_n) is characterized competely by the following random measure on $]0,\infty] \times E$:

$$\mu(]0,t] \times \Gamma) = \sum_{n \geq 1} I_{\{T_n \leq t\}} I_{\{X_n \in \Gamma\}}, \quad \Gamma \in \mathcal{E}, \ t > 0 .$$

When E consists of one point only, then, in studying such multivariate point processes, it is convenient to consider the counting process $N_t = \mu(]0,t] \times E)$. In particular, if $E = \{1\}$, the difference $T_{n+1} - T_n$, $n \geq 1$, are independent and have exponential distri butions, the counting process N_t is a Poisson process.

The aim of this paper is to obtain necessary and sufficient conditions for the absolute continiity of distributions corresponding to random measures μ.

2. We will assume that X_n takes values in a measurable space $(E_\Delta, \mathcal{E}_\Delta)$, where $E_\Delta = E \cup \Delta$, E is a Lusin space (i.e. E is Borel subset of a compact metric space) and Δ is a point.

Put $\mathcal{F}_t^\mu = \sigma\{\mu(]0,s]\times\Gamma \;;\; s\le t \;,\; \Gamma\in\mathcal{E}\}$.

The family ($\mathcal{F}_t^\mu$), $t\ge 0$, is an incressing right-continuous (see [1], compare with Lemma 1 and remark to it in [2]), at least if $(\Omega, \mathcal{F})$ is "rich" enough.

Let $\mathcal{P}$ be the σ - algebra of predictable sets of $\Omega\times]0,\infty]$ generated by the mappings $(\omega, t) \rightsquigarrow Y_t(\omega)$ which are $\mathcal{F}_t$ - measurable in ω and left-continuous in t. Let P be a probability measure on $(\Omega, \mathcal{F})$.

<u>Definition.</u> A random measure $\nu(\omega; dt, dx)$ on $[0,\infty[\times E$ is called a predictable projection (or a compensator) of a random measure μ with respect to P, if, for any nonnegative $\mathcal{P}\times\mathcal{E}$ - measurable function $X(t,x)$, the process $\int_0^t\int_E X(s,x)\,\nu(ds,dx)$, $t\ge 0$ is predictable and

$$M\int_0^\infty\int_E X(t,x)\mu(dt,dx) = M\int_0^\infty\int_E X(t,x)\nu(dt,dx) \quad . \tag{1.1}$$

According to Theorem 2.1 in [1], there exists one and only one (up to P -equivalence) predictable measure corresponding to this definition.

Put $\nu^c(]0,t]\times\Gamma) = \int_0^t\int_\Gamma I_{\{\nu(\{s\})E)=0\}}\,\nu(ds,dx)$,

$t\ge 0$, $\Gamma\in\mathcal{E}$. Remind essential properties of the measure, [1].

1) For any $\Gamma\in\mathcal{E}$, the process $\nu(]0,t]\times\Gamma)$, $t\ge 0$ is increasing, right-continuous and predictable (i.e. $\mathcal{P}$ - measurable),

2) $\nu(\{t\}\times E)\le 1$,

3) $\nu([T_\infty, \infty[\times E) = 0$,

4) $\nu(]0,T]\times E) < \infty$ for any stopping time $T<T_\infty$ (P-a.s),

5) Let $\mathcal{F}_t = \mathcal{F}_t^\mu$, ν be a predictable random mesure which satisfies properties 1) - 3). There exists a probability measure P on $(\Omega, \mathcal{F}_\infty^\mu)$ such that ν is a predictable projection of the random measure μ with respect to P .

6) If P and P' are probability measures on $(\Omega, \mathcal{F}_\infty^\mu)$, ν and ν' are predictable projections with respect to P and P' respectively such that P - a.s. (or P' - a.s.) $\nu'=(]0,t\wedge T]\times\Gamma) = \nu(]0,t\wedge T]\times\Gamma)$, where T is a stopping time relative to the family $(\mathcal{F}_t^\mu)$, $t\geq 0$, $\Gamma\in\mathcal{E}$ then the restrictions P_T and P_T' of the measures P and P' to $\mathcal{F}_T^\mu$ coincide.

It is natural to call the probability measure P on $(\Omega, \mathcal{F}_\infty^\mu)$, corresponding to a compensator ν the distribution of a random measure μ with compensator ν .

4. To formulate the main result, we will assume that $\mathcal{F}_t = \mathcal{F}_t^\mu$, $t\geq 0$ P and P' are probability measures on $(\Omega, \mathcal{F}_\infty^\mu)$ P_T and P_T' are restrictions of P and P' respectively to the σ -algebra $\mathcal{F}_T^\mu$ (T is stopping time). The random measure μ has predictable projections ν and ν' corresponding to the measures P and P' respectively.

We will use the following notation: μ_T, ν_T and ν_T' for the restrictions of random measures μ , ν and ν' to $]0,T]\times E$.

The formula

(1.2) $$d\nu' = Y d\nu \qquad (P'\text{- a.s.}) ,$$

where $Y(t,x)$ is a nonnegative and finite $\mathcal{P}\times\mathcal{E}$ - measurable function, will be used if

$$P'\{\nu'(]0,t]\times\Gamma) = \int_0^t\int_\Gamma Y(s,x)\nu(ds,dx),\ t\geq 0,\ \Gamma\in\mathcal{E}\} = 1 .$$

The similar formula $d\nu'_{|T} = Y d\nu_T$ (P'_T - a.s.) will be used also.

Theorem 1. Let $T < T_\infty$ P a.s. and P' - a.s. Then the following conditions are necessary and sufficient for absolute continuity P' with respect to P_T : P'_T - a.s.

(i) $d\nu'_T = Y d\nu_T$,

(ii) $\nu_T(\{t\}\times E) = 1 \Rightarrow \nu'_T(\{t\}\times E) = 1$,

(iii) $\nu_T(\{t\}\times E) \leq 1$,

(iv) $\nu_T(]0,T]\times E) < \infty$.

Theorem 2. Let $\nu = \nu^c P$ and P' -a.s. Then the following conditions are necessary and sufficient for absolute continuity of P' with respect to P : P' - a.s.

(i) $d\nu' = Y d\nu$,

(ii) $\int_0^\infty\int_E \frac{(Y(t,x)-1)^2}{1+|Y(t,x)-1|}\ \nu(dt,dx) < \infty$.

Let $E = \{1\}$ and P^π be the mesure corresponding to a Poisson process, i.e. the compensator ν of the random measure μ with respect to P^π is given by the formuly $\nu(]0,t]\times\{1\}) \equiv t$. Let $A_t = \nu'(]0,t]\times\{1\})$ be the compensator of a random measure μ with respect to P^N (P^π_T, P^N_T are restrictions of measures P^π, P^N to the σ-algebra $\mathcal{F}^\mu_T$). From Theorem 1 and 2, we obtain the following result.

Theorem 3. 1) The measure P^N is absolute continuous with respect to the measure P^π of a Poisson process if and only if

(i) $A_t = \int_0^t \lambda_s ds$, $t \geq 0$, P^N - a.s. .

where (λ_t , $\mathcal{F}_t^{\mu^0}$), $t \geq 0$ is a nonnegative predictable process;

(ii) $\int_0^\infty \frac{(\lambda_s - 1)^2}{1 + |\lambda_s - 1|} ds < \infty$ P^N - a.s.

2) Let $T < T_\infty$ P^π-a.s. and P^N-a.s. Then the mesure P_T^N is absolutely continuous with respect to the measure P_T^π if and only if the condition (i) is satisfied.

6. The problem of absolute continuity of measures correspinding to similar (and even more general) random processes was studied by Skorokhod, who obtained criteria of absolute continuity of measures corresponding to stochastically continuous random processes with independent increments ([3] ,Ch.9, 46). The result of Skorokhod were extended in fifferent directions. So, for exemple, necessary, sufficient, necessary and sufficient conditions of absolute continuity of measures corresponding to different types of random processes were obtained with the help of fundamental Girsanov's theorem [4] and its different generalization (see [1], [2],[5 - 17])

The problem of absolute continuity of measures corresponding to multivariate point processes was studied by Jacod [1] , who obtained some necessary conditions.Necessary and sufficient conditions of adsolute continuity of mesures counting processes with continuous compensators on finite time interval were obtained in our paper [2] .

It is necessary to note that, when the present paper was presented for publication, the authors got a preprint by Jacod and Mimin [19] , which contains the result analogous to Theorem 2 .

2. Auxiliary results

1. Let $X(t,x)$ be a $\mathcal{P} \times \mathcal{E}$ - measurable function, $|X(t,x)| < \infty$

If P - a.s.

(2.1) $$\int_0^t \int_E |X(s,x)|\,\nu(ds,dx) < \infty, \quad t < T_\infty$$

then, from (1.1), Lemma 4.4 and Proposition 5.3 in [1], it follows that the values $R_t = \int_0^t \int_E X(s,x)\mu(ds,dx)$, $\mathcal{I}_t = \int_0^t \int_E X(s,x)[\mu(ds,dx) - \nu(ds,dx)]$, $t < T_\infty$, are well defined and finite. The process $(\mathcal{I}_t, \mathcal{F}_t)$, $t \geq 0$, is a local martungale (more precisely, there exists the sequence of stopping times $S_n \uparrow T_\infty$ such that the process $(\mathcal{I}_{t \wedge S_n}, \mathcal{F}_t)$, $t \geq 0$, is a uniformly integrable martingale for each n. If $\nu = \nu^c$ and inequality (2.1) is satisfied for all $t \leq T_\infty$ P-a.s., then $|R_t| < \infty$, $|\mathcal{I}_t| < \infty$ P - a.s. for all $t \leq T_\infty$ also. If, instead of (2.1), we have the inequality

(2.2) $$M \int_0^\infty \int_E |X(t,x)|\,\nu(dt,dx) < \infty,$$

then it follows from (1.1) that the process $(\mathcal{I}_t, \mathcal{F}_t)$, $t \geq 0$, is a uniformly integrable martingale.

Let P-a.s.

$$\int_0^t \int_E X^2(s,x)\nu(ds,dx) - \sum_{s \leq t} \left(\int_E X(s,x)\nu(\{s\},dx) \right)^2 < \infty, \quad t < T_\infty.$$

In this case the values $\mathcal{I}_t$, $t < T_\infty$, are well defined and finite ([20], see also Theorem 7 in [2]) and the process $(\mathcal{I}_t, \mathcal{F}_t)$, $t \geq 0$, is a local square-integrable martingale (more precisely there exists a sequence of stopping times $S_n \uparrow T_\infty$ such that the process $(\mathcal{I}_{t \wedge S_n}, \mathcal{F}_t)$, $t \geq 0$, is a square-integrable martingale for each n). If $\nu = \nu^c$ and inequality (2.2) holds for all $t \leq T_\infty$, then $|\mathcal{I}_t| < \infty$ P-a.s. for all $t \leq T_\infty$. The inequality

$$M[\int_0^\infty \int_E X^2(t,x)\nu(dt,dx) - \sum_{t>0} (\int_E X(t,x)\nu(\{t\},dx))^2] < \infty$$

is sufficient (and necessary) for square-integrability of the martingale $(\mathcal{J}_t, \mathcal{F}_t)$, $t \geq 0$, [20], [2].

2. Put

$$\alpha_t = \nu(\{t\} \times E), \qquad a^+ = \begin{cases} \frac{1}{a}, & a \neq 0, \\ 0, & a = 0. \end{cases}$$

Let $Y(t,x)$ be a $\mathcal{P} \times \mathcal{E}$ - measurable non-negative function such that

1) $Y(t,x) < \infty$,

2) $\hat{Y}_t = \int_E Y(t,x)\nu(\{t\},dx) \leq 1$,

3) $\alpha_t = 1 \Longrightarrow \hat{Y}_t = 1$,

4) the process $\int_0^t \int_E Y(s,x)\nu(ds,dx)$, $t \geq 0$, is right-continuous.

We consider the process $(\mathcal{Z}_t(Y), \mathcal{F}_t)$, $t \geq 0$, with $\mathcal{Z}_t(Y) = \mathcal{Z}_{t \wedge \tau}(Y)$, where $\tau = \inf(t: \nu(]0,t] \times E) = \infty)$, which is defined as follows:

(2.3)
$$\mathcal{Z}_t(Y) = (\prod_{T_n \leq t} Y(T_n, X_n))(\prod_{\substack{s \leq t, \\ s \neq t \\ \alpha_s < 1}} [1 + \frac{\alpha_s - \hat{Y}_s}{1 - \alpha_s}])\exp[-(1 - Y(s,x))\nu^c(ds,dx)]$$

if $t < \tau$ and

$$\mathcal{Z}_\tau(Y) = \liminf_{t \to \tau} \mathcal{Z}_t(Y) .$$

By virtue of Proposition 4.3 in [1] the process $(\mathcal{Z}_t(Y), \mathcal{F}_t)$, $t \geq 0$, is a non-negative supermartingale, $M\mathcal{Z}_t(Y) \leq 1$, and is a local martingale having the representation

$$(2.4) \quad \mathcal{Z}_t(Y) = 1 + \int_0^t \int_E \mathcal{Z}_{s-}(Y)V(s,x)[\mu(ds,dx) - \nu(ds,dx)]$$

with

(2.5) $V(t,x) = Y(t,x) - 1 - (1 - \alpha_t)^+(\alpha_t - \hat{Y}_t)$.

Let us suppose that $M\mathcal{Z}_\infty(Y) = 1$ and let P' be a probability measure on $(\Omega, \mathcal{F})$ with $dP' = \mathcal{Z}_\infty(Y)\, dP$.
According to Theorem 4.5 in [1] the random measure ν', $d\nu' = Y d\nu$, is the compensator with respect to P' for a random measure μ .
An analogous statement is true for the restrictions of measures, P_T and P'_T, i.e. the equality $M\mathcal{Z}_T(Y) = 1$ implies that ν'_T , $d\nu'_T = Y d\nu_T$, is the compensator of μ_T with respect to P'_T .

3. We will formulate conditions for $V(t,x)$ that guarante the equality $M\mathcal{Z}_T(Y) = 1$.

Lemma 1. Let T be a stopping time and P_T -a.s.

$$\int_0^T\int_E |V(t,x)|\,\nu(dt,dx) \le C < \infty \quad .$$

Then $M\mathcal{Z}_T(Y) = 1$.

Proof. By virtue of (2.2) and (2.4) it is sufficient to show that

$$M\int_0^T\int_E \mathcal{Z}_{t-}(Y)|V(t,x)|\,\nu(dt,dx) < \infty \quad .$$

Since $\mathcal{Z}_t(Y)$ is a local martingale, there exists a sequence of stopping times $\theta_n \uparrow \infty$ such that $\mathcal{Z}_{t\wedge\theta_n}(Y)$ is a uniformly integrable martingale for each n . It follows (see IY - T- 47, V-T-27 in [21]) that

$$M\int_0^\infty\int_E \mathcal{Z}_{t-}(Y)|V(t,x)|\,\nu(dt,dx) =$$

$$= \lim_n M \int_0^{\theta_n}\int_E \mathcal{Z}_{(t\wedge\theta_n)-} |V(t,x)| \nu(dt,dx) =$$

$$= \lim_n M\mathcal{Z}_{\theta_n}(Y) \int_0^{\theta_n}\int_E |V(t,x)|\nu(dt,dx) \le C \lim_n M\mathcal{Z}_{\theta_n}(Y) = C .$$

Lemma 2. Let $\nu = \nu^c$, T be a stopping time and P_T - a.s.

$$\int_0^T\int_E (Y(t,x)-1)^2 \nu(dt,dx) \le C < \infty .$$

Then $M\mathcal{Z}_T(Y) = 1$

Proof. It is easy to see that

$$\mathcal{Z}_t^2(Y) = \mathcal{Z}_t(Y^2) \exp[\int_0^t\int_E (Y(s,x)-1)^2 \nu(ds,dx)]$$

for $t<\tau$. Defining $\mathcal{Z}_t(Y^2)$ for $t \ge \tau$ in the similar way as $\mathcal{Z}_t(Y)$, we obtain a non-negative supermartingale $(\mathcal{Z}_t(Y^2), \mathcal{F}_t)$, $t \ge 0$, with $M\mathcal{Z}_t(Y^2) \le 1$. Hence

$$M\mathcal{Z}_\sigma^2(Y) = M\mathcal{Z}_\sigma(Y^2) \exp[\int_0^\sigma\int_E (Y(s,x)-1)^2 \nu(ds,dx)] \le$$

$$\le e^C M\mathcal{Z}_\sigma(Y^2) \le e^C ,$$

where σ is a stopping time.

Let θ_n , n= 1,2,..., be the sequence of stopping times, which was used in the proof of Lemma 1. Then, by the inequality $M\mathcal{Z}_{\theta_n}^2(Y) \le e^C$ the sequence $\{\mathcal{Z}_{\theta_n}(Y),\ n = 1,2,\dots\}$ is uniformly integrable (Theorem 6, Ch. 2, §5, [22]).

Hence

$$M\mathcal{Z}_T(Y) = M \lim_n \mathcal{Z}_{T\wedge\theta_n}(Y) = \lim_n M\mathcal{Z}_{T\wedge\theta_n} = 1$$

Lemma 3. Let $\nu = \nu^c$, T be a stopping time and P - a.s.

$$\int_0^T \int_E \frac{(Y(t,x)-1)^2}{1+|Y(t,x)-1|} \nu(dt,dx) \le C < \infty$$

Then $M \mathcal{Z}_T(Y) = 1$.

Proof. Put

$$Y_1(t,x) = (Y(t,x))^{I_{\{|Y(t,x)| > \frac{1}{2}\}}} ,$$

$$Y_2(t,x) = (Y(t,x))^{I_{\{|Y(t,x)| \le \frac{1}{2}\}}} .$$

Obviously, $\mathcal{Z}_t(Y) = \mathcal{Z}_t(Y_1) \cdot \mathcal{Z}_t(Y_2)$. By assumptions, P -a.s.

$$\int_0^T \int_E |Y_1(t,x)-1| \nu(dt,dx) \le C_1 < \infty ,$$

$$\int_0^T \int_E (Y_2(t,x)-1)^2 \nu(dt,dx) \le C_2 < \infty .$$

Hence, by Lemma 1, $M \mathcal{Z}_T(Y_1)=1$. Let $\tilde{P}_T$ be the probability measure on (Ω, $\mathcal{F}_T$) with $d\tilde{P}_T = \mathcal{Z}_T(Y_1) dP_T$. Then the random measure $\tilde{\nu}_T$ with

$$\tilde{\nu}_T(]0,t] \times \Gamma) = \int_0^t \int_\Gamma Y_1(s,x) \nu(ds,dx), \ t \le T, \ \Gamma \in \mathcal{E},$$

is the compensator for a random measure μ_T with respect to $\tilde{P}_T$.

Note that P_T -a.s. (and $\tilde{P}_T$ -a.s.)

$$\int_0^T \int_E (Y_2(t,x)-1)^2 \tilde{\nu}_T(dt,dx) = \int_0^T \int_E (Y_2(t,x)-1)^2 Y_1(t,x) \nu(dt,dx) =$$

$$= \int_0^T \int_E (Y_2(t,x)-1)^2 \nu(dt,dx) \le C_2 < \infty .$$

Therefore, by Lemma 2, we obtain, $\int_{\Omega} Z_T(Y_1)d\tilde{P}_T = 1$.
Hence

$$1 = \int_{\Omega} Z_T(Y_2)d\tilde{P}_T = \int_{\Omega} Z_T(Y_2)Z_T(Y_1)dP_T = MZ_T(Y) \ .$$

Remark. In a similar way we can prove that the inequality

$$\int_0^T\int_E \frac{(Y(t,x)-1)^2}{1+|Y(t,x)-1|}\nu^c(dt,dx)+\sum_{t\le T}|V(t,x)|\nu(\{t\},dx)\le C<\infty \quad (P_T\text{-a.s.})$$

is a sufficient condition for $MZ_T(Y) = 1$.

4. Let P and P' be two probability measures on $(\Omega, \mathcal{F})$, $P'\ll P$, and ν, ν' be the compensators for a random measure μ with respect to the measures P and P' respectively. According to Theorem 4.1 in [1], $d\nu' = Y d\nu$ (P' - a.s.). Conditions 1)- 4) for $Y(t,x)$ (see §2, No.2) are true P'-a.s. only. But without loss of generality we can suppose that we chose a version of the function $Y(t,x)$, which also satisfies conditions 1) - 4) P - a.s.

Let $Y_1(t,x) = Y(t,x) \vee 1$ and $(Z_t(Y_1), \mathcal{F}_t)$, $t \ge 0$ be the supermartingale defined by (2.3).

Lemma 4. Let $\nu = \nu^c$ and

$$P'\{\inf_{t\ge 0} Z_t(Y_1) = 0\} = 0 \ .$$

Then P' - a.s.

$$\int_0^\infty\int_E (Y_1(t,x) - \ln Y_1(t,x) - 1)\,\nu(dt, dx) < \infty \ .$$

Proof. Put

$$J_t = \int_0^t\int_E \ln Y_1(s,x)[\mu(ds,dx) - \nu(ds,dx)] \ ,$$

$$R_t = \int_0^t \int_E (Y_1(s,x) - \ln Y_1(s,x) - 1)\,\nu(ds, dx).$$

Then, by (2.3), $\ln \mathfrak{Z}_t(Y_1) = \mathfrak{I}_t - R_t$. Note that trajectories of the process $\mathfrak{Z}_t(Y_1)$, $t \geq 0$ have positive jumps only. Therefore the stopping times $\theta_K = \inf\{t: \mathfrak{Z}_t(Y_1) \leq \frac{1}{K}\}$, k= 1, 2,..., have the following properties:

(2.6) $$\lim_K P'(\theta_K < \infty) = 0 \quad ,$$

(2.7) $$\ln \mathfrak{Z}_{t\wedge\theta_K}(Y_1) \geq -\ln K \quad .$$

Put $\sigma_n = \inf\{t: \int_0^t\int_E \ln Y_1(s,x)\,\nu(ds,dx) \geq n\}$ and $\sigma = \lim_n \sigma_n$.
Since $\ln(y \vee 1) \leq (y \vee 1) \leq 1 + (y \vee 0)$, we obtain for $t < T_\infty$:

$$\int_0^t\int_E \ln Y_1(s,x)\,\nu(ds,dx) \leq \int_0^t\int_E Y_1(s,x)\,\nu(ds,dx) \leq$$

$$\leq \int_0^t\int_E (1+Y(s,x))\,\nu(ds,dx) = \nu(]0,t]\times E) + \nu'(]0,t]\times E) < \infty \qquad P'\text{- a.s.}$$

It follows that $P'(\sigma = \infty) = 1$. From the definition of the stopping times σ_n and θ_K, we obtain (see (2.7)):

$$MR_{t\wedge\sigma_n\wedge\theta_K} = M\mathfrak{I}_{t\wedge\sigma_n\wedge\theta_K} - M\ln\mathfrak{Z}_{t\wedge\sigma_n\wedge\theta_K}(Y_1) =$$

$$= -M\ln\mathfrak{Z}_{t\wedge\sigma_n\wedge\theta_K}(Y_1) \leq \ln K$$

and

$$MR_{\theta_K} = \lim_{n\to\infty}\lim_{t\to\infty} MR_{t\wedge\sigma_n\wedge\theta_K} \leq \ln K \ ,$$

i.e. $P'(R_{\theta_K} < \infty) = 1$ by absolute continuity $P' \ll P$.

There $\{R_\infty = \infty\} \subseteq \{\theta_K < \infty\}$. Consequently (see (2.6))

$$P'(R_\infty = \infty) \le P'(\theta_K < \infty) \to 0, \quad K \to \infty .$$

Lemma 5. Let $\hat{\nu} = \nu^c$, $Y_2(t,x) = Y(t,x) \wedge 1$.

Then P' - a.s.

$$\int_0^\infty \int_E (Y_2(t,x) \ln Y_2(t,x) + 1 - Y_2(t,x)) \hat{\nu}(dt, dx) < \infty .$$

Proof. Put

$$\mathcal{Y}_t = \int_0^t \int_E \ln Y_2(s,x)[\mu(ds,dx) - \nu'(ds,dx)] ,$$

$$R_t = \int_0^t \int_E (Y_2(s,x) \ln Y_2(s,x) + 1 - Y_2(s,x)) \hat{\nu}(ds,dx) .$$

Then, by (2.3), $\ln \mathfrak{Z}_t(Y_2) = \mathcal{Y}_t + R_t$. According to the definition of $\mathfrak{Z}_t(Y_2)$, trajectories of the supermartingale $\mathfrak{Z}_t(Y_2)$ have negative jumps only. Therefore the stopping times $\theta_K = \inf\{t : \mathfrak{Z}_t(Y_2) > K\}$ have the following properties

(2.8) $$\lim_K P'(\theta_K < \infty) = 0 .$$

(2.9) $$\ln \mathfrak{Z}_{t \wedge \theta_K}(Y_2) \le \ln K .$$

Since $y|\ln(y \wedge 1)| = (y \wedge 1)|\ln(y \wedge 1)| \le e^{-1}$, we obtain P - a.s. (and P' -a.s.) that

$$\int_0^t \int_E |\ln Y_2(s,x)| \nu'(ds,dx) = \int_0^t \int_E Y(s,x) |\ln Y_2(s,x)| \hat{\nu}(ds,dx) =$$

$$= \int_0^t\int_E Y_2(s,x)|\ln Y_2(s,x)|\nu(ds,dx) \le e^{-1}\nu(]0,t]\times E) < \infty , \quad t < T_\infty .$$

It follows that the process ($\mathcal{J}_t$, $\mathcal{F}_t$, P'), $t \ge 0$, is a local martingale. Let a sequence $\sigma_n \uparrow T_\infty$ (P' -a.s.) of stopping times be such that the process ($\mathcal{J}_{t\wedge\sigma_n}$, $\mathcal{F}_t$, P'), $t \ge 0$, is a uniformly integrable martingale. Then, for each n= 1 , 2,...,

$$M'R_{t\wedge\sigma_n\wedge\theta_K} = -M'\mathcal{J}_{t\wedge\sigma_n\wedge\theta_K} + M'\ln \mathcal{Z}_{t\wedge\sigma_n\wedge\theta_K} \le \ln K$$

and hence

$$M'R_{\theta_K} = \lim_{n\to\infty}\lim_{t\to\infty} M'R_{t\wedge\sigma_n\wedge\theta_K} \le \ln K ,$$

i.e. we have $P'(R_{\theta_K} < \infty) = 1$.

Therefore $\{R_\infty = \infty\} \subseteq \{\theta_K < \infty\}$. Consequently (see (2.8))

$$P'\{R_\infty = \infty\} \le P'\{\theta_K < \infty\} \to 0 , \quad K \to \infty .$$

3. Proof of Theorems 1 and 2

1. Proof of Theorem 1. Necessity. Conditions (i) and(ii) were established in [1] (Th.4.1) Conditions (iii) and (iv) are satisfied P_T - a.s. Consequently, by virtue of absolute continuity $P'_T \ll P_T$, they are also satisfied P'_T - a.s.

Sufficiency. Conditions 1) - 4) (see 2. No.2) are satisfied for $Y(t,x)$ P'_T - a.s. But without loss of generality we can suppose that there exists a version of the function $Y(t,x)$ which also satisfies conditions 1) - 4) P_T -a.s.

The supermartingale ($\mathcal{Z}_t(Y)$, $\mathcal{F}^\mu_t$) , $t \ge 0$, is defined by (2.3), where $V(t,x)$ is defined by formula (2.5).Let $U_t = \int_0^t\int_E |V(s,x)|\nu(ds,dx)$.

Then (see (2.5))

$$U_t \le \int_0^t\int_E (1+Y(s,x))\,\nu(ds,dx) + \sum_{\substack{s\le t,\\ \alpha_s<1}} \frac{\alpha_s - \hat{Y}_s}{1-\alpha_s} \le$$

$$\le \nu(]0,t]\times E) + \int_0^t\int_E Y(s,x)\,\nu(ds,dx) + 2\sum_{\substack{s\le t,\\ \alpha_s\le \frac{1}{2}}} (\alpha_s + \hat{Y}_s) + 2\sum_{\substack{s\le t,\\ \frac{1}{2}<\alpha_s<1}} \frac{1}{1-\alpha_s} \le$$

$$\le 3[\nu(]0,t]\times E) + \int_0^t\int_E Y(s,x)\,\nu(ds,dx)] + 2\sum_{\substack{s\le t,\\ \frac{1}{2}<\alpha_s<1}} \frac{1}{1-\alpha_s} \,.$$

It follows that P'_T -as.

$$U_T \le 3[\nu(]0,T]\times E) + \nu'(]0,T]\times E)] + 2\sum_{\substack{s\le T,\\ \frac{1}{2}<\alpha_s<1}} \frac{1}{1-\alpha_s} < \infty \quad ,$$

since (P'_T -a.s.) $T<T_\infty$, $\nu(]0,T]\times E)<\infty$ by assumption)iv) and and the number of jumps of the process α_t , $t\le T$, of size more than 1/2 is finite.

Put $\theta = \inf\{t : U_{t\wedge T} = \infty\}$. According to Lemma 4.4 in [1] , there exists a sequence of stopping times $\theta_K \uparrow \theta$ Q - a.s., where $Q=\frac{1}{2}(P_T+P'_T)$, such that $U_{t\wedge\theta_K} \le K$ Q-a.s.

By virtue of absolute continuity of the measures P_T and P'_T with respect to the measure Q , the sequence θ_K has the same properties P_T - a.s. and P'_T -a.s. Note that $\{\theta_K < T\} \subseteq \{U_T \ge K\}$.

Consequently

(3.1) $$\lim_K P'_T\{\theta_K < T\} \le \lim_K P'_T\{U_T \ge K\} = 0 \quad .$$

By Lemma 1, the process $(\mathfrak{z}_{t\wedge\theta_\kappa}(Y), \mathcal{F}^{\mu}_t, P_T)$, $t \le T$, is a uniformly integrable martingale, Hence the mesure $\tilde{P}^{(\kappa)}_T$ with $d\tilde{P}^{(\kappa)}_T = \mathfrak{z}_{T\wedge\theta_\kappa}(Y)\, dP_T$ is a probability mesure. The random measure $\tilde{\nu}^{(\kappa)}_T$ with

$$\tilde{\nu}^{(\kappa)}_T(]0,t]\times\Gamma) = \int_0^t\int_\Gamma (Y(s,x))^{I\{\theta_\kappa \ge s\}}\, \nu(ds,dx),\ t \le T,\ \Gamma\in\mathcal{E},$$

is the compensator for a random mesure μ_T with respect to $\tilde{P}^{(\kappa)}_T$. It is easy to see that $\tilde{\nu}^{(\kappa)}_{T\wedge\theta_\kappa} = \nu'_{T\wedge\theta_\kappa}$. Therefore the restrictions of the mesures $\tilde{P}^{(\kappa)}_T$ and P_T to the σ-algebra $\mathcal{F}^{\mu}_{T\wedge\theta_\kappa}$ coincide. In particular, for any set $B\in\mathcal{F}^{\mu}_T$,

(3.2)

$$\tilde{P}^{(\kappa)}_T(B,\ \theta_\kappa \ge T) = P'_T(B,\ \theta_\kappa \ge T)$$

because $B\cap\{\theta_\kappa \ge T\} \subseteq \mathcal{F}^{\mu}_{T\wedge\theta_\kappa}$.

Let $B\in\mathcal{F}^{\mu}_T$ be a set such that $P_T(B) = 0$. Then $\tilde{P}^{(\kappa)}_T(B) = 0$, and by (3.1) and (3.2) we obtain

$$P'_T(B) = P'_T(B,\ \theta_\kappa < T) \le P'_T(\theta_\kappa < T) \to 0,\quad \kappa\to\infty .$$

The proof is completed.

2. Proof of Theorem 2. Necessity

Condition (i) was established in [1](Th.4.1). In [1] (Th. 5.1) it was also shown that P - a.s. $\mathfrak{z}_t(Y) = M(\frac{dP'}{dP} \mid \mathcal{F}^{\mu}_t)$, where $\mathfrak{z}_t(Y)$ is defined by (2.3). To establish condition (ii) note that the following inequality hold :

$$0 < C_1 \le \frac{(y-1)^2}{1+|y-1|} / (y\ln y + 1 - y) \le C_2 < \infty,\quad y\in[0,\tfrac{3}{2}],$$

$$0<C_1\leq \frac{(y-1)^2}{1+|y-1|}\Big/(y-1-\ln y)\leq C_2<\infty,\ y\in[\tfrac{1}{2},\infty[\ .$$

Therefore (ii) is equivalent to the following conditions:

(3.3) $$\int_0^\infty\int_E (Y_1(t,x)-1-\ln Y_1(t,x))\,\nu(dt,dx)<\infty \quad P'\text{- a.s.} \ ,$$

(3.4) $$\int_0^\infty\int_E (Y_2(t,x)\ln Y_2(x,t)+1-Y_2(t,x))\,\nu(dt,dx)<\infty \quad P'\text{- a.s.},$$

where $Y_1(t,x)=Y(t,x)\vee 1$, $Y_2(t,x)=Y(t,x)\wedge 1$.

Obviously $Z_t(Y)=Z_t(Y_1)\cdot Z_t(Y_2)$ and each of the process $(Z_t(Y_i),\mathcal{F}^{\mu}_t,P)$, $t\geq 0$, $i=1,2$, is a non-negative right-continuous supermartingale. It follows, [23],

(3.5) $$P\{\sup_{t\geq 0} Z_t(Y_i)<\infty\}=1 \ , \quad i=1,2.$$

Inequality (3.4) holds true by Lemma 5.

To establish inequality (3.3), it is sufficient to show, by Lemma 4, that

(3.6) $$P'(\inf_{t\geq 0} Z_t(Y_1)=0)=0 \quad .$$

By virtue of (3.5) and by absolute continuity of measure P' with respect to P, for proving (3.6) it is sufficient to establish that $P'(\inf_{t\geq 0} Z_t(Y)=0)=0$. Let $\theta=\inf\{t: Z_t(Y)=0\}$. Then $\{\inf_{t\geq 0} Z_t(Y)=0\}=\{Z_\theta=0\}$. Therefore

$$P'\{\inf_{t\geq 0} Z_t(Y)=0\}=\int_{\{Z_\theta(Y)=0\}} Z_\infty(Y)\,dP=\int_{\{Z_\theta(Y)=0\}} Z_\theta(Y)\,dP=0 \ .$$

Sufficiency. Put

$$\theta_n=\inf\{t:\int_0^t\int_E \frac{(Y(s,x)-1)^2}{1+|Y(s,x)-1|}\,\nu(ds,dx)\geq n\}$$

and $\theta = \lim_n \theta_n$. According to condition (ii), we have

(3.7) $$\lim_n P'(\theta_n < T_\infty) = 0 .$$

The process $(\mathcal{Z}_t(Y), \mathcal{F}_t^\mu, P)$, $t \geq 0$, defined by (2.3), is a non-negative supermartingale. According to Lemma 3 and by the definition of the stopping times θ_n , n = 1,2, ..., the process $(\mathcal{Z}_{t\wedge\theta_n}(Y), \mathcal{F}_t^\mu, P)$, $t \geq 0$, is a uniformly integrable martingale. Let $\tilde{P}^{(n)}$ be the probability measure with $d\tilde{P}^{(n)} = \mathcal{Z}_\theta(Y)\, dP$.

The random measure $\tilde{\nu}^{(n)}$ with

$\tilde{\nu}^{(n)}(]0,t]\times\Gamma) = \int_0^t\int_\Gamma (Y(s,x))^{I_{\{\theta_n \leq s\}}}\nu(ds,dx), t\geq 0, \Gamma\in\mathcal{E}$, is the compensator for a random measure μ with respect to $\tilde{P}^{(n)}$.

From the definition of $\tilde{\nu}^{(n)}$ it follows that $\tilde{\nu}^{(n)}_{\theta_n} = \nu'_{\theta_n}$.

Hence the restrictions of the mesures $\tilde{P}^{(n)}$ and P to the σ-algebra $\mathcal{F}^\mu_{\theta_n}$ coincide. In particular, for any set $B\in\mathcal{F}^\mu_\infty$, we have

(3.8) $$\tilde{P}^{(n)}(B, \theta_n \geq T_\infty) = P'(B, \theta_n \geq T_\infty) .$$

Let a set $B\in\mathcal{F}^\mu_\infty$ be such that $P(B)=0$. Then, by (3.7) and (3.8), we obtain

$$P'(B) = P'(B, \theta_n < T_\infty) \leq P'(\theta_n < T_\infty) \to 0 , \quad n\to\infty .$$

The proof is completed.

References

1 Jacod J., Multivariate point processes: predictable projection, Radon-Mikodym derivatives, representation of martingales, Z. Wahrscheinlichkeitstheorie verw. Gebiete 31 (1975), 235-253.

2 Kabanov Yu.M., Lipcer R.Š., Širyaev A.N., Martingale techniques in point processes theory. School-seminar on the theory of stochastic processes (Druskininkai, 25-30 Nov., 1974), Vilnius 1975, part II, 269-354. (in Russian).

3 Skorokhod A.V., Random processes with independent increments, "Nauka", Moscow, 1964.

4 Girsanov J.V., On transforming a certain class of stochastic processes by absolute continuous substitution of measures. Theor. Probability Appl. 5(1960), 285-307.

5 Ershov M.P., On the absolute continuity of measures corresponding to diffusion processes. Theor. Probability Appl. XVII, 1(1972), 173-178.

6 Lipcer R.Š., Širyaev A.N., On the absolute continuity of measures corresponding to processes of diffusion type with respect to the Wiener measure. Izv. Akad. Nauk. Ser. Mat. 36(1972), 847-889(in Russian).

7 Lipcer R.Š., Širyaev A.N., Statistics of random processes, "Nauka", Moscow, 1974. (Springer-Verlag, 1976).

8 Grigelionis B., On the absolute continuity of measures corresponding to stochastic processes; Litovsk. Mat. Sb. XI, 4(1971), 783-794 (in Russian).

9 Grigelionis B., On the structure of the densities of measures corresponding to stochastic processes, Litovsk.Mat.Sb.XIII 1(1973), 71-78(in Russian).

10 Van Shuppen, Wong E., Transformation of local margingales under a change of low, AMS, 2,5 (1974), 879-888.

11 Kailath T., Zakai M., Absolute continuity and Radon-Nikodym derivatives for certain measures relative to Wiener measure, AMS, 42, 1(1971), 130-140.

12 Kailath T., The structure of Radon-Nikodim derivatives with respect to Wiener measure and related measures, AMS, 42(1971), 1054-1067.

13 Segall A., Kailath T., Radon-Nikodim derivatives with respect to measures induced by discontinuous independent-increments process, Ann.Probability, 3, 3(1975), 449-464.

14 Boel R., Varaiya P., Wong E., Martingales on jump processes I, II, SIAM J.Control, 13, 5(1975), 999-1061.

15 Bremand P., The martingale theory of point processes over the real half line admitting an intensity, Lect.Notes Econ.and Math.Syst.107(1975), 519-542.

16 Briggs V.D., Densities for infinitely divisible random processes. J.Multivar.Analysis 5(1975), 178-205.

17 Kadota T.T., Shepp L.A., Conditions for the absolute continuity between a certain pair of probability measures, Z. Wahrscheinlichkeitstheorie verw.Gebiete 16,3(1970), 250-260.

18 Orey S., Radon-Nikodim derivatives of probability measure: martingales methods, Dpt.Found.Math.Sc.Tokyo Univ.of Education, 1974.

19 Jacod J., Memin J., Characteristiques locales et conditions de continuité absolute pour les sémi-martingales (1975), to appear.

20 Jacod J., Un théoreme de representation pour les martingales discontinues (1975), to appear.

21 Dellacherie C., Capacités et processus stochastiques. Berlin, Heidelberg, New York: Springer 1972.

22 Gikhman J., Skorokhod A., Introduction to the theory of random processes. W.B.Saunders Co., Philadelphia.

23 Meyer P.A., Probability and potential. Waltham: Blaisdell, 1966.

Steklov Mathematical Institute
Academy of Sciences of the USSR
Moscow

NORMAL NUMBERS AND ERGODIC THEORY

Teturo KAMAE
Department of Mathematics
Osaka City University
Sugimoto-cho, Osaka
Japan

In [4] and [5], the author, partly with B. Weiss, investigated normal numbers from a viewpoint of the classical notion of collective by von Mises. In this paper, applying similar tecniques used there, we prove the following result which is a generalization of a result by G. Rauzy [8] obtained in the case of 1-torus. Let G be a compact metric Abelian group and ψ be a continuous homomorphism from G onto G. By Π_ψ, we denote the set of ψ-normal numbers. We assume a kind of mixing condition called the vague specification property on (G,ψ). We also assume that there exists a subgroup K of G* (character group) such that $\psi^*(K)\subset K$ and $G^*=K\otimes F$ (direct product), where $F=\{\chi\in G^*;\ \chi^n=1$ for some non-zero integer $n\}$. Then, $x+\Pi_\psi\subset\Pi_\psi$ holds if and only if x is a ψ-deterministic number. These assumptions are satisfied in the following cases:

1. ψ is a shift on a product group $G=\Sigma^N$, and
2. G is the n-torus and ψ has no proper value with the absolute value 1.

§1. Generic points

Let X be a compact metric space and ψ be a continuous mapping from X onto X. Such a pair (X,ψ) shall be called a dynamical system. Let $N=\{0,1,2,\dots\}$. By a measure on a topological space, we always mean a probability Borel measure. For $x\in X$, let

$$V_\psi(x)=\{\text{w-}\lim_{\substack{n\in S\\ n\to\infty}}\frac{1}{n}\sum_{i=0}^{n-1}\delta_{\psi^i x}\ ;\ S \text{ is an infinite subset of } N\},$$

where δ_x is the Dirac measure at x and w-lim implies the weak limit of measures ([6]). Note that $V_\psi(x)$ is not empty and consists of ψ-invariant measures for any $x\in X$. We say that x is ψ-generic for a measure μ if $V_\psi(x)=\{\mu\}$; i.e.

$$\text{w-}\lim_{n\to\infty}\frac{1}{n}\sum_{i=0}^{n-1}\delta_{\psi^i x}=\mu.$$

Definition 1. The entropy of $x\in X$ with respect to ψ is defined and denoted by

$$h_{\psi}(x) = \sup_{\mu \in V_{\psi}(x)} h_{\mu}(\psi),$$

where $h_{\mu}(\psi)$ is the entropy of the measure preserving transformation ψ under μ (refer [1], for example).

§2. Entropy

Let G be a compact metric Abelian group. By an endomorphism of G, we mean a cotinuous homomorphism from G <u>onto</u> G. Let ψ be an endomorphism of G. Then (G,ψ) is a dynamical system, which shall be called a <u>group endomorphism</u>. A <u>homomorphism</u> from a dynamical system (X,ψ) to another dynamical system(Y,η) is a continuous mapping f from X to Y such that $f\circ\psi=\eta\circ f$.

Lemma 1. <u>Let</u> f <u>be a homomorphism from a dynamical system</u> (X,ψ) <u>into</u> (Y,η). <u>Then</u> $V_{\eta}(f(x))=V_{\psi}(x)\circ f^{-1}(=\{\mu\circ f^{-1};\ \mu\in V_{\psi}(x)\})$ <u>for any</u> $x\in X$. <u>Therefore</u>, $h_{\psi}(x)\geq h_{\eta}(f(x))$.

(Proof) Clear.

Proposition 1. <u>Let</u> (G,ψ) <u>be a group endomorphism. Then for any</u> x <u>and</u> y <u>in</u> G, <u>we have</u>

1. $h_{\psi}(x)=h_{\psi}(-x)$,
2. $h_{\psi}(x+y)\leq h_{\psi}(x)+h_{\psi}(y)$,
3. $h_{\psi}(\tau x)\leq h_{\psi}(x)$, <u>where</u> τ <u>is a continuous mapping from</u> G <u>into</u> G <u>which commutes with</u> ψ,
4. $h_{\psi^k}(x)=kh_{\psi}(x)$, <u>and</u>
5. $h_{\psi}(x-\psi^k(x))=h_{\psi}(x)$ $(k=1,2,\dots)$.

(Proof) 1 and 3 are clear from lemma 1. 2 follows from lemma 1 and the fact that $h_{\psi\times\psi}((x,y))\leq h_{\psi}(x)+h_{\psi}(y)$. 4 holds since

$$h_{\psi^k}(x) = \sup_{\mu\in V_{\psi^k}(x)} h_{\mu}(\psi^k)$$

$$= \sup_{\mu\in V_{\psi^k}(x)} \frac{1}{k}\sum_{i=0}^{k-1} h_{\mu\circ\psi^{-i}}(\psi^k)$$

$$= \sup_{\mu \in V_{\psi^k}(x)} h_{\frac{1}{k}\sum_{i=0}^{k-1} \mu \circ \psi^{-i}}(\psi^k)$$

$$= \sup_{\mu \in V_{\psi}(x)} h_{\mu}(\psi^k)$$

$$= \sup_{\mu \in V_{\psi}(x)} k h_{\mu}(\psi)$$

$$= k h_{\mu}(x).$$

To prove 5, it is sufficient to prove it in the case k=1 since $h_{\psi}(x-\psi^k(x)) = \frac{1}{k} h_{\psi^k}(x-\psi^k(x))$ and $h_{\psi}(x) = \frac{1}{k} h_{\psi^k}(x)$. Let $f(x)=x-\psi(x)$.

Then, $h_{\psi}(f(x)) \leq h_{\psi}(x)$ follows from 3 since f commutes with ψ. Now we prove $h_{\psi}(f(x)) \geq h_{\psi}(x)$. For this purpose, it is sufficient to prove

$$h_{\mu \circ f^{-1}}(\psi) \geq h_{\mu}(\psi) \tag{2.1}$$

for any ψ-invariant measure μ on G. Furthermore, as shall be shown, it is sufficient to prove (2.1) only in the case that ψ is one-to-one and μ is ergodic with respect to ψ. Let

$$\mu(.) = \int_E \nu(.)\, dP(\nu)$$

be the ergodic decomposition of μ, where P is a measure on the space E of ergodic measures on G with the weak topology. Then

$$\mu \circ f^{-1}(.) = \int_E \nu \circ f^{-1}(.)\, dP(\nu)$$

becomes the ergodic decomposition of $\mu \circ f^{-1}$. In this case, it is known [12] that

$$h_{\mu}(\psi) = \int_E h_{\nu}(\psi)\, dP(\nu)$$

and

$$h_{\mu \circ f^{-1}}(\psi) = \int_E h_{\nu \circ f^{-1}}(\psi)\, dP(\nu)$$

Therefore, (2.1) follows from the fact that $h_{\nu\circ f^{-1}}(\psi)\geq h_{\nu}(\psi)$ for any $\nu\in E$. Now, consider the product group G^Z and the shift T on G^Z; $(T\alpha)(i)=\alpha(i+1)$ for $\alpha\in G^Z$ and $i\in Z$. Let $\hat{\mu}$ be the T-invariant measure on G^Z such that

$$(2.2)\qquad \hat{\mu}(\{\alpha\in G^Z;\ \alpha(i)\in E_i \text{ for } 0\leq i\leq k\}) = \mu\Big(\bigcap_{0\leq i\leq k} \psi^{-i}E_i\Big)$$

for any Borel sets E_i $(0\leq i\leq k)$ in G. For $\alpha\in G^Z$, let $\hat{f}(\alpha)=\alpha-T\alpha$. Then it is easy to see that $h_{\mu\circ f^{-1}}(\mu)=h_{\hat{\mu}\circ\hat{f}^{-1}}(T)$. Therefore (2.1) follows from the fact that $h_{\hat{\mu}\circ\hat{f}^{-1}}(T)\geq h_{\hat{\mu}}(T)$. Thus (2.1) can be reduced to the case that ψ is one-to-one and μ is ergodic.

Now we assume them. Let us assume, further, that $h_{\mu\circ f^{-1}}(\psi)<\infty$, since otherwise, there are nothing to prove. Then, there is a finite, measurable (w.r.t. $\mu\circ f^{-1}$) partition A of G such that $\bigvee_{i=-\infty}^{+\infty}\psi^i A$ is the pointwise partition (mod $\mu\circ f^{-1}$) and

$$h_{\mu\circ f^{-1}}(\psi)=H_{\mu\circ f^{-1}}\Big(A\,\Big|\,\bigvee_{i=-\infty}^{-1}\psi^i A\Big),$$

where $H(\ |\ )$ is the conditional entropy of partitions [1]. Then it holds for $B=f^{-1}A$ that

$$(2.3)\qquad \bigvee_{i=-\infty}^{+\infty}\psi^i B=\{\{u\in G;\ f(u)=v\};\ v\in G\}\qquad (\text{mod } \mu)$$

and

$$h_{\mu\circ f^{-1}}(\psi)=H_{\mu}\Big(B\,\Big|\,\bigvee_{i=-\infty}^{-1}\psi^i B\Big).$$

Let d be a metric on G such that $d(u+w, v+w)=d(u, v)$ for any u, v and w in G. Take a countable, measurable (w.r.t. μ) partition C which satisfies the following three conditions:

1. $H(C)<\infty$,
2. $\mu(c)>0$ for any $c\in C$, and
3. for any $\varepsilon>0$, there exists $c\in C$ such that $d(c)<\varepsilon$.

For a moment, Let us assume that

$$(2.4)\qquad \Big(\bigvee_{i=-\infty}^{+\infty}\psi^i B\Big)\vee\Big(\bigvee_{i=-\infty}^{-1}\psi^i C\Big)=\text{the pointwise partition (mod } \mu).$$

Then we have

$$h_\mu(\psi)=H_\mu(B^\vee C \,|\, \bigvee_{i=-\infty}^{-1} \psi^i(B^\vee C))$$

$$=H_\mu(B\,|\, \bigvee_{i=-\infty}^{-1} \psi^i B)+H_\mu(C\,|\,(\bigvee_{i=-\infty}^{+\infty} \psi^i B)^\vee(\bigvee_{i=-\infty}^{-1} \psi^i C))$$

$$=h_{\mu\circ f^{-1}}(\psi),$$

and complete the proof of (2.1). Now, let us prove (2.4). By (2.3), there exists $\Omega_1 \subset G$ such that $\mu(\Omega_1)=1$ and that $f(u)=f(v)$ holds for any u and v in Ω_1 which belong to the same element of the partition $\bigvee_{i=-\infty}^{+\infty} \psi^i B$. Let

$\Omega_2=\{u\in G$; for any $c\in C$, there exists $i\geq 1$ such that $\psi^i u\in c\}$.
Since μ is ergodic with respect to ψ, $\mu(\Omega_2)=1$. Let $\Omega=\Omega_1\cap\Omega_2$. Then $\mu(\Omega)=1$. Take any u and v in Ω which belong to the same atom of $(\bigvee_{i=-\infty}^{+\infty} \psi^i B)^\vee(\bigvee_{i=-\infty}^{-1} \psi^i C)$. Since $f(u)=f(v)$, we have $u-\psi^i u=v-\psi^i v$ for any $i\geq 1$. For any $\varepsilon>0$, there exists $i\geq 1$ such that $\psi^i u\in c$ and $d(c)<\varepsilon$ for some $c\in C$. Since u and v belong to the same atom of $\psi^{-i}C$, $\psi^i v\in c$. Then

$$d(u,v)=d(\psi^i u,\ \psi^i v)<\varepsilon.$$

Since $\varepsilon>0$ is arbitrary, we have $u=v$, which completes the proof of (2.4).

§3. Dynamical systems with the specification property

A G-_time_ is a collection $\{I_1, I_2,\ldots\}$ of intervals in N. A G-time $\{I_1, I_2,\ldots\}$ is L-_separated_ if $|t-s|>L$ for any $t\in I_i$ and $s\in I_j$ with $i\neq j$.

Definition 2. (Slightly different definitions are given by R. Bowen [2], D. Ruelle [10] and K. Sigmund [11].) A dynamical system (X,ψ) has the _specification property_ if for any $\varepsilon>0$, there exists L such that for any L-separated G-time $\{I_1, I_2,\ldots\}$ and $x_1, x_2,\ldots$ in X, there exists x in X such that $d(\psi^i x, \psi^i x_j)<\varepsilon$ for any i and j with $i\in I_j$, where d is a metric on X.

Let (X,ψ) be a dynamical system. Let T be the shift on the product space X^N; $(T\alpha)(i)=\alpha(i+1)$ for $\alpha\in X^N$ and $i\in N$. Let

$$\Gamma_\psi=\{\alpha\in X^N;\ \alpha(i)=\psi^i(\alpha(0)) \text{ for any } i\in N\}.$$

For $S\subset N$, let

$$\sigma(S)=\lim_{n\to\infty}\frac{1}{n}|S\cap[0,n)|$$

be the density.

<u>Definition</u> 3. A dynamical system (X,ψ) has the <u>vague specification property</u> (<u>v.s.p.</u>) if for any $\alpha\in X^N$ such that $\sigma(\{i;\ T^i\alpha\in U\})=1$ for any open set U containing Γ_ψ, there exists $x\in X$ such that $\sigma(\{i;\ d(\alpha(i),\psi^i x)<\varepsilon\})=1$ for any $\varepsilon>0$.

Let (X,ψ) and (Y,η) be dynamical systems. Recall that a homomorphism from (X,ψ) to (Y,η) is a continuous mapping f from X to Y such that $f\circ\psi=\eta\circ f$.

Proposition 2. <u>A dynamical system has v.s.p. if there is a homomorphism from a dynamical system with v.s.p. onto it</u>.

(Proof) Let (X,ψ) and (Y,η) be dynamical systems and f be a homomorphism from (X,ψ) onto (Y,η). Assume that (X,ψ) has v.s.p. Let $\beta\in Y^N$ satisfy that $\sigma(\{i\in N;\ T^i\beta\in V\})=1$ for any open set V containing Γ_η, where we denote the shift on Y^N as well as X^N by the same notation T. For $k=1,2,\ldots$, let

$$V_k=\{\gamma\in Y^N;\ \rho(\eta^i\gamma(0),\ \gamma(i))<\frac{1}{k}\text{ for any } i=1,2,\ldots,k\},$$

where ρ is a metric on Y. Let

$$a_1^k=\min\{i;\ T^i\beta\in V_k\},\text{ and}$$

$$a_{n+1}^k=\min\{i;\ T^i\beta\in V_k,\ i>a_n^k+k\}\quad(n=1,2,\ldots).$$

Since

$$\bigcup_{n=1}^{\infty}\{a_n^k,\ a_n^k+1,\ldots,\ a_n^k+k\}\supset\{i\in N;\ T^i\beta\in V_k\},$$

we have $\sigma(\bigcup_{n=1}^{\infty}I_n^k)=1$, where $I_n^k=\{a_n^k,\ a_n^k+1,\ldots,\ a_n^k+k\}$. Then it is not difficult to show that there exist finite sets $M_1, M_2,\ldots$ of positive integers such that $\{I_n^k;\ n\in M_k,\ k=1,2,\ldots\}$ is a disjoint family of sets and

$$\sigma(\bigcup_{k=1}^{\infty}\bigcup_{n\in M_k}I_n^k)=1.$$

For n and k with $n\in M_k$, take $x_n^k\in f^{-1}(\beta(a_n^k))$. Define $\alpha\in X^N$ by

$$\alpha(i)=\begin{cases}\psi^{i-a_n^k}x_n^k & \text{if } i\in I_n^k \text{ and } n\in M_k\\ \text{arbitrary} & \text{else.}\end{cases}$$

Then we have $\rho(f(\alpha(i)),\ \beta(i))<\frac{1}{k}$ if $i\in\bigcup_{k=K}^{\infty}\bigcup_{n\in M_k} I_n^k$ (K=1,2,...).
Therefore,

(3.1) $\quad\sigma(\{i;\ \rho(f(\alpha(i)),\ \beta(i))<\varepsilon/2\})=1$

for any $\varepsilon>0$. Let U be any open set including Γ_ψ. Then there exists $L\in N$ and $\varepsilon>0$ such that

$$W=\{\gamma\in X^N;\ d(\psi^i\gamma(0),\ \gamma(i))<\varepsilon \text{ for any } l=1,2,\dots,L\}$$

is contained in U. Then $T^i\alpha\in W$ if $\{i,\ i+1,\dots,i+L\}$ is contained in some I_n^k with $n\in M_k$. Therefore, $\sigma(\{i;\ T^i\alpha\in U\})=1$. Since (X,ψ) has v.s.p., there exists $x\in X$ such that $\sigma(\{i;\ d(\alpha(i),\ \psi^i x)<\varepsilon\})=1$ for any $\varepsilon>0$. Let $y=f(x)$. Take any $\varepsilon>0$. Take $\delta>0$ such that $\rho(f(u),\ f(v))<\varepsilon/2$ whenever $d(u,v)<\delta$. Since $\sigma(\{i;\ d(\alpha(i),\ \psi^i x)<\delta\})=1$, we have $\sigma(\{i;\ \rho(f(\alpha(i)),\ \eta^i y)<\varepsilon/2\})=1$. Combining this with (3.1), we have $\sigma(\{i;\ \rho(\beta(i),\eta^i y)<\varepsilon\})=1$, which completes the proof.

Example 1. Let Σ be any compact metric space. Let T be the shift on Σ^N. Then the dynamical system $(\Sigma^N,\ T)$ has v.s.p.

(Proof) Assume that $\alpha\in(\Sigma^N)^N$ satisfies that $\sigma(\{i;\ \tau^i\alpha\in U\})=1$ for any open set U including Γ_T, where τ is the shift on $(\Sigma^N)^N$. Define $x\in\Sigma^N$ by $x(i)=(\alpha(i))(0)$ for any $i\in N$. Then it is easy to see $\sigma(\{i;\ d(T^i x,\ \alpha(i))<\varepsilon\})=1$ for any $\varepsilon>0$, where d is a metric on Σ^N.

Example 2: Let $r\geq 2$ be an integer. Let $\psi x=rx$ (mod Z). Then, the dynamical system $(R/Z,\psi)$ has v.s.p.

(Proof) $(R/Z,\psi)$ is a homomorphic image of the dynamical system $(\{0,1,\dots,r-1\}^N,\ T)$, where T is the shift.

Recall that a dynamical system (X,ψ) is <u>expansive</u> if there exists a constant $\delta>0$ such that for any x, $y\in X$ with $x\neq y$, there exists an integer i (i may be negative) satisfying that $d(\psi^i x,\ \psi^i y)>\delta$. where for a negative i, we denote

$$d(\psi^i x,\ \psi^i y)=\inf_{\substack{\psi^{-i}u=x\\ \psi^{-i}v=y}} d(u,v).$$

Proposition 3. <u>A dynamical system (X,ψ) has v.s.p. if it is expansive and has the specification property.</u>

(Proof) Let (X,ψ) be expansive and have the specification property. Let $\alpha\in X^N$ satisfy $\sigma(\{i;\ T^i\alpha\in U\})=1$ for any open set U containing Γ_ψ. Since (X,ψ) is expansive, there exists $\delta>0$ such that for any $\varepsilon>0$, there exists $m(\varepsilon)$ with the property that $d(x,y)<\varepsilon$ whenever $d(\psi^i x, \psi^i y)\leq\delta$ for any integer i with $|i|\leq m(\varepsilon)$. For $k=1,2,\ldots$, let

$$U_k=\{\beta\in X^N;\ d(\psi^i\beta(0), \beta(i))<\delta/2 \text{ for any } i=0,1,\ldots,k\}.$$

Then, U_k is an open set containing Γ_ψ. Let L be the constant in definition 2 for $\delta/2$. Define

$$a_1^k=\min\{i;T^i\alpha\in U_k\}, \text{ and}$$

$$a_{n+1}^k=\min\{i;T^i\alpha\in U_k,\ i>a_n^k+k+L\} \qquad (n=1,2,\ldots).$$

Then we have

$$\sigma(\bigcup_{n=1}^{\infty} I_n^k)=\frac{k+1}{k+L+1},$$

where $I_n^k=\{a_n^k, a_n^k+1,\ldots, a_n^k+k\}$. Then it is not difficult to see that we can take finite sets $M_1, M_2,\ldots$ of positive integers such that $\{I_n^k;\ n\in M_k,\ k=1,2,\ldots\}$ is a L-separated G-time satisfying

$\sigma(\bigcup_{k=1}^{\infty}\bigcup_{n\in M_k} I)=1$. Take $x_n^k\in\psi^{-a_n^k}\alpha(a_n^k)$ for any n and k with $n\in M_k$. Then

there exists $x\in X$ such that $d(\psi^i x, \psi^i x_n^k)<\delta/2$ for any i, k and n with $i\in I_n^k$ and $n\in M_k$, since (X,ψ) has the specification property. On the other hand, it holds that $d(\alpha(i), \psi^i x_n^k)<\delta/2$ for any i, k and n with $i\in I_n^k$ and $n\in M_k$. Therefore, $d(\alpha(i),\psi^i x)<\delta$ if $i\in I_n^k$ and $n\in M_k$. Take any $\varepsilon>0$. Assume that $\{i-m(\varepsilon), i-m(\varepsilon)+1,\ldots, i+m(\varepsilon)\}\subset I_n^k$ for some I_n^k with $n\in M_k$. Then we have $d(\alpha(i), \psi^i x)<\varepsilon$. It is clear that the set of i's as above has density 1, which completes the proof.

<u>Example</u> 3. It is known that the following dynamical systems are expansive and have the specification property, and therefore, have v.s.p. by proposition 3:

1. Anozov diffeomorphisms, and
2. topologically mixing subshifts of finite type on finite alphabets ([2], [11]).

Proposition 4. *Let (X,ψ) be a dynamical system with v.s.p. and (Y,η) be any dynamical system. Let ν be a $\psi\times\eta$-invariant measure on $X\times Y$. Let $y\in Y$ and $S\subset N$ satisfy*

$$\operatorname*{w-lim}_{\substack{n\in S\\ n\to\infty}} \frac{1}{n}\sum_{i=0}^{n-1} \delta_{\eta^i y} = \nu|_Y,$$

where $\nu|_Y$ is the marginal distribution of ν on Y. In addition, assume one of the following two conditions:

1. $S=N$,
2. *$\nu|_X$ is ergodic with respect to ψ.*

Then there exists $x\in X$ which is ψ-generic for $\nu|_X$ such that

$$\operatorname*{w-lim}_{\substack{n\in S\\ n\to\infty}} \frac{1}{n}\sum_{i=0}^{n-1} \delta_{(\psi^i x,\eta^i y)} = \nu.$$

(Proof) By the same notation T, we denote the shifts on X^N, Y^N and $X^N\times Y^N$, where the shift on $X^N\times Y^N$ is defined by $T(\alpha,\beta)=(T\alpha, T\beta)$ for $\alpha\in X^N$ and $\beta\in Y^N$. Let $\hat{\nu}$ be the T-invariant measure on $X^N\times Y^N$ defined by

$$\hat{\nu}(\{(\alpha,\beta)\in X^N\times Y^N;\ (\alpha(i),\ \beta(i))\in E_i \quad \text{for } i=0,1,\dots,L\})$$
$$=\nu(\bigcap_{i=0}^{L} (\psi\times\eta)^{-i} E_i)$$

for any L and Borel sets E_0, $E_1,\dots, E_L$ in $X\times Y$. Define $\beta\in Y^N$ by $\beta(i)=\eta^i y$ for any $i\in N$. Then it is clear that

$$\operatorname*{w-lim}_{\substack{n\in S\\ n\to\infty}} \frac{1}{n}\sum_{i=0}^{n-1} \delta_{T^i\beta} = \hat{\nu}|_{Y^N}.$$

Also, note that $\hat{\nu}|_{X^N}$ is ergodic with respect to T if $\nu|_X$ is ergodic with respect to ψ. Then by theorem 2 in [4], there exists $\alpha\in X^N$ which is T-generic for $\hat{\nu}|_{X^N}$ such that

$$\operatorname*{w-lim}_{\substack{n\in S\\ n\to\infty}} \frac{1}{n}\sum_{i=0}^{n-1} \delta_{T^i(\alpha,\beta)} = \hat{\nu}.$$

Since α is T-generic for the measure $\hat{\nu}|_{X^N}$ which is supported by Γ_ψ, we have $\sigma(\{i;\ T^i\alpha\in U\})=1$ for any open set U containing Γ_ψ. Therefore, there exists $x\in X$ such that $\sigma(\{i;\ d(\alpha(i),\ \psi^i x)<\varepsilon\})=1$ for any $\varepsilon>0$. It is clear that this x satisfies the required conditions.

Proposition 5. *Let (X,ψ) be a dynamical system with v.s.p. Let f be a homomorphism from (X,ψ) onto another dynamical system (Y,η). Then for any ψ-invariant measure μ on X and $y\in Y$ which is η-generic for $\mu\circ f^{-1}$, there exists $x\in X$ which is ψ-generic for μ such that $\sigma(\{i;\ \rho(\eta^i f(x),$*

$\eta^i y)<\varepsilon\})=1$ *for any* $\varepsilon>0$, *where* ρ *is a metric on* Y.

(Proof) Let $F=\{(u,v)\in X\times Y;\ f(u)=v\}$. Let ν be the $\psi\times\eta$-invariant measure on $X\times Y$ defined by $\nu(D\times E)=\mu(D\cap f^{-1}E)$ for any Borel sets D in X and E in Y. Then, y is η-generic for $\nu|_Y=\mu\circ f^{-1}$. By proposition 4, there exists $x\in X$ such that (x,y) is $\psi\times\eta$-generic for ν. Since ν is supported by F, we have

$$\sigma(\{i;\ \rho(\eta^i f(x),\ \eta^i y)\}<\varepsilon\ )$$

$$=\sigma(\{i;\ \rho(f(\psi^i x),\ \eta^i y)<\varepsilon\})$$

$$=\sigma(\{i;\ (\psi^i x,\ \eta^i y)\in U\})=1,$$

where $\varepsilon>0$ is arbitrary and $U=\{(u,v);\ \rho(f(u),\ v)<\varepsilon\}$ is an open set including F.

Corollary *Let* (X,ψ) *be a dynamical system with v.s.p. Then for any* ψ-*invariant measure* μ *on* X, *there exists* $x\in X$ *which is* ψ-*generic for* μ.

(Proof) Apply proposition 5, taking the trivial (Y,ψ); $Y=\{y\}$ and $\psi y=y$.

§4. Normal numbers and deterministic numbers

Let (G,ψ) be a group endomorphism.

Definition 4. $x\in G$ is a ψ-*normal number* if it is ψ-generic for the Haar measure on G. The set of ψ-normal numbers is denoted by Π_ψ.

Definition 5. $x\in G$ is called a ψ-*deterministic number* if $h_\psi(x)=0$.

By the ergodic theorem, if ψ is ergodic with respect to the Haar measure λ, then almost all $x\in G$ with respect to λ are ψ-normal numbers.

Example 4 (G. Rauzy [7]). Let $G=(R/Z)^N$ and T be the shift on G. Let f be an entire function such that

1. f is not a polynomial,
2. $f(R)\subset R$, and
3. $\overline{\lim_{r\to\infty}}\ \dfrac{\log\log M(r)}{\log\log r}<\dfrac{5}{4}$,

where $M(r)=\sup_{|z|=r} |f(z)|$. Then, $(f(0), f(1), f(2),\ldots) \pmod{Z} \in G$ is a T-normal number.

Example 5. (This example is due to Professor Mendes France of Bordeaux University.) Let $G=(R/Z)^{N}$ and T be the shift on G. Let f be a polynomial with real coefficients. Then, $(f(0), f(1), f(2),\ldots) \pmod{Z} \in G$ is a T-deterministic number.

(Proof) Let $x=(f(0), f(1), f(2),\ldots) \pmod{Z}$. We prove this by the induction about the degree n of the polynomial f. If n=0, then this is clear since x is a fixed point of T. Let us assume this up to n-1. Since $x-Tx=(g(0), g(1), g(2),\ldots) \pmod{Z}$, where $g(x)=f(x)-f(x+1)$ is a polynimial of degree n-1, x-Tx is a T-deterministic number. Therefore, x is a T-deterministic number by proposition 1.

Let (G,ψ) be a group endomorphism. Define ψ^* from the dual group G^* into G^* by $(\psi^*\chi)(x)=\chi(\psi x)$ for $\chi\in G^*$ and $x\in G$.

Lemma 2. Let (G,ψ) be a group endomorphism with v.s.p. Let λ be the Haar measure on G. Then, the measure preserving transformation ψ on the measure space (G,ψ) has a completely positive entropy.

(Proof) By V. A. Rohlin [9], an endomorphism on a compact metric Abelian group has a completely positive entropy with respect to the Haar measure if it is ergodic. Therefore, it is sufficient to prove that ψ on (G,λ) is ergodic. Let us suppose, to the contrary, that ψ is not ergodic with respect to λ. Note that G^* is an ortho-normal basis of $L_2(G,\lambda)$. Hence, ψ on (G,λ) has a Lebesgue spectrum and is ergodic if $\psi^{*n}(\chi)\neq\chi$ for any character $\chi\neq 1$ and n=1,2,... . Therefore, we have $\psi^{*k}(\chi)=\chi$ for some character $\chi\neq 1$ and $k\geq 1$. We may assume that $\psi^{*i}(\chi)\neq\chi$ for i=1,2,...,k-1. Let $g=\sum_{i=0}^{k-1}\psi^{*i}(\chi)$. Then g is a continuous function on G which is not constant, since $\{1, \chi, \psi^*(\chi),\ldots, \psi^{*(k-1)}(\chi)\}$ is an orthogonal system, and hence, linearly independent. Note that $g(\psi x)=g(x)$ for any $x\in G$. Take x and y in G such that $g(x)\neq g(y)$. Let

$s=\bigcup_{j=0}^{\infty}[(2j)^2, (2j+1)^2)$. Define $\alpha\in G^N$ by

$$\alpha(i)=\begin{cases}\psi^i x & i\in S\\ \psi^i y & i\notin S.\end{cases}$$

Then it is clear that $\alpha(\{i;\ T^i\alpha\in U\})=1$ for any open set $U\supset\Gamma_\psi$. Take

$z \in G$ such that $\sigma(\{i;\ d(\psi^i z, \alpha(i)) < \varepsilon\}) = 1$ for any $\varepsilon > 0$, where d is a metric on G. Then it holds that

$$\lim_{n\to\infty} \frac{1}{n} \sum_{i=0}^{n-1} |g(\psi^i z) - g(\alpha(i))| = 0.$$

But this is impossible since the left hand side is equal to $\frac{1}{2}(|g(z) - g(x)| + |g(z) - g(y)|)$.

Theorem <u>Let</u> (G,ψ) <u>be a group endomorphism with v.s.p. Assume that there exists a subgroup</u> K <u>of</u> G^* <u>such that</u> $\psi^* K \subset K$ <u>and</u> $G^* = F \otimes K$ (<u>direct product</u>), <u>where</u> $F = \{\chi \in G^*;\ \chi^n = 1$ for some integer $n \neq 0\}$. <u>Then</u>, $x \in G$ <u>is a</u> ψ-<u>deterministic number if and only if</u> $x + \Pi_\psi \subset \Pi_\psi$.

(Proof of the necessity) Let $y \in G$ be a ψ-deterministic number. Let $x \in \Pi_\psi$. Let S be any infinite subset of N. Then there exists $S' \subset S$ such that

$$\mu = \text{w-}\lim_{\substack{n\in S' \\ n\to\infty}} \frac{1}{n} \sum_{i=0}^{n-1} \delta_{\psi^i y} \tag{4.1}$$

exists. Since y is a ψ-deterministic number, $h_\mu(\psi) = 0$. Therefore, by H. Furstenberg [3], ψ on (G,λ) (λ is the Haar measure) and ψ on (G,μ) the former of which has a completely positive entropy by lemma 2 are <u>disjoint</u>; $\lambda \times \mu$ is the only $\psi \times \psi$-invariant measure ν on $G_1 \times G_2$ ($G_1 = G_2 = G$) such that $\nu|_{G_1} = \lambda$ and $\nu|_{G_2} = \mu$. Hence,

$$\text{w-}\lim_{\substack{n\in S' \\ n\to\infty}} \frac{1}{n} \sum_{i=0}^{n-1} \delta_{(\psi^i x, \psi^i y)} = \lambda \times \mu,$$

from which it follows that

$$\text{w-}\lim_{\substack{n\in S' \\ n\to\infty}} \frac{1}{n} \sum_{i=0}^{n-1} \delta_{\psi^i (x+y)} = \lambda * \mu = \lambda, \tag{4.2}$$

where * denotes the convolution of measures. Since we have proved that for any infinite subset S of N, there exists $S' \subset S$ satisfying (4.2), it holds that

$$\text{w-}\lim_{n\to\infty} \frac{1}{n} \sum_{i=0}^{n-1} \delta_{\psi^i(x+y)} = \lambda,$$

and that $x + y \in \Pi_\psi$.

To prove the sufficiency, we need the following lemma essentially due to H. Furstenberg [3].

Lemma 3. <u>For</u> $i=1,2$, <u>let</u> X_i <u>be a standard space and</u> μ_i <u>be a probability measure on</u> X_i. <u>Let</u> $\psi_i(i=1,2)$ <u>be a measure preserving transformation on</u> (X_i, μ_i). <u>Let</u> χ_i $(i=1,2)$ <u>be a complex valued, measurable function on</u> X_i <u>such that the shift</u> T <u>on</u> C^N (C <u>is the set of complex numbers) has a positive entropy with respect to</u> $\mu_i \circ f_i^{-1}$, <u>where</u> f_i <u>is a measurable mapping from</u> X_i <u>to</u> C^N <u>such that</u> $f_i(x)(n)=\chi_i(\psi_i^n x)$ <u>for any</u> $x \in X_i$ <u>and</u> $n \in N$. <u>Then there exists a</u> $\psi_1 \times \psi_2$<u>-invariant measure</u> ν <u>on</u> $X_1 \times X_2$ <u>such that</u>

1. $\nu|_{X_i} = \mu_i$ $(i=1,2)$, <u>and</u>

2. $\chi_1 \circ \pi_1$ <u>and</u> $\chi_2 \circ \pi_2$ <u>are not independent of each other under</u> ν, <u>where</u> π_i $(i=1,2)$ <u>is the projection</u> $X_1 \times X_2 \to X_i$.

(Proof) In theorem I.1 of [3], it was proved that there exists a measure τ on $C_1^N \times C_2^N$ $(C_1=C_2=C)$ such that

$$\tau|_{C_i^N} = \mu_i \circ f_i^{-1} \quad (i=1,2), \text{ and} \tag{4.3}$$

$$\tau \neq (\mu_1 \circ f_1^{-1}) \times (\mu_2 \circ f_2^{-1}). \tag{4.4}$$

But this τ fails to be T-invariant (T is the shift on $C_1^N \times C_2^N$), contradicting the statement of the theorem. Nevertheless, we can modify τ so to be T-invariant by taking a vague limit of

$$\frac{1}{n} \sum_{i=0}^{n-1} \tau \circ T^{-i}$$

when n tends to the infinity belonging to some subset of N. Thus we may assume that τ is a T-invariant measure on $C_1^N \times C_2^N$ satisfying (4.3) and (4.4). Now, we want to show the existence of $\psi_1 \times \psi_2$-invariant measure ν on $X_1 \times X_2$ satisfying the condition 1 in our lemma and that $\nu \circ (f_1 \times f_2)^{-1} = \tau$. But this was proved in the proof of proposition I.1 in [3].

(Proof of the sufficiency) Let H_1 and H_2 be the anihilators of K and F, respectively. Then, $\psi H_i \subset H_i$ $(i=1,2)$ and $G=H_1 \oplus H_2$ (direct sum). Assume that $y \in G$ is not a ψ-deterministic number. Let $y=y_1+y_2$, where $y_i \in H_i$ $(i=1,2)$. Then, at least one of y_1 or y_2 is not ψ-deterministic. Let ψ_i $(i=1,2)$ be the restriction of ψ to H_i. Since the projection $G \to H_i$ $(i=1,2)$ is a homomorphism from (G,ψ) onto (H_i, ψ_i), (H_i, ψ_i) has v.s.p. by proposition 2. For some $i=1,2$, let y_i be not a ψ_i-deterministic number. Then by lemma 4, which shall be proved later, there exists a ψ_i-normal number $x_i \in H_i$ such that x_i+y_i is not a ψ_i-

normal number. By proposition 4, there exists x_j, where $j=1-i$, such that (x_1, x_2) is $\psi_1\times\psi_2$-generic for the product of the Haar measures on H_1 and H_2; $x=x_1+x_2\in\Pi_\psi$. Since x_i+y_i is the image of $x+y$ by the projection $G\to H_i$ and is not a ψ_i-normal number, $x+y$ is not a ψ-normal number. To complete the proof, we prove the following lemma.

Lemma 4. <u>Let</u> (G,ψ) <u>be a group endomorphism with v.s.p. Assume that either</u> G^* <u>consists of non-cyclic elements or</u> G^* <u>consists of cyclic elements. Then for any</u> $y\in G$ <u>which is not a</u> ψ<u>-deterministic number, it holds that</u> $y+\Pi_\psi$ <u>is not contained in</u> Π_ψ.

(Proof) Since y is not ψ-deterministic, there exists $S\subset N$ such that

$$\mu=\text{w-}\lim_{\substack{n\in S\\ n\to\infty}}\frac{1}{n}\sum_{i=0}^{n-1}\delta_{\psi^i y}$$

exists and $h_\mu(\psi)>0$. Let $Q=\{z\in C;\ |z|=1\}$ and T be the shift on Q^N. For $\chi\in G^*$, let $f_\chi(u)(i)=\chi(\psi^i u)$ for any $u\in G$ and $i\in N$. Then, $\prod_{\chi\in G^*} f_\chi$ is an isomorphism from (G,ψ) into $(\prod_{\chi\in G^*}\Omega_\chi, \prod_{\chi\in G^*} T_\chi)$, where $\Omega_\chi=Q^N$ and $T_\chi=T$ for any $\chi\in G^*$. Hence,

$$0<h_\mu(\psi)=h_{\prod_{\chi\in G^*}(\mu\circ f_\chi^{-1})}\left(\prod_{\chi\in G^*} T_\chi\right)$$

$$\leq\sum_{\chi\in G^*} h_{\mu\circ f_\chi^{-1}}(T)$$

Therefore, there exists $\chi\in G^*$ such that $h_{\mu\circ f_\chi^{-1}}(T)>0$. In the case that G^* consists of non-cyclic elements, take any $\chi\in G^*$ such that $h_{\mu\circ f_\chi^{-1}}(T)>0$. In the case that G^* consists of cyclic elements take any $\chi\in G^*$ which satisfies that $h_{\mu\circ f_\chi^{-1}}(T)>0$ and has the minimum cycle among these elements. Since ψ on (G,λ) has a completely positive entropy and $(f_\chi(G), \mu\circ f_\chi^{-1})$ is a nontrivial measure space, we have $h_{\mu\circ f_\chi^{-1}}(T)>0$. Therefore, by lemma 3, there exists a $\psi\times\psi$-invariant measure ν on $G_1\times G_2$ $(G_1=G_2=G)$ such that $\nu|_{G_1}=\lambda$, $\nu|_{G_2}=\mu$ and that $\chi\circ\pi_1$ and $\chi\circ\pi_2$ are not independent of each other under ν, where π_i $(i=1,2)$ is the projection $G_1\times G_2\to G_i$. Since $\chi\circ\pi_1$ and $\chi\circ\pi_2$ are not independent of each other under ν, there exists $s, t\in Z$ such that

$$(4.5)\qquad \int_{G\times G}\chi^s(u)\,\chi^t(v)\,d\nu(u,v)\neq\int_G\chi^s(u)\,d\lambda(u)\cdot\int_G\chi^t(v)\,d\mu(v).$$

If s=0 or t=0, then (4.5) is not true. Therefore, $s\neq 0$ and $t\neq 0$. In this case, the right hand side of (4.5) is 0. Moreover, in the case that G* consists of cyclic elements, we can prove that t and the cycle of χ are relatively prime. Because, to the contrary, suppose that t and the cycle of χ are not reatively prime. Then by the assumption about χ, $h_{\mu\circ f^{-1}_{\chi^t}}(T)=0$. Let $B=f_{\chi^t}(G)$. Then by the disjointness (see the proof of the necessity), it holds that the unique $\psi\times T$-invariant measure θ on $G\times B$ such that $\theta|_G=\lambda$ and $\theta|_B=\mu\circ f^{-1}_{\chi^t}$ is $\lambda\times(\mu\circ f^{-1}_{\chi^t})$. Therefore $\nu\circ(I\times f_{\chi^t})^{-1}=\lambda\times(\mu\circ f^{-1}_{\chi^t})$, where I is the identical mapping on G. But this contradicts with (4.5). Thus, t and the cycle of χ are relatively prime. On the other hand, in the case that G* consists of non-cyclic elements, G is a divisible group. Let $\Omega=f_\chi(G)$. Then in the both cases, it holds that $g_t:\Omega\to\Omega$, where $g_t(\omega)=\omega^t$, is a homomorphism from (Ω,T) <u>onto</u> (Ω,T). By proposition 4, there exists $\alpha\in\Omega$ which is T-generic for $\lambda\circ f^{-1}$ such that

$$(4.6)\qquad \underset{\substack{n\in S\\ n\to\infty}}{\text{w-lim}}\ \frac{1}{n}\sum_{i=0}^{n-1}\delta_{(T^i\alpha,\,T^if_\chi(y))}=\nu\circ(f_\chi\times f_\chi)^{-1}.$$

Since g_t preserves the Haar measure $\lambda\circ f_\chi^{-1}$ on Ω, by proposition 5, there exists $\beta\in\Omega$ which is T-generic for $\lambda\circ f_\chi^{-1}$ such that $\sigma(\{i;\ d(T^i\beta^t, T^i\alpha)>\varepsilon\})=1$ for any $\varepsilon>0$, where d is a metric on Ω. Let $\gamma=\beta^s$. Then, it is easy to see that γ is also T-generic for $\lambda\circ f_\chi^{-1}$ and

$$(4.7)\qquad \sigma(\{i;\ d(T^i\gamma^t,\ T^i\alpha^s)<\varepsilon\})=1$$

for any $\varepsilon>0$. Now, by (4.5), (4.6) and (4.7), we have

$$(4.8)\qquad \lim_{\substack{n\in S\\ n\to\infty}}\frac{1}{n}\sum_{i=0}^{n-1}T^i(\gamma\cdot f_\chi(y))(0)^t$$

$$=\lim_{\substack{n\in S\\ n\to\infty}}\frac{1}{n}\sum_{i=0}^{n-1}(T^i\gamma^t)(0)\ \ (T^if_\chi(y)^t)(0)$$

$$=\lim_{\substack{n\in S\\ n\to\infty}}\frac{1}{n}\sum_{i=0}^{n-1}(T^i\alpha^s)(0)\ \ (T^if_\chi(y)^t)(0)$$

$$=\lim_{\substack{n\in S\\ n\to\infty}} \frac{1}{n}\sum_{i=0}^{n-1} (T^i\alpha)(0)^s \ (T^i f_\chi(y))(0)^t$$

$$=\int \xi(0)^s\cdot \eta(0)^t \, d(\nu\circ(f_\chi\times f_\chi)^{-1})(\xi,\eta)$$

$$=\int \chi^s(u)\chi^t(v)\,d\nu(u,v)$$

$$\neq \int \chi^s(u)\,d\lambda(u)\cdot\int \chi^t(v)\,d\mu(v)=0.$$

Since $\omega\to\omega(0)^t$ is a non-trivial character on Ω, (4.8) implies that $\gamma\cdot f_\chi(y)$ is not T-generic for the Haar measure $\lambda\circ f_\chi^{-1}$ on Ω. Since $\gamma\in\Omega$ is T-generic for $\lambda\circ f_\chi^{-1}$, by proposition 5, there exists $x\in\Pi_\psi$ such that

$$(4.9) \qquad \sigma(\{i;\ d(T^i f_\chi(x),\ T^i\gamma)<\varepsilon\})=1$$

for any $\varepsilon>0$. Then

$$\sigma(\{i;\ d(T^i f_\chi(x+y),\ T^i(\gamma\cdot f_\chi(y)))<\varepsilon\})=1$$

for any $\varepsilon>0$. Therefore,

$$\lim_{\substack{n\in S\\ n\to\infty}} \frac{1}{n}\sum_{i=0}^{n-1} \chi(\psi^i(x+y))$$

$$=\lim_{\substack{n\in S\\ n\to\infty}} \frac{1}{n}\sum_{i=0}^{n-1} (T^i f_\chi(x+y))(0)$$

$$=\lim_{\substack{n\in S\\ n\to\infty}} \frac{1}{n}\sum_{i=0}^{n-1} T^i(\gamma\cdot f_\chi(y))(0)\neq 0.$$

Since χ is a non-trivial character on G, this implies that $x+y\notin\Pi_\psi$, which completes the proof.

Example 6. Let T be the shift on a product group G^N. Then, (G^N, T) satisfies the conditions in the theorem.

Example 7. Let $G=(R/Z)^n$ and $\psi\in GL(n,Z)$. Since $\psi(Z^n)\subset Z^n$, ψ can be considered as an endomorphism on G. Assume that ψ has no proper value with the absolute value 1. Then, (G,ψ) satisfies the condition in the theorem.

REFERENCES

[1] P. Billingsley, Ergodic Theory and Information, John Wiley & Sons, New York, 1965.

[2] R. Bowen, Periodic points and measures for axiom A diffeomorphisms, Trans. Amer. Math. Soc. 154 (1971).

[3] H. Furstenberg, Disjointness in ergodic theory, minimal sets, and a problem in Diophantine approximation, Math. Systems Theory 1 (1967).

[4] T. Kamae, Subsequences of normal sequences, Israel J. Math. 16 (1973).

[5] T. Kamae & B. Weiss, Normal numbers and selection rules, Israel J. Math. 21 (1975).

[6] K. R. Parthasarathy, Probability Measures on Metric Spaces, Academic Press, New York, 1967.

[7] G. Rauzy, Fonctions entièrs et répartition modulo un II, Bull. Soc. Math. France 101 (1973).

[8] G. Rauzy, Normbres normaux et processus déterministes, to appear.

[9] V, A, Rohlin, Metric propertes of endomorphisms of compact commutative groups, Amer. Math. Soc. Transl. (2) 64 (1967).

[10] D. Ruelle, Statistical mechanics on a compact set with Z^{ν} action satisfying expansiveness and specification, Bull. Amer. Math. Soc. 78 (1972).

[11] K. Sigmund, On dynamical systems with the specification property, Trans. Amer. Math. Soc. 190 (1974).

[12] H. Totoki, Introduction to Ergodic Theory, Kyoritsu Shuppan, Tokyo, 1971 (in Japanese).

ON MULTITYPE BRANCHING PROCESSES WITH IMMIGRATION

KIYOSHI KAWAZU

1. Introduction. In this note we treat multitype branching processes with immigration with reference to the parallel phenomena corresponding to results for single type processes. We are interested in the classification of the irreducible aperiodic processes. For single type cases Foster and Williamson[11] and Yang[8] obtained an integral criterion for positive recurrences. Kaplan[9] showed the same for multitype case.

Our results and discussion are analogue of the works by Pakes[1].

Author thanks to Prof.Ogura and Prof.Pakes for their kind discussions and useful advice.

2. Definitions. Let S be the N-dimensional non-negative lattice and C be the N-dimensional non-negative unit cube. For $s = (s_1,\dots,s_N)$ in C and $j = (j_1,\dots,j_N)$ in S, we define such that $s^j = s_1^{j_1}\cdots s_N^{j_N}$. We use the natural partial order for S and C, that is, $x < x'$ means that the order holds elementwise.

Let $A^\tau(s)$, $B(s)$, $\tau = 1,\dots,N$, be probability generating functions on S and we assume that

$$A^\tau(s) = \Sigma\, a^\tau(j)s^j\ , \quad \tau = 1,\dots,N,$$

$$B(s) = \Sigma\, b(j)s^j\ , \quad s \in C.$$

$\{a^\tau(j)\}$ and $\{b(j)\}$ are probability distributions on S. We use following notations;

$$A(s) = (A^1(s),\dots, A^N(s)),$$

$$A_0(s) = s, \quad A_{n+1}(s) = A(A_n(s)),$$

$$F_n(s) = \prod_{k=0}^{n-1} B(A_k(s)),\ n = 0,1,\dots\ .$$

A Markov chain $\{Z_n\}$ on S is called, by definition, a multitype branching process with immigration if its transition function $P(i,j)$ satisfies the equation

$$(2.1) \qquad \sum_j P(i,j)s^j = A(s)^i B(s), \quad i \in S, \quad s \in C.$$

As is easily verified, the n-step transition function $P_n(i,j)$ of the process satisfies

$$(2.2) \qquad \sum_j P_n(i,j)s^j = A_n(s)^i F_n(s)\ .$$

In this note, we always assume that $b(0) < 1$ and $A(s)$ is not trivial.

It is well known that there exists a unique fixed point $q = (q_1, \ldots, q_N)$ of $A(s)$ in C which is the nearest to the origin and $q = \lim_{n\to\infty} A_n(0)$ ([7]).

We notice that by the same way as simple type cases, we can deduce the next results ([12]).

Proposition 1. If $B(0) > 0$ and $A(0) > 0$, then there exists a unique (up to constant multiple) non-trivial sequence $\{\pi(j)\}$ such that

$$(2.3) \qquad \alpha\pi(j) = \sum_k \pi(k)P(k,j), \quad j \in S,$$

where $\alpha = B(q)$. And when $\pi(k) > 0$, then

$$(2.4) \qquad \lim_{n\to\infty} \frac{P_{n+m}(i,j)}{P_n(u,k)} = \alpha^m \frac{q^i \pi(j)}{q^u \pi(k)} .$$

3. Classification. Now we assume following conditions ;

$$(3.1) \qquad \frac{\delta A^\tau}{\delta s_\nu}(1-) = m_{\tau\nu} < \infty, \quad \tau,\nu = 1,\ldots,N,$$

(3.2) the matrix $M = (m_{\tau\nu})$ is positively regular, that is, M is irreducible and aperiodic.

$$(3.3) \qquad A(s) \neq Ms .$$

By the Perron-Frobenius Theorem, the positively regular matrix M has a simple and largest (in modulus) eigenvalue ρ. And we can choose strictly positive left and right eigenvectors, $u = (u_1, \ldots, u_N)$ and $v = (v_1, \ldots, v_N)$, respectively, corresponding to ρ. We normalize u and v as follow ;

$$(3.4) \qquad |u| = \sum_\tau u_\tau = 1, \quad v\cdot u = \sum_\tau u_\tau v_\tau = 1.$$

when $\rho < 1$, $\rho = 1$ or $\rho > 1$, the process is called subcritical, critical or supercritical, respectively.

We list here some conditions below.

$$(a) \qquad t_\tau = \sum_j b(j) j_\tau , \quad t = (t_1, \ldots, t_N) < \infty .$$

$$(b) \qquad \Delta^\tau_{\nu\mu} = \frac{1}{2} \sum_j a^\tau(j) j_\nu (j_\mu - \delta_{\nu\mu}) < \infty , \quad \tau,\nu,\mu = 1,\ldots,N,$$

where $\delta_{\nu\mu}$ is the kronecker's delta function.

$$(c) \qquad T_\tau = \sum_j b(j) j_\tau \log j_\tau < \infty , \quad \tau = 1,\ldots,N.$$

(d) $\sum_j a^\tau(j) j_\nu^2 \log j_\nu < \infty$, $\tau, \nu = 1,\dots,N$.

The following results are used for our discussion.

Theorem A. (Joffe-Spitzer [5]) When $\rho \leqq 1$, $1 - A_n(s)$ converges to 0 as n tends to infinity uniformly on C.

Theorem B. ([5]) Under the assumption (b), if the process is subcritical then next equality holds uniformly in $s \in C/\{1\}$,

$$(3.5) \quad \lim_{n\to\infty} \frac{1}{n}\left[\frac{1}{v\cdot(1 - A_n(s))} - \frac{1}{v\cdot(1 - s)} \right] = Q$$

where $Q = \sum v^\tau \Delta^\tau_{\nu\mu} u_\nu u_\mu$.

Lemma C. (Ogura-Shiotani [4]) Under the assumption (b), if the process is critical, then

$$(3.6) \quad \frac{1 - A_n(s)}{v\cdot(1 - A_n(s))} = u + O\left(\frac{\log n}{n}\right) \quad \text{as } n \to \infty,$$

where $O(\log n/n)$ is uniformly in $s \in C/\{1\}$.

Lemma D. ([4]) Let the critical process satisfy the condition (d) and set

$$(3.7) \quad h_n(s) = \frac{1}{v\cdot(1 - A_n(s))} - \frac{1}{v\cdot(1 - s)} - nQ,$$

then

$$\sum_n \frac{h_n(s)}{n^2} < \infty \quad \text{for } s \in C/\{1\} .$$

Lemma E. (Pakes [2]) Let $K(x) = \sum k_j x^j$ be a probability generating function of 1-dimension. If $0 < d_n < 1$ and $1 - d_n \sim a/n$, $0 < a < \infty$ then

$$\sum_n (1 - K(d_n))/n < \infty \quad \text{iff} \quad \sum_j k_j \log j < \infty .$$

Theorem 1. If the process is supercritical, then the process is transient and if $B(0) > 0$, $A(0) > 0$, then

$$(3.8) \quad P_n(i,j) = \alpha^n \pi(j) q^j + o(\alpha^n), \quad \text{as } n \to \infty ,$$

for every i,j in S.

Proof. Since $P_n(0,0) \leqq B(q)^n$, the process is transient. Now let

$$(3.9) \quad \Psi(s) = \sum \pi(j) s^j$$

then it is easily verified that $\Psi(s)$ converges for $s \in C/\{1\}$ and satisfies the functional equation

$$(3.10) \qquad \alpha^n \Psi(s) = F_n(s)\Psi(A_n(s)) \ .$$

Choosing Ψ so that $\Psi(q) = 1$, then we obtain

$$\Sigma P_n(i,j)s^j = \alpha^n \Psi(s) A_n(s)^i / \Psi(A_n(s)) \ .$$

Since $A_n(s)$ converges to q for $0 < s < q$, the assertion follows. q.e.d.

Set $\sigma = t \cdot u/Q$, which plays an essential role.

Theorem 2. When the process is critical, under the assumptions (a) and (b), the process is recurrent if $\sigma < 1$, and transient if $\sigma > 1$. Moreover if we assume the conditions (c) and (d), then we obtain

$$(3.11) \quad P_n(i,j) = c\pi(j)n^{-\sigma} + o(n^{-\sigma}), \quad \text{as } n \to \infty \ ,$$

where c is a positive constant. Especially if $\sigma \leqq 1$, then the process is null recurrent.

Proof. The first assertion is easily verified. In fact, using Theorem B and Lemma C, we obtain

$$\begin{aligned}\frac{P_{n+1}(0,0)}{P_n(0,0)} &= B(A_n(0)) \\ &= 1 + t\cdot(A_n(0)) + o(\ |1 - A_n(0)|) \\ &= 1 - \frac{\sigma}{n} + o(\ 1/n) \ .\end{aligned}$$

And if $\sigma < 1$, the series $\Sigma P_n(0,0)$ diverges and if $\sigma > 1$, the series converges, thus proving the assertion.

To prove the latter part, we consider the following quantity which is analogous to that of Pakes [1],[2].

$$(3.12) \qquad n^\sigma P_n(0,0) = B(0) \prod_{k=1}^{n-1} (1 + 1/k)^\sigma B(A_k(0)).$$

Now as is easily verified

$$\begin{aligned}(3.13) \qquad & (1 + 1/k)^\sigma B(A_k(0)) \ - 1 \\ & = \sigma/k - t\cdot(1 - A_k(0)) + (t - t')\cdot(1 - A_k(0)) + o(1/k^2)\end{aligned}$$

where $t' = (t'_1,\ldots,t'_N)$ and $t'_\tau = \dfrac{\delta B}{\delta s_\tau}(\eta_k)$, $A_k(0) < \eta_k < 1$.

Firstly we show that first tow terms in right side of (3,13) are summable. For, using Lemma C and Lemma D, we obtain

$$\sigma/k - t.(1 - A_k(0)) = \frac{\sigma h_k(0) + \sigma/|v| + O(\log k)}{k(h_k(0) + kQ + 1/|v|)} .$$

Thus Lemma D implies the assertion. Next the third term is also summable. In fact, set

$$K_\tau(x) = \frac{\delta B}{\delta s_\tau}(x1)/t_\tau \text{ for } 0 < x < 1, \quad \tau = 1,\dots,N,$$

Then since $1 - A_k(0) = u/nQ + o(1)$ and the function $x\log x$ is convex we can apply Lemma E. Therefore the left side of (3.13) is summable implying that

$$(3.14) \qquad \lim_{n\to\infty} n^\sigma P_n(0,0) = c$$

for some constant $0 < c < \infty$. By Proposition 1,

$$P_n(i,j) = \pi(j)P_n(0,0) + o(P_n(0,0)) \quad \text{as} \quad n \to \infty ,$$

thus we obtain

$$P_n(i,j) = cn^{-\sigma}\pi(j) + o(n^{-\sigma}) \quad \text{as} \quad n \to \infty . \quad \text{q.e.d.}$$

By Theorem 2, under the assumptions (c) and (d), Karamata's Tauberian Theorem[3] implies that we can apply Darling-Kac's theorem[3].

Theorem 3. If the critical process satisfies the conditions (c) and (d), then

$$(3.15) \qquad \lim_{n\to\infty} P\{(1/U_n)\sum_{m=0}^{n} \chi(Z_m) \leqq x\} = G_{1-\sigma}(x)$$

where $G_\xi(x)$ is the Mittag-Leffler distribution, $\chi(\cdot)$ is the indicator function of the state $\{0\}$, and U_m is defined as $c\log m$ when $\sigma = 1$ and $cm^{1-\sigma}/\Gamma(1 - \sigma)$ otherwise.

The proof of this theorem is same as Pakes[1]. This means that the property of multitype irreducible processes with immigration is essentially same as one dimensional cases.

References

[1] Pakes,A.G.(1971) On the critical Galton-Watson process with immigration,J.Austral.Math.Soc.12,p.476-482.

[2] Pakes,A.G.(1971) Further results on the Critical Galton-Watson process with immigration,J.Austral.Math.Soc.p.277-290.

[3] Darling,D.A.,Kac,M.(1957) On occupation times for Markov processes Trans.Amer.Math.Soc.84,p.448-458.

[4] Ogura,Y.,Shiotani,K. On invariant measures of criticalmultitype Galton-Watson processes, To appear.

[5]Joffe,A.,Spitzer,F.(1967) On multitype branching processes with $\rho \leqq 1$, J.Math.Anal.Appl.19,p.409-430.

[6]Feller,W.(1966). An introduction to Probability Theory and its application Vol.II , NewYork Wiley.

[7] Sevastyanov,B.A.(1971) Branching Processes, Nauka,Moscow.

[8] Yang,T.S.(1972) On branching processes allowing immigration. J.Appl. Prob.9,p.24-31.

[9] Kaplan,N.(1973) The multitype Galton-Watson process with Immigration Anal.Prob.1, p.947-953.

[10] Kawazu,K.(1974) On α-invariant measures of G.W.I.,Yamaguchi Tech. Rep.Math.No.1.

[11] Foster,J.,Williamson,J.A.(1971) Limit theorems for the Galton-Watson process with time dependent immigration,Z.W.20,p.227-235.

STATISTICS OF STOCHASTIC PROCESSES WITH JUMPS

Takashi Komatsu

Statistics of diffusion type processes was studied by Shiryayev[10]. It is assumed there that the distribution of the observation process dependent on the input signal is mutually absolutely continuous to the distribution of a Wiener process. In this report, we shall attempt to replace the Wiener process by a Lévy process independent of the input signal. In the first section, we shall consider the asymptotic bihavior of statistical estimators of the unknown parameter in the smooth case. And in the second section, we shall consider a stochastic dynamical system. We shall investigate the differentiability of the density of the *a posteriori* distribution and apply it to a coding problem.

§1. Asymptotic bihavior of statistical estimators

Denote by $X_t(w)$ the value of a function $w \in W = D([0,\infty) \to R^1)$ at time t, by $\underline{\underline{W}}_t$ the σ-field generated by $(X_s; 0 \leq s \leq t)$ and $\underline{\underline{W}} = \bigvee \underline{\underline{W}}_t$. Let $(W, \underline{\underline{W}}, Q; X_t)$ be a Lévy process of the form

$$X_t = M_t + \int_0^t b(s)ds + \int_0^t \int_{|u|\leq 1} u\, J_c(ds,du) + \int_0^t \int_{|u|>1} u\, J(ds,du),$$

where M_t is a continuous martingale such that $d\langle M_t, M_t \rangle = a(t)dt$, $J(dt,du)$ a Poisson random measure with $E[J(dt,du)] = N(t,du)dt$ and $J_c(dt,du) = J(dt,du) - N(t,du)dt$. If $(W, \underline{\underline{W}}, Q'; X_t)$ is a process such that $Q'_t = Q'|_{\underline{\underline{W}}_t}$ and $Q_t = Q|_{\underline{\underline{W}}_t}$, probabilities restricted on the σ-field $\underline{\underline{W}}_t$, are mutually absolutely continuous for each t, then there are non-anticipating functions $h(t,w)$ and $\rho(t,w,u) > 0$ such that

(1.1) $\int_0^t [ah^2 + 4\int(\sqrt{\rho} - 1)^2 N](s,w)ds < \infty$ for each t

and $dQ'_t/dQ_t = \Phi_t(h,\rho;Q)$, where

$$(1.2)\qquad \log \Phi_t(h,\rho;Q) = \int_0^t h(s,w)dM_s - \frac{1}{2}\int_0^t ah^2(s,w)ds + \int_0^t\int \{\log\rho(s,w,u)J - (\rho(s,w,u) - 1)Nds\}$$

(cf. Kunita - Watanabe [6]).

Now, a family (Q^θ) of probabilities on $(W,\underline{\underline{W}})$ dependent on a parameter $\theta \in R^1$ is given. Suppose that Q_t^θ and Q_t are mutually absolutely continuous for any t and

$$(1.3)\qquad \frac{dQ_t^\theta}{dQ_t} = \Phi_t(h_\theta,\rho_\theta;Q), \quad \text{this is denoted by } \phi_t(\theta),$$

for non-anticipating functions $h_\theta = h(t,w,\theta)$ and $\rho_\theta = \rho(t,w,\theta,u) > 0$ satisfying (1,1). These functions are assumed to be continuously differentiable in θ. We may call the function

$$(1.4)\qquad I_{t,w}(\theta) = a(t)(h_\theta'(t,w))^2 + \int \frac{(\rho_\theta')^2}{\rho_\theta}(t,w,u)\, N(t,du),$$

where $h_\theta' = \partial h/\partial\theta$, $\rho_\theta' = \partial\rho/\partial\theta$, the Fisher information of (Q^θ). In fact, we have

$$(1.5)\qquad E^\theta[(\frac{\partial}{\partial\theta}\log\phi_\tau(\theta))^2] = E^\theta[\int_0^\tau I_{s,w}(\theta)ds]$$

for any stopping time τ, where $E^\theta[\cdot]$ means the expectation by Q^θ. Without loss of generality, we can suppose that the <u>true parameter</u> is at the origin 0. In this section, the following condition is always satisfied.

Condition 1. $0 < I_{t,w}(0) < \infty$ a.e. and $\int_0^\infty I_{s,w}(0)ds = \infty$.

Let $\tau(t)$ be a strictly increasing process determined by

$$(1.6)\qquad \int_0^{\tau(t)} I_{s,w}(0)\, ds = t ,$$

and define a process

$$(1.7)\qquad R^1 \ni y \rightsquigarrow Z_t(y) = \phi_{\tau(t)}(\frac{y}{\sqrt{t}})/\phi_{\tau(t)}(0).$$

Put

$$(1.8)\qquad A_{t,w}(\theta,0) = a(t)(h_\theta - h_0)^2 + 4\int(\sqrt{\rho_\theta} - \sqrt{\rho_0})^2 N(t,du)$$

$$A^1_{t,w}(\theta,0) = a(t)(h'_\theta - h'_0)^2 + \int\left(\frac{\rho'_\theta}{\sqrt{\rho_\theta}} - \frac{\rho'_0}{\sqrt{\rho_0}}\right)^2 N(t,du)$$

and introduce a condition.

Condition 2.

a) $$\limsup_{\theta\to 0}\ \sup_{t,w} \frac{A^1_{t,w}(\theta,0)}{I_{t,w}(0)} = 0 \text{ and } \limsup_{\theta\to 0}\ \sup_{t,w} \frac{\int(\rho'_\theta/\rho_\theta - \rho'_0/\rho_0)^2\rho_0 N}{I_{t,w}(0)} = 0.$$

b) There exists a function $F(x)$ such that $\lim_{|x|\to\infty} F(x)/x^2 = \infty$ and

$$\sup_{t,w} \frac{1}{I_{t,w}(0)} \int F\left(\frac{\rho'_0}{\rho_0}\right)\rho_0 N < \infty.$$

<u>Lemma 1.1. If the functions</u> h_θ <u>and</u> ρ_θ <u>satisfy Condition 1 and 2, then as</u> $t \to \infty$ <u>the finite dimensional distributions of the process</u> $(Z_t(y),Q^0)$ <u>converge to the finite dimensional distributions of the process</u> $Z(y) = \exp[yB - y^2/2]$, <u>where</u> B <u>is a normal random variable with parameter</u> $(0,1)$.

Proof. Decompose $\log Z_t(y)$ to three terms

$$\log Z_t(y) = yB_t + \frac{1}{2}\int_0^{\tau(t)} A_{s,w}\left(\frac{y}{\sqrt{t}},0\right)ds + \Delta_{t,w}(y),$$

where

$$B_t = \frac{1}{\sqrt{t}}\left[\int_0^{\tau(t)} h'_0\, dM^0_s + \int_0^{\tau(t)}\int \frac{\rho'_0}{\rho_0} J^0_c(ds,du)\right],$$

$$dM^0_t = dM_t - ah_0dt \quad\text{and}\quad J^0_c(dt,du) = J(dt,du) - \rho_0 N dt.$$

It is verified from Condition 2 that, as $t \to \infty$,

$$\Delta_{t,w}(y) \to 0,\quad \int_0^{\tau(t)} A_{s,w}\left(\frac{y}{\sqrt{t}},0\right)ds \to y^2 \text{(in probability) and}$$

$$E^0[e^{i\xi B_t}] \to e^{-\xi^2/2} \quad\text{for each } \xi. \qquad \text{Q.E.D.}$$

We shall introduce one more condition.

Condition 3.

a) $$\inf_{t,w,\theta\neq 0} \frac{A_{t,w}(\theta,0)}{I_{t,w}(0)\log(1+|\theta|^2)} > 0 ,$$

b) $$\sup_{t,w,\theta} \frac{I_{t,w}(\theta)}{I_{t,w}(0)(1+|\theta|^p)} < \infty \quad \text{for a certain constant } p \geq 0.$$

Lemma 1.2. *Under Condition 1 and 3, there is a constant* t_0 *such that*

a) $$\lim_{R\to\infty} \sup_{t\geq t_0} Q^0(\sup_y Z_t(y) > R) = 0,$$

b) $$\lim_{r\to\infty} \sup_{t\geq t_0} Q^0(\sup_{|y|>r} Z_t(y) > \varepsilon) = 0 \quad \textit{for any } \varepsilon > 0,$$

c) $$\lim_{\delta\downarrow 0} \sup_{t\geq t_0} Q^0(\sup_{|y-y'|\leq\delta} |Z_t(y)-Z_t(y')| > \varepsilon) = 0 \quad \textit{for each } \varepsilon > 0.$$

Proof. 1°. We observe that

$$E_Q[\phi_\tau(\theta)\phi_\tau(0) \exp(\tfrac{1}{8} \textstyle\int_0^\tau A_{s,w}(\theta,0)ds)] = 1 , \tag{1.9}$$

where $\tau = \tau(t)$ and $E_Q[\cdot]$ means the expectation by Q. And then

$$Q^0(Z_t(y) > y^{-2n}) \leq |y|^n E^0[\sqrt{Z_t}(y)]$$

$$\leq |y|^n \exp[- \tfrac{1}{8} \inf_w \textstyle\int_0^\tau A_{s,w}(\tfrac{y}{\sqrt{t}},0)ds] \qquad \text{(by (1.9))}$$

$$\leq |y|^n(1 + \tfrac{y^2}{t})^{-\delta t} , \quad \delta > 0 \qquad \text{(by Condition 3.a)}.$$

Since $\lim_{t\to\infty} \sup_y |y|^{3n}(1 + y^2/t)^{-\delta t} < \infty$, there are constants c_n and t_n such that, for all $n \geq 0$ and $t \geq t_n$,

$$Q^0(Z_t(y) > y^{-2n}) \leq c_n y^{-2n}. \tag{1.10}$$

2° From (1.5) we obtain the inequality

$$E_Q[(\sqrt{\phi_\tau}(\theta_1)-\sqrt{\phi_\tau}(\theta_0))^2] \leq \frac{(\theta_1-\theta_0)^2}{4} \int_0^1 E^{\theta_\alpha}[\textstyle\int_0^\tau I_{s,w}(\theta_\alpha)ds]d\alpha , \tag{1.11}$$

where $\theta_\alpha = \alpha\theta_1 + (1-\alpha)\theta_0$. If $t \geq 1$ and $m \leq y,y' \leq m+1$, then

(1.12) $E^0[(\sqrt{Z_t}(y)-\sqrt{Z_t}(y'))^2] = E_Q[(\sqrt{\phi_\tau}(\frac{y}{\sqrt{t}})-\sqrt{\phi_\tau}(\frac{y'}{\sqrt{t}}))^2]$

$\leq$ const. $(y - y')^2(1 + |m|^p)$ (by (1.11) and Condition 3.b).

3^0. From (1.10) and (1.12) the assertion of the lemma can be proved by a similar method to Ibragimov - Khas'minskii [4]. Q.E.D.

Let $\underline{\underline{C}}_0$ be the space of continuous functions $f(y)$ on R^1 such that $\lim_{|y|\to\infty} f(y) = 0$. Lemma 1.2 implies that the function $Z_t(y)$ belongs to the space $\underline{\underline{C}}_0$ for $t \geq t_0$ and the family of distributions in $\underline{\underline{C}}_0$ generated by the processes $Z_t(y)$ is tight. Combining this fact and Lemma 1.1, we obtain the following result.

Theorem 1. If Condition 1, 2 and 3 are satisfied, then the distributions in $\underline{\underline{C}}_0$ generated by the processes $Z_t(y)$ converge as $t \to \infty$ to the distribution generated by the process

(1.13) $Z(y) = e^{yB - y^2/2}$, B: a normal r. v. with parameter (0,1).

We shall denote by $\hat{\theta}_t$ the maximum likelihood estimator (M.L.E.) of the parameter θ based on $\underline{\underline{W}}_t$.

Corollary. Under Condition 1, 2, and 3, the M.L.E. exists for sufficiently large t. The distribution of the random variable $\sqrt{t}\ \hat{\theta}_{\tau(t)}$ with respect to the probability Q^0 is asymptotically normal with parameter (0,1).

Proof. Since $Z_t(y) \in \underline{\underline{C}}_0$ for $t \geq t_0$, there is a value $\hat{y}_t$ such that $Z_t(\hat{y}_t) = \max_y Z_t(y)$. The M.L.E. $\hat{\theta}_t$ is given by $\hat{\theta}_{\tau(t)} = \hat{y}_t/\sqrt{t}$. Let F_x, $x \in R^1$, be continuous functionals on $\underline{\underline{C}}_0$ defined by

$$F_x(f) = \max_{x \leq y} |f(y)| - \max_{x \geq y} |f(y)|.$$

Let Q^* be the probability for which the random variable B of (1.13) is defined. Since $Q^*(F_x(Z(\cdot)) = 0) = 0$, it follows from Theorem 1 that

$$Q^0(\sqrt{t}\ \hat{\theta}_{\tau(t)} \geq x) = Q^0(F_x(Z_t(\cdot)) \geq 0)$$

$$\rightarrow Q^*(F_x(Z(\cdot)) \geq 0) = Q^*(B \geq x) \quad \text{as } t \rightarrow \infty. \qquad \text{Q.E.D.}$$

Suppose that the a priori distribution $\Pi(d\theta) = p(\theta)d\theta$ of the parameter is given. Let $\bar{\theta}_t$ be the a posteriori mean of the parameter based on $\underline{\underline{W}}_t$. The following result is induced from Theorem 1 and (1.10) (cf. Ibragimov - Khas'minskii [5]).

Theorem 2. *Assume that the functions* h_θ *and* ρ_θ *satisfy Condition 1, 2 and 3. If* $p(\theta)$ *is bounded and continuous and if* $p(0) \neq 0$, *then the distribution of* $\sqrt{t}\ \bar{\theta}_{\tau(t)}$ *with respect to* Q^0 *tends to the normal distribution with parameter* (0,1).

§2. On the density of the a posteriori distribution

In this section, every notations are the same as in the preceeding section except that W means the space $D([0,T] \rightarrow R^1)$ of functions on $[0,T]$. The distribution Q_T of the Lévy process is simply denoted by Q. Let Z be a space of functions z on $[0,T]$ taking values $z(t) = \theta_t(z)$ in a locally compact Hausdorff space S. Denote by $\underline{\underline{Z}}_t$ the σ-field generated by $(\theta_s; 0 \leq s \leq t)$. The stochastic dynamical system to be considered here is a process $(W\times Z, \underline{\underline{W}}_T \times \underline{\underline{Z}}_T, P; (X_t, \theta_t))$ which has the following property.

a) Given a system $(\Pi^w)_{w\in W}$ of probabilities on the space $(Z, \underline{\underline{Z}}_T)$ such that if $\Gamma \in \underline{\underline{Z}}_t$, then $w \rightsquigarrow \Pi^w(\Gamma)$ is $\underline{\underline{W}}_t$-measurable.

b) Given $(\underline{\underline{W}}_t \times \underline{\underline{Z}}_t)$-non-anticipating functions $h(t,w,z)$ and $\rho(t,w,z,u)>0$ such that

$$(2.1) \qquad E_P[\textstyle\int_0^T (ah^2 + \int(\rho-1)\log\rho N)(s,w,z)ds] < \infty.$$

c) Let $Q\otimes\Pi$ be the probability on $(W\times Z, \underline{\underline{W}}_T \times \underline{\underline{Z}}_T)$ such that

$$Q\otimes\Pi(\Gamma'\times\Gamma) = \int_{\Gamma'} \Pi^w(\Gamma)Q(dw) \quad \text{for any } \Gamma \in \underline{\underline{Z}}_T \text{ and } \Gamma' \in \underline{\underline{W}}_T.$$

Then probabilities P and $Q\otimes\Pi$ are mutually absolutely continuous and

(2.2) $$dP_t/d(Q\otimes\Pi)_t = \Phi_t(h,\rho;Q),$$

where $P_t = P|_{\underline{W}_t\times\underline{Z}_t}$ and $(Q\otimes\Pi)_t = Q\otimes\Pi|_{\underline{W}_t\times\underline{Z}_t}$.

From now on, σ-fields $\underline{W}_t$ are imbeded in $\underline{W}_T\times\underline{Z}_T$. The process $(X_t, \bar{P})$, $\bar{P} = P|_{\underline{W}_T}$, is called the observation process. We shall denote the filtering $E_P[f(t,w,z)|\underline{W}_t]$ by $\pi_t(f)$ or $\pi f(t,w)$. It is easy to show that

(2.3) $$d\bar{P}_t/dQ_t = \Phi_t(\pi h,\pi\rho;Q), \quad \text{where } \bar{P}_t = P|_{\underline{W}_t}.$$

Put

(2.4) $$I(X_0^t,\theta_0^t) = E_P[\log \frac{dP_t}{d(\bar{P}\otimes\Pi)_t}],$$

where $\bar{P}\otimes\Pi$ is the probability on the space $(W\times Z, \underline{W}_T\times\underline{Z}_T)$ such that

$$\bar{P}\otimes\Pi(\Gamma'\times\Gamma) = \int_{\Gamma'}\Pi^w(\Gamma)\bar{P}(dw) \quad \text{for any } \Gamma\in\underline{Z}_T \text{ and } \Gamma'\in\underline{W}_T.$$

If Π^w is independent of $w\in W$, then the value $I(X_0^t,\theta_0^t)$ is called the mutual information between the input signal $(\theta_s; s\leq t)$ and the observation process $(X_s; s\leq t)$ up to time t. Since

(2.5) $$H(P_t) \equiv - E_P[\log(dP_t/d(Q\otimes\Pi)_t)]$$
$$= - \int_0^t E_P[\tfrac{1}{2}ah^2 + \int(1 - \rho + \rho\log\rho)N]ds,$$
$$H(\bar{P}_t) \equiv - E_{\bar{P}}[\log(d\bar{P}_t/dQ_t)]$$
$$= - \int_0^t E_{\bar{P}}[\tfrac{1}{2}a(\pi h)^2 + \int(1 - \pi\rho + \pi\rho\log(\pi\rho))N]ds,$$

the equality (Grigelionis [3])

(2.6) $$I(X_0^t,\theta_0^t) = \int_0^t E_P[\tfrac{1}{2}a(h - \pi h)^2 + \int(\rho\log\rho - \pi\rho\log(\pi\rho))N]ds$$

follows immediately from the relation $I(X_0^t,\theta_0^t) = H(\bar{P}_t) - H(P_t)$.

Let $f(x)$ be a bounded measurable function on S such that there is a bounded measurable and non-anticipating function $Lf(t,w,x)$ on $[0,T]\times W\times S$ for which the process

$$f(\theta_t) - f(\theta_0) - \int_0^t Lf(s,w,\theta_s)ds$$

is a martingale with respect to $(\underline{\underline{Z}}_t, \Pi^w)$. Grigelionis [2] showed that the filtering $\pi_t f$ satisfies the following equation.

(2.7) $$d\pi_t(f) = \pi_t(Lf)dt + \pi_t(f(h-\pi h))d\bar{M}_t + \int \pi_t(f(\frac{\rho}{\pi\rho} - 1))\bar{J}_c(dt,du),$$

where $d\bar{M}_t = dM_t - a(t)\pi_t(h)dt$ and $\bar{J}_c(dt,du) = J(dt,du) - \pi_t(\rho)N(t,du)dt$. It is immediate to show that the process $U_t(f) = \Phi_t(\pi h, \pi\rho; Q)\pi_t(f)$ satisfies the equation

(2.8) $$dU_t(f) = U_t(Lf)dt + U_t(fh)dM_t + \int U_t(f(\rho-1))J_c(dt,du).$$

We shall consider the following case.

a) $S = R^1$ and $L \equiv L_{t,w,x}$ is a differential operator such that

(2.9) $$L = \frac{1}{2} A(t,w,x)\partial^2 + B(t,w,x)\partial, \quad \partial = \partial/\partial x,$$

where $A(t,w,x) \geqq 0$ and $B(t,w,x)$ are non-anticipating functions ;

b) there are measurable functions $h_x(t,w)$ and $\rho_x(t,w,u) > 0$ such that $h(t,w,z) = h_{\theta_t(z)}(t,w)$ and $\rho(t,w,z,u) = \rho_{\theta_t(z)}(t,w,u)$.

Let C_*^n (resp. $C_*^{n,b}$) be the space of functions $f(t,x)$ on $[0,T]\times R^1$ such that $\partial^m f(t,x)$, $m \leq n$, are continuous in x and bounded in t (resp. bounded in (t,x)) and further $\lim_{x\to x_0} \sup_t |\partial^m f(t,x) - \partial^m f(t,x_0)| = 0$ for any $x_0 \in R^1$. We always assume that the following condition is satisfied.

Condition A. For almost all w (Q),

a) $A(t,w,x) \in C_*^0$ and $B(t,w,x) \in C_*^0$;

b) $$\int_0^T \sup_{|x|\leqq r} [ah_x^2 + \int(\rho_x - 1)\log\rho_x N](s,w)ds < \infty \quad \text{and}$$

$$\lim_{\delta\downarrow 0} \int_0^T \sup_{|x|\leqq r, |x-y|\leqq\delta} [a(h_x - h_y)^2 + \int(\sqrt{\rho_x} - \sqrt{\rho_y})^2 N](s,w)ds = 0$$

for each $r > 0$ and further

(2.10) $$\Psi(t,w,x) \equiv \Phi_t(h_x, \rho_x; Q) \quad \in C_*^2.$$

Equation (2.8) is regarded as a stochastic differential equation for the measure U_t. Namely, we shall say that a measure valued process $U_t(dx)$ adapted to ($\underline{W}_t$) is a solution of (2.8) if the functional $U_t : f(x) \rightsquigarrow \int f(x) U_t(dx)$ (which is again denoted by $U_t(f)$) satisfies (2.8) for each test function $f(x) \in C_K^\infty(R^1)$ and if the process $U_t(1) \equiv \int U_t(dx)$ is strictly positive and locally bounded (i.e. there is a sequence T_n of stopping times such that $T_n \uparrow$, $Q[T_n < T] \to 0$ and each stopped process $U_{t \wedge T_n}(1)$ is bounded). There is a close relation between equation (2.8) and the ones in Rozovskii [9]. Making use of ⟨the Fubini theorem⟩ for an iteration of a Lebesque's integral and a stochastic integral, it is possible to transform the equation to a partial differential equation of parabolic-type (cf. Fujisaki - Komatsu [1]).

Lemma 2.1. a) Let U_t be a solution of (2.8). If $\tilde{U}_t$ is a measure valued process associated with U_t by the relation

(2.11) $$\tilde{U}_t(f) = U_t(f\Psi^{-1}) \quad \text{for each } f \in C_K^\infty(R^1),$$

then $\tilde{U}_t(\Psi) > 0$ is locally bounded and

(2.12) $$\frac{d}{dt}\tilde{U}_t(f) = \tilde{U}_t(\Psi L \Psi^{-1} f) \quad \text{for each } f \in C_K^\infty(R^1).$$

b) Conversely, if $\tilde{U}_t$ is a measure valued process satisfying (2.12) such that $\tilde{U}_t(\Psi) > 0$ is locally bounded, then the measure valued process U_t associated with $\tilde{U}_t$ by (2.11) is a solution of (2.8).

The uniqueness of the solution of the filtering equation in a certain context is investigated by Kunita [7]. But his method is not available in our case where the operator L depends on $w \in W$. On the other hand, the following result is obtained from Lemma 2.1.

Theorem 3. Condition A is imposed here. If, for a.e. w (Q), $\sqrt{A}(t,w,x) \in C_*^{2,b}$, $B(t,w,x) \in C_*^{2,b}$ and $\Psi(t,w,x) \in C_*^{4,b}$, then the solution of equation (2.8) is uniquely determined for any initial value.

Proof. Let $\underline{F}$ be a class of functions $f(t,x)$ on $[0,T] \times R^1$ such that

$f(t,x)$ tend to 0 as $|x| \to \infty$ or $t \uparrow T$, $\partial^m f(t,x)$ $(m \leq 2)$ are bounded and continuous in (t,x) and $(\partial/\partial t)f(t,x) \in C_*^{0,b}$. It is possible to suppose that the path $w \in W$ in (2.12) is fixed. Put

$$\underline{A} = \frac{\partial}{\partial t} + \Psi L \Psi^{-1} , \qquad \underline{D} = \{f \in \underline{F};\ \underline{A}f \in \underline{F}\}.$$

Let $\tilde{U}_t$ be a solution of equation (2.12) and define a functional V_t as follows: $V_t f = \int f(t,x)\tilde{U}_t(dx)$. From (2.12), it follows immediately that

$$(2.13) \qquad \frac{d}{dt} V_t f = V_t(\underline{A}f) \quad \text{for any } f \in \underline{D}.$$

On the other hand, the following properties can be proved.

a) If $g \in \underline{F}$, then $\|g\| \leq (\lambda - \lambda_0)^{-1} \|(\lambda - A)g\|$ for $\lambda > \lambda_0$ (a certain constant), where $\|\cdot\|$ means the sup-norm.

b) For each $f \in C_*^{2,b}$ and $\lambda > \lambda_0$, there is a function $g \in \underline{F}$ such that $(\lambda - \underline{A})g = f$.

c) Let $\underline{C}$ be the space of continuous functions $f(t,x)$ on $[0,T]\times R^1$ such that $f(t,x)$ tend to 0 as $|x| \to \infty$ or $t \uparrow T$. Then $\underline{D}$ is dense in $\underline{C}$ with respect to the sup-norm $\|\cdot\|$.

Applying Hille - Yosida's semi-group theory to equation (2.13), we see that the functional V_t is uniquely determined. The assertion of the theorem follows immediately from Lemma 2.1. Q.E.D.

Roughly speaking, if the initial value $U_0(dx)$ has a smooth density $u_0(x)$ with respect to dx and if the function Ψ and the coefficients of the operator L^* (L^* stands for the adjoint of L) are bounded and smooth in the sense of $C_*^{n,b}$, then, since the equation

$$\frac{\partial \tilde{u}}{\partial t} = \Psi^{-1} L^* \Psi \tilde{u} , \quad \tilde{u}(0,x) = u_0(x)$$

has a smooth solution, the existence of the smooth density $u(t,w,x)$ of the solution $U_t(dx)$ of equation (2.8) follows from Theorem 3. But in the case when $A(t,w,x)$ is strictly positive, the existence of the smooth density $u(t,w,x)$ can be proved without the condition that the derivatives of the function Ψ and the coefficients of L^* are bounded. Namely, the following fact can be proved from Lemma 2.1. (cf. Fujisaki-Komatsu [1]).

Theorem 4. Condition A is imposed here. Let $U_t(dx)$ be a solution of equation (2.8) for an initial value $U_0(dx) = u_0(x)dx$. If, for a.e. w (Q), $A(t,w,x) > 0$, $A(t,w,x) \in C_*^3$, $B(t,w,x) \in C_*^2$, $\Psi(t,w,x) \in C_*^3$ and $u_0(x)$ is continuously differentiable up to third order, then there exists a non-anticipating function $\mathfrak{U}(t,w,x) \geq 0$ such that

$$\mathfrak{U}(t,w,x) \in C_*^2 \ , \quad (\partial/\partial t)\mathfrak{U}(t,w,x) \in C_*^0 \quad \text{and}$$

$$\int f(x)U_t(dx) = \int \mathfrak{U}(t,w,x)\Psi(t,w,x)f(x)dx \quad \text{for any } f(x) \in C_K^\infty(R^1).$$

Let β_t be a 1-dimensional Brownian motion on a certain probability space $(\Omega',\underline{\underline{G}}',P')$ and θ_0 be a random variable on the space independent of β_t. Suppose that $\sqrt{A}(t,w,x)$ and $B(t,w,x)$ are uniformly Lipshitz continuous in x. Then the equation

$$\theta_t = \theta_0 + \int_0^t \sqrt{A}(s,w,\theta_s)d\beta_s + \int_0^t B(s,w,\theta_s)ds \tag{2.14}$$

considered on the space $(W\times\Omega',\underline{\underline{W}}_T\times\underline{\underline{G}}',Q\times P')$ (direct product) has a pathwise solution θ_t^w. Let Π^w be the probability distribution on the space $(Z,\underline{\underline{Z}}_T)$ induced by the process θ_t^w. Then, for each $\Gamma \in \underline{\underline{Z}}_t$, $w \rightsquigarrow \Pi^w(\Gamma)$ is $\underline{\underline{W}}_t$-measurable and the process

$$f(\theta_t) - f(\theta_0) - \int_0^t Lf(s,w,\theta_s)ds$$

is a martingale with respect to $(\underline{\underline{Z}}_t,\Pi^w)$ for each $f(x) \in C_K^\infty(R^1)$. Suppose that the condition of Theorem 4 is satisfied. Then the a posteriori distribution $\pi_t(dx)$ has a smooth density $q_t^x = q(t,w,x)$ with respect to dx and (2.7) turns into

$$dq_t^x = L^*q_t^x dt + q_t^x\{(h_x-\pi h)d\bar{M}_t + \int(\frac{\rho_x}{\pi\rho} -1)\bar{J}_c(dt,du)\} . \tag{2.15}$$

The density $p_t^x = p(t,x)$ of the distribution of θ_t with respect to the probability P is given as follows: $p(t,x) = E_{\bar{P}}[q(t,w,x)]$. Put

$$I(X_0^t,\theta_t) = E_P[\ \log \frac{q(t,w,\theta_t)}{p(t,\theta_t)}\ ]. \tag{2.16}$$

This is the mutual information between $(X_s; s\leq t)$ and θ_t. In the case where $A(t,w,x)$ and $B(t,w,x)$ are independent of $w\in W$, there is a remakable relation between the mutual informations $I(X_0^t,\theta_0^t)$ and $I(X_0^t,\theta_t)$.

Theorem 5. Condition A is imposed. If

a) $A(t,x)\in C_*^3$, $A(t,x) > 0$ and $\partial^2 A(t,x)$ is bounded,

b) $B(t,x)\in C_*^2$, and $\partial B(t,x)$ is bounded,

c) for a.e. w, $\Psi(t,w,x)\in C_*^3$,

d) $p(0,x)\in C^3$, $p(0,x) > 0$ and $\int p(0,x)\log p(0,x)dx < \infty$,

then we have

$$(2.17)\quad I(X_0^t,\theta_0^t) - I(X_0^t,\theta_t) = \frac{1}{2}\int_0^t \{E_{\bar{P}}[F(\theta_s|X_0^s)] - F(\theta_s)\}ds$$

as long as $I(X_0^t,\theta_0^t) < \infty$, where

$$(2.18)\quad F(\theta_s|X_0^s) = \int A(s,x)\frac{(\partial q)^2}{q}(s,w,x)dx,$$

$$F(\theta_s) = \int A(s,x)\frac{(\partial p)^2}{p}(s,x)dx.$$

Proof. Though the following computations are formal, they can be justified. Put $R(x) = 1 - x + x\log x$. From (2.15) we have

$$R(q_t^x) - R(p_t^x) = \int_0^t[\log q_s^x\ L^*q_s^x - \log p_s^x\ L^*p_s^x]ds$$
$$+ \int_0^t q_s^x[\ \frac{1}{2}\ a(h_x-\pi h)^2 + \int R(\frac{\rho_x}{\pi\rho})N\ ]ds$$
$$+ [\text{a martingale w.r.t. } (\underline{\underline{W}}_t,\bar{P})],$$

Therefore, by (2.6),

$$I(X_0^t,\theta_t) = \int(E_{\bar{P}}[R(q_t^x)] - R(p_t^x))dx$$
$$= \int_0^t\{E_{\bar{P}}[\int q_s^x L(\log q_s^x)dx] - \int p_s^x L(\log p_s^x)dx\}ds + I(X_0^t,\theta_0^t).$$

Since $A(t,x)$ and $B(t,x)$ do not depend on w, we have

$$I(X_0^t,\theta_0^t) - I(X_0^t,\theta_t) - \frac{1}{2}\int_0^t \{E_{\bar{P}}[F(\theta_s|X_0^s)] - F(\theta_s)\}ds$$
$$= -\int_0^t\{E_{\bar{P}}[\int Lq_s^x dx] - \int Lp_s^x dx\}ds = 0. \qquad \text{Q.E.D.}$$

If $A(t,x) \equiv 1$, then the values $F(\theta_t|X_0^t)$ and $F(\theta_t)$ are the Fisher informations of $q(t,w,x)$ and $p(t,x)$ respectively. For an example for which (2.17) is used, we shall consider the problem: Minimize the filtering error $\Delta(t) \equiv E_P[(\theta_t - E_P[\theta_t|\underline{W}_t])^2]$ under the constraint

$$(2.19) \quad -\frac{d}{dt} H(P_t) \equiv E_P[\ \frac{1}{2} a(t)h^2 + \int(1-\rho+\rho\log\rho)N(t,du)] \leq C(t).$$

From (2.5) and (2.6), it follows that

$$I(X_0^t,\theta_0^t) \leq \int_0^t C(s)ds$$

and the equality holds only the case when $\pi h = 0$, $\pi\rho = 1$(in other words $\bar{P} = Q$) and $-(d/dt)H(P_t) = C(t)$. Let (θ_t,Π) be a Gaussian Markov process, i.e. θ_0 is a Gaussian random variable and the operator L is given as follows:

$$L = \frac{1}{2} A(t)\partial^2 + \dot{B}(t)x\partial.$$

Since $\Delta(t)E_{\bar{P}}[F(\theta_t|X_0^t)] \geq A(t)$ and $D(\theta_t)F(\theta_t) = A(t)$, where $D(\theta_t)$ is the variance of the random variable θ_t, we have from (2.17) that

$$(2.20) \quad I(X_0^t,\theta_t) \leq \int_0^t \{C(s) - \frac{1}{2} A(s)(\frac{1}{\Delta(s)} - \frac{1}{D(\theta_s)})\}ds.$$

Similarly to Liptzer [8], it is proved from (2.20) that if $\Gamma(t)$ is a solution of the ordinary differential equation

$$\frac{d\Gamma}{dt} = 2(\dot{B}(t) - C(t))\Gamma + A(t) \quad \text{with} \quad \Gamma(0) = D(\theta_0) > 0,$$

then $\Delta(t) \geq \Gamma(t)$. One of the codings (h,ρ) for which the equality $\Delta(t) = \Gamma(t)$ holds is found in the codings of the type

$$h_x(t,w) = \alpha(t)(x - \gamma(t,w)), \quad \rho_x(t,w,u) \equiv 1.$$

References

[1] M.Fujisaki, T.Komatsu: On certain equations arising in the theory of filtering, Seminar on Probability 40 (1973), 3-21 (in Japanese).

[2] B.Grigelionis: On non-linear filtering theory and absolute continuity of measures corresponding to stochastic processes, Lecture

Notes in Math. 330, Proc. U.S.S.R.-Japan Symp. on Prob., Springer Berlin, (1973), 80-94.

[3] B.Grigelionis: On mutual information for locally infinitely divisible stochastic process, Liet. Matem. Rink., XIV 4(1974), 5-11 (in Russian).

[4] I.A.Ibragimov, R.Z.Khas'minskii: Asymptotic behavior of statistical estimators I, Theory Prob. Appl. 17 (1972), 445-462.

[5] I.A.Ibragimov, R.Z.Khas'minskii: Asymptotic behavior of statistical estimators II, Theory Prob. Appl. 18 (1973), 73-91.

[6] H.Kunita, S.Watanabe: On square integrable martingales, Nagoya Math. J. 30 (1967), 209-245.

[7] H.Kunita: Asymptotic behavior of the non-linear filtering errors of Markov process, J. Multive. Anal. 1 (1971), 365-393.

[8] R.S.Liptzer: Оптимальное кодирование и декодирование при передаче Гауссовского Марковского сигнала по канлу с бесшумной обратной связью, Проблемы Передачи Информации, X 4 (1974), 3-15.

[9] B.L.Rozovskii: О стохастических дифференциальных уравнениях в частных производных, Математический Сборник 96(138)(1975), 314-341.

[10] A.N.Shiryayev: Statistics of diffusion type processes, Lecture Notes in Math. 330, Proc. U.S.S.R.-Japan Symp. on Prob., Springer Berlin, (1973), 397-411.

Department of Mathematics
Osaka City University

Evolution asymptotique des temps d'arrêt et des temps de séjour liés aux trajectoires de certaines fonctions aléatoires gaussiennes.

Norio KÔNO

Sommaire: Dans cette note, nous donnons, dans le cas d'une classe particulière de fonctions aléatoires gaussiennes à valeurs dans R^d, des résultats relatifs au comportement asymptotique de certains temps d'arrêt et certains temps de séjour liés aux trajectoires. On remarquera que dans le cas particulier du mouvement brownien dans R^d, Z. Ciesielski et S. J. Taylor [1] ont obtenu des résultats plus fins en déterminant exactement les lois des temps d'arrêt et des temps de séjour.

1. Notations:

Soit $\{X(t)\ ;\ t \in R^N\}$ une fonction aléatoire gaussienne réelle, centrée, à trajectoires continues. On lui associe la fonction aléatoire gaussienne X^d suivante, à valeurs dans R^d:

$$X^d(t) = (X_1(t),\cdots,X_d(t)) \quad \forall t \in R^N,$$

où les $\{X_i(t)\}$ sont des copies indépendantes de $\{X(t)\}$.

Posons:

$$R^0_{d,N}(r) = \sup_{0\leq\|t\|\leq r} \|X^d(t) - X^d(0)\| \quad \text{si } N \geq 2$$

$$= \sup_{0\leq t\leq r} \|X^d(t) - X^d(0)\| \quad \text{si } N = 1,$$

où $\|\cdot\|$ désigne la norme euclidienne.

De même:

$$R^u_{d,N}(r) = \sup_{\substack{0<\|s-t\|<r \\ \|s\|\leq 1,\|t\|\leq 1}} \|X^d(s) - X^d(t)\| \quad \text{si } N \geq 2$$

$$= \sup_{\substack{0<|s-t|<r \\ 0\leq s,t\leq 1}} \|X^d(s) - X^d(t)\| \quad \text{si } N = 1.$$

Nous allons étudier l'évolution asymptotique des temps d'arrêt suivants:

$$P^0_{d,N}(a) = \inf\{r \; ; \; R^0_{d,N}(r) > a\} ,$$

$$P^u_{d,N}(a) = \inf\{r \; ; \; R^u_{d,N}(r) > a\} .$$

Donnons maintenant la définition du temps de séjour que nous utilissons:

$$\forall M \in]0,+\infty] \quad T_d(x,M) = \int_0^M I_x(\|X^d(t)\|) \, dt ,$$

où I_x désigne la fonction indicatrice de l'intervalle $[0,x]$.

Enonçons maintenant les résultats annoncés:

2. Théorèmes:

Les deux premiers Théorèmes sont relatifs au cas $N = 1$.

Théorème 1:

S'il existe une fonction $\sigma : R^+ \longrightarrow R$, non triviale, vérifiant les conditions suivantes:

(i) $\forall s, t \quad E[(X(x) - X(t))^2] = \sigma^2(|s-t|)$,

(ii) σ^2 est concave dans un voisinage du 0 ,

(ii) il existe une fonction s, à croissance lente au sens de J. Karamata, $\alpha > 0$ et deux constantes réelles $c_2 \geq c_1 > 0$ telles que:

$$(1) \quad \forall x > 0 \qquad c_1 \, x^\alpha s(x) \leq \sigma(x) \leq c_2 \, x^\alpha s(x) .$$

Alors il existe deux constantes positives c_3 et c_4 telles que:

$$\overline{\lim_{a \to 0}} \frac{P^0_{d,1}(a)}{\sigma^{-1}(a) \text{ loglog } 1/a} \leq c_3 \quad \text{p.s.}$$

et:

$$\underline{\lim_{r \to 0}} \frac{R^0_{d,1}(r)}{\sigma(r/\text{loglog } 1/r)} \geq c_4 \qquad \text{p.s.}.$$

Remarque 1: Les conditions (ii) et (iii) impliquent $0 < \alpha \leq 1/2$.

Remarque 2: Dans le cas du mouvement brownien dans R^d, Z. Ciesielski et S. J. Taylor [1] ont obtenu l'énoncé-ci-dessus avec $c_3 = 2/q_d^2$ où q_d est le premier zéro positif de la fonction de Bessel $J_{d/2-1}$.

Théorème 2:

Supposons que $X(0) = 0$ p.s. et qu'il existe une fonction $\sigma : R^+ \longrightarrow R$, non triviale, telle que:

(i) $\quad s,t \quad E[(X(s) - X(t))^2] = \sigma^2(|s-t|)$,

(ii) il existe $M > 0$ tel que σ^2 soit concave sur $[0,M[$,

(iii) σ satisfait à la relation (1) avec $\alpha d > 1$.

Sous ces hypothèses, il existe, lorsque $M < +\infty$, une constante positive c_5 avec:

$$\overline{\lim_{x \to 0}} \frac{T_d(x,M)}{\sigma^{-1}(x) \operatorname{loglog} 1/x} \leq c_5 \quad \text{p.s..}$$

Si on a de plus: $\displaystyle\int_1^{\infty} \sigma(u)^{-d}\, du < +\infty$,

la relation précédente reste encore vraie si $M = +\infty$.

Remarque 3: Dans le cas particulier du mouvement brownien dans R^d, $d \geq 3$, Z. Ciesielski et S. J. Taylor [1] ont obtenu l'énoncé-ci-dessus avec $M = +\infty$ et $c_5 = 2/p_d^2$ où p_d désigne le premier zéro positif de la fonction de Bessel $J_{d/2-2}$.

Dans le cas particulier: $\sigma(x) = x^{\alpha}$ avec $\alpha d = 1$, on peut remplacer, l'énoncé précédent, la fonction:

$$x \longmapsto x^{1/\alpha} \operatorname{loglog} 1/x$$

par:

$$x \longmapsto x^{1/\alpha} \log 1/x \operatorname{loglog} 1/x .$$

Cette fonction n'est pas la meilleure posible car on sait que pour le mouvement brownien plan:

$$x \longmapsto x^2 \log 1/x \text{ logloglog } 1/x$$

convient (cf. [4]).

Nous allons donner maintenant des résultats relatifs au cas $N \geq 1$.

Théorème 3:

On suppose qu'il existe une fonction $\sigma : R^+ \longrightarrow R$ continue, vérifiant les deux propriétés suivantes:

(a) $\forall s,t \quad E[(X(s) - X(t))^2] \leq \sigma^2(\|s-t\|)$,

(b) σ est non décroissant et satisfait à la relation (1).

Sous ces hypothèses on a :

(i) $$\overline{\lim_{r\to 0}} \frac{R^0_{d,N}(r)}{\sigma(r)\sqrt{2 \text{ loglog } 1/r}} \leq c_6 \quad \text{p.s.}$$

(ii) $$\underline{\lim_{a\to 0}} \frac{P^0_{d,N}(a)}{\sigma^{-1}(a/\sqrt{2 \text{ loglog } 1/a})} \geq c_7 \quad \text{p.s.}$$

(iii) $$\overline{\lim_{r\to 0}} \frac{R^u_{d,N}(r)}{\sigma(r)\sqrt{2 N \log 1/r}} \leq c_8 \quad \text{p.s.}$$

(iv) $$\underline{\lim_{a\to 0}} \frac{P^u_{d,N}(a)}{\sigma^{-1}(a/\sqrt{2N/\alpha \log 1/a})} \geq c_9 \quad \text{p.s.}$$

où on a posé: $\sigma^{-1}(x) = \inf \{y ; \sigma(y) = x\}$.

Si de plus (1) est vérifiée avec $c_1 = c_2$, alors on peut prendre: $c_6 = c_7 = c_8 = c_9 = 1$.

Théorème 4:

Supposons pu'il existe une fonction $\sigma : R^+ \longrightarrow R$, continue, satisfaisant aux 3 conditions suivantes:

(a) $\forall s,t \quad E[(X(s) - X(t))^2] = \sigma^2(\|s-t\|)$,

(b) σ est non décroissant et satisfait à la relation (1) pour un $\alpha \in]0,1[$,

(c) il existe une constante positive c_{10} avec:

$$\forall x \in]0,1] \qquad \left|\frac{d\sigma(x)}{dx}\right| \leq c_{10}\, \sigma(x)/x \ .$$

On a alors:

(i) $$\overline{\lim_{r\to 0}} \frac{R^0_{d,N}(r)}{\sigma(r)\sqrt{2 \text{ loglog } 1/r}} \geq 1 \quad \text{p.s.}$$

(ii) il existe une constante $c_{11} > 0$ telle que:

$$\underline{\lim_{a\to 0}} \frac{P^0_{d,N}(a)}{\sigma^{-1}(a/\sqrt{2 \text{ loglog } 1/a})} \geq c_{11} \quad \text{p.s..}$$

Si de plus (1) est satisfaite avec $c_1 = c_2$, alors on peut prendre $c_{11} = 1$.

Théorème 5:

On suppose que les hypothèses (a), (b) et (c) du Théorème 4 sont satisfaites, ainsi que le suivante:

(d) il existe une constante positive c_{12} telle que:

$$\forall x \in]0,1] \quad \left|\frac{d^2\sigma(x)}{dx^2}\right| \leq c_{12}\, \sigma(x)/x^2 \ .$$

On a alors:

(iii) $$\overline{\lim_{r\to 0}} \frac{R^u_{d,N}(r)'}{\sigma(r)\sqrt{2N \log 1/r}} \geq 1 \quad \text{p.s.}$$

(iv) il existe une constante positive c_{13} telle que:

$$\underline{\lim_{a\to 0}} \frac{P^u_{d,N}(a)}{\sigma^{-1}(a/\sqrt{2N/\alpha \log 1/a})} \leq c_{13} \quad \text{p.s..}$$

Si de plus (1) est satisfaite avec $c_1 = c_2$, alors on peut prendre $c_{13} = 1$.

3. Lemmes fondamentaux:

La démonstration des Théorèmes 1, 2 et 3 repose sur les 3 lemmes suivants:

Lemma 1:

Sous les hypothèses du Théorème 1, il existe deux constantes positives c_{14} et c_{15}, indépendantes de a et r, telles que pour tout $r > 0$ et tout $a > 0$, assez petits, on ait:

$$P(P^0_{d,1}(a) > r) \leq c_{14}\ e^{-c_{15} r/\sigma^{-1}(a)} .$$

Lemme 2:

Sous les hypothèses du Théorème 2, il existe deux constantes positives c_{16} et c_{17}, indépendantes de x et de r, telles que:

$$\forall r > 0 \quad \forall x \in]0,1[\quad P(T_d(x,M) > r) \leq c_{16} e^{-c_{17} r/\sigma^{-1}(x)} .$$

Ce Lemme s'obtient en calculant les moments du $T_d(x,M)$ et en appliquant l'inégalité de Čebičev et le Lemme 2 de [3].

Lemme 3:

Soient S une partie bornée de R^N et $\{X(s)\ ;\ s \in S\}$ une fonction aléatoire gaussienne, séparable, centrée, vérifiant la condition suivante: il existe une fonction $\sigma : R^+ \longrightarrow R$, non décroissante, continue, avec:

(i) $\forall s,t \in S \quad E[(X(s) - X(t))^2] \leq \sigma^2(\|s-t\|)$,

(ii) $\int^{+\infty} \sigma(e^{-u^2})\, du < +\infty$.

Sous ces hypothèses on a:

$$\forall n \text{ entier} > 1 , \quad \forall x \geq \sqrt{2d + 4N \log n}$$

$$P(\sup_{s\in S} \|X^d(s)\| \geq x(\sqrt{\sup_{s\in S} E[X^2(s)]} + 4\int_1^\infty \sigma(d(S)\ n^{-u^2})du))$$

$$\leq (\sqrt{N}/2)^N c_d\ n^{2N} \int_{u_1^2+\cdots+u_d^2 \geq x^2} e^{-(u_1^2+\cdots+u_d^2)/2}\, du_1\cdots du_d ,$$

où $c_d = (2\pi)^{-d/2} \sum_{k=1}^{\infty} 2^{3+k(d-2)/2} e^{-(2^k-1)/2}$, et $d(S)$ désigne le diamètre de S.

Ce Lemme est une extension de l'inégalité de Fernique [2].

[1] Z. CIESIELSKI et S. J. TAYLOR. Amer. Math. Soc. 103(1962), 434-450.

[2] X. FERNIQUE. C. R. Acad. Sci. Paris, t. 258(1964), 6058-6060.

[3] M. B. MARCUS. Pacific J. Math. Vol.26 No.1(1968), 149-157.

[4] D. RAY. Trans. Amer. Math. Soc. 106(1963), 436-444.

INSTITUTE OF MATHEMATICS
Yoshida College
Kyoto University
KYOTO, JAPAN.

INSTITUT DE RECHERCHE MATHEMATIQUE AVANCEE
Laboratoire Associé au C.N.R.S.
Université Louis Pasteur
7. rue René Descartes
67084, STRASBOURG CEDEX, FRANCE.

ASYMPTOTIC ENLARGING OF SEMI-MARKOV PROCESSES WITH AN ARBITRARY STATE SPACE

V.S.Korolyuk, A.F.Turbin

I. Introduction. The paper is devoted to asymptotic enlarging of semi-Markov processes and may be considered as a generalization of the results obtained in [I],[2] .

Since [I] has been published the idea of asymptotic enlarging of stochastic processes becomes more and more popular and one can meet it in the investigation in the theory of stochastic process [2]-[5], in queueing and reliability theory [6],[7] , in stochastic automatons theory [8],[9] etc. This idea seems natural in analysis of processes describing complex systems.

In the next section neccessary concepts and definitions are introduced and Lemma I describing the class of operators involved is proved. With this lemma it is possible to apply the method of spectrum perturbed operators [I0]-[I2] to calculation of the characteristics of enlarged processes. In Section 3 asymptotic enlarging of semi-Markov processes is considered.

2. Main notions.

I. Let $(E, \mathfrak{S})$ be a measurable space with a σ-algebra $\mathfrak{S}$ containing one-point subsets and let ξ_n , $n \geq 0$, be a Markov chain with transition probabilities $P(x, \Gamma)$, $x \in E$, $\Gamma \in \mathfrak{S}$, defined on $(E, \mathfrak{S})$.

Define $\mathfrak{B} = \mathfrak{B}(E, \mathfrak{S})$ as the Banach space of real-valued $\mathfrak{S}$-measurable functions $f(x)$, $x \in E$, with the sup-norm $\|f\| = \sup_{x \in E} |f(x)|$ and let P be the linear operator in $\mathfrak{B}$ generated by the transition probabilities $P(x,\Gamma)$:

$$[Pf](x) = \int_E P(x,dy)f(y), \quad f\in \mathcal{B} \quad .$$

In what follows we suppose that the Markov chain ξ_n, $n \geq 0$, with the transition operator P satisfies the uniform ergodic hypothesis: the sequence of operators $\Pi_n = n^{-1}\sum P^k$ converges, in the uniform operator topology, to a limit Π. The operator Π, as is easily seen, is a projection operator $(\Pi^2 = \Pi)$ and will be referred to as the stationary projector of the Markov chain ξ_n, $n\geq 0$.

L e m m a I. If the Markov chain ξ_n, $n \geq 0$, satisfies the uniform ergodic hypothesis, then the operator $(I - P + \Pi)^{-1}$ exists and is bounded.

P r o o f . Suppose that there exists a vector $x_0 \in \mathcal{B}$ such that

(I) $$(I - P + \Pi)x_0 = 0 \quad .$$

The properties of Π imply

(2) $$\Pi x_0 = 0 , \quad (I-P)x_0 = 0 \quad .$$

By the uniform ergodic hypothesis, there exists an n_0 such that $\|\Pi_n - \Pi\| < 1$ for any $n \geq n_0$, i.e. the operators $I-\Pi_n+\Pi$ have the bounded inverses for any $n \geq n_0$. Moreover, since $x_0 = P^k x_0$ for any $k \geq 0$ by (2), we have

$$(I-\Pi_n+\Pi)x_0 = (I-\Pi_n)x_0 = x_0 - x_0 = 0 \quad .$$

Thus $x_0 = 0$ and $I - P + \Pi$ is invertible. Boundedness of $(I-P+\Pi)$ can be easily obtained as a consequence of existence of the n_0 and the well-known Dunford Theorem on the spectrum of linear operators.

This lemma shows that the operator $I-P$ is a simple normally solvable operator ([I2]) and hence the theory of inversion of the spectrum perturbed operators may be applied to analysis of operators of the kind $I-P+\varepsilon B$ with ε sufficiently small and B bounded.

The operator $R_0 = (I-P+\Pi)^{-1} - \Pi$ will be called the resolvent operator of a Markov chain ξ_n, $n \geq 0$, with the transition operator P. (For finite Markov chains $R_0 + \Pi$ is known as the fundamental matrix [I3]).

2. Define the basic, for the present paper, notion of enlarging.

Let U be some set which induces a partition of the space $(E, \mathfrak{S})$ in the sense that

$$(3) \qquad E = \bigcup_{u \in U} E_u \,, \quad E_{u'} \cap E_{u''} = \Phi \,, \quad E_u \in \mathfrak{S}$$

for $u' \neq u''$, $u', u'' \in U$.

Let $\mathfrak{S}$ be the least σ-algebra containing the classes E_u, $u \in U$.

The partitioning (3) generates the following equivalence relation S in E : $x \sim y$ if x and y belong to the same class E_u, $u \in U$.

The measurable space $(E, \mathfrak{S})$ and the equivalence relation S define the measurable factor-space $(\bar{E}, \bar{\mathfrak{S}}_0)$ where $\bar{E}$ is the factor-space of E relative to S and $\bar{\mathfrak{S}}$ is the corresponding factor-

-sigma-algebra. The space $(\bar{E}, \bar{\mathfrak{E}}_0)$ will be called the enlarging of the space $(E, \mathfrak{E})$ induced by the partitioning (3). There is a one-to-one correspondence between $\bar{E}$ and U inducing the σ-algebra $\mathfrak{U}$ in U, the image of $\bar{\mathfrak{E}}$.

The space $(U, \mathfrak{U})$ will be also called the enlarging of the space $(E, \mathfrak{E})$ induced by the partitioning (3) and any $\mathfrak{E}$-measurable mapping $u(x)$ of $(E, \mathfrak{E})$ onto $(U, \mathfrak{U})$. The mapping $u(x): (E, \mathfrak{E}) \to (U, \mathfrak{U})$ will be called an enlarging operation, or simply, an enlarging.

Define $\mathfrak{E}_u$ as the restriction of the σ-algebra $\mathfrak{E}$ onto E_u and let $(E_u, \mathfrak{E}_u)$ be the measurable space, corresponding to the class E_u, $u \in U$. Now let us suppose that the Markov chain ξ_n, $n \geq 0$ with transition operator P is consistent with the partitioning (3) in the sense that

$$(4) \qquad P(x, E_u) = \begin{cases} 1, & x \in E_u \\ 0, & x \bar{\in} E_u, \end{cases}$$

and the Markov chains ξ_n^u, $n \geq 0$, with state spaces $(E_u, \mathfrak{E}_u)$ and transition probabilities $P_u(x, \Gamma)$, $x \in E_u$, $\Gamma \in \mathfrak{E}_u$, are indecomposable. In other words, it is assumed that the classes E_u $u \in U$, are closed classes of states and we should speak of a totality of Markov chains $\{\xi_n^u, n \geq 0\}$, $u \in U$, with state spaces $(E_u, \mathfrak{E}_u)$.

The uniform ergodic hypothesis allows us to introduce the following measures

$$(5) \qquad \mu_{u(x)}(\Gamma) = \Pi(x, \Gamma), \ x \in E_{u(x)}, \ \Gamma \in \mathfrak{E},$$

where $\Pi(x,\Gamma)=[\Pi\chi_\Gamma](x)$ and $\chi_\Gamma(x)$ is the indicator of a set Γ.

From the definition of the measures $\mu_u(\cdot)$ one can easily obtain (we write u instend of $u(x)$ for short)

$$(6) \qquad [\mu_u P](\Gamma)=\int_E \mu_u(dy)\,P(y,\Gamma)=\mu_u(\Gamma) \quad ,$$

i.e. $\mu_u(\cdot)$, for each $u\in U$, is a stationary measure of the chain ξ_n, $n\geq 0$.

Further, if $f(x)$ is an $\mathfrak{G}$-measurable function, then $f_0(x)= =[\Pi f](x)$ is an $\mathfrak{G}$-measurable function.

Now we come to the main enlargement assumptions. Let $B(x,\Gamma)$ $x\in E$, $\Gamma\in\mathfrak{G}$, be $\mathfrak{G}$-measurable as a function of x and a charge (see [I4]) as a function of $\Gamma\in\mathfrak{G}$. Suppose that $B(x,\Gamma)$ is consistent with the transition probability $P(x,\Gamma)$ in the sense that

$$(7) \qquad P_\varepsilon(x,\Gamma)\stackrel{d}{=}P(x,\Gamma)-\varepsilon B(x,\Gamma)$$

is a transition probability of some Markov chain ξ^ε_n, $n\geq 0$, for all $\varepsilon\in(0,\varepsilon_0]$ with $\varepsilon_0>0$ fixed.

By (4) and (7) we get

$$(8) \qquad B(x,E)=0 \quad , \quad x\in E \quad ;$$

$$(9) \qquad \begin{cases} B(x,\Gamma)\leq 0 \quad , \quad x\in E_u \ , \ \Gamma\in\mathfrak{G} \ , \ \Gamma\cap E_u=\Phi \ , \\ B(x,E_u)\geq 0 \ , \quad x\in E_u \ . \end{cases}$$

The Markov chain ξ_n^ε, $n \geq 0$, with the transition probability $P_\varepsilon(x,\Gamma)$ defined by (7) is called a perturbed Markov chain, and the Markov chain ξ_n^0, $n \geq 0$ with the transition probability $P(x,\Gamma)$ in (7) is called a non-perturbed Markov chain.

Let B be the operator in $\mathcal{B}$ generated by $B(x,\Gamma)$. The operator $\Pi B \Pi$, where Π is the stationary projector of a non-perturbed Markov chain, acts non-trivially in the space $\mathcal{B}_0 = \{f \in \mathcal{B} : (I-P)f = 0\}$ which is the eigenspace of $I-P$ corresponding to the eigenvalue 0. Our next step is to describe the contraction of $\Pi B \Pi$ onto $\mathcal{B}_0$.

L e m m a 2.

$$[\Pi B \Pi \chi_\Gamma](x) = \mu_u^B(E_u)\Pi(x,\Gamma) + \int_{E \setminus E_u} \mu_u^B(dz)\Pi(z,\Gamma), \tag{10}$$

where $u = u(x) : (E, \mathfrak{E}) \to U$,

$$\mu_u^B(\Gamma) = \int_{E_u} \mu_u(dz) B(z,\Gamma), \quad \Gamma \in \mathfrak{E}. \tag{11}$$

P r o o f. Let $x \in E_u$. Then by (5)

$$[\Pi B \Pi \chi_\Gamma](x) = \int_E \int_E \Pi(x,dy) B(y,dz) \Pi(z,\Gamma) = \tag{12}$$

$$= \int_{E_u} \int_{E_u} \Pi(x,dy) B(y,dz) \Pi(z,\Gamma) +$$

$$+ \int_{E_u} \int_{E \setminus E_u} \Pi(x,dy) B(y,dz) \Pi(z,\Gamma).$$

By (11) we have

$$\int_{E_u} \int_{E_u} \Pi(x,dy) B(y,dz) \Pi(z,\Gamma) =$$

$$= \int_{E_u} \mu_u(dy) B(y, E_u) \Pi(x, \Gamma) = \mu_u^B(E_u) \Pi(x, \Gamma) \quad .$$

By the Fubini theorem

$$\int_{E_u} \int_{E \setminus E_u} \Pi(x, dy) B(y, dz) \Pi(z, \Gamma) = \int_{E \setminus E_u} \mu_u^B(dz) \Pi(z, \Gamma) \quad .$$

The one-to-one correspondence between $\bar{E}$ and $\mathcal{U}$ induces an isomorphism between the space $\mathcal{B}(\bar{E}, \bar{\mathcal{E}}_0)$ of $\bar{\mathcal{E}}$ -measurable functions on $\bar{E}$ with the sup-norm and the space $\mathcal{B}(\mathcal{U}, \mathfrak{U})$ of $\mathfrak{U}$-measurable functions on $\mathcal{U}$.

Since $\mathcal{B}_0 = \mathcal{B}(E, \mathcal{E}_0)$ is isomorphic to $\mathcal{B}(\bar{E}, \bar{\mathcal{E}}_0)$, the space $\mathcal{B}$ is isomorphic to $\mathcal{B}(\mathcal{U}, \mathfrak{U})$ as well.

By Lemma 2 the contraction of the operator $\Pi B \Pi$ onto $\mathcal{B}$ can be described as the operator in $\mathcal{B}(\mathcal{U}, \mathfrak{U})$ with the kernel

$$\bar{\mu}_u^B(\bar{\Gamma}) = \mu_u^B(u^{-1}(\bar{\Gamma})) , \qquad \bar{\Gamma} \in \mathfrak{U} \tag{I4}$$

(with bar we mark quantities corresponding to the enlarging).

3. Asymptotic enlarging of semi-Markov processes. Let $\xi^{\varepsilon}(t)$ be a regular semi-Markov process, depending on a small parameter $\varepsilon \in (0, \varepsilon_0]$ and defined on a measurable space $(E, \mathcal{E})$ by its semi-Markov kernel $Q_\varepsilon(x, \Gamma, t)$, $x \in E$, $\Gamma \in \mathcal{E}$, $t \geq 0$ (see [I5] ,[I6]).

Let, for each $\varepsilon \in (0, \varepsilon_0]$ and $t \geq 0$, $Q_\varepsilon(t)$ be the operator in $\mathcal{B}$ induced by the kernel $Q_\varepsilon(x, \Gamma, t)$ and $\tilde{Q}_\varepsilon(s)$ be its Laplace--Stiltjes transform:

$$[\tilde{Q}_\varepsilon(s)f](x) = \int_0^\infty e^{-st} Q(x, dy, dt) f(y) .$$

Now let us suppose that the partitioning (3) is defined on $(E, \mathfrak{S})$ and the following conditions are fullfilled:

A_I. $\tilde{Q}_\varepsilon(\varepsilon s)$ can be represented in the form[*)]

$$(I5) \qquad \tilde{Q}_\varepsilon(\varepsilon s) = P - \varepsilon(B + sG) + o(\varepsilon)$$

where $o(\varepsilon)$ means that

$$\| \tilde{Q}(\varepsilon s) - P + \varepsilon(B + sG) \| \xrightarrow[\varepsilon \to 0]{} 0$$

for any finite s, $\mathrm{Re}\ s \geq 0$.

A_2. Markov chain ξ_n^0, $n \geq 0$, embedded in the non-perturbed semi-Markov process $\xi^0(t)$ is consistent with the partitioning (3) in the sense (4) and satisfies the uniform ergodic hypothesis.

R e m a r k. Condition A_I implies that $m_x = E\theta_x$, where θ_x is the sojourn time of the non-perturbed process $\xi^0(t)$ in the state x, is uniformly bounded as a function of x. Besides

$$G(x, \Gamma) = [G\chi_\Gamma](x) = \int_0^\infty t\, Q_0(x, \Gamma, dt) ,$$

$$G(x, E) = G(x, E_n) = m_x$$

and without loss of generality we may assume that $m_x > 0$.

---

*) All the operators in (I5) are assumed to be bounded.

Now we are able to formulate the problem: to describe finite-dimensional distributions of the process $\bar{\eta}(t) = \lim_{\varepsilon \to 0} \bar{\eta}_\varepsilon(t)$ where $\bar{\eta}_\varepsilon(t) = u(\xi^\varepsilon(\frac{t}{\varepsilon}))$ and $u(x): (E, \mathfrak{S}) \to U$ is an enlarging.

T h e o r e m I. Let $x \in E_u$ and the conditions A_I and A_2 be fullfilled. Then

$$\lim_{\varepsilon \to 0} P\{\bar{\eta}_\varepsilon(y) = u(x),\ y \in [0,t) \mid \xi^\varepsilon(0) = x\} = \tag{I6}$$

$$= \exp\{-\lambda_{u(x)} t\} \quad ,$$

where

$$\lambda_u = \frac{\mu_u^B(E_u)}{\mu_u^G(E_u)} \quad . \tag{I7}$$

P r o o f. First note that

$$0 < \mu_u^G(E_u) \stackrel{d}{=} \int_{E_u} \mu_u(dx)\, G(x, E_u) \leq c \quad ,$$

where c is some constant, which is a consequence of condition A_I and the regularity of $\xi^0(t)$.

Let ζ_x^ε denote the sojourn time of the process $\xi^\varepsilon(t)$ in the class E_u up to the first exit from this class under the condition that the initial state was x .

It is easy to see that

$$P\{\bar{\eta}_\varepsilon(y) = u(x),\ y \in [0,t) \mid \xi(0) = x\} =$$

$$= P\{\xi^\varepsilon(\tfrac{y}{\varepsilon}) \in E_{u(x)} y \in [0,t) \mid \xi^\varepsilon(0) = x\} = P\{\varepsilon \zeta_x^\varepsilon \geq t\} \quad . \tag{I8}$$

Put $\tilde{\varphi}_x^{\varepsilon}(s) = E\exp\{-\varepsilon s \zeta_x^{\varepsilon}\}$, $\operatorname{Re} s \geq 0$. As in [17] one can show that $\tilde{\varphi}_x^{\varepsilon}(s)$ is a solution of the equation:

$$\tilde{\varphi}_x^{\varepsilon}(s) - \int_{E_u} \tilde{Q}_{\varepsilon}(x, dy, \varepsilon s)\tilde{\varphi}_y^{\varepsilon}(s) = \tilde{Q}_{\varepsilon}(x, E\setminus E_u, \varepsilon s)$$

or, in the operator notation,

(19) $$\tilde{\varphi}^{\varepsilon}(s) - \tilde{Q}_u^{\varepsilon}(\varepsilon s)\tilde{\varphi}^{\varepsilon}(s) = \tilde{Q}_{\varepsilon}(E\setminus E_u, \varepsilon s)$$

where $\tilde{Q}_u^{\varepsilon}(s)$ stands for the operator in $\mathcal{B}(E_u, \mathfrak{S}_u)$ defined by the kernel $\tilde{Q}_{\varepsilon}(x, \Gamma, s)$, $x \in E_u$, $\Gamma \in \mathfrak{S}_u$.

Using condition A_1 and the regularity of the process we get from (19)

(20) $$\tilde{\varphi}^{\varepsilon}(s) = [I_u - \tilde{Q}_u^{\varepsilon}(\varepsilon s)]^{-1}\tilde{Q}_{\varepsilon}(E\setminus E_u, \varepsilon s) =$$

$$= [I_u - P_u + \varepsilon(B_u + sG_u) + o(\varepsilon)]^{-1}(\varepsilon\tilde{B}(E_u) + o(\varepsilon)),$$

where I_u is the identity operator in $\mathcal{B}(E_u, \mathfrak{S}_u)$ and P_u, B_u, G_u are the contractions of the corresponding operators P, B, G on $\mathcal{B}(E_u, \mathfrak{S}_u)$.

Condition A_2 guarantees normal solvability of $I_u - P_u$ for any $u \in U$ as was shown in Lemma 1.

Moreover the assumed consistence of the non-perturbed Markov chain transition function $P(x, \Gamma)$ and the partitioning (3) bring us to

$$\dim N(I_u - P_u) = 1$$

where

$$\mathcal{N}(I_u-P_u)\overset{d}{=}\{f\in\mathcal{B}(E_u,\mathfrak{S}_n):(I_u-P_u)f=0\}\ .$$

Since, by our assumption, $\mu_u^G(E_u)>0$, the expansions for spretrum perturbed operators obtained in [II] (see also [I2]) and arguments similar to those of [I8] allow us to deduce from (20) that, for any fixed s, $\mathrm{Re}\ s>0$,

$$\lim_{\varepsilon\to 0}\tilde{\varphi}_x^{\varepsilon}(s)=\frac{1}{\mu_u^B(E_u)+s\mu_u^G(E_u)}[\Pi B\chi_{E_u}](x)\ .$$

But

$$[\Pi B\chi_{E_u}](x)=\int_E \Pi(x,dy)B(y,E_u)=\int_{E_u}\Pi(x,dy)B(y,E_u)=$$

$$=\int_{E_u}\mu_u(dy)B(y,E_u)=\mu_u^B(E_u)\ .$$

Thus, for $x\in E_u$ and any fixed s, $\mathrm{Re}\ s>0$,

$$\lim_{\varepsilon\to 0}\tilde{\varphi}_x^{\varepsilon}(s)=\frac{\mu_u^B(E_u)}{\mu_u^B(E_n)+s\mu_u^G(E_u)}=\frac{\lambda_u}{\lambda_u+s}\ . \tag{2I}$$

Finally by the continuity theorem for Laplace-Stilties transforms, (2I) is equivalent to (I6).

This completes the proof.

It follows from (I6) that

$$\lim_{\varepsilon\to 0}P\{\bar{\eta}_\varepsilon(y)=u(x_1),\ y\in[0,t)\,|\,\xi^\varepsilon(0)=x_1\}=$$

$$= \lim_{\varepsilon \to 0} P\{\bar{\eta}_\varepsilon(y) = u(x_2),\ y \in [0,t) \mid \xi^\varepsilon(0) = x_2\}$$

for any $x_1 \sim x_2$. Hence the enlarged process $\bar{\eta}(t)$ spends exponentially distributed time in any of its states with the exponential parameter defined by (I7).

Let D denote the multiplying operator in $\mathcal{B}$ with the kernel

$$D(x,\Gamma) = \begin{cases} m_x \ , & x \in \Gamma \\ 0 \ , & x \bar{\in} \Gamma \ , \quad x \in E \ , \quad \Gamma \in \mathfrak{S} \ , \end{cases}$$

and $G^{(-1)}$ be any operator whose contraction to $\mathcal{B}_0$ coincides with the inverse of the contraction of $\Pi G \Pi$ to $\mathcal{B}_0$ ($G^{(-1)}$ is well-defined since $\mu_u^G(E_u) > 0$).

By the same argument as above one can show that

$$(22) \qquad \lim_{t \to \infty} P\{\xi^0(t) \in \Gamma \mid \xi^0(0) = x\} = \Pi_G(x,\Gamma)$$

where $\Pi_G(x,\Gamma) = [\Pi G^{(-1)} \Pi D \chi_\Gamma](x) \overset{d}{=} [\Pi_G \chi_\Gamma](x)$.
The right-hand side of (22) is the stationary distribution of the non-perturbed semi-Markov process $\xi^0(t)$.

Now let us return to the perturbed process

Put $A = -\Pi G^{(-1)} \Pi B$ where B is the above defined perturbation operator.

L e m m a 3. Let

$$\Phi_\varepsilon(x,\Gamma,t) = P\{\xi^\varepsilon(\tfrac{t}{\varepsilon}) \in \Gamma \mid \xi^\varepsilon(0) = x\} \ .$$

In the conditions of Theorem I, for any finite $t>0$,

(23) $\lim_{\varepsilon\to 0} \Phi_\varepsilon(x,\Gamma,t) = [\Pi_G \exp\{\Pi_G A \Pi_G t\} \chi_\Gamma](x)$.

P r o o f. Put

$$\tilde{\Phi}_\varepsilon(x,\Gamma,s) = \int_0^\infty e^{-st} d_t \Phi_\varepsilon(x,\Gamma,t) .$$

It is known (see, for instance, [I5] , [I6]) that $\tilde{\Phi}_\varepsilon(x,\Gamma,s)$ is a solution of the equation:

$$\tilde{\Phi}_\varepsilon(x,\Gamma,s) = \int_E \tilde{Q}_\varepsilon(x,dy,\varepsilon s)\tilde{\Phi}_\varepsilon(y,\Gamma,s) +$$

$$+(1-\tilde{Q}_\varepsilon(x,E,\varepsilon s))\chi_\Gamma(x) .$$

or, in the operator notation,

$$\tilde{\Phi}_\varepsilon(\Gamma,s) = \tilde{Q}_\varepsilon(\varepsilon,s)\tilde{\Phi}_\varepsilon(\Gamma,s) + \tilde{T}_\varepsilon(\Gamma,\varepsilon) ,$$

where $\tilde{T}_\varepsilon(x,\Gamma,s) = (1-\tilde{Q}_\varepsilon(x,E,\varepsilon s))\chi_\Gamma(x)$.

The regularity of the process $\xi^\varepsilon(t)$ and condition A_I imply that

$$\tilde{\Phi}_\varepsilon(\Gamma,s) = [I-P+\varepsilon(B+sG)+o(\varepsilon)]^{-1}\tilde{T}_\varepsilon(\Gamma,s) .$$

Now using the expansions obtained in [I2] and arguments similar to those of [I8] we obtain, for any fixed s, $\operatorname{Re} s>0$,

(24)
$$[I-P+\varepsilon(B+sG)+o(\varepsilon)]^{-1} = \frac{1}{\varepsilon}\Pi(B+sG)^{(-1)}\Pi + o(\frac{1}{\varepsilon})$$

where the operator $(B+sG)^{(-1)}$ is defined in just the same way as $G^{(-1)}$.

Hence, noting that

$$\tilde{T}_{\varepsilon}(x, \Gamma, s) = \varepsilon s D(x, \Gamma) + o(\varepsilon)$$

by the condition A_I and combining this with (24), we get

(25) $$\lim_{\varepsilon \to 0} \tilde{\Phi}_{\varepsilon}(x,\Gamma,s) = s[\Pi(B+sG)^{(-1)} \Pi D \chi_{\Gamma}](x) .$$

Since Π is a projection operator and $\Pi G^{(-1)} \Pi G \Pi = \Pi$, it follows that

$$[\Pi(B+sG)^{(-1)} \Pi D \chi_{\Gamma}](x) = [\Pi(B+sG)^{(-1)} \Pi G^{(-1)} \Pi \Pi_G \chi_{\Gamma}](x) =$$

$$= -[\Pi(A-sI)^{(-1)} \Pi \Pi_G \chi_{\Gamma}](x) .$$

Denote

$$\tilde{\Phi}(s) = \int_0^{\infty} e^{-st} d(\Pi_G \exp\{\Pi_G A \Pi_G t\}) =$$

(26)

$$= s\Pi_G(\Pi_G A \Pi_G - sI)^{-1} .$$

Using the easily checked equalities

$$\Pi \Pi_G = \Pi_G , \quad \Pi_G \Pi = \Pi ,$$

one can show that the Taylor series on the right-hand sides of (25) and (26) at infinity neibourhood coincide. Now their analycity implies that the right-hand sides of (25) and (26) coincide for all s, $\operatorname{Re} s > 0$. Thus by the continuity theorem for Laplace--Stiltjes transforms we have completed the proof.

Consider in more detail the operator $A_\Pi = \Pi_G A \Pi_G$ in (26). Let $A_\Pi(x,\Gamma) = [A_\Pi \chi_\Gamma](x)$. First note that

$$[\Pi G^{(-1)} \Pi \chi_\Gamma](x) = \frac{1}{\mu^G_{u(x)}(E_{u(x)})} \Pi(x,\Gamma) .$$

If $x \in E_{u(x)}$, $\Gamma \in \mathfrak{S}$, then

$$A_\Pi(x,\Gamma) = -[\Pi G^{(-1)} \Pi B \Pi_G \chi_\Gamma](x) =$$

$$= -\frac{1}{\mu^G_u(E_u)} [\Pi B \Pi_G \chi_\Gamma](x) = -\frac{\mu^B_u(E_u)}{\mu^G_u(E_u)} \Pi_G(x,\Gamma) .$$

If $\Gamma \cap E_u = \Phi$, then, for $\mu^B_u(E_u) > 0$,

$$A_\Pi(x,\Gamma) = -\frac{1}{\mu^G_u(E_u)} [\Pi B \Pi_G \chi_\Gamma](x) = -\lambda_u \frac{1}{\mu^B_u(E_u)} \times$$

$$\times [\Pi B \Pi_G \chi_\Gamma](x) = -\lambda_u \frac{\int_{E \setminus E_u} \mu^B_u(dy)\, \Pi_G(y,\Gamma)}{\mu^B_u(E_u)} =$$

$$= -\lambda_u \frac{\int_{U \setminus \{u\}} \bar{\mu}^B_u(dz)\, \Pi_G(u^{-1}(z),\Gamma)}{\bar{\mu}^B_u(\{u\})} .$$

Hence by (8) and (9)

$$\bar{\Pi}(u,\bar{\Gamma}) \overset{d}{=} \begin{cases} -\dfrac{\mu_u^{-B}(\bar{\Gamma})}{\mu_u^{B}(\{u\})} , & \mu_u^{-B}(\{u\})>0 , \\ \chi_{\bar{\Gamma}}(u) , & \mu_u^{-B}(\{u\})=0 , \end{cases}$$

$$(u \bar{\in} \bar{\Gamma}, \ \bar{\Gamma} \in \mathfrak{U})$$

is a probability measure on the enlarged space $(U, \mathfrak{U})$.

Thus we get the following

L e m m a 4. The operator's $A_\Pi = \Pi_G A \Pi_G$ contraction to $\mathcal{B}_G = \{ f \in \mathcal{B} : (I - \Pi_G) f = 0 \}$ is the infinitesimal operator of some Markov process.

Now we come to the main result on enlarging of semi-Markov processes.

T h e o r e m 2. In the conditions of Theorem I

$$\lim_{\varepsilon \to 0} P\{ u(\xi^\varepsilon(\tfrac{t}{\varepsilon}) \in \bar{\Gamma} \mid \xi^\varepsilon(0) = x \} =$$

$$= P\{ \bar{\eta}(t) \in \bar{\Gamma} \mid \bar{\eta}(0) = u(x) \} ,$$

where $\bar{\eta}(t)$ is a Markov process on $(U, \mathfrak{U})$ with the infinitesimal operator $\bar{A}_\Pi$.

P r o o f. Using the definition of the enlarging operation, we get

$$P\{ u(\xi^\varepsilon(\tfrac{t}{\varepsilon}) \in \bar{\Gamma} \mid \xi^\varepsilon(0) = x \} =$$

$$= P\{\xi^{\varepsilon}(\tfrac{t}{\varepsilon}) \in u^{-1}(\bar{\Gamma}) \mid \xi^{\varepsilon}(0) = x\} = \Phi_{\varepsilon}(x, u^{-1}(\bar{\Gamma}), t).$$

By Lemma 3

$$(27) \quad \lim_{\varepsilon \to 0} \Phi(x, u^{-1}(\bar{\Gamma}), t) = [\Pi_G \exp\{\Pi_G A \Pi_G t\} \chi_{u^{-1}(\bar{\Gamma})}](x) \quad .$$

Since

$$[\Pi_G \chi_{u^{-1}(\bar{\Gamma})}](x) = \chi_{\bar{\Gamma}}(u)$$

and

$$[(\Pi_G A \Pi_G)^{\kappa} \chi_{u^{-1}(\bar{\Gamma})}](x) = [\bar{A}_{\Pi}^{\kappa} \chi_{\bar{\Gamma}}](u)$$

for any κ, decomposing the exponent in (27) into series, we get

$$[\Pi_G \exp\{\Pi_G A \Pi_G t\} \chi_{u^{-1}(\bar{\Gamma})}](x) =$$

$$= [\exp\{\bar{A}_{\Pi} t\} \chi_{\bar{\Gamma}}](u)$$

which is equivalent to the statement of the theorem.

REFERENCES

I. Korolyuk V.S., Polišyuk L.I., Tomusyak A.A., On a limit theorem for semi-Markov processes, Kibernetika, 4, I969 (in Russian).

2. Turbin A.F., Polišyuk L.I., On one case when a semi-Markov process with small parameter converges to a non-standard Markov cha-

in, Teoriya Slucainyh Protzessov, I, I973 (in Russian).

3. Gusak D.V., Korolyuk V.S., Asymptotic behavior of semi-Markov processes with splitted state space, Teoriya Veroyatnostei i Matemat. Statist., Kiev State University, 5, I97I (in Russian).

4. Korolyuk V.S., On asymptotic behavior of the sojourn time of a semi-Markov process in a subset of the state space, Ukrain. Matem. Zurnal, 2I, 6, I969 (in Russian).

5. Anisimov V.V., Enlargement of stochastic processes, Kibernetika, 3, I974 (in Russian).

6. Kovalenko I.N., Investigations on reliability of complex systems, Kiev, I975 (in Russian).

7. Korolyuk V.S., Turbin A.F., Asymptotic enlargement of complex systems. In the book: Knowledge-mathematization and science-technology progress, Kiev, I975.

8. Tsertsvadze T.N., Asymptotic enlargement of Markov chain states and automatons with random entrance effects, Doctor dissertation, Tbilisi, I97I (in Russian).

9. Rotenberg A., Asymptotic enlargement of Markov chain states, Problemy Peredači Inform. I, I974.

IO. Turbin A.F., Aplications of perturbation theory for linear operators in solving some problems related to Markov and semi-Markov processes, Teoriya Veroyatnostei i Matemat. Statist., Kiev State University,6, I972 (in Russian).

II. Plotkin Ya.D., Turbin A.F., Inversion of spectrum-perturbed linear operators,Ukrain.Matemat.Zurnal, 23, 2, I97I (in Russian).

I2. Plotkin Ya.D., Turbin A.F., Inversion of spectrum-perturbed normally solvable linear operators, Ukrain.Matemat.Zurnal, 27, 4, I975 (in Russian).

I3. Kemeny J.G., Snell J.L., Finite Markov chains, Princeton, I960.

I4. Gihman I.I., Skorohod A.V., Introduction to stochastic processes theory, Moscow, I967 (in Russian).

I5. Cinlar E., On semi-Markov processes on arbirary spaces, Proc. Cambridge Philos.Soc., 66, I969.

I6. Čerenkov A.P., Existence theorems for semi-Markov processes on arbitrary spaces, Matemat. Zametki, 3, I974 (in Russian).

I7. Korolyuk V.S., The time for a semi-Markov process to stay in a fixed set of states, Ukrain. Matemat. Zurnal, I7, 3, I965 (in Russian).

I8. Korolyuk V.S., Turbin A.F., On one method to proove limit theorems for some functionals of semi-Markov processes, Ukrain. Matemat.Zurnal, 24, 2, I972 (in Russian).

Mathematical Institute
Academy of Sciences of the Ukrainian SSR, Kiev

THE METHOD OF ACCOMPANYING INFINITELY DIVISIBLE DISTRIBUTIONS

V.M.Kruglov

The idea of the method of accompanying infinitely divisible distributions, in its simpliest form, will be demonstrated by the following example.

Let $\xi_{n1}, \xi_{n2}, \xi_{nm_n}$ $(n=1,2,\dots)$ be a double array of row-wise independent symmetric random variables.

Consider the double array of row-wise independent infinitely divisible (inf.div.) random variables $\eta_{n1}, \eta_{n2}, \dots, \eta_{nm_n}$ $(n=1,2,\dots)$ where η_{nj} is distributed according to the distribution function $\tilde{F}_{nj}$ with the characteristic function

$$\tilde{f}_{nj}(t) = exp\,(\int_{-\infty}^{\infty} (e^{itx} - 1).$$

Here F_{nj} is the distribution function of ξ_{nj} .

Suppose we are interested in some property π of the sequence $\{F_n\}$, $F_n = \overset{m_n}{\underset{j=1}{\boxed{*}}} F_{nj}$. It is often the case that the sequence $\{F_n\}$ possesses the property π if and only if the sequence $\{\tilde{F}_n\}$, $\tilde{F}_n = \overset{m_n}{\underset{j=1}{\boxed{*}}} \tilde{F}_{nj}$ does. This is very important, since we can often prove this property π for the sequence $\{\hat{F}_n\}$ more easily and, hence, can obtain the property π for the original sequence $\{F_n\}$.

Two examples of such properties are:

1) The sequence $\{F_n\}$ is weakly compact if and only if the sequence $\{\tilde{F}_n\}$ is weakly compact.

2) Fix any number $p>0$. The following two relations are

equivalent

$$\lim_{R\to\infty} \sup_n \int_{|x|>R} |x|^p dF_n(x) = 0 \quad .$$

$$\lim_{R\to\infty} \sup_n \int_{|x|>R} |x|^p d\tilde{F}_n(x) = 0 \quad .$$

Thus the method of accompanying infinitely divisible distributions consists in reducing the proof of a property $\mathcal{T}$ for the sequence $\{F_n\}$ to the equivalent problem for the sequence $\{\tilde{F}_n\}$, where $\tilde{F}_n$, $n=1,2,\dots$, are special inf.div. distributions.

It is worth noting that the method of accompanying infinitely divisible distributions can be used in infinite-dimensional spaces as well. In particular, it can be used in Abelian groups and Banach spaces.

This method seems to be most convenient in proving limit theorems for sums of independent random variables and in proving limit theorems for numerical characteristics of sums of independent random variables.

The same method can be used in more delicate problems connected with the behaviour almost sure of sums of independent random variables. But in this case we can obtain only sufficient conditions.

The method of accompanying infinitely divisible distributions is well-known because of the book [3], where this concept plays an important role. We can find the following statment in [3], if a system of random variables $\xi_{n1}, \xi_{n2}, \dots, \xi_{nm_n}$ ($n=1,2,\dots$) is infinitesimal then the sequences $\{F_n\}$ and $\{\tilde{F}_n\}$ can have only common limiting distribution functions.

In the general case we can not hope that the sequences $\{F_n\}$ and $\{\tilde{F}_n\}$ possess this property. But comparison of these two sequences $\{F_n\}$ and $\{\tilde{F}_n\}$ can give a lot of important facts. We will point out only some of them.

In the general theory of summing independent random variables due to V.M.Zolotarev,[1], this comparison allows us to formulate, in a convenient form, necessary and sufficient conditions for weak convergence of distribution functions to a given one.

Using this comparison, Yu.V.Prohorov, [2], found a number of sufficient conditions for the strong law of large numbers.

Systematic use of this method gives the possibility of constructing a satisfactory theory of summing independent random variables with values in a separable Hilbert space H (see [4],[5]).

Denote by $\mathfrak{A}$ the class of continuous functions φ on H, which satisfy the condition: for every elements x, $y \in H$ and for some number $A(\varphi) \equiv A \geq 1$,

$$\varphi(x+y) \leq A\varphi(x)\,\varphi(y) .$$

The method of accompanying infinitely divisible distributions gives necessary and sufficient conditions for convergence of φ-moments, $\varphi \in \mathfrak{A}$ of sums of independent H-valued random variables to the φ-moment of a limiting (in a weak sense) random variable (see [6],[7]).

Choosing $\varphi(x) = 2 + \|x\|^p$, $p>0$, we see that all known theorems of convergence of moments of sums of independent random variables are particular cases of our general result.

Our method gives necessary and sufficient conditions in global limit theorems in the pseudometric:

$$\vartheta(F,G)=\int_{-\infty}^{\infty}\Psi(F(u)-G(u))q(u)du \quad ,$$

where F , G are distribution functions, Ψ and q are functions from rather wide classes of functions. In particular, these classes contain functions of the form $|x|^{p}$, $p>0$ (see [8],[9]).

To prove the results, mentioned above, we make use of some properties of inf.div. distributions in a Hilbert space. We point out two of them that have, in our opinion, also independent interest (see [10], [11]).

1) The existence of the integral of φ , $\varphi\in\mathfrak{A}$, with respect to an inf.div.distribution is equivalent to the existence of the integral of φ with respect to Khinchin's spectral measure of the inf.div. distribution.

2) The existence of the integral of $\varphi_{\alpha}(x)=\exp(\alpha\|x\|\times \ln(\|x\|+1))$ for some $\alpha>0$ ($\|x\|$ is the norm of the element $x\in H$) with respect to an inf.div. distribution is equivalent to the fact that its spectral measure is concentrated in a ball of a finite radius.

As a corollary of the last fact, we have a new characterization of the normal distribution in the class of inf.div.distributions: an inf.div. distribution F is a normal distribution if and only if the integral of $\varphi_{\alpha}(x)$ with respect to F exists for all $\alpha>0$.

In conclusion we formulate two theorems that deal with random variables with values in locally compact Abelian groups and in Banach spaces.

Let G be a separable metric locally compact Abelian group. For every symmetric probability distribution F, denote by $\mathcal{U}(F)$ the set of collections of symmetric components $(F_1, F_2, \ldots)$, $\mathop{\boxed{*}}\limits_{j=1}^{\infty} F_j = F$, such that there are only finitely many $F_j \neq E(\theta)$ in every collection, where $E(\theta)$ is the probability distribution concentrated at the zero element $\theta \in H$. Denote Prohorov's metric by L (see [12]).

Let $\xi_{n1}, \xi_{n2}, \ldots, \xi_{nm_n}$, $n=1,2,\ldots$, be a double array of row-wise independent G-valued symmetric random variables. Denote by F_{nj} the probability distribution of the random variable ξ_{nj}; $F_n = \mathop{\boxed{*}}\limits_{j=1}^{\infty} F_{nj}$, where $F_{nj} = F(\theta)$ for $j > m_n$, $n = 1, 2, \ldots$.

<u>Theorem 1.</u> Let F be a symmetric probability distribution, the characteristic function of which is not equal to zero on every character of the group G.

In order that

$$\lim_{n\to\infty} L(F_n, F) = 0 ,$$

it is necessary and sufficient that the following two conditions hold.

1) There exists a sequence of collections $(G_{n1}, G_{n2}, \ldots) \in \mathcal{U}(F)$ such that

$$\lim_{n\to\infty} \sup_j L(F_{nj}, G_{nj}) = 0 .$$

2) For every character y of the group G

$$\lim_{n\to\infty} \sum_{j=1}^{\infty} [f_{nj}(y) - g_{nj}(y)] = 0 \ .$$

Here f_{nj} and g_{nj} are the characteristic functions of the probability distributions F_{nj} and G_{nj} respectively.

Now let X be a separable Banach space. We shall say that a function $\varphi(x) \geq 0$, $x \in X$, is in the class $\mathfrak{A}_0$ if it is in the class $\mathfrak{A}$ and if it satisfies the condition: for every number $\alpha > 0$,

$$\lim_{\|x\|\to\infty} \varphi(x) \exp(-\alpha \|x\|) = 0 \ .$$

Let $\xi_{n1}, \xi_{n2}, \ldots, \xi_{nm_n}$, $n = 1,2,\ldots$, be a double array of row-wise independent X -valued random variables. Let ξ be a given X -valued random variable, F being its probability distribution. Suppose that the expectations $E\varphi(\xi)$, $E\varphi(\xi_{nj})$, $1 \leq j \leq m_n$, $n = 1,2,\ldots$, are finite for some fixed function $\varphi \in \mathfrak{A}_0$.

Theorem 2. Suppose that

$$\lim_{n\to\infty} L(F_n , F) = 0 \ .$$

For the given function $\varphi \in \mathfrak{A}_0$

$$\lim_{n\to\infty} E\varphi(\xi_n) = E\varphi(\xi)$$

if and only if

$$\lim_{R\to\infty} \sup_n \sum_{j=1}^{m_n} \int_{\|x\|>R} \varphi(x) F_{nj}^{(s)}(dx) = 0 \ ,$$

where $F^{(s)}$ is the operation of symmetrization.

Note. The necessary of the condition of Theorem 2 is valid for all functions $\varphi \in \mathfrak{A}$.

We do not give the proof of the formulated theorems; they can be easily obtained by using the ideas of [4],[5],[6],[7].

REFERENCES

1 Zolotarev V.M., Théorèmes limites generaux pour les sommes de variables aléatoires indépendantes, C.R.Acad.Sci.Paris,A270 (1970), 14, 889-902.

2 Zolotarev V.M., Strong stability of sums and infinitely divisible distributions, Teorija Veroyatnostei i ee Primenen., 3(1958), 2, 153-165. (Russian).

3 Gnedenko B.V., Kolmogorov A.N., Limit distributions for sums of independent random variables, Moscow-Leningrad, 1949. (Russian).

4 Kruglov V.M., Limit theorems for sums of independent random variables with values in a Hilbert space, Teoriya Veroyatnostei i ee Primenen., 17 (1972), 2, 209-227. (Russian).

5 Kruglov V.M., Weak convergence of distributions for sums of independent Hilbert space valued random variables, Studia Scientiarum Mathematicarum Hungarica, 9 (1974), 33-44.

6 Kruglov V.M., Convergence of numeric characteristics of sums of independent random variables and global theorems, Lecture Notes in Math., Springer-Verlag, 330 (1973), 255-286.

7 Kruglov V.M., Convergence of numeric characteristics of sums of Hilbert space valued random variables, Teoriya Veroyatnistei i ee Primenen., 18 (1973), 4, 734-752. (Russian).

8 Kruglov V.M., A global limit theorem for sums of independent ran-

dom variables, Doklady Acad.Sci. SSSR, 3 (1974), 542-545. (Russian).

9 Kruglov V.M., Global limit theorems, Trudy of Leningrad Branch of Steklov Math. Inst. (Russian), to appear.

10 Kruglov V.M., On infinitely divisible distributions in Hilbert space, Matem.Zametki, 16 (1974), 4, 585-594. (Russian).

11 Kruglov V.M., Characterization of a class of infinitely divisible distributions in Hilbert space, Matem.Zametki, 16 (1974), 5, 777-782. (Russian).

12 Prohorov Yu.V., Convergence of random processes and limit theorems of probability theory, Teoriya Veroyatnostei i ee Primenen., 1 (1956), 3, 177-238. (Russian).

Department of Cybernetics
Moscow State University
Moscow

OPTIMAL STOPPING OF CONTROLLED DIFFUSION PROCESS

N.V.Krylov

1. Introduction. The theory of optimal stopping of random processes has been developed in works of Wald, Wolfowitz, Snell, Dynkin, Širyaev, Grigelionis and others (see [1]). In the author's papers [2]-[5] the problems of optimal stopping are considered in more generals settings than usually. There the game situation in a stopping problems was considered and stopping problems for controlled process. The aim of this paper is to give an extension of the results of [2],[3] on optimal stopping of a time-homogeneous controlled diffusion process to the non-homogeneous case. The possibility of this extension is based on the followingestimate.

Theorem 1. Let $(w_t, \mathcal{F}_t)$ be a d_1-dimensional Wiener process, $\sigma_t(\omega)$ be a $d \times d_1$-matrix, $b_t(\omega)$ be a d-dimensional vector, $c_t(\omega)$ be a positive number. Suppose that σ_t, b_t, c_t are bounded progressively measurable with respect to $\{\mathcal{F}_t\}$ and, for some constants R, K, $\varepsilon > 0$: $\varepsilon < \frac{1}{R}$, the following inequality

$$|b_t| \leq \varepsilon \operatorname{tr} \tfrac{1}{2} \sigma_t \sigma_t^* + K[c_t + |\sigma_t^* \lambda|^2]$$

holds for all ω, t and unit vectors λ.

Take a d-dimensional vector x, and let

$$x_t = x + \int_0^t \sigma_s \, dw_s + \int_0^t b_s \, ds \quad .$$

Denote by τ the first exit time of x_t from a ball $S_R = \{y : |y| < R\}$ (R is the same as at the beginning). Then there exists a constant N depending on R, K, ε, d only such that, for all $p \geq d$, $f \geq 0$,

$$M\int_0^\tau e^{-\varphi_t}\psi_t f(t,x_t)\,dt \leq N\left(\int |f(t,y)|^{p+1}\,dt\,dy\right)^{\frac{1}{p+1}}$$

where

$$\varphi_t = \int_0^t c_s\,ds\ ,\quad \psi_t = c_t^{\frac{p-d}{p+1}}(\det \sigma_t\sigma_t^*)^{\frac{1}{p+1}}\ .$$

There are two ways to prove this theorem. One can obtain it from the results of [6] proved with the help of the controlled processes theory. On the other hand, one can apply the following lemma of the theory of convex functions.

Let $\chi(t)$ be a nonnegative smooth function of real t which is equal to zero if $|t| \geq 1$ and equal to one if $|t| < \frac{1}{4}$. Denote by E_d a Euclidean space of dimension d,

$$\varkappa = \int_{E_d} \chi(|x|)\,dx \int_{-\infty}^{\infty} \chi(|t|)\,dt$$

$$u^{(\varepsilon)}(t,x) = \varkappa^{-1}\varepsilon^{-(d+1)}u(t,x) * [\chi(\varepsilon^{-1}|x|)\chi(\varepsilon^{-1}|t|)]$$

<u>Lemma.</u> Let $R > 0$, $h(t,x) \geq 0$, $h \in \mathcal{L}_{d+1}([0,\infty)\times S_R)$: $h(t,x) = 0$ if $(t,x) \notin [0,\infty)\times S_R$. Then there exists a bounded function $z(t,x) \leq 0$ defined on $(-\infty,\infty)\times E_d$, equal to zero if $t < 0$ and such that, for any sufficiently small $\varepsilon > 0$ and any $d\times d$ -matrix $a = (a^{ij})$ on $[0,\infty)\times S_R$

$$N(d)(\det a)^{\frac{1}{d+1}} h^{(\varepsilon)} \leq -\frac{\partial}{\partial t} z^{(\varepsilon)} + \sum_{i,j=1}^{d} a^{ij} z^{(\varepsilon)}_{x^i x^j}\ ,$$

where $N(d) > 0$. Moreover, if a vector b and a number c are such that $|b| < \frac{R}{2}c$, then, on the same set, $\sum_{i=1}^{d} b^i z^{(\varepsilon)}_{x^i} \geq c z^{(\varepsilon)}$

for ε sufficiently small. Finally, for all $t \geq 0$,

$$|z(t,x)|^{d+1} \leq N(d,R) \int_{|y|<R} \int_0^t h^{d+1}(s,y)\,ds\,dy .$$

This lemma is proved in my article in vol. 17, no. 2(1976), of Siberian Mathematical Journal. All the other results of the paper will be proved in my book on the control of diffusion type processes.

2. Notations and difinitions. E_d is a Euclidian space of dimension d , a number $T>0$, $H_T=(0,T)\times E_d$, $(w_t, \mathcal{F}_t)$ is a Wiener process in E_{d_1} , A is a separable metric space. Let $d\times d_1$ -matrix $\sigma(d,s,x)$, d -vector $b(d,s,x)$ and numbers $c^\alpha(s,x)$, $f^\alpha(s,x)$, $g(s,x)$ be defined for $s\geq 0$, $x\in E_d$, $\alpha\in A$. Let $a(\alpha,s,x)=\frac{1}{2}\sigma(\alpha,s,x)\sigma^*(\alpha,s,x)$.

Denote by $\gamma^\alpha(s,x)$ the vector of dimension $d\times d_1+d+3$ with the elements $\sigma^{ij}(\alpha,s,x)$; $i=1,\dots,d$, $j=1,\dots,d_1$; $b^i(\alpha,s,x)$, $i=1,\dots,d$; $c^\alpha(s,x)$, $f^\alpha(s,x)$, $g(s,x)$. Suppose that, for each $\alpha\in A$; $i,j=1,\dots,d$, the derivatives $\gamma^\alpha_{x^i}$, $\gamma^\alpha_{x^i x^j}$; exist and are continuous in x for each $s\in[0,T]$ and the derivative $\frac{\partial}{\partial s}\gamma^\alpha(s,x)$ exists and is continuous in s for each $x\in E_d$. Suppose also the norms of all these derivatives to be le less than $K(1+|x|)^m$ for each $\alpha\in A$; $i,j=1,\dots,d$; $(s,x)\in \bar{H}_T$ where K and $m\geq 0$ are fixed constants.

Let $\gamma^\alpha(s,x)$ be continuous in α and, for all $x,y\in E_\alpha$, $s\geq 0$, $\alpha\in A$

$$\|\sigma(\alpha,s,x)-\sigma(\alpha,s,y)\|+|b(\alpha,s,x)-b(s,\alpha,y)|\leq K|x-y| ,$$

$$\|\sigma(\alpha,s,x)\|+|b(\alpha,s,x)|\le K(1+|x|),$$

$$|c^{\alpha}(s,x)|+|f^{\alpha}(s,x)|+|g(s,x)|\le K(1+|x|)^{m} \quad .$$

Definition 1. A control α_t is an A-valued random process progressively measurable with respect to $\{\mathcal{F}_t\}$. $\mathfrak{A}$ is the set of all controls.

For $\alpha\in\mathfrak{A}$, $s\in[0,T]$, and $x\in E_d$ denote by $x_t^{\alpha,s,x}$ the solution of

$$x_t=x+\int_0^t\sigma(\alpha_\tau,s+\tau,x_\tau)\,dw_\tau+\int_0^t b(\alpha_\tau,s+\tau,x_\tau)\,d\tau \quad . \tag{1}$$

$\mathfrak{M}(T-s)$ is the set of all stopping times τ (with respect to $\{\mathcal{F}_t\}$) such that $\tau(w)\le T-s$. For $\alpha\in\mathfrak{A}$, $x\in E_d$, $s\in[0,T]$ and $\tau\in\mathfrak{M}(T-s)$ we define

$$v^{\alpha,\tau}(s,x)=M_{s,x}^{\alpha}\Big[\int_0^{\tau}e^{-\varphi_t}f^{\alpha_t}(s+t,x_t)\,dt+e^{-\varphi_\tau}g(s+\tau,x_\tau)\Big], \tag{2}$$

where the indices α,s,x mean that $x_t=x_t^{\alpha,s,x}$ and $\varphi_t=\varphi_t^{\alpha,s,x}=$ $=\int_0^t c^{\alpha_\tau}(s+\tau,x_\tau^{\alpha,s,x})\,d\tau$. Define

$$v(s,x)=\sup_{\alpha\in\mathfrak{A}}\ \sup_{\tau\in\mathfrak{M}(T-s)}v^{\alpha,\tau}(s,x).$$

It seems to be not quite natural to use controls depending on w_t (i.e. on the "noise") from the practical point of view. Therefore we will also consider other types of control.

Let $C([0,\infty),E_d)$ be the space of all E_d-valued continuous functions x_t defined on $[0,\infty)$, $\mathcal{N}_t$ be the σ-field generated by the sets $\{x_{[0,\infty)}: x_\tau\in\Gamma\}$ for $\tau\in[0,t]$ and Borel $\Gamma\subset E_d$.

<u>Definition 2.</u> An A-valued function $\alpha_t(x_{[0,\infty)}) = \alpha_t(x_{[0,t]})$ defined for $t \in [0,\infty)$, $x_{[0,\infty)} \in C([0,\infty), E_d)$ is said to be a natural control if it is progressively measurable with respect to $\{\mathcal{N}_t\}$. A natural control $\alpha_t(x_{[0,t]})$ is admissible at point (s,x) if there exists a solution (possibly non-unique) of the equation

$$x_t = x + \int_0^t \sigma(\alpha_\tau(x_{[0,\tau]}), s+\tau, x_\tau)\, dw_\tau + \int_0^t b(\alpha_\tau(x_{[0,\tau]}), s+\tau, x_\tau)\, d\tau \qquad (3)$$

which is progressively measurable with respect to $\{\mathcal{F}_t\}$. Denote by $\mathfrak{A}_{\mathcal{N}}(s,x)$ the set of all natural controls admissible at point (s,x).

For each $\alpha \in \mathfrak{A}_{\mathcal{N}}(x,s)$, we choose and fix some solution $x_t^{\alpha,s,x}$ of (3).

<u>Definition 3.</u> A natural control $\alpha_t(x_{[0,t]})$ is said to be a Markov control if $\alpha_t(x_{[0,t]}) \equiv \alpha_t(x_t)$ for some function $\alpha_t(x)$.

Denote by $\mathfrak{A}_M(s,x)$ the set of all Markov controls admissible at point (s,x).

It is easy to construct a mapping of $\mathfrak{A}_{\mathcal{N}}(s,x)$ into $\mathfrak{A}$. If $\alpha \in \mathfrak{A}_{\mathcal{N}}(s,x)$, define $\beta_t = \alpha_t(x_{[0,t]}^{\alpha,s,x})$. Obviously, $\beta = \{\beta_t\} \in \mathfrak{A}$. Since a solution of (1) is unique, $x_t^{\beta,s,x}$ coincides with $x_t^{\alpha,s,x}$. This implies

$$w_M(s,x) \equiv \sup_{\alpha \in \mathfrak{A}_M(s,x)} \sup_{\tau \in \mathfrak{M}(T-s)} v^{\alpha,\tau}(s,x) \le$$

$$\le \sup_{\alpha \in \mathfrak{A}_{\mathcal{N}}(s,x)} \sup_{\tau \in \mathfrak{M}(T-s)} v^{\alpha,\tau}(s,x) \equiv w_{\mathcal{N}}(s,x) \le w(s,x) .$$

<u>3. Randomized stopping. Main results.</u> We begin by reducing the problem of optimal stopping to a pure problem of control.

Consider the problem of optimal stopping of a Wiener process: $d_1 = d$, $\sigma(\alpha, s, x)$ is the unit matrix, $b(\alpha, s, x) \equiv 0$, $c^{\alpha}(s, x) \equiv 0$, $f^{\alpha}(s,x) \equiv f(s,x)$.

Take $s \in [0,T]$ and some nonnegative progressively measurable process z_t. With the help of z_t define the following stopping rule of $x_t = x + w_t$. Let the process x_t stop in the time integral $(t, t+\Delta t)$, conditioned that it was not stopped before, with the probability $z_t \Delta t + o(\Delta t)$. If the process is stopped at time t, we get the reward $\rho_t \equiv \int_0^t f(s+z, x_z)\,dz + g(s+t, x_t)$ (cf. the expression following the expectation sign in (2)). Since τ is less than $T-s$ in (2), we shall suppose the probability of stopping at time $T-s$ equal to 1 (conditioned the process was not stopped before). One can easily check that the expected value of ρ_t on an individual trajectory is equal to

$$\int_0^{T-s} \rho_t z_t e^{-\int_0^t z_u du} dt + \rho_{T-s} e^{-\int_0^{T-s} z_u du} =$$

$$= \int_0^{T-s} [f(s+t, x_t) + z_t g(s+t, x_t)] e^{-\int_0^t z_u du} dt + g(T, x_{T-s}) e^{-\int_0^{T-s} z_u du} .$$

The expectation of this expression is equal to

$$M\{\int_0^{T-s} [f(s+t,x_t) + z_t g(s+t, x_t)] e^{-\int_0^t z_u du} dt + g(T, x_{T-s}) e^{-\int_0^{T-s} z_u du}\} . \qquad (4)$$

It is not difficult to understand that the supremum of (4) over the set of all processes z_t must coincide with $w(s,x)$. In fact, this supremum is not less than $w(s,x)$ because immediate stopping at the moment τ can be "approximated" by randomized stopping with large intensity z_t of stopping after time τ. On the other hand, if there are some reasons showing that it is really profitable to have a possibility to stop the process with positive intensity at time t , then , for the same reasons, one can stop the process with probability one at time t . Therefore (4) is not greater than $w(s,x)$.

In the general case, define $B_n = A \times [0,n]$ and, for $\beta=(\alpha,\tau)\in B_n$, let

$$\sigma(\beta,s,x)=\sigma(\alpha,s,x),\ b(\beta,s,x)=b(\alpha,s,x),$$

$$c^{\beta}(s,x)=c^{\alpha}(s,x)+\tau,\ f^{\beta}(s,x)=f^{\alpha}(s,x)+\tau g(s,x).$$

Denote by $\mathcal{R}_n$ the set of all progressively measurable process τ_t such that $0\leq \tau_t(\omega)\leq n$. For $\beta=\beta_t=(\alpha_t,\tau_t)\in\mathfrak{A}\times\mathcal{R}_n$, define

$$x_t^{\beta,s,x}=x_t^{\alpha,s,x},\ \varphi_t^{\beta,x,s}=\varphi_t^{\alpha,s,x}+\int_0^t \tau_u\,du,$$

$$v^{\beta}(s,x)=M_{s,x}^{\beta}\Big[\int_0^{T-s} f^{\beta_t}(s+t,x_t)e^{-\varphi_t}dt+g(T,x_{T-s})e^{-\varphi_{T-s}}\Big]$$

with the same agreements about indices β,s,x as in (2). Put

$$w_n(s,x)=\sup_{\beta\in\mathfrak{A}\times\mathcal{R}_n} v^{\beta}(s,x).$$

Theorem 2. The function $w_n(s,x)$ is the reward for the problem of optimal stopping of the process under consideration if the function $g(s,x)$ is replaced by $g(s,x)\wedge w_n(s,x)$. Moreover

$$w(s,x)=\lim_{n\to\infty} w_n(s,x)=\sup_{\beta\in\bigcup_n(\mathfrak{A}\times\mathcal{R}_n)} v^{\beta}(s,x).$$

Since the last supremum is the reward function for the problem of control without stopping, this theorem allow us to use the results of [7] for investigating the problem of optimal stopping.

Theorem 3. The function $w(s,x)$ is continuous in $\bar{H}_T$,

$$w(s,x) \geq g(s,x) \quad w(T,x) = g(T,x), \; |w(s,x)| \leq N(1+|x|)^m.$$

Put

$$Q = \{(s,x) \in H_T : \inf_{|\lambda|=1} \sup_{\alpha \in A} (a(\alpha,s,x)\lambda, \lambda) > 0\}$$

Then w , in the domain Q , has two Sobolev derivatives with respect to x and one Sobolev derivative with respect to t . These derivatives are locally bounded in Q . The following expression

$$\sup_{\alpha \in A} [\sum_{i,j=1}^{d} a^{i,j}(\alpha,x,s) w_{x^i x^j}(s,x) + \sum_{i=1}^{d} b^i(\alpha,s,x) w_{x^i}(s,x) - \qquad (5)$$
$$- c^\alpha(s,x) w(s,x) + \frac{\partial}{\partial s} w(s,x) + f^\alpha(s,x)]$$

is equal (a.e.) to zero in $Q \cap \{(s,x) : w(s,x) > g(s,x)\}$ and is negative (a.e.) in Q .

Theorem 4. For $\varepsilon > 0$,

$$w(s,x) \leq \sup_{\alpha \in \mathfrak{A}} M^\alpha_{s,x} [\int_0^{\tau_\varepsilon} f^{\alpha_t}(s+t,x_t) e^{-\varphi_t} dt + g(s+\tau_\varepsilon, x_{\tau_\varepsilon}) e^{-\varphi_{\tau_\varepsilon}}] + \varepsilon$$

where $\tau_\varepsilon = \tau_\varepsilon^{\alpha,s,x}$ is the first exit time of the process $(s+t, x_t^{\alpha,s,x})$ from the domain

$$Q_\varepsilon = \{(s,x) \in \bar{H}_T : w(s,x) > g(s,x) + \varepsilon\}$$

It holds also true for $\varepsilon = 0$ if A is a singleton.

Theorem 5. $w(s,x) \equiv w_N(s,x)$ always. $w(s,x) \equiv w_M(s,x)$ if a)

$\inf \sup_{\alpha\in A} (a(\alpha,s,x)\lambda,\lambda) > 0$ for all s,x or b) $\sigma(\alpha,s,x)$ does not depend on x.

4. Explicit formulas for optimal boundary. Global "smoothness" of $w(s,x)$ and the properties of (5) sometimes enable us to write explicit formulas for the boundary of the domain Q_0. By Theorem 4 it is optimal to stop the process $(s+A, x_t^{\alpha,s,x})$ on this boundary if A is a singleton.

Consider an example (cf.[8]).

Let w_t be a one-dimensional Wiener process and

$$w(s,x) = \sup_{\tau\in \mathfrak{M}(1-s)} M(1-s-\tau)(x+w_\tau).$$

Using the invariance properties of w_t and decreasing in s of $g(s,x) \equiv (1-s)x$ one can prove without difficulty that

$$Q_0 = \{(s,x): s\in[0,1),\ x < c\sqrt{1-s}\}$$

for some constant c. Thus the problem is to find this constant c.

Note that Q from Theorem 3 coincides with $(0,1)\times(-\infty,\infty)$. Consequently, by Theorem 3,

$$\left(\frac{\partial}{\partial s}+\frac{1}{2}\frac{\partial^2}{\partial x^2}\right)w(s,x) = h(s,x) = \begin{cases} \left(\frac{\partial}{\partial s}+\frac{1}{2}\frac{\partial^2}{\partial x^2}\right)g(s,x), & x > c\sqrt{1-s} \\ 0, & x < c\sqrt{1-s} \end{cases}$$

Applying Theorem 1, it is easy to prove that the Itô formula for $w(s+t, x+w_t)$ holds true. By this formula

$$w(s,x) = -M\int_0^{1-s} h(s+t, x+w_t)\,dt = \int_0^{1-s}\frac{1}{\sqrt{2\pi t}}\int_{c\sqrt{1-s-t}-x}(x+y)e^{-\frac{y^2}{2t}}\,dy\,dt.$$

Now

$$\omega(s, c\sqrt{1-s}) = g(s, c\sqrt{1-s})$$

since $\omega(s,x) = g(s,x)$ on the boundary of Q_0. Setting $s=0$, we have

$$g(0,c) = c = \int_0^1 \frac{1}{\sqrt{2\pi t}} \int_{c\sqrt{1-t}-c}^{\infty} (c+y) e^{-\frac{y^2}{2t}} \, dy \, dt$$

This is an equation for c. One can prove that it has one and only one solution.

REFERENCES

I. Ширяев А.Н., Статистический последовательный анализ, Москва, I969.

2. Крылов Н.В., Об управлении решением стохастического интегрального уравнения, Теория вероятн. и ее применен., 17, I (I972), III-I27.

3. Крылов Н.В., Об управлении решением стохастического интегрального уравнения при наличии вырождения, Известия АН СССР, сер. матем., 36, I (I972), 248-26I.

4. Крылов Н.В., Задача с двумя свободными границами для эллиптического уравнения и оптимальная остановка марковского процесса, ДАН СССР, I94, 6 (I970), I263-I265.

5. Крылов Н.В., Управление марковскими процессами и пространства ,
Известия АН СССР, сер. матем., 35, I (I97I), 224-255.

6. Крылов Н.В., Некоторые оценки плотности распределения стохастического интеграла, Известия АН СССР, сер. матем., 38, I (I974), 228-248.

7. Крылов Н.В., Об уравнении Беллмана, Труды школы-семинара по теории случайных процессов, I, Вильнюс, I975, 20I-234.

8. Мирошниченко Т.П., Оптимальная остановка интеграла от винеровского процесса, Теория вероятн. и ее применен., 20, 2 (1975).

Department of Mathematics and Mechanics

Moscow State University

Moscow

ADDITIVE ARITHMETIC FUNCTIONS AND BROWNIAN MOTION

J.Kubilius

Consider a sequence of real numbers $\{f(p)\}$ defined for all successive primes p. Let, for all positive integers m,

$$f(m) = \sum_{p|m} f(p) \tag{I}$$

where the sum is extended over all prime divisors of m, or

$$f(m) = \sum_{p} f(p)\delta_{pm}$$

with

$$\delta_{pm} = \begin{cases} 1 & \text{if } p|m \\ 0 & \text{otherwise.} \end{cases}$$

The function $f(m)$ is called strongly additive. It has the properties: (i) $f(m_1 m_2) = f(m_1) + f(m_2)$ for all coprime m_1, m_2, (ii) $f(p^\alpha) = f(p)$ for all prime powers p^α. Conversely, from these properties it follows that $f(m)$ satisfies (1).

Some functions of this kind play an important role in number theory. Values of an additive function depend on the multiplicative structure of the argument. Therefore their distribution is very complicated. For this reason it is natural to consider the distribution of values of additive functions from the point of view of probability theory. For example, we may study the frequence of positive integers $m \leq n$ for which $f(m) < A_n + B_n x$ where A_n and B_n are suitably chosen normalizing constants. At present there are many results giving sufficient conditions for the convergence of this frequency to a distribution function (see for references [6,9,II]).

I shall mention just one of them. Let us denote by $\nu_n\{\dots\}$ the frequency of positive integers $m \leq n$ satisfying the conditions in the braces.

Suppose that $f(p)$ are not identically 0. Let

$$B_n^2 = \sum_{p \leq n} \frac{f^2(p)}{p}, \quad f_n(p) = \frac{f(p)}{B_n},$$

$$\xi_{np}(m) = f_n(p)(\delta_{pm} - \frac{1}{p}), \; \zeta_n(m) = \sum_{p \leq n} \xi_{np}(m).$$

If, for every fixed $\varepsilon > 0$,

(2) $$\sum_{\substack{p \leq n \\ |f_n(p)| > \varepsilon}} \frac{f_n^2(p)}{p} \longrightarrow 0$$

as $n \to \infty$ (an analogue of the Lindeberg condition), then

(3) $$\nu_n\{\zeta_n(m) < x\} \longrightarrow \Phi(x)$$

as $n \to \infty$, where $\Phi(x)$ is the standard normal distribution function. If the sequence $\{f(p)\}$ satisfies the Lindeberg condition (2), then B_n is a slowly increasing function (in the sense of Karamata) of $\ln n$ (or, what is the same, $B_n \sim B_{[\sqrt{n}]}$). If B_n is such a function, then (2) is also necessary for (3) to hold.

Condition (2) holds true if

(4) $$\max_{p \leq n} |f_n(p)| = \mu_n \longrightarrow 0$$

as $n \to \infty$.

It is of interest to investigate the simultaneous asymptotic behaviour of the partial sums

$$\zeta_{nq}(m) = \sum_{p \le q} \xi_{nq}(m)$$

when q takes prime values, $q = 2,3,5,\dots;\ q \le n$. Denote

$$t_{nq} = \sum_{p \le q} \frac{f_n^2(q)}{q}$$

Consider in the (t,x)-plane the points with the coordinates $(t_{nq}, \zeta_{nq}(m))(q \le n)$. Obviously all of them lie between the two straigt lines $t=0$ and $t=1$. One may ask: what is the number of such points belonging to a region of this strip? In 1955 [10] I proved the following theorem.

Let $\psi_1(t)$ and $\psi_2(t)$ by two real functions defined in the interval $[0,1]$, continuously differentiable and subject to the conditions $\psi_1(t) < \psi_2(t),\ \psi_1(0) < 0 < \psi_2(0)$. Suppose, further, that (4) holds true. Then

(5) $$\nu_n\{\psi_1(t_{nq}) < \zeta_{nq}(m) < \psi_2(t_{nq}),\ q \le n\}$$

converges to the probability

$$W(\psi_1, \psi_2) = P(\psi_1(t) < w(t) < \psi_2(t),\ 0 \le t \le 1)$$

that the standard Brownian motion process $w(t)$, starting at the point 0, does not reach the boundaries $\psi_1(t)$ and $\psi_2(t)$.

Changing a little the proof of the theorem, it is easy to see that the same is true if the Lindeberg condition (2) holds. Further, the continuous differentiability may be replaced by the Lipschitz condition: for a certain $c > 0$ and for any $t', t'' \in [0,1]$,

(6) $$|\psi_j(t') - \psi_j(t'')| \le c\,|t' - t''| \quad (j = 1,2).$$

This theorem led us to the conjecture that the sequence $\zeta_{nq}(m)$ may be used for simulation of the Brownian motion process. Let us take a polygonal curve $\zeta_n(t,m)$, $0 \le t \le 1$, which we obtain by connecting the points $(0,0), (t_{n2}, \zeta_{n2}(m)), \ldots, (t_{nq}, \zeta_{nq}(m)), \ldots$ $(q \le n)$ linearly. There are other possible choices for $\zeta_n(t,m)$. For instance, we could take

$$\zeta_n(t,m) = \sum_{\substack{p \le n \\ t_{np} \le t}} \xi_{np}(m) .$$

From the above theorem one can deduce easily that the frequency of $m \le n$ for which the curve lies between $\psi_1(t)$ and $\psi_2(t)$

$$\nu_n\{\psi_1(t) < \zeta_n(t,m) < \psi_2(t),\ 0 \le t \le 1\} \tag{7}$$

converges to the same probability $W(\psi_1, \psi_2)$.

P.Billingsley [3] proved that if the numbers $f(p)$ are bounded, then $\zeta_n(t,m)$ tends in distribution to the standard Brownian motion $w(t)$. G.Jogesh Babu [I] and W.Philipp [I3] proved this statement with the Lindeberg condition (2). The theorem is true for both the choices of $\zeta_n(t,m)$ if we consider the convergence of the corresponding probability measures in $C[0,1]$ and $\mathcal{D}[0,1]$. Moreover, if B_n is a Karamata function of $\ln n$ then (2) is also necessary.

The aim of this paper is to estimate the rate of convergence for the frequencies (5) and (7).

Theorem. Let the functions $\psi_1(t)$ and $\psi_2(t)$ satisfy Lipschitz condition (6). Suppose that the sequence $\{f(p)\}$ satisfies (4). Then

$$\nu_n\{\psi_1(t_{nq}) < \zeta_n(t_{nq},m) < \psi_2(t_{nq});\ q \le n\} = W(\psi_1,\psi_2) + \frac{B\mu_n \ln(1/\mu_n)}{\ln\ln(1/\mu_n)}$$

and, if $\zeta_n(t,m)$ is a polygonal curve in $C[0,1]$,

$$\nu_n\{\psi_1(t)<\zeta_n(t,m)<\psi_2(t),\ 0\le t\le 1\}=W(\psi_1,\psi_2)+\frac{B\mu_n\ln(1/\mu_n)}{\ln\ln(1/\mu_n)}$$

Here (and in the sequel) B denotes a function which is bounded by an absolute constant or a constant depending on c, $\psi_1(0)$, $\psi_2(0)$ only.

The proof of this theorem consists of two parts. At first we reduce the problem to an analogous one for suitably chosen independent random variables. Then we may solve it directly or apply known results from the theory of summation of independent random variables.

I°. Consider the probability space $\{\Omega_n, \mathcal{A}_n, \nu_n\}$. Here $\Omega_n=\{1,2,\dots,n\}$, $\mathcal{A}_n$ is the algebra of all subsets of Ω_n, the probability measure $\nu_n(A)$ is the frequency of elements of A,

$$\nu_n(A)=\nu_n\{m\in A\}.$$

Obviously all real-valued functions defined on Ω_n are $\mathcal{A}_n$-measurable and therefore may be considered as random variables. However this model is of no use because the random variables $\xi_{np}(m)$ are dependent in a very complicated manner.

To overcome this difficulty we shall change the probability space. Let $z=z(n)$ be a function of n such that $z(n)\to\infty$, $\ln z(n)/\ln n\to 0$ as $n\to\infty$. Denote by E_p the set of positive integers $m\le n$ divisible by p. Let $\mathbb{B}_n$ be the smallest set algebra containing Ω_n and all E_p with $p\le z$. Every $A\in\mathbb{B}_n$ has the form

$$(8) \qquad \bigcup_{k \in K} E_k$$

where K is a set of square-free integers k having no prime factors greater than z ,

$$E_k = \bigcap_{p \le z} E_p(\delta_{pk}),\ E_p(1) = E_p\,,\ E_p(0) = \Omega_n \setminus E_p\,.$$

Obviously for different k the sets E_k are desjoint.

We shall now approximate the measure ν_n by another measure P_n. Define first

$$P_n(E_p(\delta)) = \begin{cases} \frac{1}{p} & \text{if } \delta = 1\,, \\ 1 - \frac{1}{p} & \text{if } \delta = 0\,. \end{cases}$$

For all $A \in \mathbb{B}_n$, we put

$$P_n(A) = \sum_{k \in K} \prod_{p \le z} P_n(E_p(\delta_{pk}))$$

whenever A is of the form (8). The set function $P_n(A)$ is non--negative, additive and $P_n(\Omega_n) = 1$.

We have two new probability spaces $\{\Omega_n, \mathbb{B}_n, \nu_n\}$ and $\{\Omega_n, \mathbb{B}_n, P_n\}$. It is easy to prove that $\delta_{pm}\ (p \le z)$ as functions of m (and, consequently, the functions $\xi_{np}(m)$) are independent with respect to the second space. Let $p_1, \dots, p_s$ be primes not exceeding z and let $\delta_j\ (j = 1, \dots, s)$ take values 0 or 1. We need show that

$$(9) \qquad P_n(\delta_{p_1 m} = \delta_1, \dots, \delta_{p_s m} = \delta_s) = \prod_{j=1}^{s} P_n(\delta_{p_j m} = \delta_j)\,,$$

For this aim we note that

$$\{m \leq n, \delta_{p_1 m} = \delta_1, \dots, \delta_{p_s m} = \delta_s\} = \bigcup_k{}' E_k$$

where the union is taken over all possible k satisfying the conditions $\delta_{p_1 k} = \delta_1, \dots, \delta_{p_s k} = \delta_s$. Hence it follows:

$$P_n(\delta_{p_1 m} = \delta_1, \dots, \delta_{p_s m} = \delta_s) = \sum_k{}' \prod_{p \leq z} P_n(E_p(\delta_{pk})) =$$

$$= \prod_{j=1}^{s} P_n(E_{p_j}(\delta_j)) \cdot \sum_k{}' \prod_{\substack{p \leq z \\ p \neq p_j (j=1,\dots,s)}} P_n(E_p(\delta_{pk})).$$

The first product on the right-hand side is the same as in the equality (9), and the sum equals

$$\prod_{\substack{p \leq z \\ p \neq p_j (j=1,\dots,s)}} \{P_n(E_p(1)) + P_n(E_p(0))\} = 1.$$

We need estimate the difference between the measures ν_n and P_n for all $A \in \mathbb{B}_n$. We shall make use of two arithmetic lemmas.

Lemma 1. For any u, v and $y \geq 1, 3 \leq v \leq u$,

$$\nu_y\{\prod_{\substack{p|m \\ p \leq v}} p \geq u\} = B\lambda^{-c_1 \lambda}$$

where $\lambda = \ln u / \ln \max(v, \ln u)$, $c_1 > 0$ and B is bounded by an absolute constant.

Lemma 2. Let u and v be any numbers, $3 \leq v \leq u$. Let Q_1 and Q_2 be two sets of different primes not exceeding v. Suppose that

$$\prod_{p \in Q_1} p \leq \sqrt{u}.$$

Then the frequency of positive integers $m \leq u$, divisible by all primes from Q_1 and not divisible by any of primes from Q_2 , equals

$$\prod_{p \in Q_1} \frac{1}{p} \cdot \prod_{p \in Q_2} \left(1 - \frac{1}{p}\right) \cdot \left(1 + B\lambda^{-c_2\lambda}\right)$$

with the same λ as in Lemma 1. The constansts $c_2 > 0$ and that bounding B are absolute.

These lemmas are some improvement of those given in the author's book ([9] , Lemma 1.2,I.6). Their proofs may be found in [2] ,[8] .

Now we return to the proof of the theorem. Let the set A has the form (8). Then

$$\nu_n(A) = \sum_{k \in K} \nu_n(E_k).$$

The set E_k consists of positive integers $m \leq n$ satisfying the conditions

$$m \equiv 0 \bmod \prod_{\substack{p \mid k \\ p \leq z}} p \,, \quad (m, \prod_{\substack{p \nmid k \\ p \leq z}} p) = 1.$$

In case $k < \sqrt{n}$ we have from Lemma 2 (with $v = z$, $u = n$)

$$\nu_n(E_k) = (1 + BR) \prod_{p \leq z} P_n(E_p(\delta_{pk}))$$

where

$$R = \varkappa^{-c_3\varkappa}, \quad \varkappa = \frac{\ln n}{\ln \max(z, \ln n)}$$

and $c_3 = \min(c_1, c_2)$. In case $k \geq \sqrt{n}$ we use Lemma 1 (with $v = z$, $u = \sqrt{n}$, $y = n$). We obtain

$$\nu_n\Big(\bigcup_{k\geq\sqrt{n}} E_k\Big) = BR.$$

Therefore

$$\nu_n(A) = (1+BR)\sum_{\substack{k\in K\\ k<\sqrt{n}}} P_n(E_p(\delta_{pk})) + BR.$$

It remains to estimate the error if we omit the condition $k<\sqrt{n}$. By Lemma 2 (with $v=r$ and $u\to\infty$)

$$\sum_{k\geq\sqrt{n}} \prod_{\substack{p|k\\ p\leq r}} \frac{1}{p} \prod_{\substack{p\nmid k\\ p\leq r}} \Big(1-\frac{1}{p}\Big)$$

is the asymptotic density of positive integers m for which $\delta_{pm}=\delta_{pk}$ for all $p\leq r$ and at least one $k\geq\sqrt{n}$. By Lemma 1 (with $v=r$, $u=\sqrt{n}$ and $y\to\infty$) this density equals BR. Thus,

(IO) $$\nu_n(A) = (1+BR)\sum_{k\in K}\prod_{p\leq r} P_n(E_p(\delta_{pk})) + BR = P_n(A)+BR$$

uniformly for all $A\in\mathbb{B}_n$

2°. Denote

$$\sigma_n^2 = \sum_{p\leq r}\frac{f_n^2(p)}{p}\Big(1-\frac{1}{p}\Big).$$

Instead of $\xi_{np}(m)$, $\zeta_{nq}(m)$, t_{nq} we shall consider the functions

$$\hat{\xi}_{np}(m) = \frac{\xi_{np}(m)}{\sigma_n},\ \hat{\xi}_{nq}(m) = \sum_{p\leq q}\hat{\xi}_{np}(m),\ \hat{t}_{nq} = \sum_{p\leq q} E\,\hat{\xi}^2_{np}.$$

With respect to the space $\{\Omega_n, \mathrm{B}_n, \mathrm{P}_n\}$ the variables $\hat{\xi}_{np}$ $(p \le z)$ are independent and such that

$$\mathrm{E}\,\hat{\xi}_{np} = 0, \ \sum_{p \le z} \hat{\xi}_{np}^{2} = 1, \ \mathrm{P}_n \{ |\hat{\xi}_{np}| > \frac{\mu_n}{\sigma_n} \} = 0 .$$

By a theorem of B.V.Gnedenko, V.S.Korolyuk and A.V.Skorokhod [7] we have

$$\mathrm{P}_n \{ \psi_1(\hat{t}_{nq}) < \hat{\xi}_{nq}(m) < \psi_2(\hat{t}_{nq}), \ q \le z \} =$$

(II)

$$= W(\psi_1, \psi_2) + B\left(\frac{\mu_n}{\sigma_n} + \max_{p \le z} \mathrm{E}\,\hat{\xi}_{np}^{2} + \gamma_n \ln \frac{1}{\gamma_n}\right)$$

where

$$\gamma_n = \sum_{p \le z} \mathrm{E}\,\hat{\xi}_{np}^{4}$$

and the constant bounding B depends only on c, $\psi_1(0)$, $\psi_2(0)$.

Let us take

$$z = \exp\left(\frac{c_1 \ln n \cdot \ln n(1/\mu_n)}{2 \ln(1/\mu_n)}\right).$$

From (4) it follows

$$1 = \sum_{p \le n} \frac{f_n^2(p)}{p} \le \mu_n^2 \sum_{p \le n} \frac{1}{p} = B\mu_n^2 \ln\ln n , \ \mu_n^{-2} = B \ln\ln n .$$

Hence

$$z > \frac{c_4 \ln n \cdot \ln_4 n}{\ln_3 n} > \ln\ln n$$

for sufficiently large n and

$$\ln z = v(\ln n),\ \varkappa = \frac{\ln n}{\ln z} = \frac{2\ln 1/\mu_n}{c_3 \ln\ln 1/\mu_n},\ \varkappa^{-c_3\varkappa} = B\mu_n .$$

Further

$$\sigma_n^2 = \sum_{p\le n} \frac{f_n^2(p)}{p} - \sum_{z<p\le n} \frac{f_n^2(p)}{p} - \sum_{p\le z} \frac{f_n^2(p)}{p^2} = 1 + B\mu_n^2 \ln\frac{\ln n}{\ln z} + B\mu_n^2 = 1 + B\mu_n^2 \ln\varkappa ,$$

$$\max_{p\le z} E\hat{\xi}_{np}^2 = \max_{p\le z} \frac{f_n^2(p)}{\sigma_n^2 p}\left(1-\frac{1}{p}\right) = B\mu_n^2 ,$$

$$\gamma_n \le \frac{\mu_n^2}{\sigma_n^2} \sum_{p\le z} E\hat{\xi}_{np}^2 = \frac{\mu_n^2}{\sigma_n'^2} = B\mu_n^2 .$$

These estimates, (10) and (11) give us

(12) $$\nu_n\{\psi_1(\hat{t}_{nq}) < \hat{\xi}_{nq}(m) < \psi_2(\hat{t}_{nq}),\ q\le z\} = W(\psi_1,\psi_2) + B\mu_n .$$

It remains to replace $\hat{t}_{nq}$ and $\hat{\xi}_{nq}(m)$ by t_{nq} and $\xi_{nq}(m)$ and let q run over all primes $\le n$. For this aim we note that

(13) $$\hat{t}_{nq} = \frac{1}{\sigma_n^2} \sum_{p\le q} \frac{f_n^2(p)}{p}\left(1-\frac{1}{p}\right) = \frac{1}{\sigma_n^2}(t_{nq} + B\mu_n^2) = t_{nq} + B\mu_n^2 \ln\varkappa$$

and

(14) $$\xi_{nq}(m) = \sigma_n \hat{\xi}_{nq}(m) = \hat{\xi}_{nq}(1 + B\mu_n^2 \ln\varkappa)$$

for $q\le z$. For $z\le q\le n$,

$$\xi_{nq}(m) = \sigma_n \hat{\xi}_{nq'}(m) + \rho$$

where q' is the largest $p\le z$,

$$\rho = \sum_{r<p\le q} \xi_{np}(m) = \sum_{r<p\le q} f_n(p)\delta_{pm} - \sum_{r<p\le q} \frac{f_n(p)}{p} .$$

It is obvious that

$$|\rho| \le \mu_n \sum_{r<p\le q} \delta_{pm} + B\mu_n \ln \frac{\ln n}{\ln r} .$$

We note that the sum to estimate is the number of prime divisors of m which all are between r and q. This number ℓ satisfies the inequality $r^{\ell} \le n$; therefore it does not exceed

$$\frac{\ln n}{\ln r} = \varkappa .$$

Thus,

$$\xi_{nq}(m) = \hat{\xi}_{nq'}(m)(1 + B\mu_n^2 \ln \varkappa) + B\mu_n \varkappa \tag{I5}$$

for $r < q \le n$.

From (I3),(I4) and (I5) we have

$$\nu_n\{\psi_1(\hat{t}_{nq}) + C_5\mu_n\varkappa < \hat{\xi}_{nq}(m) < \psi_2(\hat{t}_{nq}) - C_5\mu_n\varkappa,\ q \le r\} \le$$

$$\le \nu_n\{\psi_1(t_{nq}) < \xi_{nq}(m) < \psi_2(t_{nq}),\ q \le n\} \le \tag{I6}$$

$$\le \nu_n\{\psi_1(\hat{t}_{nq}) - C_5\mu_n\varkappa < \hat{\xi}_{nq}(m) < \psi_2(\hat{t}_{nq}) + C_5\mu_n\varkappa,\ q \le r\}$$

with some sufficienly large positive constant C_5. We need estimates for $W(\psi_1 + C_5\mu_n\varkappa, \psi_2 - C_5\mu_n\varkappa) - W(\psi_1, \psi_2)$ and $W(\psi_1 - C_5\mu_n\varkappa, \psi_2 + C_5\mu_n\varkappa) - W(\psi_1, \psi_2)$. We may obtain it in the same way as in ([9] , lemma

7.I) or use a result of Ya.Yu.Nikitin [I2] . These differences are $B\mu_n æ$. From (I2) and (I6) we conclude that

$$\nu_n\{\psi_1(t_{nq}) < \zeta_{nq}(m) < \psi_2(t_{nq}),\ q \le n\} = W(\psi_1, \psi_2) + B\mu_n æ .$$

Hence it follows also that

$$\nu_n\{\psi_1(t) < \zeta_n(t,m) < \psi_2(t),\ 0 \le t \le 1\} = W(\psi_1, \psi_2) + B\mu_n æ .$$

The theorem is proved.

In conclusion I mention some results which either follows from the theorem proved or may be obtained by the same methods.

1°. For $x > 0$,

$$\nu_n\{\max_{q \le n} \zeta_{nq}(m) < x\} = \sqrt{\tfrac{2}{\pi}} \int_0^x e^{-u^2/2} du + \frac{B\mu_n \ln(1/\mu_n)}{\ln\ln(1/\mu_n)}$$

uniformly in x .

2°. For $x > 0$,

$$\nu_n\{\max_{q \le n} |\zeta_{nq}(m)| < x\} = \frac{1}{\sqrt{2\pi}} \sum_{k=-\infty}^{\infty} \int_{-x}^{x} e^{-(u-2kx)^2/2} du + \frac{B\mu_n \ln(1/\mu_n)}{\ln\ln(1/\mu_n)}$$

uniformly in x .

3°. $$\nu_n\{\psi_1(\tfrac{\ln\ln q}{\ln\ln n})\sqrt{\ln\ln n} < \sum_{\substack{p|m\\ p\le q}} 1 - \ln\ln q < \psi_2(\tfrac{\ln\ln q}{\ln\ln n})\sqrt{\ln\ln n},\ 3 \le q \le n\} =$$

$$= W(\psi_1, \psi_2) + \frac{B\ln_3 n}{\sqrt{\ln\ln n}\ \ln_4 n} .$$

4°. For any fixed positive integer a,

$$\nu_n\{\psi_1(\frac{B_q^2}{B_n^2})\sqrt{2}\,B_n < \sum_{\substack{p|m\\ p\le q}} f(p) - \sum_{\substack{p|m+a\\ p\le q}} f(p) < \psi_2(\frac{B_q^2}{B_n^2})\sqrt{2}\,B_n\,,\ q\le n\} =$$

$$= W(\psi_1, \psi_2) + \frac{B\mu_n \ln(1/\mu_n)}{\ln\ln(1/\mu_n)}\,.$$

The constant bounding B may depend upon $c, \psi_1(0), \psi_2(0)$.

5°. Let $\{f^{(k)}(p)\}\ (k=1,\dots,s)$ be sequences of real numbers which are not identically 0. Let $a_1,\dots,a_s$ be different fixed non--negative integers. Denote

$$A_{\ell k} = \sum_{p\le \ell} \frac{f^{(k)}(p)}{p}\,,\quad B_{\ell k}^2 = \sum_{p\le \ell} \frac{f^{(k)}(p)^2}{p} \qquad (k=1,\dots,s)\,.$$

Assume that

$$\max_{1\le k\le s} (B_{nk}^{-1} \max_{p\le n} |f^{(k)}(p)|) = \lambda_n \to 0$$

as $n\to\infty$. Then

$$\nu_n\{\psi_1(\frac{1}{s}\sum_{k=1}^{s} \frac{B_{qk}^2}{B_{nk}^2})\sqrt{s} < \sum_{k=1}^{s} B_{nk}^{-1} (\sum_{\substack{p|m+a_k\\ p\le q}} f^{(k)}(p) - A_{qk}) < \psi_2(\frac{1}{s}\sum_{k=1}^{s} \frac{B_{qk}^2}{B_{nk}^2}\sqrt{s}, q\le n\} =$$

$$= W(\psi_1, \psi_2) + \frac{B\lambda_n \ln(1/\lambda_n)}{\ln\ln(1/\lambda_n)}$$

where B is bounded by a constant depending on $c, \psi_1(0), \psi_2(0)$, $a_1,\dots,a_s$.

REFERENCES

I. G.Jogesh Babu, Probabilistic methods in the theory of additive arithmetic functions, Ph.D.dissertation, The Indian Statistical Institute, Calcutta, 1973.

2. M.B.Barban, A.I.Vinogradov, On the number-theoretic basis of probabilistic number theory (Russian), Dokl. Akad. Nauk SSSR 154 (1964), 495-496.

3. P.Billingsley, Additive functions and Brownian motion, Notices Amer. Math. Soc., 17(1970), 1050, Abstract # 681-A9.

4. P.Billingsley, Prime numbers and Brownian motion, Amer. Math. Monthly 80(1973), 1099-1115.

5. P.Billingsley, The probability theory of additive arithmetic functions, Ann. of Probability 2(1974), 749-791.

6. J.Galambos, Distribution of arithmetical functions. A survey, Ann. Inst. H.Poincaré, Sect. B 6(1970), 281-305.

7. B.V.Gnedenko, V.S.Korolyuk, A.V.Skorokhod, Asymptotic expansions in probability theory, Proc. Fourth Berkeley Sympos. Math. Statistics and Probability, Vol. 2(1961), 153-170.

8. H.Halberstam, H.-E.Richert, Sieve Methods, Academic Press,London, New York, San Francisco, 1974.

9. J.Kubilius, Probabilistic Methods in the Theory of Numbers, Amer. Math. Soc. Translations, Vol. 11, Providence, R.I., 1964.

10. J.Kubilius, An analogue of A.N.Kolmogorov's theorem on Markov processes in the theory of prime numbers (Russian), Dokl. Akad. Nauk SSSR 103(1955), 361-363.

11. J.Kubilius, Probabilistic methods in the theory of distributions

of arithmetic functions (Russian), Aktualnyje problemy analitičeskoj teorii čisel, Izd. "Nauka i technika", Minsk, 1974, 81-118.

12. Ya.Yu.Nikitin, Estimates connected with staying of the Wiener process between boundaries, and their applications (Russian), Liet. Mat. rink. 15(1975), No. 1, 199-206.

13. W.Philipp, Arithmetic functions and Brownian motion, Proc. Sympos. Pure Math., Vol. 24, Amer. Math. Soc., 1973, 233-246.

Vilnius University

Vilnius

ASYMPTOTIC BEHAVIOR OF THE FISHER INFORMATION CONTAINED IN ADDITIVE STATISTICS

Z.M.Landsman, S.H.Siraždinov

1. INTRODUCTION

Let $\mathcal{P}=\{P_\theta, \theta\in\Theta\}$ be a family of distributions defined on a a measurable space $(\mathcal{X},\mathcal{A})$ depending on the scalar parameter $\theta\in\Theta$, Θ is an interval of R^1. We denote by L^2_θ the space of random variables defined on $(\mathcal{X},\mathcal{A})$ with finite second moment with respest to P_θ and with standard inner product

$$(\varphi,\psi)_\theta = E_\theta(\varphi\psi)$$

and the corresponding norm $\|\cdot\|_\theta$. Suppose that the distributions P_θ are absolutely continuous with respect to μ and the densities $p(x,\theta)=\frac{dP_\theta}{d\mu}$ are differentiable with respect to θ μ- almost everywhere and

$$\mathcal{J}(x,\theta)=\mathcal{J}(x)=\frac{\partial}{\partial\theta}\log p(x,\theta)\in L^2_\theta .$$

Put

$$I(\theta)=\|\mathcal{J}(x,\theta)\|^2_\theta . \tag{1.1}$$

In the classical theory of estimation, $\mathcal{J}(x,\theta)$ and $I(\theta)$ are called respectively the informant (score) and the Fisher information about the parameter θ, that corresponds to random observation with the density $p(x,\theta)$.

Let H be the linear subspace of L^2_θ generated by the basic random variables $1, \varphi_1(x), \dots, \varphi_k(x)$.

Suppose that

$$\mathfrak{I}(x;H;\theta)=\hat{E}_\theta(\mathfrak{I}(x,\theta)\mid H)$$

where $\hat{E}_\theta(\cdot\mid H)$ is the projection of L^2_θ onto H, and

$$I(H;\theta)=\left\|\mathfrak{I}(x;H;\theta)\right\|^2_\theta$$

In Kagan [1] it was shown that, under certain conditions of regularity, $I(H;\theta)$ is an analogue of the Fisher information about parameter θ, contained in H *). In the same article it was shown that the formula

$$I(\underbrace{H\otimes\ldots\otimes H}_{n};\theta)=nI(H;\theta)$$

holds for the tensor product $\underbrace{H\otimes\ldots\otimes H}_{n}$. In this paper, we shall show, that $nI(H;\theta)$ is the main term of the Fisher information, as $n\to\infty$ contained in the statistics

$$T_1=\sum_{i=1}^{n}\varphi_1(x_i),\quad\ldots,\quad T_k=\sum_{i=1}^{n}\varphi_k(x_i)$$

where $x_1,\ldots,x_n$ are independent random variables with values in $(\mathfrak{X},\mathfrak{A})$ each of these variables has the distribution P_θ. This result could be foreseen because the random vector $T=(T_1,\ldots,T_k)$ has asymptotically normal distribution. But the strict proof requires

*) The construction in [1], in fact, is more general and does not require regulization of H in the form of the space of random variables.

introducing certain conditions on the family $\mathcal{P}$ and some modification of the arguements used in local limit theorems.

Special attention is paid to the case when $\mathcal{X} = R^1$, $\mathcal{O}$ is the σ - algebra of Borel sets, μ is the Lebesgue measure. If in this situation $\varphi_1(x)=x, \dots, \varphi_K(x) = x^K$, the conditions become more visual. Further simplification takes place if θ is a location parameter. The result we get here is used in the proof of asymptotic optimality of polynomial and modified polinominal Pitman estimators for the class of equivariant estimators depending on sample moments.

We take this opportunity to express our gratitude to A.M.Kagan for attracting our attention to the described problem and useful remarks.

2. FORMULATION OF BASIC THEOREMS

$$\varphi(x) = (\varphi_1(x), \dots, \varphi_K(x)),$$

$$\pi_j(\theta) = E_\theta \varphi_j \qquad (j = 1, \dots, K),$$

$$\pi_{ij}(\theta) = E \varphi_i \varphi_j \qquad (i,j = 1, \dots, K),$$

$$\bar{\Lambda}(\theta) = \left\| \pi_{ij} - \pi_i \pi_j \right\|_1^K = \left\| \lambda_{ij}(0) \right\|_1^K$$

For $t = (t_1, \dots, t_K)$, $u = (u_1, \dots, u_K)$ put

$$(t, u) = \sum_{p=1}^{K} u_\ell t_\ell , \quad |t| = \sqrt{\sum_{\ell=1}^{K} t_\ell^2}$$

We need the following conditions.

$$E_\theta |\varphi(x)|^{2+\delta} < \infty , \qquad (2.1)$$

$$E_\theta | \mathcal{J}(x,\theta)|^{2+\delta} < \infty \qquad (2.2)$$

for some $\delta > 0$, where δ may depend on θ .

Let Δ be a bounded closed interval, contained in Θ . Put

$$F(x;\Delta) = \sup_{\theta\in\Delta} \left| \frac{\partial p(x,\theta)}{\partial\theta} \right|$$

and let us introduce a regularity condition (compare[3, ch.32]):

for $\theta \in \Theta$, such a $\Delta \ni \theta$ exists that

$$\int_{\mathfrak{X}} (|\varphi(x)|^2 + 1)\, F(x;\Delta)\, d\mu < \infty . \tag{2.3}$$

We remind that the Fisher information, contained in the statistics $T=(T_1, \dots, T_k)$ with the distribution given by the density $p_T(u,\theta)$ with respect to some measure ν, is the quantity

$$I_T(\theta) = \| \mathcal{J}_T(T,\theta) \|_\theta^2$$

where $\mathcal{J}_T(u,\theta) = \frac{\partial}{\partial\theta} \log p_T(u,\theta)$.

Let

$$f(t,\theta) = E_\theta e^{i(t,\varphi)} .$$

Let us introduce the following conditions.

There exists a natural N_0 such that

$$\int_{R^k} |t|\, |f(t,\theta)|^{N_0} dt \tag{2.4}$$

is convergent uniformly with respect to $\theta \in \Delta$.

The densities $p_T(u,\theta)$ are regular in the sense of Rao [4,p.284] for $n \geq N_0$,that is

$$\frac{d}{d\theta} \int_A p_T(u,\theta)\, du = \int_A \frac{\partial p_T(u,\theta)}{\partial\theta}\, du \tag{2.5}$$

for any Borel set $A \subset R^{\kappa}$ and $\theta \in \Delta$.

THEOREM 1. Suppose that conditions (2.1) - (2.5) are fulfilled and $\bar{\Lambda}(\theta)$ is a non-singular matrix. Then

$$I_T(\theta) = n\left\{ I(H;\theta) + O(1) \right\}$$

as $n \to \infty$.

We notice, in connection with theorem 1, that the condition

$$|\bar{\Lambda}(\theta)| = \det \bar{\Lambda}(\theta) > 0 \tag{2.6}$$

obviously holds, if the measure μ has no atoms, so that $1, \varphi_1(x), \ldots, \varphi_\kappa(x)$ are linearly independent.

Now let $\mathfrak{X} = R^1$, $\mathfrak{A}$ be the σ-algebra of Borel sets, and μ be the Lebesgue measure. If $\varphi_1(x), \ldots, \varphi_\kappa(x)$ have derivatives of any order, the densities $p_T(u,\theta)$ (with respect to the Lebesgue measure in R^{κ}) satisfy condition (2.5).

In fact consider the continuosly differentiable mapping $T: R^n \to R^{\kappa}$, given by formulas

$$T_1 = \sum_{i=1}^{n} \varphi_1(x_i)$$

$$\cdots\cdots\cdots\cdots$$

$$T_\kappa = \sum_{i=1}^{n} \varphi_\kappa(x_i)$$

If $M \subset R^n$ is the set where the rank of the Jacobi matrix $T'(X)$ ($X = (x_1, \ldots, x_n)$) is less that κ (the set of critical points) from Sards theorem (see ,[5 , Theorem 3.1]) $\mu_\kappa(T(M)) = 0$*), where μ_κ

*) The existence of the derivatives of the functions $\varphi_1(x), \ldots, \varphi_\kappa(x)$ of any order is used only here and, probably, it may be dispensed with.

is the Lebesgue measure in R^{κ}. Denote $G = R^n \setminus M$. The conditions of Lemma 1, [6], are satisfied in this set since G is open. Therefore the $n-\kappa$ functions $t_1(X), \dots, t_{n-\kappa}(X)$ can be found such that the mapping $\xi : G^* \to R^n$, given by the formulas

$$T_1 = \sum_{i=1}^{n} \varphi_1(x_i) ,$$
$$\dots\dots\dots$$
$$T_\kappa = \sum_{i=1}^{n} \varphi_\kappa(x_i) ,$$
$$Y_1 = t_1(X) ,$$
$$\dots\dots\dots$$
$$Y_{n-\kappa} = t_{n-\kappa}(X) ,$$

of the open set $G^* \subset G$ onto $\xi(G^*)$ is one-to-one and bicontinuous. If $X \in G^*$ the Jacobian $\det \xi'(X) \neq 0$ and $\mu_n(\overline{G \setminus G^*}) = 0$. So the densify of the vector ξ in the domain $\xi(G^*)$ is

$$p_\xi(T, Y, \theta) = \frac{p(x_1,\theta) \dots p(x_n,\theta)}{|\det \xi'(X)|} \tag{2.7}$$

where $x_1, \dots, x_n$ must be expressed by

$$(T, Y) = (T_1, \dots, T_\kappa, Y_1, \dots, Y_{n-\kappa}) \in \xi(G^*) .$$

If we denote

$$\sup_{\theta \in \Delta} \left| \frac{\partial}{\partial \theta} [p(x_1,\theta) \dots p(x_n,\theta)] \right| = \Psi(x_1, \dots, x_n ; \Delta),$$

from the regularity condition (2.3) for $p(x,\theta)$ it follows that

$$\int_{R^n} \Psi(x_1, \dots, x_n ; \Delta) dX < \infty .$$

Now returning to (2.7), we notice that, for $(T,Y) \in \xi(G^*)$ and $\theta \in \Delta$,

$$\left| \frac{\partial p_\xi(T, Y, \theta)}{\partial \theta} \right| \leq \frac{\Psi(\xi^{-1}(T,Y) ; \Delta)}{|\det \xi'(\xi^{-1}(T,Y))|} = \Psi_1(T, Y, \Delta) .$$

Using appropriate transformations of coordinates in multiple integrals, we get

$$\iint\limits_{\xi(G^*)} \Psi_1(T,Y;\Delta)\,d\mu_n < \infty \,.$$

Hence (2.5) follows for Borel sets $A \subset T(G^*)$ and, consequently, for any Borel set $A \subset R^k$, because $\mu_k(A \cap (T(R^n) \setminus T(G^*))) = 0$. Taking into account the said above, from Theorem 1 follows

<u>THEOREM 2.</u> Let $\mathcal{X} = R^1$, $\mathcal{Ol}$ be the σ- algebra of Borel sets, μ be the Lebesgue measure, the functions $\varphi_1(x), \dots, \varphi_k(x)$ have derivatives of any order. If for $\theta \in \Theta$ conditions (2.1)-(2.4)are fulfilled, then

$$I_T(\theta) = n\{I(H;\theta) + o(1)\} \qquad (n \to \infty)\,.$$

Now suppose that, in the case of Theorem 2,

$$\varphi_j(x) = x^j, \quad j = 1, \dots, k\,.$$

Conditions (2.1) and (2.3) turn into

$$\int_{-\infty}^{\infty} |x|^{2k+\delta} p(x,\theta)\,dx < \infty\,, \tag{2.8}$$

$$\int_{-\infty}^{\infty} x^{2k} F(x;\Delta)\,dx < \infty\,. \tag{2.9}$$

Let

$$\sup_{\theta \in \Delta} \left| \frac{\partial p(x,\theta)}{\partial x} \right| = F_1(x,\Delta)\,,$$

and let

$$\int_{-\infty}^{\infty} F_1(x,\Delta)\,dx < \infty \tag{2.10}$$

for some Δ .

The following theorem considers the asymptotic behaviour of the Fisher information, contained in the sample moments

$$(a_1=\frac{1}{n}\sum_{i=1}^{n} x_i \ , \ \dots \ , \ a_K=\frac{1}{n}\sum_{i=1}^{n} x_i^K) = a$$

THEOREM 3. Suppose that conditions (2.8),(2.2),(2.9) and (2.10) are fulfilled, then

$$I_a(\theta)=n(I(H;\theta)+o(1)) \qquad (n\to\infty).$$

The special case when θ is a location parameter will be considered in §4 where the results obtained will be used for constricting asymptotically optimal equivariant estimators.

3. PROOFS OF THE THEOREMS

From the condition (2.3) it follows that $\pi_\ell(\theta)$ are differentiable. Let $\pi'(\theta)=(\pi_1'(\theta), \dots, \pi_K'(\theta))$ and denote by $\bar\Lambda_{ij}(\theta)$ the algebraic complement of the term $\lambda_{ij}(\theta)$ in the determinant of the matrix $\bar\Lambda(\theta)$,

$$q(u,\theta)=(2\pi)^{-\frac{K}{2}} \, |\bar\Lambda(\theta)|^{-\frac{1}{2}} \exp(-\frac{1}{2}\sum_{\ell,m=1}^{K} \frac{\bar\Lambda_{\ell m}}{|\bar\Lambda(\theta)|} u_\ell u_m) .$$

Let $p_{z_n}(u,\theta)$ be the probability density of the random vector

$$z_n=\frac{1}{\sqrt{n}}\sum_{i=1}^{n}(\varphi(x_i)-\pi(\theta)) .$$

The following lemma is known in the one-dimensional case (for example see [7, ch.VI,Theorem 7]). We give its proof specially for our multidimensional case.

LEMMA 1. If, for some natural N_0 ,

$$\int_{R^\kappa} |t| \, |f(t,\theta)|^{N_0} dt < \infty \tag{3.1}$$

and $\bar{\Lambda}(\theta)$ is a non-singular matrix, then, for any ℓ, $1 \le \ell \le \kappa$,

$$\max_{u \in R^\kappa} \left| \frac{\partial p_{z_n}(u,\theta)}{\partial u_\ell} - \frac{\partial g(u,\theta)}{\partial u_\ell} \right| = o(1), \quad n \to \infty .$$

PROOF. By properties of positive definite quadratic forms there exist $C(\theta) > 0$, $M(\theta) > 0$, $C_1(\theta) > 0$, $M_1(\theta) > 0$ ж) such that

$$C(\theta)|u|^2 \le \sum_{\ell,m=1}^{\kappa} \lambda_{\ell m} u_\ell u_m \le M(\theta)|u|^2, \tag{3.2}$$

$$C_1(\theta)|u|^2 \le \sum_{\ell,m}^{\kappa} \frac{\bar{\Lambda}_{\ell m}}{|\bar{\Lambda}|} u_\ell u_m \le M_1(\theta) \tag{3.3}$$

since $\bar{\Lambda}(\theta)$ is non-singular.

Put $f_{z_n}(t,\theta) = E_\theta e^{i(t,z_n)}$. By (3.1) the formula

$$p_{z_n}(u,\theta) = \frac{1}{(2\pi)^\kappa} \int_{R^\kappa} e^{-i(u,t)} f_{z_n}(t,\theta)\, dt \tag{3.4}$$

hold for $n \ge N_0$ (see [8],p.197) and it may be differentiated with respect to u_ℓ. So, according to the traditional scheme, it may be written as follows

$$\left| \frac{\partial p_{z_n}(u,\theta)}{\partial u_\ell} - \frac{\partial g(u,\theta)}{\partial u_\ell} \right| \le \int_{R^\kappa} |t| \left| f_{z_n}(t,\theta) - \exp\left(-\frac{1}{2}\sum_{\ell,m=1}^{\kappa} \lambda_{\ell m} t_\ell t_m\right) \right| dt \le \tag{3.5}$$

$$\le \mathcal{Y}_1 + \mathcal{Y}_2 + \mathcal{Y}_3 + \mathcal{Y}_4 \, ,$$

ж) Here and in the sequel constants may depend on θ but not on n .

$$\mathcal{J}_1=\int_{|t|<T} |t|\,|f_{t_n}(t,\theta)-\exp(-\tfrac{1}{2}\sum_{\ell,m=1}^{K}\lambda_{\ell m}t_\ell t_m)|\,dt, \tag{3.6}$$

$$\mathcal{J}_2=\int_{|t|\geq T} |t|\exp(-\tfrac{1}{2}\sum_{\ell,m=1}^{K}\lambda_{\ell m}t_\ell t_m)\,dt,$$

$$\mathcal{J}_3=\int_{T\leq|t|<\delta\sqrt{n}} |t|\,|f_{z_n}(t,\theta)|\,dt, \quad \mathcal{J}_4=\int_{\delta\sqrt{n}\leq|t|} |t|\,|f_{z_n}(t,\theta)|\,dt.$$

By the integral limit theorem

$$\mathcal{J}_1\to 0, \quad n\to\infty. \tag{3.7}$$

Choosing an arbitrary $\varepsilon>0$, it is easy to see that, for large T ,

$$\mathcal{J}_2\leq\int_{|t|\geq T}|t|\exp(-\tfrac{1}{2}C(\theta)|t|^2)\,dt<\varepsilon. \tag{3.8}$$

Denote $\xi=\varphi(x)-\pi(\theta)$, $f_\xi(t,\theta)=E_\theta e^{i(t,\xi)}$ and for the characteristic functions we have

$$f_{z_n}(t,\theta)=f_\xi^n(\tfrac{t}{\sqrt{n}},\theta)=e^{-i(t,\sqrt{n}\,\pi(\theta))}f^n(\tfrac{t}{\sqrt{n}},\theta).$$

Using the Taylor formula, we get

$$f_\xi(u,\theta)=1-\tfrac{1}{2}\lambda_{\ell m}(\theta)u_\ell u_m+o(|u|^2). \tag{3.9}$$

Hence choosing an appropriate δ and taking into accout (3.2), it may be written as follows

$$|f_\xi(\tfrac{t}{\sqrt{n}},\theta)|\leq 1-\tfrac{1}{4}C(\theta)\frac{|t|^2}{n} \tag{3.10}$$

for $|t|<\delta\sqrt{n}$. Hence, for large T ,

$$\mathcal{J}_3\leq\int_{T\leq|t|<\delta\sqrt{n}}|t|\exp(-\tfrac{1}{4}C(\theta)|t|^2)\,dt\leq\int_{|t|\geq T}|t|\exp(-\tfrac{1}{4}C(\theta)|t|^2)\,dt<\varepsilon. \tag{3.11}$$

From condition (3.1) for $|u| \geq \delta$ we have

$$|f(u,\theta)| \leq e^{-p(\theta)} \quad , \quad p(\theta) > 0 \tag{3.12}$$

and

$$J_4 = (\sqrt{n})^{K+1} \int_{|u|\geq\delta} |u| \, |f(u,\theta)|^n du \leq (\sqrt{n})^{K+1} e^{-p(\theta)(n-N_0)} \int_{R^K} |u| |f(u,\theta)|^{N_0} du \to 0 \quad (n \to \infty) \tag{3.13}$$

From (3.5),(3.7),(3.8) and (3.13) Lemma 1 follows.

The following lemma concerns the differentiation of the density of the vector z_n with respect to the parameter.

<u>LEMMA 2.</u> Let conditions (2.3),(2.4) be fulfilled and $\bar{\Lambda}(\theta)$ be a non-singular matrix, then, for $n \geq N_0 + 1$,

$$\max_{u \in R^K} \left| \frac{\partial p_{z_n}(u,\theta)}{\partial \theta} \right| \leq K(\theta) < \infty . \tag{3.14}$$

<u>PROOF</u>. As in Lemma 1, from condition (2.4) it follows that, for $n \geq N_0$, formula (3.4) holds. By (2.3) $f_\xi(u,\theta)$ may be differentiated with respect to θ and, if

$$\ell(u,\theta) = \int_{\mathfrak{X}} e^{i(u,\varphi(x)-\pi(\theta))} \frac{\partial p(x,\theta)}{\partial \theta} d\mu , \tag{3.15}$$

then

$$\frac{\partial f_\xi(u,\theta)}{\partial \theta} = \ell(u,\theta) - i(u,\pi'(\theta)) f_\xi(u,\theta) . \tag{3.16}$$

From (2.3) it follows that, for $j = 1, \ldots, \kappa$,

$$\pi_j'(\theta)=\int_{\mathfrak{X}}\varphi_j(x)\frac{\partial p(x,\theta)}{\partial\theta}\,d\mu \quad ,$$
$$\frac{d}{d\theta}\int_{\mathfrak{X}}\varphi_j^2(x)\,p(x,\theta)\,d\mu=\int_{\mathfrak{X}}\varphi_j^2\frac{\partial p(x,\theta)}{\partial\theta}\,d\mu \quad , \tag{3.17}$$
$$\frac{d}{d\theta}\int_{\mathfrak{X}}p(x,\theta)\,d\mu=\int_{\mathfrak{X}}\frac{\partial p(x,\theta)}{\partial\theta}\,d\mu=0 \ ,$$

the derivatives

$$\frac{\partial\ell(u,\theta)}{\partial u_j}=i\int_{\mathfrak{X}}(\varphi_j(x)-\pi_j(\theta)\exp(i(u,\varphi(x)-\pi(\theta)))\,\frac{\partial p(x,\theta)}{\partial\theta}\,d\mu \ ,$$

$$\frac{\partial^2\ell(u,\theta)}{\partial u_s\,\partial u_m}=-\int_{\mathfrak{X}}(\varphi_s(x)-\pi_s)(\varphi_m(x)-\pi_m)\exp(i(u,\varphi(x)-\pi(\theta)))\,\frac{\partial p(x,\theta)}{\partial\theta}\,d\mu$$

exist and are continuous with respect to u . Using the Taylor expantion of $\ell(u,\theta)$ and taking into account (3.17), we have

$$\ell(u,\theta)=i\sum_{m=1}^{K}\pi_m'(\theta)u_m-\frac{1}{2}\sum_{m,s=1}^{K}\lambda_{ms}'(\theta)u_m u_s+o(|u|^2) \ . \tag{3.18}$$

Returning to (3.16), we get

$$\left|\frac{df_{z_n}(t,\theta)}{\partial\theta}\right|=n\left|f_\xi\left(\frac{t}{\sqrt{n}},\theta\right)\right|^{n-1}\left|\frac{\partial f_\xi(\frac{t}{\sqrt{n}},\theta)}{\partial\theta}\right|\leq \tag{3.19}$$

$$\leq n\left|f\left(\frac{t}{\sqrt{n}},\theta\right)\right|^{N_0}\left[\int_{\mathfrak{X}}F(x;\Delta)\,d\mu+\frac{|t|}{\sqrt{n}}K\int_{\mathfrak{X}}|\varphi|\,F(x;\Delta)\,d\mu\right]$$

for $n\geq N_0+1$ and $\theta\in\Delta$.

So, thanks to conditions (2.3) and (2.4), from (3.19) it follows that

$$\int_{R^K}\left|\frac{\partial f_{z_n}(t,\theta)}{\partial\theta}\right|dt$$

uniformly converges with respect to $\theta\in\Delta$, and then (3.4) may be differentiated with respect to θ . Hence

$$\left|\frac{\partial p_{z_n}(u,\theta)}{\partial\theta}\right|\leq\int_{R^K}\left|\frac{\partial f_{z_n}(t,\theta)}{\partial\theta}\right|=J_1+J_2 \tag{3.20}$$

where

$$\mathcal{J}_1 = \int_{|t|<\delta\sqrt{n}} \left| \frac{\partial f_{\tilde{z}_n}(t,\theta)}{\partial\theta} \right| dt , \quad \mathcal{J}_2 \int_{|t|\geq\delta\sqrt{n}} \left| \frac{\partial f_{\tilde{z}_n}(t,\theta)}{\partial\theta} \right| dt . \tag{3.21}$$

From properties of quadratic forms we get

$$\left| \sum_{m,s=1}^{\kappa} \lambda'_{ms}(\theta) t_m t_s \right| \leq G(\theta) |t|^2 . \tag{3.22}$$

Choosing an appropriate from (3.16),(3.18), (3.9) and taking into account (3.2) and (3.22), we get the following estimate

$$\left| \frac{\partial f_{\xi}(\frac{t}{\sqrt{n}},\theta)}{\partial\theta} \right| \leq K_1(\theta) |t|^2 \tag{3.23}$$

for $|t|<\delta\sqrt{n}$.

From (3.10), for $|t|<\delta\sqrt{n}$, we have

$$\left| f_{\xi}\left(\frac{t}{\sqrt{n}},\theta\right) \right|^{n-1} \leq \exp\left(-\frac{1}{8} C(\theta) |t|^2\right) \quad (n \geq 2). \tag{3.24}$$

Now it is easy to estimate $\mathcal{J}_1$ using (3.23) and (3.24):

$$\mathcal{J}_1 \leq K_1(\theta) \int_{|t|<\delta\sqrt{n}} |t|^2 e^{-\frac{1}{8}C(\theta)|t|^2} dt \leq K_1(\theta) \int_{R^{\kappa}} |t|^2 e^{-\frac{1}{8}C(\theta)|t|^2} dt = K_2(\theta) < \infty . \tag{3.25}$$

From condition (2.4) and from (3.21), (3.15),(3.16),(3.12) we get

$$\mathcal{J}_2 = n \cdot n^{\frac{\kappa}{2}} \int_{|u|\geq\delta} |f(u,\theta)|^{n-1} \left| \frac{\partial f_{\xi}(u,\theta)}{\partial\theta} \right| du \leq \tag{3.26}$$

$$\leq n^{\frac{\kappa}{2}+1} e^{-p(\theta)(n-1-N_0)} \Big(\int_{\mathfrak{X}} F(x;\Delta) d\mu \int_{R^{\kappa}} |f(u,\theta)|^{N_0} du +$$

$$+ |\pi'(\theta)| \int_{R^{\kappa}} |u| \, |f(u,\theta)|^{N_0} du \Big) \leq K_3(\theta) < \infty .$$

Substituting (3.25) and (3.26) into (3.20), we get the proof of Lemma 2.

Denote

$$\Lambda(\theta) = \begin{Vmatrix} 0 & 0 & \pi_1' & \cdots & \pi_K' \\ 0 & 1 & \pi_1 & \cdots & \pi_K \\ \pi_1' & \pi_1 & \pi_{11} & \cdots & \pi_{1K} \\ \cdots & \cdots & \cdots & \cdots & \cdots \\ \cdots & \cdots & \cdots & \cdots & \cdots \\ \pi_K' & \pi_K & \pi_{K1} & \cdots & \pi_{KK} \end{Vmatrix}$$

The relations between elements of the matrices $\bar{\Lambda}(\theta)$ and $\Lambda(\theta)$ are established by the following lemma which is proved by direct calculation.

LEMMA 3.

$$|\bar{\Lambda}(\theta)| = \Lambda_{11}(\theta), \tag{3.27}$$

$$\sum_{\ell,m=1}^{K} \pi_\ell'(\theta)\, \pi_m'(\theta)\, \bar{\Lambda}_{\ell m}(\theta) = -|\Lambda(\theta)| . \tag{3.28}$$

Proof of Theorem 1. As was shown in Kagan [1], if condition (2.6) is fulfilled, then

$$I(H;\theta) = \frac{-|\Lambda(\theta)|}{\Lambda_{11}(\theta)} . \tag{3.29}$$

The regularity condition (2.5) (see [4,p.284]) guarantees the following relation*)

$$J_T(T,\theta) = \frac{\partial}{\partial\theta} \log p_{z_n}(z_n,\theta) = E_\theta\Big(\sum_{i=1}^{n} J(x_i) \,|\, T\Big) . \tag{3.30}$$

---

*) Condition (2.5) is needed only for (3.30) to hold true.

As $\frac{\partial z_n}{\partial \theta} = -\sqrt{n}\, \pi'(\theta)$,

$$\frac{1}{\sqrt{n}} J_T(T,\theta) = \left[\frac{-\sum_{\ell=1}^{k} \pi_\ell'(\theta) \frac{\partial p_{z_u}(u,\theta)}{\partial u_\ell} + \frac{1}{\sqrt{n}} \frac{\partial}{\partial \theta} p_{z_n}(u,\theta)}{p_{z_n}(u,\theta)} \right]_{u=z_n} . \tag{3.31}$$

Put

$$\eta_n(u,\theta) = -\sum_{\ell=1}^{k} \pi_\ell'(\theta) \frac{\partial p_{z_u}(u,\theta)}{\partial u_\ell} + \frac{1}{\sqrt{n}} \frac{\partial p_{z_n}(u,\theta)}{\partial \theta} ,$$

$$\Psi_n(u,\theta) = p_{z_n}(u,\theta) , \tag{3.32}$$

$$\eta(u,\theta) = -\sum_{\ell=1}^{k} \pi_\ell'(\theta) \frac{\partial g(u,\theta)}{\partial u_\ell} = g(u,\theta) \sum_{\ell,m=1}^{k} \frac{\bar{\Lambda}_{\ell m}}{|\bar{\Lambda}|} \pi_\ell'(\theta) u_m ,$$

$$\Psi(u,\theta) = g(u,\theta) .$$

By properties of conditional expectations, condition (2.2) and an inequality for absolute values of sums of independent variables (see [7, p.79 ,addition 16]), from (3.30) we get

$$E_\theta \left| \frac{1}{\sqrt{n}} J_T(T,\theta) \right|^{2+\delta} \le \frac{1}{n^{\frac{2+\delta}{2}}} E_\theta \left| \sum_{i=1}^{n} J(x_i) \right|^{2+\delta} \le C_1 E_\theta |J(x)|^{2+\delta} = C_2 < \infty . \tag{3.33}$$

Here and in the sequel the dependence of constants on θ is sometimes omitted for simplicity.Using the property of bilinear forms that corresponds to positive-definite quadratic forms, and using the same inequality for absolute moments of sums of independent random variables, inequality (C_z) ([9], p.168) and condition (2.1),from (3.32) and (3.3) we get

$$E_\theta \left| \frac{\eta(z_n,\theta)}{\psi(z_n,\theta)} \right|^{2+\delta} \le M_1^{2+\delta}(\theta) |\pi'(\theta)|^{2+\delta} E_\theta |z_n|^{2+\delta} \le \tag{3.34}$$

$$\le M_1^{2+\delta}(\theta) |\pi'(\theta)|^{2+\delta} C_3 \sum_{m=1}^{k} E_\theta |\varphi_m(x) - \pi_m|^{2+\delta} = C_4 < \infty .$$

Denote

$$Y_n = \frac{1}{\sqrt{n}} \mathcal{J}_T(T,\theta) - \frac{\eta(z_n,\theta)}{\psi(z_n,\theta)} = \frac{\eta_n(z_n,\theta)}{\psi_n(z_n,\theta)} - \frac{\eta(z_n,\theta)}{\psi(z_n,\theta)} = R_n(z_n,\theta). \qquad (3.35)$$

From (3.33),(3.34) and (3.35) we have

$$E_\theta |Y_n|^{2+\delta} \le C_5 < \infty . \qquad (3.36)$$

Now we consider

$$E_\theta |Y_n|^2 = \int_{R^k} R_n^2(u,\theta) p_{z_n}(u,\theta)\,du = L_1 + L_2 \qquad (3.37)$$

where

$$L_1 = \int_{|u|<B} R_n^2(u,\theta) p_{z_n}(u,\theta)\,du, \quad L_2 = \int_{|u|\ge B} R_n^2(u,\theta) p_{z_n}(u,\theta)\,du. \qquad (3.38)$$

From (3.32) and (3.3) it follows that

$$|\eta(u,\theta)| \le M_2 < \infty, \qquad (3.39)$$

$$|\Psi(u,\theta)| \le M_3 < \infty,$$

and, for $|u| < B$,

$$\Psi(u,\theta) \ge C(B) > 0. \qquad (3.40)$$

From Lemmas 1 and 2

$$\max_{u\in R^k} |\eta_n(u,\theta) - \eta(u,\theta)| \le \sum_{\ell=1}^{k} \pi_\ell'(\theta) \max_{u\in R^k} \left| \frac{\partial q(u,\theta)}{\partial u_\ell} - \frac{\partial p_{z_n}(u,\theta)}{\partial u_\ell} \right| + \qquad (3.41)$$

$$+ \frac{1}{\sqrt{n}} K(\theta) \to 0 \qquad (n \to \infty).$$

Since the local limit theorem follows from (2.4),

$$\max_{u\in R^k} |\Psi_n(u,\theta) - \Psi(u,\theta)| \to 0 \qquad (n \to \infty), \qquad (3.42)$$

and from (3.41) and (3.42) we have

$$\max_{u\in R^k} |\eta_n(u,\theta)-\eta(u,\theta)| \le \frac{\varepsilon}{M_2+M_3}\;\frac{C^2(B)}{2} \quad , \tag{3.43}$$

$$\max_{u\in R^k} |\Psi_n(u,\theta)-\Psi(u,\theta)| \le \frac{\varepsilon}{M_2+M_3}\;\frac{C^2(B)}{2} \quad ,$$

$$|\Psi_n(u,\theta)| \ge \frac{1}{2} C(B) \qquad \text{if} \quad |u| < B \tag{3.44}$$

for arbitrary $\varepsilon > 0$ and $n \ge N(\varepsilon)$ .Substituting (3.39),(3.40),(3.43) into (3.38),we get

$$L_1 < \varepsilon^2 \tag{3.45}$$

for $n \ge N(\varepsilon)$. Denote $A_n = \{u;\ |R_n(u,\theta)| \ge R\}$ and by A_n^c the complement of A_n . For estimation of L_2 we use (3.36):

$$L_2 = \int_{[\{|u|\ge B\}\cap A_n]} |R_n(u,\theta)|^2 p_{z_n}(u,\theta)du + \int_{[\{|u|\ge B\}\cap A_n^c]} |R_n(u,\theta)|^2 p_{z_n}(u,\theta)du \le$$

$$\le \frac{1}{R^\delta} E_\theta |Y_n|^{2+\delta} + R^2 \int_{|u|\ge B} p_{z_n}(u,\theta)\,du \le \frac{C_5}{R^\delta} + \frac{R^2}{B^2}\sum_{\ell=1}^{k}(\pi_{\ell\ell} - \pi_\ell^2) \ .$$

By choosing R and B appropriately, we get

$$L_2 < \delta \ . \tag{3.46}$$

Now substituting (3.45) and (3.46) into (3.37),we get

$$E_\theta |Y_n|^2 \to 0 \ , \quad n \to \infty \ .$$

Hence from the Minkowski inequality it follows that

$$I_T(\theta) = n\left(E_\theta \left|\frac{\eta(z_n,\theta)}{\psi(z_n,\theta)}\right|^2 + o(1)\right), \quad (n\to\infty) \ . \tag{3.47}$$

Using the row expansion of the determinant $|\bar{\Lambda}(\theta)|$ from (3.32) we obtain

$$E_\theta\left(\frac{\eta(Z_n,\theta)}{\psi(Z_n,\theta)}\right)^2 = E_\theta\left(\sum_{\ell,m=1}^{k}\frac{\bar{\Lambda}_{\ell m}}{|\bar{\Lambda}|} Z_n^{(m)}\pi_\ell'(\theta)\right)^2 = \frac{\sum_{i,j=1}^{k}\pi_i'\pi_j'\bar{\Lambda}_{ij}}{|\bar{\Lambda}(\theta)|}, \tag{3.48}$$

where

$$Z_n^{(m)} = \frac{1}{\sqrt{n}}\sum_{i=1}^{n}(\varphi_m(x_i) - \pi_m(\theta)).$$

Using Lemma 3, from (3.47) and (3.48) we have

$$I_T(\theta) = n\left(\frac{-|\Lambda(\theta)|}{\Lambda_{11}(\theta)} + o(1)\right), \qquad n \to \infty \tag{3.49}$$

Comparing (3.29) with (3.49), we complete the proof of Theorem 1.

Theorem 2 follows from Theorem 1.

Proof of Theorem 3. Taking into account the above proved, it suffices to show that condition (2.10) implies (2.4).

Denote

$$\alpha_\ell(\theta) = \int_{-\infty}^{\infty} x^\ell p(x,\theta)\,dx, \quad \bar{\alpha}_\ell(\theta) = \int_{-\infty}^{\infty} |x|^\ell p(x,\theta)\,dx,$$

$$V(\theta) = \int_{-\infty}^{\infty}\left|\frac{\partial p(x,\theta)}{\partial x}\right| dx.$$

In the considered case, $\varphi(x) = (x, x^2, \ldots, x^k)$ Hence (see [10]), for $|t| \geq t_0 > 1$,

$$|f(t,\theta)| \leq R(\theta)|t|^{-\gamma} \tag{3.50}$$

where

$$R(\theta) \max\left(\max_{2\leq j\leq k}\left(C''_{k+1-j}\left(V(\theta) + \sum_{\nu=k+2-j}^{k}\nu\bar{\alpha}_{\nu-1}(\theta)\right)\right),\; C'_k V(\theta)\right), \tag{3.51}$$

and $\gamma = 1/(1+\frac{1}{k})^{k-1} k!$ C'_{k+1-j} $(j = 2, \ldots, k)$ and C'_k are such as in [10], and, what is important, they are not dependent upon θ.

Taking into account (3.50) and (3.51), it is easy to see that, from conditions (2.9) and (2.10)*), condition (2.4) follows for

$$N_0 \geq \frac{k+2}{\gamma} = (k+2)(1+\frac{1}{k})^{k-1} k!$$

4. THE CASE OF A LOCATION PARAMETER

Let

$$\varphi(x) = (x, x^2, \ldots, x^k) \quad \text{and} \quad p(x,\theta) = p(x-\theta).$$

Then the conditions of Theorem 3 can be formulated in terms of $p(x)$.

THEOREM 4. Let the conditions

$$\int_{-\infty}^{\infty} |x|^{2k+\delta} p(x)\,dx < \infty , \tag{4.1}$$

$$\int_{-\infty}^{\infty} \left| \frac{p'(x)}{p(x)} \right|^{2+\delta} p(x)\,dx < \infty \tag{4.2}$$

be fulfilled. Then

$$I_a(0) = n\,I(H;0)\,(1+o(1)) \quad (n \to \infty).$$

PROOF. It is easy to see, that in the considered situation conditions (2.8) and (2.2) follow from (4.1) and (4.2) and

$$\int_{-\infty}^{\infty} |p'(x)|\,dx < \infty . \tag{4.3}$$

---

*) Instead of (2.10) it would be sufficient to require $V(\theta)$ to be continuous on Δ.

As in the proof of Theorem 3, condition (2.4) follows from (4.1), (4.3) and (3.50). Conditions (2.5) and (2.6) are automatically fulfilled. The further considerations fully repeat the proof of Theorem 1 with the only difference that instead of Lemma 2 the following Lemma 2′ should be used

LEMMA 2′. Let

$$\int_{-\infty}^{\infty} |x|^{2k} p(x)dx < \infty , \tag{4.4}$$

and let condition (2.4) be fulfilled. Then

$$\max_{u\in R^k} \left| \frac{\partial p_{z_n}(u,\theta)}{\partial\theta} \right| \le K(\theta) < \infty$$

for $n \ge N_0 + 1$.

PROOF. Here we indicate the nessesary changes in the proof of Lemma 2. In our case

$$\pi_\ell(\theta) = \alpha_\ell(\theta) = \int_{-\infty}^{\infty} (x+\theta)^\ell p(x)dx \;, \quad \ell = 1, \ldots, k \,,$$

$$\pi_{\ell m}(\theta) = \alpha_{\ell+m}(\theta) \,, \quad \ell, m = 1, \ldots, k \;,$$

$$f_\xi(t,\theta) = \int_{-\infty}^{\infty} \exp\left(i \sum_{\ell=1}^{k} t_\ell \left((x+\theta)^\ell - \alpha_\ell(\theta)\right)\right) p(x)dx \;.$$

According to (4.4), $\alpha_\ell(\theta)$ and $f_\xi(t,\theta)$ may be differentiated with respect to θ and, in contrast to the general case, the density must be not differentiated. Hence

$$\alpha_\ell'(\theta) = \ell\alpha_{\ell-1}(\theta) \tag{4.5}$$

and if

$$\lambda_\ell(u,\theta) = \ell \int_{-\infty}^{\infty} x^{\ell-1} \exp\left(i \sum_{j=1}^{\kappa} u_j (x^j - \alpha_j)\right) p\,(x-\theta)\,dx \ , \ \ell = 1, \dots, \kappa \ ,$$

$$\lambda(u,\theta) = (\lambda_1(u,\theta), \ \dots \ , \lambda_\kappa(u,\theta)) \ , \tag{4.6}$$

then

$$\frac{\partial f_\xi(u,\theta)}{\partial\theta} = i\,(\lambda(u,\theta), u) - i\,(u, \alpha'(\theta))\, f_\xi(u,\theta) \ . \tag{4.7}$$

Since condition (4.4) allows to express $\lambda(u,\theta)$ by the total differential formula with respect to $u_1, \dots, u_\kappa$, and $f_\xi(u,\theta)$ in the form (3.9), here (3.23) and (3.25) also hold. Taking into account (4.5),(4.6), and (4.7) conditions (4.4) and (2.4) allow to differentiate with respect to θ in (3.4). We obtain the estimate (3.26) using relations (4.6) and (4.7) instead of (3.15) and (3.16). Lemma 2, and hence Theorem 4 are proved.

Now we turn to the problem of estimating the parameter θ from the sample $x_1, \dots, x_n$ with the probability density $p(x-\theta)$.

Let

$$\bar{x} = \frac{1}{n} \sum_{i=1}^{n} x_i \quad , \quad m_j = \frac{1}{n} \sum_{i=1}^{n} (x_i - \bar{x})^j \ , \ j = 2, \dots, \kappa$$

$$\mu_j = \int_{-\infty}^{\infty} x^j p(x)\,dx \quad , \quad j = 1, \dots, \kappa \ .$$

Without loss of generality we assume $\mu_1 = 0$.

In [2] the polynomial and modified polynomial Pitman estimators

$$t_n^{(\kappa)} = \bar{x} - \hat{E}_0(\bar{x} \mid \Lambda_\kappa) \ , \ \tau_n^{(\kappa)} = A_0 + \sum_{j=2}^{\kappa} A_j m_j \tag{4.8}$$

were considered, where Λ_K is the set of all the polynomials of $(x_1-\bar{x}, \dots, x_n-\bar{x})$ of degree $\leq K$, the coefficients $A_0, A_1, \dots, A_K$ are as in [2] . In [2] if was shown that these estimators are asymptotically optimal in the class of polynomial and mean-square consistent estimators of the parameter θ of degree $\leq K$.

We consider the class of equivariant estimators of the type

$$q_n(x_1, \dots, x_n) = \bar{x} + \Phi(m_2, \dots, m_K), \tag{4.9}$$

where Φ is a measure function. Theorem 4 allows to prove the asymptotical optimality of the estimators $t_n^{(K)}$ and $\tau_n^{(K)}$ for this class.

THEOREM 5. Let conditions (4.1) and (4.2) be fulfilled, then the polynomial $t_n^{(k)}$ and modified polynomial $\tau_n^{(k)}$ Pitman estimators are asymptotically optimal for the class of equivariant estimators of type (4.9).

PROOF. First notice that the sample moments $m_2, \dots, m_K$ are expressed by the moments $a_1, \dots, a_K$. Further, the regularity conditions, sufficient for the Rao-Cramér inequality to hold, is fulfilled here. By this inequality

$$E_\theta(q_n-\theta)^2 \geq \frac{1}{I_a(\theta)} . \tag{4.10}$$

Now we use Theorem 4. From (4.10), taking into account that in our case $I(H;\theta)>0$ (see [2]), we obtain

$$E_\theta(q_n-\theta)^2 = E_0\, q_n^2 \geq \frac{1+o(1)}{nI(H;\theta)} \qquad (n\to\infty). \tag{4.11}$$

From (4.11) and from Theorems 1 and 5 in [2] we have, as $n \to \infty$,

$$E_\theta (q_n - \theta)^2 \geq E_\theta (\tau_n^{(k)} - \theta)^2 (1 + o(1)),$$

$$E_\theta (q_n - \theta)^2 \geq E_\theta (t_n^{(k)} - \theta)^2 (1 + o(1)).$$

REFERENCES

1 Kagan A.M., The Fisher information, contained in a finite-dimensional linear space and a correct variant of the method of moments, Problemy peredači informatsii, 1975 (in Russian).

2 Kagan A.M., Klebanov L.B., Fintusal S.M., Asymptotic behaviour of polynomial Pitman estimators, Zapiski naučnyh seminarov LOMI, 43, (1974), 30-39 (in Russian).

3 Cramér H., Mathematical methods in statistics, Moscow, 1948 (in Russian).

4 Rao C.R., Linear statistical inference and its applicationsm Moscow, 1968 (in Russian).

5 Sternberg S., Lectures on Differential Geometry, Moscow, 1970. (in Russian).

6 Lehmann E.L., Scheffe H., On the problem of similar regions, Proc. Nat.Acad.Sci., 33 (1947), 382-386.

7 Petrov V.V., Sums of independent random variables, Moscow, 1972, (in Russian).

8 Prohorov Yu.V., Rozanov Yu.A., Probability Theory, Moscow, 1973. (in Russian).

9 Loève M., Probability Theory, Moscow, 1962. (in Russian).

10 Sadikova S.M., Some inequalities for characteristic functions, Teoriya Veroyatnostei i ee Primeneniya, 11, 3, (1966), 500-506. (in Russian).

Mathematical Institute Academy of
Sciences of the Uzbek SSR
Tashkent

NONLINEAR FUNCTIONALS OF GAUSSIAN STATIONARY PROCESSES AND THEIR APPLICATIONS

Gisiro Maruyama
Department of Mathematics
College of General Education
University of Tokyo
Komaba, Meguro
Tokyo, Japan

Suppose we are given a flow $\{T_t\}$, i.e. a group of automorphisms on a probability space $(\Omega, \mathfrak{B}, P)$. To a real function $f \in L^2(P)$, it will be said that the central limit theorem (CLT) is applicable when

(i) $v(T) = v_f(T) =$ variance of $\int_0^T f(T_t\omega)\, dt \to \infty$, as $T \to \infty$

(ii) the probability law of

$$\frac{1}{\sqrt{v(T)}}\left(\int_0^T f(T_t\omega)dt - \int f dP\right)$$

approaches for $T \to \infty$, to the normal law $N(0, 1)$.

By writing

(1) $v(T) \asymp T, \quad T \to \infty$

is meant an almost linearity in the gorwth of $v(T)$, i.e. the existence of such $c_1, c_2 > 0$ that

$$c_1 T < v(T) < c_2 T.$$

In this paper we are mainly interested in the class of f for which v_f satisfy (1) and the CLT is applicable, and the set of such f will be denoted by $G(T_t)$. In this case, CLT implying the law of large numbers

$$P(T^{-1} \int_0^T f(T_t\omega)dt \to \int f dP) = 1,$$

if there exists a dense subset $\mathcal{L} \subset L^2(P)$ with $\mathcal{L} \subset G(T_t)$, $\{T_t\}$ must be ergodic.

Let

$$\xi(t, \omega) = \int_{-\infty}^{\infty} e^{i\lambda t} d\beta(\lambda)$$

be a real-valued Gaussian stationary process, where $d\beta(\lambda)$ is a usual complex Gaussian random measure, $E|d\beta(\lambda)|^2 = d\sigma(\lambda)$, $d\sigma(\lambda)$ a continuous measure, so that the correlation function is given by

$$\mathcal{R}(t) = \int_{-\infty}^{\infty} e^{i\lambda t} d\sigma(\lambda).$$

Consider the space $L^2(\xi)$ of L^2-functions measurable with respect to the σ-algebra $\sigma(\xi_t, -\infty < t < \infty)$. Then every $f \in L^2(\xi)$ is represented as

$$f(\omega) = c + \sum_{n \geq 1} I_n(\omega),$$

where c is a non-random constant, and $I_n(\omega) = I_n(c_n, \omega)$ is written in the form

(2) $I_n(\omega) = \int_{-\infty}^{\infty}\cdots\int c_n(\lambda_1,\ldots,\lambda_n)d\beta(\lambda_1)\cdots d\beta)\lambda_n)$,

$c_n \in L^2(d\sigma^n = d\sigma(\lambda_1) \times\cdots\times d\sigma(\lambda_n))$

$$\sum_{n\geq 1}\int_{-\infty}^{\infty}\cdots\int |c_n|^2 d\sigma^n < \infty$$

and in the r ight-hand side of (2), integration is performed off diagonals, i.e. the actual domain of integration should be

$$D = \{\lambda_i \pm \lambda_j \neq 0 \quad \text{for} \quad 1 \leq i \neq j \leq n\}.$$

Moreover c_n can be taken a symmetric function of $\lambda_1,\cdots,\lambda_n$ and $\overline{c_n}(\lambda_1,\ldots,\lambda_n) = c_n(-\lambda_1,\ldots,-\lambda_n)$ if f is real-valued.

In connection with the flow $\{T_t\}$ generated by ξ_t, we are going to set up a systematic way of finding out as many essentially different functions as possible from $L^2(\xi)$, which belong to $G(T_t)$.

Suppose first that $d\sigma(\lambda)$ is absolutely continuous, $d\sigma(\lambda) = f(\lambda)d\lambda$, and above all things, we restrict ourselves to the standard case:

(C_0) $f(\lambda)$ is bounded.

In general, there exists a $c > 0$ such that $f^{2k*}(\lambda) \geq c$ for $k \geq 1$, around $\lambda = 0$,

whereas if

(C_1) there exists a $d > 0$ such that $f(\lambda) \geq d > 0$ around $\lambda = 0$,

then $f^{k*}(\lambda) \geq c$ for $k \geq 1$, around $\lambda = 0$.

Moreover under (C_0) $f^{k*}(k \geq 2)$ is bounded, continuous, and belongs to $L^p(1 \leq p \leq \infty)$.

We will next discuss about conditions under which $v_f(T) \asymp T$ is realized for $f = I_n$, $n \geq 2$. One can write

$$v_n(T) = v_f(T) = \int_0^{\infty}\left(\frac{\sin\lambda T/2}{\lambda/2}\right)^2 \varphi_n(\lambda)d\lambda \quad \text{for} \quad f = I_n,$$

and setting

$$\Phi_n(h) = \Phi(c_n, h) = \int_0^h \varphi_n(\lambda)d\lambda,$$

one easily concludes that there exist constants $c_1, c_2 > 0$ such that

(3) $c_1 \varliminf_{h\downarrow 0} \frac{1}{h}\Phi_n(h) \leq \varliminf_{T\to\infty} v_n(T)/T \leq \varlimsup_{T\to\infty} v_n(T)/T \leq c_2 \varlimsup_{h\downarrow 0} \frac{1}{h}\Phi_n(h)$.

Then the condition

(4) $\Phi_n(h) \asymp h, \quad h \downarrow 0$

is sufficient for (1) with $f = I_n$.

Especially, when

(C_2) $c_n(\lambda_1,\cdots,\lambda_n)$ is bounded,

then

(5) $\varlimsup_{h\downarrow 0} \frac{1}{h}\Phi_n(h) < \infty$,

so that in this case (4) is equivalent to

(6) $\varliminf_{h\downarrow 0} \frac{1}{h}\Phi_n(h) > 0$.

We will exemplify sufficient conditions for $n = 3$ which guarantee (6). Either the conditions

(6.1) $c_3(\lambda+u+v, u, v) \to c_0(u, v) \; (L^2(du dv))$, as $\lambda \downarrow 0$, and $\int_{-\infty}^{\infty}\int f(u)f(v)f(u + v)|c_0(u, v)|^2 du dv > 0$,

or

the set of (u, v)-functions

(6.2) $\mathcal{H} = \{\frac{1}{h}\int_0^h |c_3(\lambda+u+v, u, v)|^2 d\lambda, \quad h > 0\}$

dose not contain zero as a limit element for $h \downarrow 0$ under the topology $\sigma(L_\mu^1, L_\mu^\infty)$, $d\mu = f(u)f(v)f(u + v)dudv$

are sufficient for (6). When (C_2) is valid a necessary condition for (6) is

(7) $$f^{n*}(0) > 0.$$

We may relax (C_2) in several ways. For instance write

$c_{nk}(\lambda_1, \cdots, \lambda_n) = c_n(\lambda_1, \cdots, \lambda_n)$ when $|c_n(\lambda_1, \cdots, \lambda_n)| \leq k$, $= 0$ othewise,

and suppose

(C_3) $\lim_{k\to\infty} \overline{\lim}_{h\downarrow 0} \frac{1}{h}\Phi(c_n - c_{nk}, h) = 0$,

then under (6), (1) is true. Next suppose

(C_4) $\underline{\lim}_{h\downarrow 0} \frac{1}{h}\Phi(c_{nk}, h) > 0$ for some $k > 0$,

which is implied by the condition that there exists a constant $\gamma > 0$ such that

(C_5) $$|c_n(\lambda_1, \cdots, \lambda_n)| \geq \gamma,$$

then (C_3), (C_4) in turn together imply (1).

After these analysis we may formulate the following theorem providing sufficient conditions for the validity of CLT.

Theorem 1. Under either the conditions

(a) (C_3) and (6) ; (b) (C_3) and (C_4) ;

or (c) (C_3) and (C_5),

$I_n \in G(T_t)$ $(n \geq 2)$.

The main ideas of the proof depend on the method of moments and basic facts on the Gaussian measure $d\beta(\lambda)$.

Theorem 2. Suppose the above sufficient conditions are satisfied for I_k, $1 \leq k \leq N$, then

$$I_k(T_t\omega)/\sqrt{v_k(T)}, \quad 1 \leq k \leq N$$

are asymptotically independent.

Theorem 3. Let

$$g(\omega) = \sum_{k=1}^{N} I_k$$

be the Nth partial sum of the series representing f and suppose that

(8) $$\underline{\lim}_{T\to\infty} \frac{1}{T} v_g(T) > 0$$

and each c_k, $1 \leq k \leq N$, satisfies (C_3), then $g \in G(T_t)$.

Define

$$p(f) = \overline{\lim}_{T\to\infty} \frac{v_f(T)}{T}$$

on the space of f with $v_f(T) = O(T)$. p is a semi-norm on this space.

The following theorem provides us with a sufficient condition for $f \in L^2(\xi)$ to be an element of $G(T_t)$.

Theorem 4. Let $f \in L^2(\xi)$, and each c_k satisfy (C_3), (8). Suppose

$$\lim_{n\to\infty} p(f - \sum_{k=1}^{n} I_k) = 0$$

then $f \in G(T_t)$.

At this stage we may raise a question whether there exists a dynamical system whose entropy is zero, but sufficiently many functions belong to $G(T_t)$. We will construct a dynamical system which serves as an answer to the question by means of [1], which rests on a Gaussian stationary process with spectral measure of such a special character that it is continuous and singular with respect to Lebesgue measure but the correlation function tends to zero as $|t| \to \infty$. According to Wiener-Wintner [2], there exists a continuous singular measure $\sigma(d\lambda)$ on $[-\pi, \pi]$ symmetric about the origin, and

$$\mathcal{R}(t) = \int_{-\infty}^{\infty} e^{i\lambda t} d\sigma(\lambda) = O(|t|^{-1/2+\varepsilon}),$$

where $\varepsilon > 0$ can be assigned as small as we may like. Then the flow $\{T_t\}$ generated by the corresponding stationary process ξ_t has zero entropy. Let $d\beta(\lambda)$ be the corresponding Gaussian measure and put

$$I_k(\omega) = \int_{-\infty}^{\infty} \cdots \int d\beta(\lambda_1) \cdots d\beta(\lambda_k), \quad 1 \le k < \infty.$$

Theorem 5. If $0 < \varepsilon < 1/8$, then

(a) $I_k(\omega) \in G(T_t)$ for $3 \le k < \infty$,

(b) $\frac{1}{\sqrt{v_k(T)}} \int_0^T I_k(T_t\omega)dt, \quad 3 \le k < \infty$

are asymptotically independent.

Starting from this we would be able to find more general functions belonging to $G(T_t)$ of the form

$$f(\omega) = \sum_{k=3}^{\infty} \int_{-\infty}^{\infty} \cdots \int c_n(\lambda_1, \cdots, \lambda_n) d\beta(\lambda_1) \cdots d\beta(\lambda_n).$$

Along our line, we would also be able to relax the boundedness restriction (C_0) for $f(\lambda)$.

Reference

[1] G. Maruyama, A singular flow with countable Lebesgue spectrum, J. Math. Soc. Japan, 19(1967), 359-365.

[2] N. Wiener - A. Wintner, Fourier-Stieltjes transforms and singular infinite convolutions, Amer. Journ. Math., 60(1938), 513-522.

STATIONARY MATRICES OF PROBABILITIES FOR STOCHASTIC SUPERMATRIX

E.A.Morozova and N.N.Čencov

0. Introduction. A classical probability distribution P on a finite set $\Omega = \{\omega_1, \dots, \omega_n\}$ may be given by the vector $(p_1, \dots, p_n)$ of the probabilities of elementary events $p_i = P\{\omega_i\}$:

$$p_i \geq 0 , \ \forall i \ ; \quad \sum_{i=1}^{n} p_i = 1 \ . \tag{0.1}$$

A square stochastic matrix $\Pi = ({}_i\pi_k)_{i,k=1}^{n}$ describes Markov random transitions on the set Ω, $\Pi(\omega_i \to \omega_k) = {}_i\pi_k$. It satisfies the conditions

$$ {}_i\pi_k \geq 0 , \ \forall i, k \ ; \quad \sum_k {}_i\pi_k = 1 , \ \forall i \ . \tag{0.2}$$

The following problem arises: To find all Π - stationary probability distributions, i.e. vectors $\vec{p}$ of the probabilities of elementary events satisfying the equations

$$\vec{p}\,\Pi = \vec{p} \ : \ \sum_i p_i \ {}_i\pi_k = p_k \ , \quad \forall k \ . \tag{0.3}$$

A solution of this problem has been given by A.N.Kolmogorov [1] in terms of communicating states classes. We shall attempt to transfer the concept of A.N.Kolmogorov to non-commutative probability theory.

1. Probability matrices. In non-commutative probability theory there appears a square self-adjoint matrix $P = (p^i_j)^n_{i,j=1}$, having compex elements, in general , instead of an n - dimensional vector of probabilities. It has to satisfy the conditions

$$P = P^* : \quad p^i_j = \overline{p^j_i} \ , \quad \forall i, j \ ; \tag{1.1}$$

$$P \geq 0 : \sum_{i,j} \xi_i \, p^i_j \, \xi_j \geq 0 \ , \quad \forall (\xi_1, \dots, \xi_n) \in \mathbb{C}^n \ ; \tag{1.2}$$

$$\operatorname{tr} P := \sum_{i=1}^{n} p^i_i = 1 \ . \tag{1.3}$$

Such matrices and their infinite-dimensional analogues were introduced by von Neumann [2]. Following [2] one calls them density matrices. We shall call them here probability matrices to emphasize analogies between classical and non-commutative theories.

If we restrict ourselves to consider diagonal matrices $P = \mathrm{diag}(p_1^1, \ldots, p_n^n)$ only, we should come to description of a scheme, isomorphic to a classical one: for any diagonal probability matrix P there correspond a vector $\vec{p}$: $p_i = p_i^i, \forall i$. When considering a family of commuting matrices P, one may reduce them simultaneously to the diagonal form by an orthogonal transformation. Such a family is also isomorphic to a classical one. Therefore we can speak about classical (i.e. commutative) probability theory and about general (non-commutative) theory. General theory differs from classical one not only by its technical tools but also by a complicated logic of events, cf. [3].

An intermediate case of block-diagonal matrices with foxed sizes of blocs is possible:

$$P = \begin{pmatrix} P_1^1 & 0 & \cdots & 0 \\ 0 & P_2^2 & \cdots & 0 \\ \cdots & \cdots & \cdots & \cdots \\ 0 & 0 & \cdots & P_m^m \end{pmatrix} \tag{1.4}$$

$$(P_\kappa^\kappa)^* = P_\kappa^\kappa \geq 0 \ , \quad \forall \kappa \ ; \tag{1.5}$$

$$\mathrm{tr}\, P = \sum_{\kappa=1}^{m} \mathrm{tr}\, P_\kappa^\kappa = 1 \tag{1.6}$$

Such schemes were sistematically considered by Wick, Wightman and Wigner [4].

In principle, we can consider probability matrices having elements of different number nature, cf.[5]. We may, e.g., restrict ourselves to the field $\mathbb{R}$ of real numbers. We can take quaternion self-adjoint matrices or blocks; see [6], where

$$\overline{\alpha_0 + \alpha_1 \mathbf{i} + \alpha_2 \mathbf{j} + \alpha_3 \mathbf{K}} = \alpha_0 - \alpha_1 \mathbf{i} - \alpha_2 \mathbf{j} - \alpha_3 \mathbf{K} \quad .$$

We may also consider self-adjoint blocks of size two on hypercomplex numbers, with ν imaginary units, where ν is any natural number.

2. Events. In non-commutative theory, there is an n-dimensional vector space $\mathcal{H}$ with sesquilinear scalar product

$$\langle x | y \rangle = \overline{\langle y | x \rangle}$$

$$\langle \lambda x | \mu y \rangle = \bar{\lambda} \mu \langle x | y \rangle \ , \quad \forall \lambda, \mu \in \mathbb{C}$$

which corresponds to a set of all n elementary events. $\mathcal{H}$ is called the abstract space of the considered stochastic object. We assume that an orthonormal basis is fixed in $\mathcal{H}$:

$$|x_1\rangle \ , \ \ldots \ , \ |x_n\rangle \ ; \ \langle x_i | x_j \rangle = \delta_{ij} \ , \quad \forall i, j \quad .$$

Then any matrix P is a matrix of some self-adjoint operator, probability operator (density operator).

For a block scheme, we form the union

$$\Omega = \mathcal{H}_1 \cup \ldots \cup \mathcal{H}_m \ ; \quad \mathcal{H} = \bigoplus_{K=1}^{m} \mathcal{H}_K \quad , \tag{2.1}$$

of subspaces $\mathcal{H}_K$ (coherent sectors), and we take the union of sector bases as a basis of $\mathcal{H}$. In particular, Ω is the union of coordinate axes in a purely diagonal scheme.

In classical theory, any subset of set Ω of all elementary events is an event. Analogously in the general theory, events are subspaces $\mathcal{F}$ of the abstract space $\mathcal{H}$. In the block scheme (2.1), events are only subspaces of the form

$$\mathcal{F} : \mathcal{F} = \bigoplus_{K=1}^{m} \mathcal{F}_m ; \; H_K(\mathcal{F}) = \mathcal{F}_K \subseteq \mathcal{H}_K , \; \forall K . \tag{2.2}$$

Here and in what follows we denote by E the orthoprojector onto a subspace $\mathcal{E}$:

$$E(\mathcal{H}) = \mathcal{E} , \; E = E^* , \; E^2 = E \quad . \tag{2.3}$$

In accordance with the classical theory, coordinate subspaces are all the possible events in a diagonal scheme. We deliberately abstain from introducing the elementary event concept.

The probability of an event $\mathcal{F}$ may be defined in two ways:

$$P\{\mathcal{F}\} := \operatorname{tr} PF = \operatorname{tr} FPF \quad . \tag{2.4}$$

Their equivalence follows from the following property of the trace: $\operatorname{tr} AB = \operatorname{tr} BA$ and from property (2.3) of the orthoprojector. If $P\{\mathcal{F}\} > 0$, there exists a conditional probability distribution with the operator

$$P^{\mathcal{F}} := \frac{1}{P\{\mathcal{F}\}} FPF \quad . \tag{2.5}$$

We refer to our survey [7] for more detailed presentation of elementary theory foundation. We have to note that the first equality in (2.4) corresponds to the definition of $P\{\mathcal{F}\}$ as the mean value of the operator F .

3. Stochastic supermatrices. Every stochastic matrix gives an affine mapping Π of the simplex of all probability distributions into itself:

$$\vec{q}\,\Pi = \vec{r} : \Sigma\, q_i \; {}_i\pi_\kappa = r_\kappa , \quad \forall \kappa \quad . \tag{3.1}$$

Stationary distributions, see (0.3), are fixed points of this mapping. Conversely, any affine mapping of the simplex of all probability vectors (0.1) into itself is given by a stochastic matrix.

DEFINITION 3.1. A stochastic supermatrix of size n is four-indexized matrix $\Pi = ({}^i_j\pi^\kappa_\ell)^n_{i,\kappa,j,\ell=1}$ with the following properties:

$$\Pi = \Pi^* : {}^i_j\pi^\kappa_\ell = \overline{{}^j_i\pi^\ell_\kappa} , \quad \forall\, i,j,\kappa,\ell : \tag{3.2}$$

$$\Pi \geqslant 0 : \sum_{i,j,\kappa,\ell} \xi_i \overline{\xi_j}\, {}^i_j\pi^\kappa_\ell\, \eta_\kappa \overline{\eta_\ell} \geqslant 0 , \quad \forall \vec{\xi}, \vec{\eta} \in \mathbb{C}^n \tag{3.3}$$

$$\mathrm{tr}\,\Pi = \mathbb{1} : \sum_\kappa {}^i_j\pi^\kappa_\ell = \delta^i_j \quad , \forall\, i,j \tag{3.4}$$

THEOREM 3.1. A stochastic supermatrix of size n gives an affine (into itself) mapping of the convex set $\mathcal{P}$ of all probability matrices of size n determined by conditions (1.1)-(1.3). Conversely, any affine mapping of $\mathcal{P}$ into itself is given by some stochastic supermatrix.

SUPPLEMENT. A stochastic supermatrix Π defines an $\mathbb{R}$ - linear mapping of the space $\mathcal{Y}$ of all self-adjoint matrices A into itself by the formula

$$A \longrightarrow A\Pi = B : \; b^{\kappa}_{\ell} = \sum_{i,j=1}^{n} a^{i}_{j}\, {}^{i}_{j}\Pi^{\kappa}_{\ell} \;, \quad \forall \kappa, \ell \;. \tag{3.5}$$

Moreover, non-negative matrices turn into non-negative ones under this mapping.

Proof is given, for example, in [7], see Theorems 6.1 and 6.2.

THEOREM 3.2. The set of all stochastic supermatrices of size n is convex.

Proof is evident.

THEOREM 3.3. A composition of two stochastic supermatrices $\Pi * \Phi$

$$\Pi * \Phi = Ш : \; {}^{i}_{j}Ш^{\kappa}_{\ell} = \sum_{\alpha,\beta} ({}^{i}_{j}\Pi^{\alpha}_{\beta}) \cdot ({}^{\alpha}_{\beta}\varphi^{\kappa}_{\ell}) \;, \quad \forall i, j, \kappa, \ell \;. \tag{3.6}$$

is also a stochastic supermatrix.

Proof. Changing the order of summation, we find that

$$P(\Pi * \Phi) = (P\Pi)\Phi = Q \in \mathfrak{P} \;, \qquad \forall P \in \mathfrak{P} \;.$$

LEMMA 3.4. Any unitary transformation $U = ({}^{i}u^{j})^{n}_{i,j=1}$ of the space $\mathcal{H}$ determines a stochastic supermatrix Π by the formula

$${}^{i}_{j}\Pi^{\kappa}_{\ell} = {}^{i}u^{\kappa} \cdot \overline{{}^{j}u^{\ell}} \;, \qquad \forall i, j, \kappa, \ell . \tag{3.7}$$

Proof. We have $UU^{*} = 1$ for unitary U, where $U^{*} = ({}^{i*}_{\,}u^{j}) : {}^{i*}u^{j} = \overline{{}^{j}u^{i}}$, $\forall i, j$, see for detail [8].

A convex hull of all supermatrices of form (3.7) does not exhaust the set of all stochastic supermatrices. Similarly, there is a stochastic supermatrix, which is determined by any antiunitary transformation V, see [9], - an antilinear transformation of the space $\mathcal{H}$ with a property

$$\langle Vx | Vy \rangle = \overline{\langle x | y \rangle} \ .$$

LEMMA 3.5. An antiunitary transformation V of the space $\mathcal{H}$ with a basis $\{|x_i\rangle\}$ defined by

$$|x\rangle = \sum_i \xi_i |x_i\rangle \xrightarrow{V} \sum_i \bar{\xi}_i |x_i\rangle = |Vx\rangle \tag{3.8}$$

determines a stochastic supermatrix Φ :

$${}^i_j\Phi^\kappa_\ell = \delta^i_\ell \, \delta^\kappa_j \quad , \quad \forall \, i,j,\kappa,\ell \, , \tag{3.9}$$

which maps each self-adjoint matrix P into the transposed one:

$$P\Phi = P^T : \ p^{T i}_j = p^j_i = \overline{p^i_j} \ , \quad \forall \, i,j \ . \tag{3.10}$$

Proof. For the supermatrix (3.9) the property (3.10) is evident. Let us consider the orthoprojector $F = |x\rangle\langle x|$ onto the straight line $\mathfrak{F}$ having $|x\rangle$ as the direction vector, see [10] , and the similar orthoprojector $G = |Vx\rangle\langle Vx|$ onto $\mathfrak{G} = V(\mathfrak{F})$. In view of (3.8) their matrices are connected according to (3.10):

$$(f^i_j) = (\bar{\xi}_i \xi_j)^n_{i,j=1} \ ; \quad (g^i_j) = (\xi_i \bar{\xi}_j)^n_{i,j=1} \ .$$

Any self-adjoint matrix P may be decomposed into an $\mathbb{R}$ - linear combination of orthoprojectors.

We have to note that there are properties of stochastic matrices which make no sense for supermatrices. Let a probability vector $\vec{q}$ and a stochastic matrix Π be given. Then one might speak about the joint distribution of a pair (ω', ω'') of positions before and after a Markov jump:

$$p_{i\kappa} = P\{\omega_i , \omega_\kappa\} = q_i \cdot {}_i\Pi_\kappa \ , \quad \forall i,\kappa \ , \tag{3.11}$$

where $q_i = \sum_\kappa p_{i\kappa} ; \ z_\kappa = \sum_i p_{i,\kappa}$, see (3.1). For a supermatrix, the joint distribution need not exist.

4. The problem of Kolmogorov. We pose the following problem: Let a stochastic supermatrix Π be given . It is required to describe all Π-stationary probability distributions, i.e. to describe the set $\mathfrak{P}_\Pi$ of all probability matrices P satisfying the system of equations:

$$P\Pi = P : \sum_{i,j} p^i_j \cdot {}^i_j\Pi^\kappa_\ell = p^\kappa_\ell \ , \quad \forall \kappa,\ell \ . \tag{4.1}$$

Let us consider some examples. Let $\Pi_0 = ({}^{i}_{j}\Pi^{\kappa}_{\ell}) = (\delta^{i\kappa}\delta_{j\ell})$. This supermatrix maps any matrix into itself, and $\mathfrak{P}_\Pi = \mathfrak{P}$. The transposition supermatrix (3.9) replaces the elements of self-adjoint matrix by the adjont, see (3.10). For Φ , the set $\mathfrak{P}_\Phi$ is a subset of $\mathfrak{P}$ consisting of all (symmetric) matrices with real elements.

LEMMA 4.1. Let a unitary transformation U_κ have the matrix

$$ {}_i\overset{(\kappa)}{u}_j = \varepsilon_i \delta^{ij} \ ; \ \varepsilon_i = -1, \ \forall i \le \kappa, \ \varepsilon_i = 1 \ , \ \forall i > \kappa \ , \tag{4.2}$$

and let Π_κ be the supermatrix determined by U_κ in (3.7). Then the supermatrix $\Pi = \frac{1}{2}(\Pi_0 + \Pi_\kappa)$ acts according to the rule

$$P = \begin{pmatrix} P^1_1 & P^2_1 \\ P^1_2 & P^2_2 \end{pmatrix} \longrightarrow P\Pi = \begin{pmatrix} P^1_1 & 0 \\ 0 & P^2_2 \end{pmatrix} \ , \tag{4.3}$$

where $P_1^1 = (p_j^i)_{i,j=1}^{\kappa}$.

Proof. The supermatrix Π_κ changes the signs of elements p_j^i when $i \leq \kappa < j$ or $j \leq \kappa < i$ .The other elements remain unchanged.

THEOREM 4.2. For the decomposition Ω of the space $\mathcal{H}$ into a sum of coherent sectors, given by (3.1), there exists a supermatrix $Ш_\Omega$, leaving fixed all the probability matrices of the corresponding block-diagonal form (1.4) and only them.

COROLLARY. The Kolmogorov problem for a collection of all probability matrices of a fixed block-diagonal construction is reduced to the Kolmogorov problem for the whole collection $\mathfrak{P}$.

Proof. Let $\kappa(1), \ldots, \kappa(m) = n$ be bounds of a basis decomposition into sector bases. Then

$$Ш_\Omega = 2^{-m}(\Pi_0 + \Pi_{\kappa(1)}) * \ldots * (\Pi_0 + \Pi_{\kappa(m)}), \tag{4.4}$$

by Lemma 4.1.

If Π is an affine mapping of the collection of all probability matrices of a given block- diagonal construction into itself, then its fixed points coincide with $(Ш_\Omega * \Pi)$ - stationary probability matrices.

The non-commutative field $\mathbb{Q}$ of quaternions, by the rule

$$\varepsilon + \alpha \mathbf{i} + \beta \mathbf{j} + \gamma \mathbf{K} \to \begin{pmatrix} \varepsilon + \alpha\sqrt{-1} & -\beta - \gamma\sqrt{-1} \\ -\beta + \gamma\sqrt{-1} & \varepsilon - \alpha\sqrt{-1} \end{pmatrix} \tag{4.5}$$

is associated with an algebra of complex matrices of size two, isomorphic to $\mathbb{Q}$, see [11] . Therefore, with any quaternion matrix A of size s one can associate a complex matrix B of size $2s$, where

$$a_{k\ell} \longrightarrow \begin{pmatrix} b_{2k-1,2\ell-1} & b_{2k-1,2\ell} \\ b_{2k,2\ell-1} & b_{2k,2\ell} \end{pmatrix} ,$$

in accordance to (4.5). We shall call B a quasi-quaternion matrix.

THEOREM 4.3. Let a unitary transformation U of an abstract space $\mathcal{H}$ of dimension $n=2\nu$ be given by the correspondence

$$|x_{2k-1}\rangle \xrightarrow{U} -|x_{2k}\rangle , \quad |x_{2k}\rangle \xrightarrow{U} |x_{2k-1}\rangle, \quad \forall k, \tag{4.6}$$

and let Π be the supermatrix corresponding to U by (3.7). Now let Φ be the transposition supermatrix (3.9), and Π be the unit supermatrix (see 4.2)).

Then the stochastic supermatrix $Ш = \frac{1}{2}(\Pi_0 + \Pi * \Phi)$ maps $\mathfrak{P}$ into a collection $\mathfrak{P}_Q \subset \mathfrak{P}$ of all quasi-quaternion probability matrices. All quasi-quaternion matrices are $Ш$-fixed.

Proof. Let us compute the transformation of the left upper block of matrix $P = (p_j^i)$. Applying successively the operations Π and Φ to the matrix (p_j^i) and then averaging, one arrives at

$$PШ = \frac{1}{2} \begin{pmatrix} p_1^1 + p_2^2 & 0 & p_1^3 + p_4^2 & p_1^4 - p_3^2 & \cdots \\ 0 & p_1^1 + p_2^2 & p_2^3 - p_4^1 & p_2^4 + p_3^1 & \cdots \\ p_3^1 + p_2^4 & p_3^2 - p_1^4 & p_3^3 + p_4^4 & 0 & \cdots \\ p_4^1 - p_2^3 & p_4^2 + p_1^3 & 0 & p_3^3 + p_4^4 & \cdots \\ \cdots & \cdots & \cdots & \cdots & \cdots \end{pmatrix}$$

It is seen, that the operation $Ш$ is idempotent. The block

$$\begin{pmatrix} p_1^3 + p_4^2 & p_1^4 - p_3^2 \\ p_2^3 - p_4^1 & p_2^4 - p_3^1 \end{pmatrix} = \begin{pmatrix} \overline{p_3^1 + p_2^4} & \overline{p_4^1 - p_2^3} \\ -(p_4^1 - p_2^3) & p_3^1 + p_2^4 \end{pmatrix}$$

is of quasi-quaternion form(4.5).

In a similar way one can obtain blocks which are isomorphic to matrices of size two over hypercomplex numbers. One can construct correspondences of type (4.5) by tensor products of (Hermitian) Pauli matrices, used in (4.5) to represent the imaginary units.

We have described various possible types of stationary matrices construction. Evidently these types may be combined: one block can be of real type, the other of quaternion one and so on. It will be proved that with a corresponding choice of coherent sector decomposition of $\mathcal{H}$ such a construction gives a complete solution of the posed problem.

5. Family of all stationary probability vectors and its construction. The homogeneous system (0.3) with matrix (0.2) is well-known to have a non-trivial solutions. It is most convenient to carry out the arguement for this statement in such a way. A stochastic matrix Π defines, by (3.1), an affine mapping of the simplex of all probability distributions into itself. The simplex is convex and compact. Consequently the mapping has fixed points.

Let us stude the family of solutions. Our approach [12] will be somewhat different from traditional. First let us enlarge the problem and take all the solutions $(x_1,\dots,x_n)$ of the system (0.3), not only of probability kind.

DEFINITION 5.1. The positive, respectively the negative, part of a vector $(x_1,\dots,x_n)$ are defined as the vectors $(x_1^+,\dots,x_n^+)$ and $(x_1^-,\dots,x_n^-)$, where

$$x^+ = max\{x,0\}, \quad x^- = -min\{x,0\}, \quad x = x^+ - x^- . \tag{5.1}$$

DEFINITION 5.2. The set $\Omega(\vec{x}) = \{\omega_i : x_i \neq 0\}$ will be called the support of the vector $(x_1, \dots, x_n)$. The subsets $\Omega^+(\vec{x}) = \{\omega_i : x_i > 0\}$ and $\Omega^-(\vec{x}) = \{\omega_i : x_i < 0\}$ will be called the positive, respectively the negative part of the support,

$$\Omega^+(\vec{x}) \cup \Omega^-(\vec{x}) = \Omega(\vec{x}) \subseteq \Omega \quad .$$

REMARK. For short we shall identify the set of elementary events and the set of their indeces.

LEMMA 5.1. Let $(x_1, \dots, x_n)$ be a solution of the system (0.3). Then

$$\sum_{\kappa \in \Omega^\pm} {}_i\Pi_\kappa = 1 \quad , \qquad \forall i \in \Omega^\pm(\vec{x}) \quad , \tag{5.2}$$

$${}_j\Pi_\kappa = 0 \quad , \quad \forall j \in \Omega^\mp(\vec{x}) \quad , \quad \forall \kappa \in \Omega^\pm(\vec{x}) \quad . \tag{5.3}$$

COROLLARY. The matrix Π maps a linear vector space $\{\vec{y} : \Omega(\vec{y}) \subseteq \Omega^+(\vec{x})\}$ into itself.

Proof. Putting x_i^+ and x_i^- into (0.3), we have

$$x_\kappa = \sum_{i \in \Omega^+} x_i^+ \, {}_i\Pi_\kappa - \sum_{j \in \Omega^-} x_j^- \, {}_j\Pi_\kappa \quad . \tag{5.4}$$

All the x_i^+ and x_j^- are strongly positive. Their coefficients ${}_i\Pi_\kappa$ and ${}_j\Pi_\kappa$ are non-negative and the sums over κ are normalized (see (0.2)). Therefore

$$\sum_{\kappa \in \Omega^+} x_\kappa^+ = \sum_{\kappa \in \Omega^+} x_\kappa = \sum_{i \in \Omega} x_i^+ \sum_{\kappa \in \Omega^+} {}_i\Pi_\kappa -$$

$$- \sum_{j \in \Omega^-} x_j^- \sum_{\kappa \in \Omega^+} {}_j\Pi_\kappa \leq \sum_{i \in \Omega^+} x_i^+ \quad .$$

The left-hand side and the right-hand side of the inequality have the same value. Therefore, the rejected non-positive terms are equal to zero, see (5.3), and the sums estimated from above by 1 are equal to 1 see (5.2). The corollary follows from (5.3).

THEOREM 5.2. Let $(x_1, \dots, x_n)$ be a solution of the system (0.3). Then $(x_1^+, \dots, x_n^+)$ and $(x_1^-, \dots, x_n^-)$ are solutions of the system (0.3).

Proof. For $\kappa \in \Omega^+(\vec{x})$ we have by (5.4):

$$x_\kappa^+ = x_\kappa = \sum_{i \in \Omega^+} x_i^+ \, {}_i\Pi_\kappa - \sum_{i \in \Omega^-} x_i^- \, {}_i\Pi_\kappa +$$

$$+ \sum_{i \bar{\in} \Omega^+} x_i^+ \, {}_i\Pi_\kappa = \sum_{i=1}^{n} x_i^+ \, {}_i\Pi_\kappa \quad ,$$

where only the first term of the middle expression differs from zero. The second term is equal to zero by (5.3), and all $x_i^+ = 0$ in the third term because $i \bar{\in} \Omega^+$. When $\kappa \bar{\in} \Omega$,

$$\sum_{i=1}^{n} x_i^+ \, {}_i\Pi_\kappa = \sum_{i \in \Omega^+} x_i^+ \, {}_i\Pi_\kappa = 0 \quad ,$$

in view of the second statement in (5.3).

THEOREM 5.3. The supports of solutions of the system (0.3) form a Boolean ring of sets.

Proof. As $\Omega(\vec{x}) = \Omega(\vec{x}^+ + \vec{x}^-)$, it is sufficient to consider only non-negative solution supports. In that case

$$\Omega(\vec{x} + \vec{y}) = \Omega(\vec{x}) \cup \Omega(\vec{y}) \quad .$$

According to Theorem 5.2 a vector $\vec{z}(t)$

$$\vec{z}(t) : \; z_i = (t x_i - y_i)^- \; , \quad \forall i \; ; \quad t > 0 \; , \tag{5.5}$$

will be a solution of the system (0.3), if $\vec{x}$ and $\vec{y}$ are. Let us denote $\Omega' = \Omega(\vec{y}) \setminus \Omega(\vec{x})$. Then

$$\left\{\begin{array}{ll} z_i(t) = y_i \ , & \forall i \in \Omega' \\ z_j(t) = 0 \ , & \forall j \bar{\in} \Omega' \end{array}\right\} \quad \forall t > \max_{i \in \Omega(x)} \frac{y_i}{x_i} \ . \tag{5.6}$$

Therefore $\Omega(\vec{z}(t)) = \Omega(\vec{y}) \setminus \Omega(\vec{x})$ for a sufficiently large t .

The structure of finite Boolean rings is well-known. The union of all the sets of the ring is its maximal set. The latter may be decomposed into minimal disjoint ones (atoms).Any set of the ring is a union of some of its atoms.

THEOREM 5.4. Let Π be a stochastic matrix. Then there exist $m \geq 1$ Π-stationary probability vectors $\vec{q}^{(1)}, \dots, \vec{q}^{(m)}$ with disjoint supports $\Omega_i = \Omega(\vec{q}^{(i)})$; $\Omega_i \cap \Omega_j = \emptyset$, $\forall i \neq j$; $\bigcup_i \Omega_i \subseteq \Omega$.

Vectors $\vec{q}^{(i)}$ form a fundamental system of solutions. In particular, the stationary probability vectors are all described by the formula

$$\vec{p} = \sum_{k=1}^{m} z_k \vec{q}^{(k)} \ , \quad \forall z_k \geq 0 \ , \quad \sum_{k=1}^{m} z_k = 1 \ . \tag{5.7}$$

Proof. Every atom of the ring of the supports has its stationary probability vector. This probability vector is unique, since the difference of the different probability vectors has non-trivial positive and negative parts, with smaller supports.

We refer to [12] for more detailed presentation of the theory. It has be used there in constructing main concepts of mathematical statistics in a systematic way.

6. The positive and the negative parts of self-adjoint operators. Let us attempt to solve the non-commutative Kolmogorov problem in the same way as the classical one with necessary supplements and generalizations.

As a probability matrix is anologous to a probability vector, so an analogue to a probability mesure is a probability operator (density operator). The same probability mesure may correspond to different vectors of elementary event probabilities: a vector depends on indexing elementary events. One has to transpose the vector coordinates after re-indexing the latter. In a similar way we shall consider any probability operator $\mathcal{P}$ and any affine mapping Π by using the most suitable coordinate system. The passage from one coordinate system to another must be determined by unitary matrix.

It follows from Lemma 3.4 that unitary matrix U generates a mapping $Ш_U$ of the collection $\mathfrak{P}$ of all probability matrices into itself. As $U^{-1} = U^*$ is a unitary matrix as well, the mapping $Ш_U$ is invertible: $Ш_U^{-1} = Ш_{U^*}$ .Under changing the coordinates,

$$\Pi \to Ш_{U^*} \Pi Ш_U \quad , \qquad \forall \Pi \quad . \tag{6.1}$$

We shall begin with definitions of the positive and the negative parts of operator.It is well-known that any self-adjoint operator $A = A^*$ on a finite-dimensional $\mathcal{H}$ has a spectral decomposition:

$$A = \sum_{\kappa=1}^{h(A)} \alpha_\kappa E_\kappa \; ; \qquad E_\kappa = E_\kappa^2 = E_\kappa^* , \quad \forall \kappa \quad . \tag{6.2}$$

$$\alpha_\kappa \neq \alpha_\ell \quad , \quad E_\kappa E_\ell = 0 \quad , \qquad \forall \kappa \neq \ell \; , \tag{6.3}$$

$$1 = \sum_{K=1}^{h(A)} E_K \, . \tag{6.4}$$

Any E_K is an orthoprojecttor on the eigen-space $\mathcal{E}_K(A)$ of the operator A, α_K is the corresponding eigen-value.

DEFINITION 6.1. We shall call the positive, respectively the negative part of a self-adjoint operator A with spectral decomposition (6.2) the operators

$$A^+ = \sum_{K:\alpha_K>0} \alpha_K E_K = \sum_{K=1}^{h(A)} \alpha_K^+ E_K \quad , \tag{6.5}$$

$$A^- = -\sum_{K:\alpha_K<0} \alpha_K E_K = \sum_{K=1}^{h(A)} \alpha_K^- E_K \quad . \tag{6.6}$$

REMARK. By virtue of (5.1)

$$A = A^+ - A^- \quad . \tag{6.7}$$

DEFINITION 6.2. The subspace

$$\mathcal{C}_A = \bigoplus_{K:\alpha_K\neq 0} E_K(\mathcal{H}) = \bigoplus_{K:\alpha_K\neq 0} \mathcal{E}_K(A) \tag{6.8}$$

will be called the support of the operator A. The subspaces

$$\mathcal{C}_A^+ = \bigoplus_{K:\alpha_K>0} \mathcal{E}_K(A), \quad \mathcal{C}_A^- = \bigoplus_{K:\alpha_K<0} \mathcal{E}_K(A) \, , \tag{6.9}$$

will be called the supports of the positive, respectively the negative part of A ,

$$\mathcal{C}_A^+ \oplus \mathcal{C}_A^- = \mathcal{C}_A \quad . \tag{6.10}$$

LEMMA 6.1. The operators A^+, A^- and the orthoprojectors $\mathcal{C}_A^+$, $\mathcal{C}_A^-$ commute.

Proof. The orthoprojectors E_K of their decompositions commute in view of (6.3).

DEFINITION 6.3. An $\mathbb{R}$ - linear space γ of self-adjoint operators will be called splitted, if $A^{+}\in\gamma$ and $A^{-}\in\gamma$ when $A\in\gamma$.

LEMMA 6.2. If a subspase $\mathcal{F}$ is the support of a self-adjoint operator A, and F is the orthoprojector onto $\mathcal{F}$, then

$$FAF = FA = AF = F\ . \tag{6.11}$$

Proof. Two orthoprojectors $F \geqslant E$ commute, and $FE = EF = E$. One has to apply this reasoning to the spectral decomposition (6.2) of the operator A.

LEMMA 6.3. The operators $A^{\pm}$ and orthoprojectors $C^{\pm}$ admit the representations

$$A^{\pm} = \frac{1}{2\pi i}\int_{\gamma_{\pm}} \zeta\,(\zeta\mathbf{1} - A)^{-1}\,d\zeta\ , \tag{6.12}$$

$$C^{\pm} = \frac{1}{2\pi i}\int_{\gamma_{\pm}} (\zeta\mathbf{1} - A)^{-1}\,d\zeta\ , \tag{6.13}$$

where the contour γ_{+} (resp. γ_{-}) in the complex plane $\mathbb{C}$ contains all the positive (resp.negative) points of the spectrum $\{\alpha_{k}\}$ of the operator A, and only them.

REMARK. If the contour γ in (6.13) contains only spectrum points $\alpha_{j(1)}, \ldots, \alpha_{j(k)}$, then $C^{\vee} = E_{j(1)} \oplus \ldots \oplus E_{j(k)}$.
Similary,

$$A^{\vee} = \sum_{k} \alpha_{j(k)} E_{j(k)}\ . \tag{6.14}$$

Proof.See, for example,[13] ,especially §5.9.

It is well-known that there is an order relation of self-adjoint operators:

$$A \geq B \Leftrightarrow \langle x|A|x\rangle \geq \langle x|B|x\rangle\ , \quad \forall |x\rangle \in \mathcal{H}\ . \tag{6.15}$$

But it is well-known also that the order structure is not a lattice: two operators (two quadratic forms) have vrai upper bound (resp.vrai lower bound), iff one of them majorizes another.

Therefore in studying operator-given measures and integrals with respect to them, one should use the concept of splittability. It is the key for our constructions.

<u>7. Stationary matrices for a stochastic supermatrix.</u> Let us enlarge the problem and find all self-adjoint matrices A satisfying the system of equations (4.1). More precisely, let us find all self-adjoint operators A , invariant under mapping Π :

$$A^* = A = A\Pi : \sum_{i,j} a^i_j\, {}^{i}_{j}\pi^{\kappa}_{\ell} = a^{\kappa}_{\ell}\ , \quad \forall \kappa,\ell\ . \tag{7.1}$$

THEOREM 7.1. The set γ_{Π} of all solutions of the system (7.1) is a linear space.Any solution is proportional to a normalized one, $A = (\mathrm{tr}\,A)\cdot P$.

Proof is evident.

Let us remind a property of non-negative matrices, which we shall often use.

LEMMA 7.2. If a matrix $B \geq 0$,then $b^i_i \geq 0\ , \forall i$. Moreover, if $b^m_m = 0$ for some m ,then

$$b^m_\ell = 0,\ \forall \ell : \quad b^\kappa_m = 0\ , \quad \forall \kappa\ .$$

Proof. By definition (6.11) the form

$$\langle x|B|x\rangle = \sum_{i,j} \xi_i \bar{\xi}_j b^i_j \geq 0 , \quad \forall |x\rangle \in \mathcal{H} .$$

Putting $\xi_i = 0, \ \forall j \neq m, \kappa ; \ \xi_m = 1$ and $\xi_\kappa = t$ or $\xi = t\sqrt{-1}$, one arrives at the reduced forms

$$t^2 b^\kappa_\kappa + t \cdot 2 \mathrm{Re}\, b^\kappa_m \ , \quad t^2 b^\kappa_\kappa + t \cdot 2\, \mathrm{Im}\, b^\kappa_m \ .$$

If $b^\kappa_m \neq 0$, at least one of them changes the sign at $t = 0$.

THEOREM 7.3. Bidiagonal elements of the stochastic supermatrix Π are non-negative,

$$ {}^j_j\Pi^\kappa_\kappa \geq 0 \ , \quad \forall j, \kappa \ . \tag{7.2}$$

If ${}^j_j\Pi^m_m = 0$,then

$$ {}^j_j\Pi^m_\ell = 0 = {}^j_j\Pi^\kappa_m \ , \quad \forall \kappa, \ell \ . \tag{7.3}$$

COROLLARY. The bidiagonal elements ${}^j_j\Pi^\kappa_\kappa$ of a stochastic supermatrix form a stochastic matrix, ${}_j\tilde{\Pi}_\kappa = {}^j_j\Pi^\kappa_\kappa$.

Proof. Let us consider the non-negative form $\langle x|A|x\rangle = \xi_j \bar{\xi}_j = |\xi_j|^2$. The supermatrix Π maps its matrix A into a matrix B of the quadratic form

$$\sum_{\kappa,\ell} b^\kappa_\ell \eta_\kappa \bar{\eta}_\ell = \sum_{\kappa,\ell} {}^j_j\Pi^\kappa_\ell \eta_\kappa \bar{\eta}_\ell \ . \tag{7.4}$$

The initial matrix A is non-negative (even probability matrix). By Theorem 3.1 the matrix $B \geq 0$ also .From here by Lemma 7.2 we deduce (7.2) and (7.3).Finally by (3.4) the sum of diagonal elements is normalized.

Let us use the proved corollary to study diagonal stationary matrices.Let a Π - stationary matrix A be diagonal,

$A = \mathrm{diag}(a_1^1, \dots, a_n^n)$. Since it is self-adjoint, $a_j^j \in \mathbb{R}$, $\forall j$. In just the same way as in §5, let us denote

$$\Omega^+(A) = \{ j : a_j^j > 0 \} , \quad \Omega^-(A) = \{ j : a_j^j < 0 \} . \tag{7.5}$$

It is more convenient to speak here about the set (of indeces) of the coordinate axes rather than about the corresponding subspace-supports.

Let us pay attention to the possible existence of

$$\Omega^0(A) = \{ j : a_j^j = 0 \} .$$

LEMMA 7.4. Let A be a solution of the system (7.1) and let the matrix A be diagonal. Then

$$\sum_{\kappa \in \Omega^+} {}_i^i\Pi_\kappa^\kappa = 1 , \qquad \forall i \in \Omega^\pm(A) \tag{7.6}$$

$${}_j^j\Pi_\ell^\kappa = 0 , \quad \forall j \in \Omega^\mp(A) , \quad \forall \kappa \in \Omega^\pm(A) . \tag{7.7}$$

COROLLARY.

$${}_j^j\Pi_\ell^\kappa = 0 , \quad \forall j \in \Omega^\mp(A) , \quad \forall \kappa, \ell \in \Omega^\pm(A) . \tag{7.8}$$

Proof. The lemma is similar to Lemma 5.1. Moreover, it follows from Lemma 5.1 by corollary from Lemma 7.3. We repeat the calculation commenting it in more detail.

If one puts $(a_i^i)^+$ and $(a_j^j)^-$ into (7.1), then

$$a_\kappa^\kappa = \sum_{i \in \Omega^+} (a_i^i)^+ \, {}_i^i\Pi_\kappa^\kappa - \sum_{j \in \Omega^-} (a_j^j)^- \, {}_j^j\Pi_\kappa^\kappa . \tag{7.9}$$

All the other terms are equal to zero, because the corresponding a_j^i are zeros. The values $(a_i^i)^+$ and $(a_j^j)^-$ in (7.9) are all strongly positive. The coefficients ${}_i^i\Pi_\kappa^\kappa$ and ${}_j^j\Pi_\kappa^\kappa$ at them are non-negative and their sums are normalized, $\sum_{\kappa} {}_i^i\Pi_\kappa^\kappa = 1$, $\forall i$, by Theorem 7.3. Therefore

$$\sum_{\kappa\in\Omega^+}(a_\kappa^\kappa)^+ = \sum_{\kappa\in\Omega^+} a_\kappa^\kappa = \sum_{i\in\Omega^+}(a_i^i)^+ \sum_{\kappa\in\Omega^+} {}_i^i\Pi_\kappa^\kappa - \tag{7.10}$$

$$-\sum_{j\in\Omega^-}(a_j^j)^- \sum_{\kappa\in\Omega^+} {}_j^j\Pi_\kappa^\kappa \leq \sum_{i\in\Omega^+}(a_i^i)^+ ,$$

where we have used (7.9) and after that changed the order of summation. The left-hand side and the right-hand side of the inequality are equal. Hence, the rejected non-positive terms are equal to zero. Since $(a_j^j)^- > 0$, $\forall j \in \Omega^-$, then the coefficients at them are zeros. The sum of non-negative ${}_j^j\Pi_\kappa^\kappa$ is equal to zero, iff all the summands are zeros. It implies (7.7).

Moreover, we are given 1 as an upper bound of the sums $\sum_{\kappa\in\Omega^+} {}_i^i\Pi_\kappa^\kappa$, $\forall i \in \Omega^+$. For the equality in (7.1) it is neccesary that all these sums coincide with their upper bound. It implies (7.6).

(7.8) follows from (7.7) by Theorem 7.3.

THEOREM 7.5. Let A be a solution of the system (7.1). Then so are A^+ and A^- .

COROLLARY. The set of all solutions of the system (7,1) is the splitted linear space od self-adjoint operators.

Proof. It is sufficient to verify the statement of the theorem only for diagonal matrices A . In fact, each unitary coordinate transformation U reducing A to a diagonal form U^*AU maps the supermatrix Π into the unitarily equivalent supermatrix (6.1). Furthermore a solution of the system (7.1) is converted into a solution of the transformed system.

Let us use the relation derived above. When $\kappa \in \Omega^+(A)$, we have by (7.9)

$$(a^\kappa_\kappa)^+ = a^\kappa_\kappa = \sum_{i\in\Omega^+} (a^i_i)^+ \, {}^i_i\pi^\kappa_\kappa - \sum_{j\in\Omega^-} (a^j_j)^- \, {}^j_j\pi^\kappa_\kappa + \tag{7.11}$$

$$+ \sum_{i\bar{\in}\Omega^+} (a^i_i)^+ \, {}^i_i\pi^\kappa_\kappa = \sum_{i=1}^{n} (a^i_i)^+ \, {}^i_i\pi^\kappa_\kappa \, .$$

Only the first sum of the middle expression is non-zero. The second sum is zero by (7.7). The added third sum is equal to zero in view of $(a^i_i)^+ = 0$, $\forall i \bar{\in} \Omega^+$. Thus, we have calculated the part of diagonal elements of the matrix $B = A^+\Pi$. It is found that

$$b^\kappa_\kappa = (a^\kappa_\kappa)^+ \, , \qquad \forall \kappa \in \Omega^+ \, .$$

Now, for $\kappa \bar{\in} \Omega$,

$$b^\kappa_\kappa = \sum_{i=1}^{n} (a^i_i)^+ \, {}^i_i\pi^\kappa_\kappa = \sum_{i\in\Omega^+} (a^i_i)^+ \, {}^i_i\pi^\kappa_\kappa = 0 = (a^\kappa_\kappa)^+ \, ,$$

according to the second formula in (7.7) with lower signs.

Up to this point, we have repeated the proof of Theorem 5.2. Let us now calculate the non-diagonal elements of the matrix $B = A^+\Pi$. Let $\kappa, \ell \in \Omega^+$. Then

$$b^\kappa_\ell = \sum_{i=1}^{n} (a^i_i)^+ \, {}^i_i\pi^\kappa_\ell = \sum_{i\in\Omega^+} a^i_i \, {}^i_i\pi^\kappa_\kappa = \sum_{i\in\Omega^+} a^i_i \, {}^i_i\pi^\kappa_\ell +$$

$$+ \sum_{j\in\Omega^0} a^j_j \, {}^j_j\pi^\kappa_\ell + \sum_{j\in\Omega^-} a^j_j \, {}^j_j\pi^\kappa_\ell = a^\kappa_\ell = 0 \, .$$

All the summands with $j \in \Omega^0$ are zeros in view of $a^j_j = 0$. All the summands with $j \in \Omega^-$ are equal to zero in view of ${}^j_j\pi^\kappa_\ell = 0$ according to (7.8).

Matrix $B \geqslant 0$ as the Π -image of a matrix $A^+ \geqslant 0$, see supplement to Theorem 3.1. Since $b^\kappa_\kappa = 0$, $\forall \kappa \in \Omega^+$, then $b^\kappa_\ell = 0 = b^\kappa_\kappa$, $\forall \kappa \in \Omega^+$, $\forall \ell$. Hence, the matrix B is diagonal coinciding with A^+ .

<u>8. Logics of subspaces.</u> According to the classical model of §5, we have to know what a structure is formed by the supports of solutions of (7.1). The cardinal question is: what a subspace structure is analogous to a Boolean ring or a Boolean algebra. In their original work on quantum logic, see [5], Birkhoff and von Neumann had shown that such a structure must be an orthomodular lattice. These desired structures have been found in a series of our papers [14]-[16]. Their operation system is found to be essentialy richer than the lattice one.

DEFINITION 8.1. Let a subspace $\mathfrak{F}$ be not orthogonal to $\mathfrak{G}$. We call the contact of $\mathfrak{F}$ with $\mathfrak{G}$ a subspace $\mathfrak{F} \mid\cap \mathfrak{G}$:

$$\mathfrak{F} \mid\cap \mathfrak{G} = \left[\lim_{n \to \infty} \left(\frac{1}{\rho^2} FGF \right)^n \right] (\mathfrak{F}) \quad , \qquad (8.1)$$

where F and G are orthoprojectors on $\mathfrak{F}$ and on $\mathfrak{G}$, $\rho^2 = \| FGF \|$ is the maximal eigen-value of the self-adjoint operator FGF .

When $\mathfrak{F} \perp \mathfrak{G}$, then $FGF = 0$, and vice versa. We put $\mathfrak{F} \mid\cap \mathfrak{G} = 0$ in this case.

It is evident, that $\mathfrak{F} \mid\cap \mathfrak{G} \subseteq \mathfrak{F}$ always. If $\mathfrak{F} \cap \mathfrak{G} \neq 0$, then $\rho^2 = 1$ and $\mathfrak{F} \mid\cap \mathfrak{G} = \mathfrak{F} \cap \mathfrak{G}$, and vice versa.

DEFINITION 8.2. Non-orthogonal subspaces $\mathfrak{F}$ and $\mathfrak{G}$ will be called isocline, if

$$FGF = \rho^2 F \quad , \quad GFG = \rho^2 G \quad , \qquad (8.2)$$

i.e., if the angle formed by any vector $|x\rangle \in \mathfrak{F}$ and its projection $G|x\rangle$ is equal to $\arccos\rho$ independently of $|x\rangle$ and similarly for $|y\rangle \in \mathfrak{Y}$.

LEMMA 8.1. Contacts $\mathfrak{F} \mid\cap \mathfrak{Y}$ and $\mathfrak{Y} \mid\cap \mathfrak{F}$ of two non-orthogonal subspaces $\mathfrak{F}$ and $\mathfrak{Y}$ are isocline subspaces. Conversely, if $\mathfrak{E}$ and $\mathfrak{R}$ are isocline, then $\mathfrak{E} = \mathfrak{E} \mid\cap \mathfrak{R}$, $\mathfrak{R} = \mathfrak{R} \mid\cap \mathfrak{E}$.

Proof. See [14].

LEMMA 8.2. There is a canonic isometry I of two isocline subspaces $\mathfrak{F}$ and $\mathfrak{Y}$:

$$\mathfrak{F} \ni |x\rangle \to \rho^{-1} G|x\rangle = |y\rangle \to \rho^{-1} F|y\rangle = |x\rangle \ . \tag{8.3}$$

Proof. See [14], Theorem 1.1.

DEFINITION 8.3. We shall call the subspace

$$\mathcal{K}_{a:b} := \{|z\rangle = a|x\rangle + bI|x\rangle \ , \ \forall |x\rangle \in \mathfrak{F}\} \tag{8.4}$$

coherent $a:b$-linear combination of isocline subspaces $\mathfrak{F}$ and $\mathfrak{Y}$.

LEMMA 8.3. Coherent linear combination of isocline subspaces $\mathfrak{F}$ and $\mathfrak{Y}$ with all possible $a,b \in \mathbb{R}$ form an $\mathbb{R}$-bundle of isocline subspaces $\mathfrak{E}_\psi$, $-\pi/2 < \psi \leq \pi/2$,

$$\mathfrak{E}_\psi := \{|y\rangle = \cos\psi \cdot |x\rangle + \sin\psi J|x\rangle \ , \ \forall |x\rangle \in \mathfrak{F}\} \ , \tag{8.5}$$

where J is the isometry $\mathfrak{F}$ onto $\mathfrak{F}' = (\mathfrak{F} + \mathfrak{Y}) \ominus \mathfrak{F}$ generated by the isometries $\mathfrak{F}$ onto $\mathfrak{Y}$ and $\mathfrak{Y}$ onto $\mathfrak{F}'$ by (8.3). Now, $\mathfrak{Y} = \mathfrak{E}_\varphi$ where $\varphi = \arccos\rho$.

Proof. See [16], Theorem 2.6.

DEFINITION 8.4. We call by $\mathbb{R}$-quasi-logic any lattice $\mathfrak{L}$ of subspaces of space $\mathcal{H}$, stable with respect to operations:

(I) proper orthogonal subtraction (of a contained subspace out of a containing);

(II) vector addition of subspaces;

(III) isolation of contact;

(IV) coherent $\mathbb{R}$ -linear combination of isocline subspaces.

A quasi-logic $\mathcal{L} \ni \mathcal{H}$ will be called a logic. A quasi-logic is an anlogue of a Boolean ring, a logic is that of a Boolean algebra.

LEMMA 8.4.

$$\mathcal{F} \cap \mathcal{G} = (\mathcal{F} + \mathcal{G}) - [(\mathcal{F} - \mathcal{G}) + (\mathcal{G} - \mathcal{F})] \quad , \tag{8.6}$$

$$\mathcal{F} \ominus \mathcal{G} = \mathcal{F} \cap (\mathcal{F} - \mathcal{G}) \ , \tag{8.7}$$

where the operation of subtraction is defined as

$$\mathcal{R} - \mathcal{E} = (\mathcal{R} + \mathcal{E}) \ominus \mathcal{E} \quad . \tag{8.8}$$

COROLLARY. It is possible to replace the operation(I) by subtraction (8.8) in Definition 8.4 of quasi-logic.

Proof. See [16], Lemma 1.3. Let us note that, in logics, it is natural to take the operation of orthocomplement:

$$\mathcal{F} \to \mathcal{F}^{\perp} := \mathcal{H} \ominus \mathcal{F} \ , \tag{8.9}$$

instead of the operation(I).

9. Logic $\mathcal{L}\mathcal{Y}$ of supports for splitted operator space $\mathcal{Y}^{\nu}$. We have to consider the list of quasi-logic operations and to establish that any operation applied to an $\mathcal{Y}$-operator support leads again to a support.

LEMMA 9.1. The support of an $\mathcal{Y}$-operator is the support of some non-negative $\mathcal{Y}$-operator.

Proof. By formula (6.10)

$$\mathcal{C}_A = \mathcal{C}_A^+ \oplus \mathcal{C}_A^- = \mathcal{C}_{A^+} \oplus \mathcal{C}_{A^-} = \mathcal{C}_{A^+ + A^-} \quad . \tag{9.1}$$

LEMMA 9.2. The vector sum of two supports is also the support of an $\mathcal{Y}$-operator.

Proof. Let $A, B \geq 0$. Then

$$\mathcal{C}_A + \mathcal{C}_B = \mathcal{C}_{A+B} \quad , \tag{9.2}$$

because $A+B \geq 0$ and

$$(\mathcal{C}_{A+B})^\perp = (\mathcal{C}_A)^\perp \cap (\mathcal{C}_B)^\perp$$

is an annulled subspace for $A+B$.

THEOREM 9.3. Let a subspace $\mathcal{F}$ be the support of an $\mathcal{Y}$-operator $P \geq 0$, and a non-orthogonal to $\mathcal{F}$ subspace $\mathcal{G}$ be the support of an $\mathcal{Y}$-operator $Q \geq 0$.

Then a subspace $\mathcal{G}' = \mathcal{F} - \mathcal{G} = (\mathcal{F} + \mathcal{G}) \ominus \mathcal{G}$ is also the support of the $\mathcal{Y}$-operator

$$R := (\mathbb{1} - G)P(\mathbb{1} - G) = \lim_{t \to \infty} [tQ - P]^- \quad , \tag{9.3}$$

where G is the orthoprojector into $\mathcal{G}$, $F \vee G$ is the orthoprojector onto $\mathcal{F} + \mathcal{G}$, and

$$R = ((F \vee G) - G)\, P\, ((F \vee G) - G) \quad .$$

Proof. Formula (9.3) is analogous to (5.6). Only in (5.6) the limit value was attained for all sufficiently small t. Now, let us try to explicitly calculate $[tQ - P]^-$ and its limit with the help of contour integrals. Since all the considered operators annul on $(\mathcal{F}+\mathcal{G})^{\perp}$, one may suppose $\mathcal{F}+\mathcal{G}=\mathcal{H}$, $F \vee G = \mathbb{1}$.

Note that the eigen-subspaces of the operators $tQ-P$ and $Q-\frac{1}{t}P$ coincide. Hence, their supports also coincide. For sufficiently small t the orthoprojector

$$G_t^{\vee} = \frac{1}{2\pi i} \int_{\gamma_+} (\zeta \mathbb{1} - Q + \frac{1}{t} P)^{-1} d\zeta = G + O(\tfrac{1}{t}) \qquad (9.5)$$

is close to the orthoprojector G onto the support of Q, see Lemma 6.3 and its corollary. Here the contour γ_+ is taken in lie in the right half-plane and to contain all the positive part of the spectrum of $Q \geq 0$, intersecting the real axe at $\zeta = \frac{1}{2}\varkappa_{min}$, where $\varkappa_{min}$ is the minimal positive eigen-value of Q.

Let us transform the integral (6.12) to prove (9.5) in such a way:

$$G_t^{\vee} = \frac{1}{2\pi i} \int_{\gamma_+} (\zeta \mathbb{1} - Q)^{-1} [\mathbb{1} + \tfrac{1}{t} P (\zeta \mathbb{1} - Q)^{-1}]^{-1} d\zeta =$$

$$= \frac{1}{2\pi i} \int_{\gamma_+} (\zeta \mathbb{1} - Q)^{-1} d\zeta + O(\tfrac{1}{t}) .$$

We are not able still to affirm that $G_t^{\vee}$ is the orthoprojector onto all the positive part of the support of the operator $tQ-P$. We don't know yet if the contour γ_+ contains all the positive part of the spectrum of the perturbed operator $Q-\frac{1}{t}P$ as well. But necessarily $G_t^{+} \geq G_t^{\vee}$, $\mathbb{1}-G_t^{+} \leq \mathbb{1}-G_t^{\vee}$, and all these four operators commute, see Lemma 6.3, with each other and each with orthoprojectors of the spectral decomposition of $tQ-P$.

Let us choose a contour γ_- containing the interval $(-a, \frac{1}{2}\varkappa_{min})$

of the real axe, $-a<0$. Beginning with some t the contours γ_+ and γ_- contain together all the spectrum of $Q=\frac{1}{t}P$. We suppose that $\zeta=\frac{1}{2}\varkappa_{min}$ is the only common point of γ_+ and γ_- .

Let us transform the integral (6.13) in the same way as (6.12). We take now two first terms in the expansion of (6.13):

$$-R_t=\frac{t}{2\pi i}\int_{\gamma_-}\zeta(\zeta\mathbb{1}-Q+\frac{1}{t}P)^{-1}d\zeta=\frac{t}{2\pi i}\int_{\gamma_-}\zeta(\zeta\mathbb{1}-Q)^{-1}d\zeta-$$

$$-\frac{t}{2\pi i}\int_{\gamma_-}\zeta(\zeta\mathbb{1}-Q)^{-1}\frac{1}{t}P(\zeta\mathbb{1}-Q)^{-1}d\zeta+O(\tfrac{1}{t}) .$$

The first term of the epansion is the zero operator, see (6.14), because the contour γ_- contains the only point $\varkappa=0$ from the spectrum of $Q\geqslant 0$. Therefore

$$R_t=\frac{1}{2\pi i}\int_{\gamma_-}\zeta(\zeta\mathbb{1}-Q)^{-1}P(\zeta\mathbb{1}-Q)^{-1}d\zeta+O(\tfrac{1}{t}) . \qquad (9.6)$$

We shall calculate this integral as follows. The operator R_t is obtained from the spectral decomposition of $tQ-P$ by ommitting the summands with the orthoprojectors $E_\kappa(t)$ entering in $G_t^{\vee}$. Hence, $(\mathbb{1}-G_t^{\vee})R_t(\mathbb{1}-G_t^{\vee})=R_t$. After multiplying the two sides of (9.6) by $\mathbb{1}-G_t^{\vee}$, bringing the factors into thw sign of integral, using (9.5) and omitting the terms of order $1/t$, we obtain

$$R_t=\frac{1}{2\pi i}\int_{\gamma_-}(\mathbb{1}-G\ )P(\mathbb{1}-G)\zeta^{-1}d\zeta+O(\tfrac{1}{t})=R+O(\tfrac{1}{t}) , \qquad (9.7)$$

see (9.3), because

$$(\mathbb{1}-G)Q=0=Q(\mathbb{1}-G);\ (\mathbb{1}-G)(\zeta\mathbb{1}-Q)^{-1}=\zeta^{-1}(\mathbb{1}-G) .$$

Let us verify that the subspace $\mathcal{G}' = (\mathcal{F} + \mathcal{G}) \ominus \mathcal{G}$ is the carrier of the non-negative operator $R = (1 - G) P (1 - G)$. We remind the convention that $\mathcal{H} = \mathcal{F} + \mathcal{G}$, $F \vee G = 1$. The operator R vanishes on the ortho-complement of $\mathcal{G}'$, i.e. on the subspace $\mathcal{G}$. By Lemma 1.1 from [15], the subspace $\mathcal{G}' = (1 - G)(\mathcal{F})$. Therefore, if $|0\rangle \neq |y\rangle \in \mathcal{G}'$, then $|y\rangle = (1 - G)|x\rangle$, where $|x\rangle \in \mathcal{F}$ and $\langle y|(1 - G) P (1 - G)|y\rangle = \langle x|(1 - G) \cdot P (1 - G)|x\rangle =$
$= \langle y|P|y\rangle > 0$, because $|y\rangle$ is not orthogonal to the support $\mathcal{F}$ of the operator P, since $\langle y|x\rangle > 0$ strongly.

Thus, there is an essential positive part of the spectrum of $tQ - P$, which is close to the spectrum of tQ, and the support of which is close to $\mathcal{G}$. Further, there is an essential negative part, which is close to the spectrum of $-R$, and the support of which is close to $\mathcal{G}'$. Since $\mathcal{G} \oplus \mathcal{G}' = \mathcal{G} + \mathcal{F} = \mathcal{H}$, the spectrum of $tQ - P$ is exhausted by those parts. In the general case, when $\mathcal{F} + \mathcal{G} \subset \mathcal{H}$, the spectrum is only added by the constant subspace $(\mathcal{G} + \mathcal{F})^{\perp}$ being annulled.

The operator $[tQ - P]^{-} \in \mathscr{Y}$ according to linearity and splittability of $\mathscr{Y}$. We have proved that its limit exists. The latter belongs to $\mathscr{Y}$ since any linear subspace of finite-dimensional operators is closed.

THEOREM 9.4. Under the conditions of Theorem 9.3 the support $\mathcal{G}''$ of the $\mathscr{Y}$-operator GPG is located in $\mathcal{G}$, i.e. $\mathcal{G}'' \subseteq \mathcal{G}$.

The subspace $\mathcal{F}$ contains the support $\mathcal{F} \cap \mathcal{G}$ of the $\mathscr{Y}$-operator $(F \wedge G) P (F \wedge G)$, where $F \wedge G$ is the orthoprojector onto the contact $\mathcal{F} \cap \mathcal{G}$ of the subspace $\mathcal{F}$ with $\mathcal{G}$.

Proof. Let us apply Theorem 9.3 to the pair $\mathcal{F}$ and $\mathcal{G}'$ of the supports of P and R , see (9.3). We shall get the $\mathcal{Y}$-operator GPG on $\mathcal{G}'' = (\mathcal{F} + \mathcal{G}') \ominus \mathcal{G}'$ since $\mathcal{G} \oplus \mathcal{G}' = \mathcal{F} + \mathcal{G}$, $GVF - G' = G$. The specific expression of the operator R is of no importance, it does not enter in the answer. Since $\mathcal{G}' \subseteq \mathcal{G} + \mathcal{F}$, we have $\mathcal{G}' + \mathcal{F} \subseteq \mathcal{G} + \mathcal{F}$, $\mathcal{G}'' \subseteq \mathcal{G}$. The inclusion may be strong.

We succeeded in constructing the $\mathcal{Y}$-operator GPG on $\mathcal{G}'' \subseteq \subseteq \mathcal{G}$ by the $\mathcal{Y}$-operator P with $\mathcal{F}$ as the support under the only condition that $\mathcal{G}$ is the support of some $\mathcal{Y}$-operator Q . Let $P_1 = GPG$, $\mathcal{F}_1 = \mathcal{G}''$, $Q_1 = P$, $\mathcal{G}_1 = \mathcal{F}$, There exists an $\mathcal{Y}$-operator

$$FGPGF = (FGF)P(FGF) , \tag{9.8}$$

with the support $\mathcal{G}_1' \subseteq \mathcal{F}$. The condition of non-orthogonality of $\mathcal{G}''$ and $\mathcal{F}$ is fulfilled, otherwise $FGF = 0$, what is equivalent to $\mathcal{F} \perp \mathcal{G}$, see (8.1).

Let $\rho = \rho(\mathcal{F}, \mathcal{G})$. By Definition 8.1

$$F \wedge G = \lim_{n \to \infty} (\rho^{-2} FGF)^n \tag{9.9}$$

To complete the proof of the theorem one need only iterate the passage from P to $(\rho^{-2} FGF)P(\rho^{-2} FGF)$ according to (9.8).

THEOREM 9.5. Let isocline subspaces $\mathcal{F}$ and $\mathcal{G}$ with the angle $\varphi = \arccos \rho$ between each other be the supports of $\mathcal{Y}$-operator $P \geq 0$ and $GPG \geq 0$ respectively.

Then every subspace $\mathcal{E}_\psi$ of the bundle of the isocline subspaces (8.5), passing through $\mathcal{F} = \mathcal{E}_0$ and $\mathcal{Y} = \mathcal{E}_\psi$ is the support of an $\mathcal{Y}$-operator $P_\psi = \frac{1}{\cos^2\psi} E_\psi P E_\psi$:

$$P_\psi = \mu_0(\psi)P + \mu_1(\psi)GPG + \mu_2(\psi)(1-G)P(1-G) , \qquad (9.10)$$

where $\mu_0(\psi)$, $\mu_1(\psi)$ and $\mu_2(\psi)$ are some trigonometrical expressions.

REMARK. If P is a probability operator, then P_ψ may be called the superposition of coherent probability distributions P and $P_\varphi = \rho^{-2} GPG$, see [16], Definition 0.2.

Proof. By Theorem 9.3 the subspace $\mathcal{Y}'$ is the support of the $\mathcal{Y}$-operator $(1-G)P(1-G)$. Let us take the basis in $\mathcal{F}$ formed by eigen-vectors of the operator P , and the corresponding basis in $\mathcal{F}' = (\mathcal{F} + \mathcal{Y}) \ominus \mathcal{F}$ by (8.5). According to Lemma 8.2 the operator $\cos^{-1}\psi \cdot E_\psi$ determines an isometry of $\mathcal{F}$ onto $\mathcal{E}_\psi$ - the rotation of $\mathcal{F}$ at angle ψ . Therefore, in the chosen basis of the subspace $\mathcal{F} + \mathcal{Y}$, the matrix of the operator P_ψ is constructed from the diagonal blocks

$$\begin{pmatrix} \pi_k \cos^2\psi & \pi_k \sin\psi\cos\psi \\ \pi_k \sin\psi\cos\psi & \pi_k \sin^2\psi \end{pmatrix} \qquad (9.11)$$

where π_k is the eigen-value of P corresponding to the vector $|x_k\rangle$ and to $|z_k\rangle = J|x_k\rangle$, connected with the former by (8.5).

A block of such a form may be decomposed into the blocks

$$\begin{pmatrix} 1 & 0 \\ 0 & 0 \end{pmatrix}, \begin{pmatrix} \cos^2\varphi & \sin\varphi\cos\varphi \\ \sin\varphi\cos\varphi & \sin^2\varphi \end{pmatrix}, \begin{pmatrix} \sin^2\varphi & -\sin\varphi\cos\varphi \\ -\sin\varphi\cos\varphi & \cos^2\varphi \end{pmatrix}$$

corresponding to $\psi = 0$, $\psi = \varphi$, $\psi = \varphi - \pi/2$, $0 < \varphi < \pi/2$, i.e. to $\mathfrak{F}$, $\mathfrak{G}$, $\mathfrak{G}'$, see [16]. Therefore, the decomposition (9.10) exists. $P_\psi \in \mathscr{Y}$ by linearity of $\mathscr{Y}$.

THEOREM 9.6. Let $\mathscr{Y}$ be a splittable linear space of self-adjoint operators, acting in a finite-dimensional unitary space $\mathcal{H}$.

Then the suppots of all the operators $A \in \mathscr{Y}$ form a certain $\mathbb{R}$ -quasi-logic $\mathfrak{L}_{\mathscr{Y}}$ of subspaces of the space $\mathcal{H}$.

Proof. The above statements 9.2 - 9.5 show that a support lattice is stable with respect to all $\mathbb{R}$ -quasi-logic operations, see Definition 8.4 and Lemma 8.4.

<u>10. The structure of splittable linear subspaces of self-adjoint operators.</u> It follows from Theorem 9.6 that our problem is at least partially redused to classifying all the $\mathbb{R}$ -quasi-logics of subspaces. As proved in our paper [16], this problem is equivalent to classification of all finite-dimensional special Jordan algebras given by Jordan, von Neumann and Wigner in their famous work [17]. Even the most reduced version of the proof of this statement is too long to give it here.

Following [16] we shall briefly explain all the facts necessary to us. Structures of finite-dimensional $\mathbb{R}$ -quasi-logics (we shall call them elementary Jordan quasi-logics) generalize, to a certain degree, those of Boolean rings.

The sum of all the subspaces of a quasi-logic $\mathfrak{L}$ is its maximal subspace $\mathcal{M} = \mathcal{M}_{\mathfrak{L}}$. If $\mathcal{M}_{\mathfrak{L}} = \mathcal{H}$, then $\mathfrak{L}$ is an $\mathbb{R}$ -logic.

The class of minimal $\mathfrak{L}$ - subspaces in $\mathfrak{L}$ can be distinguished. We shall say that an $\mathfrak{L}$ - space $\mathfrak{F}$ is minimal if it has no proper $\mathcal{L}$ - subspaces (i.e. differing from 0 and from $\mathcal{F}$). Since the whole space $\mathcal{H}$ is finite-dimensional and the dimension of successively strongly imbedded subspaces is strongly decreasing, any $\mathcal{L}$ - subspace contains a minimal one.

Minimal $\mathcal{L}$ - subspaces are similar to atoms of a Boolean ring: if they are intersecting, then they coincide. Any $\mathcal{L}$ - subspace may be decomposed into an orthogonal sum of $\mathcal{L}$ -minimal ones (the decomposition is, in general, non-unique).

Two different minimal $\mathcal{L}$ - spaces are either orthogonal or contacting. There are no presedents of the latter possibility in Boolean algebras. By their minimality, contacting minimal spaces $\mathcal{F}$ and $\mathcal{G}$ must respectively coincide with their contacts. Hence $\mathcal{F}$ and $\mathcal{G}$ are isocline spaces by Lemma 8.1.

Let $\mathcal{V}$ be splittable linear space of self-adjoint operators, $\mathcal{L}_{\mathcal{V}}$ be the quasi-logic of the supports of $\mathcal{V}$ -operators. We shall now clarify what are operators supported by a minimal $\mathcal{L}_{\mathcal{V}}$ - space.

THEOREM 10.1. Any minimal $\mathcal{L}$ - space $\mathcal{F}$ supports only one probability $\mathcal{V}$ - operator $Q = Q_{\mathcal{F}}$ and only one family $\{\lambda Q_{\mathcal{F}}\}$ of non-normalized $\mathcal{V}$-operators.

Proof. Let $\operatorname{tr}(Q'_{\mathcal{F}} - Q''_{\mathcal{F}}) = 0$. Either $Q'_{\mathcal{F}} = Q''_{\mathcal{F}}$ or $(Q'_{\mathcal{F}} - Q''_{\mathcal{F}})^+ \neq 0$, $(Q'_{\mathcal{F}} - Q''_{\mathcal{F}})^- \neq 0$, $0 \subset \mathcal{C}_{(Q'-Q'')} \subset \mathcal{F}$. The latter contradicts minimality of $\mathcal{F}$.

THEOREM 10.2. If an $\mathcal{L}_{\mathcal{V}}$ -space $\mathcal{K}$ is decomposed in an orthogonal sum of minimal ones $\mathcal{K} = \bigoplus_i \mathcal{F}_i$, then any $\mathcal{V}$-operator A with the support, contained in $\mathcal{K}$, is of the form

$$A = \sum_{i=1}^{s} \lambda_i Q_{\mathfrak{F}_i} ; \quad \lambda_i \in \mathbb{R} \ , \ \forall i \ , \tag{10.1}$$

The probability $\mathscr{Y}$-operators have all the $\lambda_i \geqslant 0$, $\sum_i \lambda_i = 1$.

Proof. By orthogonality $\mathfrak{F}_i \perp \mathfrak{F}_j$, $F_i F_j = 0$, $i \neq j$, ew have $A = \sum F_i A F_i$, because A commutes with the orthoprojector $\sum_i F_i$ on its support. Then Theorem 10.1 is applicable to $F_i A F_i$.

THEOREM 10.3. If minimal $\mathfrak{L}_{\mathscr{Y}}$-spaces $\mathfrak{F}$ and $\mathfrak{Y}$ are contacting, i.e. $\rho(\mathfrak{F}, \mathfrak{Y}) > 0$, then their correspondent probability $\mathscr{Y}$-operators $Q_{\mathfrak{F}}$ and $Q_{\mathfrak{Y}}$ are dependent:

$$Q_{\mathfrak{Y}} = \rho^{-2}(\mathfrak{F}, \mathfrak{Y})\, G Q_{\mathfrak{F}} G \ . \tag{10.2}$$

Proof. By Theorem 9.4 the operator $GQ_{\mathfrak{F}}G$ differs from zero, belongs to $\mathscr{Y}$, and its support $\mathfrak{Y}'' \subseteq \mathfrak{Y}$. Hence $Q_{\mathfrak{Y}} = \lambda G Q_{\mathfrak{F}} G$ by Theorem 10.1. The normalization can be found from (9.10).

DEFINITION 10.1. We shall call minimal $\mathfrak{L}$-spaces $\mathfrak{F}$ and $\mathfrak{Y}$ coherent if either they contact or they are orthogonal, but each of them contacts the same third minimal one.

LEMMA 10.4. The relation of coherence is reflexive, symmetric and transitive, i.e. is an equivalence relation.

COROLLARY. The system of all the minimal $\mathfrak{L}$-spaces is distributed into classes of coherence. The minimal spaces of different classes are orthogonal.

Proof. For detailed proof see [14], Theorem 4.14, Lemmas 4.11 and 4.12. We shall only outline the key point of arguments. Let minimal $\mathfrak{F}_1$, $\mathfrak{F}_2$, $\mathfrak{F}_3$ be three orthogonal straight lines in an ordinary Euclidean space. Let $\mathcal{F}_1$ and $\mathcal{F}_2$ be coherent by line $\mathcal{G}_{12} \subset (\mathcal{F}_1 \oplus \mathcal{F}_2)$, $\mathcal{F}_2$ and $\mathcal{F}_3$ by $\mathcal{G}_{23} \subset (\mathcal{F}_2 \oplus \mathcal{F}_3)$. Then the plane $\mathcal{G}_{12} + \mathcal{G}_{23}$ and the plane $\mathcal{F}_3 \oplus \mathcal{F}_1$ have neccessarily a common line $\mathcal{G}_{31}$, because two different planes having the origin as a common point are intersecting by a line. The line $\mathcal{G}_{31}$ belongs to the quasi-logic by condtruction and coherents $\mathcal{F}_3$ and $\mathcal{F}_1$. If $\mathcal{F}_j$ are multi-dimensional, the isocline minimal $\mathcal{G}_{12}$ and $\mathcal{G}_{23}$ induce some coherent orthonormal bases by (8.5). Therefore, we can decompose the whole multi-dimensional configuration into an orthogonal sum of the considered form.

DEFINITION 10.2. The sum $\mathcal{N}$ of all coherent minimal $\mathcal{L}$-spaces of the same class will be called the factor space.

LEMMA 10.5. The minimal $\mathcal{L}$ - space $\mathcal{M}$ can be decomposed into an orthogonal sum of factor ones, $\mathcal{M} = \bigoplus_{j=1}^{m} \mathcal{N}_j$.

Proof is evident (compare [14], Theorem 4.16).

THEOREM 10.6. To fully determine a splittable linear space $\mathcal{Y}$ of self-adjoint operators acting on $\mathcal{H}$, it is sufficient to find an $\mathbb{R}$-quasi-logic $\mathcal{L}_{\mathcal{Y}}$ of supports and to define certain probability $\mathcal{Y}$-operator Q_j on every specimen $\mathcal{F}_j$ of minimal $\mathcal{L}$-spaces each chosen out of each factor $\mathcal{L}$-space $\mathcal{N}_j$.

REMARK. In general, the operator Q_j must satisfy a certain condition of symmetry determined by the logic $\mathcal{L} \cap \mathcal{N}_j$.

Proof follows from Theorems 10.2 and 10.3. The probability $\mathcal{Y}$-operator Q on any minimal $\mathcal{F} \subset \mathcal{N}_j$ is determined by Q_j with the help of (10.2) maximally in two steps. Since the different chains of contacting minimal spaces have to lead to the same Q, a symmetry conditions appears.

LEMMA 10.7. Let $\mathcal{L} \cap \mathcal{N}$ be the logic of minimal $\mathcal{L}$ -subspaces of a factor $\mathcal{L}$ -space $\mathcal{N}$. Then $\mathcal{L} \cap \mathcal{N}$ can be isomorphic only to lattice of all straight lines of either a finite-dimensional real, complex or quaternionic space (of arbitrary dimension), or a plane (two-dimensional space) under hypercomplex numbers with several imaginary units.

Proof follows from the classification of Jordan algebras given in [17] and the connection between Jordan algebras and Jordan logics mentioned above. A direct but too detailed proof see in [16].

THEOREM 10.8. Let $\mathcal{Y}$ be a splittable linear space of self-adjoint matrices. Any $\mathcal{Y}$ -matrix A may be described by the formula:

$$A = \bigoplus_{j=1}^{m} \lambda_j R_j \oplus Q_j \quad , \quad \lambda_i \in \mathbb{R} \; , \; \forall j \quad ; \tag{10.3}$$

$$R_j : R_j^* = R_j \geqslant 0 \; ; \quad \mathrm{tr}\, R_j = 1 \quad , \quad \forall j \quad , \tag{10.4}$$

The dimension of R_j and the number nature ($\mathbb{R}$, $\mathbb{C}$, $\mathbb{Q}$, $\mathbb{H}$) of its elements are determined by the logic $\mathcal{L}_{\mathcal{Y}} \cap \mathcal{N}_j$ of the factor $\mathcal{N}_j$.

For the probability $\mathcal{Y}$-matrices

$$\lambda_j \geqslant 0 \quad , \forall j \; ; \quad \sum_j \lambda_j = 1 \; . \tag{10.5}$$

REMARK. The $\mathcal{Y}$ -matrices Q_j satisfy the conditions of symmetry permitting to correctly define the tensor product $R_j \oplus Q_j$.

Proof. For all the mentioned number systems, a self-adjoint operator R_j has the spectral decomposition (6.2), where any orthoprojector $E_K^{(j)} = \sum_i L_i^{(Kj)}$, $\mathcal{E}_K^{(j)} = \bigoplus_i \mathcal{L}_i^{(Kj)}$, and all the $\mathcal{L}_i$

are straight lines in $\mathbb{R}^s$, $\mathbb{C}^s$, $\mathbb{Q}^s$ or $\mathbb{H}^s$ respectively. Having taken a fixed system of homogeneous coordinates in $\mathcal{N}_j$, we obtain the possibility to identify the orthoprojectors $L_i^{(\kappa j)}$ and the orthoprojectors $F_i^{(\kappa j)}$ on the corresponding minimal subspaces. After that we calculate the $Q_i^{(\kappa j)}$ supported by $\mathcal{F}_i^{(\kappa j)}$. As a final result we shall take the expansion

$$R_j = \sum_\kappa \rho_\kappa^{(j)} E_\kappa^{(j)} = \sum_\kappa \rho_\kappa^{(j)} \sum_i L_i^{(\kappa j)} \longrightarrow \sum_\kappa \rho_\kappa^{(j)} \sum_i Q_i^{(\kappa j)} . \qquad (10.6)$$

The latter is correctly defined by the remark to Theorem 10.6.

11. The description of Π - stationary probability distribution. We are now able to give the solution of our problem.

THEOREM 11.1. Let Π be a stochastic supermatrix. Then there exist m types of stationary probability matrices $Q_1, \dots, Q_m$, $m \geq 1$.

Any stationary probability matrix P is described by the formula

$$P = \sum_{j=1}^{m} z_i R_j \otimes Q_j \quad ; \qquad (11.1)$$

$$R_j : R_j^* = R_j \geq 0 \quad ; \quad \operatorname{tr} R_j = 1 \quad ; \quad \forall j \; ; \qquad (11.2)$$

$$z_j : z_j \geq 0 , \quad \forall j \; ; \quad \sum_{j=1}^{m} z_j = 1 . \qquad (11.3)$$

The dimension of R_j and the number nature of its elements are determined by the structure of the logic $\mathcal{L}_j$, being a factor of a quasi-logic of Π - stationary matrix supports.

A matrix Q_j in case of $\mathbb{Q}$-numbers or $\mathbb{H}$-numbers must satisfy additional conditions of symmetry.

For supermatrices Π of corresponding structure there are realized the factors of the types of $\mathbb{R}^s$, $\mathbb{C}^s$, $\mathbb{Q}^s$ with any natural s, and those of $\mathbb{H}^2_\nu$ with any number ν of imaginary units (some of mentioned types coincide).

Proof. The basic statement follows from Theorems 9.6 and 10.8. The realizability of the mentioned factors follows from the results of §4. By the fixed point theorem applied to the convex set $\mathfrak{P}$ of all the probability matrices on $\mathcal{H}$ which is mapped by the super-matrix Π into itself, $\mathfrak{P}\Pi \subseteq \mathfrak{P}$, there exists at least one Π-stationary probability matrix.

From the proved results it follows that there are two non-commutative analogues of a communicating states class of a Markov chain. The first one is a minimal subspace: there exist only one Π-stationary probability distribution concentrated on it. The second one is the class of all the minimal subspaces $\mathcal{F}$ of a factor space: they are coherent one another, each of them supports its own probability distribution $Q_{\mathcal{F}}$, but the latter are connected by (10.1). Roughly speaking, all the $Q_{\mathcal{F}}$ of one class are coherent and admit superpositions, see Theorem 9.5.

We left unanswered the important question: how must one define Q_j to satisfy the conditions of symmetry. This question is beyond the scope our paper. We shall only outline the answer. One must construct a $\mathbb{C}$-quasi-logic $\mathfrak{L}'$ (a logic of von Neumann), generated by the $\mathbb{R}$-quasi-logic $\mathfrak{L}$ of supports (a Jordan logic), $\mathfrak{L}' \supseteq \mathfrak{L}$. The restriction KQ_jK of the matrix Q_j onto an $\mathfrak{L}'$-minimal subspace $\mathcal{K} \subset \mathcal{F}_j$ will be the desired.

REFERENCES

1 Kolmogorov A.N., Markov chains with denumerable set of possible states, Bull. of Moscow State Univ. (A), 1:3 (1937).

2 Neumann J.V., Mathematische Grundlagen der Quantenmechanik, Berlin, Springer, 1932.

3 Feynman R.P., The Concept of Probability in Quantum Mechanics, Proc. 2nd Berkeley Symp. Math. Statist. Probab., Berkeley, Univ. Cal. Press, 1951, 533-541.

4 Wick G.C., Whigtman A.S., Wigner E.P., The intrinsic parity of elementary particles, Phys. Rev., 88:1 (1952), 101-105.

5 Birkhoff G., Neumann J.V., The logic of quantum mechanics, Ann. Math., 37, (1936), 823-835.

6 Finkelstein D., Jauch J.M., Schiminovich S., Speiser D., Foundation of quaternion quantum mechanics, J. Math. Phys., 3:2 (1962), 207-220.

7 Morozova E.A., Čencov N.N., Probability matrices and stochastic supermatrices, preprint N 84 Inst. Appl. Math. Acad. Sci. USSR, Moscow, 1973.

8 Bellman R., On a generalization of classical probability theory, I, Markoff chains, Proc. Nat. Acad.Sci., USA, 39 (1953), 1075-1077.

9 Wigner E.P., Group Theory and its Application to the Quantum Mechanics of Atomic Spectra, N.Y., Acad. Press, 1959.

10 Dirac P.A.M., The Principles of Quantum Mechanics, 4 ed., Oxford, Clarendon Press, 1958.

11 Chevalley C., Theory of Lie Groups, I, Princeton, 1947.

12 Čencov N.N., Statistical decisions and optimal inferences, Moscow, 1972.

13 Hille E., Phillips R.S., Functional Analysis and Semi-Groups, Amer.Math.Soc.Coll.Publ.,31, Providence, 1957.

14 Morozova E.A., Čencov N.N., Unitary equivariants of a family of subspaces, preprint N 52, Inst.Appl.Math.Acad.Sci.USSR, 1974.

15 Morozova E.A., Čencov N.N., Elementary Jordan logics, preprint N 113, Inst.Appl.Math.Acad.Sci.USSR, 1975.

16 Morozova E.A., Čencov N.N., To the theorem of Jordan-von Neumann-Wigner, preprint Inst.Appl.Math.Acad. Sci.USSR, 1975.

17 Jordan P., von Neumann J.V., Wigner E., On an algebraic generalization of the quantum mechanical formalism, Ann.Math., 35:1 (1934), 29-64.

Department of Mathematics
and Mechanics
Moscow State University
Moscow

Institute of Applied
Mathematics
Academy of Sciences of the
USSR
Moscow

AN ESTIMATE OF THE REMAINDER TERM IN THE MULTIDIMENSIONAL CENTRAL LIMIT THEOREM

S.V.Nagaev

0. Introduction. Let $\xi_1, \xi_2, \dots, \xi_n$ be independent identically distributed random variables in R^K with $E\xi_1 = 0$. Assume that ξ_1 has the unit covariance matrix and $\beta_3 \equiv E|\xi_1|^3 < \infty$ (here and in the sequel $|u|$, $u = (u_1, \dots, u_K)$, denotes the norm of u, i.e. $|u|^2 = \sum_1^K u_i^2$).

Put $\Delta_n = \sup_{A \in \mathfrak{C}} |\Phi(A) - P(\frac{1}{\sqrt{n}} \sum_1^K \xi_i \in A)|$,

where $\mathfrak{C}$ is the class of convex Borel sets, Φ is the standard Gaussian law.

It follows from V.V. Sazonov's results [1] that

$$\Delta_n < CK^{5/2} \beta_3 n^{-1/2},$$

where C is an absolute constant.

We strengthen this estimate.

Theorem. There exists an absolute constant C such that

$$\Delta_n < CK\beta_3 n^{-1/2}. \tag{0}$$

Naturally the question arises whether the last estimate is the best one in the sense of dependence on the dimension K.

We shall show that, for any K, there exists a sequence $\xi_1, \xi_2, \dots, \xi_n, \dots$ in R^K such that, for sufficiently large n,

$$\Delta_n > C_0 \frac{E|\xi_1|^3}{\sqrt{n}}$$

where the constant C_0 does not depend on K. It means that the estimate (0) is rather close to the optimal one.

Let $\eta_1, \eta_2, \dots, \eta_n$ be independent identically distributed random variables in R^1 and

$$V(y) = P(\eta_1 < y).$$

Assume that

$$V(y) = \Phi\left(\frac{y+a}{\sigma}\right)(1-p) + p\delta(y-x_0),$$

where $\delta(y)$ is the indicator of the half-line $\{y : y > 0\}$ and parameters x_0, p, a, σ satisfy the following conditions

$$x_0 > 0, \quad p > 0, \quad a > 0, \quad \sigma > 0,$$

$$p x_0 = a(1-p), \quad p x_0^2 = 1/2,$$

$$\frac{1}{2} + (\sigma^2 + a^2)(1-p) = 1.$$

It is not difficult to see that

$$E\eta_1 = 0, \; E\eta_1^2 = 1, \; E\eta_1^3 = \frac{x_0}{1} - (3a\sigma^2 + a^3)(1-p),$$

$$E|\eta_1|^3 \le 2(a+\sigma)^3 + x_0/2, \; E\eta_1^4 \le \frac{1}{2}x_0^2 + 3(a+\sigma)^4.$$

Put

$$F_n(y) = P\left(\frac{1}{\sqrt{n}}\sum_1^n \eta_i < y\right), \qquad v(t) = Ee^{it\eta_1}.$$

According to Osipov's theorem on asymptotic expansions (see, e.g., [4] , p. 197, Theorem 1) (C is an absolute constant, $\delta = 1/12 E|\eta_1|^3$)

$$|F_n(0) - 1/2 - \frac{E\eta_1^3}{6\sqrt{2\pi n}}| \le C(n^{-1/2}\int_{|y|>\sqrt{n}} |y|^3 dV(y) +$$

$$+ n^{-1}E\eta_1^4 + (\sup_{|t|\ge\delta}|v(t)| + \frac{1}{2n})^n n^6),$$

Assume now that $x_0 = \sqrt{n}/\ln n$, $n \ge 3$. Then

$$\int_{|y|>\sqrt{n}} |y|^3 dV(y) = O(e^{-n/4}).$$

Taking into account that, for $x_0 > \frac{2}{3}$,

$$\sigma^2 + a^2 < 2,$$

we have the estimate

$$\eta^{-1} E\eta_1^4 = O(\ln^{-2} n).$$

Further

$$|v(t)| < 1 + (1-p)(e^{-\sigma^2 t^2/2} - 1).$$

Hence

$$|v(t)| < 1 - \tfrac{1}{2}(1 - e^{-1/4})$$

for $|t| \geq 1$, $p \leq 1/2$, $x_0 \geq 2$, and

$$|v(t)| < 1 - \frac{t^2}{4} e^{-1/4}$$

for $|t| \leq 1$, $p \leq 1/2$, $x_0 \geq 2$ $(n \geq 75 \Rightarrow x_0 \geq 2)$.

Consequently

$$\left(\sup_{|t| \geq \delta} |v(t)| + \frac{1}{2n}\right)^n = O\left(\left(1 - \frac{\delta^2}{4} e^{-1/4}\right)^n\right) =$$

$$= O\left(\exp\left\{-\frac{e^{-1/4}}{24} \ln^2 n\right\}\right).$$

Thus

$$F_n(0) - 1/2 - \frac{E\eta_1^3}{6\sqrt{2\pi n}} = O(\ln^{-2} n).$$

It is not difficult to verify that

$$\frac{1}{\sqrt{n}} E\eta_1^3 > \frac{1}{3\ln n}$$

for sufficiently large n . It means that, for large n ,

$$F_n(0) - 1/2 > \frac{E\eta_1^3}{7\sqrt{2\pi n}} .$$

Since

$$E\eta_1^3 > E|\eta_1|^3 - 3(a+\sigma)^3 ,$$

this leads to the inequality

$$F_n(0) - 1/2 > \frac{E|\eta_1|^3}{8\sqrt{2\pi n}} .$$

which is valid for sufficiently large n .

Now define ξ_i as $(\eta_i, \xi_{i1}, \dots, \xi_{i(k-1)})$, where η_i and $\xi_i' \equiv (\xi_{i1}, \dots, \xi_{i(k-1)})$ are independent and ξ_i' has the distribution Φ in R^{k-1} .

Then

$$E|\xi_1'|^3 = O(k^{3/2}),$$

and

$$E|\xi_1|^3 < 4E|\eta_1|^3 + 4E|\xi_1'|^3 .$$

Hence, for sufficiently large k ,

$$F_n(0) - 1/2 > \frac{E|\xi_1|^3}{33\sqrt{2\pi n}} ,$$

if $k < c_1 n^{1/3} / \ln^{2/3} n$, where c_1 is some constant.

In proving his results, V.V. Sazonov used the method of convolutions. Our method is rather the method of characteristic functions.

The survey on results concerning the multidimensional limit theorem is contained, e.h., in [2] (see also [3]).

<u>1. Notation. Auxiliary results</u>. First make some general remarks concerning our notation.

Symbols x, y, t, h, v, u will denote elements of R^k

The symbol x_j denotes the j-th coordinate of x . The integral

$$\int_{-\infty}^{\infty} \dots \int_{-\infty}^{\infty} f(x)\, dx_1 \dots dx_k$$

will be denoted, for the sake of brevity, by

$$\int f(x)\, dx .$$

The dimension of the space is denoted by k .

We shall denote absolute constants by one and the same symbol C .

$c(\gamma)$ stands for a constant depending on γ .

Using the symbol O , we mean that corresponding constant is an absolute one.

Let $F(u)$ be the distribution function of a random variable

ξ_1, $F_n(u) = F^{*n}(u\sqrt{n})$.

Without loss of generality we may assume that $F(u)$ has the density, $p(u)$.

Let $p_n(u)$ be the density of $F_n(u)$. Put $A_h = \{x : (h,x) \le 1\}$.

Introduce now the following notation

$$R_1(h) = \int_{A_h} e^{(h,x)}\, dF(x),\ m(h) = \int_{A_h} x e^{(h,x)} dF(x)/R_1(h),$$

$$f_h(t) = \int_{A_h} e^{(h+it,x)}\, dF(x)/R_1(h),\ \bar{f}_h(t) = f_h(t) e^{-i(t,m(h))},$$

$$\beta_3(u) = \int |(u,x)|^3\, dF(x).$$

Let

$$R_2(h,\gamma) = \int_{A'_h} e^{\gamma(h,x)} dF(x).$$

where $A'_h = \{x : |x| \le \sqrt{n}\} - A_h$.

We formulate now auxiliary results on which the proof of the main theorem is based.

Lemma 1. If $|h| \le 1/4$, then

$$R_1(h) = \exp\{\frac{|h|^2}{2} + \theta(3\beta_3(h) + 10|h|^4)\},\ |\theta| \le 1.$$

Lemma 2. There exists a constant C such that $\forall t,\ h \in R^k$

$$|(t,m(h)) - (t,h)| < C(|(t,h)||h|^2 + \beta_3^{1/3}(t)\beta_3^{2/3}(h)).$$

Lemma 3. Let $\varphi(\lambda)$, $\lambda \in R^1$, be any real measurable function.

Then $\forall \omega > 0$, $0 < \alpha \le 1$

$$\int_{|t|\le\omega} \beta_3^{\alpha}(t) e^{\varphi(|t|)} dt \le \beta_3^{\alpha} (\int_{|t|\le\omega} e^{\varphi(|t|)} dt)^{1-\alpha} \times (\int_{|t|\le\omega} |t|^3 e^{\varphi(|t|)} dt)^{\alpha}.$$

Lemma 4. For any $0 < \alpha \le 1$ and $\gamma \ge 0$,

$$\int |t|^{\gamma} \beta_3^{\alpha}(t) e^{-|t|^2/2} dt \le C(\gamma)\, k^{\gamma/2} (\sqrt{2\pi})^k \beta_3^{\alpha}.$$

Lemma 5. For any $\alpha_1 \geq 0,\ \alpha_2 \geq 0,\ \gamma_1 \geq 0,\ \gamma_2 \geq 0,\ \omega > 0.$

$$\int\limits_{|t|\leq\omega} |t|^{\gamma_1} \beta_3^{\alpha_1}(t) e^{-|t|^2/2 + \gamma_2 \beta_3^{\alpha_2}(t)}\, dt \leq \beta_3^{\alpha_1} \int |t|^{\gamma_1} |t_1|^{3\alpha_1} e^{-|t|^2/2 + \gamma_2 \beta_3^{\alpha_2} |t_1|^{3\alpha_2}}\, dt +$$

$$+ c(\gamma_1)(2\pi)^{\kappa/2} \kappa^{\gamma_1/2} (\beta_3^{\alpha_1 + m_0 \alpha_2} (e^{\gamma_2} - 1)\delta(m_0) + \beta_3^{\alpha_1}),$$

where $m_0 < \frac{1-\alpha_1}{\alpha_2}$, $m_0 + 1 \geq \frac{1-\alpha_1}{\alpha_2}$, $\delta(j) = 1$ for $j > 0$, $\delta(0) = 0$.

Lemma 6. Let $\gamma \geq 0$ and $\eta < \frac{1}{2}$. Then

$$\int |t|^{\gamma} e^{-|t|^2(1-\eta)/2}\, dt < c(\gamma)(\sqrt{2\pi})^{\kappa} e^{\kappa\eta} \kappa^{\gamma/2}.$$

Lemma 7. For any $\alpha > 0$ and $\gamma > 0$,

$$\int\limits_{|t|\leq\sqrt{\frac{n}{8\alpha}}} |t|^{\gamma} e^{-|t|^2/2 + \alpha\frac{|t|^4}{n}}\, dt < c(\gamma) e^{\frac{\alpha e^4 (\kappa+\gamma)^2}{n}} (2\pi)^{\kappa/2} \kappa^{\gamma/2}.$$

Lemma 8. For $|h| \leq \frac{1}{50\beta_3}$, $\kappa \leq n$,

$$\int\limits_{|t|\leq\omega_0} |f_h^n(\tfrac{t}{\sqrt{n}})|\, dt \leq c(\sqrt{2\pi})^{\kappa},$$

where $\omega_0 = \frac{\sqrt{n}}{50\beta_3}$.

Lemma 9. For $|h| \leq \frac{1}{50\beta_3}$, $\kappa \leq n$,

$$\int\limits_{|t|\leq\omega_0} |\bar{f}_h^n(t) - e^{-|t|^2/2}|\, dt \leq c\mathcal{G}(\beta_3, n, h, \kappa)(2\pi)^{\kappa/2},$$

where

$$\mathcal{G}(\beta_3, n, h, \kappa) = \beta_3 n^{-1/2} + \beta_3^{2/3} \beta_3^{1/3}(h) + \kappa |h|^2.$$

Lemma 10. Let $\gamma > 0$ and $\alpha \geq 2$. Then

$$\int R_2^{\alpha}(\tfrac{x}{\sqrt{n}}, \gamma) e^{-|x|^2/2}\, dx \leq c(\sqrt{2\pi})^{\kappa} 2^{3\alpha}(1 +$$

$$+ (\tfrac{\alpha\gamma}{\sqrt{n}})^{3(\alpha-1)}) e^{\frac{\alpha^2\gamma^2}{2}} \beta_3^{\alpha} n^{-\frac{3}{2}\alpha}.$$

Lemma 11. Let P be the distribution of a random variable η in R^{κ} with independent components and $E\eta = 0$, $\mathcal{B}$ be the σ-algebra of measurable sets in R^{κ}.

Then

$$\sup_{B\in\mathcal{B}} |\Phi * P(B) - \Phi(B)| < cE|\eta|^2.$$

Lemma 12. Let P and P_1 be any distributions in R^k,

$$\alpha = P_1(x : |x| > \varepsilon), \quad \alpha > \frac{1}{2} .$$

Then

$$\sup_{A\in\mathcal{C}} |P(A) - \Phi(A)| \le \frac{1}{2\alpha - 1}\left(\sup_{A\in\mathcal{C}} (P-\Phi) * P_1(A) + \frac{4\varepsilon\alpha}{\sqrt{2\pi}}\right).$$

2. Proof of the main theorem. Without loss of generality we may assume that

(1_1) $k^5 < \sqrt{n}$,

(1_2) $|\xi_i| < \sqrt{n}/16ek^{1/3}$,

(1_3) $k\beta_3/\sqrt{n} < \dfrac{1}{16e}$.

Let

$$H(\lambda) = \frac{10}{33\pi}\left(\frac{\sin\frac{\lambda}{6}}{\frac{\lambda}{6}}\right)^6, \quad \lambda\in R^1,$$

$$q(x) = \prod_1^k H(x_j) .$$

Put

$$q_n(x) = \left(\frac{\omega_0}{\sqrt{k}}\right)^k q\left(\frac{\omega_0 x}{\sqrt{k}}\right),$$

$$q_n^x(u) = q_n(u)\chi^x(u) .$$

where $\chi^x(u)$ be the indicator of the set $\{u : (x,u) \le 1\}$, ω_0 as above stands for $\sqrt{n}/50\beta_3$.

Let further

$$\vartheta_n(t) = \int e^{i(t,u)} q_n(u)\,du .$$

Notice that $\vartheta_n(t) = 0$ if $\max\limits_{1\le j\le k} |t_j| \ge \dfrac{\omega_0}{\sqrt{k}}$.

Consequently

(2) $\vartheta_n(t) = 0$

for $|t| \geq \omega_0$. Obviously

$$P_n(u) = \sum_0^n C_n^j P_{1x}^{*(n-j)} * P_{2x}^{*j}(u),$$

where

$$P_{1x}(u) = (\sqrt{n})^k p(u\sqrt{n}) \chi^x(u) ,$$

$$P_{2x}(u) = (\sqrt{n})^k p(u\sqrt{n}) (1 - \chi^x(u)) .$$

Consequently

(3) $$P_n * q_n^x(u) = \sum_0^n C_n^j P_{1x}^{*(n-j)} * P_{2x}^{*j} * q_n^x(u) .$$

Put

$$P_{jxy}(u) = e^{(u,y)} P_{jx}(u) \quad , \quad j = 1,2 .$$

$$q_{ny}^x(u) = e^{(u,y)} q_n^x(u) .$$

It is not difficult to see that

(4) $$P_{1x}^{*(n-j)} * P_{2x}^{*j} * q_n^x(u) = e^{-(u,y)} P_{1xy}^{*(n-j)} * P_{2xy}^{*j} * q_{ny}^x(u) .$$

Obviously

(5) $$P_{1xx}^{*n} * q_{nx}^x = P_{1xx}^{*n} * (q_n^x - q_n) + P_{1xx}^{*n} * q_n + P_{1xx}^{*n} * (q_{nx}^x - q_n^x) .$$

Further

(6) $$\int e^{i(t,u)} P_{1xx}(u)\,du = (\sqrt{n})^k \int_{(x,u) \leq 1} e^{(x+it,u)} p(u\sqrt{n})\,du =$$

$$\int_{(\frac{x}{\sqrt{n}},u) \leq 1} \exp\{(\frac{x}{\sqrt{n}} + \frac{it}{\sqrt{n}}, u)\} p(u)\,du =$$

$$= R_1(\frac{x}{\sqrt{n}}) f_{x/\sqrt{n}}(\frac{t}{\sqrt{n}}) .$$

In the sequel h will denote $x/\sqrt{n}$.

It is not difficult to see that

(7) $$P_{1xx}^{*n} * (q_n - q_n^x)(x) = \int_{(x,u) \geq 1} P_{1xx}^{*n}(x-u) q_n(u)\,du \leq$$

$$\leq \int P_{1xx}^{*n}(x-u)(x,u)^2 q_n(u)\,du .$$

Using (6) and the inversion formula, we obtain

(8) $$\int P_{1xx}^{*n}(x-u)(x,u)^2 q_n(u)\,du = (2\pi)^{-k} R_1^n(\frac{x}{\sqrt{n}}) \times$$

$$\times \int f^{n}_{x/\sqrt{n}}\left(\frac{t}{\sqrt{n}}\right) v^{(2)}_{nx}(t) e^{-i(t,x)} dt ,$$

where

$$v^{(2)}_{nx}(t) = \int e^{i(t,u)} (x,u)^2 q_n(u)\, du .$$

Obviously

$$v^{(2)}_{nx} = -\left(2 \sum_{i<j} x_i x_j \frac{\partial v_n}{\partial t_j} \frac{\partial v_n}{\partial t_i} + \sum_j \frac{\partial^2 v_n}{\partial t_j^2} x_j^2\right).$$

The following estimates hold

$$\left|\frac{\partial v_n}{\partial t_j}\right| < c k^{1/2} n^{-1/2} \beta_3 , \quad \left|\frac{\partial^2 v_n}{\partial t_j^2}\right| < C k \beta_3^2 / n .$$

On the other hand,

$$2 \sum_{i<j} x_i x_j \frac{\partial v_n}{\partial t_i} \frac{\partial v_n}{\partial t_j} = \left(\sum_i x_i \frac{\partial v_n}{\partial t_i}\right)^2 - \sum_i x_i^2 \left(\frac{\partial v_n}{\partial t_i}\right)^2 .$$

Consequently

$$\left|\sum_{i<j} x_i x_j \frac{\partial v_n}{\partial t_i} \frac{\partial v_n}{\partial t_j}\right| < |x|^2 \left(\max_i \frac{\partial v_n}{\partial t_i}\right) < C |x|^2 k \beta_3^2 / n .$$

Thus

$$(9) \qquad |v^{(2)}_{nx}(t)| < c |x|^2 k \beta_3 / n .$$

From (8), (9) and Lemma 8 it follows that

$$(10) \quad \int p^{*n}_{1xx}(x-u)(x,u)^2 q_n(u)\, du < C k n^{-1} \beta_3^2 |x|^2 R_1^n(h), \quad |x| \le \omega_0 .$$

Inequalities (7) and (10) yield

$$(11) \quad p^{*n}_{1xx} * (q_n - q^x_n)(x) < c(\sqrt{2\pi})^{-k} k \beta_3^2 n^{-1} R_1^n(h) |x|^2 , \quad |x| \le \omega_0 .$$

Further by the Hölder inequality and (10), we get

$$|p^{*n}_{1xx} * (q^x_{nx} - q^x_n)(x)| = \left| \int_{(x,u) \le 1} p^{*n}_{1xx}(x-u)(e^{(x,u)} - 1) q_n(u)\, du \right| \le$$

$$\le e \int_{(x,u) \le 1} |(x,u)|\, p^{*n}_{1xx}(x-u) q_n(u)\, du \le$$

$$\le e\left(\int (x,u)^2 p^{*n}_{1xx}(x-u)\, q_n(u)\, du\right)^{1/2} \left(\int p^{*n}_{1xx}(x-u)\, q_n(u)\, du\right)^{1/2} .$$

Applying (6), the inversion formula and Lemma 8, we conclude that

$$|\int p_{1\alpha x}^{*n}(x-u)\,q_n(u)\,du| =$$

$$= \left|\frac{R_1^n(h)}{(2\pi)^k}\int f_h^n\left(\frac{t}{\sqrt{n}}\right)e^{-i(t,x)}\vartheta_n(t)dt\right| \le C(2\pi)^{-k/2}R_1^n(h).$$

From the two last inequalities and estimate (10) it follows

$$(12)\quad |p_{1\alpha x}^{*n} * (q_{nx}^{x} - q_n^{x})(x)| \le c\beta_3 k^{1/2} n^{-1/2}|x|\,R_1^n(h)(2\pi)^{-k/2},\quad |x| \le \omega_0.$$

Applying (6) and the inversion formula, it is easy to verify that

$$(2\pi)^k p_{1\alpha x}^{*n} * q_n(x) = R_1^n(h)\left(\int \bar f_h^n\left(\frac{t}{\sqrt{n}}\right)\vartheta_n(t)\cdot \exp\{i(t,\sqrt{n}\,m(h)-x)\}dt\right) =$$

$$= R_1^n(h)\int_{|t|\le\omega_0} \exp\{i(t,\sqrt{n}\,m(h)-x)-|t|^2/2\}\,dt +$$

$$+ R_1^n(h)\int \vartheta_n(t)\left(\bar f_h^n\left(\frac{t}{\sqrt{n}}\right)-e^{-|t|^2/2}\right)\exp\{i(t,\sqrt{n}\,m(h)-x)\}\,dt +$$

$$+ R_1^n(h)\int_{|t|\le\omega_0} (\vartheta_n(t)-1)\exp\{i(t,\sqrt{n}\,m(h)-x)-|t|^2/2\}dt = R_1^n(h)(I_1+I_2+I_3).$$

Applying Lemmas 2 and 3, we get

$$I_1 = \int_{|t|\le\omega_0} e^{-|t|^2/2}dt + O\left(\frac{1}{\sqrt{n}}\beta_3^{2/3}(x)\int e^{-|t|^2/2}\beta_3^{1/3}(t)\,dt +\right.$$

$$\left. + \sqrt{n}\,|h|^2\int |(t,h)|\,e^{-|t|^2/2}dt\right) =$$

$$= (2\pi)^{k/2}\left(1 + O\left(\frac{\beta_3^{1/3}\beta_3^{2/3}(x)}{\sqrt{n}} + \sqrt{n}\,|h|^3 + \beta_3\sqrt{\frac{k}{n}}\right)\right).$$

By Lemma 9

$$|I_2| \le C(2\pi)^{k/2}\, g(\beta_3, n, h, k).$$

Taking into account (1_1), we have

$$|I_3| \le c\,\frac{\beta_3 k}{n}\int |t|^2 e^{-|t|^2/2}dt = O\left(\frac{(2\pi)^{k/2}}{\sqrt{n}}\beta_3\right).$$

Thus

$$(13)\quad p_{1\alpha x}^{*n} * q_n(x) = (2\pi)^{-k/2} R_1^n(h)(1 + O(\beta_3^{1/3}\beta_3^{2/3}(x)/\sqrt{n} +$$

$$+\, \mathcal{G}(\beta_3, n, h, \kappa) + \sqrt{n}\,|h|^3 + \beta_3\sqrt{\tfrac{\kappa}{n}}\,)) .$$

From (11) - (13) it follows

$$p_{1xx}^{*n} * q_{nx}^{x}(x) = (2\pi)^{-\kappa/2} R_1^n(h)\big(1 + O(\beta_3^{1/3}\beta_3^{2/3}(x)/\sqrt{n} \;+$$

$$+\, \mathcal{G}(\beta_3, n, h, \kappa) + \sqrt{n}\,|h|^3 + \beta_3\sqrt{\tfrac{\kappa}{n}} + \frac{|x|^2\kappa\beta_3^2}{n} + \frac{|x|\,\kappa^{1/2}\beta_3}{\sqrt{n}}\,)\big), \quad |x| \le \omega_0 .$$

Hence by (4) and Lemmas 3-7 we deduce

$$(14) \quad \int\limits_{|x|\le\omega_0} | p_{1x}^{*n} * q_n^x(x) - e^{-|x|^2/2}/(\sqrt{2\pi})^\kappa \,|\,dx = O\Big(\frac{\kappa\beta_3}{\sqrt{n}}\Big) .$$

Consider now the convolution

$$p_{1x}^{*(n-j)} * p_{2x}^{*j} * q_n^x(u)$$

for $u = x$, $0 < j \le n$. First estimate $p_{1x}^{*m} * q_n^x(u)$.

Applying (4), we obtain

$$(15) \quad \begin{aligned} p_{1x}^{*m} * q_n^x(u) &= e^{-(x,u)} \int\limits_{(x,l)\le 1} p_{1xx}^{*m}(u-v)\, q_{nx}^x(v)\,dv \le \\ &\le e^{(1-(x,u))} \int p_{1xx}^{*m}(u-v)\, q_n(v)\,dv . \end{aligned}$$

On the other hand, by the inversion formula

$$p_{1xx}^{*m} * q_n(u) = (2\pi)^{-\kappa} R_1^m(h) \int\limits_{|t|\le\omega_0} e^{-i(t,u)} f_h^m\Big(\frac{t}{\sqrt{n}}\Big) \vartheta_n(t)\,dt \le$$

$$\le (2\pi)^{-\kappa} R_1^m(h) \int\limits_{|t|\le\omega_0} \Big| f_h^m\Big(\frac{t}{\sqrt{n}}\Big)\Big|\,dt .$$

By Lemma 8

$$\int\limits_{|t|\le\omega_0} \Big| f_h^m\Big(\frac{t}{\sqrt{n}}\Big)\Big|\,dt = \Big(\sqrt{\frac{n}{m}}\Big)^\kappa \int\limits_{|t|\le\omega_0\sqrt{\frac{m}{n}}} \Big| f_n^m\Big(\frac{t}{\sqrt{m}}\Big)\Big|\,dt <$$

$$< c(2\pi)^{\kappa/2}\Big(\sqrt{\frac{n}{m}}\Big)^\kappa , \quad |h| < \frac{1}{50\beta_3} , \quad m \le n .$$

Thus

$$(16) \quad p_{1xx}^{*m} * q_n^x(u) < c(2\pi)^{-\kappa/2} R_1^m(h)\Big(\sqrt{\frac{n}{m}}\Big)^\kappa e^{-(x,u)} .$$

According to Lemma 1

(17) $$R_1^m(h) \le e^{|x|^2/2 + \tau(x,n)} ,$$

where $\tau(x,n) = 3\beta_3(x)\, n^{-1/2} + 10\, \frac{|x|^4}{n}$. From (15) - (17) it follows

(18) $$P_{1x}^{*m} * q_n^x(u) < C(2\pi)^{-K/2} e^{-|x|^2/2} \left(\sqrt{\tfrac{n}{m}}\right)^K \times$$
$$\times \left(e^{2(x,x-u)} + e^{2\tau(x,n)}\right).$$

Estimate ow the integral

$$I_j \equiv \int_{|x| \le \omega_0} P_{1x}^{*(n-j)} * P_{2x}^{*j} * q_n^x(x)\, dx .$$

Evidently

$$P_{1x}^{*(n-j)} * P_{2x}^{*j} * q_n^x(x) =$$
$$= n^{K/2} \int_{\min_{1\le i\le j}(x,u^i) \ge 1} P_{1x}^{*(n-j)} * q_n^x\left(x - \sum_1^j u^i\right) \prod_1^j p(\sqrt{n}\, u^i)\, du^1 \dots du^j .$$

Consequently

(19) $$I_j = \int_{A_1} + \int_{A_2} = I_{1,j} + I_{2,j} ,$$

where

$$A_1 = \{(x, u^1, \dots, u^j) : |x| \le \omega_0 ,\ \min_{1\le i\le j} (x, u^i) \ge 1 ,$$
$$|u_*| \ge K^{1/3} e/j \},\ A_2 = \{(x, u^1, \dots, u^j) : |x| \le \omega_0 ,$$
$$\min_{1 \le i \le j} (x, u^i) \ge 1 ,\ |u_*| < K^{1/3} e/j \},$$

here u_* is any u_ℓ satisfying the condition $u_\ell = \min_{1\le i\le j} |u_i|$.

It is easy to see that

$$I_{1,j} \le n^{K/2} \int_{|u_*| \ge K^{1/3} e/j} \prod_1^j p(\sqrt{n}\, u^i)\, du^1 \dots du^j \int_{(x,u_*) \ge 1} P_{1x}^{*(n-j)} *$$
$$* q_n^x\left(x - \sum_1^j u^i\right) dx .$$

Since

$$P_{1x}^{*(n-j)} * q_n^x(u) \le n^{K/2} \int p^{*(n-j)}(\sqrt{n}(u-v))\, q_n(v)\, dv ,$$

it is clear that

$$\int_{(x,u_*) \ge 1} P_{1x}^{*(n-j)} * q_n^x\left(x - \sum_1^j u^i\right) dx \le 1 .$$

Consequently

$$(20) \qquad I_{1,j} \le P^j\left(\frac{|\xi_1|}{\sqrt{n}} > \kappa^{1/3} e/j\right) \le \left(\frac{j^3 \beta_3}{\kappa e^3 n^{1/2}}\right)^j .$$

Estimate now the integral $I_{2,j}$. According to (18)

$$I_{2,j} \le [c(2\pi)^{-\kappa/2} n^{\kappa/2} \int_{A_2} e^{-|x|^2/2 + 2(x, \sum_1^j u^i)} \prod_1^j p(\sqrt{n}\, u^i) dx\, du^1 \dots du^j +$$

$$(21) \qquad + c(2\pi)^{-\kappa/2} n^{\kappa/2} \int_{A_2} e^{-|x|^2/2 + 2z(x,n)} \prod_1^j p(\sqrt{n}\, u^i) dx\, du^1 \dots du^j] \times$$

$$\times \left(\sqrt{\frac{n}{n-j}}\right)^\kappa = c(2\pi)^{-\kappa/2} \left(\sqrt{\frac{n}{n-j}}\right)^\kappa (I'_{2,j} + I''_{2,j}).$$

Further by (1_2)

$$I'_{2,j} \le \int_{\substack{0 \le |u^i| \le \frac{1}{16 e \kappa^{1/3}} \\ |u_*| < \kappa^{1/3} e/j}} n^{\kappa/2} \prod_1^j p(u^i \sqrt{n}) du^1 \dots du^j \int_{(x, u_*) \ge 1} e^{-|x|^2/2 + 2(x, \sum_1^j u^i)} dx .$$

Clearly

$$\int_{(x,u_*) \ge 1} e^{-|x|^2/2 + 2(x, \sum_1^j u^i)} dx = e^{2|\sum_1^j u^i|^2} \int_{(x,u_*) \ge 1} e^{-\frac{|x - 2\sum_1^j u^i|^2}{}} dx .$$

It is not difficult to verify that

$$(2\pi)^{-\kappa/2} \int_{(x,u_*) \ge 1} e^{-\frac{|x - \sum_1^j u^i|^2}{2}} dx = P\{(\eta + 2\sum_1^j u^i, u_*) \ge 1\},$$

where η is a normal random variable in R^κ with mean 0 and the unit covariance matrix. Evidently

$$(\eta + 2\sum_1^j u^i, u_*) \ge 1 \Rightarrow (\eta, u_*) \ge 1 - 2(\sum_1^j u^i, u_*) .$$

Therefore if $|u_*| < \kappa^{1/3} e/j$, $|u^i| \le 1/16 e \kappa^{1/3}$, then $(\eta, u_*) \ge \frac{7}{8}$

This leads to the estimate

$$e^{2|\sum_1^j u^i|^2} P\{(\eta+2\sum_1^j u^i, u_*) \geq 1\} \leq e^{-1/4|u_*|^2}.$$

Thus

$$e^{2|\sum_1^j u^i|^2} \int_{(x,u_*)\geq 1} e^{-|x|^2/2+2(x,\sum_1^j u^i)} dx \leq (2\pi)^{K/2} e^{-1/4|u_*|^2},$$

if $|u_*| < K^{1/3} e/j$, $|u^i| < 1/16eK^{1/3}$.

It means that

$$(22) \qquad I'_{2,j} \leq (2\pi)^{K/2} E e^{-1/4\xi_*^2},$$

where $\xi_* = \frac{1}{\sqrt{n}} \min_{1\leq i\leq j} |\xi_i|$. From the inequality

$$e^{-1/4\xi_*^2} < j^{\frac{3}{2}(j+1)} \xi_*^{3j}$$

it follows

$$(23) \qquad E e^{-1/4\xi_*^2} \leq j^{\frac{3}{2}(j+1)} E \xi_*^{3j}.$$

Further

$$(24) \qquad E\xi_*^{3j} = j\int_0^\infty z^{3j}(1-Q(z))^{j-1} dQ(z).$$

where

$$Q(z) = P\left(\frac{|\xi_1|}{\sqrt{n}} < z\right).$$

It is not difficult to see that

$$(25) \qquad z^3(1-Q(z)) \leq \frac{\beta_3}{(\sqrt{n})^3}.$$

From (22) - (26) it follows

$$(26) \qquad I'_{2,j} \leq j^{\frac{3}{2}(j+1)} \left(\frac{\beta_3}{n^{3/2}}\right)^j (2\pi)^{K/2}.$$

Estimate now the integral $I''_{2,j}$. It is easy to verify that

$$I''_{2,j} \leq \int_{|x|\leq\omega_0} e^{-|x|^2/2+2\tau(x,n)} P^j((\xi_1,x)\geq\sqrt{n})dx \leq$$

$$\leq \frac{1}{n^{3j/2}} \int_{|x|\leq\omega_0} \beta_3^j(x) e^{-|x|^2/2+2\tau(x,n)} dx \leq$$

$$\leq \frac{1}{n^{3j/2}} \Big(\int\limits_{|x|\leq\omega_0} \beta_3^{2j}(x) e^{-|x|^2/2} \, dx \Big)^{1/2} \Big(\int\limits_{|x|\leq\omega_0} e^{-|x|^2/2 + 4\tau(x,n)} dx \Big)^{1/2} =$$

$$= \frac{1}{n^{3j/2}} \, I_{3,j}^{1/2} \, I_{4,j}^{1/2} .$$

By Lemma 5

$$I_{3,j} \leq \beta_3^{2j} \int\limits_{|x|\leq\omega_0} x_1^{6j} e^{-|x|^2/2} dx \leq (2\pi)^{\frac{K-1}{2}} \beta_3^{2j} \int x_1^{6j} e^{-x_1^2/2} dx_1 =$$

$$= (2\pi)^{\frac{K-1}{2}} \beta_3^{2j} \, \Gamma(3j + \tfrac{1}{2}) \, 2^{3j - \frac{1}{2}} .$$

On the other hand, Lemmas 5 and 7 allow to obtain the estimate

$$I_{4,j} < C(2\pi)^{K/2},$$

Thus

(27) $$I''_{2,j} \leq C(2\pi)^{K/2} 2^{3j/2} ((3j)!)^{1/2} n^{-\frac{3}{2}j} \beta_3^j .$$

From (19), (20), (26) and (27) it follows

(28) $$I_j \leq \Big(\sqrt{\frac{n}{n-j}}\Big)^K \Big[\Big(\frac{j^3 \beta_3}{K e \, n^{3/2}} \Big)^j + C \Big(\frac{2^{3/2} \beta_3}{n^{3/2}} \Big)^j ((3j)!)^{1/2} \Big] .$$

If $j \leq n/2$, then

(29) $$\Big(\frac{n}{n-j} \Big)^K < e^{2Kj/n} \leq e^{2j} .$$

Evidently

(30) $$C_n^j \leq \frac{n^j}{j!}$$

From (28) - (30) it follows by (1_3)

(31) $$\sum_1^K C_n^j \int\limits_{|x|\leq\omega_0} P_{1x}^{*(n-j)} * P_{2x}^{*j} * q_n^{x}(x) dx < \frac{C_K \beta_3}{\sqrt{n}} .$$

Now deduce for the integral

$$I_j \equiv \int\limits_{|x|\leq\omega_0} P_{1x}^{*(n-j)} * P_{2x}^{*j} * q_n^{x}(x) dx$$

another estimate which is more precise than (28) for $j > \kappa$.

Putting in (4) $y = y^j = x/j^2$ and taking into account that

$$p_{1xy^j}(u) \leq p_{1y^jy^j}(u) \quad \text{and} \quad q^{(x)}_{ny^j}(u) \leq q^{y^j}_{ny^j}(u) ,$$

we obtain in view of (16)

$$p_{1x}^{*(n-j)} * p_{2x}^{*j} * q_n^{x}(x) \leq e^{-|x|^2/j^2} p_{1y^jy^j}^{*(n-j)} * p_{2xy^j}^{*j} * q_{ny^j}^{y^j}(x) \leq$$

$$\leq e^{-|x|^2/j^2} \sup_u p_{1y^jy^j}^{*(n-j)} * q_{ny^j}^{y^j}(u) \int p_{2xy^j}^{*j}(u)\, du \leq$$

$$\leq C e^{-|x|^2/j^2} R_1^{n-j}\left(\frac{y^j}{\sqrt{n}}\right) R_2^j(h, j^{-2})(2\pi)^{-\kappa/2}\left(\sqrt{\frac{n}{n-j}}\right)^{\kappa} .$$

We used here the equality

$$\int p_{2xy^j}^{*j}(u)du = [(\sqrt{n})^{\kappa} \int_{(x,u)\geq 1} e^{(u,y^j)} p(u\sqrt{n})du]^j = R_2^j(h, j^{-2}) .$$

By (17),

$$R_1^{n-j}(hj^{-2}) \leq e^{|x|^2/j^4 + \tau(x/j^2, n)} .$$

Thus

$$p_{1x}^{*(n-j)} * p_{2x}^{*j} * q_n^{x}(x) \leq C e^{-|x|^2/2j^2 + \tau(x/j^2, n)} \times R_2^j(h, j^{-2})(2\pi)^{-\kappa/2}\left(\sqrt{\frac{n}{n-j}}\right)^{\kappa} .$$

Hence using the Hölder inequality, we conclude that

$$(2\pi)^{\kappa/2}\left(\sqrt{\frac{n-j}{n^j}}\right)^{\kappa} I_j \leq$$

$$\leq C\left(\int e^{-|x|^2/2j^2 + 2\tau(x/j^2, n)} dx\right)^{1/2}\left(\int e^{-|x|^2/2j^2} R_2^{2j}(h, j^{-2})dx\right)^{1/2} = C(I_j' \, I_j'')^{1/2} .$$

Applying Lemmas 5 and 7, it is not difficult to obtain the estimate

$$I_j' \leq j^{\kappa} \int e^{-|x|^2/2 + \tau(x,n)} dx < C(2\pi)^{\kappa/2} j^{\kappa} .$$

On the other hand, according to Lemma 8

$$I_j'' = j^{\kappa} \int e^{-|x|^2/2} R_2^{2j}(h, j^{-2}) dx \leq$$

$$\leq C(2\pi)^{\kappa/2} 2^{6j} j^{\kappa}\left(\frac{\beta_3}{n^{3/2}}\right)^{2j} .$$

Thus

(32) $$I_j \le C\left(\sqrt{\tfrac{n}{n-j}}\right)^{\kappa} 2^{3j} j^{\kappa} (\beta_3 n^{-3/2})^{j}.$$

Hence in view of (29) and (30),

(33) $$\sum_{j=\kappa+1}^{[\frac{n}{2}]} C_n^j I_j \le C \sum_{j=\kappa+1}^{[\frac{n}{2}]} 2^{3j} e^{2j} j^{\kappa-j} (\beta_3/\sqrt{n})^{j} = O\left(\frac{\beta_3}{\sqrt{n}}\right).$$

On the other hand, by (29) and (32)

$$\sum_{[\frac{n}{2}]+1}^{n-1} C_n^j I_j \le C \sum_{[\frac{n}{2}]+1}^{n-1} 2^{3j} e^{j} n^{\kappa/2} j^{\kappa-j} \left(\frac{\beta_3}{\sqrt{n}}\right)^{j}.$$

Note that

$$n^{\kappa/2}\left(\frac{n}{2}\right)^{\kappa-\frac{n}{2}} < 2^{n/2}\, n^{\frac{3}{2}n^{1/10}-n/2} < C.$$

From two last estimates and the condition (1_3) it follows

(34) $$\sum_{[\frac{n}{2}]+1}^{n-1} C_n^j I_j = O\left(\frac{\beta_3}{\sqrt{n}}\right).$$

In conclusion, estimate the integral

$$I_n \equiv \int_{|x|\le\omega_0} p_{2x}^{*n} * q_n^{x}(x)\,dx .$$

It is easy to see that

$$p_{2x}^{*n} * q_n^{x}(x) \le \sup_u q_n^{x}(u) \int p_{2x}^{*n}(u)\,du \le$$

$$\le C(50\beta_3)^{-\kappa}\left(\frac{n}{\kappa}\right)^{\kappa/2}\left(\frac{\beta_3(x)}{n^{3/2}}\right)^{n}.$$

Since $\beta_3(x) \le \beta_3 |x|^3$, it is clear that

$$\int_{|x|\le\omega_0} \beta_3^n(x)\,dx \le 2^{\kappa}\omega_0^{3n+\kappa}\beta_3^n .$$

From the two last estimates and the conditions (1_1) and (1_3) it follows

(35) $$I_n = O\left(\frac{\beta_3}{\sqrt{n}}\right).$$

Let $S_n = \{x : |x| \leq \omega_0\}$. From (3), (14), (31), (34) and (35) we deduce

(36) $$\sup_{B \in \mathcal{B}} |F_n * Q_n^x(BS_n) - \Phi(BS_n)| < \frac{C K \beta_3}{\sqrt{n}},$$

where $\mathcal{B}$ is the σ-algebra of measurable sets in R^K, Q_n^x is the distribution corresponding to q_n^x.

Evidently

(37) $$p_n * q_n^x(x) = p_n * q_n(x) + p_n * (q_n^x - q_n)(x).$$

Put $z_n(x) = p_n * (q_n - q_n^x)(x)$. It is clear that

$$z_n(x) = \int_{(x,u) \geq 1} p_n(x-u) q_n(u) du \leq \int p_n(x-u)(x,u)^2 q_n(u) du.$$

Further

$$\int p_n(x-u)(x,u)^2 q_n(u) du =$$

$$= \sum_{i,j} x_i x_j \int p_n(x-u) u_i u_j q_n(u) du.$$

It is not difficult to see that

$$\int p_n(x-u) x_i x_j dx = u_i u_j, \quad i \neq j,$$

$$\int p_n(x-u) x_i^2 dx = 1 + u_i^2.$$

Consequently

$$\int dx \int p_n(x-u)(x,u)^2 q_n(u) du =$$

$$= \sum_i \int u_i^2 q_n(u) du + \sum_i \int u_i^4 q_n(u) du +$$

$$+ \sum_{i \neq j} \int u_i^2 u_j^2 q_n(u) du = O\left(\frac{K^2 \beta_3^2}{n}\right).$$

It means that

(38) $$\int z_n(x) dx = O\left(\frac{K \beta_3}{\sqrt{n}}\right).$$

Since $z_n(x) \geq 0$, it follows from (36) - (38) that

$$\sup_{B \in \mathcal{B}} |F_n * Q_n(BS_n) - \Phi(BS_n)| < C\kappa\beta_3/\sqrt{n},$$

$$Q_n(B) = \int_B q_n(u)du.$$

Using the Chebyshev inequality, we have

$$\int_{|x| \geq \omega_0} P_n * q_n(x)dx < C\sqrt{\kappa} \quad _3/\sqrt{n}.$$

Thus

$$(39) \quad \sup_{B \in \mathcal{B}} |F_n * Q_n(B) - \Phi(B)| < \frac{C\kappa\beta_3}{\sqrt{n}}.$$

On the other hand, by Lemma 11

$$(40) \quad \sup |\Phi * Q_n(B) - \Phi(B)| < \frac{C\kappa\beta_3}{\sqrt{n}}.$$

From (39) and(40) we deduce

$$\sup_{B \in \mathcal{B}} (\Phi - F_n) * Q_n(B) < C\frac{\kappa\beta_3}{\sqrt{n}}.$$

To complete the proof it remains to apply Lemma 12.

REFERENCES

1. V.V. Sazonov, On a bound for the rate of convergence in the multidimensional central limit theorem, Proc. Sixth Berkeley Symp. on Math. Stat. and Prob., Berkeley and Los Angeles, University of California Press. Vol. 2, 1971, 563-582.

2. V.V. Sazonov, Sur les estimations de la rapidité de convergence dans le théoreme limite central (cas de dimension finie et infinie). Actes Congrés Intern. Math., 2,

Paris, Gauthier-Villar, 1971, 577-588.

3. V.I. Rotar', A non-uniform estimate of the speed of convergence in the multidimensional central limit theorem, Teoriya Veroyatnostei i ee Primeneniya, 15, 4(1970), 647-665. (Russian)

4. V.V. Petrov, Sums of independent random variables, Moscow, 1972 (Russian)

Mathematical Institute
Academy of Sciences of the USSR
Novosibirsk

A REMARK ON THE NON-LINEAR DIRICHLET PROBLEM OF BRANCHING MARKOV PROCESSES

By Masao NAGASAWA and Koehei UCHIYAMA
TOKYO INSTITUTE OF TECHNOLOGY

1. Let $\mathbf{A}$ be the Dynkin's characteristic operator for a continuous strong Feller process (X_t, P_x) on a locally compact Hausdorff space with a countable open base. We will call the process a base process. Let D be an open subset of the state space and ψ be a bounded measurable function on the boundary ∂D of D. We assume $\|\psi\| \leqq 1$. Then the non-linear Dirichlet problem of a branching Markov process is

$$(1)\qquad \begin{aligned} &\mathbf{A}u(x)+c(x)\Big(\sum_{n=0}^{\infty} q_n(x)u(x)^n - u(x)\Big) = 0 \quad \text{in } D, \\ &u(b) = \psi(b) \quad \text{on } \partial D,^{(*)} \end{aligned}$$

where c is a nonnegative bounded continuous function on D ($c=0$ on D^c) and q_n is a bounded continuous function on D ($q_n=0$ on D^c) satisfying

$$\sum_{n=0}^{\infty} |q_n(x)| = 1, \quad \text{for } x \in D.$$

For simplicity, we assume $q_n(x) \geqq 0$ in the following. Let $\bar{X}_t$ be the stopped process of the base process at the boundary of D. Let $(\mathbf{X}_t, \mathbf{P}_{\mathbf{x}})$ be the branching Markov process on

$$\mathbf{S} = \bigcup_{n=0}^{\infty} \bar{D}^n \cup \{\Delta\},^{(**)}$$

determined by $\bar{X}_t$, c and q_n (cf. [1]).

For a bounded measurable function f on $\bar{D}$, we define $\hat{f}$ on $\mathbf{S}$ by

$$(2)\qquad \begin{aligned} &\hat{f}(\mathbf{x}) = f(x_1)\cdots f(x_n), \quad \text{when } \mathbf{x} = (x_1,\cdots,x_n), \\ &\hat{f}(\delta) = 1, \\ &\hat{f}(\Delta) = 0. \end{aligned}$$

Then if $\|f\| \leqq 1$, $\hat{f}$ is bounded on $\mathbf{S}$.

Let τ be the killing time of the base process by means of the multiplicative functional $\exp(-\int_0^t c(X_s)ds)$ and T be the first hitting time of X_t to the boundary ∂D, and we assume

$$(3)\qquad P_x[T < \tau] \geqq \varepsilon > 0, \quad \text{for all } x \in \bar{D}.$$

(*) The equality should be understood in a suitable sense.

(**) $\bar{D}^n$ is the n-fold cartesian product of $\bar{D}$ and $\bar{D}^0 = \{\delta\}$, where δ is an extra point. Δ is another extra point.

THEOREM. *For a boundary function* ψ, *set*

$$f = \begin{cases} \psi & \text{on } \partial D, \\ 0 & \text{in } D. \end{cases} \tag{4}$$

Then under the assumption (3), *there exists*

$$u(x) = \lim_{t\to\infty} \mathbf{E}_x[\hat{f}(\mathbf{X}_t)], \quad x \in \bar{D}, \tag{5}$$

which is a solution of the non-linear Dirichlet problem (1) *satisfying*

$$\lim_{\substack{x\in D \\ x\to b\in\partial D}} u(x) = \psi(b),$$

if b *is a regular point*(*) *of the boundary* ∂D *and if* ψ *is continuous at* b (cf.[2]).

Instead of (4), let us take

$$f = \begin{cases} \psi & \text{on } \partial D, \\ g & \text{in } D, \end{cases} \tag{6}$$

as an initial data in (5), where g is a measurable function on D with $\|g\| \leqq 1$. If $\|g\| < 1$, the limit in (5) exists and does not depend on the choice of the initial value g on D. In precise, let n_t^D be the number of particles in D at t, then

$$u(x) = \lim_{t\to\infty} \mathbf{E}_x[\hat{f}(\mathbf{X}_t);\ X_s^i \in \partial D \text{ for all } i \text{ or } \mathbf{X}_s=\delta \text{ at some } s<\infty] \tag{7}$$
$$+\lim_{t\to\infty} \mathbf{E}_x[\hat{f}(\mathbf{X}_t);\ n_s^D \uparrow \infty \text{ when } s\uparrow\infty].$$

Therefore, the second term vanishes when $\|g\| < 1$, and the first term does not depend on g. However, if

$$\mathbf{P}_x[n_t^D \uparrow \infty \text{ when } t\uparrow\infty] > 0,$$

at some point $x_0 \in D$, then the second term does not vanish, if we take e.g., $g \equiv 1$.

Take $\psi \equiv 1$, for simplicity. If we take $f_1 \equiv 1$ on $\bar{D}$, then

$$u_1(x) = \lim_{t\to\infty} \mathbf{E}_x[\hat{f}_1(\mathbf{X}_t)] = 1, \quad \text{for all } x \in \bar{D},^{(**)}$$

while if we take $f_0 = 1$ on ∂D (= 0 in D),

$$u_0(x_0) = \lim_{t\to\infty} \mathbf{E}_{x_0}[\hat{f}_0(\mathbf{X}_t)] < 1.$$

u_0 is known to be the extinction probability of the branching Markov process with absorbing boundary (cf.[3],[4]).

(*) The regularity is for the base process.

(**) We assume the branching Markov process does not explode.

2. The solution of the non-linear Dirichlet problem is not unique, as is seen in §1, but how many solution do we have? To discuss the problem, let us take the simplest case of one dimensional branching Brownian motion, $\bar{D} = [-\ell,\ell]$, and $\psi \equiv 1$. The problem in this case is

$$(8)\qquad \begin{aligned} &u''(x) + c(h(u) - u)(x) = 0, \quad x \in (-\ell,\ell), \quad c;\text{constant} > 0,\\ &u(-\ell) = u(\ell) = 1, \end{aligned}$$

where

$$h(u) = \sum_{n=0}^{\infty} q_n u^n, \quad q_n \geq 0, \quad \sum_{n=0}^{\infty} q_n = 1.$$

We assume

$$1 < h'(1)$$

and $h(u)$ is analytic on R^1.

We will prove, for example, when $h(u) = u^2$, <u>there is a critical length</u> $\ell_o = \pi/2\sqrt{c}$, <u>and if</u> $\ell > \ell_o$, <u>the number of solution of</u> (8) <u>is "approximately"</u> (cf. (14) and (15) in precise)

$$n \sim 2[\ell/\ell_o],^{(*)} \qquad \ell \gg \ell_o,$$

where $[a]$ denotes the greatest integer <u>strictly</u> less than a.

To solve (8) put

$$(9)\qquad C(u) = 2c(u - h(u)).$$

Then we can write the equation (8) as

$$(10)\qquad 2u'' = C(u).$$

Introducing

$$f(u) = \int_0^u C(r)dr$$

and taking the value b of $u(x)$ at $u'(x) = 0$ as a parameter, we have

$$(11)\qquad (u')^2 = f(u) - f(b).$$

Therefore the formal solution for (8) is

$$(12)\qquad x = \int^u \frac{du}{\sqrt{f(u)-f(b)}}.$$

Put

$$(13)\qquad F(b) = \int_0^1 \frac{du}{\sqrt{B}}, \qquad B = \frac{f(1+(b-1)u) - f(b)}{(b-1)^2}.$$

LEMMA 1. (1) $F'(b) < 0$, (ii) $F(1) = \dfrac{1}{\sqrt{c(h'(1)-1)}}\dfrac{\pi}{2}$, <u>and</u> (iii) $\lim_{b\to q} F(b) = \infty$, <u>where</u> $0 \leq q < 1$ <u>is the root of</u>

$$h(u) - u = 0.$$

(*) We count the trivial solution $u \equiv 1$.

<u>PROOF</u>. Because

$$B(u,b) = -\int_u^1 f'(1+(b-1)r)(b-1)^{-1}dr = \int_u^1 C(1+(b-1)r)(b-1)^{-1}dr$$

$$= -\int_u^1 dr \int_0^r C'(1+(b-1)s)ds,$$

we have

$$\frac{\partial B}{\partial b} = -\int_u^1 dr \int_0^r C''(1+(b-1)s)\cdot s ds.$$

Since

$$F'(b) = -\frac{1}{2}\int_0^1 \frac{B'}{B^{3/2}} du, \text{ and } \quad C''(u) = -2ch''(u) < 0,$$

we have $B' > 0$ and hence $F'(b) < 0$, proving (i).

Becuase $B(u,1) = c(h'(1)-1)(1-u^2)$, we have

$$F(1) = \frac{1}{\sqrt{c(h'(1)-1)}} \int_0^1 \frac{du}{\sqrt{1-u^2}},$$

proving (ii). Because $C(u) \sim u-q$, $f(u)-f(q) \sim (u-q)^2$. Therefore, $1/\sqrt{B(u,q)} \sim 1/(1-u)$, implying (iii) by monotone convergence theorem.

Fig.1.

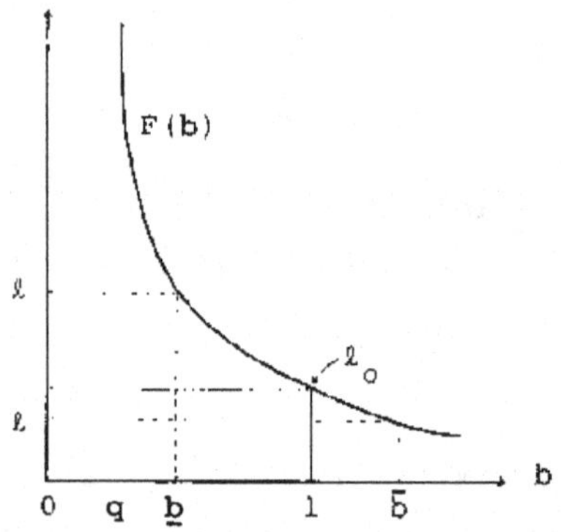

Put

$$\ell_o = \pi/2\sqrt{c(h'(1)-1)},$$

then this is the critical length of the domain as will be seen in the following.

Case 1. When $\ell < \ell_o$, we have the solution $u^0 \geqq 1$ as in Fig.2.

Fig.2.

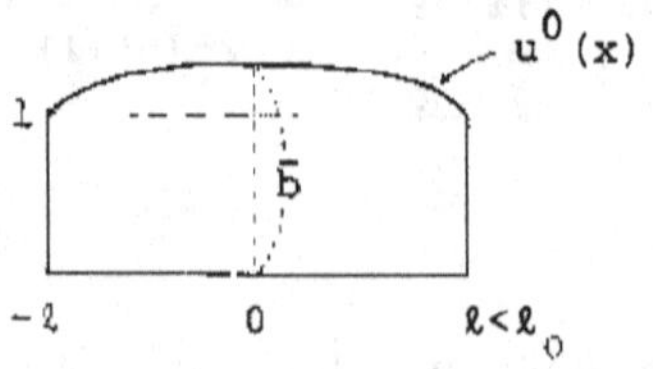

Because

$$\int_1^{\bar{b}} \frac{du}{\sqrt{f(u)-f(\bar{b})}} = F(\bar{b}),$$

to have the solution u^0, $\ell = F(\bar{b})$. It is clear that such $\bar{b} > 1$ exists when and only when $\ell < \ell_o$ (cf.Fig.1).

Case 2. When $\ell = \ell_o$, the solution of (8) is unique, i.e., we have just the trivial solution $u \equiv 1$.

Case 3. When $\ell > \ell_o$, we have the solution u_0 as in Fig.3.

Fig.3.

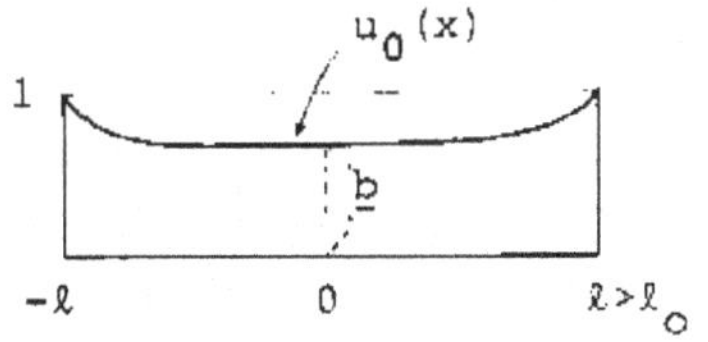

Because

$$\int_{\underline{b}}^{1} \frac{du}{\sqrt{f(u)-f(\underline{b})}} = F(\underline{b}),$$

$\underline{b}$ is determined by $\ell = F(\underline{b})$. It is clear that such $q < \underline{b} < 1$ exists when and only when $\ell > \ell_o$ (cf.Fig.1).

Let $u_0(x) = \underline{b}+\varepsilon$, then

$$|x| = \int_{\underline{b}}^{\underline{b}+\varepsilon} \frac{du}{\sqrt{f(u)-f(\underline{b})}} \longrightarrow \infty, \quad \text{if } \underline{b} \longrightarrow q,$$

where ε is arbitrary and hence $u_0(x) \longrightarrow q$, $x \in R^1$. This is understandable because the efect of absorbing boundary decreases when $\ell \to \infty$ and the extinction probability u_0 converges to that of the branching Brownian motion without boundary.

Now, if the length ℓ of the domain is large enough, we have, for example, the solution of two nodes as in Fig.4.

Fig.4.

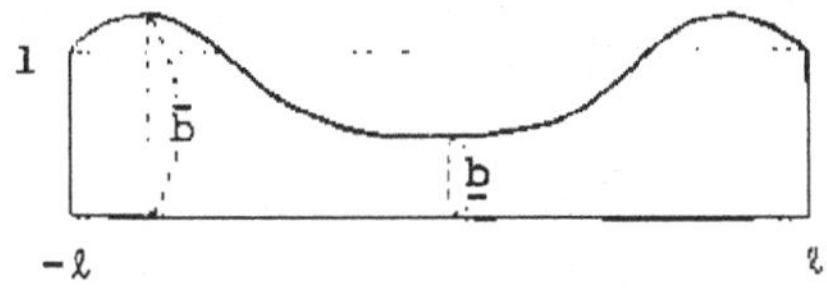

The sufficient condition for existence of such solution is

$$3F(1) < \ell.$$

In general, if

$$(n+1)F(1) < \ell,$$

there exist solutions with nodes up to n. However, it may happen that the solution of the same nodes is not unique for a given length of the domain. For example, take the solution of Fig.4, then

$$F(\underline{b}) + 2F(\bar{b}) = \ell .$$

Put

$$K_3(\underline{b}) = F(\underline{b}) + 2F(\bar{b}).$$

Since $f(\bar{b}) = f(\underline{b})$, $\bar{b}$ is a function of $\underline{b}$, $q < \underline{b} \leq 1$, satisfying

LEMMA 2.

(i) $$0 \geq \frac{d\bar{b}}{d\underline{b}} \geq -1, \quad \text{and} \quad \frac{d\bar{b}}{d\underline{b}} = -1 \quad \text{at } \underline{b} = 1,$$

(ii) $$\bar{b} \to 1, \text{ when } \underline{b} \to 1.$$

PROOF. It is clear that $\bar{b} \to 1$, when $\underline{b} \to 1$. By differentiating $f(\underline{b}) = f(\bar{b})$ with respect to $\underline{b}$, $\bar{b}' = d\bar{b}/d\underline{b} = f'(\underline{b})/f'(\bar{b})$. We have $-f'(\underline{b})/(\bar{b}-\underline{b}) = (f(\bar{b})-f(\underline{b})-f'(\underline{b})(\bar{b}-\underline{b}))/(\bar{b}-\underline{b})^2 \sim (1/2)f''(1)$, and similarly $f'(\bar{b})/(\bar{b}-\underline{b}) \sim (1/2)f''(1)$. Since $f''(1) \neq 0$, we have $\bar{b}' \to -1$, when $\underline{b} \to 1$. To show $\bar{b}' \geq -1$, assume the contrary; $\inf \bar{b}' < -1$. Since $\bar{b}'|_{\underline{b}=q} = 0$, there is a point b_o, $q < b_o < 1$, where $\bar{b}'$ attains the minimum. Since $\bar{b}'' = d^2\bar{b}/d\underline{b}^2 = (f''(\underline{b})-f''(\bar{b})(\bar{b}')^2)/f'(\bar{b})$ and f'' is monotone, it follows that $\bar{b}'' < 0$ at b_o, but this contradicts our setting of b_o.

By Lemma 2 and by the equation

$$K_3'(\underline{b}) = F'(\underline{b}) + 2F'(\underline{b})\frac{d\bar{b}}{d\underline{b}},$$

we have

$$K_3'(1) = -F'(1) > 0.$$

Clearly $\lim_{b \to q} K_3(b) = \infty$, on the other hand. This means if we take ℓ like in Fig.5, we can choose two different values of $\underline{b}$.

Fig.5.

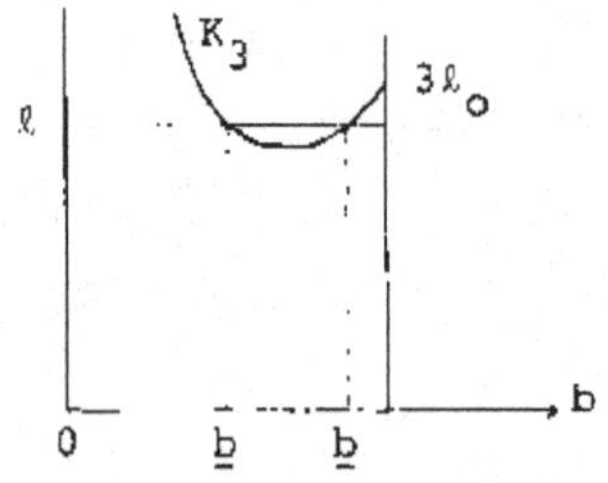

Put $$F_1 = F, \quad F_{2n}(\underline{b}) = nF(\underline{b})+nF(\bar{b}), \quad F_{2n+1}(\underline{b}) = (n+1)F(\underline{b})+nF(\bar{b}),$$

$$K_{2n}(\underline{b}) = F_{2n}(\underline{b}), \quad K_{2n+1}(\underline{b}) = nF(\underline{b})+(n+1)F(\bar{b}).$$

Then, for a given length ℓ of the domain, the number of solutions is the number of crossing of F_n and K_n with ℓ as in Fig.6.(*)

(*) We distinguish  and .

Fig.6.

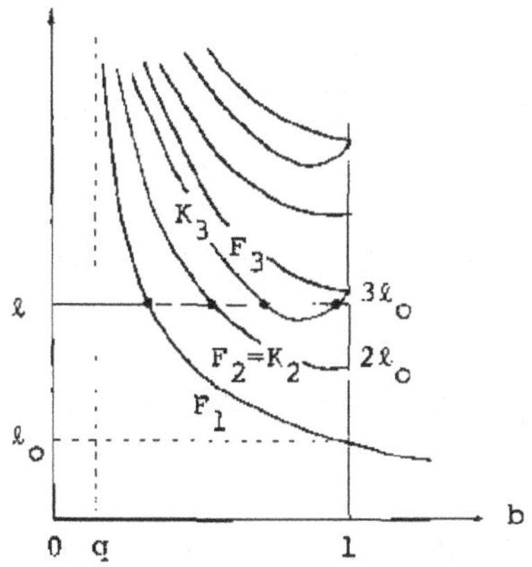

PROPOSITION. <u>The number of solutions is bounded from below as</u>

(14) $$2[\ell/\ell_o] \leqq n, \qquad \ell > \ell_o = \pi/2\sqrt{c(h'(1)-1)}.$$

<u>In the special case of</u> $h(u) = u^2$,

(15) $$2[\ell/\ell_o] \leqq n \leqq 2[\ell/\ell_o]+2, \qquad \ell > \ell_o = \pi/2\sqrt{c},$$

where [a] denotes the greatest integer <u>strictly</u> less than a.

We have the upper bound of (15) because F_n and K_n are convex in this case.

References.

[1] N.Ikeda-M.Nagasawa-S.Watanabe, Branching Markov processes I,II,III, Jounal Math. Kyoto Univ. 8(1968)233-278,365-410,9(1969)95-160.

[2] M.Nagasawa, A probabilistic approach to non-linear Dirichlet problem, Séminaire de Probabilités, Univ. de Strasbourg, to appear.

[3] B.A.Sevast'yanov, Branching stochastic processes for particles diffusing in a bounded domain with absorbing boundaries, Theory of probability and Appl. 3(1958)111-126(English translation).

[4] S.Watanabe, On the branching process for Brownian particles with an absorbing boundary, Journal Math. Kyoto Univ. 4(1965)385-398.

SOME REMARKS ON STOCHASTIC OPTIMAL CONTROLS

Makiko Nisio

Department of Mathematics, Kobe University

§1. Introduction. In the theory of stochastic optimal control of diffusion type, Bellman equation is useful. Under some condition, the solvability of this equation guarantees the existence of an optimal markovian policy, and some responses for a markovian policy are Markov [5]. This fact motivates to introduce a semi-group (non-linear semi-group) of stochastic control of diffusion type.

Let us introduce some preliminary definitions and notations.

We take for the control region Γ the convex σ-compact subset of R^k.

Given a stochastic process $X(t)$, $t \geq 0$, $\sigma_{uv}(X)$ denotes the least σ algebra generated by $\{X(t),\ t \in [u, v]\}$.

Let B be an n-dimensional Brownian motion on a probability space $\Omega = \Omega(B, P)$ and U be B-non-antiapative, i.e., $\sigma_{t+}(B, U) \equiv \bigcap_{\varepsilon>0} \sigma_{0,t+\varepsilon}(B, U)$ is independent of $\sigma_{t\infty}(dB)$, for every $t > 0$. We call U an admissible control, if, for some compact subset $F(= F(U))$ of Γ, $U(t, \omega) \in F$ for any $(t, \omega) \in [0\ \infty) \times \Omega$. To be more precise, the triple (B, U, Ω) is called an admissible system. We denote the set of all admissible systems by A.

Let $\alpha(x, u)$ be a symmetric $n \times n$ matrix, $\gamma(x, u)$ and n-vector, $c(x, u)$ non-negative and $f(x, u)$ real. Hereafter we assume that they are all bounded and continuous on $R^n \times \Gamma$, such that

(1.1) $|h(x, u) - h(y, v)| \leq K|x - y| + \rho(|u - v|)$, $h = \alpha, \gamma, c, f$

where ρ is concave, strictly increasing and continuous on $[0\ \infty)$ and $\rho(0) = 0$.

(1.2) $|h(x, u)| \leq b$.

By virtue of (1.1), the stochastic differential equation

$$dX(t) = \alpha(X(t), U(t))dB(t) + \gamma(X(t), U(t))dt$$
$$X(0) = x$$

has a unique solution X for any admissible system (B, U, Ω). Moreover $X(t)$ is $\sigma_{0t}(B, U)$-measurable. The solution X for (B, U, Ω) is called the respense for U.

Let C be the space of all bounded an uniformly continuous functions on R^n. C is a Banach space with the sup norm, $\|h\| = \sup_{x\in R^n} |h(x)|$.

In §2, we introduce a contractive semi-group on C, whose generator relates Bellman equation. In §3, we recall some results on Bellman equation and the optimal stopping problem will be considered in §4.

§2. Semi-group. We define an operator Q_t, $t \geq 0$, on C as follows,

$$(2.1)\qquad Q_t\psi(x) = \sup_A E_x \int_0^t e^{-\int_0^s c(X(\theta),\, U(\theta))d\theta} f(X(s),\, U(s))ds + e^{-\int_0^t c(X(\theta),\, U(\theta))d\theta}\psi(X(t)),$$

where X is the response for U, and x is the starting point of X. For simplicity, we denote the inside of expectation by $I(t, U, \psi)$, i.e.

$$(2.2)\qquad I(t, U, \psi) = \int_0^t e^{-\int_0^s c(X\ U)} f(X\ U)ds + e^{-\int_0^t c(X\ U)}\psi(X(t)).$$

<u>Prop. 1</u>. $Q_t\psi \in C$, i.e. Q_t is a non-linear operator from C into C.

<u>Proof</u>. Let X and Y be the responses for U starting at x and y respectively. Hereafter K_i stands for a constant independent of an admissible system, starting point and time. By a usual method we can see

$$(2.3)\qquad E|X(t)|^2 \leq K_1 t + |x|$$

$$(2.4)\qquad E|X(t) - Y(t)|^2 \leq |x - y|^2 e^{2K_2 t}$$

and

$$(2.5)\qquad E|X(t) - X(s)|^2 \leq K_3(|t - s| + |t - s|^2).$$

$$\begin{aligned} &E_x I(t, U, \psi) - E_y I(t, U, \psi) \\ &= E\int_0^t \left[e^{-\int_0^s c(X\ U)} f(X\ U) - e^{-\int_0^s c(Y\ U)} f(Y\ U)\right] ds \\ &+ E\left(e^{-\int_0^t c(X\ U)}\psi(X(t)) - e^{-\int_0^t c(Y\ U)}\psi(Y(t))\right) \equiv J_1 + J_2, \end{aligned}$$

$$\begin{aligned} |J_1| &\leq E\int_0^t |f(X(s)U(s)) - f(Y(s)U(s))| ds \\ &+ b\, E\int_0^t \left|e^{-\int_0^s c(X\ U)} - e^{-\int_0^s c(Y\ U)}\right| ds \leq K_4 |x - y| e^{K_2 t}. \end{aligned}$$

For J_2, we remark that for any positive ε, there exists a positive $\delta \equiv \delta(\varepsilon)$, such that

(2.6) $|\psi(z) - \psi(z')| < \varepsilon$ if $|z - z'| < \delta$

Hence, we have

$$E|\psi(X(t)) - \psi(Y(t))| \leq \varepsilon + 2\|\psi\|P(|X(t) - Y(t)| > \delta)$$
$$\leq \varepsilon + 2\|\psi\|E|X(t) - Y(t)|/\delta \leq \varepsilon + 2\|\psi\| \, |x - y|e^{K_2 t}/\delta.$$

Since $|J_2| \leq E|\psi(X(t)) - \psi(Y(t))| + \|\psi\|E\left|e^{-\int_0^t c(X\ U)} - e^{-\int_0^t c(Y\ U)}\right|$,

there exists a positive $\Delta = \Delta(\varepsilon, t, \psi)$, such that

(2.7) $|J_1 + J_2| < \varepsilon$ for $|x - y| < \Delta$.

Therefore, we have, for $|x - y| < \Delta$,

$$|Q_t\psi(x) - Q_t\psi(y)| \leq \sup_A |E_x I(t, U, \psi) - E_y I(t, U, \psi)| < \varepsilon,$$

namely $Q_t\psi$ is uniformly continuous.

On the other hand

$$|Q_t\psi(x)| \leq bt + \|\psi\|.$$

Hence $Q_t\psi \in C$. Q. E. D.

Prop. 2.

(i) $Q_0\psi = \psi$ i.e. $Q_0 =$ identity operator

(ii) continuity in t, $\|Q_t\psi - Q_s\psi\| \to 0$ for $t \to s$.

(iii) contractive, $\|Q_t\psi - Q_t\tilde{\psi}\| \leq \|\psi - \tilde{\psi}\|$.

(iv) monotone, $Q_t\psi(x) \leq Q_t\tilde{\psi}(x) \ \forall x$, if $\psi(x) \leq \tilde{\psi}(x) \ \forall x$.

(v) semi-group, $Q_{t+s}\psi(x) = Q_t(Q_s\psi)(x) = Q_s(Q_t\psi)(x)$.

i.e. $Q_{t+s} = Q_t \circ Q_s = Q_s \circ Q_t$.

Namely, Q_t is a contractive monotone semi-group of C. Semi-group property means the so-called Bellman principle. So, we call Q_t Bellman semi-group.

Proof. (i) and (iv) are clear.

(ii). Recalling (2.5) and (2.6), we have a positive $\Delta = \Delta(\varepsilon, \psi)$, such that, for $|t - s| < \Delta$,

$$|E_x I(t, U, \psi) - E_x I(s, U, \psi)|$$
$$\leq b|t - s| + \|\psi\|b|t - s| + E_x|\psi(X(t)) - \psi(X(s))| < \varepsilon.$$

Hence we have (ii).

(iii) Using a well-known inequality;

$|\sup \xi_\lambda - \sup \eta_\lambda| \leq \sup|\xi_\lambda - \eta_\lambda|$, we see

$$|Q_t\psi(x) - Q_t\tilde{\psi}(x)| \leq \sup_A E_x |I(t, U, \psi) - I(t, U, \tilde{\psi})| \leq \|\psi - \tilde{\psi}\|.$$

This completes the proof of (iii).

Lemma (approximation). $\forall (B, U, \Omega) \in A$, there exists a sequence of admissible system, (B, U_n, Ω), such that U_n has continuous paths and

$$U_n(t, \omega) \longrightarrow U(t, \omega) \qquad \tilde{\forall}(t, \omega).$$

Moreover, its response X_n tends to the response for U, say X, in Prohorov metric on $C^n[0\ \infty)$.

Proof. Let F be a compact set, such that $U(t, \omega) \in F, \forall (t, \omega)$. We may assume that F is convex, because the convex hull of compact set is compact. Define U_n by

$$U_n(t, \omega) = 2^n \int_{t-\frac{1}{2^n}}^{t} U(s, \omega)ds.$$

Then $U_n \in F$ and the first half is clear.

For any $T > 0$, the response on $[0, T]$ is a $C^n[0\ T]$-valued random variable and, by the boundedness of α and γ, $\{X_n\}$ is totally bounded in Prohorov topology on $C^n[0\ T]$.

On the other hand, U_n on $[0\ T]$ can be regarded as a $L^2[0\ T]$-valued random variable. So $U_n(\cdot\ \omega)$ tends to $U(\cdot\ \omega)$ in $L^2[0\ T]$. Hence $\{U_n\}$ is also totally bounded in Prohorov topology on $L^2[0\ T]$. Therefore $\{(B, U_n, X_n)\}$ is also totally bounded. So, on a probability space, we can construct a suitable version of (B, U_n, X_n), say $(\tilde{B}_n, \tilde{U}_n, \tilde{X}_n)$, such that, for some subsequence

$$\begin{aligned} &\tilde{B}_{n'}(t) \longrightarrow \tilde{B}(t) && \text{uniformly on } [0\ T] \\ (2.8)\quad &\tilde{X}_{n'}(t) \longrightarrow \tilde{X}(t) && \text{uniformly on } [0\ T] \\ &\tilde{U}_{n'}(t) \longrightarrow \tilde{U}(t) && \text{in } L^2[0\ T] \end{aligned}$$

with probability 1.

Define $W_k(t; \tilde{U})$ and $W_{k,\ell}(t; \tilde{U})$ by

$$W_k(t; \tilde{U}) = 2^k \int_{t-\frac{1}{2^n}}^{t} \tilde{U}(s)ds$$

and

$$W_{k,\ell}(t; \tilde{U}) = W_k(\tfrac{1}{2^\ell}[2^\ell t]; \tilde{U})$$

where $[\cdot]$ means the integer part of $\cdot$. Then

$$\lim_{k\uparrow\infty} \lim_{\ell\uparrow\infty} E\int_0^T |W_{k,\ell}(t; \tilde{U}) - \tilde{U}(t)|^2 dt = 0.$$

Recalling the definition of U_n, we can take two large integers $k = k(m)$ and $\ell = \ell(m)$, such that

$$(2.9)\quad E\int_0^T |W_{k,\ell}(t;\tilde{U}_{n'}) - \tilde{U}_{n'}(t)|^2 dt < 2^{-m} \qquad \forall n'$$

and

$$E\int_0^T |W_{k,\ell}(t;\tilde{U}) - \tilde{U}(t)|^2 dt < 2^{-m}.$$

Moreover, we may assume

$$(2.10)\quad E\int_0^T |\tilde{X}_{n'}(s) - \tilde{X}_{n'}(2^{-\ell}[2^\ell s])|^2 ds < 2^{-m} \qquad \forall n'$$

and

$$E\int_0^T |\tilde{X}(s) - \tilde{X}(2^{-\ell}[2^\ell s])|^2 ds < 2^{-m}.$$

By virtue of (2.8), we have

$$(2.11)\quad W_k(t, \tilde{U}_{n'}) \longrightarrow W_k(t, U) \qquad \forall t \in [0\ T]$$

with probability 1. Hence

$$(2.12)\quad E\left|\int_0^T \alpha(\tilde{X}(s), \tilde{U}(s))d\tilde{B}(s) - \int_0^T \alpha(\tilde{X}(2^{-\ell}[2^\ell s], W_{k,\ell}(s;\tilde{U}))d\tilde{B}(s)\right|$$

$$\leq K\int_0^t E|\tilde{X}(s) - \tilde{X}(2^{-\ell}[2^\ell s])|^2 ds + E\int_0^t \rho(|\tilde{U}(s) - W_{k,\ell}(s;\tilde{U})|)ds$$

$$\leq K\,2^{-m} + \text{2nd term} \qquad \text{by (2.10).}$$

Since ρ^{-1} is convex and increasing, we see

$$\rho^{-1}(\tfrac{1}{t}\text{ 2nd term}) \leq \frac{1}{t}E\int_0^t \rho^{-1}[\rho(|\tilde{U}(s) - W_{k,\ell}(s;\tilde{U})|)]ds$$

$$= \frac{1}{t}E\int_0^t |\tilde{U}(s) - W_{k,\ell}(s;\tilde{U})|ds \leq \frac{1}{t}2^{-m}, \qquad \text{by (2.9).}$$

Therefore,

$$(2.13)\quad \text{2nd term} \leq t\,\rho(\tfrac{1}{t}2^{-m}).$$

For n', we have the same evaluation. On the other hand, (2.8) and (2.11) tell us that

$$(2.14)\quad \int_0^t \alpha(\tilde{X}_{n'}(2^{-\ell}[2^\ell s]), W_{k,\ell}(s;\tilde{U}_{n'}))dB_{n'}(s)$$

$$\longrightarrow \int_0^t \alpha(\tilde{X}(2^{-\ell}[2^\ell s]), W_{k,\ell}(s;\tilde{U}))d\tilde{B}(s)$$

with probability 1. Combining (2.12),(2.13) and (2.14) for n', we can see, for any $t \geq 0$,

$$\int_0^t \alpha(\tilde{X}_{n'}(s), \tilde{U}_{n'}(s))d\tilde{B}_{n'}(s) \to \int_0^t \alpha(\tilde{X}(s), \tilde{U}(s))d\tilde{B} \qquad \text{in Proba.}$$

For the drift term, the convergence, $\int_0^t \gamma(\tilde{X}_{n'}(s), \tilde{U}_{n'}(s))ds$ $\to \int_0^t \gamma(\tilde{X}(s), \tilde{U}(s))ds$ with proba.1, is clear. So, $\tilde{X}$ is the solution for $(\tilde{U}, \tilde{B})$. This means the latter half of Lemma.

Proof of (v). For $(B, U, \Omega) \in A$, we have a sequence of (B_k, U_k, Ω_k), such that U_n has continuous paths and

$$E_x I(t+s, U, \psi) = \lim_k E_x I(t+s, U_k, \psi).$$

Since (X_k, U_k) has continuous paths, we may assume $\Omega_k = C^{2n}[0\ \infty)$.

(2.15) $E_x I(t+s, U_k, \psi)$

$$= E_x \int_0^t e^{-\int_0^s c(X_k, U_k)} f(X_k, U_k)ds$$

$$+ e^{-\int_0^t c(X_k, U_k)} E\Big(\int_t^{t+s} e^{-\int_t^\theta c(X_k, U_k)} f(X_k, U_k)d + e^{-\int_t^{t+s} c(X_k, U_k)} \psi(X_k(t+s))/\sigma_{t+}(B_k U_k)\Big).$$

Since there exists a regular conditional probability w. r. to $\sigma_{t+}(B_k, U_k)$, $\{B_k(\theta) - B_k(t), U_k(\theta), \theta \geq t\}$ becomes again an admissbile system defined on the condiitonal probability space. Hence the term of the conditional expectation of (2.15) is not bigger than $Q_s\psi(X_k(t))$. Namely we have

(2.16) $E_x I(t+s, U_k, \psi) \leq E_x I(t, U_k, Q_s\psi)$.

Tending k to ∞, we get

(2.17) $E_x I(t+s, U, \psi) \leq E_x I(t, U, Q_s\psi)$.

So,

(2.18) $Q_{t+s}\psi(x) \leq Q_t(Q_s\psi)(x)$.

Now we shall derive the converse inequality. Recalling (2.6) and (2.7), we have a positive $\delta = \delta(\varepsilon)$, such that, for $|x - y| < \delta$,

$$|\psi(x) - \psi(y)| < \varepsilon \qquad \text{and}$$

(2.19) $|E_x I(t, U, \psi) - E_y I(t, U, \psi)| < \varepsilon \quad \forall (B, U, \Omega) \in A$.

We apply a measurable partition $\{A_i\ i = 1, 2, \cdots\}$ on R^n such that, dia $(A_i) < \delta$ and any compact set of R^n can be covered by finitely many A_i. Fix $x_k \in A_k$ arbitrarily. Then we have an ε-optimal system, say (B_k, U_k, Ω_k). i.e.

(2.20) $E_{x_k} I(t, U_k, \psi) \geq Q_t\psi(x_k) - \varepsilon$.

For $y \in A_k$, the following evaluation holds, by (2.19) and (2.20)

$$E_y I(t,\ U_k,\ \psi) \geq E_{x_k} I(t,\ U_k,\ \psi) - \varepsilon \geq Q_t\psi(x_k) - 2\varepsilon \geq Q_t\psi(y) - 3\varepsilon,$$

i.e. $(B_k,\ U_k,\ \Omega_k)$ is a 3ε-optimal system for $y \in A_k$. Fix $(B_0, U_0, \Omega_0) \in A$ arbitrarily and define $(B,\ U,\ \Omega)$ as follows, Ω is the product probability space of $(\Omega_0,\ \Omega_1,\ \Omega_2,\ \cdots)$. Therefore $(B_k,\ U_k),\ k = 0,\ 1,\ 2,\ \cdots$ can be regarded as a independent system defined on Ω. Let X_i be the response for U_i on Ω. Take a compact set A such that $P_x(X_0(s) \in A) > 1 - \varepsilon$, and its covering by $\{A_i\}$, say $A \subset \bigcup_{i=1}^{N} A_i$. We define U and B by

$$U(\theta) = \begin{cases} U_0(s) & \theta < s \\ \sum_{k=1}^{N} U_k(\theta - s)\chi_{A_k}(X_0(s)) + U_{N+1}(\theta - s)\chi_{A^*}(X_0(s)), & \theta \geq s \end{cases}$$

and

$$dB(\theta) = \begin{cases} dB_0(\theta) & \theta < s \\ \sum_{k=1}^{N} dB_k(\theta - s)\chi_{A_k}(X_0(s)) + dB_{N+1}(\theta - s)\chi_{A^*}(X_0(s)), & \theta \geq s \end{cases}$$

where $A^* = \bigcup_{k=N+1}^{\infty} A_k$. Then we can prove that $(B,\ U,\ \Omega)$ is an admissible system and its response X is following

$$X(\theta) = \begin{cases} X_0(\theta) & \theta < s \\ \sum_{k=1}^{N} X_k(\theta - s)\chi_{A_k}(X_0(s)) + X_{N+1}(\theta - s)\chi_{A^*}(X_0(s)), & \theta \geq s \end{cases}$$

where X_k is the response for U_k starting at $X_0(s)$. Therefore

$$\begin{aligned} & E_x I(t+s,\ U,\ \psi) \\ &= E_x \int_0^s e^{-\int_0^\theta c(X_0 U_0)} f(X_0 U_0)\, d\theta \\ &+ e^{-\int_0^s c(X_0 U_0)} \Big[\sum_{k=1}^{N} \chi_{A_k}(X_0(s)) I(t, U_k, \psi) + \chi_{A^*}(X_0(s)) I(t, U_{N+1}, \psi)\Big] \\ &\geq E_x \int_0^s e^{-\int_0^\theta c(X_0 U_0)} f\, d\theta \\ &+ e^{-\int_0^s c(X_0 U_0)} [(Q_t\psi(X_0(s)) - 3\varepsilon)\chi_A(X_0(s)) - (\|f\|\ t + \|\psi\|)\varepsilon]. \end{aligned}$$

Therefore, we have, for any admissible control U_0,

$$Q_{t+s}\psi(x) \geq E_x I(t+s,\ U,\ \psi) \geq E_x I(t,\ U_0,\ Q_s\psi) - K_4\varepsilon.$$

Hence

$$Q_{t+s}\psi(x) \le Q_t(Q_s\psi)(x).$$

This is our wanted inequality. Q. E. D.

Prop. 3. (genrator of Q_t). Define an elliptic differential operator L^u $(u \in \Gamma)$ by

$$(2.21) \quad L^u = a_{ij}(x, u)\frac{\partial^2}{\partial x_i \partial x_j} + \gamma_i(x, u)\frac{\partial}{\partial x_i} - c(x, u)$$

where $a = \frac{1}{2}\alpha^2$.

$$G\psi(x) \equiv \lim_{t \downarrow 0} \frac{1}{t}(Q_t - I)\psi(x)$$

and $\mathcal{D}(G) \equiv \{\psi \in C, \lim \frac{1}{t}(Q_t - I)\psi$ exists in $C\}$.

Then $\mathcal{D}(G) \supset \{\psi \in C;\ \psi_i, \psi_{ij} \in C\} \equiv C^2$ and

$$G\psi(x) = \sup_{u \in \Gamma} L^u\psi(x) + f(x, u) \qquad \text{for } \psi \in C^2,$$

where the subscript i means the derivative w. r. to x_i.

For the operator $A\psi \equiv \sup_{u \in \Gamma} L^u\psi + f(x, u)$, it is still open whether our assumptions (1. 1) and (1.2) imply the conditions of Th. 1 of [1].

Proof. For $\forall (B, U, \Omega) \in A$ and $\forall \psi \in C^2$, a formula of stochastic differential tells us that

$$E_x e^{-\int_0^t c(X,\, U)} \psi(X(t)) - \psi(x) = E_x \int_0^t e^{-\int_0^s c(X,\, U)} L^{U(s)}\psi(X(s))ds.$$

Therefore

$$(2.22) \quad Q_t\psi(x) - \psi(x) = \sup_A [E_x I(t, U, \psi) - \psi(x)]$$

$$= \sup_A E_x \int_0^t e^{-\int_0^s c(X,\, U)} [L^{U(s)}\psi(X(s)) + f(X(s), U(s))]ds.$$

We denote the inside of expectation by $J(t, U, \psi)$ for simplicity.

$$(2.23) \quad |a_{ij}(X(s), U(s))\psi_{ij}(X(s)) - a_{ij}(x, U(s))\psi_{ij}(x)|$$

$$\le |a_{ij}(X(s), U(s)) - a_{ij}(x, U(s))|\,\|\psi_{ij}\| + b|\psi_{ij}(X(s)) - \psi_{ij}(x)|$$

$$\le K\,|X(s) - x|\,\|\psi_{ij}\| + b|\psi_{ij}(X(s)) - \psi_{ij}(x)|.$$

Since $\psi_{ij} \in C$, there exists $\delta = \delta(\varepsilon; \psi) > 0$, such that

$$|\psi_{ij}(x) - \psi_{ij}(y)| < \varepsilon \qquad \text{for } |x - y| < \delta.$$

$$E|\psi_{ij}(X(s)) - \psi_{ij}(s)| \le \varepsilon + 2\|\psi_{ij}\|P_x(|X(s) - x| > \delta)$$

$$\le \varepsilon + 2\|\psi_{ij}\|E_x|X(s) - x|/\delta.$$

We apply the same calculation for γ, c and f, and we have a positive $\Delta = \Delta(\varepsilon, \psi)$, such that for $t < \Delta$,

(2.24) $|E_x J(t, U, \psi) - J(t, U, \psi)_{X=x}| < \varepsilon t.$

On the other hand,

$$M(t) \mathrel{\overline{\overline{\cdot}}} \sup_{A} E_x \int_0^t L^{U(s)}\psi(x) + f(x, U(s))ds$$

$$\leq E_x \int_0^t \sup_{u\in\Gamma}(L^u\psi(x) + f(x, u)ds = \sup_{u\in\Gamma}(L^u\psi(x) + f(x, u))t$$

conversely,

$$M(t) \geq \sup_{u\in\Gamma} E_x \int_0^t (L^u\psi(x) + f(x, u))ds = \sup_{u\in\Gamma}(L^u\psi(x) + f(x, u))t.$$

Hence, we have

(2.25) $$\frac{M(t)}{t} = \sup_{u\in\Gamma} L^u\psi(x) + f(x, u).$$

Combining (2.22) ∿ (2.25), we complete the proof. Q. E. D.

§3. Bellman equation. The section is the résume of Krylov [4], [6] and Nisio [8], with little modifications. Prop.3 of §2 tells us that Bellman equation relates to the semi-group Q_t.

1. Parabolic type.

Let ϕ be in C^2. If $Q_t\phi$ is differentiable in t, then $V(t, x) \mathrel{\overline{\overline{\cdot}}} Q_t\phi(x)$ satisfies (3.1)

(3.1) $$\frac{\partial V}{\partial v} = GV, \qquad V(0, x) = \phi(x).$$

But the condition which guarantees the differentiability is still open. Bellman equation is a little strong form of (3.1), i.e. G becomes the special operator.

(3.2) $$\begin{cases} \frac{\partial V}{\partial t}(t, x) = \sup_{u\in\Gamma} L^u V(t, x) + f(x, u), & \text{a. e. in } (0\ \infty) \times R^n \\ V(0, x) = \phi(x) & \text{on } R^n. \end{cases}$$

This is Bellman equation of parabolic type.

Besides conditions in §1, we assume (A·1) ∿ (A·3),

(A·1) α, γ, c and f are C^2-functions of x, uniformly in u.

(A·2) α is uniformly positive definite, i.e. for a positive μ,

$$\sum_{ij} \alpha_{ij}(x, u)\theta_i\theta_j \geq \mu|\theta|^2, \qquad \forall\ x, u, \theta.$$

(A·3) ϕ is a C^2-function.

By (A·3), we have

$$e^{-\int_0^t c(X, U)}\phi(X(t)) - \phi(x) = \int_0^t e^{-\int_0^s c(X, U)} L^{U(s)}\phi(X(s))ds + \text{martingale}$$

Hence $V(t, x)$ is Lipschitz continuous in t, uniformly in U. We shall show that $V \in W^{1.2}_{p.\mathrm{loc}}$ for any large P. Let ξ be the response for $u(\in \Gamma)$, namely constant control. Then ξ is a diffusion. Fix T arbitrarily and define R_λ by

$$(3.3) \qquad R_\lambda g(s, x) = E_{sx}\int_s^T e^{-\lambda(t-s)-\int_s^t c(\xi,\ u)} g(t, \xi(t))dt$$

for a bounded Borel function g.

Setting $W(t, x) = V(T - t, x)$ and $\tilde{I}(s, t, U, \psi)$

$$= \int_s^t e^{-\int_s^\theta c(X(\tau),\ U(\tau))d\tau} f(X(\theta),\ U(\theta))d\theta + e^{-\int_s^t c(X(\theta),\ U(\theta))d\theta} \psi(t,\ X(t)),$$

we have, by the time homogeneity of coefficients,

$$(3.4) \quad W(s, x) = \sup_A E_{sx}\tilde{I}(s,\ T,\ U,\ \phi)$$

and

$$(3.5) \quad W(s, x) = \sup_A E_{sx}\tilde{I}(s,\ t,\ U,\ W).$$

Moreover, we can calculate

$$|\lambda R_\lambda W(s,\ x) - W(s,\ x)| \leq \frac{K_5}{\lambda} \qquad \forall (s,\ x) \in [0\ T] \times R^n.$$

Hence $g_\lambda \equiv \lambda(\lambda R_\lambda W - W)$ is bounded and some subsequence, say g_{λ_i}, converges weakly in $L_p(\bar{S}_\rho)$, where S_ρ is the open sphere with center 0 and redius ρ. Putting $g = \lim g_{\lambda_i}$, we have

$$W(s, x)$$

$$= E_{sx}\int_s^{\sigma\wedge T} e^{-\int_s^t c(\xi,\ u)} (-g(t,\ \xi(t))dt + e^{-\int_s^{\sigma\wedge T} c(\xi,\ u)} W(\sigma \wedge T,\ \xi(\sigma \wedge T))$$

where σ is the hitting time to $\partial S(\rho)$. On the other hand, the following parabolic equation has a unique solution H which is continuous on $[0\ T] \times \overline{S(\rho)}$ and belongs to $W^{1,2}_{p,\mathrm{loc}}$ in $(0\ T)\times S(\rho)$,

$$\begin{cases} \dfrac{\partial H}{\partial S} + L^u H = g & \text{a.e. in } (0\quad T) \times S(\rho) \\ H = W & \text{on } [0\ T) \times \partial S(\rho) \\ H(T\quad \cdot) = \phi & \text{on } S(\rho). \end{cases}$$

The probabilistic representation of H means "$W = H$". Since $V(t, x) = W(T - t, x)$ in $[0\ T] \times R^n$, V is in $W^{1,2}_{p,\mathrm{loc}}$. Furthermore we can show that V is a unique solution of (3.2).

2. Elliptic Type.

Suppose (A·4) is satisfied.

(A·4) $c(x, u) \geq \lambda \geq 0$.

Then, limit of $Q_t\phi$ exists when $t \to \infty$, and

$$(3.6)\quad \lim_{t\to\infty} Q_t\phi(x) = \sup_A E_x \int_0^\infty e^{-\int_0^t c(X,\, U)} f(X,\, U)dt \equiv V(x).$$

We assume (A·5) besides (A·1) ∿ (A·3).

(A·5) λ of (A·4) is large.

Then $V \in W^2_{p.\mathrm{loc}} \cap C$ for any large p and satisfies Bellman equation on R^n, i.e.

$$(3.7)\quad \sup_{u\in\Gamma}(L^u V(x) + f(x,\, u)) = 0 \qquad \text{a.e. in } R^n.$$

Moreover V is a unique solution of $C \cap W^2_{n.\mathrm{loc}}$.

If Γ is compact, then there exists a Γ-valued Borel function; (Markovian policy), $u(x)$, such that any solution ξ of

$$(3.8)\quad d\xi(t) = \alpha(\xi(t),\, u(\xi(t)))dB(t) + \gamma(\xi(t),\, u(\xi(t)))dt$$

is an optimal trajectory. i.e. $u(x)$ is an optimal Markovian policy.

Let σ be the hitting time of the boundary of the unit sphere D. Define $\tilde{V}$ by

$$(3.9)\quad \tilde{V}(x) = \sup_A E_x \int_0^\sigma e^{-\int_0^t c(X,\, U)} f(X,\, U)dt + e^{-\int_0^\sigma c(X,\, U)} \phi(X(\sigma))$$

where ϕ is a $C_{2,\delta}(\partial D)$ function.

In the general dimension, the regularity of $\tilde{V}$ is still open. So, the solvability of Dirichlet problem of Bellman equation is open. We have the following approximate result. For any $\varepsilon > 0$, the ε-approximate Bellman equation

$$(3.10)\quad \begin{cases} \left|\sup_{u\in\Gamma} L^u W(x) + f(x,\, u)\right| < \varepsilon & \text{a.e. in } D \\ |W - \phi| < \varepsilon & \text{on } \partial D \end{cases}$$

has a solution W in $C(\bar{D}) \cap W^2_{p.D}$ for any large p. Moreover

$$\sup_{x\in D}|W(x) - \tilde{V}(x)| = 0(\varepsilon)$$

and there exists an ε-optimal Markovian policy, u: $D \to \Gamma$ i.e. any solution ξ of (3.8) up to the hitting time σ, gives an ε-optimal value,

$$E_x I(\sigma,\, u(\xi),\, \phi) \geq \tilde{V}(x) - \varepsilon \qquad \forall x \in D.$$

In two 2-demensional case, we can drop the smoothness of coefficients.

Suppose α, γ, c and f are bounded borel and continuous in u, uniformly in x. We assume that c is non-negative and (A·2). Then there exists the unique solution $V \in C_1(D) \cap W^2_{2.D}$ of Bellman equation, for $\phi \in C_{0,\delta}(\partial D)$

$$(3.11) \qquad \begin{cases} \sup L^u V + f(x, u) = 0 & \text{a.e. in } D \\ V = \phi & \text{on } \partial D \end{cases}$$

If Γ is compact, then we can take a Markovian policy $u(\cdot)$, like (3.8). Although, we cannot show the existence of the response for an admissible control, $u(\cdot)$ is optimal in some sense.

§4. Optimal stopping. For $(B, U, \Omega) \in A$, $[0, \infty]$-valued random variable τ on Ω is called (B, U)-stopping time, if $(\tau \leq t) \in \sigma_{t+}(B, U)$ $\forall t \geq 0$. We denote the set of all (B, U)-stopping time by $S(B, U)$.

We assume (A·4). Let ϕ be in C and define $v(x)$ by

$$(4.1) \quad v(x) = \sup_A \sup_{S(B,U)} E_x \int_0^\tau e^{-\int_0^t c(X, U)} f(X, U)dt + e^{-\int_0^\tau c(X, U)} \phi(X(\tau))$$

$$= \sup_A \sup_{S(B,U)} E_x I(\tau, U, \phi).$$

<u>Prof. 4</u>. $v \in C$.

<u>Proof</u>. For $(B, U, \Omega) \in A$, we denote its response starting at x and y by X and Y respectively. For $\tau \in S(B, U)$, we have

$$\left|\int_{\tau \wedge T}^{\tau} e^{-\int^t c(X, U)} f(X, U)dt\right| \leq \int_T^\infty e^{-\lambda t} \|f\| dt = \|f\| e^{-\lambda T}$$

Hence, recalling the calculation of Prop. 1, we can see

$$(4.2) \quad E\left|\int_0^\tau e^{-\int_0^t c(X, U)} f(X, U)dt - \int_0^\tau e^{-\int_0^t c(Y, U)} f(Y, U)dt\right|$$

$$\leq 2\|f\| e^{-\lambda T} + E\int_0^T \left|e^{-\int_0^t c(X, U)} f(X, U) - e^{-\int_0^t c(Y, U)} f(Y, U)\, dt\right| < \varepsilon$$

if $|x - y| < \delta$, where a positive δ does not depend on an admissible system and stopping time. On the other hand, the following evalution is well-known,

$$(4.3) \quad E|X(\tau \wedge T) - Y(\tau \wedge T)|^2 \leq |x - y|^2 e^{2K_2 T} \quad \forall (B,U,\Omega) \in A, \forall \tau \in S(B,U).$$

Hence, we have a positive $\delta = \delta(\varepsilon)$ such that for $|x - y| < \delta$

$$
\begin{aligned}
&E\left|e^{-\int_0^\tau c(X,\ U)}\phi(X(\tau)) - e^{-\int_0^\tau c(Y,\ U)}\phi(Y(\tau))\right| \\
&\leq E\left|e^{-\int_0^\tau c(X,\ U)}\phi(X(\tau)) - e^{-\int_0^{\tau\wedge T} c(X,\ U)}\phi(X(\tau\wedge T))\right| \\
&+ E\left|e^{-\int_0^\tau c(Y,\ U)}\phi(Y(\tau)) - e^{-\int_0^{\tau\wedge T} c(Y,\ U)}\phi(Y(\tau\wedge T))\right| \\
&+ E\left|e^{-\int^{\tau\wedge T} c(X,\ U)}\phi(X(\tau\wedge T)) - e^{-\int_0^{\tau\wedge T} c(Y,\ U)}\phi(Y(\tau\wedge T))\right| \\
&\leq 2\|\phi\|e^{-\lambda T} + \text{the 3rd term} \leq \varepsilon \qquad \forall\,\tau \in S(B,\ U)\forall(B,\ U,\ \Omega) \in A.
\end{aligned}
$$

Namely, $E_x I(\tau,\ U,\ \phi)$ is uniformly continuous in x, uniformly in $(\tau,\ B,\ U,\ \Omega)$. So, v is also uniformly continuous. Since v is bounded, we have Prop. 4. Q. E. D.

According to [2] and [3], $\psi \in C$ is called Q-excessive, if

(4.4) $\quad Q_t\psi(x) \leq \psi(x) \qquad \forall t \geq 0.$

Recalling Prop. 2, we remark that

$$Q_{t+s}\psi(x) \leq Q_t\psi(x) \leq \psi(x) \quad \text{and} \quad Q_t\psi \to \psi, \text{ as } \ t \downarrow 0.$$

Therefore, if (4.4) holds for $t \in [0,\ \delta)$, then it holds at any t.

Prop. 5. v is Q-excessive.

Proof. We shall apply a similar method as Krylov [4]. Put $\gamma_n(\theta) = n\chi_{[t-\frac{i}{n},t]}(\theta)$. Then, for any fixed positive i,

$$\int_0^\infty e^{-\int_0^s c(X,\ U)+\gamma_n(\theta)d\theta} f(X,\ U)ds$$

$$\longrightarrow \int_0^t e^{-\int_0^t c(X,\ U)} f(X,\ U)ds + e^{-i}\int_t^\infty e^{-\int_0^s c(X,\ U)} f(X,\ U)ds$$

and

$$\int_0^\infty e^{-\int_0^s c(X,U)+\gamma_n d\theta}\gamma_n(s)v(X(s)) \to (1 - e^{-i})e^{-\int_0^t c(X,U)}v(X(t)).$$

Tending i to ∞, limit random variable becomes $I(t,\ U,\ v)$. Put $R(B,\ U)$ = set of all bounded process γ such that $\gamma(t)$ is $\sigma_t(B,\ U)$-measurable. Then, according to Krylov,

$$v(x) = \sup_A \sup_{R(B,U)} E_x\int_0^\infty e^{-\int_0^t c(X,\ U)+\gamma d\theta}\big(f(X,\ U) + (t)v(X(t))\big)dt$$

Therefore, we have, $E_x I(t,\ U,\ v) \leq v(x). \qquad \forall\,(B,\ U).$
This implies $Q_t v(x) \leq v(X)$. Q. E. D.

Cor. v is a Q-excessive majorant of ϕ.

Prop. 6. v is the least Q-excessive majorant of ϕ.

Proof. Let S be a Q-excessive majorant of ϕ. Put

$$(4.5) \quad \xi(t) = \int_0^t e^{-\int_0^s c(X,\ U)} f(X,\ U)ds + e^{-\int_0^t c(X,\ U)} S(X(t))$$

$$(= I(t,\ U,\ S)).$$

Then ξ is $\sigma_{t+}(B,\ U)$-super martingale. For the proof, it is enough to show

$$(4.6) \quad E(\xi(t)/\sigma_{s+}(B,\ U)) \leq \xi(s).$$

$$(4.7) \quad E(\xi(t)/\sigma_{s+}(B,\ U)) = \int_0^s e^{-\int_0^\theta c(X,\ U)} f(X,\ U)d\theta$$

$$+ e^{-\int_0^s c(X,\ U)} E\Big(\int_s^t e^{-\int_s^\theta c(X,\ U)} f(X,\ U)d\theta + e^{-\int_s^t c(X,\ U)} S(X(t)/\sigma_{s+}(B,\ U))\Big).$$

Recalling the proof of semi-group property of Q, we have "part of the conditional expectation of (4.7)" $\leq Q_{t-s}S(X(s))$. Therefore, by the assumption of S, we see (4.6), i.e.

$$E(\xi(t)/\sigma_{s+}(B,\ U)) \leq \text{1st term} + e^{-\int_0^s c(X,\ U)} Q_{t-s}S(X(s))$$

$$\leq \text{1st term} + e^{-\int_0^s c(X,\ U)} S(X(s)) = \xi(s).$$

Hence, $\forall \tau \in S(B, U)$, $E_x\xi(\tau) \leq S(x)$.
On the other hand, by virtue of "$\phi(x) \leq S(x)$", we get

$$(4.8) \quad \xi(t) \geq I(t,\ U,\ \phi)$$

So, $\forall (B,\ U,\ \Omega) \in A$. $\quad \tau \in S(B,\ U)$.

$$E_x I(\tau,\ U,\ \phi) \leq E_x\xi(\tau) \leq S(x).$$

This implies $v(x) \leq S(x)$. Q. E. D.

Set $D \equiv \{x \in R^n,\ v(x) > \phi(x)\}$. Then D is open.

Prop. 7. For each x, there exists a small open neighborhood of x, say W_0, such that, for any open neighborhood W of x, $W \subset W_0$, v satisfies the equation,

$$(4.9) \quad v(x) = \max\{\phi(x),\ Q_\sigma c(x)\}$$

where σ is the hitting time to ∂W and Q_σ is defined similarly, i.e.,

$$Q_\sigma v(x)$$

$$= \sup_A E_x \int_0^\sigma e^{-\int_0^s c(X(\theta),\ U(\theta))d\theta} f(X(s),\ U(s))ds + e^{-\int_0^\sigma c(X(\theta),\ U(\theta))d\theta} v(X(\sigma)).$$

Proof. Applying the calculation of Prop. 5, we can show that for any open neighborhood W of x,

(4.10) $v(x) \geq Q_\sigma v(x)$.

Hence, for $x \notin D$, (4.9) holds for any open neighborhood of x. For $x \in D$, we take $\varepsilon \in (0, v(x) - \phi(x))$ and fixed it. An open neighborhood W_0 of x, which contained in $\{y; v(y) > \phi(y) + \varepsilon\}$ is a required one. Indeed Krylov showed $v(x) = Q_\sigma v(x)$. Since $\phi(x) < v(x)$, we have (4.9).

Q. E. D.

Prop. [4]. Under the condition (A·1) $\sim$ (A·4), $v \in C_{1,B} \cap W^2_{p.\mathrm{loc}}$ for any large p, and (v, D) satisfies the following equality-inequality equation;

$$(4.11) \quad \begin{cases} v \geq \phi \quad \text{on } R^n & v = \phi \quad \text{on } D^c \\ \sup\limits_{u \in \Gamma} L^u v(x) + f(x, u) \leq 0 & \text{a. e. in } R^n \\ \sup\limits_{u \in \Gamma} L^u v(x) + f(x, u) = 0 & \text{a. e. in } D \end{cases}$$

Conversely, if $\tilde{v} \in C_{1,B} \cap W^2_{n.\mathrm{loc}}$ and $\tilde{D} \equiv \{x \in R^n; \tilde{v}(x) < \phi(x)\}$ satisfy (4.11), then $\tilde{v} = v$ and $\tilde{D} = D$.

If Γ is compact, then a Borel function $u; D \to \Gamma$ such that $\sup\limits_{u \in \Gamma} L^u v(x) + f(x, u) = L^{u(x)} v(x) + f(x, u(x))$, is an optimal Markovian policy and the hitting time to ∂D is an optimal stopping time.

References

[1] M. G. Crandall and T. M. Liggett, Generation of semi-groups of non-linear transformations on general Banach spaces, Amer. J. Math., 18 (1971), 265-278.

[2] E. B. Dynkin and A. A. Yushkevich, Markov processes, Theorems and Problems, Chap. 3, Plenum Press, 1969 (English transl.).

[3] B. I. Grigelionis and A. N. Shiryaev, On Stefan problem and Optimal stopping rules for Markov processes, Th. Prob. Appl. 11 (1966), 541-558.

[4] N. V. Krylov, Control of a solution of a stochastic integral equation Th. Prob. Appl., 17 (1972), 114-431.

[5] ————, On the selection of a Markov process from a system of processes and the construction of quasi-diffusion processes, Math. USSR. Izv., 7 (1973), 691-708.

[6] ————, On Bellman's equation, Proc. School-Seminar on the theory of random processes, Part I, Vilnius, 1974, 203-235.

[7] H. P. McKean, Stochastic integrals, Acad. Press. 1969.

[8] M. Nisio, Remarks on stochastic optimal controls, Jap. J. Math., 1 (1975), 159-183.

On stationary linear processes with Markovian property

Yasunori Okabe

§1. Introduction.

In this paper we shall discuss the Markovian property of real strictly stationary linear process $\mathbb{X} = (X(t)\ ;\ t \in \mathbb{R})$ on a probability space $(\Omega, \mathbb{F}, P)$ such that $X(t)$ is continuous in the mean and its expectation is zero. Furthermore we suppose that $\mathbb{X}$ is purely non-deterministic in the linear sense. It then follows from [1] that there exists a real L^2-function E vanishing on the negative axis and a temporally homogeneous Lévy process $(Z(t)\ ;\ t \in \mathbb{R})$ such that, for any $t \in \mathbb{R}$,

(1.1) $$X(t) = \frac{1}{\sqrt{2\pi}} \int_{\mathbb{R}} E(t-s)dZ(s)$$

and

(1.2) the closed linear hull of $\{X(s)\ ;\ s < t\}$ is equal to the closed linear hull of $\{Z(s_1)-Z(s_2)\ ;\ s_1, s_2 < t\}$.

For any open set D in $\mathbb{R}$ we define the sub-σ-field $\mathbb{F}(D)$ of $\mathbb{F}$ by

(1.3) $$\mathbb{F}(D) = \sigma(\{X(t)\ ;\ t \in D\})$$

and then for any $t \in R$ three sub-σ-fields $\mathbb{F}^-(t)$, $\mathbb{F}^+(t)$ and $\partial\mathbb{F}(t)$ by

(1.4) $$\mathbb{F}^-(t) = \mathbb{F}((-\infty, t)),\quad \mathbb{F}^+(t) = \mathbb{F}((t, \infty)) \quad\text{and}\quad \partial\mathbb{F}(t) = \bigcap_{\varepsilon>0} \mathbb{F}((t-\varepsilon, t+\varepsilon)).$$

Definition 1.1. We say that $\mathbb{X}$ has the Markovian property if the future field $\mathbb{F}^+(t)$ is independent of the past field $\mathbb{F}^-(t)$ under the condition that the germ field $\partial\mathbb{F}(t)$ is known $(t \in \mathbb{R})$.

We denote by h the Fourier inverse transform of E :

(1.5) $$h(\lambda) = \frac{1}{2\pi} \int_{\mathbb{R}} e^{it\lambda} E(t)dt \qquad (\lambda \in \mathbb{R}).$$

Recently we have proved

Theorem 1.1 ([5]). If h is a reciprocal of an entire function of infra-exponential type, then $\mathbb{X}$ has the Markovian property.

The following Theorem 1.2 is well known.

Theorem 1.2 ([3]). When the Lévy process $(Z(t)\ ;\ t \in \mathbb{R})$ is a Brownian motion, the conditions (i) and (ii) are equivalent :

(i) $\mathbb{X}$ has the Markovian property.

(ii) h is a reciprocal of an entire function of infra-exponential type.

In this paper we shall investigate the process $\mathbb{X}$ for which the Lévy process $(Z(t)\ ;\ t \in \mathbb{R})$ is a Poisson process and a necessary condition in order that X has the Markovian property. The detailed proofs of the theorems stated in this paper will be appear elsewhere.

§2. R. K. H. S. $\mathbb{K}$.

Let k be the correlation function of $\mathbf{X}$:

(2.1) $$k(t-s) = E(X(t)X(s)) \qquad (t,\ s \in \mathbb{R}).$$

It is easy to see

(2.2) $$k = \frac{1}{2\pi} E * \check{E}.$$

We denote by $\mathbb{K}$ the real Hilbert space with k in (2.1) reproducing kernel. Then we have

Proposition 2.1. There exists a unique unitary operator K from $\mathbb{K}$ onto $L^2(\mathbb{R})$ such that

(i) $$u = \frac{1}{\sqrt{2\pi}} E * Ku \qquad (u \in \mathbb{K}),$$

(ii) $$K(k(\cdot - t)) = \frac{1}{\sqrt{2\pi}} E(t - \cdot) \qquad (t \in \mathbb{R}).$$

In [4] we have characterized the Markovian property of $\mathbf{X}$ in terms of the operator K.

Theorem 2.1 ([4]). When the Lévy process $(Z(t)\ ;\ t \in \mathbb{R})$ is a Brownian motion, the conditions (i) and (ii) are equivalent :

(i) $\mathbf{X}$ has the Markovian property.

(ii) The operator K is local in the sense that, for $u \in \mathbb{K}$, if $u = 0$ in (a, b), then $Ku = 0$ in (a, b).

§3. R. K. H. S. $\mathcal{F}$.

We set $\mathcal{D}_T = \{\varphi \in C_0^\infty(\mathbb{R})\ ;\ \varphi(\mathbb{R}) \subset \mathbb{R}\}$ and define a generalized process $X(\varphi)$ by

(3.1) $$X(\varphi) = \int_{\mathbb{R}} X(t)\varphi(t)dt \qquad (\varphi \in \mathcal{D}_T).$$

Let $C(\varphi)$ be the characteristic functional of $X(\varphi)$:

(3.2) $$C(\varphi - \psi) = E(e^{iX(\varphi-\psi)}) \qquad (\varphi, \psi \in \mathcal{D}_T).$$

We denote by $\mathcal{F}$ the complex Hilbert space with $C(\cdot - \psi)$ reproducing kernel. Similarly as in (1.3) and (1.4), we define for any open set D in $\mathbb{R}$ the closed subspace $\mathcal{F}(D)$ of $\mathcal{F}$ by

(3.3) $$\mathcal{F}(D) = \text{the closed linear hull of } \{C(\cdot - \psi)\ ;\ \operatorname{supp} \psi \subset D\}$$

and then for any $t \in \mathbb{R}$ three closed subspaces $\mathcal{F}^-(t)$, $\mathcal{F}^+(t)$ and $\partial\mathcal{F}(t)$ of $\mathcal{F}$ by

(3.4) $$\mathcal{F}^-(t) = \mathcal{F}((-\infty, t)), \quad \mathcal{F}^+(t) = \mathcal{F}((t, \infty)) \quad \text{and} \quad \partial\mathcal{F}(t) = \bigcap_{\varepsilon>0} \mathcal{F}((t-\varepsilon, t+\varepsilon)).$$

Then we can show

Lemma 3.1. There uniquely exists a unitary operator T from $L^2(\Omega, \mathbb{F}(\mathbb{R}), P)$ onto $\mathcal{F}$ such that

(i) $(TY)(\varphi) = E(Y\cdot e^{iX(\varphi)})$,

(ii) $T(L^2(\Omega, \mathbb{F}(D), P)) = \mathcal{F}(D)$,

(iii) $T(L^2(\Omega, \partial\mathbb{F}(t), P)) = \partial\mathcal{F}(t)$.

Using this Lemma 3.1 we can prove

Proposition 3.1. The process X has the Markovian property if and only if, for any $t \in \mathbb{R}$,

$$\partial\mathcal{F}(t) = \mathrm{Proj}_{\mathcal{F}^-(t)} \mathcal{F}^+(t). \tag{3.5}$$

Definition 3.1. We say that the space $\mathcal{F}$ has the Markovian property if condition (3.5) holds.

By virtue of Proposition 3.1, we have reduced the problem of characterizing the Markovian property of the process X to the one of characterizing the Markovian property of the space $\mathcal{F}$.

§4. Decomposition of $\mathcal{F}$.

Since $(Z(t)\ ;\ t \in \mathbb{R})$ is a Poisson process, we have

$$E(e^{i\lambda(Z(t)-Z(s))}) = e^{(t-s)\alpha(\lambda)} \qquad (s < t), \tag{4.1}$$

where $\alpha(\lambda) = e^{i\lambda}-1-i\lambda \quad (\lambda \in \mathbb{R})$.

It is easy to see

$$C(\varphi) = \exp\{\int_{\mathbb{R}} \alpha(\frac{\check{E}*\varphi}{\sqrt{2\pi}}(t))dt\} \qquad (\varphi \in \mathcal{D}_r). \tag{4.2}$$

Lemma 4.1. (i) For any $\xi \in \mathcal{D}_r$, there exists a limit $D_\xi C(\cdot) = \lim_{\varepsilon \downarrow 0} \frac{C(\cdot+\varepsilon\xi)-C(\cdot)}{\varepsilon}$ in $\mathcal{F}$.

(ii) $D_\xi C(\varphi) = C(\varphi)\int_{\mathbb{R}} \alpha'(\frac{\check{E}*\varphi}{\sqrt{2\pi}}(t))\frac{\check{E}*\xi}{\sqrt{2\pi}}(t)dt$ and $\|D_\xi C\|_{\mathcal{F}} = \|\frac{\check{E}*\xi}{\sqrt{2\pi}}\|_{L^2}$.

We define the closed subspaces $\mathcal{F}_0$ and $\mathcal{F}_1$ of $\mathcal{F}$ by

$$\mathcal{F}_0 = \text{the linear hull of } \{C(\cdot)\} \tag{4.3}$$

and

$$\mathcal{F}_1 = \text{the closed linear hull of } \{D_\xi C\ ;\ \xi \in \mathcal{D}_r\}. \tag{4.4}$$

Similarly as in Lemma 4.1, we find that there exists a limit $D_{\xi_1}\cdots D_{\xi_n} C$ in $\mathcal{F}$ for any $\xi_j \in \mathcal{D}_r$ $(j = 1,\cdots, n)$. Then we define inductively the closed subspaces $\mathcal{F}_n$ $(n = 2, 3,\cdots)$ by

$$\mathcal{F}_n = \text{the closed linear hull of } \{\mathrm{Proj}_{(\sum_{k=0}^{n-1} \oplus\mathcal{F}_k)^-} D_{\xi_1}\cdots D_{\xi_n} C\ ;\ \xi_j \in \mathcal{D}_r,\ j = 1,\cdots, n\}. \tag{4.5}$$

Using the same consideration in [2], we have

Proposition 4.1. For any $n \in \mathbb{N}$,

$$\mathcal{F}_n = \{C(\varphi)\int_{\mathbb{R}^n} \alpha'(\frac{\check{E}*\varphi}{\sqrt{2\pi}}(t_1))\cdots\alpha'(\frac{\check{E}*\varphi}{\sqrt{2\pi}}(t_n))f(t_1,\cdots,t_n)dt_1\cdots dt_n\ ;$$

$$f \in L^2(\mathbb{R}^n) \ \text{symmetric}\}$$

and

$$\|C(\cdot)\int_{\mathbb{R}^n} \alpha'(\frac{\check{E}*\cdot}{\sqrt{2\pi}}(t_1))\cdots\alpha'(\frac{\check{E}*\cdot}{\sqrt{2\pi}}(t_n))f(t_1,\cdots,t_n)dt_1\cdots dt_n\|_{\mathcal{F}} = \sqrt{n!}\|f\|_{L^2(\mathbb{R}^n)}.$$

Next we shall obtain the reproducing kernels of the spaces $\mathcal{F}_n$. For any $\varphi,\psi \in \mathcal{D}_r$, we define $k_n(\varphi,\psi)$ by

$$k_n(\varphi,\psi) = C(\varphi)C(-\psi)\frac{1}{n!}\{\int_{\mathbb{R}} \alpha'(\frac{\check{E}*\varphi}{\sqrt{2\pi}}(t))\overline{\alpha'(\frac{\check{E}*\psi}{\sqrt{2\pi}}(t))}dt\}^n. \tag{4.6}$$

Using the following identity

$$\alpha(\lambda-\mu)-\alpha(\lambda)-\alpha(\mu) = \alpha'(\lambda)\overline{\alpha'(\mu)} \qquad (\lambda,\mu \in \mathbb{R}), \tag{4.7}$$

we have

$$C(\varphi-\psi) = \sum_{n=0}^{\infty} k_n(\varphi,\psi) \qquad (\varphi,\psi \in \mathcal{D}_r). \tag{4.8}$$

Then we can show

Lemma 4.2. $k_n(\cdot,\psi)$ is a reproducing kernel of the space $\mathcal{F}_n$.

Since the spaces $\mathcal{F}_n$ are orthogonal, Lemma 4.2 implies that

$$C(\cdot-\psi) = \sum_{n=0}^{\infty} k_n(\cdot,\psi) \ \text{ in } \mathcal{F} \text{ for any } \psi \in \mathcal{D}_r. \tag{4.9}$$

Therefore we obtain

Theorem 4.1. (i) $\mathcal{F} = \sum_{n=0}^{\infty} \mathcal{F}_n$,

(ii) $\langle u, k_n(\cdot,\psi)\rangle_{\mathcal{F}} = (\mathrm{Proj}_{\mathcal{F}_n} u)(\psi) \qquad (u \in \mathcal{F},\ \psi \in \mathcal{D}_r)$.

§5. Markovian property of the spaces $\mathcal{F}_n$.

Similarly as in (3.3) and (3.4), we define the closed subspaces $\mathcal{F}_n(D)$, $\mathcal{F}_n^-(t)$, $\mathcal{F}_n^+(t)$ and $\partial\mathcal{F}_n(t)$. Furthermore we say that the spaces $\mathcal{F}_n$ have the Markovian property if

$$\partial\mathcal{F}_n(t) = \mathrm{Proj}_{\mathcal{F}_n^-(t)} \mathcal{F}_n^+(t) \ \text{ for any } t \in \mathbb{R} \quad (n \in \mathbb{N}_*).$$

We can prove the following important

Lemma 5.1. For any open set D in $\mathbb{R}$ and $\psi_j \in \mathcal{D}_r$ with $\mathrm{supp}\,\psi_j \subset D$ $(1 \le j \le n)$,

$$C(\cdot)\int_{\mathbb{R}} \alpha'(\frac{\check{E}*\cdot}{\sqrt{2\pi}}(t))\overline{\alpha'(\frac{\check{E}*\psi_1}{\sqrt{2\pi}}(t))}dt\cdot\int_{\mathbb{R}} \alpha'(\frac{\check{E}*\cdot}{\sqrt{2\pi}}(t))\overline{\alpha'(\frac{\check{E}*\psi_2}{\sqrt{2\pi}}(t))}dt\cdot\ \cdots$$

$$\cdot \int_{\mathbb{R}} \alpha'(\frac{\check{E}*\cdot}{\sqrt{2\pi}}(t))\overline{\alpha'(\frac{\check{E}*\psi_n}{\sqrt{2\pi}}(t))}dt \in \mathcal{F}_n(D).$$

In particular, $k_n(\cdot, \psi) \in \mathcal{F}_n(D)$ for any $\psi \in \mathcal{A}_r$ with $\operatorname{supp} \psi \subset D$.

By virtue of Lemma 5.1, we find from Theorem 4.1 (1) that

Theorem 5.1. For any open set D in $\mathbb{R}$

$$\mathcal{F}(D) = \sum_{n=0}^{\infty} \mathcal{F}_n(D).$$

Immediately from Theorem 5.1, we have

Theorem 5.2. The space $\mathcal{F}$ has the Markovian property if and only if the space $\mathcal{F}_n$ has the Markovian property for any $n \in \mathbb{N}$.

Remark 5.1. The space $\mathcal{F}_0$ has the Markovian property.

Moreover, using Lemma 5.1 again, we can prove

Theorem 5.3. If the space $\mathcal{F}_1$ has the Markovian property, then every space $\mathcal{F}_n$ has the Markovian property.

Therefore, we see from Theorems 5.2 and 5.3 that the Markovian property of the space $\mathcal{F}$ is equivalent to the one of the space $\mathcal{F}_1$.

§6. Local property of the operator S.

Similarly as in Proposition 2.1, we find from Lemma 4.1 that

Proposition 6.1. There exists a unique unitary operator S form $\mathcal{F}_1$ onto $L^2(\mathbb{R}^1)$ such that

(i) $$u(\varphi) = C(\varphi)\int_{\mathbb{R}} \alpha'(\frac{\check{E}*\varphi}{\sqrt{2\pi}}(t))Su(t)dt,$$

(ii) $$S(D_\xi C) = \frac{\check{E}*\xi}{\sqrt{2\pi}}.$$

Remark 6.1. It is easy to see from (2.2) that $K^{-1}\circ S$ is a unitary operator from $\mathcal{F}_1$ onto $\mathbb{K}$ such that

(6.1) $$(K^{-1}\circ S)(D_\xi C) = k*\xi.$$

Noting that the representation kernel E in (1.1) is canonical, we find from Lemma 2.12 in [4] that

Lemma 6.1. (i) $(K^{-1}\circ S)(\mathcal{F}_1^-(t)) = \mathbb{K}^-(t)$,

(ii) $\mathcal{F}_1^-(t) = \{u \in \mathcal{F}_1 ; Su = 0 \text{ in } (t, \infty)\}$.

Using the following identity

(6.2) $$\alpha'(\lambda+\mu) - \alpha'(\lambda) = e^{i\lambda}\alpha'(\mu) \qquad (\lambda,\mu \in \mathbb{R}),$$

we can show

Lemma 6.2. (i) $\mathcal{F}_1$ is equal to the closed linear hull of $\mathcal{F}_1^-(t) \cup \mathcal{F}_1^+(t)$ $(t \in \mathbb{R})$,

(ii) $\mathcal{F}_1^-(t)$ is equal to the closed linear hull of $\mathcal{F}_1^-(s) \cup \mathcal{F}_1((s, t))$ $(s < t)$,

(iii) $\mathcal{F}_1$ is equal to the closed linear hull of $\mathcal{F}_1^-(s) \cup \mathcal{F}_1((s, t)) \cup \mathcal{F}_1^+(t)$

$(s < t)$.

Similarly as in Theorem 2.1 in [4], we see from Lemma 6.2 that

Theorem 6.1. If the space $\mathcal{F}_1$ has the Markovian property, then the operator S is local in the sense that, for $u \in \mathcal{F}$, if $u(\varphi) = 0$ for any $\varphi \in \mathcal{D}_r$ supp $\varphi \subset (a, b)$, $Su = 0$ in (a, b).

Remark 6.2. We define the closed subspaces $\mathcal{F}^{*-}(t)$ by

(6.3) $$\mathcal{F}^{*-}(t) = \mathcal{F}^-(t) \ominus \mathcal{F}_0 \qquad (t \in \mathbb{R}).$$

Then we can show

(6.4) $$\bigcap_{t \in \mathbb{R}} \mathcal{F}^{*-}(t) = \{0\}.$$

This implies that $\mathbb{X}$ is purely non-deterministic in the nonlinear sense ([2]).

References

[1] T. Hida and N. Ikeda, Note on linear processes, Journal of Mathematics of Kyoto University, 1(1961), 75-86.

[2] T. Hida and N. Ikeda, Analysis on Hilbert space with reproducing kernel arising from multiple Wiener integral, Proceedings of the Fifth Berkeley Symposium on Mathematical Statistics and Probability, Berkeley and Los Angeles, University of California Press, 1966, 117-143.

[3] N. Levinson and H. P. McKean, Jr., Weighted trigonometrical approximation on $\mathbb{R}^1$ with application to the germ field of stationary Gaussian noise, Acta Mathematica, 112(1964), 99-143.

[4] Y. Okabe, Stationary Gaussian processes with Markovian property and M.Sato's hyperfunctions, Japanese Journal of Mathematics, 41(1973), 66-122.

[5] Y. Okabe, On the structure of splitting fields of stationary Gaussian processes with Markovian property, to appear in Journal of Mathematical Society of Japan, 28(1976).

Department of Mathematics
Faculty of Science
Nagoya University,
Nagoya, Japan

SOME LIMIT THEOREMS FOR THE MAXIMUM OF NORMALIZED SUMS OF WEAKLY DEPENDENT RANDOM VARIABLES

Hiroshi Oodaira

1. Introduction. In [2] Darling and Erdős proved the following theorem:

Theorem A. Let $X_1, X_2, \ldots$ be independent random variables with mean 0, variance 1 and a uniformly bounded third absolute moment. Put $S_k = X_1+X_2+\cdots+X_k$ and let $U_n = \max_{1\le k\le n} (S_k/k^{1/2})$. Then

$$\lim_{n\to\infty} P\{ U_n < (2 \log_2 n)^{1/2} + \log_3 n/(2(2 \log_2 n)^{1/2}) + x/(2 \log_2 n)^{1/2} \}$$
$$= \exp(-e^{-x}/2\pi^{1/2}), \quad -\infty < x < \infty,$$

where $\log_2 n = \log\log n$ and $\log_3 n = \log\log\log n$.

The purpose of this note is to remark that Theorem A can be extended to a large class of weakly dependent random variables (Theorem 1), e.g., ϕ-mixing random sequences, and also to prove similar results for certain classes of relatively strongly dependent random variables (Theorems 2 and 3).

Darling-Erdős' method of proof is first to prove Theorem A in the special case of independent standard Gaussian random variables, by converting the problem to finding the limit distribution of the maximum of the Ornstein-Uhlenbeck process, and then to apply the invariance principle of Erdős-Kac.

Our method of proof for dependent random variables is essentially the same as Darling-Erdős' and is described briefly as follows. Recently the almost sure invariance principle has been extended to various classes of weakly dependent random variables, using the techniques of the Skorohod embedding and of martingale approximation due to Gordin (see, e.g., [4], [9]), i.e., for a wide class of weakly dependent random variables X_j, it is shown that

$$(1) \qquad |S_n - W(n)| = O(n^{(1/2)-\varepsilon}) \quad \text{a.s., for some } \varepsilon > 0,$$

where S_n is the partial sum of X_j, which are possibly redefined on a richer probability space, and $W(t)$, $t \ge 0$, is the Brownian motion. The almost sure invariance principle (1) makes it possible to translate immediately fluctuation results, e.g., the iterated logarithm law, from $W(t)$ to S_n (see, e.g., [9]). In the present sutuation it permits us to replace U_n by $\max_{0<c\le s\le n} (W(s)/s^{1/2})$. Then, by a simple time scale change, the limit distribution is shown to be the same as the limit distribution of the maximum of the Ornstein-Uhlenbeck process and the results of Pickands [10] are applied to conclude the proof. The details of proof of this part are given in Section 2.

Notice that the above method suggests a possible generalization of the result to random sequences with stronger dependence. Indeed, the limit theorem of the maximum of the Ornstein-Uhlenbeck process is a very special case of Pickands' result [10], which, in turn, has recently been generalized further ([11], [12]), and the simple relationship between the Brownian motion and the Ornstein-Uhlenbeck process used in the proof holds for the more general class of semi-stable Gaussian processes ([5]). Thus, if we can show, for certain random sequences $\{S_n\}$, that

$$|S_n - X(n)| = O(n^{\beta-\varepsilon}) \quad \text{a.s., for some } \varepsilon > 0, \tag{2}$$

where $X(t)$, $t \geq 0$, is a semi-stable Gaussian process and β is the order of $X(t)$, then a limit theorem similar to Theorem A can be obtained for "relatively strongly dependent" random variables, i.e., random sequences that can be approximated by a semi-stable Gaussian process in the sense of (2). In Sections 3 and 4 we shall give some results in this direction.

2. Limt theorem for weakly dependent random variables. Let

$$A(t) = (2 \log t)^{-1/2}$$

and

$$B(t) = (2 \log t)^{1/2} + (1/2)(2 \log t)^{-1/2}(\log_2 t - \log \pi).$$

We shall prove the following

Theorem 1. Let $X_1, X_2, \ldots$ be random variables with mean 0 and finite $(2+\delta)$-moment, $\delta > 0$, such that $\lim (ES_n^2/n) = 1$, where S_n are partial sums of X_j. Suppose that (1) holds for S_n. Put $U_n = \max_{1\leq k\leq n} (S_k/k^{1/2})$. Then

$$\lim_{n\to\infty} P\{ (A(\log n))^{-1}[U_n - B(\log n)] < x \} = \exp(-e^{-x}/2), \quad -\infty < x < \infty.$$

Proof. First we note that the law of the iterated logarithm holds for $\{X_j\}$ by assumption (1), and hence

Lemma 1. For any x and for almost every (a.e.) ω, there is an integer $N = N(x, \omega)$ such that

$$\max_{1\leq k\leq (\log n)^3} (S_k/k^{1/2}) < B(\log n) + x\cdot A(\log n), \quad \text{for all } n \geq N.$$

As an immediate consequence we obtain

$$\lim_{n\to\infty} P\{ (A(\log n))^{-1}[\max_{1\leq k\leq (\log n)^3} (S_k/k^{1/2}) - B(\log n)] > x \} = 0$$

for all x. It hence suffices to show that

$$\lim_{n\to\infty} P\{ (A(\log n))^{-1}[\max_{(\log n)^3<k\leq n} (S_k/k^{1/2}) - B(\log n)] < x \} = \exp(-e^{-x}/2).$$

Now we have the following lemmas.

Lemma 2. $D_1(n) = (A(\log n))^{-1}\cdot\left|\max_{(\log n)^3<k\le n}(S_k/k^{1/2}) - \max_{(\log n)^3<k\le n} W(k)/k^{1/2}\right|$

$\to 0$ a.s. as $n \to \infty$.

Proof of the lemma. Using (1), we have

$$D_1(n) \le (A(\log n))^{-1} \max_{(\log n)^3<k\le n} |S_k - W(k)|/k^{1/2}$$

$$= O(\log_2 n\cdot(\log n)^{3((1/2)-\varepsilon)}\cdot(\log n)^{-(3/2)})$$

$$= O((\log n)^{-2\varepsilon}).$$

Lemma 3. $D_2(n) = (A(\log n))^{-1}\cdot\left|\max_{(\log n)^3<k\le n}(W(k)/k^{1/2}) - \max_{(\log n)^3<k\le n}(W(t)/t^{1/2})\right|$

$\to 0$ a.s. as $n \to \infty$.

Proof of the lemma.

$$D_2(n) \le (2\log_2 n)^{1/2} \max_{(\log n)^3<k\le n}\ \max_{k-1\le t\le k} |(W(k)/k^{1/2}) - (W(t)/t^{1/2})|$$

$$\le (2\log_2 n)^{1/2} \max_{(\log n)^3<k\le n} \{\max_{k-1\le t\le k} |W(k)-W(t)|/k^{1/2}$$

$$+ \max_{k-1\le t\le k} |W(t)|/(k(k-1))^{1/2}\}.$$

Hence, using the well-known estimate for the tail probabilities of Brownian motion, we obtain, for any $\varepsilon > 0$,

$$P\{ D_2(n) > \varepsilon \} < \text{Const.}\cdot n^{-(\varepsilon/16)\log n + 1}.$$

The lemma is proved by applying the first Borel-Cantelli lemma.

Thus it is enough to prove that

$$\lim_{n\to\infty} P\{(A(\log n))^{-1}[\max_{(\log n)^3<t\le n}(W(t)/t^{1/2}) - B(\log n)] < x\} = \exp(-e^{-x}/2).$$

Let now

$$(3) \qquad Y(t) = e^{-t/2}W(e^t).$$

Then $Y(t)$ is the Ornstein-Uhlenbeck process with covariance function $r(h) = e^{-|h|/2}$. Put $n = e^T$ and $t = e^S$. Then it is sufficient to show that

$$\lim_{T\to\infty} P\{(A(T))^{-1}[\max_{\log T^3<s\le T} Y(s) - B(T)] < x\} = \exp(-e^{-x}/2).$$

We have, by Theorem 5.1 of Pickands [10],

Lemma 4. For any x and for a.e. ω, there is a $T_0 = T_0(x,\omega)$ such that

$$\max_{0\le s\le \log T^3} Y(s) < B(T) + x\cdot A(T), \quad \text{for all } T \ge T_0.$$

Therefore it suffices to prove that

$$\lim_{T\to\infty} P\{(A(T))^{-1}[\max_{0\le s\le T} Y(s) - B(T)] < x\} = \exp(-e^{-x}/2).$$

But this is a special case of Theorem 4.4 of Pickands [10]. This concludes the proof of the theorem.

In [9] Philipp and Stout show that the assumption (1) (the almost sure invariance principle) is satisfied for a large class of weakly dependent random variables, and so Theorem 1 holds for those random sequences. The following corollary is an example.

Corollary. Let $X_1, X_2, \ldots$ be a strictly stationary ϕ-mixing sequence centered at expectations with finite $(2+\delta)$-moments for some $\delta > 0$ and mixing coefficients $\phi(n)$ satisfying $\Sigma_n \phi^{1/2}(n) < \infty$. Assume that $\lim ES_n^2/n = 1$. Then the conclusion of Theorem 1 holds.

3. Limit theorem for a class of non-stationary random sequences. Let $X_0, X_1, X_2, \ldots$ be independent identically distributed (i.i.d.) random variables with mean 0, variance 1 and a finite $(2+\delta)$-moment for some $\delta > 0$. Set

$$S_n = \sum_{k=0}^{n-1} (n-k)^{\alpha} X_k, \quad \alpha \ge 0,$$

$$U_n = \max_{1\le k\le n} (S_k/s_k),$$

where $s_n^2 = ES_n^2$. Let

$$A(t) = (2\log t)^{-1/2},$$

$$B_\alpha(t) = (2\log t)^{1/2} + (2\log t)^{-1/2}\log((1/\pi)(C_\alpha/2)^{1/2}) \qquad \text{for } \alpha > 1/2,$$

where $C_\alpha = (2\alpha+1)/(8(2\alpha-1))$,

$$= (2\log t)^{1/2} - (2\log t)^{-1/2}\log(2\pi\log t)^{1/2}\sigma^{-1}((2\text{ lot } t)^{-1/2}) \qquad \text{for } \alpha = 1/2,$$

where $\sigma(s) = s(2\log(1/s))^{1/2}$,

$$= (2\log t)^{1/2} - (2\log t)^{-1/2}\{((1/2)-(1/(2\alpha+1)))\log_2 t$$
$$- \log((4\pi)^{-1/2} C_\alpha^{1/(2\alpha+1)} H_{2\alpha+1})\} \qquad \text{for } 0 \le \alpha < 1/2,$$

where $0 < C_\alpha = 1 + (2\alpha+1)\int_1^\infty \{u^{2\alpha} - (u^2-1)^\alpha\}du < \infty$,

$$0 < H_\gamma = \lim_{T\to\infty} T^{-1}\int_0^\infty e^s P(\sup_{0<t<T} Z(t) > s)\,ds < \infty,$$

and $Z(t)$ is a Gaussian process with mean $-|t|^\gamma$ and covariance function $|s|^\gamma + |t|^\gamma - |s-t|^\gamma$.

We prove the following limit theorem for U_n.

<u>Theorem 2.</u> $\lim_{n\to\infty} P\{(A(\log n))^{-1}[U_n - B_\alpha(\log n) < x\} = \exp(-e^{-x}), \ -\infty < x < \infty.$

<u>Proof.</u> We may and do assume that X_k are embedded in a Brownian motion $W(t)$, $t \geq 0$, so that

$$\left|\sum_{k=0}^{n-1} X_k - W(n)\right| = o(n^{(1/2)-\varepsilon}) \quad \text{a.s.} \quad \text{for some } \varepsilon > 0,$$

by a theorem of Breiman [1]. Define the Gaussian process $X(t)$ by

$$X(t) = \int_0^t (t-u)^\alpha W(du), \quad t \geq 0.$$

Then we have the following

<u>Lemma 5.</u> $|S_n - X(n)| = o(n^{\alpha+(1/2)-\varepsilon})$ a.s. for some $\varepsilon > 0$.

<u>Proof of the lemma.</u> Put

$$S_n^* = \sum_{k=0}^{n-1} (n-k)^\alpha [W(k+1)-W(k)].$$

Then

$$|S_n - S_n^*| = \left|\sum_{k=1}^{n} \{(n-k+1)^\alpha - (n-k)^\alpha\}[\sum_{j=0}^{k-1} X_j - W(k)]\right| \leq n^\alpha \cdot \max_{1\leq k\leq n} \left|\sum_{j=0}^{k-1} X_j - W(k)\right|$$
$$= o(n^{\alpha+(1/2)-\varepsilon}) \quad \text{a.s.}.$$

Write $X(n) = \sum_{k=0}^{n-1} \int_k^{k+1} (n-u)^\alpha W(du)$. Then it is easily seen that

$$\begin{aligned} E(S_n^* - X(n))^2 &= O(n^{2\alpha-1}) && \text{for } \alpha > 1/2 \\ &= O(\log n) && \text{for } \alpha = 1/2 \\ &= \text{bounded} && \text{for } 0 \leq \alpha < 1/2. \end{aligned}$$

Using the first Borel-Cantelli lemma, we obtain

$$|S_n^* - X(n)| = O(n^{\alpha+(1/2)-\varepsilon}) \quad \text{a.s.},$$

and hence the lemma.

The law of the iterated logarithm holds for $X(t)$ ([7]), and so for S_n, by Lemma 5 (see also [8]). Hence we need only show that

$$(4) \quad \lim_{n\to\infty} P\{(A(\log n))^{-1}[\max_{(\log n)^3 < k \leq n} (S_k/s_k) - B_\alpha(\log n)] < x\} = \exp(-e^{-x})$$

(see Section 2). Note that $s_n^2 \sim (2\alpha+1)^{-1} n^{2\alpha+1}$ and $EX(t)^2 = (2\alpha+1)^{-1} t^{2\alpha+1}$. Using Lemma 5 and Fernique-Marcus-Shepp's results ([3], [6]) on the supremum of Gaussian processes, it can be shown, as in Lemmas 2 and 3, that $\max(S_k/s_k)$ in (4) may be replaced by max $(2\alpha+1)^{1/2}X(t)/t^{\alpha+(1/2)}$, and so it suffices to show that

$$(5)\quad \lim_{n\to\infty} P\{(A(\log n))^{-1}[\max_{(\log n)^3<t\le n}(2\alpha+1)^{1/2}X(t)/t^{\alpha+(1/2)} - B_\alpha(\log n)] < x\} = \exp(-e^{-x}).$$

Now the Gaussian process $X(t)$ is semi-stable of order $\alpha + (1/2)$. If we put

$$Y(t) = (2\alpha+1)^{1/2}e^{-(\alpha+(1/2))t}X(e^t),$$

then $Y(t)$ is a stationary Gaussian process with mean 0 and covariance function

$$r(h) = (2\alpha+1)e^{-h/2}\cdot\int_0^1(1-u)^\alpha(1-ue^{-h})^\alpha du, \quad h \ge 0,$$

and it can be shown that, as $h \downarrow 0$,

$$\begin{aligned} r(h) &= 1 - \{(2\alpha+1)/(8(2\alpha-1))\}h^2 + o(h^2) && \text{for } \alpha > 1/2 \\ &= 1 + (1/4)h^2\log h + o(h^2\log(1/h)) && \text{for } \alpha = 1/2 \\ &= 1 - (1 + C_\alpha)(h/2)^{2\alpha+1} + o(h^{2\alpha+1}) && \text{for } 0 \le \alpha < 1/2. \end{aligned}$$

Put $n = e^T$ and $t = e^s$ in (5). Then (5) is the same as

$$\lim_{T\to\infty} P\{(A(T))^{-1}[\max_{\log T^3<s\le T} Y(s) - B_\alpha(T)] < x\} = \exp(-e^{-x}).$$

As in Lemma 4, Theorem 5.1 of [10] can be applied to obtain

$$\lim_{T\to\infty} P\{(A(T))^{-1}[\max_{0\le s\le \log T^3} Y(s) - B_\alpha(T)] > x\} = 0,$$

so that we have only to show that

$$\lim_{T\to\infty} P\{(A(T))^{-1}[\max_{0\le s\le T} Y(s) - B_\alpha(T)] < x\} = \exp(-e^{-x}).$$

But this follows from Theorem 2.1 of Pickands [11] and Theorem 2.1 of Qualls and Watanabe [12]. The proof of the theorem is complete.

4. Limit theorem for a class of stationary random sequences. Let ξ_k, $k = 0, \pm1, \ldots$ be i.i.d. random variables with mean 0 and variance 1. Let

$$X_j = \xi_{j-1} + 2^{-\alpha}\xi_{j-2} + 3^{-\alpha}\xi_{j-3} + \cdots\cdots, \quad 1/2 < \alpha < 1,$$

$$S_n = X_1 + X_2 + \cdots + X_n, \text{ and}$$

$$U_n = \max_{1\le k\le n}(S_k/s_k),$$

where $s_n^2 = ES_n^2$. Then

Theorem 3. If $E\xi_k^{2i} < \infty$ for some integer $i > 3/(2\alpha-1)$, then

$$\lim_{n\to\infty} P\{(A(\log n))^{-1}[U_n - B_\alpha(\log n)] < x\} = \exp(-e^{-x}), \quad -\infty < x < \infty,$$

where $A(t) = (2 \log t)^{-1/2}$,

$$B_\alpha(t) = (2 \log t)^{1/2} - (2 \log t)^{-1/2}\{((1/2)-(1/(3-2\alpha)))\log_2 t - \log((4\pi)^{-1/2}H_{3-2\alpha})\}$$

and H_γ is the same as in Section 3.

Proof. We can assume that $\xi_0, \xi_1, \xi_2, \ldots.$ and $\xi_{-1}, \xi_{-2}, \ldots.$ are respectively embedded in two independent Brownian motions $W(t)$ and $W'(t)$, $t \geq 0$, so that

$$\Big|\sum_{k=0}^{n-1} \xi_k - W(n)\Big| = O(n^{1/4}(\log n)^{1/2}(\log_2 n)^{1/4}) \quad \text{a.s., and}$$

$$\Big|\sum_{k=-n}^{-1} \xi_k - W'(n)\Big| = O(n^{1/4}(\log n)^{1/2}(\log_2 n)^{1/4}) \quad \text{a.s.,}$$

by Theorem 1.5 of Strassen [13]. Define

$$X(t) = (1-\alpha)^{-1}\int_0^t (t-u)^{1-\alpha}W(du) + (1-\alpha)^{-1}\int_0^\infty\{(t+u)^{1-\alpha}-u^{1-\alpha}\}W'(du), \quad t \geq 0.$$

Then $X(t)$ is a Gaussian process with mean 0 and covariance function $D\{s^\gamma+t^\gamma-|s-t|^\gamma\}$, $s, t \geq 0$, where $\gamma = 3-2\alpha$ and $D = \gamma^{-1} + \int_0^\infty\{(1+v)^{1-\alpha}-v^{1-\alpha}\}^2 dv$.

Lemma 6. $|S_n - X(n)| = o(n^{(\gamma/2)-\varepsilon})$ a.s. for some $\varepsilon > 0$.

Proof of the lemma. Choose $\varepsilon > 0$ sufficiently small so that $\varepsilon < (2\alpha-1)/6$ and $i > (5-4\alpha)/(2\alpha-1-6\varepsilon)$, and then choose a number δ such that $1+(1+2\varepsilon i)/((2\alpha-1)i) < \delta < (6-4\alpha-4\varepsilon)/(5-4\alpha)$. Put $N = n^\delta$ and let

$$T_N = \sum_{k=N}^{\infty} \Big(\sum_{j=1}^{n} (j+k)^{-\alpha}\Big)\xi_{-k}.$$

Then $ET_N^2 = O(n^{2+\delta(1-2\alpha)})$. Since $ET_N^{2i} <$ const.$\cdot(ET_N^2)^i$, we obtain $|T_N| = o(n^{(\gamma/2)-\varepsilon})$, by Chebyshev's inequality and the first Borel-Cantelli lemma. Now let

$$S_n^* = \sum_{k=0}^{n-1} \Big(\sum_{j=k+1}^{n} (j-k)^{-\alpha}\Big)[W(k+1)-W(k)] + \sum_{k=1}^{\infty} \Big(\sum_{j=1}^{n} (j+k)^{-\alpha}\Big)[W'(k)-W'(k-1)]$$

and

$$T_N^* = \sum_{k=N}^{\infty} \Big(\sum_{j=1}^{n} (j+k)^{-\alpha}[W'(k)-W'(k-1)].$$

Then, just as above, $|T_N^*| = o(n^{(\gamma/2)-\varepsilon})$ a.s., and

$$\begin{aligned}|(S_n-T_N)-(S_n^*-T_N^*)| &\leq \Big\{\sum_{j=1}^{n} j^{-\alpha}\Big\}\cdot \max_{1\leq k\leq n}\Big|\sum_{j=0}^{k-1}\xi_j - W(k)\Big| \\ &\quad + \Big\{\sum_{j=1}^{N-1}\big((1+j)^{-\alpha}-(n+1+j)^{-\alpha}\big)+\sum_{j=1}^{n}(j+N)^{-\alpha}\Big\}\cdot \max_{1\leq k\leq N}\Big|\sum_{j=1}^{k}\xi_{-j}-W'(k)\Big| \\ &= O(n^{(5/4)-\alpha}(\log n)^{1/2}(\log_2 n)^{1/4}) \\ &\quad + O(n^{\delta((5/4)-\alpha)}(\log n^\delta)^{1/2}(\log_2 n^\delta)^{1/4}) \\ &= o(n^{(\gamma/2)-\varepsilon}).\end{aligned}$$

Since $E(S_n^* - X(n))^2$ are bounded, we have $|S_n^* - X(n)| = o(n^{(\gamma/2)-\varepsilon})$ a.s.. The lemma is proved.

The rest of the proof is quite similar to that of Theorem 2. We need only note that if we put $Y(t) = X(e^t)/(2De^{\gamma t})^{1/2}$, then $Y(t)$ is a stationary Gaussian process with mean 0 and covariance function

$$r(h) = (1/2)\{e^{\gamma h/2} + e^{-(\gamma h/2)} + |e^{h/2} - e^{-h/2}|^{\gamma}\}, \quad h \geq 0,$$
$$= 1 - (1/2)h^{\gamma} + o(h^{\gamma}) \quad \text{as } h \downarrow 0.$$

Theorem 3 follows from Theorem 5.1 of [10] and Theorem 2.1 of [11].

References

[1] L. Breiman: On the tail behavior of sums of independent random variables, Z. Wahrscheinlichkeitstheorie verw. Geb. 9 (1967), 20-25.

[2] D. A. Darling and P. Erdös: A limit theorem for the maximum of normalized sums of independent random variables, Duke Math. J. 23 (1956), 143-156.

[3] X. Fernique: Régularité de processus gaussiens, Inventiones Math. 12 (1971), 304-320.

[4] C. C. Heyde and D. J. Scott: Invariance principle for the law of the iterated logarithm for martingales and processes with stationary increments. Ann. Probability 1 (1973), 428-436.

[5] J. Lamperti: Semi-stable stochastic processes, Trans. Amer. Math. Soc. 104 (1962), 62-78.

[6] M. B. Marcus and L. A. Shepp: Sample behavior of Gaussian processes, Proc. Sixth Berkeley Sympos. Math. Statist. Prob. 2, Univ. of California Press (1972), 423-441.

[7] H. Oodaira: On Strassen's version of the law of the iterated logarithm for Gaussian processes, Z. Wahrscheinlichkeitstheorie verw. Geb. 21 (1972), 288-299.

[8] H. Oodaira: The log log law for certain dependent random sequences, Proc. Second Japan-USSR Sympos. Prob. Theory, Lecture Notes in Math. 330, Springer-verlag (1973), 355-369.

[9] W. Philipp and W. Stout: Almost sure invariance principles for partial sums of weakly dependent random variables, Mem. Amer. Math. Soc. 2 No.161 (1975).

[10] J. Pickands III: Maxima of stationary Gaussian processes, Z. Wahrscheinlichkeitstheorie verw. Geb. 7 (1967), 190-223.

[11] J. Pickands III: Asymptotic properties of the maximum in a stationary Gaussian process, Trans. Amer. Math. Soc. 145 (1969), 75-86.

[12] C. Qualls and H. Watanabe: Asymptotic properties of Gaussian processes, Ann. Math. Statist. 43 (1972), 580-596.

[13] V. Strassen: Almost sure behavior of sums of independent random variables and martingales, Proc. Fifth Berkeley Sympos. Math. Statist. Prob. 2, Univ. of California Press (1967), 315-343.

Department of Applied Mathematics
Faculty of Engineering
Yokohama National University
Minami-Ku, Yokohama
Japan

NON-UNIFORM ESTIMATE IN THE CENTRAL LIMIT THEOREM IN A SEPARABLE HILBERT SPACE

V.I.Paulauskas

I. Introduction

Throughout the paper H stands for a real separable Hilbert space with inner product $(\cdot,\cdot)$ and norm $\|\cdot\|$ .Without loss of generality, we can assume that $H=\ell_2$ where ℓ_2 is the well-known Hilbert space of sequences $x=(x_1,\dots,x_n,\dots)$ with $\|x\|^2=\sum_{i=1}^{\infty} x_i^2<\infty$, since all separable real Hilbert spaces are isometricaly isomorphic to ℓ_2 .

Let $\xi_1,\xi_2,\dots,\xi_n,\dots$ be a sequence of independent random variables with values in H (H-r.v.). We assume that $E\xi_i=0$, $E\|\xi_i\|^3<\infty$, $i=1,2,\dots$ .Let F_i and T_i stand for the distribution and the covariance operator of H r.v. ξ_i respectively. Since $E\|\xi_i\|^2<\infty$, the operator T_i is a trace-class operator (see e.g. [14]); let $Tr\, T_i$ denote the trace of T_i . Put $B_n^2=\sum_{i=1}^{n} Tr\, T_i$ and let $\mu(0,T)$ denote the zero mean Gaussian measure on H with the covariance operator T . Denote the distribution of normed sum $B_n^{-1}\sum_{i=1}^{n}\xi_i$ by $\bar F_n$ and consider a natural problem about the closeness between the distributions $\bar F_n$ and $\mu(0,\bar T_n)$, where $\bar T_n=B_n^{-2}\sum_{i=1}^{n} T_i$.

In [8], as a consequence of a more general result, the following estimate was obtained:

$$\sup_r |\bar F_n(S_r)-\mu(0,\bar T_n)| \le C(\bar T_n)\Big(B_n^{-3}\sum_{i=1}^{n}\nu_3(i)\Big)^{1/4} , \qquad (1)$$

where $S_r\{x\in H: \|x\|<r\}$, $\nu_3(i)=\int_H \|x\|^3\,|F_i-\mu(0,T_i)|(dx)$.
Here and in what follows C stands for an absolute constant and $C(\cdot)$ denotes constants depending on parameters in the parentheses,

not the same in different places. When we want the constant to be distinguished, we shall supply it with an index.

Of course we can assume that $B_n^{-3}\sum_{i=1}^{n}\nu_3(i)\to 0$ as $n\to\infty$, and in this case it is easy to see that the estimate (1) can be regarded as the main step in constructing estimates of the remainder term in the central limit theorem (c.l.t.) (by c.l.t. we mean the weak convergence of the distributions of sums of independent H - r.v. to a fixed Gaussian distribution). There are two possibilities. One can consider new H - r.v. $\xi_{i,n}=R_n\xi_i$ where $R_n: H\to H$ is a bounded operator such that $T=R_n\bar{T}_n R_n^*$ (R_n^* is the conjugate operator) is some fixed trace-class operator. Then, by using (1), we can obtain an estimate of the quantity

$$\sup_r|\tilde{F}_n(S_r)-\mu(0,T)(S_r)| \qquad (2)$$

where $\tilde{F}_n$ is the distribution of the sum $B_n^{-1}\sum_{i=1}^{n}\xi_{i,n}$, and the estimate will be expressed by means of pseudomoments of H-r.v. $\xi_{i,n}$ which, in turn, can be estimated by $\nu_3(i)$ and some characteristics of the operator R_n.

The second possibility is the following. We can require the sequence of covariance operators $\bar{T}_n$ to satisfy some conditions, then the sequence $\mu(0,\bar{T}_n)$, $n=1,2,\dots$, converges weakly to some Gaussian distribution $\mu(0,T)$ (for details see [6]) and, in order to obtain an astimate of (2), we must construct an estimate of the quantity

$$\sup_r|\mu(0,\bar{T}_n)(S_r)-\mu(0,T)(S_r)|\ .$$

Obviously, if H-r.v. ξ_i, $i=1,2,\dots$, are independent and identically distributed then we can put $B_n^2=n$, $\bar{T}_n=T$ and formula (1) gives an estimate of the remainder term in c.l.t.

In this paper, an improved and extended version of [9], we deal with generalizations of (1) in two directions.

The simple fact that, if an H-r.v. ξ is such that $E\|\xi\|^3<\infty$, then $P\{\|\xi\|>\tau\}\le\tau^{-3}E\|\xi\|^3$, suggests that, as in the finite-dimensional case, the quantity

$$\tau^{\beta}\,|\bar{F}_n(S_\tau)-\mu(0,\bar{T}_n)(S_\tau)|$$

must be bounded for all τ with some positive β. Up to the present non-uniform estimates (which show how the remainder term depends on the location of the set with respect to the origin) had been obtained in limit theorems in the finite-dimensional case. In the second section, first non-uniform estimates in c.l.t. for H-r.v. are given.

By applying these results in Section 3 we obtain a non-uniform estimate of the remainder term in the ω^2-criterion.

The second problem we are dealing with is an extension of the class of sets for which we can construct a uniform estimate of type (1). In Section 4 we show that, in order to construct an estimate of $|\bar{F}_n(B)-\mu(0,\bar{T}_n)(B)|$, the set B must possess two properties: (i) for every $\varepsilon>0$, the Gaussian measure of the ε-neighbourhood of the boundary ∂B is majorized by some constant multiplied by ε; (ii) the boundary ∂B must be smooth in some sense.

2. Non-uniform estimates

The aim of this section is to prove the following

Theorem 1. Let ξ_i, $i=1,2,\dots n$, be independent H-r.v. with $E\xi_i=0$, $E\|\xi_i\|^3<\infty$, $i=1,\dots,n$, and let

$$\frac{\sqrt{j}^{-1}\sum_{i=1}^{j}E\|\xi_i\|^3}{\left(\sum_{i=1}^{j}E\|\xi_i\|^2\right)^{3/2}}\leq A,\quad j=1,2,\dots,n. \tag{3}$$

Then, for all $n\geq 1$ and $r>0$,

$$|\bar{F}_n(S_r)-\mu(0,\bar{T}_n)(S_r)|\leq C(\bar{T}_n,A)\left(B_n^{-3}\sum_{i=1}^{n}\nu_3(i)\right)^{1/4}(1+r^3)^{-1}. \tag{4}$$

Here we use the same notation as in the introduction. Obviously, condition (3) is fulfilled with $A=C_1C_2^{-3/2}$ if $E\|\xi_i\|^3\leq C_1$ and $E\|\xi_i\|^2\geq C_2$ for all $i=1,2,\dots,n$. The following result is an obvious consequence of Theorem 1.

Corollary. Let ξ_i, $i=1,2,\dots,n$, be independent identically distributed H - r.v. with $E\xi_i=0$, the covariance operator T and $\beta_3=E\|\xi_1\|^3<\infty$. Then, for all $n\geq 1$ and $r>0$,

$$\left|P\left\{\left\|\frac{1}{\sqrt{n}}\sum_{i=1}^{n}\xi_i\right\|>r\right\}-P\{\|\eta\|>r\}\right|\leq C(T,\beta_3)\,\frac{\nu_3^{1/4}}{n^{1/8}(1+r^3)}, \tag{5}$$

where η is an H-r.v. with the distribution $\mu(0,T)$.

To prove the theorem we need two lemmas.

Lemma 1. Let $\mu\equiv\mu(0,T)$. Then, for all $\varepsilon>0$ and $k\geq 0$,

$$\sup_{r\geq 0} r^k\mu(S_{r+\varepsilon}\setminus S_r)\leq C(T,k)\varepsilon. \tag{6}$$

Proof. Let η be an H - r.v. with the distribution $\mu(0,T)$, and let $\zeta=\|\eta\|$ be a non-negative random variable with the density $p(x)$. From [4] it is known that $p(x)\leq C(T)$, whence (6) follows in the case $k=0$ Thus we must prove (6) only for $k>0$ and $r>1$ .We have

$$r^k\mu(S_{r+\varepsilon}\setminus S_r)=r^kP\{r<\zeta\le r+\varepsilon\}\le\int_r^{r+\varepsilon}y^kp(y)\,dy. \tag{7}$$

It is well-known (see e.g. [14]) that, for all $k>0$,

$$E\zeta^k=\int_0^\infty y^kp(y)\,dy\le C(T,k).$$

Also it is possible to prove that $p'(y)$ exists for all $y>0$; hence it follows that

$$\sup_{y>0} y^kp(y)\le C(T,k). \tag{8}$$

Estimate (6) results from (7) and (8).

In the proof of Theorem 1 we shall use the following estimate, which is a consequence of a more general result [15] .

Lemma 2. If $\xi_1,\xi_2,\dots,\xi_n$ are independent symmetrically distributed H r.v. and $E\|\xi_i\|^m<\infty$, $i=1,2,\dots,n$, $m\ge 1$ then

$$E\|\xi_1+\xi_2+\dots+\xi_n\|^m\le\left(\frac{(2k)!}{k!}\right)^{\frac{m}{2k}}E(\|\xi_1\|^2+\dots+\|\xi_n\|^2)^{m/2}, \tag{9}$$

where k is the least integer satisfying the inequality $m\le 2k$.

Proof of Theorem 1. We use the method of Trotter, which was utilized for estimating the remainder term in c.l.t. in H by J. Kuelbs and Kurtz (see [4] and also [8]), and some ideas of V.V.Sazonov who obtained a non-uniform estimate in a finite-dimensional Euclidean space by means of convolutions [13] .

Define a sequence of functions $f_n : R_n \to [0,1]$, $n = 1,2,\ldots$, such that f_n is monotonically increasing, $f_n(x) = 0$ for $x \le 0$, $f_n(x) = 1$ for $x \ge \varepsilon_n$, $f_n^{(3)}(x)$ is continuous and

$$|f_n^{(3)}(x)| \le C\varepsilon_n^{-3} \chi_{(0,\varepsilon_n)}(x) \tag{10}$$

where χ_A is the indicator function of a set A and ε_n is some sequence of real numbers tending to zero (in the sequel we shall assume that $\varepsilon_n \le 1$).Put $f_{n,t}(x) = f_n(x - t + \varepsilon_n)$ and $g_{n,t}(x) = f_{n,t}(\|x\|)$, $x \in H$. From (10) we can derive the following estimates

$$|f_n^{(i)}(x)| \le C\varepsilon_n^{-3} |x|^{3-i} \chi_{(0,\varepsilon_n)}(x) \,, \quad i = 1,2,3, \tag{11}$$

$$|f_n^{(i)}(x)| \le C\varepsilon_n^{-i} \chi_{(0,\varepsilon_n)}(x) \,, \quad i = 1,2,3 \,. \tag{12}$$

Now we shall prove the following "inequality of smoothing":

$$\sup_{r \ge 0} r^3 |\bar{F}_n(S_r) - \mu(0,\bar{T}_n)(S_r)| \le \tag{13}$$

$$\le \sup_{r \ge 0} r^3 \left| \int_H g_{n,r}(x)(\bar{F}_n - \mu(0,\bar{T}_n))(dx) \right| + C(\bar{T}_n)\varepsilon_n \,.$$

Consider two cases.

a) $\bar{F}_n(S_r^c) > \mu_n(S_r^c)$ where A^c denotes the complement of a set A and μ_n , for short, denotes $\mu(0,\bar{T}_n)$. Now from (6) with $k=3$ and from the fact that $\varepsilon_n \le 1$ it follows:

$$r^3 |\bar{F}_n(S_r) - \mu_n(S_r)| = r^3(\bar{F}_n(S_r^c) - \mu_n(S_r^c)) \le \tag{14}$$

$$\le r^3 \int_H g_{n,r}(x)(\bar{F}_n - \mu_n)(dx) + r^3 \mu_n(S_r \setminus S_{r-\varepsilon_n}) \le$$

$$\leq \tau^3 \int_H g_{n,\tau}(x)(\overline{F}_n - \mu_n)(dx) + C(\overline{T}_n)\varepsilon_n .$$

b) $\overline{F}_n(S_\tau^c) \leq \mu_n(S_\tau^c)$. In this case

$$\tau^3 |\overline{F}_n(S_\tau) - \mu_n(S_\tau)| \leq \tau^3 \int_H (1 - g_{n,\tau+\varepsilon_n}(x))(\overline{F}_n - \mu_n)(dx) +$$

$$+ \tau^3 \mu_n(S_{\tau+\varepsilon_n} \setminus S_\tau) \leq -\tau^3 \int_H g_{n,\tau+\varepsilon_n}(x)(\overline{F}_n - \mu_n)(dx) + \tag{15}$$

$$+ C(\overline{T}_n)\varepsilon_n \leq (\tau+\varepsilon_n)^3 \left| \int_H g_{n,\tau+\varepsilon_n}(x)(\overline{F}_n - \mu_n)(dx) \right| + C(\overline{T}_n)\varepsilon_n .$$

From (14) and (15) the inequality (13) results, so we only have to estimate the quantity

$$\mathcal{Y}(\tau) = \tau^3 \left| \int_H g_{n,\tau}(x)(\overline{F}_n - \mu_n)(dx) \right| .$$

Let us introduce some notation:

$$H_i(A) = (\overline{F}_{i,n} - \mu_{i,n})(A),$$

$$K_i = F_{1,n} * \dots * F_{i-1,n} * \mu_{i+1,n} * \dots * \mu_{n,n} , \quad i = 1, 2, \dots, n ,$$

$$h_{i,\tau}(y) = \int_H g_{n,\tau}(y+z) K_i(dz), \quad V_{i,\tau} = \int_H h_{i,\tau}(y) H_i(dy),$$

where $F_{i,n}$ stands for the distribution of H-r.v. $\xi_{i,n} = B_n^{-1}\xi_i$, $\mu_{i,n} = \mu(0, B_n^{-2}, T_i)$ and $F_{0,n}$, and $\mu_{n+1,n}$ denote the degenerate distributions concentrated at zero element of H . Then, by means of the well-known identical expansion of the difference $\overline{F}_n - \mu_n$ (see e.g. [8]), we obtain

$$\mathcal{Y}(\tau) \leq \sum_{i=1}^{n} |V_{i,\tau}| . \tag{16}$$

In the sequel the differentation wil be in the sense of Frechet and, if $f: H \to R$, then $\mathcal{D}^{(i)} f(a)$ denotes the i - th derivative at point $a \in H$ and $\mathcal{D}^{(i)} f(a)(y)^i$, $y \in H$, denotes the corresponding multilinear form. For the sake of brevity put $\mathcal{D}^{(1)} f(a)(y)^1 = \mathcal{D} f(a)(y)$ and, if $f(x) = \|x\|$, then we shall write simply $\mathcal{D}_a^{(i)}$ instead of $\mathcal{D}^{(i)} \|a\|$.

The detailed computations of the derivatives of the functions $g_{n,\tau}$ and $h_{i,\tau}$ will not be given here (the reader finds it in [8]). Here we shall use these results without reference. Let

$$m_{i,\tau}(y) = h_{i,\tau}(y) - h_{i,\tau}(0) - \mathcal{D} h_{i,\tau}(0)(y) - \frac{1}{2} \mathcal{D}^{(2)} h_{i,\tau}(0)(y)^2$$

Then

$$V_{i,\tau} = \tau^3 \int_H m_{i,\tau}(y) H_i(dy) = V'_{i,\tau} + V''_{i,\tau} , \tag{17}$$

$$V'_{i,\tau} = \tau^3 \int_{S_1} m_{i,\tau}(y) H_i(dy), \quad V''_{i,\tau} = \tau^3 \int_{S_1^c} m_{i,\tau}(y) H_i(dy),$$

since the moments of the first and the second order of summands agree with the corresponding moments of the Gaussian distribution. By means of Taylor's formula, we obtain

$$V'_{i,\tau} = \tau^3 \int_{S_1} \mathcal{D}^{(3)} h_{i,\tau}(\theta y)(y)^3 H_i(dy),$$

where $\mathcal{D}^{(3)} h_{i,\tau}(\theta y) = \int_H \mathcal{D}^{(3)} g_{n,\tau}(\theta y + z) K_i(dz)$ with some real θ, $|\theta| < 1$. Further, if we estimate the above integral in the same manner as in [8], we to the conclusion that

$$|\mathcal{D}^{(3)} h_{i,\tau}(\theta y)(y)^3| \le C \varepsilon_n^{-3} \|y\|^3 \int_{\tau \le |\theta y + z| < \tau + \varepsilon_n} K_i(dz) . \tag{18}$$

So we must show that, for all $y \in S_1$ and all $\tau > 1$ (it is obvious that we need consider only such τ)

$$u_{i,\tau} = \tau^3 \int\limits_{\tau \le \|\theta_y + z\| < \tau + \varepsilon_n} K_i(dz) \le C(A). \tag{19}$$

First we note that

$$\{z \in H : \|z+u\| > \tau\} \subset \{z \in H : \|z\| > \tau - \|u\|\}$$

and, since $\|\theta_y\| \le 1$, the (19) is proved if we show that, for all $\tau > 0$,

$$\tilde{U}_{i,\tau} = \tau^3 \int\limits_{\|z\| \ge \tau} K_i(dz) \le C(A). \tag{20}$$

If ξ is an H-r.v., let ξ^* stand for the symmetrized H-r.v. with the distribution $\mu * \bar{\mu}$, where $\bar{\mu}(A) = \mu(-A)$, μ being the distribution of ξ. Let $\zeta_{i,n}$ stand for an H r.v. with the distribution $\mu(0, B_n^{-2} \sum_{j=i+1}^{n} T_j)$ which is symmetric itself. Due to (10) and the obvious inequality $E\|\xi_{j,n}^*\|^3 \le 8E\|\xi_{j,n}\|^3$ we have

$$P\{\|\sum_{j=1}^{i-1} \xi_{j,n}^* + \zeta_{i,n}\| > \tau^{-3} E\|\sum_{j=1}^{i-1} \xi_{j,n}^* + \zeta_{i,n}\|^3 \le$$

$$\le C\tau^{-3}(E\|\sum_{j=1}^{i-1} \xi_{j,n}^*\|^3 + E\|\zeta_{i,n}\|^3) \le$$

$$\le C\tau^{-3}(\sqrt{i-1} \sum_{j=1}^{i-1} E\|\xi_{j,n}\|^3 + E\|\zeta_{i,n}\|^3).$$

Since $E\|\zeta_{i,n}\|^2 \le 1$, the quantity $E\|\zeta_{i,n}\|^3$ is majorized by an absolute constant, thus having in mind the assumption (3), we get, for all $i = 1, 2, \dots, n$,

$$P\{\|\sum_{j=1}^{i-1} \xi_{j,n}^* + \zeta_{i,n}\| > \tau\} \le C(A)\tau^{-3}.$$

The inequality for the above symmetrized H-r.v., in the standard way (see e.g. Lemma 2 and Corollary 3 from [15]), yields the required inequality:

$$\int_{\|z\|>\tau} K_i(dz) = P\{\|\sum_{j=1}^{i-1} \xi_{j,n} + \zeta_{i,n}\| > \tau\} \leq C(A)\tau^{-3}$$

and the latter implies (20) and (19). From (19) we get

$$|V'_{i,\tau}| \leq C(A)\varepsilon_n^{-3} \int_{S_1} \|y\|^3 |H_i(dy) \leq C(A)\varepsilon_n^{-3} B_n^{-3} \nu_3(i). \tag{21}$$

We estimate the second term in (17) as follows:

$$|V''_{i,\tau}| \leq \sum_{j=1}^{5} |V''_{i,\tau,j}|,$$

where

$$V''_{i,\tau,1} = \tau^3 \int_{S_1^c \cap S_{\tau-\varepsilon_n}} h_{i,\tau}(y) H_i(dy),$$

$$V''_{i,\tau,2} = \tau^3 \int_{S^c_{\tau-\varepsilon_n}} h_{i,\tau}(y) H_i(dy), \quad V''_{i,\tau,3} = \tau^3 \int_{S_1^c} h_{i\tau}(0), H_i(dy),$$

$$V''_{i,\tau,4} = \tau^3 \int_{S_1^c} \mathcal{D} h_{i,\tau}(0)(y) H_i(dy), \quad V''_{i,\tau,5} = \frac{1}{2}\tau^3 \int_{S_1^c} \mathcal{D}^{(2)} h_{i,\tau}(0)(y)^2 H_i(dy).$$

The obvious inequality

$$\int_{S_1^c \cap S_{\tau-\varepsilon_n}} |H_i(dy)| \leq \int_H \|y\|^3 |H_i(dy)| = B_n^{-3} \nu_3(i)$$

along with (19) yields

$$|V''_{i,\tau,1}| = |\int_{S_1^c \cap S_{\tau-\varepsilon_n}} \tau^3 \int_H g_{n,\tau}(y+z) K_i(dz) H_i(dy)| \leq$$

$$\leq \int_{S_1^c \cap S_{\tau-\varepsilon_n}} \tau^3 \int_{\tau-\varepsilon_n \leq \|y+z\|} K_i(dz) |H_i(dy)| \leq \int_{S_1^c \cap S_{\tau-\varepsilon_n}} \tau^3 \int_{\tau-\varepsilon_n-\|y\| \leq \|z\|} K_i(dz) |H_i(dy)| \leq$$

$$\leq 9 \int_{S_1^c \cap S_{\tau-\varepsilon_n}} (\tau-\varepsilon_n-\|y\|)^3 \int_{\|z\|>\tau-\varepsilon_n-\|y\|} K_i(dz)\,|H_i(dy)| +$$

$$+ 9 \int_{S_1^c \cap S_{\tau-\varepsilon_n}} \|y\|^3 |H_i(dy)| + 9 \int_{S_1^c \cap S_{\tau-\varepsilon_n}} |H_i(dy)| \leq C(A) B_n^{-3} \nu_3(i) . \tag{22}$$

The trivial estimate $|h_{i,\tau}(y)| \leq 1$ gives

$$|V''_{i,\tau,2}| \leq \tau^3 \int_{S^c_{\tau-\varepsilon_n}} |H_i(dy)| \leq C\,\nu_3(i) B_n^{-3} . \tag{23}$$

We have

$$\tau^3 h_{i,\tau}(0) = \tau^3 \int_H g_{n,\tau}(z) K_i(dz) \leq \tau^3 \int_{\|z\|>\tau-\varepsilon_n} K_i(dz) \leq C(A) ,$$

whence it follows that

$$|V''_{i,\tau,3}| \leq \tau^3 \left| \int_{S_1^c} h_{i,\tau}(0) H_i(dy) \right| \leq C(A) B_n^{-3} \nu_3(i) . \tag{24}$$

Further we use the formulae of differentiation (see [3] or [8] for details) and estimates (12), (19) to obtain

$$|V''_{i,\tau,4}| < \int_{S_1^c} \tau^3 \int_{\tau-\varepsilon_n \leq \|z\| < \tau} f'_{n,\tau}(\|z\|) K_i(dz) \|y\|\,|H_i(dy)| \leq \tag{25}$$

$$\leq C(A) \varepsilon_n^{-1} \nu_3(i) B_n^{-3} .$$

Analogously we get

$$|V''_{i,\tau,5}| \leq C(A) \varepsilon_n^{-2} \nu_3(i) B_n^{-3} . \tag{26}$$

From (16),(17),(21)-(26) we obtain

$$|V_{i,\tau}| \leq C(A) \varepsilon_n^{-3} \nu_3(i) B_n^{-3} , \tag{27}$$

$$\mathcal{J}(\tau) \leq C(A) \varepsilon_n^{-3} B_n^{-3} \sum_{i=1}^{n} \nu_3(i) .$$

Denoting an H-r.v. with the distribution $\mu(0,T_i)$ by η_i and using assumption (3), we get

$$L_{3,n} = B_n^{-3}\sum_{i=1}^{n}\nu_3(i) \le B_n^{-3}\sum_{i=1}^{n}E\|\xi_i\|^3 + B_n^{-3}\sum_{i=1}^{n}E\|\eta_i\|^3 \le$$

$$\le An^{-1/2} + \frac{C\sum_{i=1}^{n}(E\|\xi_i\|^2)^{3/2}}{(\sum_{i=1}^{n}E\|\xi_i\|^2)^{3/2}} \le C_1(A)\,.$$

Now putting $\varepsilon_n = L_{3,n}^{1/4}\, C_1^{-1/4}(A) \le 1$, from (1),(13) and (27) we obtain statement (4) of the theorem.The proof of Theorem 1 is complete.

The order of the remainder term in (4) with respect to r is the right one, but it could not be said about n . For this reason the following Esseen type non-uniform estimate is of some interest . Using the notation of the theorem 1 and denoting $\Delta_n = \sup_r|\overline{F}_n(S_r)-\mu_n(S_r)|$, we shall prove the following result.

<u>Theorem 2</u>. Let ξ_i , $i=1,2,\dots,n$, be independent H-r.v. with $E\xi_i=0$, $E\|\xi_i\|^2<\infty$ and covariance operator T_i . Let $\overline{T}_n = B_n^{-2}\sum_{i=1}^{n}T_i$, where $B_n^2=\sum_{i=1}^{n}\mathrm{Tr}\,T_i$ and $\lambda_i^{(n)}$, $i=1,2,\dots$, denote the eigenvalues of the operator $\overline{T}_n$, $\lambda_1^{(n)} \ge \lambda_2^{(n)} \ge \dots$. If $\Delta_n \le \frac{1}{2}$ for $n>n_0$, then, for all $n>n_0$ and all $r\ge 0$

$$|\overline{F}_n(S_r)-\mu(0,\overline{T}_n)(S_r)| \le \min\{\Delta_n, \frac{C_2(\overline{T}_n)\Delta_n\log\frac{1}{\Delta_n}}{1+r^2}\}\,. \tag{28}$$

with $C_2(\overline{T}_n) = 4\lambda_1^{(n)}[\prod_{i=1}^{\infty}(1-\frac{\lambda_i^{(n)}}{2\lambda_1^{(n)}})^{-1/2}+3]$

<u>Proof</u>. Let us denote $\zeta_n = \|B_n^{-1}\sum_{i=1}^{n}\xi_i\|$, $\zeta=\|\eta\|$, where η is an H-r.v. with the distribution $\mu(0,\overline{T}_n)$, $G_n(r)=P\{\zeta_n<r\}$, $G(r)=P\{\zeta<r\}$.

We shall often use the following fact ([2] p.161): if ξ is an H-r.v. with the distribution F and $\gamma=\|\xi\|$ has the distribution $\overline{F}$, then, for any Borel function f on $[0,\infty)$,

$$\int_H f(\|x\|)F(dx) = \int_0^\infty f(u)\overline{F}(du)$$

under the assumption that one of these integrals exists.

By this formula we have

$$\int_0^\infty \tau^2 d\,G(\tau) = \int_H \|x\|^2 \overline{F}_n(dx) = \mathrm{Tr}\ \overline{T}_n = 1 = \int_0^\infty \tau^2 d\,G(\tau).$$

Assume that $a \geq 2\sqrt{\lambda_1^{(n)}}$, a number to be determined later. In a similar way as in [1] , we can obtain the following inequality

$$(1+\tau^2)\,|\,G_n(\tau) - G(\tau)\,| \leq 3a^2\Delta_n + \int_0^\infty \tau^2 d\quad(\tau) \tag{29}$$

for all $\tau \geq 0$. Now from [14], p. 88, it follows that

$$\int_0^\infty \tau^2 d\,G(\tau) \leq \sup_{\tau \geq a}(\tau^2 e^{-\alpha\tau^2}) \int_0^\infty e^{\alpha\tau^2} d\,G(\tau) \leq \tag{30}$$

$$\leq \sup_{\tau \geq a}(\tau^2 e^{-\alpha\tau^2}) \int_H e^{\alpha\|x\|^2} \mu(0,\overline{T}_n)(dx) = \frac{\sup_{\tau \geq a} \tau^2 e^{-\alpha\tau^2}}{\prod_{i=1}^\infty (1-2\alpha\lambda_i^{(n)})^{1/2}}$$

for every $\alpha < \frac{1}{2\lambda_1^{(n)}}$. We can choose $a = \sqrt{4\lambda_1^{(n)} \log(1/\Delta_n)}$, $\alpha = (4\lambda_1^{(n)})^{-1}$, and simple computations along with the assumption $\Delta_n \leq \frac{1}{2}$ show that inequality (28) results from (29) and (30).

<u>Remark.</u> It is easy to see that the assumption $\mathrm{Tr}\ \overline{T}_n = 1$ can be omitted.

3. On ω^2-criterion

Let $\xi_1, \xi_2, \ldots$ be independent random variables with the same continous distribution function $F(x)$. Define

$$\omega_n^2 = n\int_{-\infty}^{\infty} [\,F_n(x) - F(x)\,]^2\, dF(x)$$

where

$$F_n(x) = \frac{1}{n}\sum_{i=1}^{n}\delta(x-\xi_i)\,,\quad \delta(x)=\begin{cases}1\,, & x>0\,,\\ 0\,, & x\leq 0\,.\end{cases}$$

It is well-known that the distribution of ω_n^2 is independent of the function $F(x)$ and that

$$\Delta_n \equiv \sup_x \Delta_n(x) \to 0 \qquad \text{as} \qquad n\to\infty\,,$$

where $\Delta_n(x) = | P\{\omega_n^2 < x\} - S(x) |$, $S(x)$ is the distribution function with the characteristic function $s(t) = \prod_{j=1}^{\infty}(1-\frac{2it}{(\pi j)^2})^{-1/2}$.

There is a number of papers concerned with estimating Δ_n ,and the best estimate is due to A.I. Orlov (see [7] and references there)*:

$$\Delta_n \leq C(\varepsilon)\, n^{-1/2+\varepsilon}\,, \quad \varepsilon>0\,. \tag{31}$$

In [11] and [12] it was shown that the problem of estimating Δ_n may be considered as the problem of estimating the remainder term in the c.l.t. for sums of independent identically distributed H r.v. Namely, it was shown, that

$$\Delta_n(x) = | P\{\| Z_n\|^2 < x\} - P\{\|\eta\|^2 < x\}|\,,$$

where $Z_n = n^{-1/2}\sum_{i=1}^{n}\zeta_i$, ζ_i ($i=1,2,\ldots,n$) being independent iden-

---

* At this symposium Yu.V.Borovskih announced the following estimate $\Delta_n \leq C n^{-1}$.

tically distributed ℓ_2-r.v. with zero mean and the diagonal covariance matrix T with elements $(\pi j)^{-2}$, $j=1,2,\dots$,on the diagonal; η is a Gaussian ℓ_2-r.v. with zero mean and the same diagonal covariance matrix T . The ℓ_2-r.v. ζ_1 has a special form (see [12] for detail) but for our purpose, it is sufficient to note that $\|\zeta_1\| \le c$ so that $E\|\zeta_1\|^3 < c$. Thus applying Theorems 1 and 2 from the previous section and estimate (31), we are able to formulate the following result.

Proposition 1. For all $n \ge 1$ and $x \ge 0$,

$$\Delta_n(x) \le C_3 n^{-1/6}(1+x^{3/2})^{-1},$$

$$\Delta_n(x) \le C_4(\varepsilon) n^{-1/2+\varepsilon} \min\left(1, \frac{\log n}{1+x}\right).$$

It is not very difficult to obtain the numerical value of the absolute constant C_3 .

4. Uniform estimates for some class of sets

In this section we shall deal with the second problem: for what sets can we estimate the remainder term in c.l.t.? In [8] estimates, uniform over all balls or ellipsoids with a fixed centre, were given. Note that al though the method used there is, in some sense, similar to Bergstrom's method of convolutions, which is very useful for estimates in the c.l.t. in the finite- dimensional case, there is an essential difference between these two methods: the absence of mathematical induction in [8] . This does not allow us to obtain the right order of the remainder term with respect to the number of summands but, on the other hand, it allows us to consider the difference $|\bar{F}_n(B) - \mu(B)|$ for a separate Borel set B, where, as in section 2, $\bar{F}_n$ stands for the distribution of $B_n^{-1}\sum_{i=1}^{n} \xi_i$, and $\mu_n = \mu(0,T)$. So we shall look for conditions on the set B in order to estimate the above difference.

If we look through the proof of the main theorem in [8] or the proof in section 1, we shall see that there are two essential assumptions about the set for which we estimate the remainder term.

The first one is the existence of a function $h_\varepsilon : H \to R$ with the properties: (i) $0 \le h_\varepsilon(x) \le 1$ for all $x \in H$, $h_\varepsilon(x) = 1$ for $x \in B$ and $h_\varepsilon(x) = 0$ for $x \in (B_\varepsilon)^c$, where $B_\varepsilon = \{ y \in H : \|x-y\| \le \varepsilon, x \in B$,(ii) $h_\varepsilon \in C^3$ (C^3 denotes the class of functions which are three times continously differentiable) and the derivatives of h_ε admit estimates by negative powers of ε.

The second one is the following. Let μ be a zero mean Gaussian distribution, then there must exist a constant $C(\mu, B)$ such that, for all $\varepsilon > 0$,

$$\mu((\partial B)_\varepsilon) \le C(\mu, B)\varepsilon , \tag{32}$$

where $(\partial B)_\varepsilon = (B_\varepsilon \setminus B) \cup (B \setminus B_{-\varepsilon})$, $B_{-\varepsilon} = ((B^c)_\varepsilon)^c$.

This assumption is rather natural and can be kept in the final formulation of the result. So our aim is to discuss the first condition. If there were no need of estimating the derivates of the function h_ε, the answer to our problem would be obtained immediately by means of the partition-of-unit theorem in a separable Hilbert space [5]. And, since it seems to us impossible to obtain good estimates of the derivatives of this function with the help of the partition of unit (for detail see [9]), we present here one method of constructing, for some sets B , a function h_ε with properties (i) and (ii). The idea of the construction is the following. For a set B with a smooth boundary, we construct a map $A: H \to H$ of the class C^3 such that the set AB is a ball in H , and, for balls, we can get the required function as in section 1 .

We shall consider a set B with the following four properties:

(i) B is connected and $0 \in B$.

(ii) every ray tx, $t>0$, $\|x\|=1$ intersects the boundary ∂B of the set B at one point.

For example, convex sets, containing zero, satisfy these two conditions.

Now we define a functional, depending on the set B under consideration, as follows

$$d_B(x) = \sup \{t>0: \frac{tx}{\|x\|} \in B \}, \quad x \neq 0 . \tag{33}$$

The functional d_B means the distance between the origin of coordinates and the boundary ∂B in direction to x . It is not defined only for point $x=0$, and since $d_B(x) = d_B(kx)$, $k>0$, $x \neq 0$, it is sufficient to defined it on the unit sphere in H: $S_1 = \{x \in H: \|x\| = 1\}$

(iii) $$\inf_{x \in S_1} d_B(x) = m_1 > 0 , \quad \sup_{x \in S_1} d_B(x) = m_2 < \infty . \tag{34}$$

(iv) $$d_B \in C^3, \ \|\mathcal{D}^{(i)} d_B(x)\| \leq M_i \|x\|^{-i}, \quad i = 1,2,3, \tag{35}$$

where the norm is understood as the usual norm in the Banach space $\mathcal{L}(\underbrace{H,\dots,H}_{i};R)$(for details see [3] or [8]). For the sake of convienence let us denote the class of Borel sets with properties (i)-(iv) by $\mathcal{B}_s$ and the class of Borel sets which satisfy (32) by $\mathcal{B}_{\zeta^u}$.

Lemma 3. Let $B\in\mathcal{B}_{\zeta}$, and define the operator $A: H \to H$ by the equality

$$Ax = \frac{m_1(x)}{d_B(x)} \quad . \tag{36}$$

Then: a) $AB = S_{m_1}$; (37)

b) there exist sets $\tilde{B}$ and $\tilde{\tilde{B}}$, $B\supset\tilde{B}\supset B_{-\varepsilon}$, $B\subset\tilde{\tilde{B}}\subset B_{\varepsilon}$,

such that

$$A\tilde{B} = S_{m_1 - L\varepsilon} , \quad A\tilde{\tilde{B}} = S_{m_1 + L\varepsilon} , \quad L = \frac{m_1}{m_2} ; \tag{38}$$

c) $A\in C^3$ and

$$\|\mathcal{D}^{(i)}A(x)\| \le L_i \|x\|^{-i+1} , \quad i = 1,2,3 , \tag{39}$$

where L_1 depends on M_i $(i=1,2,3)$ m_1 and m_2 .

Proof. a) foolows from the definition of A , since, if $x\in B$, then $\|x\| \le d_B(x)$ and, if $x\in\partial B$, then $\|x\| = d_B(x)$;Now let $x\in B_{-\varepsilon}$ then

$$\|Ax\| = \frac{m_1\|x\|}{d_B(x)} \le \frac{m_1 d_{B_\varepsilon}(x)}{d_B(x)} \le \frac{m_1(d_B(x)-\varepsilon)}{d_B(x)} \le m_1 - L\varepsilon ,$$

since, by the definition of the set $B_{-\varepsilon}$, $d_B(x) \ge d_{B_{-\varepsilon}}(x) + \varepsilon$.

Thus $AB_{-\varepsilon} \subset S_{m_1 - L\varepsilon}$, and this along with the relation $AB = S_{m_1} \supset S_{m_1 - L\varepsilon}$ implies the existence of a set $\tilde{B}$, $B\supset\tilde{B}\supset B_{-\varepsilon}$, such that $A\tilde{B} = S_{m_1 - L\varepsilon}$.

Analogously we prove the second part of b). In order to complete the proof of the lemma, we give expressions for the derivatives of A ; the computations, rather tedious, are ommited. We have

$$\mathcal{D}A(x)(h_1) = \frac{m_1 h_1}{d_B(x)} - \frac{m_1 x (\mathcal{D}d_B(x), h_1)}{d_B^2(x)} ,$$

$$\mathcal{D}^{(2)}A(x)(h_1,h_2) = 2m_1 x\,(\mathcal{D}d_B(x), h_1)(\mathcal{D}d_B(x), h_2)\, d_B^{-3}(x) -$$

$$- m_1 [x\mathcal{D}^{(2)}d_B(x)(h_1,h_2) + (\mathcal{D}d_B(x), h_1)\, h_1 + (\mathcal{D}d_B(x), h_2)\, h_1]\, d_B^{-2}(x) ,$$

$$\mathcal{D}^{(3)}A(x)(h_1,h_2,h_3) = -m_1 [x\mathcal{D}^{(3)}d_B(x)(h_1,h_2,h_3) +$$

$$+\mathcal{D}^{(2)}d_B(x)(h_1,h_2)h_3 + \mathcal{D}^{(2)}d_B(x)(h_1,h_3)h_2 + \mathcal{D}^{(2)}d_B(x)(h_2,h_3)h_1]\, d_B^{-2}(x) +$$

$$+ 2m_1 \{(\mathcal{D}d_B(x), h_1)(\mathcal{D}d_B(x), h_2)h_3 + (\mathcal{D}d_B(x), h_1)(\mathcal{D}d_B(x), h_3)h_2 +$$

$$+ (\mathcal{D}d_B(x), h_2)(\mathcal{D}d_B(x), h_3)h_1 + x[\mathcal{D}^{(2)}d_B(x)(h_1,h_2)(\mathcal{D}d_B(x), h_3) +$$

$$+\mathcal{D}^{(2)}d_B(x)(h_1,h_3)(\mathcal{D}d_B(x), h_2) + \mathcal{D}^{(2)}d_B(x)(h_2,h_3)(\mathcal{D}d_B(x), h_1)]\}\, d_B^{-3}(x) -$$

$$-6m_1 x(\mathcal{D}d_B(x),h_1)(\mathcal{D}d_B(x),h_2)(\mathcal{D}d_B(x),h_3)d_B^{-4}(x),\quad x\neq 0,\ h_i\in H\ .$$

From these formulas estimates (39) are easily derived, and the proof of the lemma is complete.

Theorem 3. Let ξ_i, $i=1,2,\dots,n$, be independent H-r.v. with $E\xi_i=0$ and $E\|\xi_i\|^3<\infty$, $i=1,2,\dots,n$. Let $B\in\mathfrak{B}_S\cap\mathfrak{B}_\mu$, then, for all $n\geq 1$,

$$|\bar{F}_n(B)-\mu(0,\bar{T}_n)(B)|\leq C(\bar{T}_n,B)(B_n^{-3}\sum_{i=1}^{n}\nu_3(i))^{1/4}, \qquad (40)$$

where $C(\bar{T}_n,B)$ depends on constants M_i, m_1, m_2 and $C(\mu,B)$ from (32).

We only sketch the proof of Theorem 3, since it goes along the lines of the proof of Theorem 1 in [8]. Define the functions $f_{i,\varepsilon}: R\to[0,1]$ $(i=1,2)$ with the properties

$$f_{1,\varepsilon}(u)=\begin{cases}1, & u\leq m_1\\ 0, & u\geq m_1+L\varepsilon,\end{cases}\qquad |f_{1,\varepsilon}^{(3)}(u)|\leq C(L,\varepsilon)^{-3}\chi_{(m_1,m_1+L\varepsilon)}(u),$$

$$f_{2,\varepsilon}(u)=f_{1,\varepsilon}(u+L\varepsilon),$$

and put $g_{i,\varepsilon}(x)=f_{i,\varepsilon}(\|Ax\|)$, $x\in H$ where $A: H\to H$ is from (36). Then from (38) we have

$$g_{1,\varepsilon}(x)=\begin{cases}1, & x\in B,\\ 0, & x\in \tilde{\tilde{B}}^c,\end{cases}\qquad g_{2,\varepsilon}(x)=\begin{cases}1, & x\in \tilde{B},\\ 0, & x\in B^c,\end{cases}$$

where $B\supset\tilde{B}\supset B_{-\varepsilon}$, $B\subset\tilde{\tilde{B}}\subset B_{\varepsilon}$. These properties of the functions $g_{i,\varepsilon}$ along with condition (32) yield that, for every $\varepsilon>0$,

$$|\bar{F}_n(B)-\mu_n(B)|\le \max_{i=1,2}\Big|\int_H g_{i,\varepsilon}(x)(\bar{F}_n-\mu_n)(dx)\Big|+C(\mu_n,B)\varepsilon. \qquad (41)$$

Both integrals in (41) are estimated in the same way as in [8] (see Section 1 also): we use the identical expansion for $\bar{F}_n-\mu_n$ and in the term $\int_H h_{j,\varepsilon,i}(y)H_i(dy)$ expand the integrand $h_{j,\varepsilon,i}(y)=\int_H g_{j,\varepsilon}(y+z)K_i(dz)$ by Taylor's formula. All these computations are omitted and we confine ourselves to the computatin and the estimation of the third derivative of $g_{1,\varepsilon}$, since now $g_{1,\varepsilon}$ is more complicated and is the superposition of three functions. Let us denote $\varphi: H\to R$, $\varphi(x)=\|x\|$, $k=\varphi\circ A$, then $g_{1,\varepsilon}=f_{1,\varepsilon}\circ\varphi\circ A=f_{1,\varepsilon}\circ k$, where $f\circ g$ denotes the composition of the mappings f and g). Then, by differentiating the compound function, we have

$$\mathfrak{D}^{(3)}g_{1,\varepsilon}(v)(y)^3=f'_{1,\varepsilon}(\|Av\|)\,\mathfrak{D}^{(3)}k(v)(y)^3+ \qquad (42)$$

$$+3f^{(2)}_{1,\varepsilon}(\|Av\|)\mathfrak{D}k(v)(y)\,\mathfrak{D}^{(2)}k(v)(y)^2+f^{(3)}_{1,\varepsilon}(\|Av\|)(\mathfrak{D}k(v)(y))^3.$$

In the same way we obtain the derivatives $\mathfrak{D}^{(i)}k(v)$, $i=1,2,3$; for example,

$$\mathfrak{D}^{(2)}k(v)(y)^2=(\mathfrak{D}\varphi(Av),\mathfrak{D}^2A(v)(y)^2)+\mathfrak{D}^2\varphi(Av)(\mathfrak{D}A(v)(y),\mathfrak{D}A(v)(y)).$$

Putting $\mathcal{D}^{(i)}k(v)(y)^i$ into (42) and using estimate (11), we get

$$|\mathcal{D}^{(3)}g_{1,\varepsilon}(v)(y)^3 = f'_{1,\varepsilon}(\|Av\|)\mathcal{D}^{(3)}k(v)(y)^3 +$$

$$+\|Av\| \, \|\mathcal{D}A(v)\| \, \|\mathcal{D}^{(2)}A(v)\| + \|Av\|^2 \|\mathcal{D}^{(3)}A(v)\|] .$$

Since $\|Av\| \le \|v\|$ and we have (39), it follows that

$$\|\mathcal{D}^{(3)}g_{1,\varepsilon}(v)\| \le C(M_i,\ i=1,2,3;\ m_1, m_2)\varepsilon^{-3}\chi_{(m_1, m_1+2\varepsilon)}(\|Av\|) . \qquad (43)$$

The remaining steps in the proof can be made without essential changes, so Theorem 3 is proved.

Examples. 1. Let $E: H \to H$ be a linear operator with the property $C_1\|x\| \le \|Ex\| \le C_2\|x\|$ and let $B=\{x: \|Ex\| < r\}$. Then $d'_B(x) = r\|x\|(\|Ex\|)^{-1}$, and we can verify that $B \in \mathfrak{B}_s$. But this example is not interesting, since we have

$$P\{Z_n \in B\} - P\{\eta \in B\} = P\{\|EZ_n\| \le r\} - P\{\|\tilde{\eta}\| < r\} = P\{\|\tilde{Z}_n\| \le r\} - P\{\|\tilde{\eta}\| \le r\} ,$$

where $Z_n = B_n^{-1}\sum_{i=1}^{n}\xi_i$, η is a Gaussian H r.v. $\tilde{Z}_n = B_n^{-1}\sum_{i=1}^{n}E\xi_i$; $\tilde{\eta} = E\eta$ is a Gaussian H r.v. So we arrived at the case considered in [8] .

2. A large and interesting subclass of $\mathfrak{B}_s$ can be obtained in the following way. Let $f: H \to R$ be the functional with the properties

a) $f(x) \ge 0$ for all $x \in H$, $f(tx) = tf(x)$, $t \ge 0$

b) $\inf\limits_{\|x\|=1} f(x) = N_1$, $\sup\limits_{\|x\|=1} f(x) = N_2$;

c) $f \in C^3$, $\|\mathcal{D}^{(i)}f(x)\| \le \dfrac{N_3}{\|x\|^{i-1}}$, $i = 1,2,3$. (44)

Define $B = B(r) = \{x \in H : f(x) \le r\}$, then $d_B(x) = r\|x\|(f(x))^{-1}$.

It is easy to verify, that B satisfies conditions (i)-(iii) with $m_1 = \frac{r}{N_2}$, $m_2 = \frac{r}{}$. To establish (iv) one must find the derivatives $\mathcal{D}^{(i)} d_B(x)$. Since the computations are rather long, we give here only formulas for the two first derivates:

$$(\mathcal{D} d_B(x), h) = r\left[\frac{(x,h)}{\|x\| f(x)} - \frac{\|x\|(\mathcal{D} f(x), h)}{f^2(x)}\right], \tag{45}$$

$$\begin{aligned}\mathcal{D}^{(2)} d_B(x)(h_1, h_2) = r\{ & f^{-2}(x)\|x\|^{-1}[(h_1,h_2)f(x) - (x,h_1)(\mathcal{D} f(x), h_2) - \\ & -(x,h_2)(\mathcal{D} f(x), h_1)] - \|x\|^{-3} f^{-1}(x)(x,h_1)(x,h_2) - \\ & - f^{-2}(x)\|x\|\mathcal{D}^{(2)} f(x)(h_1,h_2) + 2 f^{-3}(x)\|x\|(\mathcal{D} f(x), h_1)(\mathcal{D} f(x), h_2)\} .\end{aligned} \tag{46}$$

The functional $d_B(x)$,for every $B \in \mathcal{B}_s$, has a property usefull when we want to verify the computations. Namely, since $d_B(tx) = d_B(x)$ for every $t > 0$, it follows that $(\mathcal{D} d_B(x), h) = 0$ for $h = kx,\ k > 0$. So we can check (45), and, for this purpose, we note, that from a) it follows that $(\mathcal{D} f(x), kx) = k f(x)$. Then

$$(\mathcal{D} d_B(x), kx) = r\left[\frac{(x,kx)}{\|x\| f(x)} - \frac{\|x\|(\mathcal{D} f(x), kx)}{f^2(x)}\right] = r\left[\frac{k\|x\|}{f(x)} - \frac{k\|x\|}{f(x)}\right] = 0.$$

From (45),(46) we can establish (35) with

$$M_i = C \frac{r}{N_1} \max\left[1 + \frac{N_3}{N_1},\ \left(\frac{N_3}{N_1}\right)^i\right], \quad i = 1,2,3 .$$

One can note that, if instead of a) we assume the more restrictive condition to be satisfied:

a$'$) the functional $f(x)$ is convex,

then we get a convex set B (recall that a non-negative functional $f(x)$ is convex if $f(x+y) \le f(x) + f(y)$ for all $x, y \in H$ and $f(t,x) = t f(x)$ for all $t \ge 0$).

Also it is interesting to note that despite the fact that the derivatives of $d_B(x)$ depend on τ , the derivatives of A , defined in Lemma 3 are independent of τ , Namely we have

$$\| \mathcal{D}^{(i)} A(x) \| \leq C(N_1, N_2, N_3) \| x \|^{-i+1} , \quad i = 1,2,3 ,$$

whence it follows that the right-hand side of estimate (43) is independent of τ .

References

1 C.G.Esseen, Acta Math., 77, 1-125(1945).

2 P.Halmos, Measure Theory, Moscow , 1953 (Russian translation).

3 A.Kartan, Differencial calculus, Moscow, 1971 (Russian translation).

4 J.Kuelbs,T.Kurtz, Ann.of Probab., 2, 3, 387-407(1974).

5 S.Lang,Introduction to the theory of differentiable manifolds, Moscow, 1967 (Russian translation)

6 Nguen Zui Tien,Bull.of the Acad. of Sci. of the Georg.SSR,69, 3, 541-543(1973) (in Russian).

7 A.I. Orlov,Teor.Veroyat.i primen., 19, 4(1974) (in Russian).

8 V.I.Paulauskas.Liet.Matem.Rink., 15, 3(1975) (in Russian).

9 V.I.Paulauskas, Liet.Matem.rink.,16, 1(1976)(in Russian).

10 V.I.Paulauskas, Submitted to Teor.Veroyat. i Primen.

11 V.V.Sazonov, Sankhya, Ser.A, 30,2, 205-210 (1968).

12 V.V.Sazonov, Veroyat, i Primen., 14,4 (1969)(in Russian).

13 V.V.Sazonov,Proc.Nat.Acad.Sci.USA,71,1,118-121 (1974).

14 N.N.Vakhaniya, Probability distributions in linear spaces, Tbilisi, 1971(in Russian).

15 V.V.Yurinskii, Mathematics and Statistics (Essays in Honour of Harald Bergstrom), 101-121,Goteborg, 1973.

Vilnius State University
Department of Mathematics
Vilnius

GENERALIZED DIFFUSION PROCESSES

N.I.Portenko

0. The following definition of a homogeneous diffusion process in an m - dimensional Euclidean space R^m is generally accepted.

Definition 1. A gomogeneous Markov process (in a wide sense) with transition probability $P(t,x,\Gamma)$, $t>0$, $x\in R^m$, $\Gamma\in\mathcal{B}^m$ ($\mathcal{B}^m$ is the σ-algebra of Borel subsets in R^m)is called a diffusion process at a point $x\in R^m$ if the following conditions are satisfied:

1) for any $\varepsilon>0$,

$$\lim_{t\downarrow 0}\frac{1}{t}\int_{|y-x|>\varepsilon} P(t,x,dy)=0 ;$$

2) for some $\varepsilon>0$, there exist the limits:

$$\lim_{t\downarrow 0}\frac{1}{t}\int_{|y-x|<\varepsilon}(y-x)P(t,x,dy),$$

$$\lim_{t\downarrow 0}\frac{1}{t}\int_{|y-x|<\varepsilon}(y-x,\theta)P(t,x,dy)$$

for all $\theta\in R^m$. Here (z,θ) is the inner product in R^m .

It is worth mentioning that condition 1) and the existence of the limits in condition 2) for some $\varepsilon>0$ imply their existence for any $\varepsilon>0$ and these limits do not depend on ε .

If a process is a diffusion process at every point $x\in R^m$, then it is called a diffusion process. In this case there exist a function $a(\cdot): R^m \to R^m$ and a function $b(\cdot): R^m\to L^+(R^m)$ ($L^+(R^m)$ is the set of all linear symmetric non-negatively definite operators in R^m) such that, for any $x \in R^m$, the first of the limits in condition 2) coincides with $a(x)$ and the second one with $(b(x)\theta,\theta)$. The function $a(x)$ is called the drift coefficient and $b(x)$ is called the diffusion operator.

In constructing of a diffusion process with given drift coefficient and diffusion operator one can usually use either the analytical method or the method of stochastic differential equations. Both the methods have lead in the long run to existence theorems for diffusions processes. Moreover they have resulted in a generalization of the concept of diffusion process itself ([1] , [2]). It became possible to construct processes with properties very close to those of diffusion processes for which, however, the limits in condition 2) may be locally unbounded and even generalized functions. In this connection it seems quite natural to give the following definition of a generalized diffusion process.

For a homogeneous Markov process with transition probability $P(t,x,\Gamma)$, $t>0$, $x \in R^m$, $\Gamma \in \mathcal{B}^m$, let us define

$$a_t^{(\varepsilon)}(x) = \frac{1}{t}\int\limits_{|y-x|<\varepsilon} (y-x)P(t,x,dy) \quad ,$$

$$(b_t^{(\varepsilon)}(x)\theta,\theta) = \frac{1}{t}\int\limits_{|y-x|<\varepsilon} (y-x,\theta)^2 P(t,x,dy),$$

$$c_t^{(\varepsilon)}(x) = \frac{1}{t}\int\limits_{|y-x|>\varepsilon} P(t,x,dy) \ ,$$

where $\theta, x \in R^m, \varepsilon>0$, $t>0$. Let $C_0(R^m)$ be the set of all real continuous finite functions on R^m and $\overset{s}{L}(R^m)$ be the set of all linear symmetric operators acting in R^m .

D e f i n i t i o n 2 . A homogeneous Markov process (in a wide sense) with transition probability $P(t,x,\Gamma)$, $t>0$, $x\in R^m$, $\Gamma\in\mathcal{B}^m$, is called a generalized diffusion process if the following conditions are satisfied:

1) for any $\varepsilon>0$ and for any $\varphi\in C_0(R^m)$,

$$\lim_{t\downarrow 0}\int_{R^m}\varphi(x)c_t^{(\varepsilon)}(x)dx=0 ;$$

2) there exist linear functionals $A(\varphi)$ and $B(\varphi)$, $\varphi\in C_0(R^m)$, taking values in R^m and $L^s(R^m)$ respectively, such that, for some $\varepsilon>0$ and for all $\varphi\in C_0(R^m)$,

$$\lim_{t\downarrow 0}\int_{R^m}\varphi(x)a_t^{(\varepsilon)}(x)dx=A(\varphi),$$

$$\lim_{t\downarrow 0}\int_{R^m}\varphi(x)b_t^{(\varepsilon)}(x)dx=B(\varphi).$$

It is evident that, if $\varphi(x)\geq 0$ for all $x\in R^m$, then $(B(\varphi)\theta,\theta)\geq 0$ for all $\theta\in R^m$. As in the case of diffusion processes, one can prove that the limits in condition 2) exist for all $\varepsilon>0$ and do not depend on ε.

In the present paper we formulate some existence theorems for generalized diffusion processes under the assumption that the diffusion operator is a sufficiently regular function and the drift coefficient either satisfies some integrability condition or is a generalized function. We also prove that the processes considered are solutions of stochastic differential equations.

1. Let us suppose that we are given a function $b(\cdot)$: $R^m\longrightarrow L^+(R^m)$ satisfying the following conditions:

a) $$\sup_{x \in R^m} (\| b(x) \| + \| b^{-1}(x) \| < \infty ,$$

where $\| b(x) \|$ is the norm of the operator $b(x)$ and $b^{-1}(x)$ is its inverse;

b) $$\| b(x) - b(y) \| < L \, |x - y|^{\alpha} ,$$

where L and α are positive constants, $0 < \alpha \leq 1$ and $|x|$ is the norm of $x \in R^m$.

Under these conditions, there exists a fundamental solution $g(t,x,y)$, $t>0$, $x, y \in R^m$, of the equation

$$\frac{\partial u}{\partial t} = \frac{1}{2} sp(b(x) \cdot u''_{xx}) \, .$$

where u''_{xx} is the second derivative operator of $u(t,x)$ with respect to x. As was shown in [3], the function $g(t,x,y)$ satisfies the following conditions:

1) $g(t,x,y)$ is non-negative, continuous with respect to t, x, y, continuously differentiable with respect to x and, for $t \in [0,T]$, $x, y \in R^m$,

$$g(t,x,y) \leq K_T \cdot t^{-\frac{m}{2}} \exp\{-\mu \frac{|y-x|^2}{t}\} ,$$

$$|\nabla_x g(t,x,y)| \leq K_T \cdot t^{-\frac{m+1}{2}} \exp\{-\mu \frac{|y-x|^2}{t}\} ,$$

where $\nabla_x g$ is the first derivative vector of g with respect to x, K_T and μ are positive constants and $K_T < \infty$ for $T < \infty$;

2) for all $t>0$, $s>0$, $x,z\in R^m$,

$$\int_{R^m} g(t,x,y)g(s,y,z)dy = g(t+s,x,z);$$

3) for all $t>0$, $x\in R^m$,

$$\int_{R^m} g(t,x,y)dy = 1 .$$

For a function $a(x): R^m \longrightarrow R^m$ let us define the norm

$$\|a\|_p = \Big(\int_{R^m} |a(x)|^p dx\Big)^{1/p}, \quad p>1 .$$

Lemma 1. If a function $b(x): R^m \longrightarrow L^+R^m$ satisfies conditions a) - b) and a function $a(\cdot): R^m \longrightarrow R^m$ is such that $\|a\|_p < \infty$ for some $p>m$, then the equation

$$G(t,x,y) = g(t,x,y) + \int_0^t ds \int_{R^m} g(s,x,z)(\nabla_z G(t-s,z,y), a(z))dz \qquad (1)$$

has a unique solution $G(t,x,y)$, $t>0$, $x,y\in R^m$, satisfying conditions 1) - 3) with the same constant μ and a constant K'_T depending only on p, $\|a\|_p$, T, K_T and μ.

The proof is elementary.

Put

$$P^{a,b}(t,x,\Gamma) = \int_\Gamma G(t,x,y)dy, \ t>0, \ x\in R^m, \ \Gamma\in\mathcal{B}^m,$$

where $G(t,x,y)$ is the solution of equation (1). Conditions 1) - 3) imply that $P^{a,b}(t,x,\Gamma)$ is the transition probability of some homogeneous Markov process.

T h e o r e m 1 . Suppose that we are given a function $a(\cdot): R^m \longrightarrow R^m$ with $\|a\|_p < \infty$ for some $p > m$ and a function $b(\cdot): R^m \longrightarrow L^+(R^m)$ satisfying conditions a) - b). Then the above constructed homogeneous Markov process with transition probability $P^{a,b}(t, x, \Gamma)$ is a generalized diffusion process such that

$$A(\varphi) = \int_{R^m} a(x)\varphi(x)dx, \; B(\varphi) = \int_{R^m} b(x)\varphi(x)dx, \quad \varphi \in C_0(R^m).$$

Besides, for any $\varepsilon > 0$,

$$\lim_{t \downarrow 0} \sup_{x \in R^m} c_t^{(\varepsilon)}(x) = 0, \quad \lim_{t \downarrow 0} \sup_{x \in R^m} \| b_t^{(\varepsilon)}(x) - b(x) \| = 0,$$

where $c_t^{(\varepsilon)}(x)$ and $b_t^{(\varepsilon)}(x)$ are constructed using transition probability $P^{a,b}(t, x, \Gamma)$ in the same way as in No. 0 .

The proof of this theorem is given in [2] .

Let Ω be the space of all continuous functions $\omega(t)$, $t \in [0, \infty)$, with values in R^m . Let $\mathcal{M}_t$, $t \in [0, \infty)$, be the minimal σ - algebra of subsets of Ω containing all the subsets of the form

$$C_{s_1, s_2, \dots, s_k}(\Gamma_1, \dots, \Gamma_k) = \{\omega(\cdot): x(s_1, \omega) \in \Gamma_1, \dots, x(s_k, \omega) \in \Gamma_k\}$$

where $k = 1, 2, \dots; \quad 0 \le s_1 \le \dots < s_k \le t; \; \Gamma_1, \dots, \Gamma_k \in \mathcal{B}^m$.

Further, let us define the function $P_x^{a,b}$, $x \in R^m$, by

$$P_x^{a,b}\{C_{0, s_1, \dots, s_k}(\Gamma_0, \Gamma_1, \dots, \Gamma_k)\} =$$

$$= \chi_{\Gamma_0}(x) \int_{\Gamma_1} P^{a,b}(s_1, x, dy_1) \int_{\Gamma_2} P^{a,b}(s_2 - s_1, y_1, dy_2) \dots \int_{\Gamma_k} P^{a,b}(s_k - s_{k-1}, y_{k-1}, dy_k),$$

where χ_Γ is the indicator of Γ; $0 < s_1 < \ldots < s_k$; $\Gamma_0, \Gamma_1, \ldots \ldots \Gamma_k \in \mathcal{B}^m$; $k=1,2,\ldots$ It follows from Theorem 1 that the function $P_x^{a,b}$ can be extended to a probability measure on the σ - algebra $\mathcal{M}_\infty$ and $P_x^{a,b}\{x(0)=x\}=1$. Thus the generalized diffusion process $(x(t), \mathcal{M}_t, P_x^{a,b})$ turns out to be continuous.

T h e o r e m 2 . Let us suppose that functions $a(x)$ and $b(x)$ satisfy conditions of Theorem 1, and let $(x(t), \mathcal{M}_t, P_x^{a,b})$ be the above constructed continuous generalized diffusion process. Then the process

$$\xi(t) = x(t) - x(0) - \int_0^t a(x(s))\, ds$$

is a continuous square-integrable martingale with respect to $(\mathcal{M}_t, P_x^{a,b})$ such that

$$\langle \xi, \xi \rangle_t = \int_0^t b(x(s))\, ds \quad .$$

This theorem is also proved in [2].

The next theorem demonstrates that in many cases the measure $P_x^{a,b}$ can be obtained by an absolutely continuous change of the measure $P_x^{0,b}$.

T h e o r e m 3 . If the conditions of Theorem 1 are fulfilled, then, for $m \geq 2$ and $p > m$ and also for $m=1$ and $p \geqslant 2$, the restrictions of the measures $P_x^{a,b}$ and $P_x^{0,b}$ on the σ -algebras $\mathcal{M}_T$ are equivalent for any $T < \infty$. If $m=1$ and $1<p<2$ then, in general, the measures $P_x^{a,b}$ and $P_x^{0,b}$ can be non-equivalent on any σ - algebra $\mathcal{M}_T$.

P r o o f . First we want to show that, for $m=1$ and

$\|a\|_p < \infty$ for some $p \in (1,2)$, the measures $P_x^{a,b}$ and $P_x^{a,b}$ can be, in general, non-equivalent on any σ - algebra $\mathcal{M}_T$

Put $b(x) \equiv 1$. Then $P_0^{a,1}$ is the Wiener measure. If, for some $N < \infty$

$$\int_{-N}^{N} (a(x))^2 \, dx = \infty \quad ,$$

then (see [5]), for any $T > 0$,

$$P_0^{a,1} \{ \int_0^T (a(x(s)))^2 ds < \infty \} < 1 \, .$$

Hence, the restrictions of the measures $P_0^{a,1}$ and $P_0^{a,1}$ on the σ - algebra $\mathcal{M}_T$ are non-equivalent since (see [6]) they are equivalent if and only if

$$\int_0^T (a(x(s)))^2 \, ds < \infty \quad (mod\ P_0^{a,1},\ mod\ P_0^{a,1}).$$

Now let us prove the first statement of the theorem. Let $f(\cdot): R^m \longrightarrow R^m$ be any measurable function. Since $(x(t), \mathcal{M}_T, P_x^{a,b})$ is a homogeneous Markov process with transition probability density $G(t, x, y)$,

$$M_x^{a,b} \{ \int_s^T | f(x(\tau)|^2 d\tau \mid \mathcal{M}_s \} = M_{x(s)}^{a,b} \int_0^{t-s} | f(x(\tau)) |^2 d\tau =$$

$$= \int_0^{t-s} d\tau \int_{R^m} | f(y) |^2 G(t, x(s), y) dy \, ,$$

where $M_x^{a,b} \eta$ denotes the expectation of a random variable η with respect to the measure $P_x^{a,b}$. Suppose that the function f is such that $\|f\|_p < \infty$ for some $p \geq 2$ if $m = 1$ and $\|f\|_p < \infty$ for some $p > m$ if $m \geq 2$. Then applying the

Hölder inequility to the last integral and using the estimate of the function $G(t,x,y)$ from condition 1), one gets:

$$\int_{R^m} |f(y)|^2 G(\tau, x(s), y)\, dy \leq$$

$$\leq \|f\|_p^2 \Big(\int_{R^m} (G(\tau, x(s), y))^{\frac{p}{p-2}} dy \Big)^{\frac{p-2}{p}} \leq const \cdot \|f\|_p^2 \cdot \tau^{-\frac{m}{p}}.$$

Therefore

$$M_x^{a,b} \{ \int_0^T |f(x(\tau))|^2 d\tau \mid \mathcal{M}_s \} \leq const \cdot \|f\|_p^2 \cdot (t-s)^{\frac{p-m}{p}} \quad . \quad (2)$$

By Lemma 1 [4] this implies

$$M_x^{a,b} \exp \{ \lambda \int_0^T |f(x(\tau))|^2 d\tau \} < \infty$$

for all real $\lambda, x \in R^m$ and $T < \infty$, and, on the other hand, this implies that the stochastic integrals

$$\int_0^T (f(x(\tau)), d\xi(\tau))$$

with $\xi(t)$ described in Theorem 2, are well-defined with respect to the measure $P_x^{a,b}$. In particular, the stochastic integral

$$\int_0^t (b^{-1}(x(s))\, a(x(s)), dx(s))$$

is well-defined with respect to the measure $P_x^{a,b}$. And besides, one can prove that

$$M_x^{a,b} \exp \{ \tfrac{1}{2} \int_0^t (b^{-1}(x(s))\, a(x(s)), a(x(s)))\, ds \} < \infty \quad . \quad (3)$$

Now put

$$R_t = \exp\{\int_0^t (b^{-1}a, dx(s)) - \frac{1}{2}\int_0^t (b^{-1}a, a)ds\} .$$

Then from (3) and [7] it follows that the process $(R_t, \mathcal{M}_t, P_x^{a,b})$ is a martingale such that $R_t > 0$ and $M_x^{a,b} R_t = 1$ Thus one can defined measure Q_x on σ - algebra $\mathcal{M}_T$, $T<\infty$ by the formula

$$dQ_x = R_T dP_x^{a,b} .$$

It is not difficult to verify that the process

$$\xi(t) = x(t) - x(0) - \int_0^t a(x)s))ds , \quad 0 \le t \le T ,$$

is a square-integrable martingale with respect to the measure Q_x such that

$$\langle \xi, \xi \rangle_t = \int_0^t b(x(\tau))d\tau , \quad 0 \le t \le T .$$

By Theorem 5 of [4] the measure Q_x is the restriction of the measure P_x to the σ - algebra $\mathcal{M}_T$.

In analogous way one can prove that

$$dP_x^{0,b} = S_T \, dP_x^{a,b}$$

on σ - algebra $\mathcal{M}_T$ for any $T < \infty$, where $S_T = R^{-1}$. This completes the proof.

2. Now we want to show that the results analogous to those of No.1 are valid even in the case $m=1$ and $\| a \|_1 < \infty$. Suppose, for simplicity, that $b(x) \equiv 1$.

Let we are given a sequence of continuous bounded functions

$a_n(\cdot): R^1 \longrightarrow R^1$, $n = 1, 2, \dots$. Put $P_x^{(n)} = P_x^{a_n, 1}$ (for definitions of the measures $P_x^{a,b}$ see No. 1). The measure $P_x^{(n)}$ is the unique measure on the space $(\Omega, \mathcal{M}_\infty)$ such that $P_x^{(n)}\{x(0) = x\} = 1$ and the process

$$\xi_n(t) = x(t) - x(0) - \int_0^t a_n(x(s))\,ds$$

is a continuous square-integrable martingale with respect to $(\mathcal{M}_t, P_x^{(n)})$ with the characteristic $\langle \xi_n, \xi_n \rangle_t = t$.

Denote

$$A_n(x) = \int_0^x a_n(y)\,dy, \quad x \in R^1, \quad n = 1, 2, \dots .$$

L e m m a 2 . If the continuous bounded functions $a_n(\cdot): R^1 \longrightarrow R^1$ satisfy the condition:

$$\sup_n \sup_{x \in R^1} |A_n(x)| < \infty , \tag{4}$$

then, for any $x \in R^1$, the sequence of the restrictions of the measures $P_x^{(n)}$ to the σ- algebra $\mathcal{M}_T$, $T < \infty$, is weakly compact.

P r o o f. Put

$$f_n(x) = \int_0^x e^{-2A_n(y)}\,dy, \quad x \in R^1, \quad n = 1, 2, \dots .$$

By (4), $f_n(x)$ satisfy the inequalities

$$c|x-y| \le |f_n(x) - f_n(y)| \le C|x-y|, \ x, y \in R^1, \ n = 1, 2, \dots ,$$

where c and C are positive constants. Ito's formula implies that $f_n(x(t)) - f_n(x(0))$ is a square-integrable martingale with respect to $(\mathcal{M}_t, P_x^{(n)})$ whose characteristic is

$$\int_0^t f_n'^2(x(s))\,ds \quad .$$

Since the functions f_n' are uniformly bounded,

$$M_x^{(n)}|x(t)-x(s)|^4 \le C^{-1} M_x^{(n)}|f_n(x(t))-f_n(x(s))|^4 \le const\cdot(t-s)^2 \quad ,$$

where $0\le s\le t<\infty$, $M_x^{(n)}$ is the expectation symbol with respect to the measure $P_x^{(n)}$. This inequality provides the proof.

It is clear that the assumption on continuouty and boundedness of functions $a_n(x)$ in Lemma 2 is unnecessary. It can be replaced by a more generalone; for instance: for each n, there exists a $p_n>1$ (it is possible that $p_n=\infty$) such that $\|a_n\|_{p_n}<\infty$. In this case the function $f_n(x)$, in general, has no second derivative. However, we can apply Ito's formula to it for it is justified by the inequalities in condition 1) No.1. One can understand that condition (4) does not guarantee that these inequalities are valid uniformly in n. The next lemma shows that, under some additional assumptions, such uniformly estimates can be obtained.

L e m m a 3 . Suppose that the sequence of continuous bounded functions $a_n(\cdot): R^1 \longrightarrow R^1$, $n=1,2,\ldots$ is such that

$$\sup_n \|a_n\|_1<\infty,\quad \lim_{\varepsilon\downarrow 0}\ \sup_{n,x}\int_{|y-x|<\varepsilon}|a_n(y)|\,dy=0 \quad . \tag{5}$$

Then there exists $\lambda_0>0$ such that

$$M_x^{(n)}\int_0^\infty e^{-\lambda t}|\varphi(x(t))|\,dt \le C_\lambda\|\varphi\|_1 \tag{6}$$

for every $\lambda \geq \lambda_0$ and every $\varphi(\cdot): R^1 \longrightarrow R^1$ with $\|\varphi\|_1 < \infty$, where C_λ is a certain bounded function of λ.

P r o o f. Put

$$q(\lambda) = \sup_{n,x} \int_{R^1} e^{-|y-x|\sqrt{2\lambda}} |a_n(y)| dy, \quad \lambda \geq 0.$$

It is clear that

$$q(\lambda) \leq e^{-\varepsilon\sqrt{2\lambda}} \sup_n \|a_n\|_1 + \sup_{n,x} \int_{|y-x|<\varepsilon} |a_n(y)| dy, \qquad \varepsilon > 0.$$

By (5) there exists a $\lambda_0 \geq 0$ such that $\sup_{\lambda \geq \lambda_0} q(\lambda) < 1$.

Now, for any bounded continuous functions $\varphi(\cdot): R^1 \longrightarrow R^1$, put

$$u_n(x,\lambda) = M_x^{(n)} \int_0^\infty e^{-\lambda t} |\varphi(x(t))| dt.$$

Equation (1) implies that the function $u_n(x,\lambda)$ is the unique continuous bounded solution of the equation

$$u_n(x,\lambda) = \frac{1}{\sqrt{2\lambda}} \int_{R^1} \exp\{-|y-x|\sqrt{2\lambda}\} |\varphi(y)| dy +$$

$$+ \frac{1}{\sqrt{2\lambda}} \int_{R^1} \exp\{-|y-x|\sqrt{2\lambda}\} \frac{\partial u_n(y,\lambda)}{\partial y} a_n(y) dy,$$

having the continuous and bounded derivative. The method of successive approximations, applied to the last equation, gives:

$$u_n(x,\lambda) \leq \frac{1}{\sqrt{2\lambda}} \cdot \frac{1}{1-q(\lambda)} \|\varphi\|_1 .$$

Thus the lemma is proved for continuous and bounded functions φ. By the routine limit procedure, one can obtain the proof in the general case.

C o r o l l a r y 1 . Since (5) implies (4), under the the assumptions of Lemma 3, the sequence of measures $P_x^{(n)}$, for any $x \in R^m$, is compact with respect to the weak convergence of their restrictions to the σ - algebras $\mathcal{M}_T$, $T < \infty$, This means that any subsequence of this sequense contains a subsequence $P_x^{(n)}$ such that, for any $T < \infty$, the restrictions of the measures $P_x^{(n_k)}$ on σ - algebra $\mathcal{M}_T$ are weakly convergent. If P_x is the limiting measure, in the described sense, for a subsequence $P_x^{(n_k)}$, then by passing to the limit in (6) for this subsequence, we obtain:

$$M_x \int_0^\infty e^{-\lambda t} |\varphi(x(t))| \, dt \le C_\lambda \|\varphi\|_1 \quad , \qquad (6^1)$$

where M_x is the expectation symbol with respect to the measure P_x and $\varphi(\cdot): R^1 \longrightarrow R^1$ is a continuous bounded function. By the usual limiting procedure, this inequality can be extended to all functions $\varphi(x)$, $\|\varphi\|_1 < \infty$. Thus every limiting measure for the sequence $P_x^{(n)}$ satisfies inequality (6^1).

C o r o l l a r y 2 . If in (6) $\varphi(x) = \chi_\Gamma(x)$, where Γ is a Borel set of finite Lebesgue measure ($mes\,\Gamma < \infty$), then

$$\int_0^T P^{(n)}(t,x,\Gamma)dt \le e^{\lambda T} \int_0^\infty e^{-\lambda t} P^{(n)}(t,x,\Gamma)dt \le e^{\lambda T} C_\lambda \, mes\,\Gamma ,$$

where $\lambda \ge \lambda_0$, $P_n(t,x,\Gamma) = P_x^{(n)}\{x(t) \in \Gamma\}$, $T < \infty$. Hence, for every $T < \infty$, there exists a constant H_T such that

$$\int_0^T P^{(n)}(t,x,\Gamma)dt \le H_T \cdot mes\,\Gamma , \; x \in R^1, \; n = 1, 2, \ldots \; .$$

T h e o r e m 4 . Suppose that a sequence of continuous bounded functions $a_n(\cdot): R^1 \longrightarrow R^1, n=1,2,\ldots$ satisfies conditions (5) and there exists a function $a(\cdot): R^1 \longrightarrow R^1$ with $\|a\|_1 < \infty$ such that

$$\lim_{n\to\infty} \int_{R^1} a_n(x)\varphi(x)dx = \int_{R^1} a(x)\varphi(x)dx \tag{7}$$

for any $\varphi \in C_0(R^1)$. Then, for any $x \in R^1$, there exists a probability measure P_x on $(\Omega, \mathcal{M}_\infty)$ such that the sequence of the restrictions of the measures $P_x^{(n)}$ to the σ-algebra $\mathcal{M}_T$, $T<\infty$, is weakly convergent to the restriction of the measure P_x to the σ-algebra $\mathcal{M}_T$. In addition, $P_x\{x(0)=x\}=1$ and the process

$$x(t)-x(0)-\int_0^t a(x(s))ds$$

is a square-integrable martingale with respect to $(\mathcal{M}_t, P_x)$ whose characteristic is t .

P r o o f. Let $x \in R^1$ be fixed. By Lemma 2 there exists a subsequence $P_x^{(n_k)}$ $(n_k \longrightarrow \infty)$ of the sequence of measures $P_x^{(n)}$ and a probability measure P_x on $(\Omega, \mathcal{M}_\infty)$ such that, for any $T<\infty$, the restrictions of the measures $P_x^{(n_k)}$ to the σ-algebra $\mathcal{M}_T$ are weakly convergent to the restriction of the measure P_x to the σ-algebra $\mathcal{M}_T$.

Further, (7) implies (in the notations of Lemma 2) that, for any $y \in R^1$,

$$\lim_{n\to\infty} A_n(y) = A(y) = \int_0^y a(z)dz .$$

This means that $f_n(y) \longrightarrow f(y)$ for any $y \in R^m$ where

$$f(y)=\int_0^y \exp\{-2A(z)\}\,dz \ .$$

Moreover, since the functions $f_n(y)$ are uniformly Lipschitzian, one can assume that, for the chosen subsequence $n_k \to \infty$, $f_{n_k}(y)$ is locally uniformly convergent to $f(y)$.

Now let us show that the process $f(x(t)) - f(x(0))$ is a square-integrable martingale with respect to $(\mathcal{M}_T, P_x)$ whose characteristic is

$$\int_0^t f'^2(x(s))\,ds .$$

Let $0 \leqslant s \leqslant t$ and let η be a continuous bounded functional of paths $x(\tau)$, $0 \leqslant \tau \leqslant s$. It is sufficient to show that the following relations are true:

$$M_x \eta\,[f(x(t)) - f(x(s))] = 0 \quad ,$$

$$M_x \eta\,[f(x(t)) - f(x(s))]^2 = M_x \eta \int_s^t f'^2(x(\tau))\,d\tau \ .$$

Since $f_{n_k}(x(t)) - f_{n_k}(x(0))$ is a martingale with respect to $(\mathcal{M}_t, P_x^{(n_k)})$,

$$M_x \eta [f(x(t)) - f(x(s))] = \lim_{n_k \to \infty} M_x^{(n_k)} \eta [f(x(t)) - f(x(s))] =$$

$$= \lim_{n_k \to \infty} M_x^{(n_k)} \eta [f_{n_k}(x(s)) - f(x(s))] + \lim_{n_k \to \infty} M_x^{(n_k)} \eta [f(x(t)) - f_{n_k}(x(t))] \ .$$

Both the limits on the right are equal to zero. If follows that the supports of the restrictions of the measures $P_x^{(n_k)}$ to the σ-algebra

$\mathcal{M}_T$ are concentrated on a compact inside a set of arbitrarily small measure, uniformly in n_K. On these compacts,

$f_{n_K}(x(\tau)) \longrightarrow f(x(\tau))$ uniformly with recpect to $\tau \in [s, t]$

Thus the first equality in (8) has been proved.

Similarly,

$$M_x \eta [f(x(t)) - f(x(s))]^2 = \lim_{n_K \to \infty} M_x^{(n_K)} \eta [f(x(t)) - f(x(s))]^2 =$$

$$= M_x \eta \int_0^t f'^2(x(\tau)) d\tau + \lim_{n_K \to \infty} M_x^{(n_K)} \eta \int_s^t [f'_{n_K}(x(\tau)) - f'^2(x(\tau))] d\tau .$$

Up to an arbitrarily small number $\varepsilon > 0$, the last limit is equal to

$$I = \lim_{n_K \to \infty} M_x^{(n_K)} \eta \int_s^t [f'^2_{n_K}(x(\tau)) - f'^2(x(\tau))] \chi_{C_\varepsilon}(x(\tau)) d\tau ,$$

where C_ε depends on ε only and $\chi_{C_\varepsilon}(x(\tau))$ is the indicator of the event $\{|x(\tau)| \le C_\varepsilon\}$. To estimate I we use the Markov property of the process $(x(t), \mathcal{M}_t, P_x^{(n_K)})$ and Corollary 2 of Lemma 3 Since $f'_n(y) \longrightarrow f'(y)$ for all $y \in R^1$, we get

$$|I| \le const \cdot \overline{\lim_{n_K \to \infty}} M_x^{(n_K)} \int_0^{t-s} d\tau \int_{|y| \le C_\varepsilon} |f'^2_{n_K}(y) - f'^2(y)| P^{(n_K)}(\tau, x(s), dy) \le$$

$$\le const \cdot \lim_{n_K \to \infty} \int_{|y| \le C_\varepsilon} |f'^2_{n_K}(y) - f'^2(y)| dy = 0 .$$

Thus the second equality in (8) has been also proved.

We have shown that any measure P_x, limiting for some subsequence of the sequence $P_x^{(n)}$, is such that $P_x\{x(0) = x\} = 1$ and the process $f(x(t) - f(x(0))$ is a square-integrable martingale with respect to $(\mathcal{M}_t, P_x)$ whose characteristic is

$$\int_0^t f'^2(x(s))\,ds.$$

By the uniqueness theorem in [8] , such a measure is unique. This and Lemma 2 imply that the sequence $P_x^{(n)}$ itself converges in the above sense to the measure P_x . For x was arbitrary in the previous arguements, P_x , with the described properties, exist for every $x \in R^1$.

Further, if the function $a(x)$ is continuous, then $f^{-1}(x)$ is twice continuously differentiable. Applying Ito's formula to the process $f^{-1}(\xi(t))$ with $\xi(t) = f(x(t))$, one can easily prove that the process

$$x(t) - x(0) - \int_0^t a(x(s))\,ds$$

is a square-integrable martingale with respect to $(\mathcal{M}_t, P_x)$ whose characteristic is t . In general, if $a(x)$ satisfies the single condition $\|a\|_1 < \infty$, the derivative $(f^{-1})''(x)$ exists only for almost every $x \in R^1$, Nevertheless the arguement based on Lemma 3 is sufficient to apply Ito's formula to the process $f^{-1}(\xi(t))$ even in this case.

In fact, let $g_n(x)$ be a sequence of twice continuously differentiable functions such that $\|g_n'' - (f^{-1})''\|_1 \longrightarrow 0$ as $n \longrightarrow \infty$ and $g_n \longrightarrow f^{-1}$, $g_n' \longrightarrow (f^{-1})$ locally uniformly.Then it follows by Ito's formula that the process

$$g_n(f(x(t))) - g_n(f(x)) - \frac{1}{2}\int_0^t g_n''(f(x(s))\cdot f'^2(x(s))\,ds$$

is a square-integrable martingale with respect to $(\mathcal{M}_T, P_x)$ whose characteristic is

$$\int_0^t g_n'^2(f(x(s)))\, f'^2(x(s))\, ds.$$

Since $(f^{-1})''(f(x))f'^2(x) = 2a(x)$, by Corollary 1 of Lemma 3 we get:

$$M_x \sup_{0\le t\le T} \Big|\int_0^t [\tfrac{1}{2} g_n''(f(x(s)))f'^2(x(s)) - a(x(s))]\,ds\Big| \le$$

$$\le const\cdot e^{\lambda T} M_x \int_0^\infty e^{-\lambda t} |g_n''(f(x(s))) - (f^{-1})(f(x(s)))|\,ds \le$$

$$\le const\cdot \int_{R^1} |g_n''(f(y)) - (f^{-1})''(f(x))|\,dy \le const\cdot \|g_n'' - (f^{-1})''\|_1 \to 0$$

for some $\lambda > 0$. Similarly, one can prove that

$$M_x \sup_{0\le t\le T} \Big|\int_0^t [g_n'^2(f(x(s)))\, f'^2(x(s)) - 1]\,ds\Big| \longrightarrow 0$$

as $n \longrightarrow \infty$. Hence, the process

$$x(t) - x(0) - \int_0^t a(x(s))\,ds$$

is a square-integrable martingale with respect to $(\mathcal{M}_t, P_x)$ and its characteristic is t. This completes the proof.

C o r o l l a r y 1. To every function $a(\cdot): R^1 \longrightarrow R^1$ with $\|a\|_1 < \infty$ there corresponds a unique, for every $x \in R^1$, measure P_x^a, defined on $(\Omega, \mathcal{M}_\infty)$ with the following properties:

1) for any $\lambda \ge \lambda_0 \ge 0$ and for all $\varphi(\cdot): R^1 \longrightarrow R^1$ with

$\|\varphi\|_1 < \infty$,

$$M_x^a \int_0^\infty e^{-\lambda t} |\varphi(x(t))| dt \le C_\lambda \|\varphi\|_1 ,$$

where C_λ is a certain function of λ, $0 < C_\lambda < \infty$, and M_x^a is the expectation symbol for the measure P_x^a ;

2) $P_x^a\{x(0)=x\} = 1$ and the process

$$x(t) - x(0) - \int_0^t a(x(s))\, ds$$

is a square-integrable martingale with respect to $(\mathcal{M}_t, P_x^a)$ whose characteristic is t .

In fact, the measure, constructed in Theorem 5, satisfies conditions 1) - 2), If there would exist an other measure with properties 1) - 2), then, applying Ito's formula to the process $f(x(t))$ where

$$f(x) = \int_0^x e^{-2A(y)} dy , \quad A(x) = \int_0^x a(y) dy , \quad x \in R^1 ,$$

we could come to the existence of two different measures such that the process $f(x(t)) - f(x(0))$ is a square-integrable martingale with characteristic

$$\int_0^t f'^2(x(s))\, ds$$

with respect to both of them. But this contradicts the uniqueness theorem in [8] .

C o r o l l a r y 2 . If the function $a(\cdot): R^1 \longrightarrow R^1$ satisfies the condition $\|a\|_1 < \infty$, then the process $(x(t), \mathcal{M}_t, P_x^a)$ is a homogeneous Markov process.

From theorem 4 one can easily get

T h e o r e m 5 . If a function $a(\cdot): R^1 \quad R^1$ id such that $\|a\|_1 < \infty$, then the process $(x(t), \mathcal{M}_t, P_x^a)$ is a generalized diffusion process with

$$A(\varphi) = \int_{R^1} a(x)\varphi(x)\,dx,\ B(\varphi) = \int_{R^1} \varphi(x)\,dx,\quad \varphi \in C_0(R^1).$$

3. The following theorem, proved in [2] , demonstrates that there exist generalized diffusion processes with drift coefficients being generalized functions.

T h e o r e m 6. There exists a generalized diffusion process such that

$$A(\varphi) = c\varphi(0),\ B(\varphi) = \int_{R^1} \varphi(x)\,dx,\quad \varphi \in C_0(R^1)$$

and, for any $\varepsilon > 0$,

$$\lim_{t \downarrow 0} \sup_{x \in R^1} C_t^{(\varepsilon)}(x) = 0 ,$$

where $C_t^{(\varepsilon)}(x)$ is defined in No.0 and c is a constant, $0 \le c \le 1$

The corresponding semigroup is defined by the formula

$$T_t\varphi(x) = \int_{R^1} (2\pi t)^{-1/2} \exp\{-\frac{(y-x)^2}{2t}\}\varphi(y)\,dy +$$

$$+ c\int_0^\infty (2\pi t)^{-1/2} \exp\{-\frac{(y+|x|)^2}{2t}\}\,[\varphi(y) - \varphi(-y)]\,dy .$$

Let P_x, $x \in R^1$, be a measure on the space $(\Omega, \mathcal{M}_\infty)$ such that, for all measurable bounded functions $\varphi(\cdot): R^1 \longrightarrow R^1$

$$M_x\,\varphi(x(t)) = T_t\,\varphi(x) .$$

where T_t is defined by the formula (9) and M_x corresponds to to integrating with respect to the measure P_x . Then the process $(x(t), \mathcal{M}_t, P_x)$ is a continuous generalized diffusion process with the unit diffusion coefficient and the drift coefficient $c\delta(x)$, where $\delta(x)$ is Dirac's δ- function.

Further, denote

$$f_t(x)=\frac{c}{\sqrt{2\pi}}\int_0^t e^{-\frac{x^2}{2s}}\frac{ds}{\sqrt{s}}, \quad t\geq 0, \quad x\in R^1 .$$

There exists a homogeneous continuous non-negative functional η_t of the process $(x(t), \mathcal{M}_t , P_x)$ such that

$$M_x\eta_t = f_t(x) , \; t\geq 0 , \; x\in R^1 .$$

In addition,

$$\eta_t = \underset{h\downarrow 0}{\text{l.i.m.}} \int_0^t \frac{f_h(x(s))}{h} ds .$$

Since, for $\varphi\in C_0(R^1)$,

$$\lim_{h\downarrow 0}\int_{R^1}\varphi(x)\frac{f_h(x)}{h}dx = c\varphi(0) ,$$

it is natural to write

$$\eta(t)=\int_0^t c\delta(x(s))ds .$$

T h e o r e m 7. The process

$$x(t)-x(0)-\int_0^t c\delta(x(s))ds$$

is a continuous square-integrable martingale with respect to $(\mathcal{M}_t, P_x)$ whose characteristic is t .

This theorem was also proved in [2] .

Finally, using the arguement similar to that of Theorem 4, one gets

T h e o r e m 8 . Let a sequence of non-negative measurable functions $a_n(\cdot): R^1 \longrightarrow R^1$ be such that, for any continuous bounded function $\varphi(\cdot): R^1 \longrightarrow R^1$,

$$\lim_{h \quad \infty} \int_{R^1} \varphi(x)\, a_n(x)\, dx = a\varphi(0) ,$$

where a is a non-negative constant. Let $P_x^{(n)}$, $x \in R^1$, be a measure on $(\Omega, \mathcal{M}_\infty)$ such that $P_x^{(n)}\{x(0)=x\}=1$ and the process

$$x(t) - x(0) - \int_0^t a_n(x(s))\, ds$$

is a continuous square-integrable martingale with respect to $(\mathcal{M}_t, P_x^{(n)})$ with the characteristic t . Then, for all $x \in R^1$, the restrictions of the measures $P_x^{(n)}$ to σ-algebras $\mathcal{M}_t$, $T < \infty$, are weakly convergent to the corresponding restrictions of the measures P_x determined by the semigroup (9) with the constant $c = th\, a = \dfrac{e^{a} - e^{-a}}{e^{a} + e^{-a}}$.

REFERENCES

1 Krylov N.V., On quasi-diffusion processes, Theoriya Veroyatn. i Primen., XI,3 (1966), 424–443 (in Russian).

2 Portenko N.I., Diffusion processes with irregular drift, Trudy Shkoly-Seminara po Teorii Sluchainyh Protsessov (Druskininkai, November 25-30, 1974), part II, 127-146, Vilnius,1975 (in Russian).

3 Friedman A., Partial differential equations of parabolic type, Prentice-Hall, Inc., Englewood Cliffs., N.Y.1964.

4 Portenko N.I., Diffusion processes with unbounded drift coefficient, Teoriya Veroyatn. i Primen., XX, 1(1975),29-39 (in Russian).

5 Skorohod A.V., Slobodenyuk N.P., Limit theorems for random walks, Kiev, 1970 (in Russian).

6 Lipcer R.Š ., Shiryaev A.N., Statistics of stochastic processes, Moscow, 1974 (in Russian).

7 Novikov A.A., On a certain identity for stochastic integrals, Theoriya Veroyatn.Primen., XYII, 4(1972), 761-765 (in Russian).

8 Stroock D.W.,Varadhan S.R.S., Diffusion processes with continuous coefficients, I, Communications of Pure and Applied Mathematics, 22(1969), 345–400.

Mathematical Institute, Academy of
Sciences of the Ukrainian SSR, Kiev

SEMIFIELDS AND PROBABILITY THEORY

T.A. SARYMSAKOV

The aim of this work is to give an algebraic exposition of probability theory. To this end we introduce and systematically use the concept of semifield. A.N. Kolmogorov, in his famous axiomatics of probability theory begins with the concept of an elementary event. This enabled him to build the whole theory on the basis of set-theoretical conceptions. In fact we deal with events but not with elementary events, that is why V.I. Glivenko suggested another approach, in which the basic concept is the algebra of events.

But in developing further probability theory, we have to deal not only with events but also with random variables. We build an algebraic object - semifield, which describes, in axiomatical form, classes of measurable functions defined on a measure space. This enables as to turn to set-theoretical conceptions, if necessary. The fact that a random variable is not a function but a class of functions is completely in accordance with the needs of probability theory in which such classes arise all the same, for instance, when conditional expectations are introduced. We emphasize that the choice of a measure space by means of which we represent elements of the semifield is not unique but there is no reason to prefer any of such spaces.

Axioms of semifields and their properties are studied in § I.

We note that the class of semifields considered in § I is much wider than the class of topological semifields studied in [I], [3] and, on the other hand, is a subclass of the class of K_σ -spaces, introduced and investigated by L.V.Kantorovich, B.Z. Vulikh and others.

In § 2 we prove a theorem on representation of a semifield by classes of equivalent functions. One may consider probability distributions as linear functionals on a semifield, that are non-negative on non-negative elements of the semifield and are equal to 1 on the unit element.

Such functionals are studied in § 3. More general functionals with values in an arbitrary semifield are also introduced there (we call them stochastic operators).

In § 4 the convergence of a sequence of elements of the semifield almost everywhere is defined and connection of this convergence with (O)-convergence is established. In § 5 we establish connection between semifields and probability theory.

In § 6 we consider ergodic properties of Markov chains. In the last section we prove an analogue of the strenghened law of large numbers of A.N.Kolmogorov for a probability measure with values in a semifield.

§ I. Semifields and their properties. Examples.

Let E be a partially ordered ring. We say that a seguence $\{x_n\} \subset E$ monotonically increases (decresses) and converges to an element $x \in E$ and write $x_n \uparrow x$ ($x_n \downarrow x$) if $x_1 \leq x_2 \leq \ldots$ ($x_1 \geq x_2 \geq \ldots$) and $\bigvee_{n=1}^{\infty} x_n = x$ ($\bigwedge_{n=1}^{\infty} x_n = x$) (where $\vee M$ ($\wedge M$) means the supremum (infimum) of M. A sequence $\{x_n\}$ is called (O)-convergent to element x ($x_n \xrightarrow{(o)} x$) if there exist two sequenses $\{y_n\}$, $\{z_n\}$ such that $\{y_n\}$ increases, $\{z_n\}$ decreases, $y_n \leq x_n \leq z_n$

and $\bigvee\limits_{n=1}^{\infty} y_n = x = \bigwedge\limits_{n=1}^{\infty} z_n$.

For an arbitrary subset $A \subset E$, $\bar{A}$ will denote the set of all $x \in E$ for which there is such a sequence $\{x_n\}$ in A that $x_n \xrightarrow{(o)} x$.

We shall consider a partially ordered ring E , which satisfies the following condition.

(C.C) For every upper bounded countable set $M \subset E$, there exist $\vee M$.

Definition 1. The partially ordered ring E , satisfying condition (C.C) and having more than one element, is called a semifield if there is such a subset K , that

I° $\bar{K}$ coincides with the set of all non-negative elements of E ;

2° $K + \bar{K} \subset K$; $K \cdot K \subset K$;

3° $E = K - K$;

4° if $\alpha, \beta \in K$, then the equation $\alpha x = \beta$ has at least one solution which is an element of K .

The simplest example of a semifield is R_1 - the set of all real numbers and K is the set of all positive numbers.

Let us consider other examples of semifields.

1) Semifield R^Δ . Let Δ be any set. Denote by R^Δ the set of all real-valued functions on Δ . R^Δ is a commutative associative ring with respect to the usual addition and multiplication. We introduce partial ordering in R^Δ : $f_1 \leq f_2$ if $f_1(q) \leq f_2(q)$ for all $q \in \Delta$. Then R^Δ becomes a partially ordered ring with condition (C.C). By K we denote the set of all positive functions. $\bar{K}$ is the set of all non-negative functions. It is obvious that axioms I°-4° are fulfilled.

2) Semifield $S(L)$. Let us consider the set of all Lebesgue-measurable functions on $[0,1]$ and identify those which are equal almost everywhere. Denote by $S(L)$ the set of all of equivalence classes. This is a partially ordered ring with respect to the usual addition, multiplication and partial order, and it satisfies condition (C.C). Define K as the set of all positive a.e. (almost everywhere) functions. Then $\bar{K}$ is the set of all a.e. non-negative functions.

3) Semifield $S(F)$. Let F be a σ-algebra in a set Δ. Denote by $S(F)$ the set of all functions from R^{Δ}, measurable with respect to F. If K is the set of all positive functions from $S(F)$ the latter is a semifield with respect to operations induced by R^{Δ}.

4) Semifield $M(F)$. It is the set of all bounded functions from $S(F)$. consists of all positive functions $f \in M(F)$, for which $\frac{1}{f} \in M(F)$.

We shall give some properties of semifields without proving them. Every semifield has a ring unit $\mathbb{1}$; K concides with the set of all non-negative invertible elements. The family of all idempotents ∇ (i.e. elements $x \in E$ for which $x^2 = x$) forms a σ-complete Boolean algebra. There is a unique linearly-ordered ring in E, which contains $\mathbb{1}$ and is isomorfhic to the ring of real numbers. That is why every semifield is a linear space over the field of real numbers. For every $x \in E$, there is an idempotent e_x such that: $xe_x = x$, $ye_x = \theta$ for every element y which is disjoint with x (i.e. $|x| \wedge |y| = \theta$, where $|x| = x \vee \theta + (-x) \vee \theta$ as usual). Such an idempotent will be called a support of the element x.

By means of supports it is easy to define analogues of Lebesgue sets $\{x<\lambda\}$, $\{x>\lambda\}$ taking into consideration that

$$e_{(x-\lambda\cdot 1)_-} = \{x<\lambda\} \ ; \ e_{(x-\lambda\cdot 1)_+} = \{x>\lambda\}$$

This enables us to interpolate every element of a semifield by linear combinations of idempotents. More precisely: an element $x\in E$ is called simple if $x=\sum_{i=1}^{n}\lambda_i e_i$, where $e_i\in\nabla$ and $e_i\wedge e_j=\theta$ $(i\neq j)$, λ_i are real numbers $(i=1,2,\dots,n)$. Every element of a semifield E is an (0)-limit of a sequence of simple elements. If $x\geq\theta$, this sequence may the chosen increasing.

§ 2. Representation of a semifield by classes of equivalent functions

A subring E_1 of a semifield E is called a subsemifield if it contains an axis of E and is a semifield with respect to the partial ordering induced from E. A subsemifield E_1 will be called solid if $|x|\leq|y|$ and $y\in E_1$ implies $x\in E_1$. It is easy to verify that, if a subring E_1 of a semifield E contains 1 and $|x|\leq|y|$, $y\in E_1$ implies $x\in E_1$, the E_1 is a solid subsemifield of E.

Theorem 1. (Representation of semifields) Every semifield E may be obtained from some solid subsemifield of the semifield $S(F)$ by identifying those functions f and g for which $\{q: f(q)\neq g(q)\}\in\mathcal{J}$ where $\mathcal{J}$ is a σ-ideal in F.

We shall use the following result.

Theorem (Loomis L.H.) Let ∇ be a σ-complete Boolean algebra. Then there exist a set Δ and a σ-algebra F of subsets of Δ and a σ-homomorphism $\varphi: F\to\nabla$ such that $\varphi(F)=\nabla$ and, if $\varphi(A)=\varphi(B)$ then $A\triangle B\in\mathcal{J}$, where $\mathcal{J}$ is some σ-ideal in F.

Let ∇ - be the algebra of idempotents of a semifield E and φ

be a σ-homomorphism from Loomis' theorem. We put $\psi(f)=\sum\lambda_i\varphi(e_i)\in E$ for a simple function $f=\sum_{i=1}^{n}\lambda_i e_i\in S(F)$. Obviously,

I) $\psi(f+g)=\psi(f)+\psi(g)$, $\psi(f\cdot g)=\psi(f)\cdot\psi(g)$,

2) $\psi(af)=a\psi(f)$,

3) $\psi(f)\geq\psi(g)$ if $f\geq g$,

4) $\psi(f\vee g)=\psi(f)\vee\psi(g)$; $\psi(f\wedge g)=\psi(f)\wedge\psi(g)$,

5) $|\psi(f)|=\psi(|f|)$,

where f, g are simple functions from $S(F)$, a is a real number. It is easy to show that, if f_n, f are simple functions and $f_n\uparrow f$ or $f_n\downarrow f$, then $\psi(f_n)\longrightarrow\psi(f)$. Now we extend the map ψ to a subring $S_1\subset S(F)$. Let $f\in S(F)$, $f\geq\theta$. There exists an increasing seguence $\{f_n\}$ of non-negative simple functions such that $f=\bigvee_{n=1}^{\infty}f_n$ If $\bigvee_{n=1}^{\infty}\psi(f_n)$ exists, then we put $\psi(f)=\bigvee_{n=1}^{\infty}\psi(f_n)$. Let us show that $\psi(f)$ is well-defined, i.e. $\psi(f)$ does not depend on the choice of a sequence of simple functions $f_n\uparrow f$. Let $f_n\uparrow f$, $g\uparrow f$ and $\psi(f)=\bigvee_{n=1}^{\infty}\psi(f_n)$. Clearly $(f_n\wedge g_m)\uparrow g_m$ as $n\to\infty$, hence $\psi(f_n\wedge g_m)\uparrow\psi(g_m)$ and $\psi(f)\geq\psi(g_m)$,i.e. $x=\bigvee_{m=1}^{\infty}\psi(g_m)\leq$ $\leq\psi(f)$ exists. By symmetry, $x\geq\psi(f)$; so $x=\psi(f)$.

We extend ψ to all functions $f\in S(F)$ for which $\psi(f_+)$ and $\psi(f_-)$ are defined by putting $\psi(f)=\psi(f_+)-\psi(f_-)$. We denote the set of such functions by S_1. S_1 is a solid subsemifield of $S(F)$ The map $\psi: S_1\to E$ has the following properties:

1') $\psi(f+g)=\psi(f)+\psi(g)$, $\psi(f\cdot g)=\psi(f)\cdot\psi(g)$,

2') $\psi(af)=a\psi(f)$,

3') $\psi(f)\geq\psi(g)$ if $f\geq g$,

4') $\psi(f\vee g)=\psi(f\vee g)=\psi(f)\vee\psi(g)$, $\psi(f\wedge g)=\psi(f)\wedge\psi(g)$,

5') if $f_n \uparrow f$ or $f_n \downarrow f$, then $\psi(f_n) \xrightarrow{(o)} \psi(f)$,

6') if $\psi(f) = \psi(g)$, then $\{q : f(q) \neq g(q)\} \in \mathcal{J}$.

Properties 1' - 4' easily follow from properties 1-4 of the map ψ for simple functions. Let us prove 5'. It is sufficient to consider the case $f_n \uparrow f$, $f_n \in S_1$, $f_n \geq \theta$. For every f_n there exists such a sequence of simple functions $\{f_m^{(n)}\}$ that $f_m^{(n)} \uparrow f_n$ as $m \to \infty$. Let us put $g_n = \bigvee_{k=1}^{n} \bigvee_{m=1}^{n} f_m^{(n)}$; obviously $g_n \uparrow f$. Hence $\psi(g_n) \to \psi(f)$. But $g_n \leq f_n \leq f$ and $\psi(g_n) \leq \psi(f_n) \leq \psi(f)$ hence $\psi(f_n) \xrightarrow{(o)} \psi(f)$.

Property 6' easily follows from the corresponding property of the map φ.

To complete the proof we must show that $\psi(S_1) = E$. Let x be a non-negative bounded* element in E, i.e. $\theta \leq x \leq K \cdot 1$ for some natural K. There exists a sequence of non-negative simple elements $\{x_n\} \subset E$ such that $x_n \uparrow x$. We denote by f_n a simple non-negative function from S_1 such that $\psi(f_n) = x_n$. We may suppose that $f_n \leq f_{n+1}$, $n = 1, 2, \ldots$. There exists $f = \bigvee_{n=1}^{\infty} f_n \in S(F)$. It follows, from the construction, that $\psi(f) = x$.

Consider now x, a non-negative element of E. It is easy to see, that $x = \bigvee_{n=1}^{\infty} x\{n \leq x < n+1\}$ where $y_n = x \cdot \{n \leq x < n+1\}$ is a bounded element now; besides y_n are disjoint. Denote by g_n a

---

* An element a of a semifield E is called bounded if there exists a number $K \geq 0$ such that $|a| \leq K \cdot 1$.

bounded function from S_1 for which $\psi(g_n) = y_n$. We can assume that $\{g_n\}$ are disjoint. Hence $f = \bigvee_{n=1}^{\infty} g_n$ exists in $S(F)$. It is clear that $\psi(f) = x$.

Finally we note that, if $x \in E$, then $x = x_+ - x_-$, and hence there exists such an $f \in S_1$ that $\psi(f) = \psi(f_+) - \psi(f_-) = x_+ - x_-$. This completes the proof.

We shall call a semifield universal if any its countable disjoint subset is bounded. We note that, for any σ-algebra ∇, there exists, up to an isomorphism, the only universal semifield the algebra of idempotents of which is isomorphic to ∇. We shall denote this semifield by $S(\nabla)$

Moreover every semifield E with the algebra of idempotents ∇ is isomorphic to any solid subsemifield of the universal semifield $S(\nabla)$.

Now we shall define Borel functions on elements of a semifield. We need one more notion.

We shall call a subsemifield E_1 correct if $\vee M \in E_1$ for every countable set $M \subset E_1$ which is upper bounded in E. It is clear that, if E_1 is a correct subsemifield, then $\wedge M \in E_1$ for every countable set $M \subset E_1$ which is lower bounded in E. Further we shall consider only correct subsemifields without mentioning this.

It is not difficult to understand that the intersection of subsemifields is again a semifield in E. Therefore every set $M \subset E$ has the smallest subsemifield containing M. This subsemifield will be denoted by $E(M)$.

Theorem 2. There is a transformation $\varphi(x)$ of a semifield E which corresponds to a bounded Borel function φ on R such that

1) for every fixed x, the set $\varphi(x)$ coincides with the set of bounded elements of the subsemifield $E(x)$;

2) $\varphi(\psi(x)) = (\varphi(\psi))(x)$ for any bounded Borel functions φ and ψ.

To prove it we shall use Theorem 1. In view of this theorem x can be considered as the class of equivalent functions on some set Δ. which are measurable with respect to a σ-algebra F. If φ is a Borel function, then $\varphi(f)$ is measurable with respect to F if f is measurable with respect to F; if f and g are equivalent, then so are $\varphi(f)$ and $\varphi(g)$. Therefore it has sense to speak about the class of functions $\varphi(x)$ for every bounded Borel function φ. Property 2) is obvious. If φ is the indicator of set $(-\infty, \lambda)$ then $\varphi(x) = \{x < \lambda\} \in \in E(x)$. It follows that $\varphi(x) \in E(x)$ for every bounded Borel function φ. Conversely, if y is a non-negative bounded element in $E(x)$ then there exists such a sequence of simple elements $\{x_n\} \subset E(x)$ that $x_n \uparrow y$. Since we may assume $x_n = \varphi_n(x)$ where φ_n are simple Borel functions, $y = \varphi(x) = (\bigvee_{n=1}^{\infty} \varphi_n)(x)$ where φ is a bounded Borel function.

§ 3. Linear functionals and measures on semifields

It would be natural to consider a linear functional on a semifield to be such a real-valued function μ which is continuous with respect to (O)-convergence and satisfies the condition:

$$\mu(ax + by) = a\mu(x) + b\mu(y)$$

where $x, y \in E$ and a, b are real numbers.

But in the most interesting semifields (for example, in $S(L)$) there are only trivial examples of such functions. So we have to consider functions existing not on the whole semifield but on its subset.

Definition 2. A function μ on $\bar{K}$ with values in the extended semiaxis $[0,+\infty]$ is called a positive linear functional if

1) for every $x, y \in \bar{K}$ and real numbers $a, b > 0$,

$$\mu(ax+by) = a\mu(x) + b\mu(y);$$

2) if $x_n \uparrow x$, then $\mu(x_n) \uparrow \mu(x)$.

A positive linear functional can be extended to some elements $x \in \bar{K}$. We call $x \in E$ μ summable if $\mu(|x|) < +\infty$ and denote by E_μ the set of all μ-summable elements of E. We put, for $x \in E_\mu$, $\mu(x) = \mu(x_+) - \mu(x_-)$. The map $\mu: E_\mu \to (-\infty,+\infty)$, obviously, satisfies the conditions:

a) $\mu(ax+by) = a\mu(x) + b\mu(y)$; a, b are real numbers; $x, y \in E_\mu$;

b) $\mu(x_n) \to 0$ if $x_n \downarrow \theta$, $x_n \in E_\mu$, $n = 1, 2, \dots$;

c) if $x_n \xrightarrow{(o)} x$ and $\bigvee_{n=1}^{\infty} |x_n| \in E_\mu$, then $\mu(x_n) \to \mu(x)$.

Let μ be a positive linear functional. Consider its restriction m on the set ∇ of idempotents of the semifield E. It is easy to verify that $m(\bigvee_{e \in Q} e) = \sum_{e \in Q} m(e)$ for every finite or countable set Q of disjount elements. So, for every positive linear functional, there is a corresponding measure on the σ-algebra ∇. The converse is also true: for every measure m on ∇, there is a unique positive linear functional on E which coinsides with m on ∇.

Further we shall suppose that $\mu(\mathbb{1}) < +\infty$ so that $\nabla \subset E_\mu$.

By $M(E)$ we shall designate the set of all positive functionals on E .

We shall call a functional strictly positive if it positive on all the elements from $\bar{K}\setminus\{\theta\}$. The following theorem gives a description of the set $M(E)$ for semifields of countable type, i.e. for semifields every disjount set of elements of which is at most countable.

Theorem 3. In any semifield E of countable type there exists the largest idempotent e_0 with the property: $\mu(e_0)=0$ for all $\mu\in\mathfrak{M}(E)$. On the semifield $(1-e_0)E$ there exists a strictly positive functional μ_0 . All positive functionals on the semifield E are absolutely continious with respect to the functional defined by the formula:

$$\tilde{\mu}_0(x)=\mu_0((1-e_0)x).$$

The proof of Theorem 3 is based on the following lemma.

Lemma. Let E be a semifield of countable type. If, for every nonzero idempotent e , there exists a $\mu\in\mathfrak{M}(E)$ such that $\mu(e)>0$, then, in E , there exists a strictly positive functional.

Let an idempotent e of the semifield E belong to the set M if there exists a measure m on the σ-algebra of idempotents ∇ the restriction of which on $[\theta,e]=\{g\in\nabla: g\leq e\}$ is strictly positive. We shall prove that M is minorant in ∇ , i.e., for every $g\in\nabla$, there exists an $e\in M$ such that $e\leq g$.

Let $g\in\nabla$; then, in view of the conditions of the lemma, there exists a measure m_1 such that $m_1(g)\neq 0$. Since ∇ is σ-complete and of countable type, ∇ is a complete Boolean algebra [2].Hence there exists a $g_1 = V\{e \in \nabla : m_1(e) = 0\}$. Moreover there exists a countable set $\{e_n\} \subset \{e \in \nabla : m_1(e) = 0\}$ such that $Ve_n = g_1$ [2], and hence $m_1(g_1) = 0$. Let $e = g \wedge Cg_1$. It is clear that $e \neq \theta$ (otherwise $g \leq g_1$ and $m_1(g) = 0$) and $e \in M$. Thus M is minorant in ∇, therefore there exists a countable subset $\{e_n\} \subset M$ for which $Ve_n = \mathbb{1}$, [2]. It is not difficult to see that the measure $m = \sum_{n=1}^{\infty} \frac{1}{2^n m_n(\mathbb{1})} m_n$ is strictly positive on ∇ (here m_n is a measure on ∇ which is strictly positive on $[\theta, e_n]$). There exists a strictly positive functional on E corresponding to this measure.

Now we turn to the proof or Theorem 3.

Let an idempotent e belong to the set H if $\mu(e) = 0$ for all $\mu \in \mathfrak{M}(E)$. In the same way as in Lemma, we get $\vee H = e_0$ and $e_0 \in H$. The first part of The theorem 3 is proved.

To prove the second part it is sufficient to check (by Lemma) that, for any nonzero idempotent $e \leq \mathbb{1} - e_0$, there exists a $\mu \in \mathfrak{M}(E)$ such that $\mu(e) > 0$. But it is evident, because otherwise we would have $e \leq e_0$ and therefore $e = \theta$.

Now let us analyse the more general situation when the values of functionals and measures are in an arbitrary semifield E_1. All the definitions remains, only instead of the term "functional" we shall use the word "operator". Convergence of numbers here is replaced by (O)-convergence in E. If we wish a positive operator T to be defined on the whole $\overline{K}$, the point $+\infty$ should be added to E_1 putting $+\infty + a = +\infty$ for any $a \in E_1$ and $+\infty \cdot c = +\infty$ for any positive number c. Instead of this we can suppose that T is defined only on some subset $\overline{K}_T \subset \overline{K}$, which is closed with respect to addition and multiplication by positive numbers. We assume ∇ to belong to the domain of T. As in the case $E_1 = R$, to each positive operator there corresponds a measure on ∇ with values in the semifield E_1. The converse is also true.

<u>Theorem 4.</u> If m is an arbitrary measure on ∇ with values in a semifield E_1, then there exists a positive operator T with values in the same semifield, which coincides with m on ∇.

It fact, for a simple element $x = \sum_{i=1}^{n} \lambda_i e_i$ from E, let us put $T(x) = \sum_{i=1}^{n} \lambda_i m(e_i) \in E_1$. It is evident, that

a) $T(x+y) = Tx + Ty$,

b) $T(a \cdot x) = a \cdot Tx$,

c) $Tx \geq Ty$ if $x \geq y$,

d) $|T(x)| \leq T(|x|)$,

where x, y are simple elements of E, a is a real number.

Now we show that, if $x, x_n \in E$, $n = 1, 2, \dots$ are simple elements and $x_n \downarrow x$ or $x_n \uparrow x$, then $Tx_n \xrightarrow{(o)} Tx$. Let first $x_n \downarrow \theta$ and $x_1 = \sum_{i=1}^{n} a_i e_i$. Put $t = \max_{1 \leq i \leq n} a_i$; then $x_n \leq t \cdot \mathbb{1}$. If we take an arbitrary $\varepsilon > 0$, it is evident that $\theta \leq x_n \leq t\{x_n > \varepsilon\} + \varepsilon \cdot \mathbb{1}$ and hence $\theta \leq T(x_n) \leq t \cdot m(\{x_n > \varepsilon\}) + \varepsilon \cdot m(\mathbb{1})$. Since $\{x_n > \varepsilon\} \downarrow \theta$, we have $m(\{x_n > \varepsilon\}) \xrightarrow{(o)} \theta$. Therefore passing in the above inequality to the (o)-limit, we have

$$\theta \leq \bigwedge_{n=1}^{\infty} Tx_n \leq \varepsilon \cdot m(\mathbb{1}).$$

Since ε is arbitrary, $Tx_n \downarrow \theta$. If now $x_n \downarrow x$, then $(x_n - x) \downarrow \theta$, and, by the proved above, $T(x_n - x) \to \theta$, that is $Tx_n \xrightarrow{(o)} Tx$. If $x_n \uparrow x$, then $(-x_n) \downarrow (-x)$, and consequently $Tx_n \xrightarrow{(o)} Tx$.

Now we extend the operator T to some subset $\bar{K}_T \subset \bar{K}$, which is under addition and multiplication by positive numbers. Let $\overline{K}_T$ be the set of all such $x \in \overline{K}$ for which there exists a sequence of simple elements $x_n \uparrow x$ such that Tx_n (o)-converges to some element of the selifield E_1, which we shall define by Tx. As in the proof of Theorem 1, it is easy to check that Tx is defined correctly; it does not depend on the choice of the sequence of simple elements $x_n \uparrow x$.

The operator T, in the usual way, can be extendel to all those $x \in E$ for which $T(x_+)$ and $T(x_-)$ are defined. Let the set of $x \in E$, for which Tx exists, be denoted by E_T. It is obvious that E_T is a linear subset in E containing the set of bounded elements.

It should be added that the constructed above mapping $T: E_T \to E_1$ has the following properties:

a') $T(x+y) = Tx + Ty$;

b') $T(ax) = aTx$, a is a real number;

c') $Tx \geq Ty$ if $x \geq y$;

d') $|Tx| \leq T(|x|)$;

e') if $x_n \uparrow x$ or $x_n \downarrow x$, then $Tx_n \xrightarrow{(o)} Tx$ where $x, y, x_n \in E_T$, $n = 1, 2, \ldots$.

Properties a')-d') immediately follow from properties a)-d), and property e') can be proved in the same way as property 5') in Theorem 1.

Thus T is a positive operator on E with values in the semifield E_1. Its restriction on ∇ evidently coincides with the measure m.

§ 4. Convergence almost everywhere and its connection with (O)-convergence

Let E be a semifield, ∇ be the Boolean algebra of its idempotents and $m: \nabla \to E_1$ be a measure on ∇, with values in a semifield E_1. A sequence $\{x_n\}$ is said to converge almost everywhere with respect to m to an element x : $x_n \to x$ (a.e. m) if $m(\bigvee_{k\geq n}\{|x_n - x| \geq \varepsilon\}) \xrightarrow{(o)} \theta$, for every $\varepsilon > 0$.

We show that the (o)-convergence always implies the almost everywhere convergence for any measure m.

If fact, if $x_n \to x$ and $v_n = |x_n - x|$, then $v_n \to \theta$, and therefore there exists a decreasing sequence $\{u_n\}$ such that $v_n \leq u_n$ and $\wedge u_n = \theta$. Let $m: \nabla \to E_1$ be a measure on ∇ with values in E_1. We show that $v_n \to \theta$ (a.e. m). We take an arbitrary $\varepsilon > 0$ and put $a_n = \{u_n > \varepsilon\}$. Since $u_{n+1} \leq u_n$, $a_{n+1} \leq a_n$. Let $a = \bigwedge_n a_n$; it is evident that $\varepsilon a_n \geq \varepsilon a$, and therefore $\wedge u_n a_n \geq \wedge \varepsilon a_n \geq \varepsilon a$. Hence $\theta = \wedge u_n \geq \varepsilon a \geq \theta$. Consequently $a = \theta$, that is $a_n \downarrow \theta$. But then $m(a_n) \to \theta$. Since $\bigvee_{k\geq n}\{v_k > \varepsilon\} \leq \bigvee_{k\geq n} a_k = a_n$, $v_n \to \theta$ (a.e. m). From this we get $x_n \to x$ (a.e. m).

Now let us find out under what conditions the (o)-convergence is implied by the convergence almost everywhere. We suppose that $m: \nabla \to E_1$ is a strictly positive measure, that is $m(e) = \theta$ if and only if $e = \theta$. If the measure m is not strictly positive, then, defining classes of equivalent with respect to m elements (elements x and y are equivalent with respect to m, if $m(e_{|x-y|}) = \theta$ where $e_{|x-y|}$ is the support of the element $|x-y|$), we get the semifield with a strictly positive measure.

Theorem 5. If $x_n \to x$ (a.e. m) and $\bigvee_n |x_n|$ exists, then $x_n \xrightarrow{(o)} x$.

Let $v_n = |x_n - x|$; it is evident that $v_n \to \theta$. Let us show that $v_n \xrightarrow{(o)} \theta$. Choose an arbitrary $\varepsilon > 0$ and put $a_n = \{u_n > \varepsilon\}$ where $u_n = \bigvee_{k \geq n} v_k$ (this supremum exists since $v_k \leq |x| + \bigvee_n |x_n|$). It is not difficult to notice that $a_n = \bigvee_{k \geq n} \{v_k > \varepsilon\}$. Put $a = \bigwedge_n a_n$; then $a_n \downarrow a$ and $m(a_n) \downarrow m(a)$. But $v_n \to \theta$ (a.e. m). Therefore $m(a_n) \xrightarrow{(o)} \theta$, that is $m(a) = \theta$, and, since m is strictly positive, $a = \theta$, that is $a_n \downarrow \theta$. It follows that

$$u_n = u_n a_n + u_n(1 - a_n) \leq u_n a_n + \varepsilon \cdot 1 .$$

Hence $u = \bigwedge_n u_n \leq \varepsilon \cdot 1$. Now ε being arbitrary, we get $u = \theta$, that is $u_n \downarrow \theta$. But then $v_n \to \theta$ and therefore $x_n \to x$.

Theorem 6. If E is a universal semifield, then the convergence $x_n \to x$ (a.e. m) implies the existence of $\bigvee_n |x_n|$.

As above for $v_n = |x_n - x|$, we have $m(\bigvee_{k \geq n} \{v_k \geq 1\}) \xrightarrow{(o)} \theta$. Let $b_n = \bigvee_{k \geq n} \{v_n \geq 1\}$ and $b = \bigwedge_n b_n$. Since $m(b_n) \downarrow m(b)$ and m is strictly positive, then $b = \theta$. Put $e_1 = 1 - b_1$, $e_n = b_{n-1} - b_n$, $n = 2, 3, \ldots$. It is easy to see that $e_n \wedge e_m = \theta$ if $n \neq m$. Since $b = 0$, we have $\bigvee_n e_n = 1$. Let $z_1 = 1$, $z_n = \bigvee_{k=1}^{n} v_k$, $n = 2, 3, \ldots$. The elements $(z_n e_n)$ being disjoint and E being a universal semifield, there exists $\bigvee_n (z_n e_n) = z$. Now $v_k = \bigvee_n v_k e_n \leq \bigvee_{n=1}^{k-1} v_k e_n + z$. We have $e_n \wedge \{v_k \geq 1\} = \theta$ if $1 \leq n \leq k-1$, and therefore $v_n e_n < 1$. So $v_k \leq z + 1$, $k = 1, 2, \ldots$. But $|x_n| \leq v_k + |x| \leq z + 1 + |x|$. Therefore $\bigvee_n |x_n|$ exists.

It follows from Theorems 5 and 6 that, in a universal semifields with a strictly positive measure m, the convergence almost everywhere coincides with the (O)-convergence.

§ 5. Basic concepts of probability theory

A positive linear functional P will be called a distribution or a probability functional if $P(\mathbb{1}) = \mathbb{1}$. While in the Kolmogorov axioms every probability concept is interpreted in terms of set theory, we shall give their algebraic interpretation in terms of the theory of semifields.

Elements of semifields are interpreted as random variables, and the corresponding value of the distribution P as their expectations. Events are described by idempotents; their probabilities as the values of the measure P. $\mathbb{1}$ corresponds to the certain event and to the impossible event. The event, opposite to e, is $Ce = \mathbb{1} - e$.

The incompatibility of events means that the corresponding idempotents are disjoint. The inequality $e_1 \leq e_2$ means that the event e_1 implies e_2. Hence, if $\{e_\alpha\}$ is a system of events, then their intersection is described by $\bigwedge_\alpha e_\alpha$, while $\bigvee_\alpha e_\alpha$ means their union.

The distribution function of a random variable x is a function of real argument defined by the formula: $F(\lambda) = P\{x < \lambda\}$. If the expectation of a random variable $f(x)$ (where f is a Borel function) exists, then it is expressed by distribution function $F(\lambda)$ of the element x as follows: $Pf(x) = \int f(\lambda)dF(\lambda)$.

Let P be a probability functional on a semifield E. Subsemifields E_1 and E_2 of the semifield E are called independent if $P(x\cdot y)=P(x)P(y)$ for all $x\in E_1$, $y\in E_2$. Random variables $x, y\in E$ are called independent if the subsemifields $E(x)$ and $E(y)$ are independent. It is easy to show that the independence of x and y is necessary and sufficient for the equality $P[\varphi(x)\cdot\psi(y)] = P[\varphi(x)]\cdot P[\psi(y)]$ for all bounded Borel functions φ and ψ. In particular, the independence of x and y implies that of $\varphi(x)$ and $\psi(y)$.

The conditional expectation is introduced as usual. Let E_1 be a subsemifield of E. An element $x_1\in E_1$ is called the conditional expectation of an element $x\in E$ if, for every bounded element $a\in E_1$, we have $P(ax) = P(ax_1)$. In particular, $P(x) = P(x_1)$. If E is a universal semifield, the conditional expectation exists for every $x\in E$.

§ 6. Markov chains

In this section we introduce homogeneous Markov chains and, under some conditions, we prove an ergodic theorem and a theorem on the regularity of the chain.

Let E be a semifield and M be the set of all bounded elements of E. By $\mathfrak{M}(E)$, as above, we denote the set of all distributions on E. A map $T: M \to M$ is called a stochastic operator if

1) $T(ax+by)=aTx+bTy$; $x,y\in M$; a,b are real numbers

2) $Tx\geq\theta$ if $x\geq\theta$;

3) $T\mathbb{1} = \mathbb{1}$;

4) if $x_n\downarrow\theta$, then $Tx_n\downarrow\theta$.

A stochastic operator T defines a map from $\mathfrak{M}(E)$ into $\mathfrak{M}(E)$ by the formula $(T\mu)(x) = \mu(T(x))$, $x\in M$.

Let each pair of non-negative real numbers $s \leq t$ correspond to a stochastic operator $T(s,t)$ such that, for $s < t < u$, we have $T(s,t)T(t,u) = T(s,u)$ (Kolmogorov-Chapman equation); then we whall say that $T(s,t)$ is the Markov transition functions or the transition operator of a Markov chain. For every $\mu \in \mathfrak{M}(E)$ and every idempotent g of the semifield E, the real number $P(s,\mu,t,g) = \mu T(s,t) g$ is called the probability for the chain to be in the state g at moment given the distribution μ at moment s. In view of the Kolmogorov-Chapman equation transition operators are expressed by $T(t,t+1)$, i.e. the one-step transition operators. We say that a Markov chain is homogeneous if $T(t,t+1) = T$ is independent on t. For a homogeneous chain we have $T(s,t) = T^{t-s}$. The probability $\mu T^t g$ is denoted by $P(t,\mu,g)$.

Our aim will be to study the asymptotic behaviour of these transition probabilities as $t \to \infty$.

Theorem 7 (Ergodic theorem). The limit $\lim_{m\to\infty} (\mu T^m - \nu T^m) = \theta$ for each $\mu, \nu \in \mathfrak{M}_1(E)$ if the following condition is satisfied:

(A) there exist such $g_o \in \nabla$, $\mu_o \in \mathfrak{M}(E)$ and a real $\delta > 0$ that $\mu_o(g_o) > 0$ and, for each $\nu \in \mathfrak{M}_1(E)$, there exists such a m that $(\nu T^m)(e) \geq \delta \mu(e)$ for each $m \geq m_o$ and for each idempotent $e \leq g_o$.

Condition (A) is a generalization of the well-known Bernstein condition for Markov chains with a finite number of states. Namely, there is such a state, transition to which is possible, with positive probability, from any other state.

For the proof of Theorem 3 we need the following

Lemma. Let a stochastic operator T satisfy condition (A). Then, for every $\lambda, \nu \in \mathfrak{M}_1(E)$, there are such elements $w \in \mathfrak{M}(E)$, $u, v \in \mathfrak{M}_1(E)$ and a natural number m that

$$\nu T^m - w = (1-c)u, \quad \lambda T^m - w = (1-c)v$$

where $c = \delta\mu_0(g_0)$ (δ, μ_0, g_0 are from condition (A)). Without loss of generality we may suppose that $c < 1$

Proof. In view of condition (A), there exists such a number m_0 that $\nu(T^m e) \geq \delta\mu_0(e)$ and $\lambda(T^m e) \geq \delta\mu_0(e)$ when $m \geq m_0$ and $e \leq g_0$, $e \in \nabla$. Put $\omega(x) = \delta\mu_0(x g_0)$, $x \in M$. Obviously $\omega \in \mathfrak{M}(E)$ and

$$\nu T^m(\mathbb{1}) - \omega(\mathbb{1}) = 1 - \delta\mu_0(g_0) = 1 - c,$$

$$\lambda T^m(\mathbb{1}) - \omega(\mathbb{1}) = 1 - \delta\mu_0(g_0) = 1 - c.$$

Let $u = \dfrac{\nu T^m - w}{1-c}$, $v = \dfrac{\lambda T^m - w}{1-c}$. Then $u, v \in \mathfrak{M}_1(E)$ and $\nu T^m - w = (1-c)u$, $\lambda T^m - w = (1-c)v$.

Now we turn to the proof of Theorem 7.

Let $\lambda, \nu \in \mathfrak{M}_1(E)$. By Lemma, there exist such $u_1, v_1 \in \mathfrak{M}_1(E)$, a number c $(0 < c < 1)$ and a natural number m_1 that

$$\nu T^{m_1} - \lambda T^{m_1} = (1-c)(u_1 - v_1).$$

Further, for $u_1, v_1 \in \mathfrak{M}_1(E)$ we can find such $u_2, v_2 \in \mathfrak{M}_1(E)$ and natural m_2 that $u_1 T^{m_2} - v_1 T^{m_2} = (1-c)(u_2 - v_2)$. Hence

$$\nu T^{m_1+m_2} - \lambda T^{m_1+m_2} = (1-c)^2(u_2 - v_2).$$

Repeating this process, we shall have after t steps:

$$\nu T^{m_1+m_2+\dots+m_t} - \lambda T^{m_1+m_2+\dots+m_t} = (1-c)^t (u_t - v_t) .$$

Every natural number K can be represented as $K = m_1 + m_2 + \dots + m_t + z$ where $z < m_{t+1}$. Then $\nu T^k - \lambda T^k = (1-c)^t (u_t - v_t) T^z$, where $u_t, v_t \in m_1(E)$.

Let $x \in M$, then there exists such a natural n that $|x| \le n \cdot \mathbb{1}$. Hence

$$|\nu(T^k x) - \lambda(T^k x)| \le$$

$$\le (1-c)^t (u_t(T^z |x|) + v_t(T^z |x|) \le 2 \cdot n (1-c)^t .$$

The proof is completed since $t \to \infty$ as $K \to \infty$.

The limit $\lim_{m\to\infty} \nu T^m$ may not exist. The following theorem gives a condition under which this limit exists.

Theorem 8. Let condition (A) be satisfied. The limit $\lim_{m\to\infty} \nu T^m$ exist for each $\nu \in m_1(E)$ and does not depend on ν if the following conditions are satisfied.

(B) For each $\nu \in m_1(E)$

$$\sum_{t=0}^{\infty} (t+1)\, \nu(u^t T g) < \infty$$

where $u = T \circ (\mathbb{1} - g)$ $\quad ((T \circ g)(x) = T(gx))$.

(C) For each $\nu \in m_1(E)$ such that $\nu(\mathbb{1} - g_0) = 0$, and for each $e \in V$, there exists such a natural number $K_0 = K_0(e, \nu)$ that $\nu(T^{K_0} e) > \alpha e > 0$.

The following proposition will be used in the proof of this

theorem.

Proposition. Let a stochastic operator T satisfy conditions (A), (B), (C). Then, for each $\varepsilon>0$, there exists such a natural m that $R(n,\nu,n+m,g)<\varepsilon$ (here we need that $\mu_0(g)>0$) where

$$R(n,\nu,n+m,g)=$$

$$=\sum_{\ell=n+m-1}^{\infty}\nu(u^{\ell}T_g)+\sum_{k=1}^{n-1}P(k,\nu,g)\sum_{\ell=m+n-k-1}\nu^{(k)}(u^{\ell}T_g)$$

and

$$\nu^{(k)}=\frac{\nu T^{k}\circ g}{\nu(T^{k}g)}\ .$$

The proof of this proposition will be omitted.

Let us take any $\varepsilon>0$. For given ν, g and the ε, we shall choose a number m according to the above proposition. Put

$$\underline{\lim}_{n\to\infty}P(n,\nu,g)=a\ ,\quad \overline{\lim}_{n\to\infty}P(n,\nu,g)=b\ ,$$

and let

$$\lambda_{k,0}=\frac{\nu T^{k}\circ g_0}{\nu T^{k}(g_0)}\ ,\quad \lambda_{k,1}=\frac{(\nu T^{k}\circ(1-g_0))T\circ g_0}{(\nu T^{k}\circ(1-g_0))(T(g_0))}\ ,\ \dots\ ,$$

$$\lambda_{k,t}=\frac{(\dots(((\nu T^{k}\circ(1-g_0)T\circ(1-g_0)T\dots)T\circ(1-g_0)T\circ g_0}{(\dots(((\nu T^{k}\circ(1-g_0))T\circ(1-g_0))T\dots)T\circ(1-g_0))(Tg_0)}$$

Let us denote by $A(\kappa,\nu,t,g_0)$ the number

$$(\dots(((\nu T^{\kappa}\circ(1-g_0))T\circ(1-g_0))T\dots)T\circ(1-g_0)(T(g_0)).$$

This number may be interpreted as the probability for the chain with the initial distribution ν to arrive at g_0 at the $(\kappa+t)$-th step for the first time after step $\kappa-1$. By Theorem 3, we have for each $\lambda\in\mathfrak{M}_1(F)$:

$$\underline{\lim}_{n\to\infty} P(n,\lambda,g) = \underline{\lim}_{n\to\infty} P(n,\nu,g) = a,$$

$$\overline{\lim}_{n\to\infty} P(n,\lambda,g) = \overline{\lim}_{n\to\infty} P(n,\nu,g) = b.$$

Moreover, as m is fixed, we can choose such numbers κ_0, $n \geqslant m+\kappa_0$ and $n' > n+\kappa_0$ that, for $i=0,1,\dots,m$, we have

$$b+\varepsilon < P(\tau,\lambda_{\kappa,i},g) < a+\varepsilon \qquad (\tau>\kappa_0),$$

$$P(n,\lambda_{\kappa,i},g) < a+\varepsilon, \quad P(n',\nu,g) > b-\varepsilon,$$

$$P(n'-n,\nu,g_0) > \delta\mu_0(g_0)$$

where δ is taken from condition (A). Let $n'-n=\kappa$, then

$$P(n',\nu,g) \leqslant P(\kappa,\nu,g_0)P(n,\lambda_{\kappa,0},g) + A(\kappa,\nu,1,g_0)P(n-1,\lambda_{\kappa,1},g_0) + \dots$$

$$\dots + A(\kappa,\nu,n,g_0)P(0,\lambda_{\kappa,n},g) + R(\kappa,\nu,n',g_0).$$

where $P(0,\lambda_{\kappa,n},g)=1$ and $R(\kappa,\nu,n',g_0)$ is the probability, under the initial distribution ν, not to get at idempotent g_0 between the κ-th and n'-th steps. We have $R(\kappa,\nu,n',g_0) < \varepsilon$, hence

$$b-\varepsilon < P(\kappa,\nu,g_0)(a+\varepsilon) + (b+\varepsilon)\sum_{s=1}^{m} A(\kappa,\nu,s,g_0) +$$

$$+\sum_{s=m+1}^{n} A(\kappa,\nu,s,g_0)\cdot P(n-s,\lambda_{\kappa,s},g) + \varepsilon.$$

Since $\sum_{s=m+1}^{n} A(\kappa,\nu,s,g_0) < \varepsilon$, we have

$$b-\varepsilon < P(\kappa,\nu,g_0)(a+\varepsilon)+(1-P(\kappa,\nu,g_0))(b+\varepsilon)+2\varepsilon =$$

$$= -P(\kappa,\nu,g_0)(b-a)+b+3\varepsilon .$$

Hence $P(\kappa,\nu,g_0)(b-a) < 4\varepsilon$ or $b-a < 4\varepsilon/\delta\mu_0(g_0)$. As ε is arbitrary, we have $b=a$.

In conclusion of this section, note that if a stochastic operator T satisfies conditions (A), (B), (C), then there exists such a unique $\lambda \in \mathcal{M}_1(E)$ that $\lambda T(g) = \lambda(g)$ for every $g \in \nabla$. This follows directly from Theorem 8 .

§ 7. The strong law of large numbers

In probability theory laws of large numbers are of greet importance. In our case (i.e. in the case of E_1-valued measures it is supposed that $P(\mathbb{1}) = \mathbb{1}$) these laws hold if all random variables are centred, i.e. they have zero expectations. Moreover, the proofs are the same as in the classical case. We shall give as an example the proof of the Kolmogorov theorem (strong law of large numbers) which is based on the Hajek-Renyi inequality.

The Hajek-Renyi inequality.

Let $\{x_n\}$ be a sequence of independent random variables with $M(x_\kappa) = \theta$ and existing $M(x_\kappa^2)$, $\kappa = 1,2,\ldots$. Let $\{c_\kappa\}$ be a non-increasing sequence of positive numbers, then for each m and n $(m > n)$ and $\varepsilon > 0$, we have

$$P(\{\bigvee_{\kappa=n}^{m} c_\kappa |s_\kappa| \geq \varepsilon\}) \leq \frac{1}{\varepsilon^2}\left(c_n^2 \sum_{\kappa=1}^{n} M(x_\kappa^2) + \sum_{\kappa=n+1}^{m} c_\kappa M(x_\kappa^2)\right)$$

where $S_k = \sum_{i=1}^{k} x_i$.

The series $\sum_{k=1}^{\infty} x_k$ of elements of a semifield is called convergent, if the sequence $\{\sum_{k=1}^{n} x_k\}$ of partial sums converges to some element x which is called the sum of the series.

We give now (without proof) an analogue of the well-known Kronecker lemma for our semifield case.

The Kronecker lemma. If a series $\sum_{n=1}^{\infty} x_n$ converges in the semifield and b_n is a monotone increasing sequence of real numbers, $b_n \uparrow \infty$, $b_n > 0$, then $\frac{1}{b_n} \sum_{k=1}^{n} b_k x_k \xrightarrow{(o)} \theta$ $(n \to \infty)$.

If for a sequence $\{x_n\}$ of random variables and a sequence $\{b_n\}, b_n \uparrow \infty$, $b_n > 0$ $(n = 1, 2, \dots)$, we have

$$\frac{\sum_{m=1}^{n} x_m}{b_n} \longrightarrow \theta \qquad (\text{a.e. } P),$$

then we say that the sequence $\{x_n\}$ satisfies the strong law of large numbers.

Theorem 9. (Kolmogorov) Let $b_n \uparrow \infty$, $b_n > 0$. If for a sequence $\{x_n\}$ of independent random variables the series $\sum_{n=1}^{\infty} \frac{\mathcal{D} x_n}{b_n^2}$ converges, then this sequence satisfies the strong law of large numbers.

Proof. Let $\varepsilon > 0$. We put $a_m = \bigvee_{k=1}^{m} \{ | \frac{S_k}{b_k} | \geq \varepsilon \}$, where $S_k = \sum_{n=1}^{k} x_n$. By the Hajek-Rényi inequality

$$P(a_m) \leq \frac{1}{\varepsilon^2} \left(\frac{1}{b_n^2} \sum_{k=1}^{n} M(x_k^2) + \sum_{k=n+1} \frac{M(x_k^2)}{b_k^2} \right) .$$

Since $a_m \uparrow \left(\bigvee_{k=n}^{\infty} \{ | \frac{S_k}{b_k} | \geq \varepsilon \} \right)$,

$$P\Big(\bigvee_{k\geq n}\Big\{\Big|\frac{S_k}{b_k}\Big|\geq\varepsilon\Big\}\Big)\leq\frac{1}{\varepsilon^2}\Big(\frac{1}{b_n^2}\sum_{k=1}^{n}M(x_k^2)+\sum_{k=n+1}^{\infty}\frac{M(x_k^2)}{b_k^2}\Big).$$

But the series $\sum_{k=1}^{\infty}\frac{M(x_k^2)}{b_k^2}$ converges, hence we have $\sum_{k=n+1}^{\infty}\frac{M(x_k^2)}{b_k^2}\xrightarrow{(o)}\theta$, and by the Kronecker lemma $\frac{1}{b_n^2}\sum_{k=1}^{n}\mathcal{D}(x_k)\xrightarrow{(o)}\theta$. So $P\Big(\bigvee_{k\geq n}\Big\{\Big|\frac{S_k}{b_k}\Big|\geq\varepsilon\Big\}\Big)\xrightarrow{(o)}\theta$, i.e. the sequence $\{x_n\}$ satisfies the strong law of large numbers.

REFERENCES

I. Антоновский М.Я., Болтянский В.Г., Сарымсаков Т.А., Очерк теории топологических полуполей. Успехи математических наук, 21,4(130) (1966), 185-218.

2. Владимиров Д.А., Булевы алгебры. Москва, "Наука", 1969.г.

3. Сарымсаков Т.А., Топологические полуполя и теория вероятностей, Ташкент, "ФАН", 1969 г.

Department of Physics and Mathematics

Tashkent State University

Tashkent

Convergence to diffusion processes for a class of Markov chains related to population genetics

Ken-iti Sato

Karlin and McGregor introduced in [5] a class of Markov chains (finite state, discrete time) induced by direct product branching processes, with the intention of unified treatment of Markov chains in population genetics. We study in Section 1 the problem of convergence for their Markov chains in the case of d types $(d \geq 2)$ of identical fertility (that is, selection does not occur) with mutation and migration allowed. In order to give Markov chain models involving various kinds of selection forces, we make in Section 2 a generalization of induced Markov chains of Karlin and McGregor by weakening of the branching property, and study the problem of convergence to diffusion processes for this class of Markov chains. Section 3 contains some comments on proof of our results, and Section 4 contains additional comments on Ethier's recent work. Our results have connection with diffusion approximation to genetics model Markov chains, which is not given a rigorous justification but is a powerful tool in population genetics (see Crow and Kimura [1]).

§1. Induced Markov chains with mutation and migration.

For each positive integer N, let $\{Z^{(N)}(n) = (Z_1^{(N)}(n),\ldots,Z_d^{(N)}(n));\ n = 0,1,\ldots\}$ be a d type branching process with stationary immigration. That is, $\{Z^{(N)}(n)\}$ is a Markov chain taking values in $\mathbb{Z}_+^d$ (the set of d-dimensional lattice points with nonnegative coordinates) and there exist generating functions $f_{N,p}(s_1,\ldots,s_d)$, $p = 1,\ldots,d$, and $g_N(s_1,\ldots,s_d)$ of distributions in $\mathbb{Z}_+^d$ such that, for any $j = (j_1,\ldots,j_d)$ and $k = (k_1,\ldots,k_d)$ in $\mathbb{Z}_+^d$,

$$P(Z^{(N)}(n+1) = k \mid Z^{(N)}(n) = j\,) = \text{coefficient of } s_1^{k_1}\cdots s_d^{k_d} \text{ in } g_N(s_1,\ldots,s_d)\prod_{p=1}^{d} f_{N,p}(s_1,\ldots,s_d)^{j_p} .$$

We make the following assumptions.

Assumption 1.1. (i) $f_{N,p}$ is of the form

$$f_{N,p}(s_1,\dots,s_d) = \sum_{n=0}^{\infty} c_n \Big(\sum_{q=1}^{d} a_{pq}^{(N)} s_q\Big)^n \quad \text{for } N \geq 1,\ 1 \leq p \leq d \ ,$$

where c_n and $a_{pq}^{(N)}$ satisfy the following conditions.

(ii) $\{c_n\}$ is a probability distribution on $\mathbb{Z}_+$ independent of N and p with $c_0 > 0$ and with maximum span 1 (that is, there is no pair of $\gamma > 1$ and δ such that $\sum_n c_{n\gamma+\delta} = 1$) . Let

$a = \sum_{n=0}^{\infty} nc_n$ (mean), $f(w) = \sum_{n=0}^{\infty} c_n w^n$ (generating function), $M(w) = \sum_{n=0}^{\infty} c_n e^{nw}$ (moment generating function), $F(w) = M(w)e^{-w}$, $b = \sup\{w;\ M(w) < \infty\}$. Then, one of the following holds:

(a) $1 < a \leq +\infty$,

(b) $a = 1$ and $b > 0$,

(c) $a < 1$ and $\lim_{w \to b-} F'(w) > 0$.

(iii) $\{a_{pq}^{(N)}\}$ is of the form $a_{pq}^{(N)} = a_{pq}/N\ (p \neq q)$ and $a_{pp}^{(N)} = 1 + (a_{pp}/N)$ for all sufficiently large N , where $\{a_{pq}\}$ is independent of N and satisfies $a_{pq} \geq 0\ (p \neq q)$, $a_{pp} \leq 0$, $\sum_{q=1}^{d} a_{pq} = 0$.

We remark that if $a < 1$, $b > 0$ and $\lim_{w\to b-} M(w) = \infty$, then (c) holds. It is easy to prove that (ii) implies the existence of a unique $\beta \in (-\infty, b)$ such that $F'(\beta) = 0$. Let $\sigma^2 = F''(\beta)/F(\beta)$. Then $\sigma^2 > 0$. If we define an associated distribution $\{\hat{c}_n\}$ of $\{c_n\}$ by $\hat{c} = c_n e^{n\beta}/M(\beta)$, then $\{\hat{c}_n\}$ has mean 1 and variance σ^2 .

Assumption 1.2. g_N is independent of N , that is,

$$g_N(s_1,\dots,s_d) = g(s_1,\dots,s_d) = \sum_{k\in \mathbb{Z}_+^d} b_k s_1^{k_1}\cdots s_d^{k_d} ,$$

and g satisfies $g(e^{\beta+\epsilon},\dots,e^{\beta+\epsilon}) < \infty$ for some $\epsilon > 0$.

Assumption 1.1 implies that reproduction of offspring by one individual of type p is made in two steps ---- first it produces independently a random number of children of the same type p according to the distribution $\{c_n\}$, and then, each child mutates to type $q (q \neq p)$ with probability α_{pq}/N. Assumption 1.2 implies that the immigration probability is independent of N.

Let $\{\hat{b}_k ; k \in \mathbb{Z}_+^d\}$ be an associated distribution of $\{b_k; k\in \mathbb{Z}_+^d\}$ defined by $\hat{b}_k = b_k e^{(k_1+\dots+k_d)\beta}/g(e^\beta,\dots,e^\beta)$, let $(\mu_1,\dots,\mu_d)$ be the mean of $\{\hat{b}_k\}$, that is, $\mu_p = \sum_{k\in\mathbb{Z}_+^d} k_p \hat{b}_k$, and let $\mu_p' = (\sum_{q=1}^d \mu_q) - \mu_p$.

Let $\mathcal{J}^{(N)}$ be the set of points $j = (j_1,\dots,j_d) \in \mathbb{Z}_+^d$ such that $|j| = n$, where we define $|j| = \sum_{p=1}^d j_p$. Following Karlin and McGregor, we define the induced Markov chain $\{X^{(N)}(n) = (X_1^{(N)}(n),\dots,X_d^{(N)}(n)); n = 0,1,\dots\}$ as a Markov chain on $\mathcal{J}^{(N)}$ with one-step transition probability $P_{jk}^{(N)}$ defined by

$$(1.1)\qquad P_{jk}^{(N)} = P(Z^{(N)}(n+1) = k \mid Z^{(N)}(n) = j,\ Z^{(N)}(n+1) \in \mathcal{J}^{(N)}) .$$

Since the sum of components is N, we can consider the induced Markov chain as a Markov chain on $(d-1)$-dimensional state space. We normalize and interpolate this chain as follows:

$$(1.2)\qquad Y^{(N)}(t) = (\tfrac{1}{N} X_1^{(N)}(n),\dots, \tfrac{1}{N} X_{d-1}^{(N)}(n)) \qquad \text{for } t = \tfrac{n}{N},$$

$$(1.3)\qquad Y^{(N)}(t) = (n+1-Nt)\, Y^{(N)}(\tfrac{n}{N}) + (Nt-n)\, Y^{(N)}(\tfrac{n+1}{N})$$

$$\text{for } \tfrac{n}{N} \le t \le \tfrac{n+1}{N} .$$

Let $\mathbb{K} = \{x = (x_1,\dots,x_{d-1}) \in R^{d-1} ; x_1 \ge 0,\dots,x_{d-1} \ge 0 , 1 - \sum_{p=1}^{d-1} x_p \ge 0\}$.

Then $\{Y^{(N)}(t)\ ;\ 0 \le t < \infty\}$ is a continuous time parameter process taking values in $\mathbb{K}$. Let Ω be the space of continuous paths $\omega : [0,\infty) \to \mathbb{K}$, endowed with the topology of uniform convergence on compact subsets of $[0,\infty)$. Let $\mathcal{M}$ be the topological σ-algebra of Ω and $\mathcal{M}_t$ be the σ-algebra generated by $x(s,\omega) = \omega(s)$, $s \le t$. Define $(d-1)\times(d-1)$-matrix $a(x) = (a_{pq}(x))$ and $(d-1)$-vector $b(x) = (b_p(x))$ on $\mathbb{K}$ by

(1.4) $a_{pp}(x) = \sigma^2 x_p(1-x_p)$,

(1.5) $a_{pq}(x) = -\sigma^2 x_p x_q \qquad (p \ne q)$,

(1.6) $b_p(x) = \sum_{q=1}^{d-1} x_q \alpha_{qp} + (1-\sum_{q=1}^{d-1} x_q)\alpha_{dp} + (1-x_p)\mu_p - x_p\mu'_p$.

Consider the following martingale problem: given $x \in \mathbb{K}$, to find a probability measure P_x on $(\Omega,\mathcal{M})$ satisfying $P_x(x(0) = x) = 1$ such that $(M_\theta(t), \mathcal{M}_t, P_x;\ 0 \le t < \infty)$ is a martingale for each $\theta \in R^{d-1}$, where $M_\theta(t)$ is given by

$$M_\theta(t) = \exp\{\langle\theta, x(t)-x(0)\rangle - \int_0^t \langle\theta, b(x(s))\rangle ds - \frac{1}{2}\int_0^t \langle\theta, a(x(s))\theta\rangle ds\} .$$

Here $\langle\ ,\ \rangle$ denotes inner product. We call this problem the martingale problem on $\mathbb{K}$ for a, b starting from x or, for short, martingale problem $(\mathbb{K},a,b,x)$. We have proved the following results.

Theorem 1.1. *For each* $x \in \mathbb{K}$, *the martingale problem* $(\mathbb{K}, a,b,x)$ *has a solution.*

Theorem 1.2. *Suppose that for each* $x \in \mathbb{K}$, *the solution* P_x *of the martingale problem* $(\mathbb{K}, a,b,x)$ *is unique. Let* $Y^{(N)}(0) = x^{(N)}$ *(non-random), and let* $P^{(N)}$ *be the probability measure on* $(\Omega,\mathcal{M})$ *which the process* $\{Y^{(N)}(t);\ 0 \le t < \infty\}$ *induces. If* $x^{(N)} \to x$, *then* $P^{(N)}$ *converges to* P_x *in the sense of* weak convergence of probability measures on the topological space Ω.

We have conjectured that the uniqueness holds for the martingale problem on $\mathbb{K}$ for a,b. Though we have not succeeded to prove it, we have given the proof in some special cases as follows.

Theorem 1.3. Let $d = 2$. Then, for each x, the solution of the martingale problem $(\mathbb{K}, a,b,x)$ is unique, and hence the conclusion of Theorem 1.2 holds.

Theorem 1.4. Suppose that if $q > p$ and $q' > p$ then $\alpha_{qp} = \alpha_{q'p}$. Then, for each $x \in \mathbb{K}$, the solution of the martingale problem $(\mathbb{K}, a,b,x)$ is unique, and the conclusion of Theorem 1.2 holds.

If the martingale problem has a unique solution P_x, then $(x(t), \mathcal{M}_t, P_x; x \in \mathbb{K})$ is a strong Markov process. By Theorem 1.3, the process $\{Y^{(N)}(t); 0 \leq t < \infty\}$ converges, in case of $d-1 = 1$, to a diffusion process on $[0,1]$ with backward Kolmogorov equation

$$\frac{\partial u}{\partial t} = \frac{\sigma^2}{2} x(1-x) \frac{\partial^2 u}{\partial x^2} + (-x(\alpha_{12} + \mu_2) + (1-x)(\alpha_{21} + \mu_1))\frac{\partial u}{\partial x} .$$

In Feller's boundary classification into four types, the boundary 0 is pure exit, regular, or pure entrance according as $\alpha_{21} + \mu_1$ is 0, positive and $< \sigma^2/2$, or $\geq \sigma^2/2$, respectively. The nature of the boundary 1 is determined similarly by $\alpha_{12} + \mu_2$. In case the boundary is regular, the limiting diffusion has reflecting boundary condition there. In case of pure exit boundary, it is a trap. In case of pure entrance boundary, the limiting diffusion starting from there immediately enters the interior and does not come back.

The result for $d-1 = 1$ in case $\alpha_{12} > 0$, $\alpha_{21} > 0$, $\mu_1 = \mu_2 = 0$ is asserted in [5] without proof. The fact that the induced Markov chains defined above include various genetics models is shown in Karlin [4]. Asymptotic behavior of eigenvalues as $N \to \infty$ is discussed in [6] .

§2. Induced Markov chains with selection.

For each positive integer N , let $\{Z^{(N)}(n) = (Z_1^{(N)}(n),\ldots,Z_d^{(N)}(n))$; $n = 0,1,\ldots\}$ be a Markov chain taking values in $\mathbb{Z}_+^d$ and let $f_{N,j}(s_1,\ldots,s_d)$ be the generating function of one-step transition probability from j .

Assumption 2.1. For all sufficiently large N , $f_{N,j}$ is of the form

$$f_{N,j}(s_1,\ldots,s_d) = \prod_{p=1}^{d} \left(\sum_{n=0}^{\infty} c_n s_p^n \right)^{j_p(1 + N^{-1}\gamma_p(|j|^{-1}j))}$$

$$\text{for } 0 \leq s_p \leq 1 \quad (p = 1,\ldots,d) \quad ,$$

where $\{c_n\}$ satisfies (ii) of Assumption 1.1 and $\gamma_p(x)$, $p = 1,\ldots,d$, are continuous functions of x defined on $\{x \in R^d ; x_1 \geq 0 , \ldots, x_d \geq 0 , \sum_{p=1}^{d} x_p = 1\}$.

This assumption implies that individuals of a type p reproduce their children of the same type p independently of each other according to a common distribution (fertility), but fertility may vary with types and with the proportion of types in the present generation.

Define the induced Markov chain $\{ X^{(N)}(n); n = 0,1,\ldots\}$ as a Markov chain on $J^{(N)}$ with one-step transition probability (1.1) , and define the continuous time parameter process $\{ Y^{(N)}(t); 0 \leq t < \infty\}$ by (1.2), (1.3) . Let $a(x) = (a_{pq}(x))$ be the $(d-1)\times(d-1)$ - matrix defined on $\mathbb{K}$ by (1.4), (1.5) and let $c(x) = (c_p(x))$ be a $(d-1)$-vector defined on $\mathbb{K}$ by

$$c_p(x) = x_p\{\gamma_p(x) - \sum_{q=1}^{d-1} x_q \gamma_q(x) - (1 - \sum_{q=1}^{d-1} x_q) \gamma_d(x)\} ,$$

where $\gamma_r(x) = \gamma_r(x_1,\ldots,x_{d-1}, 1 - \sum_{q=1}^{d-1} x_q)$ for $r = 1,\ldots,d$. We have proved, for this new induced Markov chain and its normalization and interpolation, results parallel to those in Section 1.

Theorem 2.1. For each $x \in K$, the martingale problem (K, a, c, x) has a solution.

Theorem 2.2. The statement of Theorem 1.2 is true with b replaced by c.

Theorem 2.3. Let $d = 2$ and suppose that $\gamma_1(x_1, 1-x_1) - \gamma_2(x_1, 1-x_1)$ is Lipschitz continuous. Then, for each x, the martingale problem (K, a, c, x) has a unique solution P_x. The probability measures $P^{(N)}$ converge to P_x, provided that $x^{(N)} \to x$.

Again we have conjectured that the solution of the martingale problem (K, a, b, x) is unique in general dimensions under some condition on continuity modulus of c, but we have given only the following result.

Theorem 2.4. Suppose that $\gamma_2(x) = \gamma_3(x) = \ldots = \gamma_d(x)$ and that $\gamma_1(x) - \gamma_2(x)$ is a function of x_1 alone and Lipschitz continuous. Then, for each $x \in K$, the martingale problem (K, a, c, x) has a unique solution P_x. If $x^{(N)} \to x$, then $P^{(N)}$ converges to P_x.

By Theorem 2.3, the process $\{ Y^{(N)}(t) ; 0 \leq t \leq \infty \}$ converges, in case of $d-1 = 1$, to the diffusion process on $[0, 1]$ with backward Kolmogorov equation

$$\frac{\partial u}{\partial t} = \frac{\sigma^2}{2} x(1-x) \frac{\partial^2 u}{\partial x^2} + x(1-x) (\gamma_1(x, 1-x) - \gamma_2(x, 1-x)) \frac{\partial u}{\partial x} .$$

The boundaries are of pure exit type and act as traps. Among various choices of γ_1 and γ_2, those which correspond to models in population genetics are linear functions. Namely,

1° If γ_1 and γ_2 are constant functions and $\gamma_1 \neq \gamma_2$, then gametric selection occurs. If $\gamma_1 > \gamma_2$, then type 1 has larger fertility than type 2.

2° Let us denote types 1 and 2 by A_1 and A_2, respectively. If zygotic selection occurs and A_1A_1, A_1A_2, A_2A_2 have relative advantages $1 + N^{-1}\lambda_1$,

$1 + N^{-1}\lambda_2$, $1 + N^{-1}\lambda_3$, respectively, then it is natural to set $\gamma_1(x,1-x) = \lambda_1 x + \lambda_2(1-x)$ and $\gamma_2(x,1-x) = \lambda_2 x + \lambda_3(1-x)$, because we get then

$$1 + N^{-1}\gamma_1(|j|^{-1}j) = (1 + N^{-1}\lambda_1)|j|^{-1}j_1 + (1 + N^{-1}\lambda_2)|j|^{-1}j_2 ,$$
$$1 + N^{-1}\gamma_2(|j|^{-1}j) = (1 + N^{-1}\lambda_2)|j|^{-1}j_2 + (1 + N^{-1}\lambda_3)|j|^{-1}j_2$$

in the expression of $f_{N,j}$.

§3 . Comments on proofs.

Let us make some comments on proofs of the theorems in Section 1 . Similar comments can be made also to the theorems in Section 2 .

Modifying the invariance principle of Stroock and Varadhan [9] , we can show the following: Suppose that

(i) $$\lim_{N\to\infty} \sup_{j\in\mathcal{J}^{(N)}} \Big| \sum_{k\in\mathcal{J}^{(N)}} (k_p - j_p) P^{(N)}_{jk} - b_p\Big(\frac{j_1}{N},\dots,\frac{j_{d-1}}{N}\Big) \Big| = 0$$

for $p = 1,\dots,d-1$;

(ii) $$\lim_{N\to\infty} \sup_{j\in\mathcal{J}^{(N)}} \Big| \frac{1}{N} \sum_{k\in\mathcal{J}^{(N)}} (k_p - j_p)(k_q - j_q) P^{(N)}_{jk} - a_{pq}\Big(\frac{j_1}{N},\dots,\frac{j_{d-1}}{N}\Big) \Big| = 0 \quad \text{for } p,q = 1,\dots,d-1 ;$$

(iii) $$\lim_{N\to\infty} \sup_{j\in\mathcal{J}^{(N)}} \frac{1}{N^{1+\varepsilon}} \sum_{k\in\mathcal{J}^{(N)}} |k-j|^{2+\varepsilon} P^{(N)}_{jk} = 0 \quad \text{for some } \varepsilon > 0 .$$

Then the sequence of the probability measures $P^{(N)}$ has a convergent subsequence, and the limit of any convergent subsequence is a solution of the martingale problem $(\mathbb{K}, a, b, x)$, provided that $Y^{(N)}(0) \to x$. If, moreover, uniqueness of the solution of the martingale problem $(\mathbb{K}, a, b, x)$ holds, then $P^{(N)}$ converges to the solution P_x . Note that existence of a solution of the martingale problem is proved if we check (i), (ii), (iii) .

We can derive the properties (i), (ii), and (iii) with $_\epsilon = 2$ from Assumptions 1.1 and 1.2. Indeed, it can be shown that the convergences in (i), (ii), (iii) are all of order $O(N^{-1})$. In the course of the proof, we have to make many estimations of coefficients of power series. The estimations are carried out by the saddle point method, which is familiar in the theory of limit theorems for large deviations (see [3]). Naturally the proof of Condition (iii) is the most complicated. This condition guarantees the limit process to be a diffusion process and this fact is by no means clear, since transition of the induced Markov chain in one step is not restricted to neighboring points.

It is known that uniqueness of the solution of the martingale problem is equivalent to uniqueness, in the sense of probability law, of the solution on $\mathbf{K}$ to a stochastic differential equation

$$(3.1) \qquad dY_p(t) = \sum_{q=1}^{d-1} a^0_{pq}(Y(t))\,dB_q(t) + b_p(Y(t))\,dt\,, \quad p = 1,\dots,d-1\,,$$

where $(B_1(t),\dots,B_{d-1}(t))$ is a $(d-1)$-dimensional Brownian motion and $a^0(x) = (a^0_{pq}(x))$ is such that $\sum_{r=1}^{d-1} a^0_{pr}(x)a^0_{qr}(x) = a_{pq}(x)$. If $d-1 = 1$, then, by a result of Yamada and S. Watanabe [10], the uniqueness follows from the Hölder continuity of a^0 with exponent $1/2$. If $d-1 \geq 2$, then it is known that Hölder continuity of a^0 with exponent $1/2$ does not imply uniqueness of the solution. But, in the present case, choosing a^0 as a lower triangular matrix, we find that $a^0_{pq}(x)$ is a function of $x_1, x_2, \dots, x_p$ and Hölder continuous with respect to x_p with exponent $1/2$ for fixed $x_1,\dots,x_{p-1}$. This proves uniqueness of the solution of (3.1) if the drift coefficients are null or, more generally, if the assumption in Theorem 1.4 is satisfied. This is based on a suggestion made by S. Watanabe.

Detailed proof will appear in [7], [8].

§4. Additional comments.

The preceding three sections consist of my report at the Tashkent Symposium in summer, 1975. On November 1, 1975, at Chicago S. N. Ethier pointed out to me that he can prove the uniqueness of the solution to the stochastic differential equation (3.1), and also the uniqueness for (3.1) with b_p replaced by c_p if c_p, $p = 1,\dots,d-1$, are of class C^5 . His proof is based on his recent results in [2] and uses the method of Girsanov. Now the uniqueness holds for the solution of the relevant martingale problems, and the assumption in Theorem 1.2 (and Theorem 2.2 if $c(x)$ is of class C^5) is always satisfied.

References

[1] J.F. Crow and M. Kimura: An introduction to population genetics theory, theory, Harper and Row, New York, 1970.

[2] S.N. Ethier: An error estimate for the diffusion approximation in population genetics, Ph.D. Thesis, University of Wisconsin, 1975.

[3] I.A. Ibragimov and Yu.V.Linnik: Independent and stationary sequences of random variables, Nauka, Moscow, 1965. (Russian)

[4] S.Karlin: A first course in stochastic processes, Academic Press, New York, 1966.

[5] S.Karlin and J.McGregor: Direct product branching processes and related Markov chains, Proc. Nat. Acad. Sci. USA, 51 (1964), 598-602.

[6] K.Sato: Asymptotic properties of eigenvalues of a class of Markov chains induced by direct product branching processes, J. Math. Soc. Japan, to appear.

[7] K.Sato: Diffusion processes and a class of Markov chains related to population genetics, Osaka J. Math., to appear.

[8] K.Sato: A class of Markov chains related to selection in population genetics, to appear.

[9] D.W. Stroock and S.R.S. Varadhan: Diffusion processes with continuous coefficients, I and II, Comm. Pure Appl. Math. 22 (1969), 345-400 and 479-530.

[10] T. Yamada and S. Watanabe: On the uniqueness of solutions of stochastic differential equations, J. Math. Kyoto Univ., 11(1971), 155-167 .

Department of Foundations of
Mathematical Sciences
Tokyo University of Education
Otsuka 3-29-1, Bunkyo-ku, Tokyo
and
School of Mathematics
University of Minnesota
Minneapolis, Minnesota 55455

variables $(\mathcal{A}(\omega)x, y) \in R(\Omega)$ satisfy the conditions:

W 1) $\forall x_1, x_2, y_1, y_2 \in X$, $\alpha_1, \alpha_2, \beta_1, \beta_2 \in R$

$$P\{(\mathcal{A}(\omega)[\alpha_1 x_1 + \alpha_2 x_2], \beta_1 y_1 + \beta_2 y_2) = \sum_{i,j} \alpha_i \beta_j (\mathcal{A}(\omega)x_i, y_j)\} = 1;$$

W 2) $(\mathcal{A}(\omega)x, y)$ is continuous in probability on $X \times X$.

Two weak random operators $\mathcal{A}_1(\omega)$ and $\mathcal{A}_2(\omega)$ are stochastically equivalent if

$$P\{(\mathcal{A}_1(\omega)x, y) = (\mathcal{A}_2(\omega)x, y)\} = 1, \quad x, y \in X.$$

We shall not distinguish stochastically equivalent operators.

Let $\mathcal{L}_W(\Omega, X)$ be the set of all weak random operators. Then

$$\mathcal{L}(\Omega, X) \subset \mathcal{L}_W(\Omega, X).$$

By $\langle x \circ y \rangle$ we denote the operator, for which

$$\langle x \circ y \rangle z = (x, z) y;$$

and let $\mathcal{L}_0(X)$ be the set of the operators $\sum_{k=1}^{n} \alpha_k \langle x_k \circ y_k \rangle$; $\alpha_k \in R$; $x_k, y_k \in X$; $n \geq 1$.

The function

$$\chi_{\mathcal{A}}(B) = E \exp\{i \operatorname{tr} \mathcal{A}(\omega) B\}, \tag{1}$$

defined for all $B \in \mathcal{L}_0(X)$, is called the characteristic function of the random operator $\mathcal{A}(\omega)$.

T h e o r e m 1. The function $\chi(B)$ defined on $\mathcal{L}_0(X)$

is the characteristic function of a weak random operator if and only if the following conditions are satisfied:

1) $\chi(B)$ is a positive-definite function:

$$\sum_{K=1}^{n} \chi(B_K - B_j)\lambda_K \lambda_j \geqslant 0;\ \forall n,\ B_i \in \mathcal{L}_0(\mathcal{X}),\ i=1,\dots,n,\ \lambda_j \in R;$$

2) $\chi(\langle x \cdot y \rangle)$ is a contituous function on $\mathcal{X} \times \mathcal{X}$.

Necessity of these conditions is evident. Sufficiency follows from the A.N.Kolmogorov theorem [1] on existence of a measure with given finite-dimensional projections.

3. Strong random operator. A strong random operator $\mathcal{A}(\omega)$ is defined if, for any $x \in \mathcal{X}$, the random variable $\mathcal{A}(\omega)x \in \mathcal{X}(\Omega)$ satisfies the conditions:

S 1) $P\{\mathcal{A}(\omega)(\alpha x + \beta y) = \alpha \mathcal{A}(\omega)x + \beta \mathcal{A}(\omega)y\} = 1,$

$$\alpha, \beta \in R,\ x, y \in \mathcal{X};$$

S 2) $\mathcal{A}(\omega)\, x$ is continuous in probability on $\mathcal{X}$.

It is evident that

$$\mathcal{L}_W(\Omega, \mathcal{X}) \supset \mathcal{L}_S(\Omega, \mathcal{X}) \supset \mathcal{L}(\Omega, \mathcal{X}).$$

Let $\mathcal{A}(\omega) \in \mathcal{L}_S(\Omega, \mathcal{X})$, $\chi_{\mathcal{A}}(B)$ be the characteristic function of $\mathcal{A}(\omega)$. Then

$$\chi_{\mathcal{A}}(\langle z \cdot x \rangle) = E\, exp\,\{i(\mathcal{A}(\omega)x, z)\}$$

is the characteristic function of the random variable $\mathcal{A}(\omega)x \in \mathcal{X}(\Omega)$.

It follows, from the Minlos-Sazonov theorem ([2]), that $\chi_{\mathcal{A}}(\langle z \circ x \rangle)$ is continuous in z in the Σ-topology in which neighbourhoods of zero are of the form $\{x : (Cx, x) < 1\}$ where C is an arbitrary kernel operator from $\mathcal{L}^{+}(X)$.

If $\mathcal{A}(\omega) \in \mathcal{L}_W(\Omega, X)$, then there exists the function

$$\Psi_{\mathcal{A}}(x) = \lim_{n \to \infty} E \exp\{-\sum_{K=1}^{n} (\mathcal{A}(\omega)x, e_k)^2\}$$

where $\{e_k\}$ is a fixed orthonormal basis in X. For $\mathcal{A}(\omega) \in \mathcal{L}_S(\Omega, X)$

$$\Psi_{\mathcal{A}}(x) = E \exp\{-\|\mathcal{A}(\omega)x\|^2\},$$

and therefore $\Psi_{\mathcal{A}}(x)$ is continuous in x.

T h e o r e m 2. Let $\mathcal{A}(\omega) \in \mathcal{L}_W(\Omega, X)$ and $\chi_{\mathcal{A}}(B)$ be the characteristic function of $\mathcal{A}(\omega)$. Continuity of the function $\chi_{\mathcal{A}}(\langle z \circ x \rangle)$ in z in the Σ-topology is necessary and sufficient for $\mathcal{A}(\omega)$ to be a strong random operator.

P r o o f. Necessity of this condition is evident.

Under the condition of the theorem the value

$$\|\mathcal{A}(\omega)x\|^2 = \sum_{K=1}^{\infty} (\mathcal{A}(\omega)x, e_K)^2, \quad x \in X,$$

is determined. This random function in x is bounded in probability on $S_1(0)$ (by $S_\delta(x_0)$ we denote the sphere with the centre at x_0 and the radius δ).

To prove this assertion it is sufficient to prove, for all $\varepsilon > 0$, existence of a sphere $S_\delta(x_0)$ and $\alpha > 0$ such that

$$\sup_{x \in S_\delta(x_0)} P\{\|\mathcal{A}(\omega)x\|^2 > \alpha^2\} < \varepsilon \quad , \tag{1}$$

for it follows from (1) that

$$\sup_{x \in S_1(0)} P\{\|\mathcal{A}(\omega)x\|^2 > \left(\frac{\alpha+\beta}{\delta}\right)^2\} < \varepsilon + P\{\|\mathcal{A}(\omega)x_0\|^2 > \beta^2\} \ .$$

If we can find an $\varepsilon > 0$ such that

$$\sup_{x \in S_\delta(x_0)} P\{\|\mathcal{A}(\omega)\|^2 > \alpha^2\} \geq \varepsilon$$

for all spheres $S_\delta(x_0)$ and $\alpha > 0$, then we can construct a sequence of spheres $S_{\delta_1}(x_1) \supset S_{\delta_2}(x_2) \supset \ldots$ such that $\delta_K \to 0$ and

$$P\{\|\mathcal{A}(\omega)x\|^2 > K\} \geq \varepsilon \ , \quad x \in S_{\delta_K}(x_K) \ .$$

Thus we have

$$P\{\|\mathcal{A}(\omega)x_0\|^2 > K\} \geq \varepsilon \qquad \forall K \tag{2}$$

if $x_0 \in \bigcap_K S_{\delta_K}(x_K)$. But (2) is impossible. Therefore (1) holds true and

$$\lim_{\alpha \to \infty} \sup_{x \in S_1(0)} P\{\|\mathcal{A}(\omega)x\|^2 > \alpha\} = 0 \ . \tag{3}$$

From (3) it follows that, for any $\varepsilon > 0$,

$$\lim_{\delta \to 0} \sup_{x \in S_\delta(0)} P\{\|\mathcal{A}(\omega)x\|^2 > \varepsilon\} = \lim_{\delta \to 0} \sup_{x \in S_1(0)} P\{\|\mathcal{A}(\omega)x\|^2 > \frac{\varepsilon}{\delta}\} = 0 \ .$$

Hence

$$\lim_{x \to x_0} P\{\|\mathcal{A}(\omega)x - \mathcal{A}(\omega)x_0\| > \varepsilon\} = \lim_{\|x - x_0\| \to 0} \sup_{z \in S_{\|x-x_0\|}(0)} P\{\|\mathcal{A}(\omega)z\| > \varepsilon\} = 0 \ .$$

4. Product of random operators. Let $\mathcal{A}(\omega), \mathcal{B}(\omega) \in \mathcal{L}_W(\Omega, X)$. If, for any $x, y \in X$, the series

$$\sum_{K=1}^{\infty} (\mathcal{B}(\omega)x, e_K)(\mathcal{A}(\omega)e_K, y)$$

converges in probability, we denote its sum by $(\mathcal{A}(\omega)\mathcal{B}(\omega)x, y)$. This expression, as a function of x and y, satisfies conditions W 1) and W 2). Thus there exists a random operator $C(\omega) \in \mathcal{L}_W(\Omega, X)$ for which the equality

$$\sum_{K=1}^{\infty} (\mathcal{B}(\omega)x, e_K)(\mathcal{A}(\omega)e_K, y) = (C(\omega)x, y)$$

holds. The operator $C(\omega)$ is said to be the product of the operators $\mathcal{A}(\omega)$ and $\mathcal{B}(\omega)$ and it is denoted by $\mathcal{A}(\omega)\mathcal{B}(\omega)$.

If $\mathcal{A}(\omega) \in \mathcal{L}_W(\Omega, X)$, then there exists an operator $\mathcal{A}^*(\omega) \in \mathcal{L}_W(\Omega, X)$ such that

$$P\{(\mathcal{A}(\omega)x, y) = (\mathcal{A}^*(\omega)y, x)\} = 1, \quad x, y \in X \quad .$$

The operator $\mathcal{A}^*(\omega)$ is called the conjugate operator to the operator $\mathcal{A}(\omega)$.

If $\mathcal{A}^*(\omega) \in \mathcal{L}_s(\Omega, X)$, $\mathcal{B}(\omega) \in \mathcal{L}_s(\Omega, X)$, then there exists $\mathcal{A}(\omega)\mathcal{B}(\omega) \in \mathcal{L}_W(\Omega, X)$. Indeed,

$$\sum_{K=1}^{\infty} |(\mathcal{B}(\omega)x, e_K)(\mathcal{A}(\omega)e_K, y)| \leq \sum_{K=1}^{\infty} |(\mathcal{B}(\omega)x, e_K)(\mathcal{A}^*(\omega)y, e_K)| \leq$$

$$\leq \left[\sum_{K=1}^{\infty} (\mathcal{B}(\omega)x, e_K)^2 \sum_{K=1}^{\infty} (\mathcal{A}^*(\omega)y, e_K)^2\right]^{\frac{1}{2}} = \|\mathcal{B}(\omega)x\| \cdot \|\mathcal{A}^*(\omega)y\| .$$

An operator $\mathcal{A}(\omega) \in \mathcal{L}_W(\Omega, X)$ is said to be a symmet-

RANDOM OPERATORS IN A HILBERT SPACE

A.V.Skorohod

This paper is devoted to general properties of random operators and operator-valued stochastic processes. We shall also consider operator-valued martingales, stochastic integral, stochastic differential equations and stochastic semigroups.

§ 1. Definitions

1. Notation. X is a separable Hilbert space with elements $x, y, z, \ldots$; $\mathfrak{A}$ is the σ-algebra of Borel sets in X, elements of $\mathfrak{A}$ are denoted by $Y, Z, U, W, \ldots$.

$\mathcal{L}(X)$ is the set of bounded linear operators in X, elements of $\mathcal{L}(X)$ are denoted by $A, B, C, \ldots$; $\mathfrak{B}$ is the σ-algebra of subsets of $\mathcal{L}(X)$ generated by

$$\{A: (Ax, y) < \lambda\}, \quad x, y \in X, \quad \lambda \in R,$$

$(\cdot,\cdot)$ is the inner product in X, R is the set of real numbers.

$\mathcal{L}^+(X)$ is the set of positive symmetric operators.

$\{\Omega, \mathcal{F}, P\}$ is a fixed probability space, $R(\Omega)$ is the space of real random variables, $X(\Omega)$ is the space of X-valued random variables, $\mathcal{L}(\Omega, X)$ is the space of measurable mappings from $\{\Omega, \mathcal{F}\}$ into $\{\mathcal{L}(X), \mathfrak{B}\}$. Elements of $\mathcal{L}(\Omega, X)$ are called bounded random operators and are denoted by $\mathcal{A}(\omega)$, $\mathcal{B}(\omega), \ldots$.

2. Weak random operators. A weak random operator $\mathcal{A}(\omega)$ is defined if, for any $x, y \in X$, the random

ric random operator if $\mathcal{A}(\omega) = \mathcal{A}^*(\omega)$. It is called positive if $P\{(\mathcal{A}(\omega)x,x)\geq 0\}=1 \quad \forall x\in X$. The set of all positive symmetric operators from $\mathcal{L}_w(\Omega,X)$ will be denoted by $\mathcal{L}_w^+(\Omega,X)$,

$$\mathcal{L}_s^+(\Omega,X)=\mathcal{L}_w^+(\Omega,X)\cap\mathcal{L}_s(\Omega,X),\ \mathcal{L}^+(\Omega,X)=\mathcal{L}_w^+(\Omega,X)\cap\mathcal{L}(\Omega,X).$$

For any $\mathcal{A}(\omega)\in\mathcal{L}_s(\Omega,X)$, the operators $\mathcal{A}^*(\omega)\,\mathcal{A}(\omega)$ and $\mathcal{A}^*(\omega)\mathcal{A}(\omega)\in\mathcal{L}_w^+(\Omega,X)$ are well-defined.

5. Bounded random operators. Let $\mathcal{A}(\omega)\in\mathcal{L}_s(\Omega,X)$. Then for all $x\in X$ there exists $\|\mathcal{A}(\omega)x\|$. Let $Z\subset X$ be a dense denumerable set. The condition

$$P\{\sup_{x\in Z\cap S_1(0)}\|\mathcal{A}(\omega)x\|<\infty\}=1 \tag{4}$$

is necessary and sufficient for $\mathcal{A}(\omega)\in\mathcal{L}(\Omega,X)$.

Theorem 3. If

$$P\{\sum_{i,k}(\mathcal{A}(\omega)e_i,e_k)^2<\infty\}=1\ , \tag{5}$$

then $\mathcal{A}(\omega)\in\mathcal{L}(\Omega,X)$.

Proof. Let $Z=\{x: x=\sum_{k=1}^{n}\alpha_k e_k,\ n\geq 1\}$, where α_k are rational real numbers. Then

$$\|\mathcal{A}(\omega)x\|^2=\sum_{k=1}^{\infty}(\mathcal{A}(\omega)x,e_k)^2=\sum_{k=1}^{\infty}(\sum_{i=1}^{n}\alpha_i(\mathcal{A}(\omega)e_i,e_k))^2\leq$$

$$\leq\sum_{k=1}^{\infty}\sum_{i=1}^{n}\alpha_i^2\sum_{j=1}^{n}(\mathcal{A}(\omega)e_j,e_k)^2\leq\sum_{i=1}^{n}\alpha_i^2\sum_{k,j}(\mathcal{A}(\omega)e_k,e_j)^2=$$

$$= \|x\|^2 \sum_{k,j=1}^{\infty} (\mathcal{A}(\omega)e_k, e_j)^2 \quad .$$

Thus (4) follows from (5).

C o r o l l a r y. If $\mathcal{A}(\omega) \in \mathcal{L}_W(\Omega, X)$, then there exists an invertible operator $C \in \mathcal{L}^+(X)$ such that $C^*\mathcal{A}(\omega)C \in \mathcal{L}(\Omega, X)$.

Indeed, let $Ce_k = \lambda_k e_k$. Then

$$\sum_{i,k} (C^*\mathcal{A}(\omega)Ce_k, e_i) = \sum_{i,k} \lambda_i^2 \lambda_k^2 (\mathcal{A}(\omega)e_k, e_i)^2 \; .$$

It is evident that there exist λ_k's such that $P\{\sum \lambda_i^2 \lambda_k^2 (\mathcal{A}e_i, e_k)^2 < \infty\} = 1$.

6. G a u s s i a n r a n d o m o p e r a t o r s. A weak random Gaussian operator $\mathcal{A}(\omega)$ is an operator for which the common distribution of the random variables $(\mathcal{A}(\omega)x_k, y_k)$ is Gaussian $\forall n > 0, x_1, \dots, x_n \in X$, $y_1, \dots, y_n \in X$. We set $a(x,y) = E(\mathcal{A}(\omega)x, y)$,

$$b(x,y,u,v) = E[(\mathcal{A}(\omega)x,y) - a(x,y)][(\mathcal{A}(\omega)u,v) - a(u,v)] .$$

These functions determine the characteristic function of $\mathcal{A}(\omega)$:

$$\chi_{\mathcal{A}}(B) = \exp\{i \sum_{k=1}^{n} a(x_k, y_k) - \frac{1}{2} \sum_{k,j} b(x_k, y_k, x_j, y_j)\}, B \in \mathcal{L}_0(X) . \quad (6)$$

(6) is the characteristic function of a weak random Gaussian operator if and only if the following conditions hold:

a) $a(x,y)$ is a continuous bilinear function,

b) $b(x,y,u,v)$ is a continuous 4-linear function for which

$$b(x,y,u,v) = b(u,v,x,y) \quad ,$$

$$b(x,y,x,y)+2b(x,y,u,v)+b(u,v,u,v)\geq 0\,,\quad x,y,u,v\in X\,.$$

Under these conditions $\mathcal{A}(\omega)\in\mathcal{L}_g(\Omega,X)$ if and only if, for all $x\in X$,

c) $$\sum_{k=1}^{\infty} b(x,e_k,x,e_k)<\infty\,.$$

The condition

d) $$\sum_{k,i=1}^{\infty} b(e_i,e_k,e_i,e_k)<\infty\,.$$

is sufficient for $\mathcal{A}(\omega)$ to belong to $\mathcal{L}(\Omega,X)$.

The operator of the Gaussian "white noise" is an example of a weak random Gaussian operator. For this operator $a(x,y)=0$, $b(x,y,u,v)=(x,u)(y,v)$. The function $b(x,y,u,v)$ is called the correlation function of a random operator. The functions

$$b(x,y,u,v)=(Cx,u)(\mathcal{D}y,v)$$

where $C,\mathcal{D}\in\mathcal{L}^+(X)$ satisfy condition b). Thus they are the correlation functions of weak random Gaussian operators.

§ 2. Convergence of random operators

1. Weak convergence of random operators. Let $\mathcal{A}_n(\omega)\in\mathcal{L}_W(\Omega,X)$, $n=0,1,\ldots$. The sequence of random operators $\mathcal{A}_n(\omega)$ is weakly convergent to $\mathcal{A}_0(\omega)$ if, for all $x,y\in X$,

$$P\text{-}\lim_{n\to\infty}(\mathcal{A}_n(\omega)x,y)=(\mathcal{A}_0(\omega)x,y) \quad . \tag{1}$$

T h e o r e m 1. Let $\mathcal{A}_n(\omega)\in\mathcal{L}_W(\Omega,X)$, $n=1,2,\ldots$, and $\forall x,y\in X$ there exist

$$P\text{-}\lim_{n\to\infty}(\mathcal{A}_n(\omega)x,y)=\beta(x,y) \quad .$$

Then there exists such a random operator $\mathcal{A}_0(\omega)\in\mathcal{L}_W(\Omega,X)$, that $\beta(x,y)=(\mathcal{A}_0(\omega)x,y)$.

P r o o f . $\beta(x,y)$ satisfies condition W 1) of § 1. To prove that it satisfies condition W 2) as well we shall establish the relation

$$\lim_{\alpha\to\infty}\ \sup_n\ \sup_{x,y\in S_1(0)} P\{|(\mathcal{A}_n(\omega)x,y)|>\alpha\}=0 \quad . \tag{2}$$

(2) is true if, for any $\varepsilon>0$, there exist a sphere $S_\delta(x_0)$ and an $\alpha>0$ such that

$$\sup_n\ \sup_{x,y\in S_\delta(x_0)} P\{|(\mathcal{A}_n(\omega)x,y)|>\alpha\}<\varepsilon \quad . \tag{3}$$

Suppose that there exists an $\varepsilon>0$ such that, for all spheres and $\alpha>0$,

$$\sup_n\ \sup_{x,y\in S_\delta(x_0)} P\{|(\mathcal{A}_n(\omega)x,y)|>\alpha\}\geq\varepsilon \quad .$$

Then we can construct a sequence of the spheres $S_{\delta_1}(x_1)\supset S_{\delta_2}(x_2)\supset\ldots$ such that $\delta_K\to 0$ and, for all $x,y\in S_{\delta_K}(x_K)$,

$$P\{|(A_{n_\kappa}(\omega)x, y)| > \kappa\} \geq \varepsilon$$

and $n_\kappa \to \infty$. If $x_0 \in \bigcap_\kappa S_{\delta_\kappa}(x_\kappa)$, then

$$P\{|(A_{n_\kappa}(\omega)x_0, x_0)| > \kappa\} \geq \varepsilon$$

for all κ . But this is impossible. Thus (3) and (2) hold true.

2. Strong convergence of random operators. A sequence of random operators $A_n(\omega) \in \mathcal{L}_s(\Omega, X)$ is said to be strongly convergent to $A_0(\omega) \in \mathcal{L}_s(\Omega, X)$ if, for any $x \in X$,

$$P\text{-}\lim_{n\to\infty} \|A_n(\omega)x - A_0(\omega)x\| = 0 \quad .$$

Theorem 2. Let $A_n(\omega) \in \mathcal{L}_s(\Omega, X)$ and $A_n(\omega)$ be weakly convergent to $A_0(\omega) \in \mathcal{L}_w(\Omega, X)$. If

$$\Psi_{A_n}(x) = E\exp\{-\|A_n(\omega)x\|^2\}$$

and the function $\inf_n \Psi_{A_n}(x) = \Psi(x)$ is continuous at $x = 0$, then $A_0(\omega) \in \mathcal{L}_s(\Omega, X)$. If, in addition,

$$\lim_{n\to\infty} E\exp\{-\|A_n(\omega)x\|^2\} = E\exp\{-\|A_0(\omega)x\|^2\} \ ,$$

then $A_n(\omega)$ is strongly convergent to $A_0(\omega)$.

Proof. For all, ℓ, m ,

$$E\exp\{-\sum_{\kappa=1}^{\ell}(A_m(\omega)x, e_\kappa)^2\} \geq \inf_n \Psi_{A_n}(x) \quad .$$

Therefore

$$E\exp\{-\sum_{K=1}^{\ell}(A_0(\omega)x,e_K)^2\}\geq \inf_n \Psi_{A_n}(x) \quad ,$$

$$P\{\sum_{K=1}^{\ell}(A_0(\omega)x,e_K)^2>\alpha\}\leq \frac{1-\inf_n \Psi_{A_n}(x)}{1-e^{-\alpha}} \quad .$$

and

$$P\{\sum_{K=1}^{\ell}(A_0(\omega)x,e_K)^2>\alpha\}\leq \frac{1-\inf_n \Psi_{A_n}(\lambda x)}{1-e^{-\lambda^2\alpha}} \quad .$$

Thus

$$P\{\sum_{K=1}^{\infty}(A_0(\omega)x,e_K)^2<\infty\}=1,\ A_0(\omega)\in \mathcal{L}_S(\Omega,X) \quad .$$

Suppose that

$$\zeta_n=1-\exp\{-\|A_n(\omega)x\|^2\},\quad n=0,1,\ldots \quad .$$

Then

$$\varliminf_{n\to\infty}\zeta_n\geq\zeta_0 \quad . \tag{4}$$

Denote $\bar{\zeta}_n=\zeta_n\vee\zeta_0$; $P\{\bar{\zeta}_n-\zeta_n<0\}=0$. For $\varepsilon>0$,

$$\lim_{n\to\infty}P\{\bar{\zeta}_n-\zeta_n>\varepsilon\}=\lim_{n\to\infty}P\{\zeta_0-\zeta_n>\varepsilon\}=0 \quad .$$

Thus $\lim_{n\to\infty}E(\bar{\zeta}_n-\zeta_n)=0$, $\lim_{n\to\infty}E(\bar{\zeta}_n-\zeta_0)=0$ since $\lim_{n\to\infty}E\zeta_n=E\zeta_0$.
From the inequalites $\bar{\zeta}_n\geq\zeta_0$, $\bar{\zeta}_n\geq\zeta_n$ it follows that

$$P\text{-}\lim\bar{\zeta}_n=\zeta_0,\ P\text{-}\lim(\bar{\zeta}_n-\zeta_n)=0,\ P\text{-}\lim\zeta_n=\zeta_0 .$$

Hence

$$P\text{-}\lim_{n\to\infty}\|A_n(\omega)x\|^2=\|A_0(\omega)x\|^2 .$$

This relation and the weak convergence $A_n(\omega)$ to $A_0(\omega)$ imply the strong convergence (this fact is known for convergence of elements in a Hilbert space).

3. Convergence of distributions corresponding to random operators.

The distributions of random operators $A_n(\omega)$ are said to be weakly convergent to the distribution of a random operator $A_0(\omega)$ if, for all $x_1, y_1, \dots, x_k, y_k \in X$, the common distribution of the random variables $(A_n(\omega)x_i, y_i)$, $i = 1, 2, \dots, k$, is convergent to the common distribution of the random variables $(A_0(\omega)x_i, y_i)$, $i = 1, \dots, k$.

The distributions of random operators $A_n(\omega) \in \mathcal{L}_S(\Omega, X)$ are said to be strongly convergent to the distribution of $A_0(\omega) \in \mathcal{L}_S(\Omega, X)$ if, for all $x_1, x_2, \dots, x_k \in X$ and all continuous bounded functions $f(y_1, \dots, y_k)$ on X^k, the equality

$$\lim_{n\to\infty} E f(A_n(\omega)x_1, \dots, A_n(\omega)x_k) = E f(A_0(\omega)x_1, \dots, A_0(\omega)x_k).$$

holds.

The distributions of the random operators $A_n(\omega)$ are weakly convergent to the distribution of $A_0(\omega)$ if and only if, for all $B \in \mathcal{L}_0(X)$,

$$\lim_{n\to\infty} \chi_{A_n}(B) = \chi_{A_0}(B) \tag{5}$$

where (B) is the characteristic function of $A_n(\omega)$. If, in addition to (5), $A_n(\omega) \in \mathcal{L}_S(\Omega, X)$ and

$$\lim_{n\to\infty} E\exp\{-\|\mathcal{A}_n(\omega)x\|^2\} = E\exp\{-\|\mathcal{A}_0(\omega)x\|^2\} \qquad \forall x\in X ,$$

then the distributions of $\mathcal{A}_n(\omega)$ are strongly convergent to the distribution of $\mathcal{A}_0(\omega)$. We shall prove a more general assertion.

T h e o r e m 3. Let $\mathcal{A}_n(\omega)\in \mathcal{L}_W(\Omega, X)$ be a sequence of random operators such that, for all $x,y\in X$, the sequence of random variables $(\mathcal{A}_n(\omega)x, y)$ is bounded in probability. Then there exists a subsequence n_k and random operators $\mathcal{B}_{n_k}(\omega)$ for which

1) $\chi_{\mathcal{A}_{n_k}}(B) = \chi_{\mathcal{B}_{n_k}}(B)$, $\quad B\in \mathcal{L}_0(X)$,

2) $\mathcal{B}_{n_k}$ are weakly convergent to a random operator $\mathcal{B}_0(\omega)$.

P r o o f. In the same way as in Theorem 1 we can establish relation (2). Choose a sequence $\{x_k\}$ dense in X. There exists a subsequence n_k for which the common distributions of the random variables $\{(\mathcal{A}_{n_k}(\omega)x_i, x_j),\ i,j = 1,\dots,m\}$, for all m, are convergent to the common distribution of random variables $\{\beta_0(x_i,x_j),\ i,j = 1,\dots,m\}$. Then we can construct random variables $\{\beta_{n_k}(x_i,x_j),\ i,j = 1,2,\dots\}$ such that

1) $$P\text{-}\lim \beta_{n_k}(x_i,x_j) = \beta_0(x_i,x_j), \quad \forall i,j ;$$

2) for all m the random variables $\{\beta_{n_k}(x_i,x_j), i,j = 1,\dots,m\}$ have the same distribution as the random variables $\{(\mathcal{A}_{n_k}(\omega)x_i, x_j);\ i,j = 1,\dots,m\}$ (this fact follows from [3], Ch.1. § 6). Suppose that

$$(\mathcal{B}_{n_K}(\omega)x_i, x_j) = \beta_{n_K}(x_i, x_j), \ (\mathcal{B}_0(\omega)x_i, x_j) = \beta_0(x_i, x_j).$$

Then, for all K, there exists

$$P\text{-}\lim_{n\to\infty} (\mathcal{B}_{n_K}(\omega)x_{i_n}, x_{j_n}) = (\mathcal{B}_{n_K}(\omega)x, y)$$

if $x_{i_n} \to x$, $x_{j_n} \to y$. For all x_i, x_j, x_ℓ, x_n,

$$P\{|\beta_0(x_i, x_j) - \beta_0(x_\ell, x_n)| > \varepsilon\} \leq$$

$$\leq \overline{\lim_{K\to\infty}} P\{|\beta_{n_K}(x_i, x_j) - \beta_{n_K}(x_\ell, x_n)| > \varepsilon\} =$$

$$= \lim_{K\to\infty} P\{|(\mathcal{A}_{n_K}(\omega)(x_i - x_\ell), x_j)| > \frac{\varepsilon}{2}\} + \lim_{K\to\infty} P\{|(\mathcal{A}_{n_K}(\omega)x_\ell, x_j - x_n)| >$$

$$> \frac{\varepsilon}{2}\} \leq \sup_n \sup_{x,y\in S_1(0)} P\{|(\mathcal{A}_n(\omega)x, y)| > \frac{\varepsilon}{2\|x_j\|\cdot\|x_i - x_\ell\|}\} +$$

$$+ \sup_n \sup_{x,y\in S_1(0)} P\{|(\mathcal{A}_n(\omega)x, y)| > \frac{\varepsilon}{2\|x_\ell\|\,\|x_j - x_n\|}\}$$

Therefore there exists

$$P\text{-}\lim_{\substack{x_i\to x\\ x_j\to y}} \beta_0(x_i, x_j) = \beta_0(x, y)$$

and

$$\beta_0(x, y) = P\text{-}\lim_{K\to\infty} (\mathcal{B}_{n_K}(\omega)x, y).$$

To complete the proof one can use Theorem 1.

§3. Martingales and stochastic integrals

1. Operator martingale. Let $\mathcal{F}_s \subset \mathcal{F}$, $s \geq 0$,

be a family of increasing σ-algebras. The function $\mathcal{A}_t(\omega)$, $t \geq 0$, with values in $\mathcal{L}_S(\Omega, X)$ is called an operator martingale if it satisfies the following conditions:

1) $(\mathcal{A}_t(\omega)x, y)$ is a local martingale with respect to $\mathcal{F}_t$; $x, y \in X$;

2) $\mathcal{A}_t(\omega)x$ is continuous as a function of t for all $x \in X$ with probability 1.

If $\mathcal{A}_t(\omega)$ is an operator martingale, $\mathcal{A}_t(\omega)x$ is a local continuous martingale in X . From the results of [4] it follows that there exists a function $\gamma_t(x)$ such that

a) $\gamma_t(x)$ is an $\mathcal{F}_t$ -measurable increasing continuous function in t ,

b) $\|\mathcal{A}_t(\omega)x\|^2 - \gamma_t(x)$ is a martingale with respect to $\mathcal{F}_t$.

$$\gamma_t(x) = P\text{-}\lim_{\lambda \to 0} \sum_{K=1}^{n} \|\mathcal{A}_{t_K}(\omega)x - \mathcal{A}_{t_{K-1}}(\omega)x\|^2,$$

where $0 = t_0 < t_1 < \dots < t_n = t$, $\lambda = \max_K (t_K - t_{K-1})$.

If we put

$$\gamma_t(x, y) = \frac{1}{2}[\gamma_t(x+y) - \gamma_t(x) - \gamma_t(y)] ,$$

then

$$\gamma_t(x, y) = P\text{-}\lim_{\lambda \to 0} \sum_{K=1}^{n} ([\mathcal{A}_{t_K}(\omega) - \mathcal{A}_{t_{K-1}}(\omega)]^* [\mathcal{A}_{t_K}(\omega) - \mathcal{A}_{t_{K-1}}(\omega)] x, y).$$

Thus there exists a random operator $\langle \mathcal{A}(\omega) \rangle_t \in \mathcal{L}_W^+(\Omega, X)$ which is the weak limit of

$$\sum_{K=1}^{n} [\mathcal{A}_{t_K}(\omega) - \mathcal{A}_{t_{K-1}}(\omega)]^* [\mathcal{A}_{t_K}(\omega) - \mathcal{A}_{t_{K-1}}] .$$

The random operator $\langle \mathcal{A}(\omega)\rangle_t$ is called the square characteristic of the operator martingale $\mathcal{A}_t(\omega)$.

2. Convergence of martingales.

Theorem 1. Let $\mathcal{A}_t^{(n)}(\omega), n=1,2,\ldots,$ be a sequence of operator martingales, $\mathcal{A}_0^{(n)}(\omega)=0$, and, for all $t>0$, $x\in X$, $\varepsilon>0$,

$$\lim_{n,m\to\infty} P\{(\langle \mathcal{A}^{(n)}(\omega)-\mathcal{A}^{(m)}(\omega)\rangle_t x,x)>\varepsilon\}=0. \tag{1}$$

Then there exists an operator martingale $\mathcal{A}_t(\omega)$ such that, for all $t>0, x\in X, \varepsilon>0$,

$$\lim_{n\to\infty} P\{\sup_{\theta\le t} \|\mathcal{A}_\theta^{(n)}(\omega)x-\mathcal{A}_\theta(\omega)x\|>\varepsilon\}=0. \tag{2}$$

Proof. Put

$$x_t(\omega)=\mathcal{A}_t^{(n)}(\omega)x-\mathcal{A}_t^{(m)}(\omega)x, \qquad \gamma_t=(x,\langle \mathcal{A}^{(n)}(\omega)-\mathcal{A}^{(m)}(\omega)\rangle_t x).$$

Let $\inf[t:\gamma_t=\beta\}$. Then $\|x_{\theta\wedge\tau}(\omega)\|^2-\gamma_{\theta\wedge\tau}$ is a martingale,

$$E\|x_{\theta\wedge\tau}(\omega)\|^2=E\gamma_{\theta\wedge\tau}\le\beta.$$

Therefore

$$P\{\sup_{\theta\le t}\|x_\theta(\omega)\|>\varepsilon\}=P\{\sup_{\theta\le t}\|x_\theta(\omega)\|>\varepsilon,\ \tau>t\}+$$

$$P\{\sup_{\theta\le t}\|x_\theta(\omega)\|>\varepsilon,\ \tau\le t\}\le P\{\sup_{\theta\le t}\|x_{\theta\wedge\tau}(\omega)\|>\varepsilon\}+P\{\gamma_t\ge\beta\}.$$

From the Kolmogorov inequality for submartingales it follows:

$$P\{\sup_{\theta\le t}\|x_{\theta\wedge\tau}(\omega)\|^2>\varepsilon^2\}\le\frac{N E\|x_{t\wedge\tau}(\omega)\|^2}{\varepsilon^2}\le\frac{\beta}{\varepsilon^2}\quad.$$

Thus

$$P\{\sup_{\theta\le t}\|x_\theta(\omega)\|>\varepsilon\}\le\frac{\beta}{\varepsilon^2}+P\{\gamma_t>\beta\},$$

that is

$$P\{\sup_{\theta\le t}\|\mathcal{A}^{(n)}_\theta(\omega)x-\mathcal{A}^{(m)}_\theta(\omega)x\|>\varepsilon\}\le\frac{\beta}{\varepsilon^2}+P\{(\langle\mathcal{A}^{(n)}(\omega)-\mathcal{A}^{(m)}(\omega)\rangle_t x, x)>\beta\}.$$

From (1) we obtain

$$\lim_{n\to\infty}P\{\sup_{\theta\le t}\|\mathcal{A}^{(n)}_\theta(\omega)x-\mathcal{A}^{(m)}_\theta(\omega)x\|>\varepsilon\}=0 \tag{3}$$

for all $\varepsilon>0$. This relation implies the existence of a process $\mathcal{A}_t(\omega)\in\mathcal{L}_s(\Omega,X)$ such that (2) holds. $\mathcal{A}_t(\omega)x$ is a continuous local martingale in X .

3. Operator stochastic integrals. Let $\mathcal{A}_t(\omega)$ be an operator martingale, $\langle\mathcal{A}(\omega)\rangle_t$ its square characteristic. We consider the set of all operator random functions $\mathcal{B}_t(\omega)$ which satisfy the following conditions:

1) $\mathcal{B}_t(\omega)\in\mathcal{L}_s(\Omega,X)$, $t\ge 0$;

2) $(\mathcal{B}_t(\omega)x, y)$ is $\mathcal{F}_t$ -measurable for fixed t and is measurable in t process;

3) the integral

$$\int_0^t\mathcal{B}^*_\theta(\omega)d_\theta\langle\mathcal{A}(\omega)\rangle_\theta\mathcal{B}_\theta(\omega) \tag{4}$$

is defined for all $t>0$.

Integral (4) is defined for step-functions $\mathcal{B}_\theta(\omega)$, i.e.

such that $\mathcal{B}_\theta(\omega)=\mathcal{B}_{\theta_K}(\omega)$, $\theta_K \le \theta < \theta_{K+1}$, $0=\theta_0<\theta_1<\ldots<\theta_n=t$, as the sum

$$\sum_{K=0}^{n-1} \mathcal{B}^*_{\theta_K}(\omega)[\langle \mathcal{A}(\omega)\rangle_{\theta_{K+1}} - \langle \mathcal{A}(\omega)\rangle_{\theta_K}]\, \mathcal{B}_{\theta_K}(\omega) \tag{5}$$

It is extended by continuity to other functions.

For all functions $\mathcal{B}_t(\omega)$,satisfying conditions 1)-3), one can construct the integral

$$\int_0^t d_\theta \mathcal{A}_\theta(\omega)\, \mathcal{B}_\theta(\omega) \tag{6}$$

with the properties:

a) the process

$$C_t(\omega)=\int_0^t d_\theta \mathcal{A}_\theta(\omega)\, \mathcal{B}_\theta(\omega)$$

is an operator martingale with the square characteristic

$$\langle C(\omega)\rangle_t = \int_0^t \mathcal{B}^*_\theta(\omega)\, d_\theta \langle \mathcal{A}(\omega)\rangle_\theta\, \mathcal{B}_\theta(\omega) \quad ;$$

b)

$$\int_0^t d_\theta \mathcal{A}_\theta(\omega)[\alpha_1 \mathcal{B}^{(1)}_\theta(\omega)+\alpha_2 \mathcal{B}^{(2)}_\theta(\omega)] = \alpha_1 \int_0^t d_\theta \mathcal{A}_\theta(\omega)\mathcal{B}^{(1)}_\theta(\omega) + \alpha_2 \int_0^t d_\theta \mathcal{A}_\theta(\omega)\mathcal{B}^{(2)}_\theta(\omega).$$

At first, we assume that $\mathcal{A}_t(\omega)\in \mathcal{L}(\Omega, X)$ and define the integral for step-functions:

$$\int_0^t d_\theta \mathcal{A}_\theta(\omega)\, \mathcal{B}_\theta(\omega) = \sum_{K=0}^{n-1} (\mathcal{A}_{\theta_{K+1}}(\omega) - \mathcal{A}_{\theta_K}(\omega))\, \mathcal{B}_{\theta_K}(\omega) \quad .$$

And then we use Theorem 1 to extend the construction of the stochastic integral for all functions $\mathcal{B}_\theta(\omega)$ satisfying conditions 1)-3).

In the general case we need the following lemma.

L e m m a 1. If $\mathcal{A}_t(\omega)$ is a continuous process in $\mathcal{L}_S(\Omega, X)$, then there exist operators $C_\varepsilon \in \mathcal{L}^+(X)$ such that $\mathcal{A}_t(\omega) C_\varepsilon \in \mathcal{L}(\Omega, X)$ and $\forall x \in X \quad \lim_{\varepsilon \to 0} \| C_\varepsilon x - x \| = 0$.

P r o o f. Suppose that

$$\xi_n = \sup_{i \le n} \sup_{t \le n} \| \mathcal{A}_t(\omega) e_i \| .$$

If λ_n are choosen so that

$$P\{ \sum_n \lambda_n^2 \xi_n^2 < \infty \} = 1$$

and

$$C_\varepsilon e_\kappa = \frac{\lambda_\kappa}{\varepsilon + \lambda_\kappa} e_\kappa ,$$

then, for $t \le \kappa$,

$$\sum_{i,j} (\mathcal{A}_t(\omega) C_\varepsilon e_i, e_j)^2 = \sum_i \| \mathcal{A}_t(\omega) C_\varepsilon e_i \|^2 \le$$

$$\le \sum_{i \le \kappa} \left(\frac{\lambda_i}{\varepsilon + \lambda_i}\right)^2 \xi_\kappa^2 + \sum_{i > \kappa} \left(\frac{\lambda_i}{\varepsilon + \lambda_i}\right)^2 \xi_i^2 < \infty$$

with probability 1.

Now we consider the stochastic integral

$$\int_0^t d_\theta [\mathcal{A}_\theta(\omega) C_\varepsilon] \mathcal{B}_\theta(\omega) .$$

For all $x \in X$, the process

$$x_\varepsilon(t,\omega) = \int_0^t d_\theta [\mathcal{A}_\theta(\omega) C_\varepsilon] \mathcal{B}_\theta(\omega) x$$

is a local martingale in X . The square characteristic of $x_\varepsilon(t,\omega) - x_\delta(t,\omega)$ is

$$\int_0^t (\mathcal{B}_\theta^*(\omega)(C_\varepsilon - C_\delta) d_\theta \langle \mathcal{A}(\omega) \rangle_\theta (C_\varepsilon - C_\delta) \mathcal{B}_\theta(\omega) x, x) . \tag{7}$$

Consequently integral (6) is defined for step-function $\mathcal{B}_\theta(\omega)$ with values in $\mathcal{L}(\Omega, X)$. Let $\mathcal{B}_\theta(\omega)$ be a step-function with values in $\mathcal{L}_S(\Omega, X)$. In the same way as in Lemma 1 we can construct operators $\mathcal{D}_\varepsilon$ for which $\mathcal{B}_\theta(\omega)\mathcal{D}_\varepsilon \in \mathcal{L}(\Omega, X)$ and $\lim_{\varepsilon \to 0} \|\mathcal{D}_\varepsilon x - x\| = 0$. Then

$$\lim_{\substack{\varepsilon \to 0 \\ \delta \to 0}} \left(\int_0^t (\mathcal{D}_\varepsilon - \mathcal{D}_\delta) \mathcal{B}_\theta^*(\omega) d_\theta \langle \mathcal{A}(\omega) \rangle_\theta \mathcal{B}_\theta(\omega) (\mathcal{D}_\varepsilon - \mathcal{D}_\delta) x, x \right) =$$

$$= \lim_{\substack{\varepsilon \to 0 \\ \delta \to 0}} \left(\int_0^t \mathcal{B}_\theta^*(\omega) d_\theta \langle \mathcal{A}(\omega) \rangle_\theta \mathcal{B}_\theta(\omega) (\mathcal{D}_\varepsilon - \mathcal{D}_\delta) x, (\mathcal{D}_\varepsilon - \mathcal{D}_\delta) x \right) = 0$$

since

$$\int_0^t \mathcal{B}_\theta^*(\omega) d_\theta \langle \mathcal{A}(\omega) \rangle_\theta \mathcal{B}_\theta(\omega) \in \mathcal{L}_W(\Omega, X), \ (\mathcal{D}_\varepsilon - \mathcal{D}_\delta) x \to 0 .$$

Thus integral (6) with properties a), b) is determined for all step-functions. Using Theorem 1,we can exted the definition of the integral to all functions satisfying conditions 1)-3).

§4. Operator Wiener process and stochastic equations.

1. An operator Wiener process is a Gaussian random process $\mathcal{B}_t(\omega)$ with values in $\mathcal{L}_S(\Omega, X)$ with stationary independent increments and $E(\mathcal{B}_t(\omega)x, y) = 0$, $x, y \in X$. Let

$$E(\mathcal{B}_t(\omega)x, y)(\mathcal{B}_t(\omega)u, v) = t \cdot b(x, y, u, v).$$

From the inclusion $\mathcal{B}_1(\omega)\, x \in X(\Omega)$ it follows that

$$b(x,y,u,x) = (\mathcal{B}^{(x)} y, v)$$

where $\mathcal{B}^{(x)}$ is a kernel operator. We denote

$$tr\,\mathcal{B}^{(x)} = \sum_{K=1}^{\infty} b(x,e_K,x,e_K) = (Bx,x) = E\|\mathcal{B}_1(\omega)x\|^2,\ B\in\mathcal{L}^+(X).$$

Let $\mathcal{F}_t$ be the σ-algebra generated by $(\mathcal{B}_\theta(\omega)x, y)$, $\theta\le t$, $x,y\in X$, $(\mathcal{B}_t(\omega)x,y)$ is a martingale with respect to $\mathcal{F}_t$.

$\mathcal{B}_t(\omega)x = x_t(\omega)$ is a Gaussian process in X with the correlation function

$$E(x_t(\omega),y)(x_\theta(\omega),u) = (t\wedge\theta)(\mathcal{B}^{(x)} y, u).$$

Therefore $x_t(\omega)$ is continuous with probability 1. Thus $\mathcal{B}_t(\omega)$ is an operator martingale. We can obtain its square characteristic from the relation

$$\langle\mathcal{B}(\omega)\rangle_t = P\text{-}\lim_{n\to\infty}\sum(\mathcal{B}_{\frac{Kt}{n}}(\omega) - \mathcal{B}_{\frac{K-1}{n}t}(\omega))^*(\mathcal{B}_{\frac{Kt}{n}}(\omega) - \mathcal{B}_{\frac{K-1}{n}t}(\omega)) = tB.$$

2. S t o c h a s t i c d i f f e r e n t i a l e q u a t i o n s. Now we want to consider an equation for an operator random process, which is a generalization of Itô's equation. Let $\mathcal{F}(t,\mathcal{A}(\omega))$ be a function on $[0,\infty)\times\mathcal{L}_s(\Omega,X)$ with values in $\mathcal{F}(t,\mathcal{A}(\omega))$. We suppose that it satisfies the following conditions:

a) $(\mathcal{F}(t,\mathcal{A}(\omega))u, v)$ is measurable with respect to the σ-algebra $\mathcal{F}_t$ if $\mathcal{A}(\omega)$ is $\mathcal{F}_t$-measurable;

b) there exists a $\gamma\in R$ such that, for all $t\ge 0$ and $\mathcal{F}_t$-measurable $\mathcal{A}(\omega)$, $\mathcal{B}(\omega)$, the inequality

$$[\mathcal{F}(t,\mathcal{A}(\omega))-\mathcal{F}(t,\mathcal{B}(\omega))]^*[\mathcal{F}(t,\mathcal{A}(\omega))-\mathcal{F}(t,\mathcal{B}(\omega))]\leq \gamma(\mathcal{A}(\omega)-\mathcal{B}(\omega))^*(\mathcal{A}(\omega)-\mathcal{B}(\omega))$$

holds.

Let $\mathcal{B}_t(\omega)$ be an operator Wiener process. We consider the equation

$$\mathcal{A}_t(\omega)=\mathcal{A}(\omega)+\int_0^t d\mathcal{B}_\theta(\omega)\mathcal{F}(\theta,\mathcal{A}_\theta(\omega)) . \tag{1}$$

If $\mathcal{A}_t(\omega)$ satisfies (1), then $\mathcal{A}_t(\omega)$ is an operator martingale. We shall show that equation (1) has a unique solution which is an operator martingale satisfying

$$\int_0^t \mathcal{A}_\theta^*(\omega)\mathcal{A}_\theta(\omega)d\theta\in\mathcal{L}_W(\Omega,X),\quad t>0 . \tag{2}$$

Denote by $\mathcal{M}(\Omega,X)$ the set of all operator martingales $\mathcal{A}_t(\omega)$ satisfying (2). For $\mathcal{A}_t(\omega)\in\mathcal{M}(\Omega,X)$,

$$C_t(\omega)=\mathcal{A}(\omega)+\int_0^t d\mathcal{B}_\theta(\omega)\mathcal{F}(\theta,\mathcal{A}_\theta(\omega))\in\mathcal{M}(\Omega,X) \tag{3}$$

as well. Thus (3) determines a transformation from $\mathcal{M}(\Omega,X)$ into $\mathcal{M}(\Omega,X)$:

$$\mathcal{F}\mathcal{A}_t(\omega)=\mathcal{A}(\omega)+\int_0^t d\mathcal{B}_\theta(\omega)\mathcal{F}(\theta,\mathcal{A}_\theta(\omega)) .$$

We have

$$\langle\mathcal{F}\mathcal{A}^{(1)}(\omega)-\mathcal{F}\mathcal{A}^{(2)}(\omega)\rangle_t=\int_0^t[\mathcal{F}(\theta,\mathcal{A}_\theta^{(1)}(\omega)-\mathcal{F}(\theta,\mathcal{A}_\theta^{(2)}(\omega))]^*B[\mathcal{F}(\theta,\mathcal{A}_\theta^{(1)}(\omega))-$$

$$-\mathcal{F}(\theta,\mathcal{A}_\theta^{(2)}(\omega))]d\theta\leq|||B|||\int_0^t[\mathcal{F}(\theta,\mathcal{A}_\theta^{(1)}(\omega))-\mathcal{F}(\theta,\mathcal{A}_\theta^{(2)}(\omega))]^*[\mathcal{F}(\theta,\mathcal{A}_\theta^{(1)}(\omega))- \tag{4}$$

$$-\mathcal{F}(\theta,\mathcal{A}^{(2)}_\theta(\omega))]d\theta \le \|\|B\|\|\gamma\int_0^t[\mathcal{A}^{(1)}_\theta(\omega)-\mathcal{A}^{(2)}_\theta(\omega)]^*[\mathcal{A}^{(1)}_\theta(\omega)-\mathcal{A}^{(2)}_\theta(\omega)]\,d\theta \quad .$$

If $\mathcal{F}\mathcal{A}^{(\kappa)}_t(\omega)=\mathcal{A}^{(\kappa)}_t(\omega)$, $\kappa=1,2$, then

$$\langle\mathcal{A}^{(1)}(\omega)-\mathcal{A}^{(2)}(\omega)\rangle_t \le \gamma_1\int_0^t[\mathcal{A}^{(1)}_\theta(\omega)-\mathcal{A}^{(2)}_\theta(\omega)]^*[\mathcal{A}^{(1)}_\theta(\omega)-\mathcal{A}^{(2)}_\theta(\omega)]\,d\theta \,. \tag{5}$$

Let $x\in X$ and $\tau=\inf[\theta:(\langle\mathcal{A}^{(1)}(\omega)-\mathcal{A}^{(2)}(\omega)\rangle_\theta x,x)>\beta]$. From (5) we have

$$(\langle\mathcal{A}^{(1)}(\omega)-\mathcal{A}^{(2)}(\omega)\rangle_{t\wedge\tau}x,x)\le\gamma_1\Big(\int_0^{t\wedge\tau}(\mathcal{A}^{(1)}_\theta(\omega)-\mathcal{A}^{(2)}_\theta(\omega))^*(\mathcal{A}^{(1)}_\theta(\omega)-\mathcal{A}^{(2)}_\theta(\omega))\,ds\,x,x\Big),$$

$$E(\langle\mathcal{A}^{(1)}(\omega)-\mathcal{A}^{(2)}(\omega)\rangle_{t\wedge\tau}x,x)\le\gamma_1\int_0^t E(\mathcal{A}^{(1)}_{\theta\wedge\tau}(\omega)-\mathcal{A}^{(2)}_{\theta\wedge\tau}(\omega)]^*[\mathcal{A}^{(1)}_{\theta\wedge\tau}(\omega)-$$

$$-\mathcal{A}^{(2)}_{\theta\wedge\tau}(\omega)]x,x)\,d\theta=\gamma_1\int_0^t E(\langle\mathcal{A}^{(1)}(\omega)-\mathcal{A}^{(2)}(\omega)\rangle_{\theta\wedge\tau}x,x)\,d\theta \quad .$$

Therefore, for all $\beta>0$, $E(\langle\mathcal{A}^{(1)}(\omega)-\mathcal{A}^{(2)}(\omega)\rangle_{t\wedge\tau}x,x)=0$. The uniqueness of the solution of the equation (1) is thus proved.

Now put $\mathcal{A}^{(0)}_t(\omega)=\mathcal{A}(\omega)$, $\mathcal{A}^{(n)}_t(\omega)=\mathcal{F}\mathcal{A}^{(n-1)}_t(\omega)$, $n=1,2,\dots$.

From (4) it follows that

$$\langle\mathcal{A}^{(n)}(\omega)-\mathcal{A}^{(n+1)}(\omega)\rangle_t\le\gamma_1\int_0^t[\mathcal{A}^{(n-1)}_\theta(\omega)-\mathcal{A}^{(n)}_\theta(\omega)]^*[\mathcal{A}^{(n-1)}_\theta(\omega)-\mathcal{A}^{(n)}_\theta(\omega)]\,d\theta \quad .$$

Let $\tau=\inf[\theta:(\langle\mathcal{A}^{(0)}(\omega)-\mathcal{A}^{(1)}(\omega)\rangle_\theta x,x)\ge\beta]$. Then, for all $t>0$,

$$E(\langle\mathcal{A}^{(1)}(\omega)-\mathcal{A}^{(2)}(\omega)\rangle_{t\wedge\tau}x,x)\le\gamma_1\int_0^t E(\langle\mathcal{A}^{(0)}(\omega)-\mathcal{A}^{(1)}(\omega)\rangle_{\theta\wedge\tau}x,x)\,d\theta<\infty$$

and consequently, for all $n\ge1$,

$$E(\langle\mathcal{A}^{(n)}(\omega)-\mathcal{A}^{(n+1)}(\omega)\rangle_{t\wedge\tau}x,x)\le\gamma_1\int_0^t E(\langle\mathcal{A}^{(n-1)}(\omega)-\mathcal{A}^{(n)}(\omega)\rangle_{\theta\wedge\tau}x,x)\,d\theta<\infty\,.$$

Therefore

$$E(\langle \mathcal{A}^{(n)}(\omega) - \mathcal{A}^{(n+1)}(\omega)\rangle_{t\wedge\tau} x, x) \leq$$

$$\leq \gamma_1^n \int\int_{0\leq\theta_1<\cdots<\theta_{n-1}\leq t}\cdots\int^t E(\langle \mathcal{A}^{(0)}(\omega) - \mathcal{A}^{(1)}(\omega)\rangle_{\theta_{n-1}\wedge\tau} x, x)\; d\theta_1 \ldots d\theta_{n-1} d\theta = \tag{6}$$

$$= \gamma_1^n E\int_0^t \frac{(t-\theta)^{n-1}}{(n-1)!} (\langle \mathcal{A}^{(0)}(\omega) - \mathcal{A}^{(1)}(\omega)\rangle_{\theta\wedge\tau} x, x) d\theta \leq \beta\gamma_1^n \frac{t^n}{n!}$$

and

$$\lim_{n,m\to\infty} P\{\sup_{\theta\leq t}(\langle \mathcal{A}^{(n)}(\omega) - \mathcal{A}^{(m)}(\omega)\rangle_\theta x, x) > \varepsilon\} =$$

$$\leq \lim_{n,m\to\infty} P\{\sup_{\theta\leq t}(\langle \mathcal{A}^{(n)}(\omega) - \mathcal{A}^{(m)}(\omega)\rangle_{\theta\wedge\tau} x, x) > \varepsilon\} + P\{\tau\leq t\} = P\{\tau \leq t\} ,$$

since, for $n < m$,

$$E(\langle \mathcal{A}^{(n)}(\omega) - \mathcal{A}^{(m)}(\omega)\rangle_{t\wedge\tau} x, x) = E(\langle \sum_{K=n}^{m-1} [\mathcal{A}^{(K)}(\omega) - \mathcal{A}^{(K+1)}(\omega)]\rangle_{t\wedge\tau} x, x) \leq$$

$$\leq \sum_{K=n}^{m-1} \frac{1}{K^2} \sum_{K=n}^{m-1} E(\langle \mathcal{A}^{(K)}(\omega) - \mathcal{A}^{(K+1)}(\omega)\rangle_{t\wedge\tau} x, x) K^2 \leq \sum_{K=n}^{m-1} \frac{1}{K^2} \sum_{K=n}^{m-1} K^2 \frac{1}{K!} t^K \beta$$

and $\lim_{\beta\to\infty} P\{\tau \leq t\} = 0$.

The existence of an operator martingale $\mathcal{A}_t(\omega)$ for which

$$\lim_{n\to\infty} P\{\sup_{\theta\leq t} |\mathcal{A}_\theta^{(n)}(\omega) x - \mathcal{A}_\theta(\omega) x| > \varepsilon\} = 0 , \quad x \in X, \quad t > 0 ,$$

follows from Theorem 1, § 3. $\mathcal{A}_t(\omega)$ is a solution of (1).

§ 5. Stochastic semigroups

1. **Linear stochastic equations.** Let $\mathcal{B}_t(\omega)$ be an operator Wiener process. We consider the stochastic equation

$$\mathcal{A}_{t_0,t}(\omega) = \mathcal{I} + \int_{t_0}^t d\mathcal{B}_\theta(\omega)\, \mathcal{A}_{t_0,\theta}(\omega) , \quad t \geq t_0 . \tag{1}$$

($\mathcal{J}$ is the identity operator). This is an equation of form (1), § 4. Existence and uniqueness of a solution of (1) follow from the results of § 4.

Let $\mathcal{F}_t^{t_0}$ be the σ-algebra generated by the random variables $\{(\mathcal{B}_\theta(\omega)x,y)-(\mathcal{B}_{t_0}(\omega)x,y),\theta\in[t_0 t], x, y\in X\}$. Then $\mathcal{A}_{t_0,t}(\omega)$ is $\mathcal{F}_t^{t_0}$-measurable. Since $\mathcal{B}_t(\omega)$ has stationary increments, the distributions of $\mathcal{A}_{t_0,t_0+h}(\omega)$ and $\mathcal{A}_{0,h}(\omega)$ are identical. We show finally that, for $t_1<t_2<t_3$.

$$P\{\mathcal{A}_{t_2,t_3}(\omega)\mathcal{A}_{t_1,t_2}(\omega)=\mathcal{A}_{t_1,t_3}(\omega)\}=1 \quad . \tag{2}$$

Suppose that $\hat{\mathcal{A}}_{t_1,t}(\omega)=\mathcal{A}_{t_2,t}(\omega)\mathcal{A}_{t_1,t_2}(\omega)$ for $t\geqslant t_2$. Then

$$\hat{\mathcal{A}}_{t_1,t}(\omega)=\mathcal{A}_{t_1,t_2}(\omega)+[\int_{t_2}^{t}d\mathcal{B}_\theta(\omega)\mathcal{A}_{t_2,\theta}(\omega)]\mathcal{A}_{t_1,t_2}(\omega)=$$
$$=\mathcal{A}_{t_1,t_2}(\omega)+\int_{t_2}^{t}d\mathcal{B}_\theta(\omega)\hat{\mathcal{A}}_{t_1,\theta}(\omega) \quad . \tag{3}$$

$\mathcal{A}_{t_1,t}(\omega)$ satisfies the equation

$$\mathcal{A}_{t_1,t}(\omega)=\mathcal{A}_{t_1,t_2}(\omega)+\int_{t_2}^{t}d\mathcal{B}_\theta(\omega)\mathcal{A}_{t_1,\theta}(\omega) \tag{4}$$

for $t\geqslant t_2$. Using the uniqueness of a solution of the stochastic equation, we obtain (2) from (3) and (4). We have thus proved that the family of the random operators $\{\mathcal{A}_{t_0,t}(\omega), 0\leqslant t_0\leqslant t<\infty\}$ is a stochastic semigroup. A general definition of such semigroups will be given below.

2. S t o c h a s t i c s e m i g r o u p s. A family of random operators $\{\mathcal{A}_{\theta,t}(\omega), 0\leqslant\theta\leqslant t<\infty\}$ is called a weak continuous stochastic semigroup if it satisfies the following conditions :

a) $A_{\theta,t}(\omega)\in \mathcal{L}_W(\Omega, X)$ and $(A_{\theta,t}(\omega)x, y)$ are continuous in t with probability 1, $(A_{\theta,\theta}(\omega)x, y) = (x, y)$;

b) $P\{A_{t_2,t_3}(\omega)A_{t_1,t_2}(\omega) = A_{t_1,t_3}(\omega)\} = 1$, $t_1 < t_2 < t_3$;

c) $A_{\theta,\theta+h}(\omega)$ and $A_{0,h}(\omega)$ are identically distributed;

d) if $\mathcal{F}_t^\theta$ is the σ-algebra generated by the random variables $\{(A_{\theta,t_1}(\omega)x, y), t_1\in[\theta, t],\ x, y\in X\}$, then the σ-algebras $\mathcal{F}_\theta^0$, $\mathcal{F}_t^\theta$, $\mathcal{F}_\infty^t$ are independent.

If condition a) is replaced by

a[1]) $A_{\theta,t}(\omega)\in \mathcal{L}_S(\Omega, X)$ and $A_{\theta,t}(\omega)x$ is continuous in t with probability 1, $A_{\theta,\theta}(\omega)x = x$,

then the cemigroup is called strongly continuous.

We consider a strongly contituous semigroup $A_{\theta,t}(\omega)$ for which $A_{\theta,t}(\omega)$, as a function of t, is an operator martingale and

$$E\, \| A_{\theta,t}(\omega)x \|^2 \leq \gamma_{\theta,t} \| x \|^2 , \tag{5}$$

where $\gamma_{\theta,t}$ is a constant. We can obtain such semigroups from equation (1).

T h e o r e m 1. Let $A_{\theta,t}$ be a strongly continuous semigroup and an operator martingale (in t) for which (5) holds. Then, for all $x, y\in X$, there exists

$$\lim_{t\to 0} \frac{1}{t} E(A_{0,t}(\omega)x - x, y)^2 = b(x, y)$$

and $b(x,y)\leq \gamma \|x\|^2 \|y\|^2, \sum b(x, e_k) = (Bx, x), B\in \mathcal{L}^+(X)$.

If

$$b(x,y,u,v)=\lim_{t\to 0}\frac{1}{t}E(A_{0,t}(\omega)x-x,y)(A_{0,t}(\omega)u-u,v) ,$$

then there exists an operator Wiener process $\mathcal{B}_t(\omega)$ with $E(\mathcal{B}_t(\omega)x,y)(\mathcal{B}_t(\omega)u,v)=b(x,y,u,v)$ such that $A_{t_0,t}(\omega)$ satisfies equation (1).

Proof. Put

$$\beta_h(x,y)=\frac{1}{h}E(A_{0,h}(\omega)x_0-x,y)^2 .$$

We have

$$\beta_{nh}(x,y)=\frac{1}{nh}E\Big(\sum_{k=0}^{n-1}[A_{kh,(k+1)h}(\omega)-\mathcal{I}]A_{0,kh}(\omega)x,y\Big)^2=$$

$$=\frac{1}{nh}\sum_{k=0}^{n-1}E([A_{kh,(k+1)h}(\omega)-\mathcal{I}]A_{0,kh}(\omega)x,y)^2=\frac{1}{n}\sum_{k=0}^{n-1}E\beta_h(A_{0,kh}(\omega)x,y)\geq$$

$$\geq\beta_h\Big(\frac{1}{n}\sum_{k=0}^{n-1}EA_{0,kh}x,y\Big)=\beta_h(x,y)$$

(we have used the convexity of β_h). Using this inequality, it is easy to check that there exists

$$\lim_{t\to 0}\beta_t(x,y)=\beta(x,y) .$$

In the same way one can obtain the inequality

$$\frac{1}{nh}E\|A_{0,nh}(\omega)x-x\|^2\geq\frac{1}{h}E\|A_{0,h}(\omega)x-x\|^2$$

and to prove the existence of

$$\lim_{t\to 0}\frac{1}{t}E\|A_{0,t}(\omega)x-x\|^2\geq\sum_{k=1}^{\infty}\beta(x,e_k) ,$$

and

$$(Bx,x)=\sum_{k=1}^{\infty}\beta(x,e_k)\le E\|\mathcal{A}_{0,1}(\omega)x-x\|^2 .$$

If $t_1<t_2<t_3$, then

$$E\|\mathcal{A}_{t_1,t_3}(\omega)x-x-[\mathcal{A}_{t_1,t_2}(\omega)x+\mathcal{A}_{t_2,t_3}(\omega)x-2x]\|^2\le$$

$$\le(t_3-t_2)E(B[\mathcal{A}_{t_1,t_2}(\omega)x-x],[\mathcal{A}_{t_1,t_2}(\omega)x-x])\le$$

$$\le|||B|||(t_3-t_2)(t_2-t_1)(Bx,x)\le|||B|||^2(t_3-t_2)(t_2-t_1)\|x\|^2.$$

Using the last inequality, we can establish the existence of the mean-square limit

$$\mathcal{B}_t(\omega)x=\lim_{n\to\infty}\sum_{k<t2^n}(\mathcal{A}_{\frac{k}{2^n},\frac{k+1}{2^n}}(\omega)-\mathcal{I})x .$$

$\mathcal{B}_t(\omega)$ is an operator Wiener process. Equation (1) follows from the equality

$$\mathcal{A}_{\frac{k}{2^n},\frac{l}{2^n}}(\omega)x=x+\sum_{j=0}^{l-1}(\mathcal{A}_{\frac{j}{2^n},\frac{j+1}{2^n}}(\omega)-\mathcal{I})\mathcal{A}_{\frac{k}{2^n},\frac{j}{2^n}}(\omega)x .$$

REFERENCES

1. Kolmogoroff A., Grundbegriffe der Wahrscheinlichkeitsrechnung, Berlin, 1933.
2. Gihman I.I., Skorohod A.V., The theory of stochastic processes, I, Spronger-Verlag, Berlin-Heidelberg-New York, 1974.
3. Skorohod A.V., Studies in the theory of random processes, Reading, Mass., Addison-Wesley, 1965.
4. Skorohod A.V., K-martingales and stochastic equations, Trudy shkoly-seminara po teorii sluchainych protsessov (Druskinin-

kai, November 25-30, 1-74), Vilnius, 1975 (in Russian).

Mathematical Institute

Academy of Sciences of the Ukrainian

SSR, Kiev

BERNOULLI SHIFTS ON GROUPS AND DECREASING SEQUENCES OF PARTITIONS

A.M.Stepin

This report is devoted to the isomorphism problem for Bernoulli shifts on countable discrete groups and related equivalence problem for decreasing sequences of σ-algebras up to MP-transformations. Motivation for the study of dynamical systems with nonclassical time arises in ergodic theory as well as in its applications to statistical physics models (cf.[1]).

First the notion of entropy for group actions is discussed. After that the extension of isomorphism for Bernoulli shifts from a subgroup to the whole group is described. The entropy equidistribution property for Bernoulli shifts on locally finite groups is proved. This fact together with a certain modification of Ornstein's construction [2] leads us to isomorphism of Bernoulli shifts with the same entropy for a wide class of countable groups. Every action with positive entropy of a countable periodic subgroup of $\mathbb{Q}/\mathbb{Z}$ has Bernoulli factor-action with the same entropy. This result is applied for evaluation of the entropy for decreasing sequences of homogeneous partitions.

1. Let $T=\{T_g, g\in G\}$ be measure-preserving action of a countable amenable group G in a probability measure space (X,μ). It is natural to define the entropy $h(T,\xi)$ of the action T relative to partition ξ with finite entropy $H(\xi)$ as the limit (if it exists)

$$\lim_{n\to\infty} |G_n|^{-1} \cdot H(\bigvee_{g\in G_n} T_g \xi), \qquad (1)$$

where $\{G_n\}$ is increasing sequence of finite subsets in G, satisfying conditions $\bigcup_n G_n = G$,

$$\lim_{n\to\infty} |G_n|^{-1} |gG_n \cap G_n| = 1, \quad g\in G;$$

here $|G_n|$ is the cardinality of G_n. It is known [3] that such a sequence always exists in amenable group G. For groups with an algebraic ordering certain approach is suggested in [4] to prove correctness of definition (1). Limit (1) exists in this case, it does not depend on the choice of sequence $\{G_n\}$ and is equal the conditional entropy

$$H(\xi | \bigvee_{g\in P} T_g \xi),$$

where P is the set of negative elements in G. Using integral representation for the entropy of action of commutative periodic group [4] B.S.Pitskel proved recently that for these groups limit (1) does not depend on the choice of averaging sequence $\{G_n\}$.

A subset P in amenable algebraically ordered group G is called the information past, if for any action T of G and any partition ξ with $H(\xi) < \infty$

$$H(T, \xi) = H(\xi | \bigvee_{g\in P} T_g \xi).$$

It was conjectured that every information past $P \subset G$ is the group past, i.e. the set of negative elements in G relative to some algebraic ordering of G. B.S.Pitskel [5] proved this conjecture for commutative groups and A.Safonov proved it for nilpotent groups (to appear).

Among papers devoted to entropy theory of M.P. group actions mention here [6] and [7].

2. Let G be a countable discrete group, (Y,ν) a Lebesgue space (i.e. complete separable probability measure space). Set $(X,\mu) = \prod_G (Y,\nu)$. Points of X are Y-valued functions (or configurations) $x(g)$ on G. Assign element $g\in G$ the transformation T_g : $x(g') \to x(gg')$, $g'\in G$. The action $T: g \to T_g$ of group G in the space (X,μ) is called a Bernoulli action and the space (Y,ν) the state space for T. We define the entropy $h(T)$ of Bernoulli action T as the entropy of the state space for T.

Theorem 1. Let G be a countable group with an element of infinite order. Bernoulli actions T and T' of group G are isomorphic if $h(T) = h(T')$.

Idea of the proof consists in extension of isomorphism from a subgroup generated by an element of infinite order to the whole group.

I shall outline the proof of Th.1 in the case when state spaces (Y,ν) and (Y',ν') are finite or countable. Consider map $\pi: X \to Y$ (resp. $\pi': X' \to Y'$), assigning to configuration x its value $x(e)$, e is the unity element of G. Denote by ξ (resp. by ξ') inverse image of the partition of Y (resp. Y') into points relative to π (resp. π'). Partition ξ is a generator for the action T, and partitions $T_g\xi$, $g\in G$, are independent. To establish isomorphism of the actions T and T' it suffices to find in X a partition η such that

a) partitions $T_{g_1}\eta$ and $T_{g_2}\eta$, $g_1 \neq g_2$, are independent;

b) η is a generator for T ; c) η has the same distribution as partition ξ' . To this end consider projection $\rho: X \to \prod_{h\in H} Y$ restricting configuration x onto a subgroup $H \subset G$, generated by an element of infinite order. The inverse image of the partition into points relative to projection ρ is invariant with respect to the restriction of T onto subgroup H . Corresponding factor-action is isomorphic to Bernoulli action S of the group H with the state space (Y, ν) . Since Bernoulli shifts with the same entropy on the group H are isomorphic [2], there exists in the space $\prod_{h\in H}(Y, \nu)$ a partition ζ such that a) partitions $S_h\zeta$, $h \in H$, are independent; b) ζ is a generator for the action S ; c) ζ has the same distribution as partition ξ' . Inverse image η of the partition ζ relative to projection ρ is the required partition in X .

Indeed, according to the construction, partition η is measurable relative to σ-algebra $\xi_H = \bigvee_{h\in H} T_h \xi$. Partition $T_g \eta$ is measurable relative to σ-algebra $\xi_{Hg^{-1}} = \bigvee_{h\in H} T_{gh}\eta$. Since σ-algebras ξ_H and $\xi_{Hg^{-1}}$ are independent if $g \bar{\in} H$, partitions η and $T_g\eta$ are independent too. Besides, partitions η and $T_h\eta$ are independent, if $h\in H$, $h \neq e$. Further, partitions $T_h\eta$, $h \in H$, generate σ-algebra ξ_H and partitions $T_{gh}\eta$, $h\in H$, generate σ-algebra $\xi_{Hg^{-1}}$. Hence η is a generator for the action T .

In fact we have proved

Theorem 2. Let G be a countable group, H - its subgroup, $f^{(1)}$ and $f^{(2)}$ - stationary random fields on G such that $f^{(i)}_g$ and $f^{(i)}_h$, $i=1,2$, are independent, if $h\in H$, $g\bar{\in} H$. Then isomorphism of the actions of the group H that are generated by the restrictions of random fields $f^{(1)}$ and $f^{(2)}$ onto H, implies isomorphism of the actions of G that correspond to random fields $f^{(1)}$ and $f^{(2)}$.

3. We shall say that a countable group is an Ornstein group if the entropy is a complete isomorphism invariant for Bernoulli actions of this group. It may be verified that a finite group G is not an Órnstein group if $|G| \geq 2$. It follows from Th.2 that a group containing an Ornstein subgroup is an Ornstein group itself. In particular, every periodic group is an Ornstein group.

Let us consider now the case of countable periodic groups. It turns out that a wide class of these groups is contained in the class of Ornstein groups [8]. Extension procedure of Th.1 will be applied in the case of periodic groups to countable locally finite subgroups instead of subgroup $\mathbb{Z}$ in Th.1.

Theorem 3. Bernoulli actions of a countable locally finite group with the same entropy are isomorphic.

The following lemmas allow us to apply Ornstein's construction [2] for the proof of Th.3.

Lemma 1. For any sequence $\{\xi_i\}_1^n$ of independent identically distributed finite partitions and any $\varepsilon > 0$, there exists a $\delta > 0$ such that, for every sequence $\{\eta_i\}_1^n$ of identically distributed finite partitions satisfying the following conditions

$$\left|\frac{1}{n} H(\eta_1 \vee \dots \vee \eta_n) - H(\xi_1)\right| < \delta, \quad |\operatorname{dist} \eta_1 - \operatorname{dist} \xi_1| < \delta,$$

$\bar{d}$ – the distance between sequences $\{\xi_i\}_1^n$ and $\{\eta_i\}_1^n$ – does not exceed ε .

Corollary. The random field on locally finite group, generated by a Bernoulli action and a finite partition, is finitely determined.

Lemma 2. Let G be a countable locally finite group, $\{G_n\}$ an increasing to G sequence of finite subgroups, T a Bernoulli action of G , ξ a Bernoulli generator for the action T , η any partition measurable relative to $\bigvee_{g \in G_K} T_g \xi$, $\mu_{n,\eta}(x)$ the measure of the element of partition $\bigvee_{g \in G_n} T_g \eta$ containing point x . Then limit

$$\lim_{n \to \infty} \frac{1}{|G_n|} \log \mu_{n,\eta}(x)$$

exists a.e. and is equal to $-h(T, \eta)$.

It is essential for the proof of Th.3 that partitions for which the preceding lemma guarantees asymptotic equidistribution property, form a dense set in the space of partitions with finite entropy.

It follows from Theorems 2 and 3 that a countable group, containing a countable locally finite subgroup, is Ornstein group. In particular, Bernoulli shifts with the same entropy on a countable commutative group are isomorphic. It was conjectured in [9] that Ornstein's isomorphism theorem for Bernoulli shifts with the same entropy admits a generalization for Bernoulli actions of approximately finite groups. It is unknown up to now, are there approximately finite or even amenable groups among those periodic groups that do not contain countable locally finite subgroups. Certain results in this direction were obtained by R.Grigorchuk [10]. It is well known [11] that amenability of a group with finite number of generators is completely characterized by spectral radius τ of transition operator for symmetric random walk on G . Let G be represented as a quotient group F/N of free group F with m generators; let the growth characteristic α of N be defined by

$$\alpha = \lim_{n \to \infty} \frac{1}{n} \log N_n \quad .$$

where N_n is the number of words of length n in the normal subgroup N . It is proved in [10] that

$$\tau = \frac{\alpha}{m} + \frac{2m-1}{m\alpha}$$

and in some cases estimates for the growth characteristic α are obtained.

4. As it was mentioned in the introduction, dynamical systems with nonclassical time appear in connection with some problems of ergodic theory. One of them is the equivalence problem for decreasing sequences of partitions up to M.P.- transformations.

Our approach to this problem is as follows. Let Γ be the inductive limit of finite cyclic groups of orders q_n . To every isomorphism T of group Γ into the group of automorphisms of Lebesgue space, there corresponds a decreasing sequence of partitions ξ_n with orbits of subgroup Γ_n as its elements. Conversely, for every decreasing sequence Ξ of partitions ξ_n with q_n-point elements of equal conditional measure, there exists an action T of group Γ generating a sequence Ξ in the above sense. Such an action is not unique. Class $\mathcal{T}(\Xi)$ of actions of the group Γ generating a sequence Ξ comprises complete information about this sequence.

Entropy $h(\Xi)$ of a decreasing sequence Ξ is defined in [10] as the least upper bound of $h(T)$ over class $\mathcal{T}(\Xi)$. Nontrivial character of this invariant is guaranteed by

Theorem 4. If the series $\sum_1^\infty q_n^{-1} \log q_{n+1}$ converges, then the oscillation of a function $h(T)$ on $\mathcal{T}(\Xi)$ does not exceed $\sum_1^\infty q_n^{-1} \log q_{n+1}$ (When $h(T)=\infty$ for some action $T\in\mathcal{T}(\Xi)$, this means that $h(S)=\infty$ for all actions $S\in\mathcal{T}(\Xi)$).

Proof. Choose $T, S \in \mathcal{T}(\Xi)$, take a finite partition η and set $\eta^n = \bigvee_{g\in\Gamma_n} T_g \eta$, $\zeta^n = \bigvee_{g\in\Gamma_n} S_g \eta^n$.
If $k>n$, we have

$$h(S,\zeta^n) = \lim_{k\to\infty} \frac{1}{q_k} H\Big(\bigvee_{g\in\Gamma_k} S_g \zeta^n\Big) \le \frac{1}{q_n} H(\zeta^n).$$

To prove the theorem it suffices to check the following estimate

$$H(\zeta^n) \le H(\eta^n) + \log p_n ,$$

where $q_n^{-1} \log p_n \le \sum_1^\infty q_n^{-1} \log q_{n+1}$.

Choose a $C\in\eta^n$. Every element of partition $\{T_g C,\ g\in\Gamma_1\}$ of the set $\bigcup_{g\in\Gamma_1} T_g C$ splits, under the action of group $\{S_g,\ g\in\Gamma_1\}$ into not more than $q_1^{q_1}$ subsets. Suppose that elements of partition $\{T_g C,\ g\in\Gamma_k\}$ of the set $\bigcup_{g\in\Gamma_k} T_g C$ are divided, under the action of group $\{S_g,\ g\in\Gamma_k\}$, into p_k subsets. Then elements of partition $\{T_g C,\ g\in\Gamma_{k+1}\}$ of the set $\bigcup_{g\in\Gamma_{k+1}} T_g C$ split, under the action of $\{S_g,\ g\in\Gamma_{k+1}\}$ into not more than $(q_{k+1}\,p_k)^{q_{k+1}/q_k}$ subsets (this is a combinatorial fact which follows from trajectory coincidence of groups $\{T_g,\ g\in\Gamma_{k+1}\}$ and $\{S_g,\ g\in\Gamma_{k+1}\}$).

Since

$$\frac{\log p_{k+1}}{q_{k+1}} \leq \frac{\log p_k}{q_k} + \frac{\log q_{k+1}}{q_k}$$

the theorem is proved.

The following result allows us to infer that $h(T)$ is constant on $\mathcal{T}(\Xi)$ from boundedness of the oscillation of this function on the class $\mathcal{T}(\Xi)$. This property gives a way of evaluation of the entropy $h(\Xi)$.

Theorem 5. Every ergodic action with positive entropy of a countable periodic subgroup of the circle has a Bernoulli factor-action with the same entropy.

It is proved in [10] that, for Bernoulli action T of a group Γ, the entropy of the action T/ξ_n of the quotient group Γ/Γ_n is equal to $q_n h(T)$. According to Theorem 5 this holds true for any action of group Γ. Hence, if there is a nonzero oscillation of the entropy on $\mathcal{T}(\Xi)$, then the oscillation of the entropy on $\mathcal{T}(\Xi/\xi_n)$ is equal $q_n\delta$. On the other hand, entropies of factor actions T/ξ_n and T'/ξ_n, where $T, T'\in\mathcal{T}(\Xi)$, differ not more than in

$$q_n\sum_{n+1}^{\infty} q_k^{-1}\log q_{k+1} - q_n \log q_n \sum_{n+1}^{\infty} q_k^{-1} \quad ;$$

the latter can be made less than $q_n\delta$ for sufficiently large n.

We establish, in conclusion, the relation between the invariant $h(\Xi)$ and the entropy invariant introduced by A.M.Vershik [12]. Let

$\Xi = \{\xi_n\}$ beadecreasing sequence of homogeneous partitions, η a finite partition, $\Phi_{n,\eta}$ a group of automorphisms of (X,μ), leaving partition η fixed and partitions $\xi_1 > \xi_2 > \dots > \xi_n$ invariant. Denote by η^n the intersection of ξ_n and the partition on ergodic components of group $\Phi_{n,\eta}$; set

$$h'(\Xi,\eta) = \inf_n q_n^{-1} H(\eta^n) \ , \quad h'(\Xi) = \sup h'(\Xi,\eta) \, .$$

Invariants h and h' coincide for sequences generated by Bernoulli actions of group Γ ; besides $h'(\Xi) \le h(\Xi)$ for all sequences Ξ and h' is nonincreasing under factorization. Now we shall prove coincidence of invariants h and h' . Let Ξ be an ergodic decreasing sequence of partitions. According to Theorem 5, there exists a Bernoulli factor-sequence of partitions Ξ_B such that $h(\Xi_B) = h(\Xi)$. We have

$$h(\Xi_B) \le h'(\Xi_B) \le h'(\Xi) \le h(\Xi) \, .$$

REFERENCES

I. D.S.Ornstein, Ergodic Theory,Randomness and Dynamical Systems, New Haven & London, Jale University Press, 1974.

2. D.S.Ornstein, Bernoulli shifts with the same entropy are isomorphic, Adv. in Math., 4, 3(1970), 337-352.

3. E.Følner, On groups with full Banach mean value, Mathematica Scandinavica, 3(1955), 243-254.

4. Б.С.Пицкель, А.М.Степин, О свойстве распределенности энтропии коммутативных групп метрических автоморфизмов, Доклады АН СССР, 198, 5 (1971).

5. Б.С.Пицкель, Об информационных будущих аменабельных групп, Доклады АН СССР, 223, 5 (1975), 1067-1070.

6. J.-P.Conze, Entropie d'un groupe abelien de transformations, Z.Wahrcheinlichkeitstheorie und verw. Geb., 25, I(1972).

7. J.-P.Thouvenot, Convergence en moyenne de l'information pour l'action de $\mathbb{Z}^2$, Z.Wahrcheinlichkeitstheor und verw. Geb., 24,2 (1972).

8. А.М.Степин, Сдвиги Бернулли на группах, Доклады АН СССР, 223, 2 (1975), 300-302.

9. А.М.Степин, Об энтропийном инварианте убывающих последовательностей измеримых разбиений, Функциональный анализ и его приложения, 5, 3 (1971), 80-84.

10. Р.И.Григорчук, Симметрические случайные блуждания на дискретных группах, Изв. АН СССР, 40 (1976).

II. H.Kesten, Full Banach mean values on countable groups, Mathematica Scandinavica, 7(1959).

I2. А.М.Вершик, Континуум попарно неизоморфных диадических последовательностей, Функциональный анализ и его приложения, 5, 3 (I97I), I6-I8.

Department of Mechanics and Mathematics

Moscow State University

Moscow

ON THE SECOND ORDER ASYMPTOTIC EFFICIENCIES OF ESTIMATORS

Kei TAKEUCHI and Masafumi AKAHIRA
University of Tokyo and University of Electro-Communications

1. Introduction

Second order efficiency of asymptotically efficient estimators has been discussed by R.A.Fisher [6], C.R.Rao [9], [10] and others in terms of the loss of information. Recently Chibisov ([4], [5]) has shown that a ML (maximum likelihood) estimator is second order asymptotically efficient in some sense. Pfanzagl ([7], [8]) obtained similar results. One of the authors established similar results in a book written in Japanese [11] in terms of the asymptotic distribution of the estimators. In this paper we shall present the outline of the discussion given in [11] and proceed further to the third order asymptotic efficiency. Further it is shown that the results can be extended to non-regular situations.

2. Notations and definitions

Let $\mathcal{X}$ be an abstract sample space whose generic point is denoted by x, $\mathcal{B}$ a σ-field of subsets of $\mathcal{X}$, and let Θ be a parameter space, which is assumed to be an open set in a Euclidean 1-space R^1. We shall denote by $(\mathcal{X}^{(n)},\mathcal{B}^{(n)})$ the n-fold direct products of $(\mathcal{X},\mathcal{B})$. For each n=1,2,..., the points of $\mathcal{X}^{(n)}$ will be denoted by $\tilde{x}_n=(x_1,\ldots,x_n)$. We consider a sequence of classes of probability measures $\{P_{n,\theta}:\theta\in\Theta\}$ (n=1,2,...) each defined on $(\mathcal{X}^{(n)},\mathcal{B}^{(n)})$ such that for each n=1,2,... and each $\theta\in\Theta$ the following holds:

$$P_{n,\theta}(B^{(n)})=P_{n+1,\theta}\quad(B^{(n)}\times\mathcal{X})$$

for all $B^{(n)}\in\mathcal{B}^{(n)}$.

An estimator of θ is defined to be a sequence $\{\hat{\theta}_n\}$ of $\mathcal{B}^{(n)}$-measurable functions $\hat{\theta}_n$ on $\mathcal{X}^{(n)}$ into Θ (n=1,2,...). For simplicity we denote an estimator as $\hat{\theta}_n$ instead of $\{\hat{\theta}_n\}$.

For an increasing sequence of positive numbers $\{c_n\}$ ($\lim_{n\to\infty} c_n=\infty$) an estimator $\hat{\theta}_n$ is called consistent with order $\{c_n\}$ (or $\{c_n\}$ - consistent for short) if for every $\varepsilon>0$ and every $\vartheta\in\Theta$, there exist a sufficiently small positive number δ and a sufficiently large number L satisfying the following :

$$\overline{\lim_{n\to\infty}} \sup_{\theta:|\theta-\vartheta|<\delta} P_{n,\theta}\{c_n|\hat{\theta}_n - \theta| \geq L\} < \varepsilon \quad .$$

The order $\{c_n\}$ of convergence of consistent estimators and its bound are discussed in [1] and [11]. In the subsequent discussions we shall deal only with the case when $c_n=\sqrt{n}$. Let $\hat{\theta}_n$ be a $\{\sqrt{n}\}$-consistent estimator.

Definition 1. $\hat{\theta}_n$ is asymptotically median unbiased (or AMU for short) if for any $\vartheta\in\Theta$ there exists a positive number δ such that

$$\lim_{n\to\infty} \sup_{\theta:|\theta-\vartheta|<\delta} \left| P_{n,\theta}\{\sqrt{n}(\hat{\theta}_n-\theta)\leq 0\} - \frac{1}{2}\right| = 0 \ ;$$

$$\lim_{n\to\infty} \sup_{\theta:|\theta-\vartheta|<\delta} \left| P_{n,\theta}\{\sqrt{n}(\hat{\theta}_n-\theta)\geq 0\} - \frac{1}{2}\right| = 0 \ .$$

Definition 2. For $\hat{\theta}_n$ asymptotically median unbiased $F_\theta(t)$ is called an asymptotic distribution of it if

$$\lim_{n\to\infty} \left| P_{n,\theta}\{\sqrt{n}(\hat{\theta}_n-\theta)\leq t\} - F_\theta(t) \right| = 0 \ .$$

Since $\hat{\theta}_n$ is a $\{\sqrt{n}\}$-consistent estimator, it follows that $F_\theta(-\infty)=0$ and $F_\theta(\infty)=1$. Let $\hat{\theta}_n$ be a AMU estimator. Then it follows that $F_\theta(0)=1/2$ for all $\theta\in\Theta$. Let θ_0 be arbitrary and fixed in Θ . Putting

$$A_n(t)=\{\tilde{x}_n:\ \hat{\theta}_n(\tilde{x}_n)-\theta_0\leq t/\sqrt{n}\}$$

we have

$$\lim_{n\to\infty} P_{n,\theta_0}\{A_n(t)\} = F_{\theta_0}(t) \ ;$$

$$\lim_{n\to\infty} P_{n,\theta_0+(t/\sqrt{n})}\{A_n(t)\} = \frac{1}{2} \ .$$

Consider the problem of testing hypothesis $H^+:\theta=\theta_0+(t/\sqrt{n})$ $(t>0)$ against $K:\theta=\theta_0$. We shall denote by $\beta_n(t,\alpha_n,\theta_0)$ the power of the most powerful level α_n test $(0<\alpha_n<1)$. Then we obtain for each $t>0$

(2.1) $$F_{\theta_0}(t)\leq \sup_{\{\{\alpha_n\}:\ \lim_{n\to\infty}\alpha_n=1/2\}} \overline{\lim_{n\to\infty}}\ \beta_n(t,\alpha_n,\theta_0) \ .$$

Denote by $\bar{\beta}_{\theta_0}(t)$ the right-hand side of (2.1).

Consider next the problem of testing hypothesis $H^-:\theta=\theta_0+(t/\sqrt{n})$ $(t<0)$ against alternative $K:\theta=\theta_0$.

In a similar way as the case $t>0$ we define $\beta_n(t,\alpha_n,\theta_0)$ and $\bar{\beta}_{\theta_0}(t)$. Then we have for each $t<0$

$$F_{\theta_0}(t) \geq 1 - \bar{\beta}_{\theta_0}(t) \quad . \tag{2.2}$$

Since θ_0 is arbitrary, the bounds of the asymptotic distributions of AMU estimators are obtained as follows :

$$F_\theta(t) \leq \bar{\beta}_\theta(t) \quad \text{for all } t>0 \quad ;$$

$$F_\theta(t) \geq 1 - \bar{\beta}_\theta(t) \quad \text{for all } t<0 \quad .$$

For any $\theta \in \Theta$ letting $\bar{\beta}_\theta(0)=1/2$ we make the following definition.

<u>Definition 3.</u> An asymptotically median unbiased estimator $\hat{\theta}_n$ is called asymptotically efficient if for each

$$F_\theta(t) = \begin{cases} \bar{\beta}_\theta(t) & \text{for all } t \geq 0 \quad ; \\ 1-\bar{\beta}_\theta(t) & \text{for all } t<0 \quad . \end{cases}$$

<u>Definition 4.</u> $\hat{\theta}_n$ is second order asymptotically median unbiased (or second order AMU for short) if for any $\vartheta \in \Theta$, there exists a positive number δ such that

$$\lim_{n\to\infty} \sup_{\theta:|\theta-\vartheta|<\delta} \sqrt{n} \left| P_{n,\theta}\left\{\sqrt{n}(\hat{\theta}_n-\theta) \leq 0\right\} - \frac{1}{2} \right| = 0 \quad ;$$

$$\lim_{n\to\infty} \sup_{\theta:|\theta-\vartheta|<\delta} \sqrt{n} \left| P_{n,\theta}\left\{\sqrt{n}(\hat{\theta}_n-\theta) \geq 0\right\} - \frac{1}{2} \right| = 0 \quad .$$

<u>Definition 5.</u> For $\hat{\theta}_n$ second order asymptotically median unbiased $F_\theta(t)+(1/\sqrt{n})G_\theta(t)$ is called a second order asymptotic distribution of it if

$$\lim_{n\to\infty} \sqrt{n} \left| P_{n,\theta}\left\{\sqrt{n}(\hat{\theta}_n-\theta) \leq t\right\} - F_\theta(t) - (1/\sqrt{n})G_\theta(t) \right| = 0 \quad .$$

Let θ_0 be arbitrary but fixed in Θ. Then we have

$$P_{n,\theta_0}\left\{A_n(t)\right\} = F_{\theta_0}(t) + \frac{1}{\sqrt{n}} G_{\theta_0}(t) + o\left(\frac{1}{\sqrt{n}}\right) \quad ;$$

$$P_{n,\theta_0+(t/\sqrt{n})}\left\{A_n(t)\right\} = \frac{1}{2} + o\left(\frac{1}{\sqrt{n}}\right) \quad .$$

Consider the problem of testing hypothesis $H^+ : \theta = \theta_0 + (t/\sqrt{n}) \ (t>0)$ against alternative $K : \theta = \theta_0$. Let $\beta_n(t, \alpha_n, \theta_0)$ denote the power of the most powerful level α_n test $(0<\alpha_n<1)$. We assume that

$$\sup_{\left\{\{\alpha_n\}: \, \alpha_n = \frac{1}{2} + o\left(\frac{1}{\sqrt{n}}\right)\right\}} \beta_n(t, \alpha_n, \theta_0)$$

$$= \beta_{\theta_0}(t) + \frac{1}{\sqrt{n}} \gamma_{\theta_0}(t) + o(\frac{1}{\sqrt{n}}) .$$

Considering the case $t<0$ in a similar way as the case $t>0$, we make the following definition.

Definition 6. A second order asymptotically median unbiased estimator $\hat{\theta}_n$ is called second order asymptotically efficient if for each $\theta \in \Theta$

$$F_\theta(t) = \begin{cases} \beta_\theta(t) & \text{for all } t \geq 0 , \\ 1-\beta_\theta(t) & \text{for all } t < 0 , \end{cases}$$

$$G_\theta(t) = \begin{cases} \gamma_\theta(t) & \text{for all } t \geq 0 , \\ -\gamma_\theta(t) & \text{for all } t < 0 , \end{cases}$$

where for each $\theta \in \Theta$ $\beta_\theta(0)=1/2$ and $\gamma_\theta(0)=0$.

In a similar and obvious way we may define the third or k-th order asymptotic efficiency of estimators.

3. Asymptotic efficiency and second order asymptotic efficiency

Let $X_1, X_2, \ldots, X_n, \ldots$ be independently and identically distributed random variables with a density function $f(x,\theta)$ satisfying (i), (ii) and (iii).

(i) $\{x : f(x,\theta) > 0\}$ does not depend on θ.

(ii) For almost all $x[\mu]$, $f(x,\theta)$ is twice continuously differentiable in θ.

(iii) For each $\theta \in \Theta$

$$0 < I(\theta) = \int \left\{\frac{\partial}{\partial\theta} \log f(x,\theta)\right\}^2 f(x,\theta)\,d\mu(x) = -\int \left\{\frac{\partial^2}{\partial\theta^2} \log f(x,\theta)\right\} f(x,\theta)\,d\mu(x) < \infty .$$

Let θ_0 be arbitrary but fixed in Θ. Consider the problem of testing hypothesis $H^+:\theta=\theta_0+(t/\sqrt{n})\,(t>0)$ against alternative $K:\theta=\theta_0$. Then the rejection region of the most powerful test is given by

$$T_n=\sum_{i=1}^{n} Z_{ni} > c$$

, where $\quad Z_{ni}=\log\left\{ f(X_i,\theta_0)\,/\,f(X_i,\ \theta_0+\frac{t}{\sqrt{n}})\right\}$.

Since

$$T_n=\sum_{i=1}^{n} Z_{ni} \sim -\frac{t}{\sqrt{n}}\sum_{i=1}^{n}\frac{\partial}{\partial\theta}\log f(X_i,\theta)$$
$$-\frac{t^2}{2n}\sum_{i=1}^{n}\frac{\partial^2}{\partial\theta^2}\log f(X_i,\theta) ,$$

If $\theta=\theta_0$, then T_n is asymptotically normal with mean $t^2 I_{\theta_0}/2$ and variance $t^2 I_{\theta_0}$ and if $\theta=\theta_0+(t/\sqrt{n})$, then T_n is asymptotically normal with mean $-t^2 I_{\theta_0}/2$ and variance $t^2 I_{\theta_0}$. Hence it follows that

$$\bar{\beta}_{\theta_0}(t)=\Phi(t\sqrt{I_{\theta_0}}) ,$$

where $\quad \Phi(u)=\int_{-\infty}^{u}\frac{1}{\sqrt{2\pi}}e^{-\frac{x^2}{2}}dx$.

From (2.1) we have

$$F_{\theta_0}(t)\le\Phi(t\sqrt{I_{\theta_0}}) \text{ for all } t>0 .$$

In a similar way as the case $t>0$, we obtain from (2.2)

$$F_{\theta_0}(t)\ge 1-\Phi(|t|\sqrt{I_{\theta_0}})=\Phi(t\sqrt{I_{\theta_0}}) .$$

Since θ_0 is arbitrary we have now established the following well known theorem.

<u>Theorem 3.1.</u> Under conditions (i), (ii) and (iii), if $\sqrt{n}(\hat{\theta}_n-\theta)$ is asymptotically normal with mean 0 and variance $1/I_\theta$, then $\hat{\theta}_n$ is asymptotically efficient.

It has been well established that under some regularity conditions the maximum likelihood estimator $\hat{\theta}_{ML}$ has the same asymptotic distributions as above, hence $\hat{\theta}_{ML}$ is asymptotically efficient.

Using Gram-Charlier expansion of the distribution of $\sum_{i=1}^{n} Z_{ni}$ we get the asymptotic series of the power of the most powerful test.

We further assume the following :

(iv) $f(x,\theta)$ is three times continuously differentiable in θ .

(v) There exist

$$J_\theta = E_\theta\left[\left\{ \frac{\partial^2}{\partial\theta^2} \log f(X,\theta) \right\}\left\{ \frac{\partial}{\partial\theta} \log f(X,\theta) \right\} \right]$$

and

$$K_\theta = E_\theta\left[\left\{ \frac{\partial}{\partial\theta} \log f(X,\theta) \right\}^3 \right]$$

and the following hold :

$$E_\theta\left[\frac{\partial^3}{\partial\theta^3} \log f(X,\theta) \right] = -3J_\theta - K_\theta \quad .$$

We denote $(\partial/\partial\theta)f(x,\theta)$, $(\partial^2/\partial\theta^2)f(x,\theta)$ and $(\partial^3/\partial\theta^3)f(x,\theta)$ by f_θ , $f_{\theta\theta}$ and $f_{\theta\theta\theta}$, respectively.
Since

$$\frac{\partial^3}{\partial\theta^3} \log f(x,\theta) = \frac{f_{\theta\theta\theta}(x,\theta)}{f(x,\theta)} - \frac{3f_{\theta\theta}(x,\theta) f_\theta(x,\theta)}{\{f(x,\theta)\}^2} + \frac{2\{f_\theta(x,\theta)\}^3}{\{f(x,\theta)\}^3}$$

$$= \frac{f_{\theta\theta\theta}(x,\theta)}{f(x,\theta)} - 3\left\{ \frac{\partial}{\partial\theta} \log f(x,\theta)\right\}\left\{ \frac{\partial^2}{\partial\theta^2} \log f(x,\theta)\right\}$$

$$- \left\{ \frac{\partial}{\partial\theta} \log f(x,\theta) \right\}^3 \quad ,$$

it follows by the last condition of (v) that

$$\int f_{\theta\theta\theta}(x,\theta)\,d\mu(x) = 0 \quad .$$

Let $t > 0$. If $\theta = \theta_0$, then

$$T_n = \sum_{i=1}^{n} Z_{ni} \sim - \frac{t}{\sqrt{n}} \sum_1^n \frac{\partial}{\partial\theta} \log f(X_i, \theta_0)$$

$$- \frac{t^2}{2n} \sum_1^n \frac{\partial^2}{\partial\theta^2} \log f(X_i, \theta_0)$$

$$- \frac{t^3}{6n\sqrt{n}} \sum_1^n \frac{\partial^3}{\partial\theta^3} \log f(X_i, \theta_0)$$

Hence it follows that

$$E_{\theta_0}(T_n) \sim \frac{t^2}{2} I + \frac{t^3}{6\sqrt{n}} (3J+K) \quad ,$$

$$V_{\theta_0}(T_n) \sim n(\frac{t^2}{n} I + \frac{t^3}{n\sqrt{n}} J) = t^2 I + \frac{t^3}{\sqrt{n}} J \quad ,$$

$$E_{\theta_0}[\left\{ T_n - E_{\theta_0}(T_n) \right\}^3] \sim - \frac{t^3}{\sqrt{n}} K \quad ,$$

where I, J and K denote I_{θ_0}, J_{θ_0} and K_{θ_0}, respectively.
Put $\theta_1 = \theta_0 + (t/\sqrt{n})$. If $\theta = \theta_1$, then

$$T_n \sim - \frac{t}{\sqrt{n}} \sum_1^n \frac{\partial}{\partial\theta} \log f(X_i, \theta_1)$$

$$+ \frac{t^2}{2n} \sum_1^n \frac{\partial^2}{\partial\theta^2} \log f(X_i, \theta_1)$$

$$- \frac{t^3}{6n\sqrt{n}} \sum_1^n \frac{\partial^3}{\partial\theta^3} \log f(X_i, \theta_1)$$

Hence it follows that

$$E_{\theta_1}(T_n) \sim - \frac{t^2}{2} I' + \frac{t^3}{6\sqrt{n}} (3J'+K')$$

, where I', J' and K' denote I_{θ_1}, J_{θ_1} and K_{θ_1} , respectively.
On the other hand we have

$$I' \sim I + \frac{t}{\sqrt{n}} \frac{\partial}{\partial\theta} I_{\theta_0}$$

$$= I + \frac{t}{\sqrt{n}} \frac{\partial}{\partial\theta} \int \left\{ \frac{\partial}{\partial\theta} \log f(x, \theta_0) \right\}^2 f(x, \theta_0) d\mu$$

$$= I + \frac{t}{\sqrt{n}} \int 2 \left\{ \frac{\partial^2}{\partial\theta^2} \log f(x, \theta_0) \right\} \left\{ \frac{\partial}{\partial\theta} \log f(x, \theta_0) \right\} f(x, \theta_0) d\mu$$

$$+ \frac{t}{\sqrt{n}} \int \left\{ \frac{\partial}{\partial\theta} \log f(x, \theta_0) \right\}^3 f(x, \theta_0) d\mu$$

$$= I + \frac{t}{\sqrt{n}} (2J+K) \quad .$$

Hence we obtain

$$E_{\theta_1}(T_n) \sim -\frac{t^2}{2} I - \frac{t^3}{6\sqrt{n}} (3J+2K) \quad .$$

Since

$$J' \sim J + \frac{t}{\sqrt{n}} \frac{\partial}{\partial\theta} J_{\theta_0} \quad ;$$

$$K' \sim K + \frac{t}{\sqrt{n}} \frac{\partial}{\partial\theta} K_{\theta_0} \quad ,$$

it follows by a similar way as above that

$$V_{\theta_1}(T_n) \sim t^2 I + \frac{t^3}{\sqrt{n}} (J+K) \quad ;$$

$$E_{\theta_1}[\{ T_n - E_{\theta_1}(T_n) \}^3] \sim -\frac{t^3}{\sqrt{n}} K \quad .$$

Letting a_n be a rejection bound, we have

$$P_{n,\theta_1} \{ T_n < a_n \} = P_{n,\theta_1} \left\{ \frac{T_n + (t^2 I/2)}{t\sqrt{I}} < \frac{a_n + (t^2 I/2)}{t\sqrt{I}} \right\} \quad .$$

Putting $c_n = \{ a_n + (t^2 I/2) \} / (t\sqrt{I})$, we obtain

$$P_{n,\theta_1} \{ T_n < a_n \}$$

$$= \Phi(c_n) - \phi(c_n) \left\{ -\frac{t^2}{6\sqrt{n}\, I} (3J+2K) + \frac{t}{2\sqrt{n}\, I} (J+K) c_n \right.$$

$$\left. - \frac{1}{6\sqrt{nI}\, I} K (c_n^2 - 1) \right\} + o\left(\frac{1}{\sqrt{n}}\right)$$

, where $\phi(u) = \Phi'(u) = \frac{1}{\sqrt{2\pi}} e^{-\frac{u^2}{2}}$.

If $P_{n,\theta_1}\{T_n < a_n\} = 1/2$, then it follows that $c_n = O(1/\sqrt{n})$ and $\Phi(c_n) = (1/2) + c_n \phi(c_n)$.

Since

$$c_n = -\frac{t^2}{6\sqrt{n}\ I}(3J+2K) + \frac{K}{6\sqrt{nI}\ I} + o(\frac{1}{\sqrt{n}})$$

, we have

$$a_n = -\frac{t^2 I}{2} - \frac{t^3}{6\sqrt{n}\ I}(3J+2K) + \frac{tK}{6\sqrt{n}\ I} + o(\frac{1}{\sqrt{n}}) \ .$$

Then we have

$$P_{n,\theta_0}\left\{T_n \geq a_n\right\}$$

$$= 1 - P_{n,\theta_0}\left\{T_n < a_n\right\}$$

$$= 1 - P_{n,\theta_0}\left\{\frac{T_n-(t^2 I/2)}{t\sqrt{I}} - c_n < -t\sqrt{I}\right\}$$

$$= 1 - \Phi(-t\sqrt{I}) + \phi(-t\sqrt{I})\left\{\frac{t^2}{6\sqrt{nI}}(3J+K) - c_n + \frac{t}{2\sqrt{n}\ I}J(-t\sqrt{I})\right.$$

$$\left. - \frac{1}{6\sqrt{nI}\ I}K(t^2 I-1)\right\} + o(\frac{1}{\sqrt{n}})$$

$$= \Phi(t\sqrt{I}) + \frac{t^2}{6\sqrt{nI}}(3J+2K)\phi(t\sqrt{I}) + o(\frac{1}{\sqrt{n}}) \ .$$

Since θ_0 is arbitrary, it follows that for each $t>0$

$$G_\theta(t) = \Phi(t\sqrt{I_\theta}) \ ,$$

$$\gamma_\theta(t) = \frac{t^2}{6\sqrt{I_\theta}}(3J_\theta+2K_\theta)\ (t\sqrt{I_\theta}) \ .$$

In a similar way as the case $t>0$, we have for each $t<0$,

$$G_\theta(t) = \Phi(-t\sqrt{I}\ ),$$

$$\gamma_\theta(t) = -\frac{t^2}{6\sqrt{I_\theta}}(3J_\theta+2K_\theta)\phi(t\sqrt{I_\theta}) \ .$$

Therefore we have now established the following which is analogous to the result of Pfanzagl [7].

<u>Theorem 3.2.</u> Under conditions (i)$\sim$(v), if

$$(3.1)\quad P_{n,\theta}\left\{\sqrt{nI_\theta}\quad(\hat{\theta}_n-\theta)\leq t\right\}$$

$$=\Phi(t)+\frac{3J_\theta+2K_\theta}{6\sqrt{n}\;I_\theta^{\frac{3}{2}}}\quad t^2\phi(t)+o\left(\frac{1}{\sqrt{n}}\right),$$

then $\hat{\theta}_n$ is second order asymptotically efficient.

(3.1) means that $\sqrt{nI_\theta}\;(\hat{\theta}_n-\theta)$ has an asymptotic distribution with mean $-\left\{(3J_\theta+2K_\theta)/(6\sqrt{n}I_\theta^{3/2})\right\}+o(1/\sqrt{n})$ and variance $1+o(1/\sqrt{n})$ and third moment $-(3J_\theta+2K_\theta)/(\sqrt{n}I_\theta^{3/2})$.

Let $X_1, X_2, \ldots, X_n, \ldots$ be independently and identically distributed random variables with an exponential distribution having the following density function $f(x,\theta)$:

$$(3.2)\qquad f(x,\theta)=\begin{cases}\frac{1}{\theta}e^{-\frac{x}{\theta}}, & x>0,\\ 0, & x\leq 0.\end{cases}$$

Since

$$\log f(X,\theta)=-\frac{X}{\theta}-\log\theta,$$

$$\frac{\partial}{\partial\theta}\log f(X,\theta)=\frac{X}{\theta^2}-\frac{1}{\theta},$$

$$\frac{\partial^2}{\partial\theta^2}\log f(X,\theta)=-\frac{2X}{\theta^3}+\frac{1}{\theta^2},$$

it follows that

$$I_\theta=\frac{1}{\theta^4}E_\theta[\,(X-\theta)^2\,]=\frac{1}{\theta^2},$$

$$J_\theta=-\frac{1}{\theta^5}E_\theta[\,(X-\theta)(2X-\theta)\,]=-\frac{2}{\theta^5}E_\theta[\,(X-\theta)^2\,]=-\frac{2}{\theta^3},$$

$$K_\theta=\frac{1}{\theta^6}E_\theta[\,(X-\theta)^3\,]=\frac{2}{\theta^3}.$$

If $\sqrt{n}(\hat{\theta}_n-\theta)/\theta$ has an asymptotic distribution with mean $1/(3\sqrt{n})$ and variance $1+o(1/\sqrt{n})$ and third moment $2/\sqrt{n}$, then $\hat{\theta}_n$ is second order asymptotically efficient.

The maximum likelihood estimator of θ is given by $\hat{\theta}_{ML}=\overline{X}=\sum_1^n X_i/n$. Putting

$$\hat{\theta}^*_{ML}=(1+\frac{1}{3n})\overline{X},$$

we have

$$E_\theta\left[\frac{\sqrt{n}\;(\hat{\theta}^*_{ML}-\theta)}{\theta}\right]=\frac{1}{3\sqrt{n}},$$

$$V_\theta\left[\frac{\sqrt{n}\ \hat{\theta}^*_{ML}}{\theta}\right] = \left(1+\frac{1}{3n}\right)^2 = 1+o\left(\frac{1}{\sqrt{n}}\right) ,$$

$$E_\theta\left[\left\{\frac{\sqrt{n}(\hat{\theta}^*_{ML}-\theta)}{\theta}\right\}^3\right] = \frac{2}{\sqrt{n}}\left(1+\frac{1}{3n}\right)^3 = \frac{2}{\sqrt{n}} + o\left(\frac{1}{\sqrt{n}}\right) .$$

Hence $\hat{\theta}^*_{ML}$ is second order asymptotically efficient.
In this case $\overline{X}$ is sufficient statistic and the distribution has monotone likelihood ratio. Consider the problem of testing hypothesis $\theta=\theta_1$ against alternative $\theta=\theta_0$, where $\theta_1=\theta_0+(t/\sqrt{n})$ $(t>0)$. Then the rejection region of the most powerful test is given by the following form :

$$\overline{X} < c .$$

On the other hand since the cumulants of $\overline{X}$ are given as follows :

$$V_\theta(\overline{X}) = \frac{\theta^2}{n} ,$$

$$E_\theta[(\overline{X}-\theta)^3] = \frac{2\theta^3}{n^2} .$$

$\sqrt{n}(\overline{X}-\theta)/\theta$ has an asymptotic distribution with mean 0 and variance 1 and third cumulant $2/\sqrt{n}$.
Using Gram-Charlier expansion we have

$$P_{n,\theta}\left\{\frac{\sqrt{n}(\overline{X}-\theta)}{\theta} < c\right\} \tag{3.3}$$

$$= \Phi(c) - \frac{1}{3\sqrt{n}}\phi(c)\ (c^2-1) .$$

Let $\theta=\theta_1 (=\theta_0+t/\sqrt{n})$ $(t>0)$. If (3.3) agrees with $1/2$ up to the order of $1/\sqrt{n}$, then the following must hold :

$$c\,\phi(c) - \frac{1}{3\sqrt{n}}\phi(c)\ (c^2-1) \doteqdot 0 .$$

Hence it follows that $c=-1/(3\sqrt{n})$. The rejection region of the level $1/2$ test of hypothesis $\theta=\theta_1$ is given by

$$\overline{X} < \left(1-\frac{1}{3n}\right)\theta_1 .$$

This agrees asymptotically with

$$\hat{\theta}^*_{ML} = (1+\frac{1}{3n})\ \bar{X}(<\theta_1)$$

up to the order of $1/\sqrt{n}$.

Assume that $\sqrt{n}(\hat{\theta}_n-\theta)$ has the asymptotic normal distribution with mean 0 and variance $1/I_\theta$, and that the cumulants of the asymptotic distribution are given as follows :

$$\sqrt{n}\ E_\theta(\hat{\theta}_n-\theta) = \frac{1}{\sqrt{n}} c_1(\theta) + o(\frac{1}{\sqrt{n}}) \ ;$$

$$n\ V_\theta(\hat{\theta}_n) = \frac{1}{I(\theta)} + \frac{1}{\sqrt{n}} c_2(\theta) + o(\frac{1}{\sqrt{n}}) \ ;$$

$$E_\theta[\ \{\sqrt{n}(\hat{\theta}_n-E_\theta(\hat{\theta}_n))\}^3\ ] = \frac{1}{\sqrt{n}} c_3(\theta) + o(\frac{1}{\sqrt{n}}) \ .$$

Further the m-th ($m\geq 4$) cumulants of $\sqrt{n}\,\hat{\theta}_n$ is assumed to be less than order $1/\sqrt{n}$. We assume that $c_1(\theta)$, $c_2(\theta)$ and $c_3(\theta)$ are continuous in θ . Put $\hat{\theta}^*_n=\hat{\theta}_n-(1/n)k(\hat{\theta}_n)$. Since $\hat{\theta}^*_n$ agrees with $\hat{\theta}_n-(1/n)k(\theta)$ up to the order of $1/n$, the cumulants of the asymptotic distribution of $\sqrt{nI_\theta}\ (\hat{\theta}^*_n-\theta)$ are given as follows :

$$\sqrt{nI_\theta}\ E_\theta(\hat{\theta}^*_n-\theta)=\sqrt{\frac{I_\theta}{n}}\left\{c_1(\theta)-k(\theta)\right\} + o(\frac{1}{\sqrt{n}}) \ ;$$

$$nI_\theta V_\theta(\hat{\theta}^*_n)=1+\frac{I_\theta c_2(\theta)}{\sqrt{n}}+ o(\frac{1}{\sqrt{n}}) \ ;$$

$$E_\theta[\{\sqrt{nI_\theta}\ (\hat{\theta}^*_n-E_\theta(\hat{\theta}^*_n))\}^3] = \frac{I_\theta^{\frac{3}{2}} c_3(\theta)}{\sqrt{n}}+ o(\frac{1}{\sqrt{n}}) \ .$$

If

$$k(\theta)=c_1(\theta)-\frac{1}{6}I_\theta c_3(\theta) \ ,$$

then in the Gram-Charlier expansion the following holds :

$$P_{n,\theta}\left\{\sqrt{n}(\hat{\theta}^*_n-\theta)\leq 0\right\} = \frac{1}{2} + o(\frac{1}{\sqrt{n}}) \ .$$

It also follows that

(3.4) $\quad P_{n,\theta}\left\{\sqrt{nI_\theta}\ (\hat{\theta}^*_n-\theta)\leq t\right\}$

$$= \Phi(t) - \frac{1}{\sqrt{n}}\phi(t)\ [\frac{c_2(\theta)}{2I_\theta}t + \frac{c_3(\theta)}{6\ I_\theta^{3/2}}t^2] + o(\frac{1}{\sqrt{n}}) .$$

Hence it is seen that (3.4) agrees with the bound (3.1) if and only if $c_2(\theta)=0$ and $c_3(\theta)=(3J_\theta+2K_\theta)/I_\theta^3$.

If $c_2(\theta)=0$, then $c_3(\theta)$ must automatically be equal to $-(3J_\theta+2K_\theta)/I_\theta^3$. Indeed since $\phi(t)t^2>0$ for all $t\neq0$, it follows that (3.4)$\gtrless$(3.1) uniformly in θ if $c_3(\theta)\gtrless-(3J_\theta+2K_\theta)/I_\theta^3$. This means that the bound (3.1) fails to hold at either positive or negative t.

Hence if $c_2(\theta)=0$, we must have $c_3(\theta)=-(3J_\theta+2K_\theta)/I_\theta^3$. This apparently seems mysterious. However it will be naturally understood from different examples that if the second moments is decided up to order of $1/\sqrt{n}$, then the third moment is done.

We shall verify the above fact by the maximum likelihood estimator. The second order asymptotic efficiency of the maximum likelihood estimator will be verified using the above fact. The continuous differentiability of the likelihood function is assumed up to the necessary order. Let $\hat{\theta}_{ML}$ be a maximum likelihood estimator.

Since

$$0 = \sum_1^n \frac{\partial}{\partial\theta}\log f(X_i, \hat{\theta}_{ML})$$

$$= \sum_1^n \frac{\partial}{\partial\theta}\log f(X_i,\theta) + \sum_1^n\left\{\frac{\partial^2}{\partial\theta^2}\log f(X_i,\theta)\right\}(\hat{\theta}_{ML}-\theta)$$

$$+\frac{1}{2}\sum_1^n\left\{\frac{\partial^3}{\partial\theta^3}\log f(X_i,\theta^*)\right\}(\hat{\theta}_{ML}-\theta)^2 ,$$

putting $T_n=n(\hat{\theta}_{ML}-\theta)$ we have

$$0 = \frac{1}{\sqrt{n}}\sum_1^n\frac{\partial}{\partial\theta}\log f(X_i,\theta) + \frac{1}{n}\left\{\sum_1^n\frac{\partial^2}{\partial\theta^2}\log f(X_i,\theta)\right\}T_n$$

$$+\frac{1}{2n\sqrt{n}}\left\{\sum_1^n\frac{\partial^3}{\partial\theta^3}\log f(X_i,\theta^*)\right\}T_n^2 .$$

Put

$$Z_1 = \frac{1}{\sqrt{n}}\sum_1^n\frac{\partial}{\partial\theta}\log f(X_i,\theta) ;$$

$$Z_2 = \frac{1}{\sqrt{n}}\sum_1^n\left\{\frac{\partial^2}{\partial\theta^2}\log f(X_i,\theta) + I_\theta\right\} ;$$

$$W = \frac{1}{n}\sum_1^n\frac{\partial^3}{\partial\theta^3}\log f(X_i,\theta^*) .$$

Then Z_1 and Z_2 have the asymptotic normal distributions with mean 0 and variance I_θ and $L_\theta(=E_\theta\{(\partial^2/\partial\theta^2) \log f(X,\theta) + I_\theta\}^2)$ and covariance J_θ , respectively and also W converges in probability to $-3J_\theta-K_\theta$. Hence it follows that

$$Z_1 + (-I_\theta + \frac{1}{\sqrt{n}} Z_2) T_n - \frac{3J_\theta + K_\theta}{2\sqrt{n}} T_n^2 \sim 0 ,$$

$$T_n \sim \frac{1}{I_\theta} Z_1 + \frac{1}{\sqrt{n}\, I_\theta^2} Z_1 Z_2 - \frac{3J_\theta + K_\theta}{2\sqrt{n}\, I_\theta^3} Z_1^2 .$$

Since convariance of Z_1 and the term of the order is 0, it follows that

$$V_\theta(T_n) = \frac{1}{I_\theta} + o(\frac{1}{\sqrt{n}}) .$$

Hence it is seen that the maximum likelihood estimator is second order asymptotically efficient. If we indeed calculate the asymptotic cumulants, they are obtained as follows :

$$E_\theta(T_n) \sim - \frac{J_\theta + K_\theta}{2\sqrt{n}\, I_\theta^2} + o(\frac{1}{\sqrt{n}}) ;$$

$$V_\theta(T_n) \sim \frac{1}{I_\theta} + o(\frac{1}{\sqrt{n}}) ;$$

$$E_\theta[\{T_n - E_\theta(T_n)\}^3] \sim - \frac{3J_\theta + 2K_\theta}{\sqrt{n}\, I_\theta^3} + o(\frac{1}{\sqrt{n}}) .$$

Hence it is shown that if the second moment is decided up to the order of $1/\sqrt{n}$, the third moment is equal to what it should be.

As another example, we consider the location parameter case. Let $X_1, X_2, \dots , X_n, \dots$ be a sequence of independently and identically distributed random variables with a density function $f(x-\theta)$. It is well known that under appropriate regularity conditions the best linear estimator $\hat{\theta}_n = \sum_{i=1}^{n} c_{in} X_{(i|n)} + c_{on}$ is asymptotically efficient, where $X_{(1|n)} < \dots < X_{(n|n)}$ and c_{in} are optimal constants as was established by Blom and others. Then the estimator $\hat{\theta}_n$ is second order asymptotically efficient. Indeed, if $U_{(i|n)}$ are order statistics from the uniform distribution, then it follows that

$$F^{-1}(X_{(i|n)}) = \theta + F^{-1}\left(\frac{i}{n+1}\right)$$

$$+ \left\{F^{-1}\left(\frac{i}{n+1}\right)\right\}' \left(U_{(i|n)} - \frac{i}{n+1}\right)$$

$$+ \frac{1}{2}\left\{ F^{-1} \left(\frac{i}{n+1}\right)\right\}'' \left(U_{(i|n)} - \frac{i}{n+1} \right)^2 + R .$$

Let $c_{on} = - \sum_{i=1}^{n} c_{in} F^{-1}(i/(n+1))$ and $\sum_{i=1}^{n} c_{in} = 1$.
Putting

$$a_{in} = \left\{F^{-1}\left(\frac{i}{n+1}\right)\right\}' ,$$

$$b_{in} = \left\{F^{-1}\left(\frac{i}{n+1}\right)\right\}'' ,$$

we have

$$\sqrt{n}(\hat{\theta}_n - \theta) = \sqrt{n} \sum_{i=1}^{n} c_{in}\, a_{in} \left(U_{(i|n)} - \frac{i}{n+1} \right)$$

$$+ \frac{\sqrt{n}}{2} \sum_{i=1}^{n} c_{in}\, b_{in} \left(U_{(i|n)} - \frac{i}{n+1}\right)^2 + \sqrt{n}\, R_{in} .$$

Then the asymptotic variance is given by

$$V(\sqrt{n}(\hat{\theta}_n - \theta))$$

$$\sim n \sum_{i=1}^{n} c_{in}^2 a_{in}^2 \frac{i(n+1-i)}{(n+1)(n+2)} + 2n \sum\sum_{i<j} c_{in}\, c_{jn}\, a_{in}\, a_{jn} \frac{i(n+1-j)}{(n+1)^2(n+2)}$$

$$+ n \sum\sum_{i\leq j} c_{in}\, c_{jn}\, a_{in}\, b_{jn} \frac{i(n+1-j)(n+1-2j)}{(n+1)^3(n+2)(n+3)}$$

$$+ n \sum\sum_{i>j} c_{in}\, c_{jn}\, a_{in}\, b_{jn} \frac{j(n+1-i)(n+1-2j)}{(n+1)^3(n+2)(n+3)}$$

Let $J(u)$ be a function sufficiently smooth and put $c_{in} + (1/n)J(i/(n+1))$
Putting $h(u) = f(F^{-1}(u))$, we have

$$(F^{-1}(u))' = h(u)^{-1}$$

$$(F^{-1}(u))'' = -h'(u)h(u)^{-2} .$$

Since

$$V(\sqrt{n}(\hat{\theta}_n - \theta))$$

$$\sim \quad 2\iint_{u<v} u(1-v)J(u)J(v)h(u)^{-1}h(v)^{-1}\,dudv$$

$$+\frac{1}{n}\iint_{u<v} u(1-v)(1-2v)J(u)J(v)h(u)^{-1}h'(v)h(v)^{-2}\,dudv$$

$$+\frac{1}{n}\iint_{u>v} v(1-u)(1-2v)J(u)J(v)h(u)^{-1}h'(v)h(v)^{-2}\,dudv$$

$$+o\left(\frac{1}{n}\right) ,$$

it follows that

$$J(u)=-h(u)h'(u)\big/ I$$

, where $I=\int\{f'(x)\}^2\big/ f(x)dx=\int\{h'(u)\}^2\,du$.

Then we have

$$V(\sqrt{n}(\hat{\theta}_n-\theta))\sim\frac{1}{I}+o\left(\frac{1}{\sqrt{n}}\right) .$$

Let $\hat{\theta}_n^*$ be a modified estimator which is second order AMU. Then it will be shown that $\hat{\theta}_n^*$ is second order asymptotically efficient. Since

$$J=-\int\left\{\frac{d^2}{dx^2}\log f(x)\right\}\left\{\frac{d}{dx}\log f(x)\right\}f(x)dx$$

$$=\frac{1}{2}\int\left\{\frac{d}{dx}\log f(x)\right\}^2 f'(x)dx$$

$$=\frac{1}{2}\int\left\{\frac{d}{dx}\log f(x)\right\}^3 f(x)dx$$

$$=\frac{1}{2}\int\{h'(u)\}^3\,du$$

$$=-\frac{K}{2} ,$$

the third asymptotic moment of the second order asymptotically efficient estimator must be $-K\big/(2n^2I^3)$.
Indeed, since

$$\mathrm{cov}\left[\left(U_{(i|n)}-\frac{i}{n+1}\right)\left(U_{(j|n)}-\frac{j}{n+1}\right),\ \left(U_{(k|n)}-\frac{k}{n+1}\right)^2\right]$$

$$= 2 \operatorname{cov}(U_{(i|n)}, U_{(k|n)}) \operatorname{cov}(U_{(j|n)}, U_{(k|n)}) + o(\frac{1}{n}) \quad ;$$

$$E[(U_{(i|n)} - \frac{i}{n+1})(U_{(j|n)} - \frac{j}{n+1})(U_{(k|n)} - \frac{k}{n+1})]$$

$$= \frac{i(n+1-j)(n+1-k)}{(n+1)^2(n+2)(n+3)} \quad , \; i \leq j \leq k \quad ,$$

it follows that

$$E[\{\hat{\theta}_n - E(\hat{\theta}_n)\}^3]$$

$$\sim \frac{6}{n^2} \iiint_{u<v<w} u(1-2v)(1-w)J(u)J(v)J(w)h(u)^{-1}h(v)^{-1}h(w)^{-1}\,dudvdw$$

$$+ \frac{3}{n^2} \int \left\{ \int (\min(u,v)-uv)J(u)h(u)du \right\}^2 J(v)h(v)^{-2}h'(v)dv + o(\frac{1}{n^2}) \quad .$$

Putting $J(u)=-h(u)h''(u)/I$, we have

$$E[\{\hat{\theta}_n - E(\hat{\theta}_n)\}^3]$$

$$\sim -\frac{1}{n^2 I^3} \int \{h'(u)\}^3 du - \frac{1}{n^2 I^3} \int h(u)h'(u)h''(u)du$$

$$= -\frac{1}{n^2 I^3} \int \{h'(u)\}^3 du + \frac{3}{2n^2 I^3} \int \{h'(u)\}^3 du$$

$$= -\frac{K}{2n^2 I^3} \quad .$$

Hence it is shown that the third asymptotic moment is $-K/(2n^2 I^3)$.

The second order asymptotic efficiency of ML estimators depends on the assumption of differentiability of the density function. The following example is to make this point clear. Suppose that X_i's are independently and identically distributed according to the double (bilateral) exponential distribution with the density function

$$f(x-\theta) = \frac{1}{2} e^{-|x-\theta|}$$

, where θ is unknown location parameter.

In this case it can be shown that the first order asymptotic bound is obtained with I=1, and it can be easily shown that the ML estimator

$$\hat{\theta}_{ML} = \underset{1 \le i \le n}{\text{med}} X_i$$

is asymptotically efficient.
The second order asymptotic bound is obtained after some algebraic manipulations (the detail is given in [11]) as follows :

$$\Phi(t) - \frac{t^2}{6\sqrt{n}} \phi(t) \text{ sgn } t \tag{3.5}$$

But the asymptotic distribution of $\hat{\theta}_{ML}$ is given by

$$P_{n,\theta}\left\{\sqrt{n}(\hat{\theta}_{ML} - \theta) \le t\right\} = \Phi(t) - \frac{t^2}{2\sqrt{n}} \phi(t) \text{ sgn } t + o(\frac{1}{\sqrt{n}}) ,$$

which differs in the second term from (3.5).

4. Third order asymptotic efficiency

We proceed to the problem of third order asymptotic efficiency. $\hat{\theta}_n$ is called third order asymptotically median unbiased if for any $\vartheta \in \Theta$ there exists a positive number δ such that

$$\lim_{n\to\infty} \sup_{\theta:|\theta-\vartheta|<\delta} n \left| P_{n,\theta}\left\{\sqrt{n}(\hat{\theta}_n - \theta) \le 0\right\} - \frac{1}{2}\right| = 0 ;$$

$$\lim_{n\to\infty} \sup_{\theta:|\theta-\vartheta|<\delta} n \left| P_{n,\theta}\left\{\sqrt{n}(\hat{\theta}_n - \theta) \ge 0\right\} - \frac{1}{2}\right| = 0 .$$

Let $\hat{\Theta}_3$ be the class of all third order asymptotically median unbiased estimators.
For all $\hat{\theta}_n \in \hat{\Theta}_3$ we shall obtain the bounds of G_θ and H_θ such that

$$\lim_{n\to\infty} n \left| P_{n,\theta}\left\{\sqrt{n}(\hat{\theta}_n - \theta) \le t\right\} - F_\theta(t) - \frac{1}{\sqrt{n}} G_\theta(t) - \frac{1}{n} H_\theta(t) \right| = 0 .$$

In a similar way as section 3 we consider the problem of testing hypothesis $\theta = \theta_0 + (t/\sqrt{n})$ against alternative $\theta = \theta_0$ and the asymptotic expansion of the power of the most powerful test of it.

The continuous differentiability of the likelihood function is assumed up to the necessary order. Putting $\psi' = (\partial/\partial\theta) \log f(x,\theta)$,

$\psi''=(\partial^2/\partial\theta^2)\log f(x,\theta)$ and $\psi^{(i)}=(\partial^i/\partial\theta^i)\log f(x,\theta)\,(i=3,4)$, we have

$$
\begin{aligned}
T_n &= \sum_{i=1}^{n} Z_{ni} \\
&= -\frac{t}{\sqrt{n}}\sum_{1}^{n}\psi'(X_i) - \frac{t^2}{2n}\sum_{1}^{n}\psi''(X_i) - \frac{t^3}{6n\sqrt{n}}\sum_{1}^{n}\psi^{(3)}(X_i) \\
&\quad -\frac{t^4}{24n^2}\sum_{1}^{n}\psi^{(4)}(X_i) + o\left(\frac{1}{n}\right),
\end{aligned}
$$

where $Z_{ni}=\log\left\{ f(X_i,\theta_0)\,/\,f(X_i,\theta_0+(t/\sqrt{n}))\right\}$.

Define

$$L=E_{\theta_0}\left\{\psi^{(3)}(X)\psi'(X)\right\},$$

$$M=E_{\theta_0}\left\{\psi''(X)^2\right\},$$

$$N=E_{\theta_0}\left\{\psi''(X)\psi'(X)^2\right\},$$

$$H=E_{\theta_0}\left\{\psi'(X)^4\right\}.$$

Since under appropriate conditions the following holds :

$$E_{\theta_0}\left\{\psi^{(4)}(X)\right\} = -4L - 3M - 6N - H,$$

the asymptotic cumulants are given as follows :

$$E_{\theta_0}(T_n) \sim \frac{t^2 I}{2} + \frac{t^3}{6\sqrt{n}}(3J+K) + \frac{t^4}{24n}(4L+3M+6N+H),$$

$$V_{\theta_0}(T_n) \sim t^2 I + \frac{t^3}{\sqrt{n}} J + \frac{t^4}{4n}(M-I^2) + \frac{t^4}{3n} L,$$

$$E_{\theta_0}\left[\left\{T_n - E_{\theta_0}(T_n)\right\}^3\right] \equiv \gamma_{\theta_0}(T_n) \sim -\frac{t^3}{\sqrt{n}} K - \frac{3t^4}{2n}(N+I^2),$$

$$E_{\theta_0}\left[\left\{T_n - E_{\theta_0}(T_n)\right\}^4\right] - 3\left\{V_{\theta_0}(T_n)\right\}^2 \equiv \delta_{\theta_0}(T_n) \sim \frac{t^4}{n}(H-3I^2).$$

If $\theta=\theta_1(=\theta_0+(t/\sqrt{n}))$, then for any measurable function g

$$E_{\theta_1}\left\{g(X)\right\}$$

$$= E_{\theta_0}[g(X)\left\{1+\frac{t}{\sqrt{n}}\frac{f'_{\theta_0}(X)}{f_{\theta_0}(X)}+\frac{t^2}{2n}\frac{f''_{\theta_0}(X)}{f_{\theta_0}(X)}+\frac{t^3 f^{(3)}_{\theta_0}(X)}{6n\sqrt{n}\, f_{\theta_0}(X)}+\dots\right\}]$$

$$= E_{\theta_0}[g(X)\left\{1+\frac{t}{\sqrt{n}}\psi'(X)+\frac{t^2}{2n}(\psi''(X)+\psi'(X)^2)\right.$$

$$+\frac{t^3}{6n\sqrt{n}}(\psi^{(3)}(X)+3\psi''(X)\psi'(X)+\psi'(X)^3)$$

$$\left.+\dots\right\}] .$$

Hence we have

$$E_{\theta_1}(T_n)\sim E_{\theta_0}(T_n)-E_{\theta_0}[\left\{\frac{t}{\sqrt{n}}\sum_1^n\psi'(X_i)+\frac{t^2}{2n}\sum_1^n\psi''(X_i)+\frac{t^3}{6n\sqrt{n}}\sum_1^n\psi^{(3)}(X_i)\right\}$$

$$\left\{\frac{t}{\sqrt{n}}\sum_1^n\psi'(X_i)+\frac{t^2}{2n}\sum_1^n(\psi''(X_i)+\psi'(X_i)^2)\right.$$

$$\left.+\frac{t^3}{6n\sqrt{n}}\sum_1^n(\psi^{(3)}(X_i)+3\psi''(X_i)\psi'(X_i)+\psi'(X_i)^3)\right\}]$$

$$+o\left(\frac{1}{n}\right)$$

$$= -\frac{t^2}{2}I-\frac{t^3}{6\sqrt{n}}(3J+2K)-\frac{t^4}{24n}(4L+3M+12N+3H) .$$

Similarly we obtain

$$V_{\theta_1}(T_n)\sim t^2 I+\frac{t^3}{\sqrt{n}}(J+K)+\frac{t^4}{12n}(4L+3M+18N+6H-3I^2) ,$$

$$\gamma_{\theta_1}(T_n)\sim -\frac{t^3}{\sqrt{n}}K-\frac{t^4}{2n}(3N+2H-3I^2) ,$$

$$\delta_{\theta_1}(T_n)\sim \frac{t^4}{n}(H-3I^2) .$$

Put $V_n=\left\{T_n-E_{\theta_1}(T_n)\right\}/(t\sqrt{I})$.

If

$$V_{\theta_1}(V_n) = \frac{1}{t^2 I}V_{\theta_1}(T_n)$$

$$= 1+\frac{1}{\sqrt{n}}\beta_1+\frac{1}{n}\beta_2+o\left(\frac{1}{n}\right) ,$$

$$\gamma_{\theta_1}(V_n) = \frac{1}{t^3 I\sqrt{I}} \gamma_{\theta_1}(T_n)$$

$$= \frac{1}{\sqrt{n}} \gamma_1 + \frac{1}{n} \gamma_2 + o(\frac{1}{n}) ,$$

$$\delta_{\theta_1}(V_n) = \frac{1}{t^4 I} \delta_{\theta_1}(T_n)$$

$$= \frac{1}{n} \delta + o(\frac{1}{n}) ,$$

then we have

$$P_{n,\theta_1}\left\{V_n \leq a\right\}$$

$$= \Phi(a) - \phi(a)\left\{ \frac{1}{2}(\frac{\beta_1}{\sqrt{n}} + \frac{\beta_2}{n})a + \frac{1}{6}(\frac{\gamma_1}{\sqrt{n}} + \frac{\gamma_2}{n})(a^2-1) \right.$$

$$+ (\frac{\delta}{24n} + \frac{\beta_1^2}{8n})(a^3-3a) + \frac{\beta_1\gamma_1}{12n}(a^4+6a^2+3)$$

$$\left. + \frac{\gamma_1^2}{72n}(a^5-10a^3-15a)\right\} + o(\frac{1}{n}) .$$

Choose a such that

$$P_{n,\theta_1}\left\{V_n \leq a\right\} = \frac{1}{2} + o(\frac{1}{n}) .$$

Since

$$\Phi(a) = \frac{1}{2} + a\phi(a) - \frac{1}{2} a^3\phi'(a) + \dots ,$$

it follows that

$$a = \frac{\beta_1}{2\sqrt{n}} a - (\frac{\gamma_1}{6\sqrt{n}} + \frac{\gamma_2}{n}) + \frac{1}{4n}\beta_1\gamma_1 + o(\frac{1}{n})$$

$$= - \frac{\gamma_1}{6\sqrt{n}} - \frac{\gamma_2}{n} + \frac{\beta_1\gamma_1}{6n} + o(\frac{1}{n}) .$$

if for $\theta = \theta_0$

$$E_{\theta_1}(V_n) - E_{\theta_0}(V_n) - a = -t\sqrt{I} + \frac{1}{\sqrt{n}}\alpha_1' + \frac{1}{n}\alpha_2' + o(\frac{1}{n}) ,$$

$$V_{\theta_0}(V_n) = \frac{1}{\sqrt{n}} \beta_1' + \frac{1}{n} \beta_2' + o(\frac{1}{n}) ,$$

$$\gamma_{\theta_0}(V_n) = \frac{1}{\sqrt{n}}\gamma_1' + \frac{1}{n}\gamma_2' + o(\frac{1}{n}) ,$$

$$\delta_{\theta_0}(V_n) = \frac{1}{n}\delta_1' + o(\frac{1}{n}) ,$$

then we have

(4.1) $$P_{n,\theta_0}\left\{ V_n \geq a \right\}$$

$$= 1 - P_{n,\theta_0}\left\{ V_n < a \right\}$$

$$= \Phi(t') + \phi(t')\left\{ \frac{1}{\sqrt{n}}\alpha_1' + \frac{1}{n}\alpha_2' - (\frac{\beta_1'}{2\sqrt{n}} + \frac{\beta_2'}{2n} + \frac{\alpha_1'^2}{2n})t' \right.$$

$$+ (\frac{\gamma_1'}{6\sqrt{n}} + \frac{\gamma_2'}{2n} + \frac{\alpha_1'\beta_1'}{2n}) (t'^2 - 1)$$

$$- (\frac{\delta_1'}{24n} + \frac{\alpha'\gamma_1'}{6n} + \frac{\beta_1'^2}{8n}) (t'^3 - 3t')$$

$$+ \frac{\beta_1'\gamma_1'}{12n} (t'^4 - 6t'^2 + 3)$$

$$\left. - \frac{\gamma_1'^2}{72n}(t'^5 - 10t'^3 + 15t') \right\} + o(\frac{1}{n}) ,$$

where $t' = t\sqrt{I}$.

Substituting in (4.1)

$$\alpha_1' = \frac{t'^2}{2I^{3/2}} (2J+K) - \frac{K}{6I^{3/2}} ,$$

$$\alpha_2' = \frac{t'^3}{12I^2} (4L + 3M + 6N + 3I^2) + \frac{t'}{6I^3} K(J+K) ,$$

$$\beta_1' = \frac{t'J}{I^{3/2}} ,$$

$$\beta_2' = \frac{t'^2}{4I^2} (M-I^2) + \frac{t'^2}{3I^2} L ,$$

$$\gamma_1' = - \frac{K}{I^{3/2}} ,$$

$$\gamma_2' = -\frac{3t'}{2I^2}(N+I^2) ,$$

$$\delta = \frac{1}{I^2}(H-3I^2) ,$$

we obtain

$$P_{n,\theta_0}\{V_n \geq a\}$$

$$= \Phi(t') + \phi(t')\, [\, \frac{t'^2}{6\sqrt{n}\, I^{3/2}}(3J+2K) - \frac{t'^5}{72nI^{3/2}}(3J+2K)^2$$

$$- \frac{t'^3}{72nI^3}\left\{(3J+2K)^2 - 3I(4L+3M+6N-H+6I^2)\right)$$

$$+ \frac{t'}{72nI^3}\left\{2K^2 + 9I(2N+H-I^2)\right\}\,]$$

$$+ o(\frac{1}{n}) .$$

Hence it follows that for $t > 0$

$$P_{n,\theta}\left\{\sqrt{nI}\,(\hat{\theta}_n-\theta) \leq t\right\}$$

$$\leq \Phi(t) - \phi(t)\left\{\frac{\beta_3''}{6\sqrt{n}} + \frac{\beta_3''}{6\sqrt{n}}(t^2-1) + \frac{\beta_3''^2}{72n}(t^5-10t^3+15t)\right.$$

$$+ \left(\frac{\beta_4''}{24n} + \frac{\beta_3''^2}{36n}\right)(t^3-3t)$$

$$\left. + \left(\frac{\beta_2''}{2n} + \frac{\beta_3''^2}{72n}\right)t\right\} + o(\frac{1}{n}) ,$$

where $\beta_3'' = -(3J+2K) / I^{\frac{3}{2}}$,

$$\beta_4'' = \frac{1}{I^3}\left\{3(3J+2K)^2 - I(4L+3M+6N-H+6I^2)\right\} ,$$

$$\beta_2'' = \frac{1}{36I^3}\left\{17(3J+2K)^2 - 2K^2 - 9I(4L+3M+8N+5I^2)\right\} .$$

Further the moments of the third order asymptotically efficient estimator must be as follows :

$$(4.2)\qquad E_\theta(\hat{\theta}_n) = \theta + \frac{\beta_3''}{6\sqrt{n}\ I^{3/2}} + o(\frac{1}{n\sqrt{n}}) ,$$

$$(4.3)\qquad V_\theta(\sqrt{n}\,\hat{\theta}_n) = I + \frac{\beta_2''}{nI} + o(\frac{1}{n}) ,$$

$$(4.4)\qquad \gamma_\theta(\sqrt{n}\,\hat{\theta}_n) = \frac{\beta_3''}{\sqrt{n}\ I^{3/2}} + o(\frac{1}{n}) ,$$

$$(4.5)\qquad \delta_\theta(\sqrt{n}\,\hat{\theta}_n) = \frac{\beta_4''}{nI^2} + o(\frac{1}{n}) .$$

Next we shall consider the example of the exponential distribution with the density (3.2) given in section 3. Since

$$\hat{\theta}_n = (1 + \frac{1}{3n})\bar{X}$$

it follows that

$$(4.6)\qquad E_\theta(\hat{\theta}_n) = \theta + \frac{\theta}{3n},$$

$$(4.7)\qquad V_\theta(\sqrt{n}\,\hat{\theta}_n) = (1+\frac{2}{3n})\theta^2 ,$$

$$(4.8)\qquad \gamma_\theta(\sqrt{n}\,\hat{\theta}_n) = \frac{2}{n}\theta^3 + o(\frac{1}{n}) ,$$

$$(4.9)\qquad \delta_\theta(\sqrt{n}\,\hat{\theta}_n) = \frac{6}{n}\theta^4 + o(\frac{1}{n}) .$$

Since $I=1/\theta^2$, $J=-2/\theta^3$, $K=2/\theta^3$, $L=6/\theta^4$, $M=5/\theta^4$, $N=-5/\theta^4$ and $H=9/\theta^4$, it follows from (4.2)$\sim$(4.5) that $\beta_3''=2$, $\beta_4''=6$ and $\beta_2''=2/3$. Then the moments of (4.2)$\sim$(4.5) are equal to those of (4.6)$\sim$(4.9), respectively. This fact should hold because $\bar{X}$ is a sufficient statistic.

We next consider the maximum likelihood estimator $\hat{\theta}_{ML}$. Putting $T_n=\sqrt{n}(\hat{\theta}_{ML}-\theta)$, we have

$$Z_1 + (-1+\frac{1}{\sqrt{n}}Z_2)T_n + \frac{1}{2\sqrt{n}}\left\{-(3J+K) + \frac{1}{\sqrt{n}}Z_3\right\}T_n^2$$

$$-\frac{4L+3M+6N+H}{6n}\quad T_n^3 \sim 0 ,$$

where $Z_3 = \frac{1}{\sqrt{n}} \sum_1^n \left\{ \psi^{(3)}(X_i) + 3J + K \right\}$.

We also obtain

$$T_n \sim \frac{1}{I} Z_1 + \frac{1}{\sqrt{n}\, I^2}\left(Z_2 Z_1 - \frac{3J+K}{2I}\right) Z_1^2$$

$$+ \frac{1}{n\, I^3} \left\{ Z_1 Z_2^2 + \frac{1}{2} Z_1^2 Z_3 - \frac{3(3J+K)}{2\, I} Z_1^2 Z_2 \right.$$

$$\left. + \frac{(3J+K)^2}{2\, I^2} Z_1^3 - \frac{4L+3M+6N+H}{6} Z_1^3 \right\}.$$

Since

$$E(Z_1^2 Z_2) \sim \frac{1}{\sqrt{n}}(N+I^2) ,$$

$$E(Z_1^3 Z_2) \sim \frac{3}{n} IJ ,$$

$$E(Z_1^4 Z_2^2) \sim \frac{3}{n}(M-I^2) + 12J^2 ,$$

it follows that

$$E(T_n) \sim - \frac{J+K}{2\sqrt{n}\, I^2} + o\left(\frac{1}{n}\right) ,$$

$$V(T_n) \sim \frac{1}{I} + \frac{7J^2+14JK+5K^2}{24I^4} - \frac{L+4N+H+I^2}{nI^3} + o\left(\frac{1}{n}\right) ,$$

$$\gamma(T_n) \sim - \frac{3J+2K}{nI} + o\left(\frac{1}{n}\right) ,$$

$$\delta(T_n) \sim \frac{12(J+K)(2J+K)}{n\, I^5} - \frac{4L+12N+3H-3I}{n\, I^4} + o\left(\frac{1}{n}\right) .$$

We define a modified maximum likelihood estimator $\hat{\theta}^*_{ML}$ as follows

$$\hat{\theta}^*_{ML} = \hat{\theta}_{ML} + \frac{K_{\hat{\theta}_{ML}}}{6nI^2_{\hat{\theta}_{ML}}} .$$

Then $\hat{\theta}^*_{ML}$ is third order AMU, that is,

$$P_{n,\theta}\left\{ \hat{\theta}^*_{ML} - \theta \leq 0 \right\} = \frac{1}{2} + o\left(\frac{1}{n}\right) .$$

Further it follows that

$$nV(\hat{\theta}^*_{ML}) = n\left(1 + \frac{1}{6n}\frac{\partial}{\partial\theta}\frac{K_\theta}{I_\theta}\right)^2 V(\hat{\theta}_{ML})$$

$$\sim V(T_n) + \frac{(3N+H)I - 2K(2J+K)}{3nI^4}$$

, where

$$\frac{\partial}{\partial\theta}K_\theta = \frac{\partial}{\partial\theta}\int\{\psi'(x)\}^3 f(x)dx$$

$$= 3\int\psi''(x)\psi'(x)^2 f(x)dx + \int\{\psi'(x)\}^4 f(x)dx$$

$$= 3N + H ,$$

$$\frac{\partial}{\partial\theta}I_\theta = 2J + K .$$

In general the maximum likelihood estimator is shown not to be third order asymptotically efficient.
It is enough to consider a special case, and we take the Cauchy distribution as an example which has the following density function $f(x,\theta)$:

$$f(x,\theta) = \frac{1}{\pi(1+(x-\theta)^2)} .$$

Let $X_1, X_2, \ldots , X_n, \ldots$ be independent identically distributed random variables with above density.
Since

$$\frac{\partial}{\partial\theta}\log f(x,0) = -\frac{\partial}{\partial x}\log f(x,0) = \frac{2x}{1+x^2} ,$$

$$\frac{\partial^2}{\partial\theta^2}\log f(x,0) = \frac{\partial^2}{\partial x^2}\log f(x,0) = -\frac{2(1-x^2)}{(1+x^2)^2},$$

$$\frac{\partial^3}{\partial\theta^3}\log f(x,0) = -\frac{\partial^3}{\partial x^3}\log f(x,0) = -\frac{4x(3-x^2)}{(1+x^2)^3},$$

it follows that $I=1/2$, $J=K=0$, $L=-3/4$, $M=7/8$, $N=-1/8$ and $H=3/8$.
Since the density is symmetric, the maximum likelihood estimator $\hat{\theta}_{ML}$ is third order AMU. The second and fourth cumulants are given as follows :

$$(4.10)\qquad V(\sqrt{n}\,\hat{\theta}_{ML}) \sim 2 + \frac{5}{n},$$

$$(4.11)\qquad \gamma(\sqrt{n}\,\hat{\theta}_{ML}) \sim \frac{66}{n}.$$

On the other hand since the values of the bound of the asymptotic distribution is given by

$$\beta_2'' = 1/4, \quad \beta_4'' = 0,$$

they are larger than the above moments (4.10) and (4.11). Hence $\hat{\theta}_{ML}$ is not third order asymptotically efficient.

5. Some further remarks

The above results can be extended to the cases (I),(II),(III),(IV).

(I) When the distribution is discrete, some modification is necessary to apply the above argument.

(II) Let X_i's be independent but not identically distributed random variables with density functions $f_i(x_i,\theta)$. Let $\hat{\theta}_n$ be a second order AMU estimator. When $c_n(\hat{\theta}_n-\theta)$ has an asymptotic distribution, we obtain the bound of the distributions of $\hat{\theta}_n$. Let θ_0 be arbitrary but fixed in Θ. Consider the most powerful test of the problem of testing hypothesis $\theta=\theta_1(=\theta_0+c_n^{-1}a,\ a>0)$ against alternative $\theta=\theta_0$. Then the test statistic of it is given by

$$(5.1)\qquad T_n = \sum_1^n \log\frac{f_i(X_i,\theta_1)}{f_i(X_i,\theta_0)}$$

$$= -\sum_1^n\left\{\frac{\partial}{\partial\theta}\log f_i(X_i,\theta_0)\right\}\cdot(c_n^{-1}a)$$

$$-\frac{1}{2}\sum_1^n\left\{\frac{\partial^2}{\partial\theta^2}\log f_i(X_i,\theta_0)\right\}\cdot(c_n^{-1}a)^2$$

$$-\frac{1}{6}\sum_1^n\left\{\frac{\partial^3}{\partial\theta^3}\log f_i(X_i,\theta_0)\right\}\cdot(c_n^{-1}a)^3$$

$$+\sum_{i=1}^n R_{in}$$

Put

$$I_i = E_{\theta_0}\left\{\frac{\partial}{\partial\theta}\log f_i(X_i,\theta_0)\right\}^2 = -E_{\theta_0}\left\{\frac{\partial^2}{\partial\theta^2}\log f_i(X_i,\theta_0)\right\}.$$

It follows that the variance of the first term of the right-hand side of (5.1) is given by $(c_n^{-1}a)^2\sum_1^n I_i$. When $c_n=\sqrt{\tilde{I}_n}$, where $\tilde{I}_n=\sum_1^n I_i$, it's variance convergence to some finite value.
If the following "Lindeberg type" condition holds : for every $\varepsilon > 0$,

$$\lim_{n\to\infty}\frac{1}{\tilde{I}_n}\sum_1^n E_{\theta_0}\left[\left\{\frac{\partial}{\partial\theta}\log f_i(X_i,\theta_0)\right\}^2 \chi_{in}(\varepsilon)\right] = 0\ ,$$

where

$$\chi_{in}(\varepsilon) = \begin{cases} 1, & \text{if } \frac{\partial}{\partial\theta}\log f_i(X_i,\theta) > \varepsilon\sqrt{\tilde{I}_n}\ , \\ 0, & \text{otherwise}\ , \end{cases}$$

then the first term of the right-hand side of (5.1) is asymptotically normal with mean 0 and variance a^2. The necessary condition for this to hold is given by $\lim_{n\to\infty}\max_{1\le i\le n} I_i/\tilde{I}_n = 0$.
If a similar condition as above holds : for every $\varepsilon > 0$,

$$\lim_{n\to\infty}\frac{1}{\tilde{I}_n}\sum_1^n E_{\theta_0}\left[\left\{-\frac{\partial^2}{\partial\theta^2}\log f_i(X_i,\theta_0)\right\}\chi^*_{in}(\varepsilon)\right] = 0\ ,$$

where

$$\chi^*_{in}(\varepsilon) = \begin{cases} 1, & \text{if } -\frac{\partial^2}{\partial\theta^2}\log f_i(X_i,\theta_0) > \varepsilon\cdot\tilde{I}_n\ , \\ 0, & \text{otherwise}\ , \end{cases}$$

then the second term of the right-hand side of (5.1) converges in probability to $a^2/2$.

If under alternative $\theta = \theta_0$, similar regularity conditions as above, then the test statistic T_n is asymptotically normal with mean $-a^2/2$ and variance a^2. Put $c_n=\sqrt{\tilde{I}_n}$. For $a>0$, the supremum of the power functions is given by $\Phi(a)$. Hence if $\sqrt{\tilde{I}_n}(\hat{\theta}_n-\theta)$ is asymptotically normal with mean 0 and variance 1, then $\hat{\theta}_n$ is an asymptotically efficient estimator.

Let $\hat{\theta}_{ML}$ be a maximum likelihood estimator.
Then

$$\begin{aligned} 0 = &\sum_1^n\left\{\frac{\partial}{\partial\theta}\log f_i(X_i,\theta)\right\}(\hat{\theta}_{ML}-\theta) \\ &+\sum_1^n\left\{\frac{\partial^2}{\partial\theta^2}\log f_i(X_i,\theta)\right\}(\hat{\theta}_{ML}-\theta)^2 \\ &+\sum_1^n R'_{in} \end{aligned}$$

Under some regularity conditions $\sqrt{\tilde{I}_n}(\hat{\theta}_{ML}-\theta)$ is asymptotically normal with mean 0 and variance 1. Hence the maximum likelihood estimator $\hat{\theta}_{ML}$ is asymptotically efficient.

In order to consider the higher order of the bound of the power functions, we must decide the orders of

$$\tilde{K}_n = \sum_1^n K_i = \sum_1^n E_{\theta_0}\left[\left\{\frac{\partial}{\partial\theta}\log f_i(X_i,\theta_0)\right\}^3\right]$$

and

$$\tilde{J}_n = \sum_1^n J_i = \sum_1^n E_{\theta_0}\left[\left\{\frac{\partial^2}{\partial\theta^2}\log f_i(X_i,\theta_0)\right\}\left\{\frac{\partial}{\partial\theta}\log f_i(X_i,\theta_0)\right\}\right].$$

Then the asymptotic moments of the test statistic T_n are formally obtained as follows :

$$E_{\theta_0}(T_n) \sim \frac{a^2}{2} - \frac{3\tilde{J}_n+\tilde{K}_n}{6\tilde{I}_n^{3/2}}a^3 ,$$

$$V_{\theta_0}(T_n) \sim a^2 + \frac{\tilde{J}_n}{\tilde{I}_n^{3/2}}a^3 ,$$

$$E_{\theta_0}\left\{T_n - E_{\theta_0}(T_n)\right\}^3 \sim -\frac{\tilde{K}_n}{\tilde{I}_n^{3/2}}a^3 .$$

Similarly we have for $\theta=\theta_1$

$$E_{\theta_1}(T_n) \sim -\frac{a}{2} - \frac{3\tilde{J}_n+2\tilde{K}_n}{6\tilde{I}_n^{3/2}}a^3 ,$$

$$V_{\theta_1}(T_n) \sim a^2 + \frac{\tilde{J}_n+\tilde{K}_n}{\tilde{I}_n^{3/2}}a^3 ,$$

$$E_{\theta_1}\left\{T_n - E_{\theta_1}(T_n)\right\}^3 \sim -(\tilde{K}_n/\tilde{I}_n^{\frac{3}{2}})a^3 .$$

If $\tilde{J}_n$ and $\tilde{K}_n$ have the same order which is less than $\tilde{I}_n^{\frac{3}{2}}$ and the residual terms are negligible, we can apply Gram-Charlier expansion to the asymptotic distribution of the statistic T_n. Put $d_n^{-1}=\tilde{K}_n/\tilde{I}_n^{3/2}$. If

$$\lim_{n\to\infty} d_n \left| P_{n,\theta}\left\{\sqrt{\tilde{I}_n}(\hat{\theta}_n-\theta)\leq 0\right\} - 1/2\right| = 0$$

uniformly in any neighborhood of θ_0, then for $a>0$

$$(5.2)\quad \overline{\lim_{n\to\infty}}\, d_n \left[P_{n,\theta_0} \left\{ \sqrt{\tilde{I}_n}(\hat{\theta}_n-\theta) \leq a \right\} - \Phi(a) - \frac{3\tilde{J}_n+2\tilde{K}_n}{6\,\tilde{I}_n^{3/2}} a^2 \phi(a) \right] \leq 0$$

For $a<0$, a similar inequality as above holds.
If the equality "=" of (5.2) holds, then $\hat{\theta}_n$ may be called second order asymptotically efficient. The necessary condition for second order asymptotic efficiency is that $\hat{\theta}_n$ has the following asymptotic moments :

$$E_\theta \left\{ \sqrt{\tilde{I}_n}(\hat{\theta}_n-\theta) \right\} \sim - \frac{3\tilde{J}_n+2\tilde{K}_n}{3\,\tilde{I}_n^{3/2}} + o\left(\frac{1}{d_n}\right) ;$$

$$V_\theta \left\{ \sqrt{\tilde{I}_n}(\hat{\theta}_n-\theta) \right\} \sim 1 + o\left(\frac{1}{d_n}\right) ;$$

$$E_\theta \left\{ \sqrt{\tilde{I}_n}(\hat{\theta}_n-\theta) \right\}^3 \sim - \frac{3\tilde{J}_n+2\tilde{K}_n}{\tilde{I}_n^{3/2}} + o\left(\frac{1}{d_n}\right) .$$

If $\hat{\theta}_n$ has a smooth asymptotic distribution and $V_\theta \left\{ \sqrt{\tilde{I}_n}(\hat{\theta}_n-\theta) \right\} \sim 1+o(1/d_n)$, then the modified estimator $\hat{\theta}_n^*$ which is second order asymptotically median unbiased is second order asymptotically efficient.
Let

$$\hat{\theta}_{ML}^* = \hat{\theta}_{ML} + \frac{\tilde{K}_n(\hat{\theta}_{ML})}{6n\tilde{I}_n(\hat{\theta}_{ML})^2} .$$

Under regularity conditions $\hat{\theta}_{ML}^*$ is second order asymptotically efficient. If the following generally hold :

$$\tilde{I}_n / n \to \bar{I} \quad (n\to\infty) ,$$

$$\tilde{J}_n / n \to \bar{J} \quad (n\to\infty) ,$$

$$\tilde{K}_n / n \to \bar{K} \quad (n\to\infty) ,$$

then it follows that $d_n=\sqrt{n}$. Since the second term of the asymptotic distribution has order of $1/\sqrt{n}$, a similar result as the i.i.d. case holds. But the fact does not always hold. For the purpose we consider the following examples.

Let $X_i=\theta z_i+U_i$, $i=1,2,\ldots$, where z_i's are constants and U_i's are independent identically distributed ramdom variables with a density $f(u)$. Putting

$$I = \int\left\{\frac{d}{du}\log f(u)\right\}^2 f(u)du = \int \frac{\{f'(u)\}^2}{f(u)} du ,$$

$$J = \int\left\{\frac{d^2}{du^2}\log f(u)\right\}\left\{\frac{d}{du}\log f(u)\right\} f(u)du$$

$$= \int \left[\frac{f''(u)f'(u)}{f(u)} - \frac{\{f'(u)\}^3}{\{f(u)\}^2}\right] du ,$$

$$K = \int\left\{\frac{d}{du}\log f(u)\right\}^3 f(u)du$$

$$= \int \frac{\{f'(u)\}^3}{\{f(u)\}^2} du ,$$

we have for each i

$$I_i = z_i^2 I , \qquad J_i = -z_i^3 J , \qquad K_i = -z_i^3 K .$$

Then it follows that

$$\tilde{I}_n = (\sum_1^n z_i^2) I , \quad J = -(\sum_1^n z_i^3) J , \quad K = -(\sum_1^n z_i^3) K .$$

By Lindeberg's condition we obtain

$$\max_{1 \le i \le n} z_i^2 \Big/ \sum_1^n z_i^2 \longrightarrow 0 \qquad (n \to \infty) .$$

Since

$$\frac{|\sum_1^n z_i^3|}{(\sum_1^n z_i^2)^{3/2}} \le \frac{\max |z_i|}{(\sum_1^n z_i^2)^{1/2}} \to 0 \qquad (n \to \infty) ,$$

it follows that

$$\tilde{J}_n \Big/ \tilde{I}_n^{\frac{3}{2}} \longrightarrow 0 \qquad (n \to \infty) ,$$

$$\tilde{K}_n \Big/ \tilde{I}_n^{\frac{3}{2}} \longrightarrow 0 \qquad (n \to \infty) .$$

Then we can take $d_n = (\sum_1^n z_i^2)^{\frac{3}{2}} \Big/ (\sum_1^n z_i^3)$.
If $\sum_1^n z_i^2/n \to m_2$ $(n \to \infty)$ and $\sum_1^n z_i^3/n \to m_3 (\neq 0)$, then it follows that $d_n = O(\sqrt{n})$. Then the second term of the asymptotic distribution has order of 1.

But letting $z_i = i$, we have

$$\tilde{I}_n \doteq n(n+1)\ (2n+1)\ I/6 \sim n^3 I/3 .$$

Hence if $\sqrt{n^3 I/3}\ (\hat{\theta}_n - \theta)$ is asymptotically normal with mean 0 and variance 1, then $\hat{\theta}_n$ is an asymptotically efficient estimator. Since

$$\tilde{J}_n = -\frac{n^2(n+1)^2}{4} J \sim -\frac{n^4}{4} J ,$$

$$\tilde{K}_n = -\frac{n^2(n+1)^2}{4} K \sim -\frac{n^4}{4} K ,$$

it follows that $d_n = O(\sqrt{n})$. Hence in this case the second term of the asymptotic distribution has also order of $1/\sqrt{n}$.

If $z_i \to 0$ $(i \to \infty)$, then it follows by Lindeberg's conditions that the following must hold :

$$\tilde{I}_n = (\sum_1^n z_i^2)I \to \infty \qquad (n \to \infty) .$$

But it is possible that $\sum_1^n z_i^2 \to \infty (n \to \infty)$ and $\sum_1^n z_i^3 < \infty$. Since $d_n = O(\tilde{I}_n^{\frac{1}{2}})$, letting $z_i = 1/\sqrt{i}$ we obtain $\tilde{I}_n = O(\log n)$. Hence $\sqrt{\log n}\ (\hat{\theta}_n - \theta)$ is asymptotically normal and the second term of the asymptotic distribution has order of $(\log n)^{-\frac{1}{2}}$.

Let X_i's be independently distributed random variables with the following density :

$$f(x_i, \theta) = \frac{1}{\theta^{p_i}\Gamma(p_i)} x^{p_i - 1} e^{-x/\theta} , \quad x > 0 ,$$

, where p_i's are known positive integer and θ is an unknown parameter. Since

$$\log f_i(X_i, \theta) = (p_i - 1)\log X_i - X_i/\theta - p_i \log\theta - \log\Gamma(p_i) ,$$

$$\frac{\partial}{\partial\theta}\log f_i(X_i, \theta) = \frac{X_i}{\theta^2} - \frac{p_i}{\theta} ,$$

$$\frac{\partial^2}{\partial\theta^2}\log f_i(X_i, \theta) = -\frac{2X_i}{\theta^3} + \frac{p_i}{\theta^2} ,$$

it follows that

$$I_i = E_\theta\left(\frac{X_i}{\theta^2} - \frac{p_i}{\theta}\right)^2 = \frac{p_i}{\theta^2} \quad ,$$

$$J_i = -E_\theta\left(\frac{X_i}{\theta^2} - \frac{p_i}{\theta}\right)\left(\frac{2X_i}{\theta^3} - \frac{p_i}{\theta^2}\right) = -\frac{2p_i}{\theta^3} \quad ,$$

$$K_i = E_\theta\left(\frac{X_i}{\theta^2} - \frac{p_i}{\theta}\right)^3 = \frac{2p_i}{\theta^3} \quad .$$

Then we have

$$\tilde{I}_n = \sum_1^n p_i/\theta^2 \ , \ \tilde{J}_n = -2\sum_1^n p_i/\theta^3 \ , \ \tilde{K}_n = 2\sum_1^n p_i/\theta^3 \ .$$

Since $d_n = (\sum_1^n p_i)^{\frac{1}{2}}$, the asymptotic distribution of the second order asymptotically efficient estimator $\hat{\theta}_n$ is given by

$$P_{n,\theta}\left\{\sqrt{\sum_1^n p_i}(\hat{\theta}_n - \theta)/\theta < a\right\} \sim \Phi(a) + \frac{1}{3\sqrt{\sum_1^n p_i}} a^2 \phi(a) \ .$$

The maximum likelihood estimator $\hat{\theta}_{ML}$ of θ is also given by

$\sum_1^n X_i / \sum_1^n p_i$. Let

$$\hat{\theta}^*_{ML} = \left(1 + \frac{1}{3\sum_1^n p_i}\right)\hat{\theta}_{ML} \ .$$

Then $\hat{\theta}^*_{ML}$ is a second order asymptotically efficient estimator. Since in this case $\sum_1^n X_i$ is a sufficient statistic, the second order AMU estimator based on it must be second order asymptotically efficient. Indeed it is easily shown that the asymptotic distribution of $\hat{\theta}^*_{ML}$ agrees with the above bound of the asymptotic distributions.

(III) Let $\{X_i\}$ be defined recursively by

$$X_i = \theta X_{i-1} + U_i \quad , \quad i = 1, 2, \ldots ,$$

where $X_0 = 0$ and $\{U_i\}$ is a sequence of independent identically distributed real random variables having density f with mean 0 and variance σ^2. Let θ_0 be arbitrary but fixed in Θ . We consider the problem of testing hypothesis $\theta = \theta_1 (= \theta_0 + c_n^{-1} a)$ against alternative $\theta = \theta_0$. Then the test statistic of the most powerful test is given by

$$\sum_{i=1}^n Z_{ni} = \sum_{i=1}^n \log \frac{f(X_i - \theta_0 X_{i-1})}{f(X_i - \theta_1 X_{i-1})} \quad .$$

By Taylor expansion we have

$$\sum_{i=1}^{n} Z_{ni} = \sum_{i=1}^{n} c_n^{-1} a\, X_{i-1} \frac{f'(U_i)}{f\,(U_i)}$$

$$-\frac{1}{2} \sum_{i=1}^{n} c_n^{-2} a^2 X_{i-1}^2 \left\{ \frac{\partial^2}{\partial U_i^2} \log f(U_i) \right\}$$

$$+ \sum_{i=1}^{n} R_{in} \quad .$$

Since for $|\theta|>1$, X_i is stochastically order of θ_i, the asymptotic distribution of $\sum_{i=1}^{n} Z_{ni}$ is not always written as an easy type. If $|\theta|< 1$, then at least X_i is asymptotically stationary. If $c_n=\sqrt{n}$, then $\sum_{i=1}^{n} Z_{ni}$ is asymptotically normal ([11]). From above it is shown that $\sqrt{n}(\hat{\theta}_n-\theta)$ is asymptotically normal with mean 0 and variance $(1-\theta^2)/I\sigma^2$, where $I=\int\{f'(u)\}^2/f(u)du$ if $\hat{\theta}_n$ is an asymptotically efficient estimator.

When $|\theta|< 1$, using asymptotic moments we can formally obtain the bound of the second order asymptotic distributions of second order AMU estimators of θ and show that a modified least squares estimator of θ is second order asymptotically efficient (Akahira [2]).

(IV) When the parameter θ is vector valued, we may decompose it to $\theta=\binom{\xi}{\eta}$ where ξ is a real number and η is a real vector, and we consider estimating ξ, while η is treated as a "nuisance" parameter. Then the second order asymptotic bound for any estimator $\hat{\xi}$ of ξ is given in a similar but somewhat complicated algebraic expression, and it can be shown that a modified ML estimator attains this bound under some regularity conditions ([3]).

References

[1] Akahira,M., "Asymptotic theory for estimation of location in non-regular cases, I : Order of convergence of consistent estimators," Rep. Stat. Appl. Res., JUSE, 22, (1975).

[2] Akahira,M., "A note on the second asymptotic efficiency of estimators in an autoregressive process," Rep. Univ. Electro-Comm. 26-1, (1975)

[3] Akahira,M. and Takeuchi,K., "On the second order asymptotic efficiency of estimators in multiparameter cases," Rep. Univ. Electro-Comm. 26-2, (1976).

[4] Chibisov,D.M., "On the normal approximation for a certain class of statistics," Proc. Sixth Berkeley Symp. on Math. Statist. and Prob. 1, (1972).

[5] Chibisov,D.M., "Asymptotic expansions for Neyman's C(α) tests," Proc. Second Japan-USSR Symp.on Probability Theory. Lecture Notes in Mathematics. Springer Verlag, Berlin, (1973).

[6] Fisher,R.A., "Theory of statistical estimation ," Proc. Camb. Phil. Soc. 22, (1925).

[7] Pfanzagl,J., "Asymptotic expansions related to minimum contrast estimators," Ann. Statist. 1, (1973).

[8] Pfanzagl,J., "On asymptotically complete classes," Proc. Summer Research Inst. of Statistical Inference for Stochastic Processes, 2, (1975).

[9] Rao,C.R.,"Asymptotic efficiency and limiting information," Proc. Fourth Berkely Symp. on Math. Statist. and Prob. 1, (1961).

[10] Rao,C.R., "Efficient estimates and optimum inference procedures in large samples," J. Roy. Statist. Soc. 24 (B), (1962).

[11] Takeuchi,K., "Tokei-teki suitei no Zenkinriron (Asymptotic Theory of Statistical Estimation)," (In Japanese) Kyoiku-Shuppan , Tokyo, (1974).

K. Takeuchi
Faculty of Economics
University of Tokyo
Hongo, Bunkyo-ku
Tokyo, Japan

M. Akahira
Department of Mathematics
University of Electro-Communications
Chofugaoka, Chofu-shi
Tokyo, Japan

ON THE RELAXED SOLUTIONS OF A CERTAIN STOCHASTIC DIFFERENTIAL EQUATION

by Masaaki TSUCHIYA

§1. Introduction

In this paper, we shall study the following stochastic differential equation:

(0) $dx(t) = dB(t) + a(x(t))d\psi(t)$,

where $\{B(t)\}$ is a one-dimensional Brownian motion and $\{\psi(t)\}$ is the local time at 0 for a reflecting Brownian motion $\{y(t)\}$ on $[0,\infty)$ which is independent of $\{B(t)\}$. This equation is one of the stochastic equations introduced by N. Ikeda [4] and Ikeda treated the equation (0) in the form of the integral equation:

(1) $$x(t) = x(0) + B(t) + \int_0^{\psi(t)} a(x(\psi^{-1}(s)))ds,$$

and he has shown that if the coefficient $a(x)$ is Lipschitz continuous, then the equation (1) has a unique solution $\{x(t)\}$ and the process $\mathbb{P}_a = \{(x(t),y(t))\}$ becomes a diffusion on the upper half plane $\overline{D} = \{(x,y);\ y \geqq 0\}$ such that it has the absorbing Brownian motion as minimal process and satisfies the boundary condition $\frac{\partial u}{\partial y} + a(x)\frac{\partial u}{\partial x} = 0$.

However, in non smooth coefficient cases, it is, in general, unknown that the equation (1) has a solution. So, we treat the equation (0) in the following form:

(2) $$x(t) = x(0) + B(t) + \int_0^{\psi(t)} \phi(s)ds,$$

where $\phi(s) \in A[x(\psi^{-1}(s))]$ a.e. s (with respect to Lebesgue measure) and $A[x]$ denotes the interval $[\underline{a}(x), \overline{a}(x)]$, setting $\underline{a}(x) = \lim_{\delta\downarrow 0} \operatorname{ess\,inf}_{|y-x|\leqq\delta} a(y)$ and $\overline{a}(x) = \lim_{\delta\downarrow 0} \operatorname{ess\,sup}_{|y-x|\leqq\delta} a(y)$. We call a solution of the equation (2) a <u>relaxed solution</u> (see §2 for the precise definition) of the equation (0) following E.D. Conway [1],[2], who has introduced the notion of the relaxed solution to Itô's stochastic differential equations.

It is easily to see that if $a(x)$ is continuous, the relaxed solutions of the equation (0) coincide with the ordinary solutions (i.e., the solutions of the equation (1)) and if $a(x) = b(x)$ a.e. x (with respect to Lebesgue measure), then the relaxed solutions of the equation (0) with the coefficient $a(x)$ coincide with the corresponding ones to the coefficient $b(x)$ and the same initial value.

In the present paper, we shall construct relaxed solutions $\{x(t)\}$ such that the processes $\{(x(t),y(t))\}$ become diffusions on $\bar{D}$, and the method is similar to S. Watanabe's one (cf. [12],[13]). We also refer to M. Motoo's works [8],[9] which motivate our work.

§2. Main result

Assumption. *In this paper, we assume that the coefficient* $a(x)$ *of the equation* (0) *is essentially bounded with respect to Lebesgue measure.*

Let $\{(B(t),\beta(t))\}$ be a two-dimensional Brownian motion on a probability space $(\Omega,F,P;F_t)$ so that it is adapted to (F_t) and $B(0) = \beta(0) = 0$.

Set $\psi(t) = -\min_{0\leqq s\leqq t}\{(y(0)+\beta(s))\wedge 0\}$ and $y(t) = y(0) + \beta(t) + \psi(t)$ $(y(0)\geqq 0)$, then it is well-known (cf. [5],[14]) that $\{y(t)\}$ is a description of the reflecting Brownian motion on $[0,\infty)$ and $\{\psi(t)\}$ is the local time at 0 for $\{y(t)\}$ and $\{(y(t),\psi(t))\}$ is a unique solution of the following stochastic differential equation:

$$dy(t) = d\beta(t) + d\psi(t),$$

where with probability one

(i) $y(t)$ and $\psi(t)$ are continuous functions of $t\in[0,\infty)$,

(ii) $y(t) \geqq 0$ for all $t\in[0,\infty)$, $\psi(t)$ is increasing and $I_{\{0\}}(y(t))d\psi(t) = d\psi(t)$ (where $I_{\{0\}}$ denotes the indicator function of the set $\{0\}$),

(iii) $y(0)$ is independent of $\{\beta(t)\}$ and

(iv) $y(t) = y(0) + \beta(t) + \psi(t)$ for all $t \geqq 0$.

Definition 2.1. *By a* relaxed solution *of the equation* (0) *on the probability space* $(\Omega,F,P;F_t)$*, we mean a process* $\{x(t)\}$ *defined on it such that*

(i) *with probability one,* $x(t)$ *is continuous in* t,

(ii) *it is adapted to* (F_t) *and*

(iii) *with probability one, it satisfies for all* $0\leqq t_1<t_2$ *and* $v\in(-\infty,\infty)$

$$v\cdot(x(t_2)-x(t_1)) \leqq v\cdot(B(t_2)-B(t_1)) + \int_{\psi(t_1)}^{\psi(t_2)} \overline{v\cdot a}(x(\psi^{-1}(s)))ds,$$

where $\psi^{-1}(t) = \inf\{s;\ \psi(s) > t\}$ *(the inverse local time).*

Remark 2.1. *For an essentially bounded function* $f(x)$, $\bar{f}(x)$ *is a bounded Borel function. So, using T.12 Chap. VII in* [7], *in Definition 2.1, we can replace the condition* (iii) *by the following condition* (iii'):

(iii') *with probability one, it satisfies for all* $0\leqq t_1<t_2$ *and* $v\in(-\infty,\infty)$

$$v\cdot(x(t_2)-x(t_1)) \leqq v\cdot(B(t_2)-B(t_1)) + \int_{t_1}^{t_2}\overline{v\cdot a}(x(s))d\psi(s).$$

Definition 2.2. *We call that a function* f(x) *satisfies a* one-sided Lipschitz condition, *if there exists a positive constant* K *such that it holds one of the following inequalities:*

(L-1) $(x-y)(f(x)-f(y)) \leqq K(x-y)^2$ *for all* $x,y\in(-\infty,\infty)$;

(L-2) $-(x-y)(f(x)-f(y)) \leqq K(x-y)^2$ *for all* $x,y\in(-\infty,\infty)$.

Our main result is the following theorem.

Theorem 2.1. *On some probability space* $(\Omega,F,P;F_t)$, *for each initial value there exist relaxed solutions* $\{x_i(t)\}$ $(i=1,2)$ *such that*

(1°) *for any relaxed solution* $\{x(t)\}$ *with the same initial value,*

$$P[x_1(t) \leqq x(t) \leqq x_2(t) \quad \text{for all } t \geqq 0] = 1,$$

(2°) *the processes* $\{\Omega,(x_i(t),y(t)),F_t,P\}$ $(i=1,2)$ *are diffusions on* $\overline{D}$ *and moreover*

(3°) *if the coefficient* a(x) *is continuous or satisfies a one-sided Lipschitz condition,*

$$P[x_1(t) = x_2(t) \quad \text{for all } t \geqq 0] = 1.$$

§3. Construction of solutions

We shall introduce the equation on the boundary $\partial D = \{(x,0);\ -\infty<x<\infty\}$. Let $\{\ell(t)\}$ is a one-dimensional symmetric Cauchy process on a probability space $(W,B,Q;B_t)$ such that it is adapted to (B_t) and $\ell(0) = 0$. Over this probability space, we consider the stochastic equation:

(3) $d\xi(t) = d\ell(t) + a(\xi(t))dt$.

Definition 3.1. *By a* relaxed solution *of the equation* (3) *on the probability space* $(W,B,Q;B_t)$, *we mean a process* $\{\xi(t)\}$ *defined on it so that*

(i) *with probability one,* $\xi(t)$ *is right continuous and has left-hand limits as a function of* t,

(ii) *it is adapted to* (B_t) *and*

(iii) *with probability one, it satisfies for all* $0\leqq t_1<t_2$ *and* $v\in(-\infty,\infty)$

$$v\cdot(\xi(t_2)-\xi(t_1)) \leqq v\cdot(\ell(t_2)-\ell(t_1)) + \int_{t_1}^{t_2}\overline{v\cdot a}(\xi(s))ds.$$

Then the following lemma is fundamental to show the theorem.

Lemma 3.1. (1°) *If on the probability space* $(\Omega,F,P;F_t)$ *there exists a relaxed solution* $\{x(t)\}$ *of the equation* (0), *setting* $B_t = F_{\psi^{-1}(t)}$, $\ell(t) = B(\psi^{-1}(t)) - B(\psi^{-1}(0))$ *and* $\xi(t) = x(\psi^{-1}(t))$, *then on the probability space* $(\Omega,F,P;B_t)$, $\{\xi(t)\}$ *becomes a relaxed solution of the equation* (3) *with initial value* $x(0)+B(\psi^{-1}(0))$. *Moreover, put*

$$X(t) \equiv X(t,\omega) = \begin{cases} x(0) + B(t,\omega) & \text{if } (t,\omega)\in\Omega_0 \\ \xi(\psi(t,\omega),\omega) & \text{if } (t,\omega)\in\Omega_1 \\ \lim_{n\to\infty} \xi(\psi(t-\frac{1}{n},\omega),\omega) & \text{if } (t,\omega)\in\Omega_2 \\ \lim_{n\to\infty} \xi(\psi(\sigma-\frac{1}{n},\omega),\omega) + B(t,\omega) - B(\sigma,\omega) & \text{if } (t,\omega)\in\Omega_3, \end{cases}$$

then

$$P[x(t) = X(t) \quad \text{for all } t \geqq 0] = 1,$$

where

$$\Omega_0 = \{(t,\omega);\ 0 \leqq t < \psi^{-1}(0,\omega)\},$$

$$\Omega_1 = \{(t,\omega);\ \psi^{-1}(\psi(t,\omega),\omega) = t\}\cap\Omega_0^c,$$

$$\Omega_2 = \{(t,\omega);\ \lim_{n\to\infty}\psi^{-1}(\psi(t-\tfrac{1}{n},\omega),\omega) = t < \psi^{-1}(\psi(t,\omega),\omega)\}\cap\Omega_0^c,$$

$$\Omega_3 = \{(t,\omega);\ \lim_{n\to\infty}\psi^{-1}(\psi(t-\tfrac{1}{n},\omega),\omega) > t\}\cap\Omega_0^c \quad \text{and}$$

$$\sigma \equiv \sigma(t,\omega) = \begin{cases} \lim_{n\to\infty}\psi^{-1}(\psi(t,\omega)-\frac{1}{n},\omega) & \text{if } (t,\omega)\in\Omega_3 \\ 0 & \text{otherwise.} \end{cases}$$

(2°) *For the probability space* $(\Omega,F,P;F_t)$, *suppose that* $F_t = F_{t+}$ *for all* $t \geqq 0$ *and* F_0 *contains all null sets of* F *and further assume that on the probability space* $(\Omega,F,P;B_t)$, *for any* $x\in(-\infty,\infty)$ *there exists a relaxed solution* $\{\xi(t,x,\omega)\}$ *of the equation* (3) *with initial value* x, *which is measurable in* (t,x,ω). *Then, on the probability space* $(\Omega,F,P;F_t)$ *the process* $\{X(t)\}$ *obtained from* $\xi(t) = \xi(t,x+B(\psi^{-1}(0)),\omega)$ *in* (1°) *becomes a relaxed solution of the equation* (0) *with initial value* x *and it holds*

$$P[X(\psi^{-1}(t)) = \xi(t) \quad \text{for all } t \geqq 0] = 1.$$

Using the results in [3],[11] and [10], we have:

Lemma 3.2. *On any probability space* $(W,B,Q;B_t)$ *provided with a one-dimensional symmetric Cauchy process* $\{\ell(t)\}$ $(\ell(0)=0)$ *which is adapted to* (B_t), *for each* $x\in(-\infty,\infty)$, *there exist the minimum relaxed solution* $\{\xi_1(t,x,\omega)\}$ *and the maximum relaxed solution* $\{\xi_2(t,x,\omega)\}$ *of the equation* (3) *with initial value* x *such that for each* $t > 0$, $\xi_i(s,x,\omega)$ $(s\in[0,t]$, $x\in(-\infty,\infty)$, $w\in W)$ $(i=1,2)$ *are* $B[0,t]\otimes B(-\infty,\infty)\otimes B_t$*-measurable, where* $B[0,t]$ *(resp.* $B(-\infty,\infty)$*) is the* σ*-field of Borel sets in* $[0,t]$ *(resp.* $(-\infty,\infty)$*).*

Moreover, if $a(x)$ *is continuous or satisfies a one-sided Lipschitz condition, then for all* $x\in(-\infty,\infty)$,

$$Q[\xi_1(t,x,w) = \xi_2(t,x,w) \quad \text{for all } t \geqq 0] = 1.$$

By Lemma 3.1 and Lemma 3.2, it follows that the statements (1°) and (3°) of Theorem 2.1 hold. To show the statement (2°), we shall consider the following probability space. Let Ω be the space of continuous mappings ω from $[0,\infty)$ into $(-\infty,\infty)\times(-\infty,\infty)$ so that $\omega(0) = (0,0)$ and set $\omega(t) = (B(t),\beta(t)) = (B(t,\omega),\beta(t,\omega))$ for each $t \geqq 0$. For each $s \geqq 0$, define the shift operator θ_s on Ω by $\theta_s\omega(t) = \omega(t+s)$.

Let the process $\{\Omega,F,(B(t),\beta(t)),F_t,P\}$ be the two-dimensional Brownian motion such that F is the P-completion of the σ-field $\sigma\{\omega(s);\ 0\leqq s<\infty\}$ and F_t is the P-completion of the σ-field $\sigma\{\omega(s);\ 0\leqq s\leqq t\}$, where $\sigma\{\omega(s);\ 0\leqq s<\infty\}$ (resp. $\sigma\{\omega(s);\ 0\leqq s\leqq t\}$) is the σ-field generated by $\omega(s)$, $0\leqq s<\infty$ (resp. $0\leqq s\leqq t$). Put for each $y\in[0,\infty)$

$$\psi(t) \equiv \psi(t,y,\omega) = -\min_{0\leqq s\leqq t}\{(y+\beta(s,\omega))\wedge 0\} \quad \text{and}$$

$$y(t) \equiv y(t,y,\omega) = y + \beta(t,\omega) + \psi(t,y,\omega).$$

If we set $B_t = F_{\psi^{-1}(t)}$ and $\ell(t) = B(\psi^{-1}(t)) - B(\psi^{-1}(0))$, then on the probability space $(\Omega,F,P;B_t)$ we can construct the minimum relaxed solution $\{\xi_1(t)\}$ and the maximum relaxed solution $\{\xi_2(t)\}$ of the equation (3) with initial value $x+B(\psi^{-1}(0))$ and $y(0) = y$, and denote by $\{x_i(t;x,y;\omega)\}$ $(i=1,2)$ the process $\{X(t)\}$ obtained from $\{\xi_i(t)\}$ in Lemma 3.1. Then the following lemma is necessary to show the statement (2°).

Lemma 3.3. *For every* $t,s,y\in[0,\infty), x\in(-\infty,\infty)$ *and* $\omega\in\Omega$, *it holds*

(1°) $\psi(t+s,y,\omega) = \psi(s,y,\omega) + \psi(t,y(s,y,\omega),\theta_s\omega)$,

(2°) $y(t+s,y,\omega) = y(t,y(s,y,\omega),\theta_s\omega)$ *and*

(3°) $x_i(t+s;x,y;\omega) = x_i(t;x_i(s;x,y;\omega),y(s,y,\omega);\theta_s\omega)$ $(i=1,2)$.

Set $z = (x,y)$ for $x\in(-\infty,\infty)$, $y\in[0,\infty)$ and

$$z_i(t,z,\omega) = (x_i(t;x,y;\omega),y(t,y,\omega)) \quad (i=1,2),$$

then by Lemma 3.3 and the result in [6] (pp 10-11), it follows that for each bounded function $f(z) = f(x,y)$ on $\overline{D}$ and finite valued (F_t)-stopping time $S = S(\omega)$,

$$\begin{aligned} E[f(z_i(t+S,z,\omega))|F_S] &= E[f(z_i(t,z_i(S,z,\omega),\theta_S\omega))|F_S] \\ &= E[f(z_i(t,\zeta,\hat{\omega}))]\big|_{\zeta=z_i(S,z,\omega)}. \end{aligned}$$

This show that the assertion (2°) of Theorem 2.1 holds.

To show the existence of the relaxed solutions $\{x_i(t)\}$ $(i=1,2)$, we use the result of Viktrovskii [11]. Another method to show the existence of a solution is based upon the approximation of the coefficient $a(x)$

by smooth functions. We can obtain the following theorem.

Theorem 3.1. *For each* $x \in (-\infty,\infty)$ *and* $y \in [0,\infty)$, *there exists a sequence* $\{a_n(x)\}$ *of smooth functions so that* $a_n(x) \to a(x)$ a.e. x *as* $n \to \infty$ *and on some probability space, with probability one, the relaxed solution* $x_n(t)$ *of the equation* (0) *with the coefficient* $a_n(x)$ *and initial value* x *converses uniformly to a relaxed solution* $x(t)$ *of the equation* (0) *with the coefficient* $a(x)$ *and initial value* x *on each compact interval, where* $y(0) = y$.

It should be noticed that in order to show the existence a solution of the equation (1) it is necessary to obtain similar inequality to the one presented by N.V. Krylov for Itô's stochastic integrals, but for the relaxed solution such an inequality is unnecessary.

The detailed proofs of Theorem 2.1 and Theorem 3.1 will be given elsewhere.

References

[1] E.D. Conway; Stocahstic equations with discontinuous drift, Trans. Amer. Math. Soc. 157(1971) 235-245.

[2] E.D. Conway; Stochastic equations with discontinuous drift II, Indiana Univ. Math. Jour. 22(1972) 91-99.

[3] A.F. Filippov; Differential equations with discontinuous right-hand side, Mat. Sb. 51(93) (1960) 99-168, (English transl.: Amer. Math. Soc. Transl. (2) 42(1964) 199-231.

[4] N. Ikeda; On the construction of two dimensional diffusion processes satisfying Wentzell's boundary conditions and its application to boundary value problems, Mem. Coll. Sci. Univ. Kyoto, Ser. A, 33 (1961) 367-427.

[5] K. Itô and H.P. McKean, Jr.; Diffusion processes and their sample paths, Springer, 1965.

[6] H.P. McKean, Jr.; Stochastic integrals, Academic Press, 1969.

[7] P.A Meyer; Probabilités et potentiel, Hermann, 1966.

[8] M. Motoo; The boundary condition with the discontinuous inclined derivative, USSR-Japan Symp. on Prob. Theory, Habarovsk, Vol.II (1967) 247-256.

[9] M. Motoo; Brownian mortions in the half plane with singular inclined periodic boundary conditions, Topics in Prob. Theory, ed. by Stroock and Varadhan, New York Univ. (1973) 163-179.

[10] H. Tanka, M. Tsuchiya and S. Watanabe; Perturbation of drift-type for Lévy processes, Jour. Math. Kyoto Univ. 14(1974) 73-92.

[11] E.E. Viktrovskii; On a generalization of the concept of integral curves for a discontinuous field of direction, Mat. Sb. 34(76) (1954) 213-248 (in Russian)

[12] S. Watanabe; Application of Poisson point processes to Markov processes, (1973) 1-35 (in Japnese).

[13] S. Watanabe; Construction of diffusion processes with Wentzell's boundary conditions by means of Poisson point processes, III USSR-Japan Symp. on Porb. Theory, Tashkent, Vol.II (1975) 311-345.

[14] S. Watanabe; Stochastic differential equations, Sangyo Tosho, 1975 (in Japanese).

Department of Mathematics
College of Liberal Arts
Kanazawa University
Kanazawa, Japan

ON LIMIT THEOREMS FOR NON-CRITICAL GALTON-WATSON PROCESSES WITH $EZ_1 \log Z_1 = \infty$

By Koehei UCHIYAMA
TOKYO INSTITUTE OF TECHNOLOGY

1. Let $Z_0 = 1, Z_1, Z_2, \cdots$ denote a non-critical Galton-Watson processes with a non-degenerate offspring distribution $p_j = P[Z_1 = j]$, $j = 0,1,2,\cdots$, having a finite mean $m = EZ_1 < \infty$. Write $f_0(s) = s$, $f_1(s) = f(s) = \sum_{j=0}^{\infty} p_j s^j$, $f_2(s) = f(f_1(s))$, $f_3 = f(f_2(s)), \cdots$, $0 \leq s \leq 1$. When $\sum_{j=0}^{\infty} p_j (j\log j) < \infty$ it is known (cf. Athreya and Ney [1]) that in case $m > 1$ Z_n grows up to infinity as m^n a.s. on $\{Z_n \to \infty$ as $n \to \infty\}$, and in case $m < 1$ $1 - f_n(0) = P[Z_n \neq 0]$ decays to zero as m^n. Stating more minutely;

(i) when $m < 1$, $\lim_{n\to\infty}(1 - f_n(0))^{-1} P[Z_n = j]$ exist for $j = 1,2,3, \cdots$ and define a probability distribution on $\{1,2,3,\cdots\}$; and furthermore $m^{-n}(1-f_n(0))$ converges to a positive constant if and only if $\sum p_j (j\log j) < \infty$; and

(ii) when $m > 1$, for some sequence of positive constants c_n, $c_n^{-1} Z_n$ converges almost surely to a proper non-degenerate random variable; and furthermore $m^{-n} c_n$ converges to a positive number if and only if $\sum p_j (j\log j) < \infty$.

If $\sum p_j (j\log j) = \infty$, we have $\lim (1 - f_n(0)) m^{-n} = 0$ in case $m < 1$ and $\lim c_n m^{-n} = 0$ in case $m > 1$. In this note we will investigate the rate of decay for $1 - f_n(0)$ or of growing up for Z_n in certain cases with $\sum p_j (j\log j) = \infty$.

2. Let us introduce a condition:

$$\sum_{j=n}^{\infty} j p_j \sim (\log n)^{-\alpha} L(\log n) \qquad \text{as } n \to \infty, \text{(*)} \tag{1}$$

where $L(x)$ is a function slowly varying at infinity and α a real constant.

<u>THEOREM</u>. <u>Let</u> $L(x)$ <u>be as above and</u> $0 \leq \alpha < 1$.

(a) <u>If</u> $m < 1$, <u>then the condition</u> (1) <u>is equivalent to</u>

$$\log (m^{-n}(1 - f_n(s))) \sim - A n^{1-\alpha} L(n) \qquad \text{as } n \to \infty \tag{2}$$

<u>for</u> $0 \leq s < 1$, <u>where</u> $A = |\log m|^{-\alpha} / m(1 - \alpha)$.

(b) <u>If</u> $m > 1$, <u>then the condition</u> (1) <u>is equivalent to</u>

$$\log \frac{Z_n}{m^n} \sim - A n^{1-\alpha} L(n) \qquad \text{as } n \to \infty \quad \text{a.s. on } \{Z_n \to \infty \text{ as } n \to \infty\}, \tag{3}$$

(*) "$a(x) \sim b(x)$ as $x \to c$" means that $\lim_{x\to c} a(x)/b(x) = 1$.

where A *is defined as in* (a).

REMARK 1. Let (1) hold with $\alpha \geq 0$. Then $\sum p_j(j\log j) = \infty$ if and only if either $0 \leq \alpha < 1$, or $\alpha = 1$ and $\int_1^\infty L(y)y^{-1}dy = \infty$. This follows from LEMMA 3 and 4 which are stated in the following section. When $\alpha = 1$, the assertions of THEOREM are also true if we replace $-An^{1-\alpha}L(n)$ by $\int_1^n L(y)y^{-1}dy/(m\log m)$ in (2) and (3).

REMARK 2. Suppose $\alpha > 0$. Then (1) follows from

$$\sum_{j=1}^{n} j^2 p_j \sim n(\log n)^{-\alpha-1}L(\log n) \qquad \text{as} \quad n \to \infty,$$

which is in turn implied by

$$\sum_{j=n}^{\infty} p_j \sim n^{-1}(\log n)^{-\alpha-1}L(\log n) \qquad \text{as} \quad n \to \infty.$$

These conditions are satisfied with $p_n \sim n^{-2}(\log n)^{-\alpha-1}L(\log n)$.

3. Before going into the proof of THEOREM we state several preliminary lemmas.

LEMMA 1. *If* $m < 1$, *then*

$$G(s) = \lim_{n\to\infty} \frac{1-f_n(s)}{1-f_n(0)} \qquad \textit{exists for} \quad 0 \leq s < 1,$$

and

$$-G'(s) = \lim_{n\to\infty} \frac{f_n'(s)}{1-f_n(0)} > 0 \qquad \textit{for} \quad 0 < s < 1.$$

PROOF. It is known that G(s) is well defined in the above formula because of the increasingness and the boundedness of the defining sequence of functions (c.f. [1]). Since $1-f_n(s)$ is analytic we get the latter formula except $G' < 0$. By the definition of G we find the functional equation $G(f(s)) = mG(s)$, which shows G(s) is not constant because $m < 1$. Therefore $-G'(s) > 0$ in the interval (0,1), because $-G'(s)$ is non-negative, analytic and non-decreasing.

LEMMA 2. $\sum p_j(j\log j) = \infty$ *if and only if* $\int_0^\infty (m - f'(1-r^t))dt = \infty$ *for some* $0 < r < 1$.

PROOF. Since $\int_0^\infty (m - f'(1-r^t))dt = \sum_{k=1}^\infty kp_k \int_0^\infty (1 - (1-r^t)^{k-1})dt$, it suffices to see that $\int_0^\infty (1 - (1-r^t)^{k-1})dt \sim \text{const.}\log k$ as $k \to \infty$. But $(-\log r)\int_0^\infty (1-(1-r^t)^k)dt = \int_0^1 (1-(1-v)^k)v^{-1}dv = 1 + \frac{1}{2} + \frac{1}{3} + \cdots + \frac{1}{k} \sim$ $\sim \log k$ as $k \to \infty$.

The next lemma is part of the standard textbook literature (see W.Feller[2] $_p422, _p237$).

LEMMA 3. *Suppose that* L(x) *is slowly varying at infinity and* u(x) *a non-negative function on* $[0\ \infty)$. *Put*

$$U(t) = \int_0^t u(x)dx \quad \textit{and} \quad U^*(t) = \int_t^\infty u(x)dx.$$

Then the condition that $u(x) \sim t^{d}L(t)$ *as* $t \to \infty$ *implies*

$$U(t) \sim (d+1)^{-1}t^{d+1}L(t) \quad \text{as } t \to \infty \quad \text{when } d+1 > 0,$$

and

$$U^{*}(t) \sim |d+1|^{-1}t^{d+1}L(t) \quad \text{as } t \to \infty \quad \text{when } d+1 < 0.$$

Moreover if $u(x)$ *is monotonic, the converse assertions also hold.*

LEMMA 4. (1) *is equivalent to*

$$(4) \qquad m - f'(1-v) \sim (-\log v)^{-\alpha}L(-\log v) \quad \text{as } v \downarrow 0$$

where α *may be an arbitrary real number.*

PROOF. Observe the equation

$$m - f(s) = \sum_{j=2}^{\infty} jp_j(1 - s^j) = (1-s)\sum_{k=0}^{\infty}\Big(\sum_{j=k+2}^{\infty} jp_j\Big)s^k$$

and then apply a Tauberian theorem to coefficients $a_k = \sum_{j=k+2}^{\infty} jp_j$.

4. We now prove THEOREM. Let $m < 1$. Suppose $m^{-n}(1 - f_n(0)) \to 0$ as $n \to \infty$. Then we can see, by LEMMA 1,2 and the equation

$$\frac{f_n'(s)}{m^n} = \prod_{k=0}^{n-1}\frac{f'(f_k(s))}{m},$$

that as $n \to \infty$

$$(5) \qquad -\log\frac{1 - f_n(s)}{m^n} \sim -\sum_{k=0}^{n-1}\log\Big(1 - \Big(1 - \frac{f'(f_k(s))}{m}\Big)\Big) \sim \sum_{k=0}^{n-1}\Big(1 - \frac{f'(f_k(s))}{m}\Big) \qquad 0 < s < 1.$$

Taking any tow numbers r and s, $0 < r < m$, $0 < s < 1$, fixed,

$$1 - m^k \leq f_k(s) \leq 1 - r^k \qquad k > k_0$$

with some integer k_0, and hence

$$(6) \qquad \int_0^n\Big(1 - \frac{f'(1-m^t)}{m}\Big)dt \geq \sum_{k=k_0}^{n}\Big(1 - \frac{f'(f_k(s))}{m}\Big) \geq \int_{k_0}^{n}\Big(1 - \frac{f'(1-r^t)}{m}\Big)dt.$$

By LEMMA 2 we get $\sum p_j(j\log j) = \infty$. Conversely, from the above argument, we see that $\sum p_j(j\log j) = \infty$ implies $\lim_{n\to\infty}(1 - f_n(s))m^{-n} = 0$. Therefore , for the proof of THEOREM these two conditions may be always assumed.

From (6) it follows that

$$(7) \qquad \sum_{k=0}^{n}\Big(1 - \frac{f'(f_k(s))}{m}\Big) \sim \int_0^n\Big(1 - \frac{f'(1-m^t)}{m}\Big)dt \qquad \text{as } n \to \infty.$$

Indeed, setting $a_n(x) = \int_0^n\Big(1 - \frac{f'(1-x^t)}{m}\Big)dt = \int_{x^n}^{1}\Big(1 - \frac{f'(1-v)}{m}\Big)\frac{dv}{v\log v}$, $0 < x < 1$, and $b_n(s) = \sum_{k=0}^{n}\Big(1 - \frac{f'(f_k(s))}{m}\Big)$, we get $a_n(r)/a_n(m) \geq$

$\geq \log m/\log r$ and hence, by (6), $1 \geq \overline{\lim}_{n\to\infty} b_n / a_n(m) \geq \underline{\lim}\, b_n / a_n(m) \geq \log m/\log r$, which leads to $\lim b_n / a_n(m) = 1$ since r can be taken arbitrarily near m.

Because $1 - f'(1-r^x)$ is a monotone function of x, the converse part of LEMMA 3 is applicable to it. Using LEMMA 4 it is seen that (1) is equivalent to

$$(8) \qquad \int_0^n \left(1 - \frac{f'(1-r^t)}{m}\right) dt \sim n^{1-\alpha} L(n) \frac{(-\log r)^{-\alpha}}{m(1-\alpha)} \quad \text{as } n \to \infty.$$

By (5) and (7), we have that (2) is also equivalent to (8). Thus (a) has been proved.

The proof of (b) is performed along the same lines as in the above. Let $m > 1$. Denote by $g(x)$ the inverse function of $f(x)$ defined for $q < x < 1$ where q is the smallest non-negative root of $f(s) = s$, and by $g_n(x)$ its n-times iteration. Let s_o be any fixed number between q and 1. Then c_n in section 1 can be taken as $(-\log g_n(s_o))^{-1}$. Thus $c_n^{-1} \sim 1 - g_n(s_o)$, and (3) is equivalent to

$$m^n \log(1 - g_n(s_o)) \sim A n^{1-\alpha} L(n) .$$

Corresponding to LEMMA 1 we have that for $q < s < 1$

$$\lim \frac{1 - g_n(s)}{1 - g_n(s_o)} = H(s) \quad \text{exists and} \quad \lim \frac{g_n'(s)}{1 - g_n(s_o)} = -H'(s) > 0 .$$

Observe that

$$g_n'(s) = 1 / \prod_{k=1}^{n} f'(g_k(s))$$

and that

$$r^k \geq 1 - g_k(s_o) \geq m^{-k} \quad \text{for } k > k_o$$

where r is an arbitrarily fixed number in $(m^{-1}, 1)$ and k_o some constant. Then imitate the proof of (a) to see that

$$\log[(1 - g_n(s_o))m^n] \sim \sum_{k=0}^{n-1} \left(1 - \frac{f'(g_n(s_o))}{m}\right)$$

$$\sim \int_0^n \left(1 - \frac{f'(1-m^t)}{m}\right) dt \quad \text{as } n \to \infty.$$

It is clear that (b) follows from these formulas as in the case $m < 1$.

References.

[1] K.B.Athreya-P.E.Ney, Branching Processes. Springer, Berlin.(1973)

[2] W.Feller, An introduction to probability theory and its application. Vol.2, Wiley, New York.(1966)

Construction of diffusion processes by means of Poisson point process of Brownian excursions.

Shinzo Watanabe

Department of Mathematics, Kyoto University

In this note, we shall show how diffusion processes with various boundary conditions are constructed from a given Poisson point process of Brownian excursions. For simplicity, we consider a case when the minimal diffusion process is a Brownian motion and when there is no discontinuity of path functions on the boundary, through our method can be applied to more general cases. Also, this note is only a summary: a full exposition with complete proofs in the most general case will be given elsewhere.

1. Poisson point process of Brownian excursions.

Let $\{X, \mathcal{B}(X)\}$ be a measurable space. By a point function p on X, we mean a map $p : D_p \subset (0, \infty) \to X$, where the domain D_p is a countable set of $(0, \infty)$. p defines a counting measure $N_p(dt, dx)$ on $(0, \infty) \times X$ by

$$N_p((0, t] \times U) = \# \{s \in D_p \ ; \ s \leq t, \ p(s) \in U\} \qquad t > 0, \ U \in \mathcal{B}(X).$$

Let Π_X be the set of all point functions on X and $\mathcal{B}(\Pi_X)$ be the smallest σ-algebra with respect to which, all $N_p((0, t] \times U)$, $t > 0$, $U \in \mathcal{B}(X)$, are measurable.

Let $(\Omega, \mathcal{F}, P)$ be a probability space. A measurable map $p : \Omega \to \Pi_X$ is called a point process on X. Let $\{\mathcal{F}_t\}_{t\in[0,\infty)}$ be a right continuous, increasing family of sub σ-fields of $\mathcal{F}$. A point process p is called $\mathcal{F}_t$-adapted if $N_p((0, t] \times U)$ is $\mathcal{F}_t$-measurable for each t, $(\forall U \in \mathcal{B}(X))$. Let $n(dx)$ be a σ-finite measure on $(X, \mathcal{B}(X))$. A point process p is called $\mathcal{F}_t$-stationary Poisson point process with characteristic measure n if it is $\mathcal{F}_t$-adapted and if, for every $U \in \mathcal{B}(X)$ such that $n(U) < \infty$, $t \leadsto N_p((0, t] \times U) - t \cdot n(U)$ is an $\mathcal{F}_t$-martingale. It is easy to see, then, that

$$(1.1) \qquad E(e^{-\sum_{i=1}^{n} \lambda_i N_p((s, t] \times U_i)} / \mathcal{F}_s) = \exp\{ (t-s) \sum_{i=1}^{n} n(U_i)(e^{-\lambda_i} - 1)\}$$

for every $t > s \geq 0$, $\lambda_i > 0$, $U_i \in \mathcal{B}(X)$, $n(U_i) < \infty$, $i = 1, 2, \cdots, n$, such that $\{U_i\}$ are disjoint. Given a σ-finite measure n, there exists a unique (in the law sense) $\mathcal{F}_t$-stationary Poisson point process with n as its characteristic measure on a suitable quadruplet $(\Omega, \mathcal{F}, P: \mathcal{F}_t)$.

Let $D = \{x = (x_1, x_2, \cdots, x_n) \ R^n \ ; \ x_1 \geq 0\}$ and $\partial D = \{x \in D \ ; \ x_1 = 0\}$.

Let

$$(1.2) \qquad W_0(D) = \{ w : [0, \infty) \ni t \leadsto w(t) \in D, \ \text{continuous}, \ w(0) = 0 \ \text{and} \ w(t \wedge \sigma(w)) = w(t) \}$$

where

(1.3) $$\sigma(w) = \inf\{\, t > 0; \quad w(t) \in \partial D\} .$$

Let

(1.4) $$K(t, x) = \sqrt{\frac{2}{\pi t^3}}\, x_1\, e^{-\frac{x_1^2}{2t}} \prod_{i=2}^{n} \frac{1}{\sqrt{2\pi t}}\, e^{-\frac{x_i^2}{2t}} \qquad t > 0 , \quad x \in \mathring{D}$$

and

(1.5) $$P^0(t, x, y) = \frac{1}{\sqrt{2\pi t}} \left(e^{-\frac{(x_1-y_1)^2}{2t}} - e^{-\frac{(x_1+y_1)^2}{2t}}\right) \prod_{i=2}^{n} \frac{1}{\sqrt{2\pi t}}\, e^{-\frac{(x_i-y_i)^2}{2t}}$$

$$t > 0, \quad x, y \in \mathring{D}.$$

It is well known that there exists a (unique) σ-finite measure Q on $W_0(D)$ (with σ-field generated by Borel cylinder sets) such that

(1.6) $$Q\{w ; w(t_1) \in E_1, w(t_2) \in E_2, \cdots, w(t_m) \in E_m, \sigma(w) > t_m\}$$
$$= \int_{E_1} K(t_1, x_1)dx_1 \int_{E_2} P^0(t_2-t_1, x_1, x_2)dx_2 \int \cdots \int_{E_m} P^0(t_m-t_{m-1}, x_{m-1}, x_m)dx_m$$
$$0 < t_1 < t_2 < \cdots < t_m, \quad E_i \in \mathcal{B}(\mathring{D}).$$

A (F_t-adapted) stationary Poisson point process on $W_0(D)$ with Q as its characteristic measure is called the Poisson point process of n-dimensional Brownian excursions.

2. Construction of diffusions.

We would like to construct on D a diffusion process which is n-dimensional Brownian motion (a diffusion process with generator $\frac{1}{2}\Delta$) before it hits the boundary ∂D and, on ∂D, satisfies the following boundary condition:

(2.1) $$\frac{1}{2} \sum_{i,j=2}^{n} a^{ij}(x) \frac{\partial^2 u}{\partial x_i \partial x_j}(x) + \sum_{j=2}^{n} b^i(x) \frac{\partial u}{\partial x_i}(x) + c(x) \frac{\partial u}{\partial x_1}(x)$$
$$= \rho(x)(\tfrac{1}{2}\Delta u)(x) , \qquad x \in \partial D,$$

where $a^{ij}(x)$, $b^i(x)$, $c(x)$ and $\rho(x)$ are functions defined on ∂D such that $\{a^{ij}(x)\}_{i,j=2}^{n}$ is non-negative definite, $c(x) \geq 0$, $\rho(x) \geq 0$ and $c(x) + \rho(x) \equiv 1$ on ∂D. Let $\{\alpha_k^i(x)\}_{i,k=2}^{n}$ be such that $a^{ij}(x) = \sum_{k=2}^{n} \alpha_k^i(x)\, \alpha_k^j(x)$ and assume that, $\alpha_k^i(x)$, $b^i(x)$, $c(x)$ and $\rho(x)$ are all Lipschitz continuous on ∂D.

Let

(2.2) $$W(D) = \{w : [0, \infty) \ni t \rightsquigarrow w(t) \in D \ : \ \text{continuous},$$
$$w(0) \in \partial D \quad \text{and} \quad w(t \wedge \sigma(w)) = w(t)\}$$

where $\sigma(w)$ is defined as in (1.3). Let, for $c \geq 0$, $T_c : W_0(D) \to W_0(D)$ be defined by

(2.3) $$(T_c w)(t) = \begin{cases} cw\left(\frac{t}{c^2}\right) & , c > 0 \\ 0 & , c = 0 \end{cases}$$

and let a map

$$\Phi : (x, w) \in \partial D \times W_0(D) \leadsto \Phi(x, w) \in W(D)$$

be defined by

(2.4) $$\Phi(x, w)(t) = x + (T_{c(x)} w)(t) \qquad (x \in \partial D, \quad t \geq 0).$$

Clearly,

(2.5) $$\Phi(x, w)(0) = x$$

and

(2.6) $$\sigma(\Phi(x, w)) = c^2(x)\sigma(w).$$

Let $\phi : (x, w) \in \partial D \times W_0(D) \leadsto \partial D$ be defined by

(2.7) $$\begin{aligned} \phi(x, w) &= \Phi(x, w)(\sigma(\Phi(x, w))) - x \\ &= c(x)\, w(\sigma(w)) . \end{aligned}$$

It is easy to see that, for $x, y \in \partial D$,

(2.8) $$\int_{W_0(D) \cap \{\sigma(w) \leq 1\}} |\phi(x, w) - \phi(y, w)|^2 \, Q(dw) = \sqrt{\frac{2}{\pi}}\, |c(x) - c(y)|^2 \leq K \cdot |x - y|^2$$

($K > 0$: const.).

Now, we construct path functions of the diffusion. We prepare, on a suitable quadruplet $(\Omega, F, P; F_t)$, the following :

(i) $G_t \subset F_0$: an increasing family of σ-fields and n-dim. G_t-Brownian motion $B(t) = (B^i(t))_{i=1}^n$,

(ii) (n-1)-dim. F_t-Brownian motion $B^*(t) = (B_*^k(t))_{k=2}^n$,

(iii) F_t-Poisson point process of n-dim. Brownian excursions, i.e. F_t-stationary Poisson point process on $W_0(D)$ with the characteristic measure Q.

Let the starting point $x \in D$ be given. We set, first of all,

(2.9) $$X^x(t) = x + B(t), \quad \text{for } t \leq \sigma,$$

where $\sigma = \inf \{ t \geq 0 ;\ x + B(t) \in \partial D \}$. Then $\xi_0 = X^x(\sigma)$ ($\in \partial D$) is F_0-measurable. Secondly, we solve the following stochastic differential equation of jump-type for the process $\xi(t)$ on ∂D :

$$(2.10)\quad \begin{cases} \xi^1(t)=0 \\ \xi^i(t)=\xi^i_0+\sum_{k=2}^{n}\int_0^t \alpha^i_k(\xi(s))\,dB^k_*(s)+\int_0^t b^i(\xi(s))ds \\ \qquad +\int_0^{t+}\int_{W_0(D)\cap\{\sigma(w)\leq 1\}}\phi^i(\xi(s-),w)[N_p(ds,dw)-ds\cdot Q(dw)] \\ \qquad +\int_0^{t+}\int_{W_0(D)\cap\{\sigma(w)>1\}}\phi^i(\xi(s-),w)\,N_p(ds,dw) \end{cases}$$

$i = 2, 3, \cdots, n.$

Thirdly, we set

$$(2.11)\qquad A(t)=\sigma+\int_0^{t+}\int_{W_0(D)}\sigma(\Phi(\xi(s-),w))\,N_p(ds,dw)+\int_0^t\rho(\xi(s))ds$$

$$=\sigma+\sum_{\substack{s\leq t\\ s\in D_p}}c^2(\xi(s-))\,\sigma(p(s))+\int_0^t\rho(\xi(s))ds.$$

Then, we can show that, with probability one, $t \rightsquigarrow A(t)$ is right continuous, strictly increasing and $\lim_{t\uparrow\infty} A(t) = \infty$. For every $t\geq 0$, there exists unique s such that $A(s-)\leq t\leq A(s)$. If $s = 0$, i.e. $0\leq t\leq\sigma$, $X^x(t)$ is already defined by (2.9). If $s>0$ and $A(s-)< A(s)$, then this implies $s\in D_p$ and we set

$$(2.12)\qquad X^x(t) = \Phi(\xi(s-), p(s))\,(t - A(s-)).$$

If $s>0$ and $A(s-) = A(s)$, then $\xi(s-) = \xi(s)$ and we set

$$(2.13)\qquad X^x(t) = \xi(s).$$

Thus we have defined the path functions $\{X^x(t)\}$ and this is the diffusion we want.

3. Extensions.

$1^0)$ Two sided boundary. Let $D^+ = \{x\in R^n;\ x_1\geq 0\}$, $D_- = \{x\in R^n;\ x_1\leq 0\}$ and $\{\partial D = x\in R^n;\ x_1 = 0\}$. Let $W_0(D^+)$ and $W_0(D^-)$ be defined in a similar way as in §2. Let Q^+ and Q^- be the σ-finite measures on $W_0(D^+)$ and $W_0(D^-)$, respectively, defined as in §2. Let $W_0 = W_0(D^+)\cup W_0(D^-)$ as a sum and Q be the σ-finite measure, which, restricted on $W_0(D^+)$ (resp. $W_0(D^-)$) coincides with Q^+ (resp. Q^-). Instead of (2.4), we define $\Phi : \partial D\times W_0 \to W = W(D^+)\cup W(D^-)$ by

$$(3.1)\qquad \Phi(x,w)(t) = \begin{cases} x+T_{c_1(x)}\,w(t), & w\in W_0(D^+) \\ x+T_{c_2(x)}\,w(t), & w\in W_0(D^-) \end{cases}.$$

where $c_1(x)$ and $c_2(x)$ are non-negative functions on ∂D. We assume that $a^{ij}(x)$, $b^i(x)$, $c(x) = c_1(x) + c_2(x)$ and $\rho(x)$ satisfy the same conditions as in §2. We do the same construction as above. Then we have a sample path $X^x(t)$ of diffusion process which is the n-dim. Brownian motion before it hits ∂D and, on ∂D, satisfies the boundary condition

$$(3.2)\quad \frac{1}{2}\sum_{i,j=2}^{n} a^{ij}(x)\frac{\partial^2 u}{\partial x_i \partial x_j}(x) + \sum_{j=2}^{n} b^i(x)\frac{\partial u}{\partial x_i}(x) + c_1(x)\frac{\partial u}{\partial x_1}(x_+) - c_2(x)\frac{\partial u}{\partial x_1}(x_-)$$
$$= \rho(x)(\tfrac{1}{2}\Delta u)(x), \qquad x \in \partial D,$$

where $\frac{\partial u}{\partial x_1}(x_+) = \lim_{x_1 \downarrow 0} \frac{\partial u}{\partial x_1}(x_1, x_2, \cdots, x_n)$ and $\frac{\partial u}{\partial x_1}(x_-) = \lim_{x_1 \uparrow 0} \frac{\partial u}{\partial x_1}(x_1, x_2, \cdots, x_n)$.

2^0) Minimal diffusion with variable coefficients. If we want to construct the diffusion process for which the minimal diffusion (i.e. the process before hitting the boundary ∂D) is more general than the Brownian motion, we define the mapping $\Phi(x, w) : (x, w) \in \partial D \times W_0(D) \rightsquigarrow \Phi(x, w) \in W(D)$ by solving a stochastic differential equation on the measure space $\{W_0(D), Q\}$ (which can be treated similarly as by the ordinary Itô calculus and $\Phi(x, w)$ is determined as a strong solution of the equation). Also, we can treat the case when jumps take place on the boundary by introducing so called Poisson point process of the Brownian excursions of the second kind. Details will appear in a future publication.

NON-ANTICIPATING SOLUTIONS OF STOCHASTIC EQUATIONS

M. P. Yershov

0. Introduction

0.1. Mathematical models of real systems, evolving in time, usually satisfy certain "physical" conditions.

One of these conditions is connected with non-reversibility of time, with a sort of "causality principle": the consequence cannot proceed its cause.

Another condition reflects the fact that it is impossible to precisely predict, "anticipate", the future if the system is sufficiently complicated.

Thus, the noise or random disturbances (ie the disturbances which are caused by such a large number of various factors that all of them cannot be taken into account deterministically, and one is forced to consider them as random) are, by their nature, to possess the property that their behaviour in the "future" cannot be precisely described on the basis of information about the "past" evolution of the system.

The same property of non-anticipation is used in such mathematical constructions as the Itô stochastic integral with respect to the Wiener process or, more generally, the stochastic integral with respect to martingales. These constructions have, in fact, arisen as tools of describing certain "physical" processes.

The causality property is, in turn, closely connected, e.g., with the notion of stopping time, which also has been motivated by "physical" considerations.

But despite the fact that these properties are fundamental and that they are intensively used in various theoretical problems and applications, there are no accurate general definitions of them in the literature.

In this paper, we will make an attempt to give such definitions and consider the unavoidable question of solving "causal" stochastic equations, so that the solution does not "anticipate" the noise and hence can be, in principle, used in mathematical models of "physical" systems.

<u>0.2.</u> Let us consider some simple although, in our opinion, characteristic examples.

<u>Example A.</u> In the literature, models of systems are often considered which are described by equations of the form

$$(0.1) \qquad \xi_n = g_n(\xi_0, \xi_1, \dots, \xi_{n-1}) + \eta_n$$
$$n = 1, 2, \dots$$

where g_n are given functions of n (one-dimensional, say) variables, ξ_n is interpreted as the state of the system at time n and $\{\eta_n\}_{n=1,2,\dots}$ is noise.

Usually (0.1) is considered as an equation with respect to $\xi = \{\xi_n\}_{n=1,2,\dots}$ with given $\eta = \{\eta_n\}_{n=1,2,\dots}$ (ξ_0 is a given "initial condition").

The "causality principle" here is the fact that the state of the system at time n does not depend (functionally) on its states at future moments.

And, in accordance with the "non-anticipation principle", it should be not possible to extrapolate $\eta_{n+1}, \dots$ from known $\xi_0, \dots, \xi_n$ better than from known $\eta_1, \dots, \eta_n$ only. In other words, only those solutions of (0.1) which do not "anticipate" the noise are considered adequate.

Of course, it should be explained how to compare different extrapolations. For this purpose, we will use, as one most often does, conditional expectations. For instance, the "non-anticipation principle" in the example under consideration may be defined by the equalities:

$$E(f(\eta)|\xi_0,\dots,\xi_n) = E(f(\eta)|\eta_1,\dots,\eta_n) \tag{0.2}$$

for each f from some set of test functions.

<u>Example B.</u> In the continuous time case, the analogue of (0.1) is obvious:

$$\xi_t = g_t(\xi_s ; 0 \leq s \leq t) + \eta_t$$
$$t \in [0, \infty)$$

where the "causality principle" and the "non-anticipation principle" have the same interpretation.

<u>Example C.</u> Let $\xi = \{\xi_t\}_{t \geq 0}$ be a continuous martingale. It can be shown)[x] that there exists a mapping A of the space of continuous functions $C = C[0, \infty)$ into itself such that

)[x] The results, implying this statement, are indicated in the author's paper [1] , sections 1.1 - 1.6.

(i) $(A\circ x)_t$, $x=\{x_s\}_{s\in[0,\infty)}\in C$, for each t, depends only on $\{x_s\}_{s\in[0,t]}$;

(ii) $A\circ\xi=\langle\xi\rangle$ (the quadratic variation) with probability 1;

(iii) the process $\eta=\{\eta_t\}_{t\in[0,\infty)}$:

(0.3) $$\eta_t=\xi_{A_t^*(\xi)},$$

where

$$A_t^*(x)=\begin{cases}\infty & \text{if } (A\circ x)_u<t \ \forall u,\\ \inf\{u:(A\circ x)_u=t\} & \text{otherwise,}\end{cases}$$

is a Wiener process with respect to the flow of σ-algebras $\{\mathcal{A}_t\}_{t\geq 0}$ ($\mathcal{A}_t$ is the σ-algebra generated by the random variables $\xi_{s\wedge A_t^*(\xi)}$, $s\geq 0$) stopped at the moment

$$(A\circ\xi)_\infty=\lim_{t\to\infty}(A\circ\xi)_t.$$

In this example, equation (0.3) (with respect to ξ) contains "causal" transformation

$$x\longmapsto\{x_{A_t^*(x)}\}_{t\geq 0}$$

and does not anticipate η in the sense that the latter is a martingale with respect to the flow of σ-algebras generated by the process with, so to say, "proper" time $A_t^*(\xi)$. When defining "non-anticipation" with the help of conditional expectations as in (0.2), here is the "translation" of the above property:

$$\forall s \qquad E(f(\eta)\mid\mathcal{A}_s)=E(f(\eta)\mid\eta_u, 0\leq u\leq s)$$

where the "coordinate" functions $f:\eta\longmapsto\eta_t$ are taken as test functions.

<u>0.3.</u> In what follows we will deal only with stochastic equations of the "standard" form

$$F \circ \xi = \eta \tag{0.4}$$

where η is a known stochastic process, F is a given transformation, ξ is a solution to be found.

Equation (0.3) is already of this form and the equations in examples 0.2 A and 0.2 B can be trivially solved with respect to the "noise". However, many other equations can be reduced to the standard form (0.4). For instance, in the case of a relatively general equation

$$F \circ (\xi, \eta) = 0$$

one can consider the extended system: $\widetilde{\xi} = ((\xi, \zeta), \zeta)$ and the new "noise": $\widetilde{\eta} = (0, \eta)$, replacing thus the original problem by that for the standard equation:

$$\widetilde{F} \circ \widetilde{\xi} = \widetilde{\eta}$$

where $\widetilde{F} = F \times$ [identical transformation].

In the last 10 - 15 years a series of new and important existence theorems have been obtained in the theory of stochastic equations. These theorems are connected with the abandoning of a scheme of sequential approximations in favour of approximating the original equation by simpler ones and checking the compactness, in a certain sense, of the set of their solutions.

However, in this approach it is necessary to change the idea of a solution, even to change the idea of stochastic equations themselves. The problem is that the compactness of the approximations can not be established in the sense of convergence of trajectories but only in the sense of convergence of distributions. Therefore, in the new approach, stochastic processes (or random elements) are considered as processes in the wide sense, ie probability measures in the sample space.

It is therefore natural to write stochastic equations in terms of distributions from the very beginning. For instance, standard equation (0.4) can be rewritten in the form

$$F \circ \mu = \nu \tag{0.5}$$

where ν is a (given) measure on the sample space ($Y, \mathcal{Y}$) of the random element η , μ is a measure to be found (a solution) on the sample space ($X, \mathcal{X}$) of ξ (note that, although ξ is not available yet, it must be, if possible, found, we are given its sample space: it is the domain of F), and F maps measures μ on (X , $\mathcal{X}$) into measureson(Y , $\mathcal{Y}$) by the formula

$$[F \circ \mu](B) = \mu(F^{-1} \circ B) \qquad \forall B \in \mathcal{Y}.$$

To each solution of problem (0.4), there corresponds, in an obvious way, a solution of problem (0.5), but not vice versa. In fact, the "initial" data in problem (0.4) are a probability space and a random element η on it, and it may turn out that there is no random element on this probability space with values in (X , $\mathcal{X}$) and the distribution μ , being a solution of problem (0.5), which is transferred by F into η . In other words, the given probability space may be insufficiently rich.

Note that, using the new approach, we are, actually, solving problem (0.5) instead of (0.4).

In this paper, we will give definitions of causality and non-anticipation for "standard" stochastic equations (0.5) and investigate the problem of existence of their non-anticipating solutions.

Concrete applications will be not dealt with here (except example 3.3). Note, however, that we have already used, in essence, the method described below in [1] to prove the existence of a non-anticipating solution of equation (0.3), ie the existence of a continuous martingale with given diffusion functional (squared variation). (Cf. also [2] , Theorems 1 and 2).

0.5. Equations of the type (0.5) were studied by the author in [3] - [6] where a number of general existence theorems were obtained. For reader's convenience, we state here two results of these papers, which will be referred to in the sequel. First, we remind two definitions.

A measurable space ($X, \mathcal{X}$) with countably generated σ-algebra $\mathcal{X}$ is said to be a Blackwell space if the image of X with respect to any measurable mapping into the Borel line is an analytic set.

A σ-algebra $\mathcal{Y}$ in a space Y is said to be ν-countably generated, where ν is a measure on ($Y, \mathcal{Y}$), if there exists a countably generated σ-algebra $\mathcal{Y}_0 \subset \mathcal{Y}$ such that its completion with respect to ν (by subsets of ν-null sets from $\mathcal{Y}_0$!) contains $\mathcal{Y}$.

Theorem A)^x. Let ($X, \mathcal{X}$) be a Blackwell space, F be a measur-

)^x Theorem 4.1, Chapter II in [6] ; it can be also obtained from Theorem 2.6 of [4] by taking into account the fact that Blackwell spaces ($X, \mathcal{X}$), under the canonical mapping onto the space of $\mathcal{X}$-atoms, turn into (up to the isomorphism of measurable spaces) analytic sets in $\mathbb{R}^1$ with the trace of the Borel σ-algebra.

able transformation of ($X, \mathcal{X}$) into a measurable space ($Y, \mathcal{Y}$) with a measure ν on it such that the σ-algebra $\mathcal{Y}$ is ν-countably generated.

The condition

$$[B \in \mathcal{Y}, \nu(B) \neq 0] \Rightarrow [F^{-1} \circ B \neq \emptyset] \tag{0.6}$$

is necessary and sufficient for the stochastic equation

$$F \circ \mu = \nu \qquad [F: (X, \mathcal{X}) \to (Y, \mathcal{Y})] \quad)^x$$

to have a solution.

Theorem A is equivalent to the following ([3] - [6])

Theorem B)[xx] . Let ($X, \mathcal{X}_0, \mu_0$) be a measure space and the σ-algebra $\mathcal{X}_0$ be μ_0-countably generated.

The measure μ_0 can be extended to any σ-algebra $\mathcal{X} \supset \mathcal{X}_0$ such that ($X, \mathcal{X}$) is a Blackwell space.

0.6. Basic notation. The sign $\subset$ always denotes non-strict inclusion.

Let μ be a measure on a measurable space ($X, \mathcal{X}$). The set function

$$\mathrm{Out}_{\mathcal{X}}\, \mu(A) = \inf_{A \subset B \in \mathcal{X}} \mu(B)$$

on the power set $\mathcal{P}(X)$ is called the outer measure corresponding to ($\mathcal{X}; \mu$).

The restriction of a measure μ onto $\mathcal{X}_0 \subset \mathcal{X}$ is denoted by $\mu | \mathcal{X}_0$.

)[x] The writing in the brackets, which will be also used in the sequel, underlines that solutions μ must have $\mathcal{X}$ as their domain.

)[xx] Independently, this theorem has been also obtained in [7] and [8] .

The σ-algebra, generated by $\mathcal{X}_0$ and all $A \subset X$ with $\mathrm{Out}_{\mathcal{X}}\,\mu(A)=0$, is termed the completion of the σ-algebra $\mathcal{X}_0$ with respect to the system ($\mathcal{X};\mu$) and is denoted by $\mathcal{X}_0(\mathcal{X};\mu)$. If $\mathcal{X}_0=\mathcal{X}$, we will simply write $\mathcal{X}_0(\mu)$ instead of $\mathcal{X}_0(\mathcal{X};\mu)$.

We will write $A\subset B$ (mod μ) if $\mu(A\setminus B)=0$.

$$[A=B \pmod{\mu}] \Longleftrightarrow [A\subset B \pmod{\mu} \ \&\ A\supset B \pmod{\mu}].$$

The inclusion $\mathcal{X}_0\subset\mathcal{X} \pmod{\mu}$ means that

$$\forall A\in\mathcal{X}_0 \quad \exists B\in\mathcal{X}: \ A=B \pmod{\mu}.$$

$$[\mathcal{X}_0=\mathcal{X} \pmod{\mu}] \Longleftrightarrow [\mathcal{X}_0\subset\mathcal{X} \pmod{\mu} \ \&\ \mathcal{X}_0\supset\mathcal{X} \pmod{\mu}].$$

1. Causal transformations. Non-anticipating solutions of stochastic equations

1.1. Our definition of causal transformations is based on formalization of properties of "physical" systems which transform accumulating information.

Let X and Y be some sets and let, to each t from a linearly ordered by a relation $\leqslant$ set T, there correspond σ-algebras $\mathcal{X}_t$ and $\mathcal{Y}_t$ of sets in X and Y respectively so that the families

$$\mathfrak{X}=\{\mathcal{X}_t\,;t\in T\}\,,\quad \mathfrak{Y}=\{\mathcal{Y}_t\,;t\in T\}$$

form flows of σ-algebras:

$$[s\leqslant t] \Longrightarrow [\mathcal{X}_s\subset\mathcal{X}_t \ \&\ \mathcal{Y}_s\subset\mathcal{Y}_t].$$

Definition. A transformation $F: X\to Y$ is called *causal with respect to the ordered pair* ($\mathfrak{X},\mathfrak{Y}$) if

$$F^{-1}\circ\mathcal{Y}_t\subset\mathcal{X}_t \qquad \forall t\in T.$$

If it is clear from the context what flows are meant, we shall simply say: *causal transformation*.

Interpreting t as time and $\mathcal{X}_t$ and $\mathcal{Y}_t$ as information accumulated respectively at the "input" and the "output" of the "transformator" F by the moment t, the causality of F means that any new "input" information, which will be available in the future with respect to the moment t, will not affect the "output" information accumulated by this moment.

In the rest of this section, F is always assumed to be a causal transformation.

1.2. Put

$$\mathcal{X} = V_t \mathcal{X}_t, \quad \mathcal{Y} = V_t \mathcal{Y}_t .$$

Let ν be a measure)[x] on $\mathcal{Y}$.

Definition. A solution μ_* of the stochastic equation

$$F \circ \mu = \nu \qquad [F : (X, \mathcal{X}) \to (Y, \mathcal{Y})]$$

is called causal with respect to ($\mathfrak{X}, \mathfrak{Y}$) if

$$\mathcal{X}_t = F^{-1} \circ \mathcal{Y}_t \quad (\mathrm{mod}\ \mu_*) \quad \forall t \in T.$$

If it is clear from the context what flows are meant, we shall simply say: a causal solution.

Remark. Any causal solution is obviously strong.

Proposition. The causality of μ_* is equivalent to the condition

$$(1.1) \qquad \mu_*(A|\mathcal{X}_t) = \mu_*(A|F^{-1} \circ \mathcal{Y}_t) \quad (\mathrm{mod}\ \mu_*) \quad \forall t \in T, \forall A \in \mathcal{X}.$$

Proof. Let μ_* be a causal solution. For any $\tilde{A} \in \mathcal{X}_t$, there exists a $\tilde{B} \in \mathcal{Y}_t$ such that, $\tilde{A} = F^{-1} \circ \tilde{B}$ $(\mathrm{mod}\ \mu_*)$ and, by the definition of conditional probabilities, $\forall A \in \mathcal{X}$,

$$\int_{\tilde{A}} \mu_*(A|\mathcal{X}_t)\, d\mu_* = \mu_*(A \cap \tilde{A}) = \mu_*(A \cap F^{-1} \circ \tilde{B}) = \int_{F^{-1} \circ \tilde{B}} \mu_*(A|F^{-1} \circ \mathcal{Y}_t)\, d\mu_*$$
$$= \int_{\tilde{A}} \mu_*(A|F^{-1} \circ \mathcal{Y}_t)\, d\mu_* \Longrightarrow (1.1)$$

)[x] Here and in what follows, as a rule, probability measures only are considered.

To prove the converse, let (1.1) be satisfied and $\tilde{A} \in \mathcal{X}_t$.

Then

$$\mathbb{1}_{\tilde{A}} = \mu_*(\tilde{A} \mid F^{-1} \circ \mathcal{Y}_t) \quad (\text{mod}\ \mu_*)$$

and hence

$$\tilde{A} = F^{-1} \circ \tilde{B} \quad (\text{mod}\ \mu_*)$$

where $\tilde{B} \in \mathcal{Y}_t$ is such that

$$F^{-1} \circ \tilde{B} = \{x \in X : \mu_*(\tilde{A} \mid F^{-1} \circ \mathcal{Y}_t) = 1\} \quad (\text{mod}\ \mu_*) \ \blacksquare$$

1.3. Definition. A solution μ_* of the stochastic equation

$$F \circ \mu = \nu \qquad [F : (X, \mathcal{X}) \to (Y, \mathcal{Y})]$$

is called non-anticipating with respect to $(\mathfrak{X}, \mathfrak{Y})$ if

$$\mu_*(F^{-1} \circ B \mid \mathcal{X}_t) = \mu_*(F^{-1} \circ B \mid F^{-1} \circ \mathcal{Y}_t) \quad (\text{mod}\ \mu_*)$$

$$\forall t \in T \quad \forall B \in \mathcal{Y}.$$

If it is clear from the context what flows are meant, we shall simply say: a non-anticipating solution.

Remark. By Proposition 1.2, any causal solution is non-anticipating.

Proposition. Any strong non-anticipating solution is causal.

Proof. Let $A \in \mathcal{X}_t$ and let $B \in \mathcal{Y}$ be such that $A = F^{-1} \circ B$ $(\text{mod}\ \mu_*)$. By the definition of conditional probabilities,

$$\mu_*(A \,\Delta\, F^{-1} \circ B \mid \mathcal{X}_t) = 0 \quad (\text{mod}\ \mu_*)$$

$$\Longrightarrow \mu_*(A \mid \mathcal{X}_t) = \mathbb{1}_A = \mu_*(F^{-1} \circ B \mid \mathcal{X}_t) \quad (\text{mod}\ \mu_*).$$

Since μ_* is non-anticipating,

$$\mathbb{1}_A = \mu_*(F^{-1} \circ B \mid F^{-1} \circ \mathcal{Y}_t) \quad (\text{mod}\ \mu_*)$$

i.e.

$$A = F^{-1} \circ \tilde{B} \quad (\text{mod}\ \mu_*)$$

where $\tilde{B} \in \mathcal{Y}_t$ is such that

$$F^{-1}\circ\tilde{B}=\{x\in X: \mu_*(F^{-1}\circ B\,|\,F^{-1}\circ \mathcal{Y}_t)=1\}\ (\mathrm{mod}\ \mu_*)\qquad\blacksquare$$

Taking Remark 1.2 into account, we come to the conclusion that the class of causal solutions coincides with the class of strong non-anticipating solutions.

1.4. Let $\mathfrak{F}$ be a class of real measurable integrable functions on $(Y, \mathcal{Y}, \nu)$.

Definition. A solution μ_* of the stochastic equation

$$F\circ\mu=\nu \qquad [F:(X,\mathcal{X})\longrightarrow(Y,\mathcal{Y})]$$

is called $\mathfrak{F}$-non-anticipating with respect to $(\mathfrak{X}, \mathfrak{Y})$ if

$$E_{\mu_*}(f\circ F\,|\,\mathcal{X}_t)=E_{\mu_*}(f\circ F\,|\,F^{-1}\circ\mathcal{Y}_t)\quad(\mathrm{mod}\ \mu_*)$$
$$\forall t\in T \quad \forall f\in\mathfrak{F}$$

where $E_{\mu_*}(f\circ F\,|\,\bullet)$ is the conditional expectation of the random variable $f\circ F$ on the probability space $(X, \mathcal{X}, \mu_*)$.

Remark. Taking for $\mathfrak{F}$ the class of indicators of sets from $\mathcal{Y}$, an $\mathfrak{F}$-non-anticipating μ_* is non-anticipating. In general, the latter property is stronger than the first.

2. Non-anticipating transformations

2.1. Let $F:X\longrightarrow Y$ be a causal with respect to $(\mathfrak{X}, \mathfrak{Y})$ transformation. Let us consider the stochastic equation

$$F\circ\mu=\nu \qquad [F:(X,\mathcal{X})\longrightarrow(Y,\mathcal{Y})] \qquad (\mathcal{X}=\vee_{t\in T}\mathcal{X}_t,\ \mathcal{Y}=\vee_{t\in T}\mathcal{Y}_t).$$

We shall be interested in the following problem:

Under what conditions does this equation have a causal, a non-anticipating or, more general, an $\mathfrak{F}$-non-anticipating solution for a specified class $\mathfrak{F}$ of real integrable functions on $(Y, \mathcal{Y}, \nu)$?

Of course, condition (0.6), necessary for the existence of a solution, holds in this problem, too. In the case of probability measures, this condition is equivalent to

$$\mathrm{Out}_{\mathcal{Y}}\, \nu(F\circ X) = 1.$$

In what follows, unless otherwise specified, we shall assume this condition to be satisfied.

2.2. Now we shall give a simple example of a solvable stochastic equation with a causal transformation, no solution of which is causal.

Example. Let X and Y be the square of a two-point set:

$$X = Y \overset{\cdot}{=} \{0,1\}^2 = \{(0,0), (0,1), (1,0), (1,1)\},$$

and let

$$T = \{1, 2\},$$

$$\mathcal{X}_1 = \mathcal{Y}_1 = \{\emptyset, \{(0,0)\}\cup\{(0,1)\}, \{(1,0)\}\cup\{(1,1)\}, \{0,1\}^2\},$$

$\mathcal{X}_2 = \mathcal{Y}_2$ be the power set of $\{0,1\}^2$,

$$\nu(\{(0,0)\}) = \nu(\{(0,1)\}) = \tfrac{1}{2},\ \nu(\{(1,0)\}) = \nu(\{(1,1)\}) = 0$$

and F be the transformation of X into Y defined by the equalities:

$$F\circ(0,0) = F\circ(0,1) = (0,0),\ F\circ(1,0) = F\circ(1,1) = (0,1).$$

The transformation F is causal (with respect to $(\mathfrak{X}, \mathfrak{Y})$):

(2.1) $$F^{-1}\circ\mathcal{Y}_1 = \{\emptyset, X\},$$

(2.2) $$F^{-1}\circ\mathcal{Y}_2 = \mathcal{X}_1.$$

The stochastic equation

$$F\circ\mu = \nu \qquad [F:(X,\mathcal{X})\to(Y,\mathcal{Y})]$$

has a solution: for instance, the one defined by

$$\mu(\{(0,0)\}) = \mu(\{(1,1)\}) = \tfrac{1}{2},$$

$$\mu(\{(0,1)\}\cup\{(1,0)\}) = 0.$$

Let $\mathfrak{F}$ consist of the only "coordinate" function

$$f\circ(y_1, y_2) = y_2$$

(one may say that this is the minimal non-trivial class in the case under consideration). Let us show that no solution μ of the equation in question is $\mathfrak{F}$-non-anticipating.

In fact, the function

$$f \circ F : X \to \{0,1\}$$

is $\mathcal{X}_1$-measurable:

$$(f \circ F)^{-1} \circ \{r\} = F^{-1} \circ f^{-1} \circ \{r\} \in F^{-1} \circ \mathcal{Y}_2 \overset{((2.2))}{=} \mathcal{X}_1 , \qquad r = 0, 1 .$$

Therefore, with μ-probability one,

$$E\{f \circ F \mid \mathcal{X}_1\} = f \circ F \circ x = x_1 ,$$

where $x = (x_1, x_2) \in X$, but

$$E\{f \circ F \mid F^{-1} \circ \mathcal{Y}_1\} \overset{((2.1))}{=} E\{f \circ F\} = 0 \cdot \nu(\{(0,0)\}) + 1 \cdot \nu(\{(0,1)\}) = \frac{1}{2} .$$

<u>2.3.</u> The absence of a non-anticipating solution in Example 2.2 is implied by the fact that the "transformator" F is "anticipating": the "output signal" at moment 2 is completely determined by the "input signal" at moment 1 (more precisely, the "input signal" anticipates the signal at the "output" of F).

We shall define now the notion of a non-anticipating transformation; for such transformations, it will be possible to prove theorems on the existence of non-anticipating solutions in a number of interesting cases.

<u>Definition.</u> A causal with respect to ($\mathfrak{X}, \mathfrak{Y}$) transformation is called <u>ν-non-anticipating (with respect to ($\mathfrak{X}, \mathfrak{Y}$))</u> if

$$(2.3)\qquad \forall t\in T\quad \forall A\in\mathcal{X}_t\quad \forall B\in\mathcal{Y}: F\circ A\subset B$$

$$\exists B'\in\mathcal{Y}_t:\ F\circ A\subset B'\quad \&\quad B'\subset B\ (\mathrm{mod}\ \nu).$$

Remark. Condition (2.3) is obviously weaker than the condition

$$(2.4)\qquad \forall t\in T\quad \forall A\in\mathcal{X}_t\qquad F\circ A\in\mathcal{Y}_t(\nu)$$

but, in general, is not weaker than

$$(2.4')\qquad \forall t\in T\quad \forall A\in\mathcal{X}_t\qquad F\circ A\in\mathcal{Y}_t(\mathcal{Y};\nu).$$

The non-anticipation property of the "transformator" F means, roughly speaking, that the set of all the possible "input signals" up to time t does not depend on whether we know the "output signal" up to time t only or on a longer time interval. In other words, one can say that the "input signal" in the past does not influence the set of all the possible "output signals" in the future, i.e. the "input" does not anticipate the future "output".

2.4. The meaning of the non-anticipation property is especially transparent in the following example.

Example. Let X as well as Y be the space of real continuous functions $x=\{x_t\}_{t\in T}$ and $y=\{y_t\}_{t\in T}$ on $T=[0,1]$ and let $\mathcal{X}_t$ and $\mathcal{Y}_t$ be the σ-algebras generated respectively by the mappings

$$\{x\mapsto x_s;\ s\le t\}\ \text{and}\ \{y\mapsto y_s;\ s\le t\}$$

into the Borel line. The causality (with respect to ($\mathfrak{X},\mathfrak{Y}$)) of F means that

$$\forall t\qquad [\pi_t\circ x=\pi_t\circ\tilde{x}]\Rightarrow[\pi_t\circ F\circ x=\pi_t\circ F\circ x']$$

where π_t is the restriction (mapping) onto $[0,t]$ (in X as well as in Y) and the ν-non-anticipation property, above defined, is equivalent to the condition

$$(2.5)\qquad \forall t\in T\ \ \forall A\in\mathcal{X}_t \quad \nu(\pi_t^{-1}\circ\pi_t\circ F\circ A\setminus F\circ A)=0,$$

i.e., for "almost all" $x\in X$,

$$F\circ\pi_t^{-1}\circ\pi_t\circ x=\pi_t^{-1}\circ\pi_t\circ F\circ x.$$

2.5. The transformation considered in Example 2.2 is not ν-non-anticipating: let us take $A=\{(0,0)\}\cup\{(0,1)\}\in\mathcal{X}_1$, then

$$F\circ A=\{(0,0)\}\notin\mathcal{Y}_1\,(\mathcal{Y}_2;\nu).$$

However, it is trivially $\tilde{\nu}$-non-anticipating for the measure $\tilde{\nu}$ on $\mathcal{Y}=\mathcal{Y}_2$ concentrated on the singleton (0,0). It is also easy to note that the measure $\tilde{\mu}$ on $\mathcal{X}=\mathcal{X}_2$ concentrated on the singleton (0,0) is a solution of the stochastic equation

$$F\circ\mu=\tilde{\nu}\qquad [F:(X,\mathcal{X})\to(Y,\mathcal{Y})].$$

This solution is causal:

$$\mathcal{X}_1=\{\emptyset,X\}=F^{-1}\circ\mathcal{Y}_1\ (\mathrm{mod}\,\tilde{\mu}),$$
$$\mathcal{X}_2=\{\emptyset,X\}=\mathcal{X}_1=F^{-1}\circ\mathcal{Y}_2\ (\mathrm{mod}\,\tilde{\mu}).$$

3. Theorem on the existence of a causal solution for a finite T

3.1. Theorem. Let T be a finite set $(T=\{t_k\}_{k=1,\dots,m}$, $t_1<\dots<t_m)$,

$$\mathfrak{X}=\{\mathcal{X}_t\}_{t\in T}\ \text{and}\ \mathfrak{Y}=\{\mathcal{Y}_t\}_{t\in T}$$

be flows of σ-algebras in X and Y respectively,

$$\mathcal{X}=\vee_{t\in T}\mathcal{X}_t=\mathcal{X}_{t_m}\ \text{and}\ \mathcal{Y}=\vee_{t\in T}\mathcal{Y}_t=\mathcal{Y}_{t_m},$$

let the σ-algebras $\mathcal{X}_t$ $(t\in T)$ be countably generated and such that $(X,\mathcal{X})$ is a Blackwell space and let the σ-algebras $\mathcal{Y}_t$ $(t\in T)$ be ν-countably generated where ν is a given measure on $(Y,\mathcal{Y})$.

Let $F : (X, \mathcal{X}) \to (Y, \mathcal{Y})$ be a $\mathcal{Y}$-non-anticipating with respect to ($\mathfrak{X}, \mathfrak{Y}$) transformation satisfying condition (0.6).

Then the stochastic equation

$$F \circ \mu = \nu \qquad [F : (X, \mathcal{X}) \to (Y, \mathcal{Y})]$$

has a causal with respect to ($\mathfrak{X}, \mathfrak{Y}$) solution.

Proof. Clearly we may, and will, assume that $t_k = k \quad (k = 1, \dots, m)$. Let us introduce, for convenience, the trivial σ-algebras $\mathcal{X}_0$ and $\mathcal{Y}_0$ in X and Y respectively.

Define inductively measures $\mu_k \ (k=0, \dots, m)$ on the σ-algebras

$$\mathcal{X}_k' = \mathcal{X}_k \vee F^{-1} \circ \mathcal{Y} \quad (k=0, \dots, m)$$

in X as follows:

$\mu_0 = F^{-1} \circ \nu$ (this definition is correct in view of condition (0.6));

if $\mu_k \quad (0 \leqslant k < m)$ such that

$$\mu_k \mid F^{-1} \circ \mathcal{Y} = \mu_0$$

and

$$\mathcal{X}_k = F^{-1} \circ \mathcal{Y}_k \ (\mathrm{mod}\ \mu_k)$$

is defined, put

$$\mu_{k+1}(A \cap F^{-1} \circ B) = M(A, B) = \int_A \mu_0 (F^{-1} \circ B \mid F^{-1} \circ \mathcal{Y}_{k+1}) \, d\tilde{\mu}_k , \tag{3.1}$$

$$A \in \mathcal{X}_{k+1}, \ B \in \mathcal{Y},$$

where $\mu_0 (\cdot \mid F^{-1} \circ \mathcal{Y}_{k+1})$ is any)^x $[F^{-1} \circ \mathcal{Y}_{k+1}](\mu_0)$-measurable (and not only (mod μ_0) $F^{-1} \circ \mathcal{Y}_{k+1}$-measurable!) variant of the conditional μ_0-probability and $\tilde{\mu}_k$ is an extension of $\mu_k \mid \mathcal{X}_k \vee F^{-1} \circ \mathcal{Y}_{k+1}$ to $\mathcal{X}_{k+1}(\tilde{\mu}_k)$ such that

)^x $M(A, B)$ does not, obviously, depend on the choice of an $[F^{-1} \circ \mathcal{Y}_{k+1}](\mu_0)$-measurable variant since $\tilde{\mu}_k \mid F^{-1} \circ \mathcal{Y}_{k+1} = \mu_0$.

$$(3.2) \qquad (\mathrm{mod}\ \tilde{\mu}_k) \qquad \mathcal{X}_{k+1} = \mathcal{X}_k \vee F^{-1}\circ\mathcal{Y}_{k+1}$$

(by the induction hypothesis)

$$= F^{-1}\circ\mathcal{Y}_k \vee F^{-1}\circ\mathcal{Y}_{k+1} = F^{-1}\circ\mathcal{Y}_{k+1}.$$

Let us justify this definition.

(i) The existence of $\tilde{\mu}_k$ with the above properties follows from Theorem 0.5 B since the σ-algebra $\mathcal{X}_k \vee F^{-1}\circ\mathcal{Y}_{k+1}$ is obviously μ_k-countably generated (for instance, the union of a sequence of sets, generating $\mathcal{X}_k$, and a sequence of sets, μ_0-generating $F^{-1}\circ\mathcal{Y}_{k+1}$, μ_k-generates $\mathcal{X}_k \vee F^{-1}\circ\mathcal{Y}_{k+1}$).

(ii) Equalities (3.1) correctly define $\mu_{k+1}(A \cap F^{-1}\circ B)$; in other words,

$$[A \cap F^{-1}\circ B = A' \cap F^{-1}\circ B'] \Rightarrow [M(A,B) = M(A',B')].$$

In fact, let us first show that

$$[A \cap F^{-1}\circ B = \emptyset] \Rightarrow M(A,B) = 0.$$

Since F satisfy (2.3), the inclusion $A \subset X \setminus F^{-1}\circ B \quad (B \in \mathcal{Y})$ implies that there exists such a $B_0 \in \mathcal{Y}_{k+1}$ that $A \subset X \setminus F^{-1}\circ B_0$ and

$$(3.3) \qquad \nu(B \setminus B_0) = 0.$$

It follows from (3.3) that

$$(\mathrm{mod}\ \mu_0) \qquad \mu_0(F^{-1}\circ B \mid F^{-1}\circ\mathcal{Y}_{k+1}) \le \mu_0(F^{-1}\circ B_0 \mid F^{-1}\circ\mathcal{Y}_{k+1}) = \mathbb{1}_{F^{-1}\circ B_0}$$

and, by our agreement to take only $[F^{-1}\circ\mathcal{Y}_{k+1}](\mu_0)$-measurable variants of $\mu_0(\cdot \mid F^{-1}\circ\mathcal{Y}_{k+1})$,

$$\mu_0(F^{-1}\circ B \mid F^{-1}\circ\mathcal{Y}_{k+1}) \le \mathbb{1}_{F^{-1}\circ B_0} \qquad (\mathrm{mod}\ \tilde{\mu}_k).$$

Since $A \subset X \setminus F^{-1}\circ B_0$, we have

$$M(A,B) \le \int_{X\setminus F^{-1}\circ B_0} \mathbb{1}_{F^{-1}\circ B_0}\, d\tilde{\mu}_k = \tilde{\mu}_k(\emptyset) = 0.$$

Now let

$$A\cap F^{-1}\circ B = A'\cap F^{-1}\circ B' \;;\quad A, A'\in \mathcal{X}_{k+1} \;;\quad B, B'\in \mathcal{Y}.$$

By definition,

$$\begin{aligned} M(A,B) &- M(A\cap A', B\cap B') \\ &= M(A,B) - M(A\cap A', B) \\ &\qquad + M(A\cap A', B) - M(A\cap A', B\cap B') \\ &= M(A\setminus A', B) + M(A\cap A', B\setminus B') \end{aligned}$$

what, by the above proved, equals zero because

$$[(A\setminus A')\cap F^{-1}\circ B]\cup[A\cap A'\cap F^{-1}\circ(B\setminus B')]$$

$$=(A\cap F^{-1}\circ B)\setminus(A'\cap F^{-1}\circ B')=\emptyset,$$

i.e.

$$M(A,B) = M(A\cap A', B\cap B').$$

Interchanging the pairs (A,B) and (A', B') in the above argument, we arrive at the required equality.

Let us return to constructing μ_{k+1}.

By (3.1) and additivity, we define an additive set function μ_{k+1} on the algebra generated by $\mathcal{X}_{k+1}\cup F^{-1}\circ\mathcal{Y}$, that is on the class of sets of the form

$$\bigcup_{n=1}^{N}(A_n\cap F^{-1}\circ B_n);\ N=1,2,\dots;$$

$$A_n\in\mathcal{X}_{k+1},\ B_n\in\mathcal{Y};\ n=1,\dots,N.$$

Namely, if $\{A_n\}$ and $\{B_n\}$ are such that

$$A_i\cap F^{-1}\circ B_i\cap A_j\cap F^{-1}\circ B_j=\emptyset,\ i\neq j,$$

we put

$$\begin{aligned}\mu_{k+1}(\textstyle\bigcup_{n=1}^{N}(A_n\cap F^{-1}\circ B_n)) &= M_N(\{A_n\cap F^{-1}\circ B_n\}_{n=1,\dots,N}) \\ &= \textstyle\sum_{n=1}^{N}\mu_{k+1}(A_n\cap F^{-1}\circ B_n).\end{aligned}$$

If members of $\{A_n\cap F^{-1}\circ B_n\}_{n=1,\dots,N}$ can intersect, we first represent $\bigcup_{n=1}^{N}(A_n\cap F^{-1}\circ B_n)$ as a finite union of disjoint sets $A'_n\cap F^{-1}\circ B'_n$,

$n=1,\dots,N'$ $(A'_n \in \mathcal{X}_{k+1}, B'_n \in \mathcal{Y})$. It is easy to see that such a representation is always possible.

To show that this definition is correct, i.e. does not depend on choosing a finite sequence with the same union, we shall make use of the equality $\mu_{k+1}(\emptyset)=0$, already proved, and of the following property implied by (3.1):

if a set $A \cap F^{-1} \circ B$ $(A \in \mathcal{X}_{k+1}, B \in \mathcal{Y})$ is equal to a finite union of pairwise disjoint sets of the form

$$A_n \cap F^{-1} \circ B_n \quad (A_n \in \mathcal{X}_{k+1}, B_n \in \mathcal{Y}; n=1,\dots,N),$$

then

$$\mu_{k+1}(A \cap F^{-1} \circ B) = \sum_{n=1}^{N} \mu_{k+1}(A_n \cap F^{-1} \circ B_n). \tag{3.4}$$

In fact, let

$$A \cap F^{-1} \circ B = \bigcup_{n=1}^{N} (A_n \cap F^{-1} \circ B_n). \tag{3.5}$$

Decompose A and B into constituents of $\{A_n\}$ and $\{B_n\}$ (without loss of generality we assume that $\bigcup_1^N A_n = A$, $\bigcup_1^N B_n = B$):

$$A = \bigcup_{(\alpha_1,\dots,\alpha_N)} \bigcap_{n=1}^{N} A_n^{(\alpha_n)}, \quad B = \bigcup_{(\alpha_1,\dots,\alpha_N)} \bigcap_{n=1}^{N} B_n^{(\alpha_n)}$$

where

$$A_n^{(1)} = A_n, \quad A_n^{(0)} = A \setminus A_n; \quad B_n^{(1)} = B_n, \quad B_n^{(0)} = B \setminus B_n; \quad n=1,\dots,N.$$

Any constituent can be, in an obvious way, identified with an N-dimensional vector $\alpha = \{\alpha_1,\dots,\alpha_N\}$ with coordinates 0 or 1. The summands in (3.5) being pairwise disjoint, it is easy to see that all the pairs of an A-constituent and the F-preimage of a B-constituent with a non-empty intersection are contained in the set of all such pairs for which the inner product of the corresponding pair of vectors is equal to 1.

From (3.1) and from the fact that $\mu_{k+1}(\emptyset)=0$ we get immediately:

$$\mu_{k+1}(A\cap F^{-1}\circ B) = \sum_{\alpha} \mu_{k+1}(\cap_{n=1}^{N} A_n^{(\alpha_n)} \cap F^{-1}\circ B)$$
$$= \sum_{\alpha,\beta} \mu_{k+1}(\cap_{n=1}^{N} A_n^{(\alpha_n)} \cap F^{-1}\circ \cap_{n=1}^{N} B_n^{(\beta_n)})$$
$$= \sum_{\alpha,\beta:(\alpha,\beta)=1} (\cdots) .$$

Performing analoguous computation for each set

$$A_i = \cup_{\alpha:\alpha_i=1} \cap_{n=1}^{N} A_n^{(\alpha_n)} \quad \text{and} \quad B_i = \cup_{\alpha:\alpha_i=1} \cap_{n=1}^{N} B_n^{(\alpha_n)},$$

we obtain $(i=1,\dots,N)$

$$\mu_{k+1}(A_i\cap F^{-1}\circ B_i) = \sum_{\substack{\alpha,\beta:(\alpha,\beta)=1\\ \alpha_i=\beta_i=1}} \mu_{k+1}(\bigcap_{n=1}^{N} A_n^{(\alpha_n)} \cap F^{-1}\circ \bigcap_{n=1}^{N} B_n^{(\beta_n)}).$$

Since

$$\sum_{\alpha,\beta:(\alpha,\beta)=1} = \sum_{i=1}^{N} \sum_{\alpha,\beta:(\alpha,\beta)=1;\,\alpha_i=\beta_i=1},$$

we arrive at (3.4).

Now, if the unions $\cup_{n=1}^{N_i}(A_n^i \cap F^{-1}\circ B_n^i)$; $i=1,2$; of pair-wise disjoint sets coincide, there exists a finite class $A_n \cap F^{-1}\circ B_n$; $n=1,\dots,N$; such that

$$\cup_{n=1}^{N}(A_n\cap F^{-1}\circ B_n) = \cup_{n=1}^{N_i}(A_n^i \cap F^{-1}\circ B_n^i)$$

and each set $A_n^i \cap F^{-1}\circ B_n^i$ $(n=1,\dots,N_i\,;\ i=1,2)$ is a union of some its subclass.

Making use of the above proved additivity property, we obtain

$$M_{N_1}(\{A_n^1 \cap F^{-1}\circ B_n^1\}_{n=1,\dots,N_1}) = \sum_{n=1}^{N} \mu_{k+1}(A_n\cap F^{-1}\circ B_n)$$
$$= M_{N_2}(\{A_n^2\cap F^{-1}\circ B_n^2\}_{n=1,\dots,N_2}).$$

Let us check now that μ_{k+1} is σ-additive on the algebra generated by $\mathcal{X}_{k+1} \cup F^{-1}\circ \mathcal{Y}$. To this end it suffices to verify that, if

$$\cup_{n=1}^{\infty}(A_n\cap F^{-1}\circ B_n) = X \ ;\ A_n\in\mathcal{X}_{k+1},\ B_n\in\mathcal{Y};$$

then

$$\sum_{n=1}^{\infty} \mu_{k+1}(A_n \cap F^{-1}\circ B_n) \geq 1.$$

Let $B_n^0 \in \mathcal{Y}_{k+1}$ be such that

$$A_n = F^{-1}\circ B_n^0 \quad (\mathrm{mod}\, \tilde{\mu}_k), \quad n=1,\dots .$$

We have

$$X \subset \bigcup_n [(F^{-1}\circ(B_n^0 \cap B_n)) \cup ((A_n \setminus F^{-1}\circ B_n^0) \cap F^{-1}\circ B_n)]$$

$$\subset [F^{-1}\circ(\bigcup_n (B_n^0 \cap B_n))] \cup [\bigcup_n (A_n \setminus F^{-1}\circ B_n^0)] = (F^{-1}\circ B) \cup A,$$

$$B \in \mathcal{Y}, \quad A \in \mathcal{X}_{k+1}; \quad \tilde{\mu}_k(A) = 0 \qquad (\Rightarrow \mu_{k+1}(A) = 0).$$

By finite additivity of μ_{k+1},

$$1 = \mu_{k+1}(X) \leq \mu_{k+1}(F^{-1}\circ B) + \mu_{k+1}(A) = \mu_{k+1}(F^{-1}\circ B).$$

But $\mu_{k+1} | F^{-1}\circ \mathcal{Y} = \mu_0$ and, by σ-additivity of μ_0, we get

$$1 = \mu_{k+1}(F^{-1}\circ B) = \mu_0(F^{-1}\circ B) \leq \sum_n \mu_0(F^{-1}\circ(B_n^0 \cap B_n)) =$$

$$= \sum_n \mu_{k+1}(F^{-1}\circ(B_n^0 \cap B_n)) = \sum_n \mu_{k+1}(A_n \cap F^{-1}\circ B_n).$$

Hence μ_{k+1} is σ-additive. It follows that μ_{k+1} can be uniquely extended to a measure on $\mathcal{X}'_{k+1}$.

Thus, starting from (3.1), we have defined a measure μ_{k+1} on $\mathcal{X}'_{k+1}$. By (3.1) and by continuity and additivity, it follows:

(3.6) $$\mu_{k+1} | F^{-1}\circ \mathcal{Y} = \mu_0$$

and, by (3.2),

(3.7) $$\mathcal{X}_{k+1} = F^{-1}\circ \mathcal{Y}_{k+1} \quad (\mathrm{mod}\, \mu_{k+1})$$

and also

(3.8) $$\mu_{k+1} | \mathcal{X}'_k = \mu_k .$$

Let us continue the procedure of constructing untill $k+1=m$. From (3.6) - (3.8), by definition, it follows that $\mu=\mu_m$ is a desired causal solution. ∎

<u>3.2.</u> The proof of Theorem 3.1 clarifies the meaning of the condition that F should be a γ-non-anticipating transformation. This

condition enables to sequentially construct measures μ_k without "peeping into the future", i.e. utilizing solely the "information" contained in the σ-algebras $\{\mathcal{Y}_i\}_{i\leq k}$.

In the case of a linear not well-ordered)^x set T , our proof fails. In fact, using any discrete procedure, we have to "run ahead", and the obtained solution will be anticipating. However, this anticipation can be made arbitrarily small in the following sense.

Theorem. Let $T=[0,1]$, $\mathfrak{X}=\{\mathcal{X}_t\}_{t\in T}$ and $\mathfrak{Y}=\{\mathcal{Y}_t\}_{t\in T}$ be flows of σ-algebras in X and Y respectively, $\mathcal{X}=\bigvee_{t\in T}\mathcal{X}_t=\mathcal{X}_1$ and $\mathcal{Y}=\bigvee_{t\in T}\mathcal{Y}_t=\mathcal{Y}_1$. Let the σ-algebras $\mathcal{X}_t$ $(t\in T)$ be countably generated and such that $(X,\mathcal{X})$ is Blackwell space, and suppose that the σ-algebras $\mathcal{Y}_t$ $(t\in T)$ be ν-countably generated. Let $F:(X,\mathcal{X})\to(Y,\mathcal{Y})$ be a ν-non-anticipating with respect to $(\mathfrak{X},\mathfrak{Y})$ transformation satisfying condition (0.6).

Then, for any $\varepsilon>0$, the stochastic equation

$$F\circ\mu=\nu \qquad [F:(X,\mathcal{X})\to(Y,\mathcal{Y})]$$

has a (strong) solution μ^ε such that

$$\forall t\in T \tag{3.9}$$
$$\mathcal{X}_t\subset F^{-1}\circ\mathcal{Y}_{(t+\varepsilon)\wedge 1} \qquad (\mathrm{mod}\ \mu^\varepsilon).$$

To prove this assertion we need only apply Theorem 3.1 to the restrictions of the flows $\mathfrak{X}$ and $\mathfrak{Y}$ to a sufficiently dense in T finite net $\{t_k\}_{k=0,\dots,m}$:

)^x Here we put aside another difficulty one is faced with even for $T=\{1,2,\dots\}$: the question of the existence of the projective limit for the consistent family $\{\mu_k\}_{k=1,2,\dots}$.

$$0=t_0<\dots<t_m=1\,,\quad t_{k+1}-t_k<\varepsilon \qquad \forall k \qquad \blacksquare$$

3.3. Example. Let $X, Y, T, \mathfrak{X}$ amd $\mathfrak{Y}$ be defined as in Example 2.4 with the difference that all continuous functions, which are points of X and Y, "start" from zero. The transformation $F: X \to Y$ maps $x=\{x_t\}_{t\in T} \in X$ into $y=\{y_t\}_{t\in T} \in Y$ by the formula

$$y_t = x_t - \int_0^t f(s,x)\,ds$$

where f is a measurable functional on $([0,1]\times X\,,\ \mathcal{B}\times\mathcal{X})$ ($\mathcal{B}$ is the Borel σ-algebra in $[0,1]$, $\mathcal{X}=\bigvee_{t\in T}\mathcal{X}_t=\mathcal{X}_1$) such that (i) for each $t\in T$, the mapping $f(t,\cdot)$ of X into the Borel line is $\mathcal{X}_t$-measurable, (ii) for each $x\in X$,

$$\int_0^1 f^2(t,x)\,dt<\infty$$

and (iii)

$$E\exp\left\{\int_0^1 f(t,w)\,dw_t-\frac{1}{2}\int_0^1 f^2(t,w)\,dt\right\}=1$$

(here $w=\{w_t\}_{t\in[0,1]}$ is a Wiener process on $[0,1]$; $\int_0^1\cdots dw_t$ is the Itô stochastic integral).
It is clear that F is causal.

Under these conditions, it was shown in [9] with the help of Girsanov's theorem that the stochastic equation

$$F\circ\mu=\nu \qquad [F:(X,\mathcal{X})\to(Y,\mathcal{Y})]$$

where ν is the Wiener measure on Y, has a solution μ_* such that, for any its representation $\xi(\omega)=\{\xi_t(\omega)\}_{t\in T}$ on a probability space ($\Omega, \mathfrak{S}, P$), the random process $W=\{W_t\}_{t\in T}\overset{\text{def.}}{=}F\circ\xi$ is a Wiener process on this probability space with respect to the flow of σ-algebras $\{\xi^{-1}\circ\mathcal{X}_t\}_{t\in T}$. Using our terminology, this precisely means that the solution μ_* is $\mathfrak{F}$-non-anticipating with respect to ($\mathfrak{X}, \mathfrak{Y}$), where $\mathfrak{F}$ is the class of "coordinate" functions on Y :

$$\mathfrak{F} = \{ y \longmapsto y_t \; ; t \in T \}.$$

Untill quite recently the so-called innovation problem (T. Kailath)was open:

Is this $\mathfrak{F}$-non-anticipating solution causal)[x]? The recent counterexample of B. Cirel'son [10] shows that this question must be answered, in general, negatively.

However, we shall prove, with the help of Theorem 3.2, that, under the above conditions, there exists a strong solution of the equation under consideration with arbitrarily small "anticipation" in the sense of (3.9)[xx]. To this end we need only check that the transformation F is ν-non-anticipating with respect to ($\mathfrak{X}, \mathfrak{Y}$).

Let π_t $(t \in [0,1])$ map any continuous function on $[0,1]$ into its restriction onto $[0,t]$ and ρ_t map any continuous function $z = \{ z_s \}_{s \in [0,1]}$ into ${}^t z = \{ {}^t z_s \}_{s \in [0,1-t]}$ by the formula

$${}^t z_s = z_{s+t} - z_t .$$

For any $x \in X$, let $F^{t, \pi_t \circ x}$ denote the transformation of

)[x] Note that, thanks to specific properties of the Wiener measure, the $\mathfrak{F}$-non-anticipation and simply non-anticipation (i.e. if $\mathfrak{F}$ is the class of indicators) are one and the same thing. Thus, by Proposition 1.3, the innovation problem can be formulated as follows: Is the solution under consideration strong?

)[xx] The author was informed by A.V. Skorohod that he proved an analogous result using quite a different method.

$\rho_t \circ X$ into $\rho_t \circ Y$ defined by

$$F^{t,\pi_t \circ x} \circ \rho_t \circ x = \rho_t \circ F \circ x .$$

Let $(\Omega, \mathfrak{S}, P)$, ξ and W be the same as above in this section. One can trivially check that, for any $A \in \mathcal{X}_t$ $(A = \pi_t^{-1} \circ \pi_t \circ A\,!)$,

$$\text{(3.10)} \qquad (\text{mod}\, P) \qquad P(\{W \in F \circ A\} \mid \pi_t \circ \xi) =$$

$$= 1\!\!1_{\pi_t \circ F \circ A}(\pi_t \circ F \circ \xi)\,[P(\{\rho_t \circ W \in F^{t,\pi_t \circ x} \circ {}^t X\})]_{x=\xi} .$$

In particular, taking $A = X$, the second co-factor on the right can be replaced by 1. Integrating, we get (2.5):

$$\nu(F \circ A) = E\, P(\{W \in F \circ A\} \mid \pi_t \circ \xi)$$
$$= E\, 1\!\!1_{\pi_t \circ F \circ A}(\pi_t \circ F \circ \xi) = E\, 1\!\!1_{\pi_t^{-1} \circ \pi_t \circ F \circ A}(W)$$
$$= \nu(\pi_t^{-1} \circ \pi_t \circ F \circ A)$$

as required.

Remark. We need, in fact, the conditions imposed on the transformation F (except those which are necessary for F to be correctly defined) only to ensure that there exists an $\mathfrak{F}$-non-anticipating solution of the equation $F \circ \mu = \nu$ (or, equivalently, existence of a "weak" solution of the Itô stochastic differential equation $\xi_t = \int_0^t f(s, \xi)ds + W_t$). If such a solution is known to exist from the very beginning, our assertion about an "ε-anticipating" solution still holds. Even the specific form of F is unessential since all we need is formula (3.10) which is obtained by making use of the fact that increments of W do not depend on the "past" of ξ. Moreover, we venture to formulate, without proof, the following statement.

3.4. Statement. Let T be a linearly ordered set, $\mathfrak{X} = \{\mathcal{X}_t\}_{t \in T}$ and $\mathfrak{Y} = \{\mathcal{Y}_t\}_{t \in T}$ be flows of σ-algebras in spaces X and Y

respectively, $\mathcal{X} = \bigvee_{t\in T} \mathcal{X}_t$ and $\mathcal{Y} = \bigvee_{t\in T} \mathcal{Y}_t$, μ be a measure on $(X, \mathcal{X})$ such that $(X, \mathcal{X}, \mu)$ is a Lebesgue space, F be a causal transformation of X into Y such that, for each t, the σ-algebra $\mathcal{Y}$ can be represented as the sum $\mathcal{Y}_t \vee \mathcal{Y}^t$ where the σ-algebra $F^{-1} \circ \mathcal{Y}^t$ does not depend (with respect to μ) on $\mathcal{X}_t$.

Then F is an $[F\circ\mu]$-non-anticipating transformation.

4. The case of not well-ordered T

4.1. To construct non-anticipating solutions of stochastic equations in the case of linearly ordered but not well-ordered T, e.g. T = the set of rationals or $T = [0, \infty)$ or $T = [0,1]$, we are going to use non-anticipating solutions for finite subsets of T as approximations of a solution for T. (Theorem 3.1 ensures the existence of a solution for any finite subset of T.) In other words, we will approximate a solution of the original problem by "ε-anticipating" ones from Theorem 3.2.

It is clear that the topology in the space of solutions, with respect to which approximation is understood, should be chosen so that the non-anticipation property in question would be invariant under passage to a limit. However, the stronger the topology is, the more restrictive compactness conditions must be imposed upon our "finite approximations". We choose the weak-star topology in the space of measures, and the purpose of all the additional restrictions as compared with the case of a finite T, is to enable its application.

4.2. Now we formulate basic assumptions which we will stick to in the rest of this section.

(I) X is an analytic metric space.

(II) T is a linear set.

(III) $\mathfrak{X} = \{\mathcal{X}_t\}_{t\in T}$ is a flow of σ-algebras in X, each σ-algebra $\mathcal{X}_t$ being generated by a family $\mathcal{G}_t$ of continuous functions $g : X \to \mathbb{R}^1$, i.e.

$$\mathcal{X}_t = \bigvee_{g\in\mathcal{G}_t} g^{-1}\circ\mathcal{B}$$

where $\mathcal{B}$ is the Borel σ-algebra in $\mathbb{R}^1$. (For instance, if $T=[0,1]$, conditions (I) and (II) are satisfied for the space C of continuous functions on T with the topology of uniform convergence, or for the Skorohod space D, in both of which the flows of σ-algebras are generated by the restrictions of functions to intervals $[0,t]$).

It is easy to see that, since X is separable (this follows from analyticity), each σ-algebra $\mathcal{X}_t$ is countably generated.

(IV) The σ-algebra $\mathcal{X}=\bigvee_{t\in T}\mathcal{X}_t$ coincides with the Borel σ-algebra in X.

Since $\mathcal{X}$, under condition (III), is always a part of the Borel σ-algebra, it follows from the properties of Blackwell spaces [11] that (IV) holds iff all the atoms of $\mathcal{X}$ are singletons.

Condition (IV) is not principal, and we assume it to be satisfied only for convenience. Moreover, one can make it be satisfied by quite a simple trick. Namely, one can always add one more point θ to T putting $\theta > t$ for any $t\in T$ (note that it can be done even if $T=[0,\infty]$) and defining $\mathcal{X}_\theta$ to be the Borel σ-algebra in X. In other words, one can imbed the original problem into its "one-point extension" with the desired property.

(V) Y is a metric space, $\mathcal{Y}=\{\mathcal{Y}_t\}_{t\in T}$ is a flow of σ-algebras in Y and $\mathcal{Y}=\bigvee_{t\in T}\mathcal{Y}_t$ is the Borel σ-algebra in Y (of course, the note in Condition (IV) concerns this condition as well). ν is a measure on $(Y,\mathcal{Y})$, and each σ-algebra $\mathcal{Y}_t$ $(t\in T)$ is ν-countably generated.

Denote by $\mathcal{F}_t$ $(t\in T)$ the set of all real ν-integrable continuous functions f on Y such that

$$E_\nu(f\,|\,\mathcal{Y}_t)=\Phi_{f,t} \quad (\mathrm{mod}\,\nu)$$

where $\Phi_{f,t}$ is a ν-integrable continuous function on Y ($E_\nu(\cdot|\cdot)$ is the conditional expectation with respect to ν).

Remark. The set $\mathcal{F}_t$ is obviously non-empty: it contains, e.g., constants. However it is desirable that this set would be rich enough. How large $\mathcal{F}_t$ is depends on the measure ν and on the σ-algebra $\mathcal{Y}_t$. For instance, if ν corresponds to a martingale with sample paths in $Y=C$ or D with the usual metrics and $\mathcal{Y}_t$'s are generated by cylinder sets over intervals $[0,t]$, $\mathcal{F}_t$ contains all the "coordinate" functions $y=\{y_u\}_{u\in T}\longmapsto y_s$ $(s\in T)$.

4.3. Theorem. Let Conditions (I) - (V) be satisfied and T be a countable set. Let $F: X\to Y$ be a continuous transformation such that $\nu(F\circ X)=1$)[x], F-preimages of compacts are compact and let F be ν-non-anticipating with respect to $(\mathfrak{X},\mathcal{Y})$.

Then the stochastic equation

$$F\circ\mu=\nu \qquad [F:(X,\mathcal{X})\to(Y,\mathcal{Y})]$$

)[x] Note that $F\circ X$ is measurable with respect to $\mathcal{Y}(\nu)$, since, by (I), it is an analytic set.

has an $\mathcal{F}$-non-anticipating with respect to ($\mathfrak{X}, \mathfrak{Y}$) solution, where

$$\mathcal{F} = \bigcap_{t \in T} \mathcal{F}_t .$$

Proof. Let $\{T_n\}_{n \geq 1}$ be an increasing sequence of finite subsets of T, which exhausts T:

$$T_1 \subset T_2 \subset \dots , \quad \bigcup_n T_n = T.$$

For convenience, let us add a point ∞ to T and to each T_n, $n = 1, 2, \dots$ ($\infty > t$ for any $t \in T$), keeping the original notation. Put $\mathcal{X}_\infty = \mathcal{X}$, $\mathcal{Y}_\infty = \mathcal{Y}$.

Now the conditions of Theorem 3.1 being satisfied in view of the made assumptions, by applying it, we can, for any n, construct a causal with respect to $(\mathfrak{X}_n, \mathfrak{Y}_n) = (\{\mathcal{X}_t\}_{t \in T_n}, \{\mathcal{Y}_t\}_{t \in T_n})$ solution μ_n of the stochastic equation

$$F \circ \mu = \nu \qquad [F: (X, \mathcal{X}) \to (Y, \mathcal{Y})].$$

Let us show that the sequence of measures $\{\mu_n\}$ is relatively $\mathbf{C}$-compact: there exist a subsequence $\{\mu_{n'}\}$ of it and a measure μ_∞ on ($X, \mathcal{X}$) such that

$$\int_X g \circ x \ \mu_{n'}(dx) \xrightarrow[n']{} \int_X g \circ x \ \mu_\infty(dx) \qquad \forall g \in \mathbf{C}$$

where $\mathbf{C} = \mathbf{C}(X)$ is the set of all real bounded continuous functions on X.

To do it, as is well-known (Prohorov [12]), it suffices to check that the measures μ_n are uniformly tight.

Let ε be a small positive number and K' be such a compact in Y that

$$\nu(K') \geq 1 - \varepsilon$$

(ν is a tight measure; this follows from the condition: $\nu(F \circ X) = 1$ and from the tightness of any Borel measure on an analytic metric space). By assumption, the set $K = F^{-1} \circ K'$ is compact in X. Hence we have

$$\mu_n(K) = \nu(K') \geq 1 - \varepsilon \quad \forall n$$

since $\mu_n \mid F^{-1} \circ \mathcal{Y} = F^{-1} \circ \nu$ for each n.

Without loss of generality we can and do suppose that the sequence $\{\mu_n\}$ itself $\mathbf{C}$-converges to μ_∞.

Since $\mathbf{C}$-convergence is invariant under continuous transformations, we have

$$\nu = F \circ \mu_n \xrightarrow[n]{\mathbf{C}(Y)} F \circ \mu_\infty$$

so that $F \circ \mu_\infty = \nu$. Thus μ_∞ is a solution of our equation. Let us check that it is $\mathcal{F}$-non-anticipating.

Let $t \in T$ $(t \neq \infty)$ be arbitrary and μ' be any measure on $(X, \mathcal{X})$ such that $F \circ \mu' = \nu$ (i.e. any solution of the equation under consideration). For any real bounded $\mathcal{Y}_t$-measurable function h on Y and any $f \in \mathcal{F}$,

$$\int_X (h \circ F \circ x)(f \circ F \circ x)\, \mu'(dx) = \int_Y (h \circ y)(f \circ y)\, \nu(dy)$$

$$= \int_Y (h \circ y) E_\nu (f \mid \mathcal{Y}_t)\, \nu(dy)$$

$$= \int_Y (h \circ y)(\Phi_{f,t} \circ y)\, \nu(dy) = \int_X (h \circ F \circ x)(\Phi_{f,t} \circ F \circ x)\, \mu'(dx).$$

Hence

$$E_{\mu'}(f \circ F \mid F^{-1} \circ \mathcal{Y}_t) = \Phi_{f,t} \circ F \pmod{\mu'}. \tag{4.1}$$

Since μ_n is causal with respect to $(\mathfrak{X}_n, \mathfrak{Y}_n)$, $n = 1, 2, \ldots$, we have, for each sufficiently large n,

$$E_{\mu_n}(f \circ F \mid \mathcal{X}_t) = E_{\mu_n}(f \circ F \mid F^{-1} \circ \mathcal{Y}_t) = \Phi_{f,t} \circ F \pmod{\mu_n}.$$

That is, for any real bounded $\mathcal{X}_t$-measurable function g on X,

$$\int_X (g \circ x)((f \circ F \circ x) - (\Phi_{f,t} \circ F \circ x))\, \mu_n(dx) = 0.$$

Taking first continuous g and passing to a limit as n goes to infinity, Condition (III) implies the same equality with μ_∞ instead of μ_n for any $\mathcal{X}_t$-measurable g.

Thus

$$E_{\mu_\infty}(f\circ F \mid \mathcal{X}_t) = \Phi_{f,t}\circ F \quad (\mathrm{mod}\ \mu_\infty),$$

and, by (4.1), it follows

$$E_{\mu_\infty}(f\circ F \mid \mathcal{X}_t) = E_{\mu_\infty}(f\circ F \mid F^{-1}\circ \mathcal{Y}_t) \quad (\mathrm{mod}\ \mu_\infty) \blacksquare$$

4.4. In order to obtain a theorem on existence of a non-anticipating solution for uncountable sets T by making use of the corresponding result in the "countable" case, one has to require that the flows $\mathfrak{X}$ and $\mathfrak{Y}$ could be consistently, in a certain sense, approximated by the flows $\mathfrak{X}' = \{\mathcal{X}_t\}_{t\in S}$ and $\mathfrak{Y}' = \{\mathcal{Y}_t\}_{t\in S}$ where S is a countable subset of T.

Let us introduce the following definitions.

($VI_{\mathcal{X}_{t_0-}}$) The flow $\{\mathcal{X}_t\}_{t\in T}$ is left-continuous at point $t_0\in T$, $t_0 \neq \inf\{s\in T\}$:

$$\mathcal{X}_{t_0-} = \vee_{s<t_0, s\in T}\, \mathcal{X}_s = \mathcal{X}_{t_0}$$

(if $t_* = \inf\{s\in T\}\in T$, we put $\mathcal{X}_{t_*-} = \mathcal{X}_{t_*}$).

($VI_{\mathcal{X}_{t_0+}}$) The flow $\{\mathcal{X}_t\}_{t\in T}$ is right-continuous at point $t_0\in T$, $t_0\neq \sup\{s\in T\}$:

$$\mathcal{X}_{t_0+} = \wedge_{s>t_0, s\in T}\, \mathcal{X}_s = \cap_{s>t_0, s\in T}\, \mathcal{X}_s = \mathcal{X}_{t_0}$$

(if $t^* = \sup\{s\in T\}\in T$, we put $\mathcal{X}_{t^*+} = \mathcal{X}_{t^*}$).

For brevity, we will write

$$(VI^{\mu}_{\mathcal{X}_{t\pm}}) \equiv (VI_{\mathcal{X}_{t\pm}(\mathcal{X};\mu)})$$

where $\mathcal{X} = \vee_{t\in T}\mathcal{X}_t$ as usual and μ is a measure on $\mathcal{X}$, and

$$(VI^{u}_{\mathcal{X}_{t\pm}}) \equiv \wedge_{\mu}\, (VI^{\mu}_{\mathcal{X}_{t\pm}})$$

where the intersection is taken by all the measures μ on $\mathscr{X}$.

Clearly, for any measure μ on $\mathscr{X}$,

$$(\mathrm{VI}_{\mathscr{X}_{t\pm}}) \Rightarrow (\mathrm{VI}^{u}_{\mathscr{X}_{t\pm}}) \Rightarrow (\mathrm{VI}^{\mu}_{\mathscr{X}_{t\pm}}).$$

Theorem. Let T be an uncountable set and F be a causal with respect to ($\mathfrak{X},\mathfrak{Y}$) transformation of X into Y . Suppose that, for any countable $T' \subset T$, the stochastic equation

$$F \circ \mu = \nu \qquad [F : (X, \mathscr{X}) \to (Y, \mathscr{Y})]$$

has an $\mathfrak{F}'$-non-anticipating solution ($\mathfrak{F}' = \bigcap_{t \in T'} \mathfrak{F}_t$).

Then this stochastic equation has an $\mathfrak{F}$-non-anticipating solution ($\mathfrak{F} = \bigcap_{t \in T} \mathfrak{F}_t$)

1) always with respect to $(\mathfrak{X}_-, \mathfrak{Y}_-) = (\{\mathscr{X}_{t-}\}_{t\in T}, \{\mathscr{Y}_{t-}\}_{t\in T})$;

2) always with respect to $(\mathfrak{X}_+, \mathfrak{Y}_+) = (\{\mathscr{X}_{t+}\}_{t\in T}, \{\mathscr{Y}_{t+}\}_{t\in T})$;

3) with respect to ($\mathfrak{X},\mathfrak{Y}$) if, for all $t \in T$ except, may be, countably many points, the following condition is satisfied:

$$(\mathrm{VI}^{u}_{\mathscr{X}_{t-}}) \vee (\mathrm{VI}^{\nu}_{\mathscr{Y}_{t+}}).$$

Proof. 1) [2)] . Take a dense in T from the left [right] countable set $S \subset T$:

$$\forall t \in T \quad \exists \{s_n\} \subset S : \quad s_n \uparrow t \qquad [s_n \downarrow t],$$

take an $[\bigcap_{t\in S} \mathfrak{F}_t]$-non-anticipating solution and use the martingale convergence theorem. (Note that in the same way one can prove the existence of an $\mathfrak{F}$-non-anticipating solution with respect to $(\{\mathscr{X}_{t\pm}\}_{t\in T}, \{\mathscr{Y}_{t\pm}\}_{t\in T})$ where the plus and the minus may alternate, but always the same sign stands at each t .)

3) Take a dense in T from the left and from the right countable set $S \subset T$ which also contains all the exceptional points. Take an $[\bigcap_{t\in S} \mathfrak{F}_t]$-non-anticipating solution. If, for instance,

($VI^{u}_{\mathcal{X}_{t-}}$) holds, then by taking $\{s_n\} \subset S$, $s_n \uparrow t$, and by using the martingale convergence theorem, we get:

$$\forall f \in \mathcal{F}, \quad E_{\mu}(f \circ F \mid \mathcal{X}_t) = E_{\mu}(f \circ F \mid F^{-1} \circ \mathcal{Y}_{t-}) \quad (\mathrm{mod}\ \mu).$$

This implies the same equalities with $\mathcal{Y}_t$ instead of $\mathcal{Y}_{t-}$. The opposite case is analogous.

4.5. Let us show that, under Condition V. 4.2, condition 3) of Theorem 4.4 is always fulfilled. Namely, all points $t \in T$, except at most countably many, are points of continuity for the flow $\{\mathcal{Y}_t(\mathcal{Y};\nu)\}_{t \in T}$.

Proposition. (A.M. Stepin) Let $T \subset \mathbb{R}^1$ be an arbitrary set and $\mathfrak{H} = \{\mathcal{H}_t\}_{t \in T}$ be a flow of separable Hilbert spaces over T:

$$[T \ni s \leq t \in T] \Leftrightarrow [\mathcal{H}_s \text{ is a subspace of } \mathcal{H}_t].$$

The number of discointinuity points of $\mathfrak{H}$ is at most countable. (A point t is called a discontinuity point of $\mathfrak{H}$ if the linear closure of $\mathcal{H}_{t-} = \bigcup_{s<t} \mathcal{H}_s$ does not coincide with $\mathcal{H}_{t+} = \bigcap_{s>t} \mathcal{H}_s$.)

Proof. The spaces $\mathcal{H}_{t+} \ominus \mathcal{H}_{t-}$, $t \in T$, are mutually orthogonal. It follows that at most countably many of them can be non-zero, since $\mathcal{H}_t$, $t \in T$, are separable. ■

Now consider the flow of Hilbert spaces $\{L^2(Y, \mathcal{Y}_t(\mathcal{Y};\nu), \nu)\}_{t \in T}$ which corresponds to the flow of σ-algebras $\{\mathcal{Y}_t(\mathcal{Y};\nu)\}_{t \in T}$. It is easy to see that these flows have the same discontinuity points. Since each $\mathcal{Y}_t$, $t \in T$, is ν-countably generated, each $L^2(Y, \mathcal{Y}_t(\mathcal{Y};\nu), \nu)$, $t \in T$, is separable. Applying the above Proposition, we arrive at the desired result.

Note. It is essential that we considered the σ-algebras $\mathcal{Y}_t(\mathcal{Y};\nu)$ but not $\mathcal{Y}_t$; otherwise we could not even consider the corresponding L^2-spaces as a flow of Hilbert spaces.

4.6. Corollary. In Theorem 4.3 one can assume T to be uncountable.

Proof. Apply Theorem 4.4 taking into account the above result.■

REFERENCES

1. Yershov M.P., The existence of a martingale with given diffusion functional, Teoriya Veroyatnostei i ee Primeneniya, 19, 4(1974), 665-687 (Russian). English translation in: Theory of Probability and its Applications, 19, 4(1974), 633-655.

2. - " - Stochastic equations, Teoriya Veroyatnostei i ee Primeneniya, 19, 3(1974), 652-654 (Russian). English translation in: Theory of Probability and its Applications, 19, 3(1974), 431-444.

3. - " - Extensions of measures. Stochastic equations Proc. 2nd Japan-USSR Sympos. on Probability Theory, August 1972, Kyoto, Springer-Verlag, ser. Lecture Notes in Math., vol. 330, 1973, 516-526.

4. - " - Extension of measures and stochastic equations, Teoriya Veroyatnostei i ee Primeneniya, 19, 3 (1974), 457-471 (Russian). English translation

in: Theory of Probability and Its Applications, 19, 3(1974), 622-624.

5. - " - Analytic sets and some applications to probability theory, Teoriya Veroyatnostei i ee Primeneniya, 19, 3(1974), 655-656 (Russian). English translation in: Theory of Probability and Its Applications, 19, 3(1974), 625-626.

6. - " - Studies in the theory of stochastic equations, A dissertation, Moscow (1974) (Russian).

7. Lubin A., Extension of measures and the von Neumann selection theorem, Proc.Amer.Math.Soc., 43, 1(1974), 118-122.

8. Landers D. & Rogge L., On the extension problem for measures, Z. Wahrscheinlichkeitstheorie verw. Gebiete, 30, 2(1974), 167-169.

9. Yershov M.P., On absolute continuity of measures corresponding to diffusion type processes, Teoriya Veroyatnostei i ee Primeneniya, 17, 1(1972), 182-187 (Russian). English translation in: Theory of Probability and Its Applications, 17, 1(1972).

10. Cirel'son B.S. An example of a stochastic differential equation having no strong solution, Teoriya Veroyatnostei i ee Primeneniya, 20, 2(1975), 427-429 (Russian). English translation in: Theory of Probability and Its Applications, 20, 2(1975).

12. Prokhorov Yu.V., Convergence of random processes and limit theorems of probability theory, Teoriya Veroyatnostei i ee Primeneniya, 1, 2(1956), 177-237 (Russian). English translation in: Theory of Probability and Its Applications, 1, 2(1956), 157-214.

Steklov Mathematical Institute
Academy of Sciences of the USSR
Moscow

A STOCHASTIC MAXIMUM PRINCIPLE IN CONTROL PROBLEMS WITH DISCRETE TIME

V.I.Arkin, L.I.Krečetov (participants in paper)

1. Introdution. Consider measurable spaces $(S_t, \mathcal{F}_t)$, $0 \leq t \leq \tau$ (t is a discrete variable) and denote $S^t = S_0 \times \ldots \times S_t$, $\mathcal{F}^t = \mathcal{F}_0 \otimes \ldots \otimes \mathcal{F}_t$. Let $P_t(\cdot|\cdot)$, $1 \leq t \leq \tau$, be Markovian transition probabilities, i.e., $P_t(A|\cdot)$ is an $\mathcal{F}_{t-1}$ - measurable random variable for any $A \in \mathcal{F}_t$ and $P_t(\cdot \mid s_{t-1})$ is a probability on $\mathcal{F}_t$ for any $s_{t-1} \in S_{t-1}$. Let a probability P^o on $\mathcal{F}_0$ be given. The probabilities P^o and $P_t(\cdot|\cdot)$, $1 \leq t \leq \tau$, generate a probability P on $\mathcal{F}^\tau$.

An $\mathcal{F}^t$ - measurable function z_t with values in R^n is referred to as the state of a system. The transition from z_t to z_{t+1} is fulfilled in accordance with the equations

$$z_{t+1}(s^{t+1}) = f^{t+1}(z_t(s^t), u_t(s^t), s_t, s_{t+1}) \quad \text{a.s.,} \tag{1}$$

where the functions f^{t+1} are given and u_t (controls) are $\mathcal{F}^t$- - measurable with values in R^m. The controls u_t are supposed to meet the constraints

$$g^{t+1}(z_t(s^t), u_t(s^t), s_t, s_{t+1}) \leq 0 \quad \text{a.s.,} \quad 0 \leq t \leq \tau - 1, \tag{2}$$

$$u_t(s^t) \in U^t(s_t) \quad \text{a.s.,} \quad 0 \leq t \leq \tau - 1, \tag{3}$$

where g^{t+1} are given and take values in R^k, $U^t : S_t \to 2^{R^m}$ are set-valued mappings and $s^t = (s_0, \ldots, s_t)$.

Let us introduce one more specific constraint:

$$u_t(s^t) = v_t(z_t(s^t),\ s_t) \qquad \text{a.s.} \qquad (4)$$

for some function $v_t : \mathcal{R}^n \times S_t \to \mathcal{R}^m$.

The function z_0 is supposed to be fixed.

The problem is to maximize the functional

$$M \sum_{t=0}^{\tau-1} \varphi^{t+1}(z_t(s^t),\ u_t(s^t),\ s_t,\ s_{t+1}) \longrightarrow max, \qquad (5)$$

where $\{z_t,\ 1 \leq t \leq \tau\}$, $\{u_t,\ 0 \leq t \leq \tau-1\}$ (or, simply, $\{z_t\}$, $\{u_t\}$) satisfy (1)-(4).

We shall make some assumptions about the mappings in question.

a) $\varphi^t : \mathcal{R}^n \times \mathcal{R}^m \times S_{t-1} \times S_t \longrightarrow \mathcal{R}$,

$$f^t : \mathcal{R}^n \times \mathcal{R}^m \times S_{t-1} \times S_t \longrightarrow \mathcal{R}^n,$$

$$g^t : \mathcal{R}^n \times \mathcal{R}^m \times S_{t-1} \times S_t \longrightarrow \mathcal{R}^k$$

are bounded and measurable with respect to $(z, u, s_{t-1}, s_t) \in \in \mathcal{R}^n \times \mathcal{R}^m \times S_{t-1} \times S_t$.

b) φ^t, f^t, g^t are twice differentiable with respect to z for all $u \in \mathcal{R}^n$ and almost all s_t , and their first- and second-order derivativs are bounded in $\mathcal{R}^n \times \mathcal{R}^m \times S_{t-1} \times S_t$.

c) U^t is measurable in the sense, that its graph belongs to $\mathcal{F}_t \otimes \mathcal{B}^m$, $0 \leq t \leq \tau-1$.

d) For every $z \in \mathcal{R}^n$ and almost all s_{t-1} the subset of $U^{t-1}(s_{t-1})$, where the condition

$$g^t(z, u, s_{t-1}, s_t) \le 0 \quad P_t(\cdot \mid s_{t-1}) - \text{ a.s.}$$

holds, is nonempty $(1 \le t \le \tau)$.

2. Auxiliary results on measurable selectors

Proposition 1. [1]. Let $(\Omega, \mathcal{M})$ be a measurable space and $(\mathcal{R}^n, \mathcal{B}^n)$ be a Euclidean space with the Borel σ-algebra. If $G \in \mathcal{M} \otimes \mathcal{B}^n$, then $\mathrm{proj}_\Omega G$ (projection G onto Ω) is a universally measurable set ($\mathrm{proj}_\Omega G \in \hat{\mathcal{M}}$, where $\hat{\mathcal{M}}$ is the universal completion of $\mathcal{M}$).

Proposition 2. [1]. Let $(\Omega, \mathcal{M})$ be a measurable space, $G \in \mathcal{M} \otimes \mathcal{B}^n$ and G_ω is the section of G at $\omega \in \Omega$. Then there exists a mapping (selector) $u : \Omega \longrightarrow \mathcal{R}^n$, such that

(a) $u(\omega) \in G_\omega$ for all $\omega \in \mathrm{proj}_\Omega G$,

(b) $u^{-1}(A) \in \hat{\mathcal{M}}$ for all $A \in \mathcal{B}^n$.

Let a probability space $(\Omega, \mathcal{M}, P_0)$ and a measurable space $(S, \mathcal{F})$ be given, and let a transition probability $\pi(ds \mid \omega)$ be defined. The probabilities P_0 and $\pi(\cdot \mid \cdot)$ generate a probability P on $\mathcal{M} \otimes \mathcal{F}$.

Let us consider the functions

$$h_1 : \Omega \times S \to \mathcal{R}^n,$$

$$h_2 : \Omega \times S \to \mathcal{R}^k,$$

$$f_1 : \mathcal{R}^m \times \Omega \times S \longrightarrow \mathcal{R}^n,$$

$$f_2 : R^m \times \Omega \times S \longrightarrow R^k ,$$

such that h_1, h_2 are $\mathcal{M} \otimes \mathcal{F}$ - measurable and f_1, f_2 are $\mathcal{B}^m \otimes \mathcal{M} \otimes \mathcal{F}$ - measurable.

Let $U : \Omega \longrightarrow 2^{R^m}$ be an $\mathcal{M}$ -measurable set-valued mapping.

<u>Proposition 3.</u>*) Suppose that, for almost all $\omega \in \Omega$, there exists a vector $u_\omega \in U(\omega)$ such that

$$h_1(\omega, s) = f_1(u_\omega, \omega, s) \quad \pi(\cdot\,|\,\omega)\text{-a.s.}$$

$$h_2(\omega, s) \leq f_2(u_\omega, \omega, s) \quad \pi(\cdot\,|\,\omega)\text{-a.s.}$$

Then there exists an $\mathcal{M}$ -measurable function $u(\cdot)$, $u(\omega) \in U(\omega)$ a.s., such that

$$h_1(\omega, s) = f_1(u(\omega), \omega, s) \qquad P\text{ - a.s.},$$

$$h_2(\omega, s) \leq f_2(u(\omega), \omega, s) \qquad P\text{ - a.s.}$$

<u>Proof.</u> Let us demonstrate that

$$A = \{ u, \omega : h_1(\omega, s) = f_1(u, \omega, s) \quad \pi(\cdot\,|\,\omega) \text{ - a.s.} \}$$

and

$$B = \{ u, \omega : h_2(\omega, s) \leq f_2(u, \omega, s) \quad \pi(\cdot\,|\,\omega) \text{ - a.s.} \}$$

---

*) Proposition 3 is a generalization of the so-called Filippov lemma see, e.g., [1]).

belong to the σ-algebra $\mathcal{B}^m \otimes \mathcal{M}$.

For some number $c > 0$, define the function

$$\alpha(u,\omega) = \int \left| (h_1(\omega,s) - f_1(u,\omega,s))_c \right| \pi(ds|\omega),$$

where $\quad (h_1 - f_1)_c = \min((h_1-f_1)^+, c) - \min((f_1-h_1)^+, c).$*)

The function α is known to be $\mathcal{B}^m \otimes \mathcal{M}$-measurable. It follows that

$$A = \{u,\omega : \alpha(u,\omega) = 0\} \in \mathcal{B}^m \otimes \mathcal{M}.$$

To prove the measurability of B, it suffices to note that

$$B = \{u,\omega : (h_2(\omega,s) - f_2(u,\omega,s))^+ = 0 \quad \pi(\cdot|\omega) - \text{a.s.}\}.$$

Now apply Proposition 2 to the set

$$A \cap B \cap \{u,\omega : u \in U(\omega)\}$$

which belongs to $\mathcal{B}^m \otimes \mathcal{M}$. We obtain a selector which one can alter on a P_0-negligible set so that the resulting function $u(\cdot)$ would be $\mathcal{M}$-measurable. The function $u(\cdot)$ is a required one.

The next theorem is of great importance in proving the main result of the paper.

<u>Theorem 1.</u> For every pair $\{z_t\}$, $\{u_t\}$, satisfying (1)-(3), there exists a pair $\{\tilde{z}_t\}$, $\{\tilde{u}_t\}$ which satisfies (1)-(4) and has the value of the functional (5) not less than that for $\{z_t\}$, $\{u_t\}$.

---

*) By definition,

$$\varphi^+ = \begin{cases} \varphi, & \varphi \geq 0 \\ 0, & \varphi < 0. \end{cases}$$

<u>Proof.</u> Let $\{z_t\}$, $\{u_t\}$ be the given pair which meets conditions (1)-(3). First, we construct a pair $\{\tilde{z}_t\}$, $\{\tilde{v}_t\}$, which satisfies constraint (1), the function $\tilde{v}_t$ being $\mathcal{F}_{\tilde{z}_t, s_t}$ -measurable and

$$\tilde{v}_t(s^t) = v_t(\tilde{z}_t(s^t), s_t) \quad \text{a.s.}$$

for some $\widehat{\mathcal{B}^n \otimes \mathcal{F}_t}$ -measurable function v_t , such that

$$g^{t+1}(z, v_t(z, s_t), s_t, s_{t+1}) \leq 0, \quad P(\cdot|s_t)\text{-a.s.}, \tag{6}$$

$$v_t(z, s_t) \in U^t(s_t)$$

for all $z \in R^n$ and almost all s_t , and the value of the functional (5) for $\{\tilde{z}_t\}$, $\{\tilde{v}_t\}$ is not less than that for $\{z_t\}$, $\{u_t\}$.

The pair $\{\tilde{z}_t\}$, $\{v_t\}$ (hence $\{\tilde{z}_t\}$, $\{\tilde{v}_t\}$) will be constructed by induction starting from the moment τ backwards. Suppose that we have already found z'_k , $k = t+2, \dots, \tau$, v_k , $k = t+1, \dots, \tau-1$, such that $z_0, \dots, z_{t+1}$, $z'_{t+1}, \dots, z_\tau$, $u_0, \dots, u_t$, $\tilde{v}_{t+1}, \dots, \tilde{v}_{\tau-1}$ satisfy (1) and

$$M \sum_{k=t+1}^{\tau-1} \varphi^{k+1}(z'_k, \tilde{v}_k, s_k, s_{k+1}) \geq M \sum_{k=t+1}^{\tau-1} \varphi^{k+1}(z_k, u_k, s_k, s_{k+1}),$$

where $z'_{t+1} = z_{t+1}$.

The function

$$\sum_{k=t}^{\tau-1} \varphi^{k+1}(z'_k, \tilde{v}_k, s_k, s_{k+1}), \quad \text{where } z'_t = z_t,\ z'_{t+1} = z_{t+1},\ \tilde{v}_t = u_t$$

is equal almost surely to some function $\tilde{\varphi}(u_t(s^t), z_t(s^t), s_t, \dots, s_\tau)$, where the mapping

$$\tilde{\varphi} : R^m \times R^n \times S_t \times \dots \times S_\tau \to R$$

is measurable with respect to $(u, z, s_t, \dots, s_\tau)$.

Let

$$\alpha^t(z, u, s_t) = \int \tilde{\varphi}(u, z, s_t, \dots, s_\tau) P_\tau(ds_\tau | s_{\tau-1}) \dots P_{t+1}(ds_{t+1} | s_t).$$

The function

$$\alpha^t : R^n \times R^m \times S_t \to R$$

is measurable with respect to (z, u, s_t) .

Fix an arbitrarily choosen version of the conditional expectation

$$\beta^t(z_t(s^t), s_t) = M(\alpha^t(z_t(s^t), u_t(s^t), s_t) \mid \mathcal{F}_{z_t, s_t}).$$

Consider the set

$$A_t = \{s^t, u : \alpha^t(z_t(s^t), u, s_t) \geq \beta^t(z_t(s^t), s_t)\}.$$

It belongs to the σ -algebra $\mathcal{F}_{z_t, s_t} \otimes \mathcal{B}^n$; hence, its projection onto S^t , by Proposition 1, belongs to $\hat{\mathcal{F}}_{z_t, s_t}$. Denote by $\hat{P}$ the completion of P .

Let us show that $\hat{P}(\mathrm{proj}_{S^t} A_t) = 1$. Since $\overline{\mathrm{proj}_{S^t} A_t} = \{s^t : \alpha^t(z_t(s^t), u, s_t) < \beta^t(z_t(s^t), s_t)$ for all $u \in R^m\}$,*)

---

*) The set $\bar{C}$ is the complement to C .

the strict inequality

$$\alpha^t(z_t(s^t), u_t(s^t), s_t) < \beta^t(z_t(s^t), s_t)$$

holds on $\overline{proj_{S^t} A_t}$. Multiply both parts of this inequality by the characteristic function $\chi_{\overline{proj_{S^t} A_t}}$, and take the conditional expextation with respect to $\mathcal{F}_{z_t, s_t}$ of both parts. The strict inequality holds almost surely on $\overline{proj_{S^t} A_t}$:

$$M(\chi_{\overline{proj_{S^t} A_t}} \alpha^t | \mathcal{F}_{z_t, s_t}) < M(\chi_{\overline{proj_{S^t} A_t}} \beta^t | \mathcal{F}_{z_t, s_t})$$

This results in

$$\chi_{\overline{proj_{S^t} A_t}} M(\alpha^t | \mathcal{F}_{z_t, s_t}) < \chi_{\overline{proj_{S^t} A_t}} \beta^t$$

a.s. on $\overline{proj_{S^t} A_t}$. But this condition is true in the case $\hat{P}(\overline{proj_{S^t} A_t}) = 0$ only. Hence $\hat{P}(proj_{S^t} A_t) = 1$.

Consider the set

$$B_t = \{s^t, u : g^t(z_t(s^t), u, s_t, s_{t+1}) \leq 0 \quad P_{t+1}(\cdot | s_t) \text{ - a.s.}\}$$

By repeating the reasoning in the proof of Proposition 3, one can show that $B_t \in \mathcal{F}_{z_t, s_t} \otimes \mathcal{B}^m$. It follows from condition d) that s_t -sections of B_t are empty almost surely.

Let

$$C_t = A_t \cap B_t \cap \{s^t, u : u \in U^t(s_t)\}.$$

The set C_t belongs to the σ-algebra $\mathcal{F}_{z_t, s_t} \otimes \mathcal{B}^m$, and its s_t -sections are not empty almost surely. Applying Proposition 2 to C_t and altering the resulting $\hat{\mathcal{F}}_{z_t, s_t}$ -measurable selector, one can obtain an $\mathcal{F}_{z_t, s_t}$ -measurable function $\hat{w}_t$, which one can consider as the composition

$$w_t(s^t) = w_t(z_t(s^t), s_t),$$

$w_t : R^n \times S_t \to R^m$ is measurable with respect to the pair (z, s_t) .

Thus, the functions z''_t , $\tilde{w}_t$ weet constraints (2), (3),(4), and

$$M \sum_{k=t}^{\tau-1} \varphi^{k+1}(z_k(s^k), u_k(s^k), s_k, s_{k+1}) \leq M \sum_{k=t}^{\tau-1} \varphi^{k+1}(z''_k(s^k), \tilde{w}_k(s^k), s_k, s_{k+1})$$

where $\mathcal{F}^k$ -measurable functions z''_k are defined recursively by the formulas

$$z''_{k+1} = f^{k+1}(z''_k, w_k(z''_k, s_k), s_k, s_{k+1}) \quad \text{a.s.}, \; t \leq k \leq \tau - 1,$$

$$z''_t = z_t .$$

Reconstruct now w_t and obtain a function, satisfying (6).

Consider the sets

$$\mathcal{D}'_t = \{z, s_t, u : g^{t+1}(z, u, s_t, s_{t+1}) \leq 0 \quad P_{t+1}(\cdot \mid s_t) - \text{a.s.}\},$$

$$\mathcal{D}''_t = \{z, s_t, u : u \in U^t(s_t)\} = R^n \times \{s_t, u : u \in U^t(s_t)\}.$$

These sets belong to the σ -algebra $\mathcal{B}^n \otimes \mathcal{F}_t \otimes \mathcal{B}^m$. Apply Proposition 2 to the set $\mathcal{D}_t = \mathcal{D}'_t \cap \mathcal{D}''_t$ with nonempty (z, s_t)--sections for all z and almost all s_t (in accordance with condition d)). Let $w'_t(z, s_t)$ be a $\widehat{\mathcal{B}_n \otimes \mathcal{F}_t}$ -measurable selector.

Consider the set

$$\mathcal{E}=\{z,s_t: g^{t+1}(z,w_t(z,s_t),s_t,s_{t+1})\leq 0 \quad P_{t+1}(\cdot|s_t)\text{-a.s.}\}\cap$$
$$\cap\{z,s_t: w_t(z,s_t)\in U^t(s_t)\}.$$

Both of the sets on the right-hand side belong to the σ-algebra $\widehat{\mathcal{B}^n\otimes\mathcal{F}_t}$. Hence, $\mathcal{E}\in\widehat{\mathcal{B}^n\otimes\mathcal{F}_t}$.

Let

$$v_t=\begin{cases} w_t & \text{in } \mathcal{E}, \\ w_t' & \text{in } R^n\times S_t\setminus\mathcal{E}. \end{cases}$$

The function v_t is $\widehat{\mathcal{B}^n\otimes\mathcal{F}_t}$ -measurable and satisfies constraints (6).

We need show that replacing w_t by v_t does not alter the value of the functional

$$M\sum_{k=t}^{\tau-1}\varphi^{k+1}(z_k'',\tilde{w}_k,s_k,s_{k+1}).$$

We define the measure Q_t on $\mathcal{B}^n\otimes\mathcal{F}_t$ to be the image of P under the measurable mapping

$$s^t\longrightarrow(z_t(s^t),s_t).$$

One can easily verify, that $Q_t(\mathcal{E})=1$. Hence, w_t and v_t may differ from each other on a Q_t -negligible set at most. But then $\tilde{w}_t$ and $\tilde{v}_t=v_t(z_t,s_t)$ can differ from each other on a P -negligible set only.

These remarks yield at once:

$$M\Big(\varphi^{t+1}(z_t(s^t), \tilde{w}_t, s_t, s_{t+1}) +$$

$$+\sum_{k=t+1}^{\tau-1} \varphi^{k+1}(z''_k, v_k(z''_k, s_k), s_k, s_{k+1})\Big) =$$

$$= M\Big(\varphi^{t+1}(z_t(s^t), \tilde{v}_t(s^t), s_t, s_{t+1}) +$$

$$+\sum_{k=t+1}^{\tau-1} \varphi^{k+1}(z''_k, v_k(z'', s_k), s_k, s_{k+1})\Big).$$

The general induction step is completed, and the functions z''_k, $t+1 \leq k \leq \tau$, v_k, $t \leq k \leq \tau-1$, are constructed.

Using the same reasoning, the first induction step may be performed.

Define the measure $\tilde{Q}_t$ on $\mathcal{B}^m \otimes \mathcal{F}_t$ as the image of P under the measurable mapping

$$s^t \longrightarrow (\tilde{z}_t(s^t), s_t).$$

Given the functions $\tilde{z}_t$, $1 \leq t \leq \tau$, v_t, $0 \leq t \leq \tau-1$, which we have constructed, alter v_t on a $\tilde{Q}_t$-negligible set, so as to obtain $\mathcal{B}^n \otimes \mathcal{F}$ - measurable functions v'_t. Let $\tilde{u}_t(s^t) = v'_t(\tilde{z}_t(s^t), s_t)$. Then the pair $\{\tilde{z}_t\}$, $\{\tilde{u}_t\}$ satisfies (1)-(4) and has the value of the functional (5) not less than that of $\{z_t\}$, $\{u_t\}$, i.e. possesses all the required properties.

3. Maximum principle

The problem (1)-(5) will be treated in the space of pairs $(\{u_t\}, \{z_t\})$, where u_t, $0 \leq t \leq \tau-1$, is $\mathcal{F}^t$ -measurable with values in R^m and $z_t \in L_\infty^n(S^t, \mathcal{F}^t, P)$.

Let $(\{z_t^*\}, \{u_t^*\})$ be a solution of the problem (1)-(5). Define the set

$$\Theta = \{ t : \operatorname{vrai} \sup_{s^t} g^t (z_{t-1}^*, u_{t-1}^*, s_{t-1}, s_t) = 0 \}.$$

Let

$$H^{t+1}(\lambda, \psi, z, u, s_t, s_{t+1}) =$$

$$= \varphi^{t+1}(z, u, s_t, s_{t+1}) - \psi f^{t+1}(z, u, s_t, s_{t+1}) -$$

$$- \lambda g^{t+1}(z, u, s_t, s_{t+1}), \qquad 0 \leq t \leq \tau - 1,$$

where $\lambda \in R^k$, $\psi \in R^n$.

<u>Theorem 2.</u> Let $\{z_t^*\}$, $\{u_t^*\}$ be a solution of the problem (1)-(5). Assume that the following conditions are satisfied.

1) The convexity condition: for every t, $0 \leq t \leq \tau - 1$, and α, $0 \leq \alpha \leq 1$, for all $z \in R^n$, $u^{(1)}$, $u^{(2)} \in U^t(s_t)$, there exists a vector $u_{s_t} \in U(s_t)$ such that

$$\varphi^{t+1}(z, u_{s_t}, s_t, s_{t+1}) \geq \alpha \varphi^{t+1}(z, u^{(1)}, s_t, s_{t+1}) + (1-\alpha) \varphi^{t+1}(z, u^{(2)}, s_t, s_{t+1}),$$

$$f^{t+1}(z, u_{s_t}, s_t, s_{t+1}) = \alpha f^{t+1}(z, u^{(1)}, s_t, s_{t+1}) + (1-\alpha) f^{t+1}(z, u^{(2)}, s_t, s_{t+1}),$$

$$g^{t+1}(z, u_{s_t}, s_t, s_{t+1}) \leq \alpha g^{t+1}(z, u^{(1)}, s_t, s_{t+1}) + (1-\alpha) g^{t+1}(z, u^{(2)}, s_t, s_{t+1}).$$

These relations are supposed to hold $P_{t+1}(\cdot | s_t)$ - a.s.

2) The regularity condition: there exist $\{z_t^o\}$, $\{u_t^o\}$, where u_t^o satisfies constraint (3), such that

$$z^0_{t+1} + z^*_t - f^{t+1}_z(z^*_t, u^*_t, s_t, s_{t+1}) z^0_t - f^{t+1}(z^*_t, u^0_t, s_t, s_{t+1}) = 0,$$

$$g^{t+1}_z(z^*_t, u^*_t, s_t, s_{t+1}) z^0_t + g^{t+1}(z^*_t, u^0_t, s_t, s_{t+1}) < -\delta < 0$$

for $t \in \theta$ on one of the sets $\{s^t : g^{t+1}(z^*_t, u^*_t, s_t, s_{t+1}) > -\frac{1}{n}\}$ at least, where $n = 1, 2, \dots$.

Then there exist mappings

$$\lambda_{t+1} : R^n \times S_t \times S_{t+1} \longrightarrow R^k, \quad 0 \le t \le \tau - 1,$$

$$\psi_{t+1} : R^n \times S_t \times S_{t+1} \longrightarrow R^n, \quad 0 \le t \le \tau - 1,$$

measurable with respect to the triple (z, s_t, s_{t+1}), $\lambda_{t+1} \ge 0$, $M\lambda_{t+1}(z^*_t, s_t, s_{t+1}) < \infty$, $M|\psi_{t+1}(z^*_t, s_t, s_{t+1})| < \infty$, such that

$$M(H^{t+1}(\lambda_{t+1}(z^*_t, s_t, s_{t+1}), \psi_{t+1}(z^*_t, s_t, s_{t+1}), z^*_t, u^*_t, s_t, s_{t+1}) | \mathcal{F}_{z^*_t, s_t}) =$$

$$\max_{u \in U^t(s_t)} M(H^{t+1}(\lambda_{t+1}(z^*_t, s_t, s_{t+1}), \psi_{t+1}(z^*_t, u, s_t, s_{t+1}), z^*_t, u, s_t, s_{t+1}) \mathcal{F}_{z^*_t, s_t}),$$

$$0 \le t \le \tau - 1.$$

Besides,

$$M(\psi_t(z^*_{t-1}, s_{t-1}, s_t) | \mathcal{F}_{z^*_t, s_t}) =$$

$$= -M(H^{t+1}_z(\lambda_{t+1}(z^*_t, s_t, s_{t+1}), \psi_{t+1}(z^*_t, s_t, s_{t+1}), z^*_t, u^*_t, s_t, s_{t+1}) | \mathcal{F}_{z^*_t, s_t}), \quad 1 \le t \le \tau - 1,$$

$$\psi_\tau = 0$$

and

$$\lambda_{t+1}(z^*_t, s_t, s_{t+1}) g^{t+1}(z^*_t, u^*_t, s_t, s_{t+1}) = 0, \quad 0 \le t \le \tau - 1.$$

4. Proof of the maximum principle

Before going into the proof, let us formulate two results from the functional analysis, which we shall make use of.

Definition. A functional $\lambda^{*s} \in L^*_\infty(\Omega,\mathcal{F},P)$ is called singular, if there exists a sequence of sets $\mathcal{E}_n$, $P(\mathcal{E}_n) \xrightarrow[n\to\infty]{} 0$, such that $<\lambda^{*s}, z> = 0$ for every $z \in L_\infty(\Omega,\mathcal{F},P)$, which is equal to zero on an $\mathcal{E}_n$ for some n.

Definition. A functional $\lambda^{*a} \in L^*_\infty(\Omega,\mathcal{F},P)$ is called absolutely continuous if it is determined by an element of $L_1(\Omega,\mathcal{F},P)$.

Proposition 4.[2]. For a functional $\lambda^* \in L^*_\infty(\Omega,\mathcal{F},P)$, the Lebesgue decomposition holds:

$$\lambda^* = \lambda^{*a} + \lambda^{*s},$$

where λ^{*a} is an absolutely continuous functional and λ^{*s} is a singular one.

Let X, Y be Banach spaces and U be an arbitrary set. Denote by $L^k_\infty(\Omega,\mathcal{F},P)$, where P is a probability on a σ-algebra $\mathcal{F}$, the Banach space of equivalence classes of measurable P-a.s. bounded functions taking values in R^k.

Let

$$\Phi : X \times U \longrightarrow R,$$

$$F : X \times U \longrightarrow Y,$$

$$G : X \times U \longrightarrow L^k_\infty(\Omega,\mathcal{F},P).$$

Consider the problem

$$\Phi(x,u) \to max, \tag{10}$$

$$F(x,u) = 0, \tag{11}$$

$$G(x,u) \leq 0, \tag{12}$$

$$u \in U. \tag{13}$$

Let $\lambda_0 \in R$, $\psi^* \in Y^*$, $\lambda^* \in L_\infty^{*k}(\Omega, \mathcal{F}, P)$. Define the Lagrange function for the problem (10)-(13):

$$\mathcal{L}(x,u,\lambda_0,\lambda^*,\psi^*) = \lambda_0 \Phi(x,u) - \langle \psi^*, F(x,u) \rangle - \langle \lambda^*, G(x,u) \rangle .$$

<u>Lemma.</u> Let x^*, u^* be a solution of the problem (10)-(13). Suppose that there exists a neighbourhood V of x^* with the following properties.

a°. For every $u \in U$, the mappings

$$x \longrightarrow F(x,u)$$

and

$$x \longrightarrow G(x,u)$$

have Frechet derivatives which are continuous at x^* .

b°. For every $x \in V$, u_1 , $u_2 \in V$ and $0 \leq \alpha \leq 1$, one can find a $u \in U$ such that the relations

$$\Phi(x,u) \geq \alpha \Phi(x,u_1) + (1-\alpha)\Phi(x,u_2),$$

$$F(x,u) = \alpha F(x,u_1) + (1-\alpha) F(x,u_2),$$

$$G(x,u) \leq \alpha G(x,u_1) + (1-\alpha) G(x,u_2)$$

hold.

Besides, the following condition is imposed on F'_x .

c°. The range of the operator $F'_x(x^*,u^*)$ coincides with Y .

Under these conditions the Lagrange multipliers $\lambda_0 \geq 0$, $\lambda_0^* \geq 0$, ψ^* exist, one of them, at least, is non-zero, which satisfy the equations

$$\mathcal{I}_x(\lambda_0,\lambda^*,\psi^*,x^*,u^*) = -\lambda_0 \Phi_x'(x^*,u^*) + F_x'(x^*,u^*)\psi^* + G_x'(x^*,u^*)\lambda^* = 0,$$

$$\mathcal{I}(\lambda_0,\lambda^*,\psi^*,x^*,u^*) = \max_{u\in U} \mathcal{I}(\lambda_0,\lambda^*,\psi^*,x^*,u)$$

where $F_x'^*$, $G_x'^*$ are conjugate to F_x' , G_x' .

Besides, the following condition holds:

$$\lambda \cdot G(x^*,u^*) = 0 \qquad \text{a.s.}$$

where $\lambda \in L_1^k(\Omega,\mathcal{F},P)$ is the absolutely continuous part of the functional λ^* . Moreover, $\lambda_0 \neq 0$, if, apart from the assumptions made, the regularity conditions hold:

d°. there exist $x_0 \in X$, $u_0 \in U$ and a constant $\delta > 0$, such that

$$F_x'(x^*,u^*)x_0 + F(x^*,u_0) = 0,$$

$$G_x'^i(x^*,u^*)x_0 + G^i(x^*,u_0) < -\delta < 0$$

on the set $\{G^i(x^*,u^*) > -\delta\}$ for all i, $1 \leq i \leq k$, such that $\text{vrai sup}\; G^i(x^*,u^*) = 0$.

Let us now turn to the proof of the maximum principle.

It follows from Theorem 1 that, if $\{z_t^*\}$, $\{u_t^*\}$ is a solution of the problem (1)-(5), then this pair is a solution of (1)-(3) and (5), i.e. gives a maximum to the functional in a wider class of contròls $\{u_t\}$. Using this fact, derive first necessary conditions for the problem (1)-(3), (5) and then, considering (4), obtain the maximum pronciple, we need.

Necessary conditions for (1)-(3), (5) will be derived from the above lemma.

State the problem (1)-(5) in terms of the lemma.

Put

$$X = Y = \prod_{t=1}^{\tau} L_{\infty}^{n}(S^t, \mathcal{F}^t, P),$$

$$x = (z_1, \ldots, z_\tau),$$

$$u = (u_0, \ldots, u_{\tau-1}),$$

$$\Phi(x,u) = M \sum_{t=1}^{\tau} \varphi^t(z_{t-1}, u_{t-1}, \cdot),$$

$$F(x,u) = \{z_t - f^t(z_{t-1}, u_{t-1}, s_{t-1}, s_t); \quad 1 \leq t \leq \tau\},$$

$$G(x,u) = \{g^t(z_{t-1}, u_{t-1}, s_{t-1}, s_t); \quad 1 \leq t \leq \tau\}.$$

First of all, verify that condiyions a°, b°, c°, d° hold for the problem (1)-(3), (5).

The existence of the conditions Fréchet derivative for the mapping

$$F : \{z_t(\cdot)\} \rightarrow \{z_t(\cdot) - f^t(z_{t-1}(\cdot), u^*_{t-1}(\cdot), \cdot), \quad 1 \leq t \leq \tau\}$$

follows from b). The matrix F'_z is nondegenerate, therefore F'_z is "onto" at every point $\{z_t\}$.

The set U for the problem (1)-(3), (5) consists of satisfying constraint (3).

The convexity condition b° follows from 1) and Proposition 3. The regularity condition follows directly from 2).

Apply the above lemma to the problem (1)-(3), (5). In accordance with this lemma, there exist functionals $\{\psi^*_t\}$, $\{\lambda^*_t\}$, $\psi^*_t \in$

$\in L_{\infty}^{*n}(S^t, \mathcal{F}^t, P)$, $\lambda_t^* \in L_{\infty}^{*k}(S^t, \mathcal{F}^t, P)$, $\lambda_t^* \geq 0$, such that, for any $\mathcal{F}^t$ -measurable functions u_t , $u_t(s^t) \in U^t(s_t)$, $0 \leq t \leq \tau - 1$,

$$M \sum_{t=0}^{\tau-1} (-\varphi^{t+1}(z_t^*, u_t, s_t, s_{t+1}) + \varphi^{t+1}(z_t^*, u_t^*, s_t, s_{t+1})) -$$

$$- \langle \psi_{t+1}^*, f^{t+1}(z_t^*, u_t^*, s_t, s_{t+1}) - f^{t+1}(z_t^*, u_t, s_t, s_{t+1}) \rangle +$$

$$+ \langle \lambda_{t+1}^*, g^{t+1}(z_t^*, u_t, s_t, s_{t+1}) - g^{t+1}(z_t^*, u_t, s_t, s_{t+1}) \rangle \geq 0.$$

Let ψ_t^{*a} and λ_t^{*a} be the absolutely continuous parts of ψ_t^* and λ_t^* respectively. The last inequality holds of ψ_t^* , λ_t^* are replaced by $\psi_t^{*a}, \lambda_t^{*a}$. We have

$$M \sum_{t=0}^{\tau-1} (-\varphi^{t+1}(z_t^*, u_t, s_t, s_{t+1}) + \varphi^{t+1}(z_t^*, u_t^*, s_t, s_{t+1}) -$$
$$- \tilde{\psi}_{t+1}(f^{t+1}(z_t^*, u_t^*, s_t, s_{t+1}) - f^{t+1}(z_t^*, u_t, s_t, s_{t+1})) + \tag{17}$$
$$+ \tilde{\lambda}_{t+1}(g^{t+1}(z_t^*, u_t, s_t, s_{t+1}) - g^{t+1}(z_t^*, u_t^*, s_t, s_{t+1})) \geq 0.$$

This inequality holds for all $\{u_t\}$ such that u_t is $\mathcal{F}^t$ -measurable and satisfies (3). Among these we shall consider only $\{u_t\}$ satisfying constraint (4).

Put $u_k = u_k^*$, $k \neq t$. Then u_t satisfies (4), where z_t is replaced by z_t^* . Therefore in the inequality for the t -th term of the sum (17), one can substitute the compositions

$$\psi_{t+1}(z_t^*(s^t), s_t, s_{t+1}) = M(\tilde{\psi}_{t+1} \mid \mathcal{F}_{z_t^*, s_t, s_{t+1}}),$$
$$\lambda_{t+1}(z_t^*(s^t), s_t, s_{t+1}) = M(\tilde{\lambda}_{t+1} \mid \mathcal{F}_{z_t^*, s_t, s_{t+1}}) \geq 0$$

for $\tilde{\psi}_{t+1}$ and $\tilde{\lambda}_{t+1}$. Here the functions

$$\psi_{t+1} : R^n \times S_t \times S_{t+1} \longrightarrow R^n,$$
$$\lambda_{t+1} : R^n \times S_t \times S_{t+1} \longrightarrow R^k$$

are measurable with respect to (z, s_t, s_{t+1}). We have

$$MH^{t+1}(\lambda_{t+1}, \psi_{t+1}, z_t^*, u_t, s_t, s_{t+1}) \leq \tag{18}$$
$$\leq MH^{t+1}(\lambda_{t+1}, \psi_{t+1}, z_t^*, u_t^*, s_t, s_{t+1})$$

for all $u_t(s^t) \in U^t(s_t)$.

Consider the $\mathcal{F}_{z_t^*, s_t} \otimes \mathcal{B}^m$-measurable function

$$h(s^t, u) = M(H^{t+1}(\lambda_{t+1}, \psi_{t+1}, z_t^*, u, s_t, s_{t+1} \mid \mathcal{F}_{z_t^*, s_t}).$$

The set

$$A = \{s^t, u : h(s^t, u) > h(s^t, u_t^*(s^t))\}$$

belongs to $\mathcal{F}_{z_t^* s_t} \otimes \mathcal{B}^m$. By Proposition 1, $proj_{S^t} A \in \hat{\mathcal{F}}_{z_t^*, s_t}$. If $\hat{P}(proj_{S^t} A) = 0$, then (7) is proved. Suppose that $\hat{P}(proj_{S^t} A) > 0$. One can apply Proposition 2 on measurable selectors to the set

$$B = A \cap \{s^t, u : u \in U^t(s_t)\}$$

from the σ-algebra $\mathcal{F}_{z_t^*, s_t} \otimes \mathcal{B}^m$. Let u' be an $\hat{\mathcal{F}}_{z_t^*, s_t}$-measurable selector of $s^t \longrightarrow B_{s^t}$. Define a function u'' as follows

$$u_t^* \quad \text{in} \quad S^t \setminus proj_{S^t} A,$$

$$u' \quad \text{in} \quad proj_{S^t} A.$$

The function u'' is $\hat{\mathcal{F}}_{z_t^*, s_t}$ - measurable, so that one can alter it on a $\mathcal{P}$ -negligible set and obtain an $\mathcal{F}_{z_t^*, s_t}$ -measurable function $u(\cdot)$. For this function

$$M H^{t+1}(\lambda_{t+1}, \psi_{t+1}, z_t^*, u_t, s_t, s_{t+1}) >$$

$$> M H^{t+1}(\lambda_{t+1}, \psi_{t+1}, z_t^*, u_t^*, s_t, s_{t+1}).$$

But this contradicts (18). Therefore,

$$M(H^{t+1}(\lambda_{t+1}, \psi_{t+1}, z_t^*, u_t^*, s_t, s_{t+1}) | \mathcal{F}_{z_t^*, s_t}) >$$

$$\geqslant M(H^{t+1}(\lambda_{t+1}, \psi_{t+1}, z_t^*, u, s_t, s_{t+1}) | \mathcal{F}_{z_t^*, s_t})$$

for all $u \in U^t(s_t)$ on a set of $\mathcal{P}$ -measure one. Statement (7) is proved.

To obtain system (8), use equation (14) which in

$$M \sum_{t=0}^{\tau-1} \left(-\varphi_z^{t+1}(z_t^*, u_t^*, s_t, s_{t+1}) z_t\right) - \langle \psi_{t+1}^*, z_{t+1} -$$

$$- f_z^{t+1}(z_t^*, u_t^*, s_t, s_{t+1}) z \rangle + \langle \lambda_{t+1}^*, g_z^{t+1}(z_t^*, u_t^*, s_t, s_{t+1}) z_t \rangle = 0.$$

Write this equation in components, substituting $\tilde{\psi}_{t+1}$, $\tilde{\lambda}_{t+1}$ for ψ_{t+1}^*, λ_{t+1}^* :

$$M((-\varphi_z^{t+1} - \tilde{\psi}_t + \tilde{\psi}_{t+1} f_z^{t+1} + \tilde{\lambda}_{t+1} g_z^{t+1}) z_t) = 0, \quad 1 \le t \le \tau - 1,$$

$$M \tilde{\psi}_\tau z_\tau = 0 .$$

These equations hold for all $z_t \in L_\infty^n(S^t, \mathcal{F}^t, P)$, therefore

$$\tilde{\psi}_t = -M((-\varphi_z^{t+1} + \tilde{\psi}_{t+1} f_z^{t+1} + \tilde{\lambda}_{t+1} g_z^{t+1}) \mid \mathcal{F}^t), \quad 1 \le t \le \tau - 1,$$

$$\tilde{\psi}_\tau = 0.$$

It follows from here

$$M(\tilde{\psi}_t \mid \mathcal{F}_{z^*_{t-1}, s_{t-1}, s_t} \mid \mathcal{F}_{z^*_t, s_t}) =$$

$$= -M(H_z^{t+1}(\tilde{\lambda}_{t+1}, \tilde{\psi}_{t+1}, z^*_t, u^*_t, s_t, s_{t+1}) \mid \mathcal{F}_{z^*_t, s_t, s_{t+1}} \mid \mathcal{F}_{z^*_t, s_t})$$

$$M(\tilde{\psi}_\tau \mid \mathcal{F}_{z^*_{\tau-1}, s_{\tau-1}, s_\tau}) = 0.$$

Now remember that $\psi_t = M(\tilde{\psi}_t \mid \mathcal{F}_{z^*_{t-1}, s_{t-1}, s_t})$ and obtain (8).

The last statement of the maximum principle follows directly from (16). This completed the proof.

References

1 M.-P.Sainte-Beuve, On the extension of von Neumann-Aumann's theorem, J.Functional Analysis,17 (1974),112-129.

2 K.Yosida, E.Hewitt, Finitely additive measures, Trans. Amer.Math. Soc., 72 (1952), 46-66.

The Central Mathematical Economical Institute

Academy of Sciences of the USSR

Moscow

SELECTION OF VARIABLES IN MULTIPLE REGRESSION ANALYSIS

Toshiro Haga
Sanyo-Kokusaku Pulp Co., Ltd.

Tadakazu Okuno
University of Tokyo

1. INTRODUCTION

The problem of selecting variables in multiple regression has received a great deal of attention. Among the more common procedures are the forward selection method, the backward elimination method and the stepwise regression of Efroymson (1960). These procedures have been discussed in Chapter 6 of Draper and Smith (1966). The present authors (1971) have devised an improved method on the stepwise regression, which is a backward stepwise regression. The backward elimination method and the backward stepwise regression are known from experience to be superior to the forward selection method and that of Efroymson, respectively.

The criterion used in these procedures is to minimize the residual sum of squares (RSS) - the sum of the squared differences between the observations and their predicted values by regression -, which is equivalent to maximizing the multiple correlation coefficient R. If the RSS were the sole criterion, then one would always use all of the variables. An additional criterion must be used if one wishes to reduce the number of variables. The degree to which these criteria are weighted is arbitrary. Allen (1971) proposed to use the mean square error of prediction (MSEP) as a criterion for selecting variables, which takes into account the values of the predictor variables associated with the future observation and eliminates the arbitrariness with the RSS. Since the values associated with the future observation can not generally be given definitely, Allen (1974) has introduced a new criterion of the prediction sum of squares (PRESS), referred to as PSS in this paper, which utilizes exclusively the observations pertaining to estimation of multiple regression.

In this paper we will demonstrate the higher validity of PSS in a simple numerical example and investigate the relation among the three criteria for selecting variables; PSS, MSEP and F-test with RSS. Comparison of the above criteria is illustrated with the use of a numerical example.

2. THE PREDICTION SUM OF SQUARES (PSS) AND ITS VALIDITY

The multiple linear regression model is

$$\underset{\sim}{y} = X\underset{\sim}{\beta} + \underset{\sim}{\varepsilon} \tag{1}$$

where $\underset{\sim}{y}$ is an $n \times 1$ vector of observable random variables, X is an $n \times (p+1)$ matrix of full rank with first column unities, $\underset{\sim}{\beta}$ is a $(p+1) \times 1$ vector of regression coefficients, and $\underset{\sim}{\varepsilon}$ is an $n \times 1$ vector of normal random variables with $\underset{\sim}{\varepsilon} \sim N(0, \sigma^2 I)$. Let x'_α be the α-th row vector of X. Then the least square predictor associated with $\underset{\sim}{x}_\alpha$ is

$$\hat{y}_\alpha = \underset{\sim}{x}'_\alpha \underset{\sim}{b} \tag{2}$$

where

$$\underset{\sim}{b} = (X'X)^{-1}X'\underset{\sim}{y} . \tag{3}$$

Then the RSS is written as

$$RSS = \sum_{\alpha=1}^{n} (y_\alpha - \hat{y}_\alpha)^2 . \tag{4}$$

According to Allen (1974), the PSS is defined as

$$PSS = \sum_{\alpha=1}^{n} (y_\alpha - \hat{y}^*_\alpha)^2 , \tag{5}$$

where

$$\hat{y}^*_\alpha = \underset{\sim}{x}'_\alpha \underset{\sim}{b}_\alpha , \tag{6}$$

$\underset{\sim}{b}_\alpha$ being obtained in a same way as $\underset{\sim}{b}$ from $(n-1)$ observations excluding the α-th observation. This implies that n calculations of different multiple regressions are required for PSS. Okuno and Takeuchi (1976) have found the expression:

$$PSS = \sum_{\alpha=1}^{n} \{(y_\alpha - \hat{y}_\alpha)/(1 - c_\alpha)\}^2 , \tag{7}$$

where

$$c_\alpha = \underset{\sim}{x}'_\alpha (X'X)^{-1} \underset{\sim}{x}_\alpha .$$

By use of this expression the calculation has become much simpler.

A simple numerical example with 5 observations is given as in Table 1 and we compare the values of RSS and of PSS by use of the two different models of the first degree polynomial and the second degree polynomial which correspond, respectively, to the multiple regressions with one and two predictor variables. The corresponding least square predictors $\hat{y}_1$ and $\hat{y}_2$ for RSS and $\hat{y}^*_1$ and $\hat{y}^*_2$ for PSS are listed in the same Table and illustrated in Fig. 1. Although the second degree polynomial model has given much smaller RSS than the first degree polynomial model, the PSS of the former is remarkably larger than that of the latter. This suggests that the simpler model, the first degree polynomial, will give better results in predicting future observations. The values of y in this numerical example were determined by the following third degree polynomial:

$$y = 10X_1 - 5X_2 + 4X_3$$
$$= 10u - 5(u^2 - 2) + 4\tfrac{5}{6}(u^3 - \tfrac{17}{5}u)$$
$$= 10 - \tfrac{4}{3}u - 5u^2 + \tfrac{10}{3}u^3$$

where, X_1, X_2 and X_3 are the orthogonal polynomials in u of the first, second and third degree, respectively. The distinction between the results obtained by the two criteria, RSS and PSS, is clearly displayed in this example. From the result it can be generally said that PSS will have the higher validity than RSS in selecting variables for the prediction of future observations.

3. THE RELATION AMONG PSS, MSEP AND F-TEST WITH RSS

The expected values of RSS and PSS, and TMSEP are (Okuno and Takeuchi, 1976):

$$\left.\begin{aligned} E[RSS] &= \sigma^2\Sigma(1 - c_\alpha) = (n - p - 1)\sigma^2 , \\ E[PSS] &= \sigma^2\Sigma\frac{1}{1 - c_\alpha} \geq \frac{n^2}{n - p - 1}\sigma^2 , \\ TMSEP &= \sigma^2\Sigma(1 + c_\alpha) = (n + p + 1)\sigma^2 , \end{aligned}\right\} \quad (8)$$

where TMSEP is defined as the total of the expected values of the squared differences between future observations of y having the same set of n values of $\underset{\sim}{x}_\alpha$ and their predicted values. From these results we have

$$E[PSS] > TMSEP > E[RSS] . \quad (9)$$

TMSEP in this sense is the infimum of E[PSS].

Okuno and Takeuchi have also introduced the multiple correlation doubly adjusted by the degrees of freedom (d.f.) as follows:

Multiple correlation

$$R^2 = 1 - \frac{RSS}{Syy}$$

Multiple correlation adjusted by d.f.

$$R^{*2} = 1 - \frac{RSS/(n - p - 1)}{Syy/(n - 1)} = 1 - \frac{n - 1}{n - p - 1}(1 - R^2) ,$$

Multiple correlation doubly adjusted by d.f.

$$R^{**2} = 1 - \left(\frac{n + p + 1}{n - p + 1} RSS\right) \Big/ \left(\frac{n + 1}{n - 1} Syy\right)$$
$$= 1 - \frac{(n + p + 1)(n - 1)}{(n + 1)(n - p - 1)} (1 - R^2) , \quad (10)$$

where Syy denotes the total sum of squares of y with (n - 1) d.f., which is RSS with $\hat{y}_\alpha = \bar{y}$ in (4). Accordingly, $1 - R^2$ is considered to be the ratio of RSS with p variables of $\underset{\sim}{x}$ to that with no $\underset{\sim}{x}$-variables. $1 - R^{*2}$ is a similar ratio of RSSs divided by the corresponding d.f.s. $1 - R^{**2}$ is analogously a ratio of estimates of TMSEP = $(n + p + 1)\sigma^2$.

Table 1 A simple numerical example

(1) Data and RSS

x_1	$x_2 = x_1^2$	y	$\hat{y}_1$	$y - \hat{y}_1$	$(y - \hat{y}_1)^2$	$\hat{y}_2$	$y - \hat{y}_2$	$(y - \hat{y}_2)^2$
-2	4	-34	-20	-14	196	-30	-4	16
-1	1	3	-10	13	169	- 5	8	64
0	0	10	0	10	100	10	0	0
1	1	7	10	- 3	9	15	-8	64
2	4	14	20	- 6	36	10	4	16
					$RSS_1 = 600$			$RSS_2 = 160$
Regression equation: $\hat{y}_1 = 10x_1$ (linear)						$\hat{y}_2 = 10x_1 - 5(x_2 - 2)$ (quadratic)		

(2) Data and PSS

x_1	$x_2 = x_1^2$	y	$\hat{y}_1^*$	$y - \hat{y}_1^*$	$(y - \hat{y}_1^*)^2$	$\hat{y}_2^*$	$y - \hat{y}_2^*$	$(y - \hat{y}_2^*)^2$
-2	4	-34	1.0	-35.0	1,225	1.0	-35.0	1,225
-1	1	3	-15.6	18.4	345	- 9.7	12.7	162
0	0	10	- 2.5	12.5	156	10.0	0.0	0
1	1	7	11.3	- 4.3	18	19.7	-12.7	162
2	4	14	29.0	-15.0	225	-21.0	35.0	1,225
					$PSS_1 = 1,969$			$PSS_2 = 2,774$

(3) Calculation of $\hat{y}_1^*$ and $\hat{y}_2^*$ for the fifth observation

x_1	x_2	y	$u_1 = 2x_1 + 1$	$u_2 = x_2 + x_1 - 1$	u_1y	u_2y
-2	4	-34	-3	1	102	-34
-1	1	3	-1	-1	- 3	- 3
0	0	10	1	-1	10	-10
1	1	7	3	1	21	7
(2)	(4)	(14)	(5)	(5)		
		-14		sum	130	-40
	$\bar{y}_{(5)}$	= -3.5			$b_1=6.5$	$b_2=-10$

$\hat{y}_1^* = -3.5 + 6.5u_1 = 3.0 + 13x_1$ (linear)

$\hat{y}_2^* = -3.5 + 6.5u_1 - 10u_2 = 13.0 + 3x_1 - 10x_2$ (quadratic)

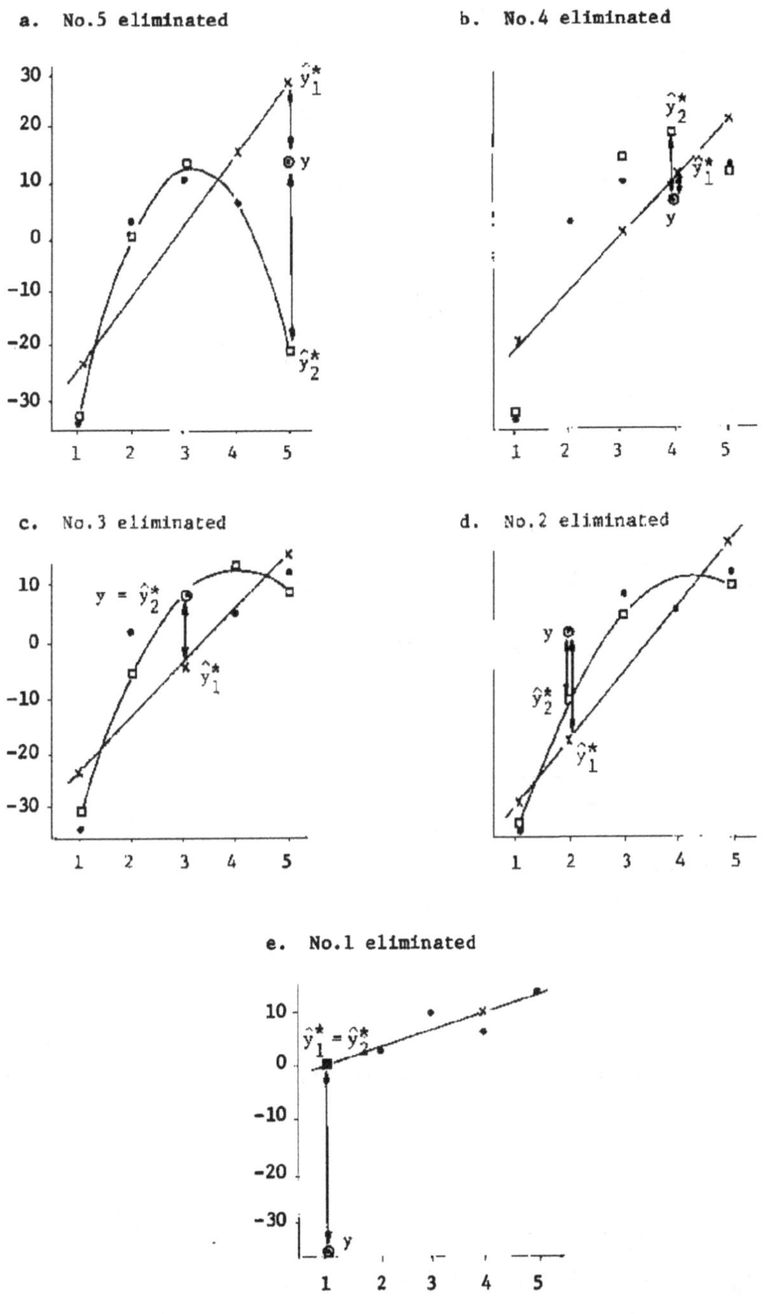

Fig. 1 Fitting linear and quadratic polynomials for PSS

The usual F-test for adding one variable to preceding (k - 1) variables or eliminating one variable from k variables is expressed by

$$F_k = \frac{(RSS)_{k-1} - (RSS)_k}{(RSS)_k/(n-k-1)} = \frac{R_k^2 - R_{k-1}^2}{(1-R_k^2)/(n-k-1)} . \tag{11}$$

Since $R_k^2 \geqq R_{k-1}^2$, F_k is always larger than or equal to zero. The criteria $R_k^{*2} \geqq R_{k-1}^{*2}$ and $R_k^{**2} \geqq R_{k-1}^{**2}$ are expressed in terms of F_k statistic as follows:

$$R_k^{*2} - R_{k-1}^{*2} = \frac{n-1}{n-k}(1-R_{k-1}^2) - \frac{n-1}{n-k-1}(1-R_k^2)$$
$$= \frac{n-1}{n-k}(R_k^2 - R_{k-1}^2) - \frac{(n-1)}{(n-k)(n-k-1)}(1-R_k^2) \geqq 0 ,$$

which is equivalent to

$$F_k = \frac{R_k^2 - R_{k-1}^2}{(1-R_k^2)/(n-k-1)} \geqq 1.0 . \tag{12}$$

And

$$R_k^{**2} - R_{k-1}^{**2} = \frac{(n+k)(n-1)}{(n+1)(n-k)}(R_k^2 - R_{k-1}^2)$$
$$- \frac{2n(n-1)}{(n+1)(n-k)(n-k-1)}(1-R_k^2)$$
$$\geqq 0 ,$$

whence

$$F_k \geqq \frac{R_k^2 - R_{k-1}^2}{(1-R_k^2)/(n-k-1)} = \frac{2n}{n+k} . \tag{13}$$

For $n \gg k$, the latter criterion implies that $F_k \geqq 2.0$. This outcome gives a sound basis for the stopping rule $F_k = 2.0$ recommended in a book by the present authors (1971).

4. EXAMPLE

The use of the above criteria for selecting predictor variables is illustrated with 30 sets of observations generated randomly from the following model:

$$\begin{aligned}
x_1 &= e_1 \\
x_2 &= 0.3e_1 + ge_2 \\
x_3 &= 0.3e_1 + 0.6ge_2 + 0.8ge_3 \\
x_4 &= x_1 + 0.5\, x_2 + 0.3\, x_3 + 0.5e_4 \\
x_5 &= 0.5x_1 + x_2 + \quad + 0.5e_5 \\
y &= x_1 + 0.5\, x_2 + 0.3\, x_3 + 0.5e ,
\end{aligned} \tag{14}$$

where $g = \sqrt{1-0.3^2} = \sqrt{0.91}$ and e_i $(i = 1, \cdots, 5)$ and e are independently distributed with $N(0,1^2)$.

Means, standard deviations and simple correlation coefficients are given in Table 2, with the information relevant to the multiple regressions for all the possible subsets of the five predictor variables. If the criterion $F_k \geq 2.0$ is adopted for selecting variables, the four methods stated in section 1 yield the following results.

Selection methods	Addition (+) and elimination (-) of variables							Set of variables selected
Step	1	2	3	4	5	6	7	Final
Forward selection	$+x_4$	$+x_5$	$+x_1$	$+x_2$				x_1 x_2 x_4 x_5
Stepwise regression	$+x_4$	$+x_5$	$+x_1$	$+x_2$	$-x_4$	$+x_3$	$-x_5$	x_1 x_2 x_3
Backward elimination / Stepwise backward	$-x_4$	$-x_5$						x_1 x_2 x_3

From the model (14) the best set of predictor variables is known to be the set of three variables (x_1, x_2, x_3), because x_4 and x_5 contribute to y nothing more than random terms in spite of the highest correlation between x_4 and y, which reflects the major part $x_1 + 0.5x_2 + 0.3x_3$ common to x_4 and y. In this example the best set is attained through the stepwise regression and the two backward selection methods. The use of the criterion with either R^{**2} or PSS leads to the same set of variables (x_1, x_2, x_3), whereas the criterion with R^{*2} leads to a different set (x_1, x_2, x_3, x_4). Fig. 2 shows the values of RSS and PSS for the best set of the variables.

Table 3 summarizes several similar results, which are obtained from different sets of observations generated randomly from the model (14). Table 3 also reveals a very high possibility in leading to the best set of variables (x_1, x_2, x_3) by use of the criterion with PSS or R^{**2}.

5. SUMMARY

One of the most important aspects of predicting a future y_0 associated with a new given value of $\underset{\sim}{x}_0$ is to select the best set of predictor variables and their functional forms to be used in multiple regression. The predictive ability of alternative prediction models must be evaluated.

Table 2 One example of randomly generated data

Variable	mean	s.d.	Simple correlation coefficient					
			x_1	x_2	x_3	x_4	x_5	y
x_1	-0.405	1.087	1	.171	.204	.769	.560	.787
x_2	-0.386	0.922		1	.751	.649	.790	.615
x_3	-0.116	1.136			1	.660	.719	.619
x_4	-0.622	1.433				1	.886	.869
x_5	-0.661	1.176					1	.784
y	-0.667	1.692						1

No. of selected variables	PSS	RSS	$100R^2$	$100R^{*2}$	$100R^{**2}$
1	35.583	31.624	61.95	60.59	59.32
2	59.366	51.695	37.80	35.58	33.50
3	58.647	51.271	38.31	36.11	34.04
4	23.967	20.333	75.53	74.66	73.84
5	37.597	32.034	61.45	60.08	58.79
1 2	14.748	11.858	85.73	84.67	83.68
1 3	15.979	13.398	83.87	82.68	81.56
2 3	21.717	17.455	78.99	77.44	75.98
1 4	21.608	17.352	79.12	77.57	76.12
2 4	58.114	46.990	43.46	39.27	35.35
3 4	25.199	19.963	75.98	74.20	72.53
1 5	38.733	32.030	61.46	58.60	55.93
2 5	24.747	20.032	75.89	74.11	72.44
3 5	38.516	31.503	62.09	59.29	56.66
4 5	26.422	20.255	75.63	73.82	72.13
1 2 3	12.317	9.774	88.24	86.88	85.61
1 2 4	15.859	11.820	85.77	84.13	82.60
1 3 4	15.447	11.858	85.73	84.08	82.54
2 3 4	17.133	13.086	84.25	82.43	80.73
1 2 5	15.650	12.598	84.84	83.09	81.45
1 3 5	22.877	16.325	80.35	78.09	75.97
2 3 5	26.883	19.901	76.05	73.29	70.70
1 4 5	40.451	31.327	62.30	57.96	53.89
2 4 5	26.913	19.950	75.99	73.22	70.63
3 4 5	27.232	20.027	75.90	73.12	70.52
1 2 3 4	12.909	9.386	88.70	86.89	85.20
1 2 3 5	12.690	9.528	88.53	86.70	84.98
1 2 4 5	16.811	11.803	85.79	83.52	81.40
1 3 4 5	18.499	12.596	84.84	82.41	80.15
2 3 4 5	28.693	19.884	76.07	72.24	68.66
1 2 3 4 5	14.166	9.329	88.77	86.43	84.24

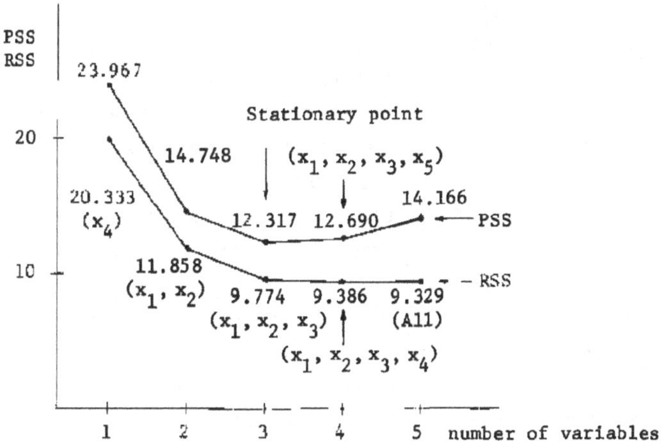

Fig. 2 Minimum values of RSS and of PSS for the number of variables involved

Table 3 Results obtained from simulation

Case	Forward selection	Stepwise regres-sion	Backward elimina-tion	Variables selected under the following criteria		
				R*2	R**2	PSS
1	(1,2,4,5)	(1,2,3)	(1,2,3)	(1,2,3,5)	(1,2,3)	(1,2,3)
2	(1,3,4,5)	(1,3,5)	(1,2,3)	(1,2,3)	(1,2,3)	(1,2,3)
3	(1,2,4)	(1,2,3)	(1,2,3)	(1,2,3)	(1,2,3)	(1,2,3)
4	(1,2,3,4)	(1,2,3)	(1,2,3)	(1,2,3,4)	(1,2,3)	(1,2,3)
5	(1,2,3,4)	(1,2,3)	(1,2,3)	(1,2,3)	(1,2,3)	(1,2,3)
6	(1,2,3,4)	(1,2,3)	(1,2,3)	(1,2,3)	(1,2,3)	(1,2,3)
7	(1,2,3,4)	(1,2,3,5)	(1,2,3,5)	(1,2,3,5)	(1,2,3,5)	(1,2,3,5)
8	(1,2,3,4)	(1,2,3,4)	(1,2,3,4)	(1,2,3,4)	(1,2,3,4)	(1,2,3,4)
9	(1,2,4)	(1,2,4)	(1,2,4)	(1,2,3)	(1,2,3)	(1,2,3)

The usual criterion for determining the best prediction function is the residual sum of squares (RSS). To make the RSS value as small as possible, a great number of potentially important variables are identified and a full model, which includes all relevant functions of these variables, is adopted. The full model was considered to be the best for prediction by many experimenters. Recently, experimenters have noticed that estimated regression coefficients in the full model very often lack reliability and that the precision in predicting future observations by use of the full model is much lower than that by use of the model with a small number of variables suitably chosen.

For the selection of such variables, we have investigated the validity of the two criteria of PSS recommended by Allen (1974) and of R^{**2} derived by Okuno and Takeuchi (1976) from the estimate of TMSEP, which is equal to infimum of E[PSS]. The latter is also found to be equivalent to the criterion with the help of the F_k statistic accompanied by the usual stopping rule. Consequently, our final recommendations are:

1) To find the set of variables having the minimum PSS when the number of observations is not very large.
2) To find the set of variables having the maximum R^{**2} or to use the equivalent criterion in the stepwise selection that F_k is larger than $2n/(n+k)$, which is approximately equal to 2.0 for $n \gg k$, when the number of observations is very large.

R^{**2} is a simple, convenient substitute for PSS and any discernible discrepancy between the two sets of variables selected by these two criteria is not found with our numerical example. The superiority of PSS to R^{**2} will, however, appear with some actual data, because the weight of residual $1-c_\alpha$ in (7) is dependent on the α-th observation.

References

1. Allen, D.M. (1971): Mean square error of prediction as a criterion for selecting variables, Technometrics 13, No.3, 469-475.
2. Allen, D.M. (1974): The relationship between variable selection and data augmentation and a method for prediction, Technometrics 16, No.1, 125-126.
3. Anderson, R.L., Allen, D.M. and Cady, F.B. (1972): Selection of predictor variables in linear multiple regression, Statistical Papers in Honour of George W. Snedecor, ed. T.A. Bancroft, Iowa State Univ. Press.
4. Draper, N.R. and Smith, H. (1966): Applied Regression Analysis, John Wiley & Sons Inc., New York.
5. Efroymson, M.A. (1960): Multiple Regression Analysis, Mathematical Methods for Digital Computers, ed. A. Ralston and H.S. Wilf, John Wiley & Sons Inc., New York.
6. Okuno, T., Haga, T. and others (1971): Multivariate Analysis, (in Japanese), Nikkagiren Shuppan.
7. Okuno, T. and Takeuchi, K. (1976): Prediction sum of squares, Akaike's information and doubly adjusted multiple correlation coefficient, to appear.

Vol. 399: Functional Analysis and its Applications. Proceedings 1973. Edited by H. G. Garnir, K. R. Unni and J. H. Williamson. II, 584 pages. 1974.

Vol. 400: A Crash Course on Kleinian Groups. Proceedings 1974. Edited by L. Bers and I. Kra. VII, 130 pages. 1974.

Vol. 401: M. F. Atiyah, Elliptic Operators and Compact Groups. V, 93 pages. 1974.

Vol. 402: M. Waldschmidt, Nombres Transcendants. VIII, 277 pages. 1974.

Vol. 403: Combinatorial Mathematics. Proceedings 1972. Edited by D. A. Holton. VIII, 148 pages. 1974.

Vol. 404: Théorie du Potentiel et Analyse Harmonique. Edité par J. Faraut. V, 245 pages. 1974.

Vol. 405: K. J. Devlin and H. Johnsbråten, The Souslin Problem. VIII, 132 pages. 1974.

Vol. 406: Graphs and Combinatorics. Proceedings 1973. Edited by R. A. Bari and F. Harary. VIII, 355 pages. 1974.

Vol. 407: P. Berthelot, Cohomologie Cristalline des Schémas de Caracteristique $p > o$. II, 604 pages. 1974.

Vol. 408: J. Wermer, Potential Theory. VIII, 146 pages. 1974.

Vol. 409: Fonctions de Plusieurs Variables Complexes, Séminaire François Norguet 1970–1973. XIII, 612 pages. 1974.

Vol. 410: Séminaire Pierre Lelong (Analyse) Année 1972–1973. VI, 181 pages. 1974.

Vol. 411: Hypergraph Seminar. Ohio State University, 1972. Edited by C. Berge and D. Ray-Chaudhuri. IX, 287 pages. 1974.

Vol. 412: Classification of Algebraic Varieties and Compact Complex Manifolds. Proceedings 1974. Edited by H. Popp. V, 333 pages. 1974.

Vol. 413: M. Bruneau, Variation Totale d'une Fonction. XIV, 332 pages. 1974.

Vol. 414: T. Kambayashi, M. Miyanishi and M. Takeuchi, Unipotent Algebraic Groups. VI, 165 pages. 1974.

Vol. 415: Ordinary and Partial Differential Equations. Proceedings 1974. XVII, 447 pages. 1974.

Vol. 416: M. E. Taylor, Pseudo Differential Operators. IV, 155 pages. 1974.

Vol. 417: H. H. Keller, Differential Calculus in Locally Convex Spaces. XVI, 131 pages. 1974.

Vol. 418: Localization in Group Theory and Homotopy Theory and Related Topics. Battelle Seattle 1974 Seminar. Edited by P. J. Hilton. VI, 172 pages. 1974.

Vol. 419: Topics in Analysis. Proceedings 1970. Edited by O. E. Lehto, I. S. Louhivaara, and R. H. Nevanlinna. XIII, 392 pages. 1974.

Vol. 420: Category Seminar. Proceedings 1972/73. Edited by G. M. Kelly. VI, 375 pages. 1974.

Vol. 421: V. Poénaru, Groupes Discrets. VI, 216 pages. 1974.

Vol. 422: J.-M. Lemaire, Algèbres Connexes et Homologie des Espaces de Lacets. XIV, 133 pages. 1974.

Vol. 423: S. S. Abhyankar and A. M. Sathaye, Geometric Theory of Algebraic Space Curves. XIV, 302 pages. 1974.

Vol. 424: L. Weiss and J. Wolfowitz, Maximum Probability Estimators and Related Topics. V, 106 pages. 1974.

Vol. 425: P. R. Chernoff and J. E. Marsden, Properties of Infinite Dimensional Hamiltonian Systems. IV, 160 pages. 1974.

Vol. 426: M. L. Silverstein, Symmetric Markov Processes. X, 287 pages. 1974.

Vol. 427: H. Omori, Infinite Dimensional Lie Transformation Groups. XII, 149 pages. 1974.

Vol. 428: Algebraic and Geometrical Methods in Topology. Proceedings 1973. Edited by L. F. McAuley. XI, 280 pages. 1974.

Vol. 429: L. Cohn, Analytic Theory of the Harish-Chandra C-Function. III, 154 pages. 1974.

Vol. 430: Constructive and Computational Methods for Differential and Integral Equations. Proceedings 1974. Edited by D. L. Colton and R. P. Gilbert. VII, 476 pages. 1974.

Vol. 431: Séminaire Bourbaki – vol. 1973/74. Exposés 436–452. IV, 347 pages. 1975.

Vol. 432: R. P. Pflug, Holomorphiegebiete, pseudokonvexe Gebiete und das Levi-Problem. VI, 210 Seiten. 1975.

Vol. 433: W. G. Faris, Self-Adjoint Operators. VII, 115 pages. 1975.

Vol. 434: P. Brenner, V. Thomée, and L. B. Wahlbin, Besov Spaces and Applications to Difference Methods for Initial Value Problems. II, 154 pages. 1975.

Vol. 435: C. F. Dunkl and D. E. Ramirez, Representations of Commutative Semitopological Semigroups. VI, 181 pages. 1975.

Vol. 436: L. Auslander and R. Tolimieri, Abelian Harmonic Analysis, Theta Functions and Function Algebras on a Nilmanifold. V, 99 pages. 1975.

Vol. 437: D. W. Masser, Elliptic Functions and Transcendence. XIV, 143 pages. 1975.

Vol. 438: Geometric Topology. Proceedings 1974. Edited by L. C. Glaser and T. B. Rushing. X, 459 pages. 1975.

Vol. 439: K. Ueno, Classification Theory of Algebraic Varieties and Compact Complex Spaces. XIX, 278 pages. 1975.

Vol. 440: R. K. Getoor, Markov Processes: Ray Processes and Right Processes. V, 118 pages. 1975.

Vol. 441: N. Jacobson, PI-Algebras. An Introduction. V, 115 pages. 1975.

Vol. 442: C. H. Wilcox, Scattering Theory for the d'Alembert Equation in Exterior Domains. III, 184 pages. 1975.

Vol. 443: M. Lazard, Commutative Formal Groups. II, 236 pages. 1975.

Vol. 444: F. van Oystaeyen, Prime Spectra in Non-Commutative Algebra. V, 128 pages. 1975.

Vol. 445: Model Theory and Topoi. Edited by F. W. Lawvere, C. Maurer, and G. C. Wraith. III, 354 pages. 1975.

Vol. 446: Partial Differential Equations and Related Topics. Proceedings 1974. Edited by J. A. Goldstein. IV, 389 pages. 1975.

Vol. 447: S. Toledo, Tableau Systems for First Order Number Theory and Certain Higher Order Theories. III, 339 pages. 1975.

Vol. 448: Spectral Theory and Differential Equations. Proceedings 1974. Edited by W. N. Everitt. XII, 321 pages. 1975.

Vol. 449: Hyperfunctions and Theoretical Physics. Proceedings 1973. Edited by F. Pham. IV, 218 pages. 1975.

Vol. 450: Algebra and Logic. Proceedings 1974. Edited by J. N. Crossley. VIII, 307 pages. 1975.

Vol. 451: Probabilistic Methods in Differential Equations. Proceedings 1974. Edited by M. A. Pinsky. VII, 162 pages. 1975.

Vol. 452: Combinatorial Mathematics III. Proceedings 1974. Edited by Anne Penfold Street and W. D. Wallis. IX, 233 pages. 1975.

Vol. 453: Logic Colloquium. Symposium on Logic Held at Boston, 1972–73. Edited by R. Parikh. IV, 251 pages. 1975.

Vol. 454: J. Hirschfeld and W. H. Wheeler, Forcing, Arithmetic, Division Rings. VII, 266 pages. 1975.

Vol. 455: H. Kraft, Kommutative algebraische Gruppen und Ringe. III, 163 Seiten. 1975.

Vol. 456: R. M. Fossum, P. A. Griffith, and I. Reiten, Trivial Extensions of Abelian Categories. Homological Algebra of Trivial Extensions of Abelian Categories with Applications to Ring Theory. XI, 122 pages. 1975.

Vol. 457: Fractional Calculus and Its Applications. Proceedings 1974 Edited by B Ross. VI, 381 pages 1975

Vol. 458: P. Walters, Ergodic Theory – Introductory Lectures. VI, 198 pages. 1975.

Vol. 459: Fourier Integral Operators and Partial Differential Equations. Proceedings 1974. Edited by J. Chazarain VI, 372 pages 1975

Vol. 460: O. Loos, Jordan Pairs. XVI, 218 pages. 1975.

Vol. 461: Computational Mechanics. Proceedings 1974. Edited by J T Oden VII, 328 pages 1975.

Vol. 462: P. Gérardin, Construction de Séries Discrètes p-adiques. »Sur les séries discrètes non ramifiées des groupes réductifs déployés p-adiques« III, 180 pages. 1975.

Vol. 463: H.-H. Kuo, Gaussian Measures in Banach Spaces VI, 224 pages. 1975.

Vol. 464: C. Rockland, Hypoellipticity and Eigenvalue Asymptotics. III, 171 pages 1975.

Vol. 465: Séminaire de Probabilités IX Proceedings 1973/74. Edité par P. A. Meyer. IV, 589 pages 1975.

Vol. 466: Non-Commutative Harmonic Analysis. Proceedings 1974 Edited by J Carmona, J. Dixmier and M. Vergne VI, 231 pages. 1975

Vol. 467: M. R. Essén, The Cos $\pi\lambda$ Theorem. With a paper by Christer Borell VII, 112 pages. 1975.

Vol. 468: Dynamical Systems – Warwick 1974 Proceedings 1973/74. Edited by A. Manning. X, 405 pages. 1975

Vol. 469: E. Binz, Continuous Convergence on C(X). IX, 140 pages 1975

Vol. 470: R. Bowen, Equilibrium States and the Ergodic Theory of Anosov Diffeomorphisms III, 108 pages. 1975.

Vol. 471: R. S. Hamilton, Harmonic Maps of Manifolds with Boundary III, 168 pages 1975.

Vol. 472: Probability-Winter School. Proceedings 1975 Edited by Z. Ciesielski, K. Urbanik, and W. A. Woyczyński. VI, 283 pages. 1975.

Vol. 473: D. Burghelea, R. Lashof, and M. Rothenberg, Groups of Automorphisms of Manifolds. (with an appendix by E. Pedersen) VII, 156 pages. 1975.

Vol. 474: Séminaire Pierre Lelong (Analyse) Année 1973/74. Edité par P. Lelong. VI, 182 pages. 1975.

Vol. 475: Répartition Modulo 1. Actes du Colloque de Marseille-Luminy, 4 au 7 Juin 1974. Edité par G. Rauzy. V, 258 pages. 1975. 1975

Vol. 476: Modular Functions of One Variable IV. Proceedings 1972. Edited by B. J. Birch and W. Kuyk. V, 151 pages. 1975.

Vol. 477: Optimization and Optimal Control. Proceedings 1974 Edited by R Bulirsch, W Oettli, and J Stoer VII, 294 pages. 1975.

Vol. 478: G. Schober, Univalent Functions – Selected Topics. V, 200 pages. 1975.

Vol. 479: S. D. Fisher and J. W. Jerome, Minimum Norm Extremals in Function Spaces. With Applications to Classical and Modern Analysis VIII, 209 pages. 1975.

Vol. 480: X. M. Fernique, J P. Conze et J Gani, Ecole d'Eté de Probabilités de Saint-Flour IV–1974. Edité par P.-L Hennequin. XI, 293 pages. 1975.

Vol. 481: M. de Guzmán, Differentiation of Integrals in R^n. XII, 226 pages. 1975.

Vol. 482: Fonctions de Plusieurs Variables Complexes II. Séminaire François Norguet 1974–1975. IX, 367 pages. 1975.

Vol. 483: R. D M. Accola, Riemann Surfaces, Theta Functions, and Abelian Automorphisms Groups. III, 105 pages. 1975.

Vol. 484: Differential Topology and Geometry. Proceedings 1974. Edited by G. P. Joubert, R. P. Moussu, and R. H. Roussarie. IX, 287 pages. 1975.

Vol 485: J. Diestel, Geometry of Banach Spaces – Selected Topics. XI, 282 pages 1975.

Vol 486: S Stratila and D. Voiculescu, Representations of AF-Algebras and of the Group U (∞). IX, 169 pages. 1975

Vol. 487: H. M. Reimann und T. Rychener, Funktionen beschränkter mittlerer Oszillation. VI, 141 Seiten. 1975.

Vol. 488: Representations of Algebras, Ottawa 1974. Proceedings 1974. Edited by V. Dlab and P. Gabriel. XII, 378 pages. 1975.

Vol 489: J. Bair and R Fourneau, Etude Géométrique des Espaces Vectoriels Une Introduction. VII, 185 pages 1975.

Vol. 490: The Geometry of Metric and Linear Spaces. Proceedings 1974 Edited by L M Kelly X, 244 pages 1975

Vol. 491: K. A Broughan, Invariants for Real-Generated Uniform Topological and Algebraic Categories X, 197 pages. 1975.

Vol. 492: Infinitary Logic: In Memoriam Carol Karp. Edited by D. W. Kueker. VI, 206 pages 1975

Vol. 493: F. W. Kamber and P Tondeur, Foliated Bundles and Characteristic Classes XIII, 208 pages. 1975.

Vol. 494: A Cornea and G Licea. Order and Potential Resolvent Families of Kernels IV, 154 pages 1975.

Vol 495: A Kerber, Representations of Permutation Groups II. V, 175 pages. 1975.

Vol. 496: L H. Hodgkin and V P Snaith, Topics in K-Theory. Two Independent Contributions III, 294 pages 1975.

Vol 497: Analyse Harmonique sur les Groupes de Lie. Proceedings 1973–75. Edité par P. Eymard et al. VI, 710 pages. 1975

Vol. 498: Model Theory and Algebra A Memorial Tribute to Abraham Robinson. Edited by D. H Saracino and V. B. Weispfenning. X, 463 pages. 1975.

Vol 499: Logic Conference, Kiel 1974. Proceedings. Edited by G. H Müller, A. Oberschelp, and K Potthoff V, 651 pages 1975.

Vol. 500: Proof Theory Symposion, Kiel 1974. Proceedings. Edited by J. Diller and G H. Müller VIII, 383 pages 1975.

Vol. 501: Spline Functions, Karlsruhe 1975. Proceedings. Edited by K. Böhmer, G. Meinardus, and W Schempp. VI, 421 pages. 1976.

Vol. 502: János Galambos, Representations of Real Numbers by Infinite Series. VI, 146 pages. 1976

Vol. 503: Applications of Methods of Functional Analysis to Problems in Mechanics. Proceedings 1975. Edited by P. Germain and B Nayroles XIX, 531 pages 1976

Vol 504: S. Lang and H F Trotter, Frobenius Distributions in GL_2-Extensions III, 274 pages 1976

Vol 505: Advances in Complex Function Theory. Proceedings 1973/74. Edited by W E Kirwan and L Zalcman. VIII, 203 pages. 1976

Vol. 506: Numerical Analysis, Dundee 1975. Proceedings. Edited by G. A. Watson X, 201 pages 1976

Vol 507: M C Reed, Abstract Non-Linear Wave Equations. VI, 128 pages 1976.

Vol. 508: E Seneta, Regularly Varying Functions. V, 112 pages 1976

Vol 509: D E. Blair, Contact Manifolds in Riemannian Geometry. VI, 146 pages. 1976

Vol. 510: V. Poènaru, Singularités C^∞ en Présence de Symétrie. V, 174 pages. 1976

Vol. 511: Séminaire de Probabilités X Proceedings 1974/75. Edité par P. A. Meyer VI, 593 pages 1976.

Vol. 512: Spaces of Analytic Functions, Kristiansand, Norway 1975. Proceedings. Edited by O. B. Bekken, B. K. Øksendal, and A. Stray. VIII, 204 pages. 1976.

Vol. 513: R. B. Warfield, Jr Nilpotent Groups. VIII, 115 pages. 1976.

Vol. 514: Séminaire Bourbaki vol. 1974/75. Exposés 453 – 470. IV, 276 pages. 1976

Vol. 515: Bäcklund Transformations. Nashville, Tennessee 1974. Proceedings. Edited by R. M. Miura. VIII, 295 pages. 1976.